Atlas of the Breeding Birds of Tennessee

Atlas of the Breeding Birds of Tennessee

Charles P. Nicholson

THE UNIVERSITY OF TENNESSEE PRESS / KNOXVILLE

Copyright © 1997 by The University of Tennessee Press / Knoxville.
All Rights Reserved. Manufactured in the United States of America.
First Edition.

Frontispiece: Blue-winged Teal. Chris Myers.

The paper in this book meets the minimum requirements of the American National Standard for Permanence of Paper for Printed Library Materials. ∞ The binding materials have been chosen for strength and durability.

♻ Printed on recycled paper.

Library of Congress Cataloging-in-Publication Data

Nicholson, Charles P.
   Atlas of the breeding birds of Tennessee / Charles P. Nicholson. — 1st ed.
     p. cm.
Includes bibliographical references and index.
ISBN 0-87049-987-4 (cloth : alk. paper)
1. Birds—Tennessee. 2. Birds—Tennessee—Geographical distribution. 3. Birds—Tennessee—Geographical distribution—Maps. 4. Birds—Habitat—Tennessee. 5. Bird populations—Tennessee.
I. Title.
QL684.T2N53 1997
598'.09768—dc21   97-4622

To Ben B. Coffey Jr., Joseph C. Howell, and James T. Tanner, Tennessee ornithologists who helped educate me and set such high standards for all of us to follow.

Contents

Preface   ix
Acknowledgments   xi

The Atlas Project
    The Atlas Project   3
Landscape and Ornithology of Tennessee
    The History of Tennessee Ornithology   11
    The Environment of Tennessee   19
    The Landscape of Tennessee   30
    Historic Changes in the Tennessee Avifauna   41
    An Overview and Analysis of Atlas Results   48
Species Accounts
    Introduction to the Species Accounts   57
    Confirmed Breeding Species   63
    Miscellaneous Species   373

Appendix 1. Taxonomic List   387
Appendix 2. Summary of Breeding Chronology   391
Appendix 3. Brown-headed Cowbird Hosts   395
Literature Cited   397
Index   423

Preface

The first attempt at a comprehensive description of Tennessee's birdlife was Samuel Rhoads's *Contributions to the Zoology of Tennessee, No. 2—Birds,* published in 1895. In the ensuing century, knowledge of the birds of Tennessee increased exponentially. So, too, did the need for detailed information on their distribution, population trends, habitat requirements, and breeding biology. This information is necessary both to satisfy our intellectual curiosity and to conserve our birdlife, as well as the other biota, in the face of the seemingly rapid homogenization of the Tennessee landscape.

To meet these needs, I and several others launched the Tennessee Breeding Bird Atlas Project, with the goal of providing a picture of the detailed distribution and relative abundance of birds nesting in Tennessee from 1986 through 1991. Our findings make up the heart of this book.

This work is designed to appeal to beginning and advanced birdwatchers, natural historians, conservationists, professional ornithologists, and ecologists. Because of the availability of several high-quality field guides, as well as space restrictions, the species accounts contain little information that will assist a beginning birdwatcher in identifying birds. The line drawings accompanying the species accounts do, however, accurately portray parts of the birds' nesting cycle. Birdwatchers in Tennessee and nearby states will find the maps useful in locating the different species, and the text of the species accounts will add to their knowledge of the biology of the species. Conservationists will find the distribution, abundance, and habitat descriptions useful; these descriptions are already being used by Tennessee's Partners In Flight bird conservation program. Ecologists will find detailed distribution and relative abundance information, useful in studies of habitat relationships and biogeography. The sixth chapter below, "An Overview and Analysis of Atlas Results," contains some cursory analyses of this sort.

While the information here provides a detailed picture, remember that it is a still snapshot. A few parts of the print are not fully developed, and the scene is moving. The results for some rare species were not sufficient for a full evaluation of their status, and more fieldwork is needed for species such as rails, bitterns, and owls. The breeding biology of many species, including some that are relatively common, is still poorly known. And, as the environment changes, our birdlife changes. It is time to begin planning the next picture.

At sunset on 14 June 1989, while walking along a gravel road high in the Tellico District of Cherokee National Forest, I marveled at a chorus of singing Veeries, occasionally interrupted by an outburst of Winter Wren song. I thought how this was a suitable reward for the alarm clock ringing at 4 A.M., chiggers, long, hot days of Atlas fieldwork, and sore wrists from hours at the computer. Later that night, while I was camped on the mountain top, a severe thunderstorm arose with gale force winds and several centimeters of rain. The rain continued off and on much of the next day, and I spent part of the next evening in a laundromat drying my tent and sleeping bag. I also wondered whether the reward of the previous evening was enough. I believe other Atlas workers experienced similar doubts along the way. The results you see here show that most of us quickly dismissed our doubts.

Acknowledgments

The Tennessee Breeding Bird Atlas project was a collaborative effort that succeeded because of the involvement of a large proportion of the professional and amateur ornithologists in Tennessee, as well as numerous other people and institutions both in Tennessee and elsewhere. These people and institutions are listed here; I sincerely apologize for any omissions.

Project Coordination

The project was directed by the Atlas Committee, composed of Paul Hamel (chairman, 1986–87), Robert Hatcher, and Charles Nicholson (chairman, 1987–93), and the regional coordinators: Fred Alsop, Linda Cartwright, Ken Dubke, Lil Dubke, Bob Ford, Pete Kalla, Richard Knight, Charles Nicholson, George Payne, Damien Simbeck, Steve Stedman, David Vogt, and Morris Williams. During the 1986–91 fieldwork period, the Atlas project newsletter was an important means of communicating with observers and donors; newsletter editors were Paul Hamel, Charles Nicholson, Donna Smith, and Steve Stedman.

Data Entry and Review

Retrieving the results of the fieldwork from the observers, editing the results, and typing them into computer files was primarily the responsibility of the regional coordinators. Others who ably assisted in this immense task were Dianne Bean, Carolyn Bullock, Robbie Hassler, Virginia Reynolds, and Martha Waldron.

The Atlas Book

Species accounts were contributed by Dianne Bean, Ralph Dimmick, Bob Ford, William Fowler, Paul Hamel, Robert Hatcher, Audrey Hoff, Ron Hoff, Richard Knight, Knox Martin, Douglas McNair, Jerry Nagle, Charles Nicholson, Lisa Petit, David Pitts, John Robinson, Damien Simbeck, Barbara Stedman, Candy Swan, Ann Tarbell, and David Vogt. Authors are identified at the ends of the accounts they wrote.

Species accounts were reviewed by David Buehler, Ralph Dimmick, Bob Ford, Linda Fowler, Paul Hamel, Jerome Jackson, Ross James, Richard Knight, Douglas McNair, David Pitts, Marian Pitts, and Burline Pullin. Other portions of the text were reviewed by Bob Ford, Betty Mason, Niki Nicholas, Milo Pyne, and Andrea Shea. Mike Baltz and especially Dianne Bean conducted bibliographic research that aided in the preparation of many species accounts.

David Vogt coordinated the work of artists Elizabeth S. Chastain and Chris Myers; their drawings, as well as David's, beautifully illustrate the species accounts.

Financial Contributors

Because the Atlas project had no fixed source of income to pay the numerous bills, the generous contributions and in-kind services of numerous individuals, clubs, agencies, businesses, and foundations were critical. Most of the individual donations were raised through direct appeals to Tennessee Ornithological Society

(TOS) members. Robert Hatcher coordinated much of the fund raising from businesses and foundations.

Individual Donors—Species Sponsors: Great Egret—In Memory of Dr. Wendell L. Whittemore; Snowy Egret—Lula C. Coffey; Bald Eagle—Dr. O. Ray Jordan; Eastern Screech-Owl—David and Rebekah Chaffin; Northern Saw-whet Owl—Dr. and Mrs. Robert H. Collier Jr.; Chimney Swift—In Honor of R. Demett Smith Jr.; Red-headed Woodpecker—Richard Figari; Black-capped Chickadee—In Memory of Dr. James T. Tanner; Tufted Titmouse—Betty Goff Cartwright; Eastern Towhee—Edward L. and Shirley Nicholson; Lark Sparrow—John W. Sellars Chapter, TOS; Song Sparrow—Louise Jackson.

Other Individual Donors: George A. Ammann, Trish Ardovino, Dianne P. Bean, Mrs. William F. Bell, Phine Britton, Jim Brooks, Janet Brown, Robert D. Brown, John Bruner, Ted Caldwell, Joyce Campion, Hazel Cassel, Howard Chitwood, Marylin Cline, Ben B. Coffey Jr., J. Wallace Coffey, Lula C. Coffey, J. Cole, Herbert and Cynthia Cragin, Marcia Davis, Judith Deaderick, Helen Dinkelspiel, Ken and Lil Dubke, Bobby DuBois, Jim and Sue Ferguson, Lucy Finch, Sarah Funkhouser, Bob Garst, Margaret Gaut, Sally Goodin, Paul Hamel, Marjorie Harper, Van Harris, Robbie Hassler, Bob Hatcher, Ann Houk, Ginger Ilardi, Ruth Luckado, Johnnie Sue Lyons, Nelle Moore, Dan Neves, Dick Newton, Charles P. Nicholson, Sara Stafford Nolan, Dr. and Mrs. William C. North, Linda Northrop, Harriet Brown Overton, George R. Payne Jr., Richard Peake, Karen Petrey, Ruth F. Pierce, Virginia Price, Joanne Routledge, Beth Simms, Norene Smith, Dave Snyder, Frances Spence, John and Mary Ann Spence, Stephen Stedman, Jim and Nancy Tanner, Ann Tarbell, Joe Wahl, Miriam Weinstein, Dan Williams, Barbara Wilson, and Terry Witt. Honoraria: Bob Hatcher, Arlo and Noreen Smith, Barbara Stedman, David Vogt, and Dr. Terry Witt. Memorials: for Charles A. Buckner, Maxie Swindell, and Mrs. Glenn Swofford.

Clubs: Chattanooga Chapter, TOS; Highland Rim Chapter, TOS; John W. Sellars Chapter, TOS; Kinsman Klay Tillers Club; Knoxville Chapter, TOS; Lake Hills Home and Garden Club; Memphis Audubon Society; Memphis Chapter, TOS; and Nashville Chapter, TOS.

Businesses and Foundations: Monsanto Corporation; for donation of proceeds from bird seed sales held to support the Atlas project, Acme Farm Supply, Nashville; Agri Feed and Supply, Knoxville; and Garr's Rental and Feed, Mt. Juliet.

Government Agencies: Tennessee Department of Conservation, Tennessee Valley Authority, Tennessee Wildlife Resources Agency, U.S. Fish and Wildlife Service, and U.S.D.A. Forest Service.

Other Contributors

Use of the Breeding Bird Survey (BBS) results in the species accounts would not have been possible without the efforts of the many volunteers who have diligently censused the routes since 1966. Sam Droege and Bruce Peterjohn of U.S. Fish and Wildlife Service and National Biological Service provided the BBS results.

Several of my co-workers in the Tennessee Valley Authority Regional Natural Heritage Program provided help and encouragement throughout the project; this assistance was critical in the latter stages. For this I especially thank Peggy Shute, Susan Jeffers, and Bill Redmond. Bill Redmond and Ralph Jordan were instrumental in arranging release time for fieldwork and manuscript preparation. David Hankins patiently tutored me on the computer system used to make most of the maps and figures.

Atlas Participants

Lastly, but not in their importance, are the many participants, most of them volunteers, who conducted field surveys:

Robert Abernathy, Sharon Adams, Doug Alexander, Fred Alsop, Bruce F. Anderson, Charles Anderson, Kenneth G. Anderson, Trish Ardovino, Mark Armstrong.

Duane Baker, Bob Barni, Dianne P. Bean, Dorothy Beck, T. Edward Beddow, Jane Beintema, Susan M. Bell, Mary Benson, George R. Beringer, Earline C. Berry, Michael A. Beuerlein, Jean Biddle, Sandra D. Bivens, Clyde Del Blum, Donald W. Blunk, L. Bordenave, Elaine Borders, Judy Boyles, Sharon Bracy, Bob Brady, R. M. Brayden, Steven Brinkley, Ben Britton, Ed Britton, Edith S. Bromley, Les Brown, Robert Browne, Carolyn H. Bullock, Danny Bystrak.

Don Caldwell, John Caldwell, Betty Reid Campbell, James M. Campbell, Linda and David Cartwright, Robert Casey, Jean Cashion, David C. Chaffin, Howard Chitwood, Richard M. Clark, Robert B. Clark III, Sally Clark, Will J. Cloyd, Ben B. Coffey Jr., J. Wallace Coffey, Lula C. Coffey, Paul W. Cole, Carol Coleman, Gene Coleman, Judy Coleman, Robert Collier, C. W. Comer, Walter Cook, Lynn Coppinger, Andrew Core, Robert Core, Steven Cottrell, Andrew Cowherd, H. E. and Cynthia Cragin, Dot T. Crawford, J. Paul Crawford, Robert L. Crawford, Ron K. Cristen, Brian L. Cross.

Robert Daly, Donald L. Davidson, Marcia L. Davis, Chris Debold, Rowan Debold, Fred O. Detlefsen, Paul H. Dietrich, Martha Dillenbeck, Ralph Dimmick, Helen B. Dinkelspiel, Michael Dinsmore, Edna Dixon, Marion Dorsey, Mary Kay Doyle, Randall L. Doyle, C. Gerald Drewry, Sam Droege, Kenneth H. Dubke, Lillian H. Dubke.

David A. Easterla, T. B. Easterla, J. Eberhardt, Dee Eiklor, John Eiklor, Glen D. Eller, Lynn Elliot, J. Ely, Becky Endres, Bob Endres.

Julie Faulk, Victor Fazio, James A. Ferguson, Susan B. Ferguson, Richard A. Figari, Barry Fleming, Gertrude L. Fleming, Robert P. Ford, Ann Forshey, Brian Foster, Linda Fowler, William B. Fowler Jr., John Froeshaurer, Louise Fuller.

Murray L. Gardler, Debbie Gillis, Sally Goodin, Katherine A. Goodpasture, Katie Greenberg, Mark A. Greene, Tom Grelen, Joe B. Guinn.

Mark S. Hackney, Tom Haggerty, William G. Haley Jr., Gordon E. Hall, Paul B. Hamel, David A. Hankins, Dorenda Hanna, Bobby Harbin, Carol Hardy, Barbara Harris, Glenn Harris, Linda B. Harris, Paul C. Harris, Van Harris, W. Mark Harris, Paul Hartigan, David Hassler, Robbie C. Hassler, Robert M. Hatcher, Lisa Hays, Annie H. Heilman, R. John Henderson, Jim Heptinstall, Marguerite Hernandez, Lois Herndon, Tim Higgs, Bob Hill, Debbie W. Hill, Dorris Hill, Audrey R. Hoff, Ronald D. Hoff, Don Holt, Jim Holt, Andrea Hopkins, J. N. and Ella Howard, Joseph C. Howell, Susan E. Hoyle, D. Huffine, Susan Humphrey.

Robert Ilardi, Virginia Ilardi, Jerry L. Ingles, Theresa Irion.

Daniel R. Jacobson, Wesley K. James, Albert M. Jenkins, Tim Jenkins, Tim Jensen, William Jernigan, Burt Jordan.

Peter I. Kalla, Alice Kaserman, Jim Keeton, John Kennedy, Paul D. Kittle, Richard L. Knight, Jon A. Koella, Brad Kovach, Lee Kramer, Roger Kroodsma.

Beth Lacy, Howard Langridge, Thomas F. Laughlin, Allyn Lay, Robert Ledbetter, Barbara Lee, Galen Lenhert, Jo Levy, Selma S. Lewis, Richard Lewis, Selena Little, Anne Lochridge, O. Bedford Lochridge, Ruth Luckado, Jonnie S. Lyons.

Margaret L. Mann, Knox Martin, Linda Mascuch, Paul G. Mascuch, Jim Mason, Cleo Mayfield, George R. Mayfield Jr., Mark H. Mayfield, Rad Mayfield, Sarah McClellan, John R. McClure, Chester A. McConnell, George W. McKinney, F. Joseph McLaughlin, Peter K. McLean, Louise and Ralph McLeroy, Barbara G. McMahan, Michael A. McMahan, Ruth V. McMillan, Douglas B. McNair, Mac McWhirter, Susan McWhirter, Nickey Medley, C. Metz, P. Metz, Lynn Miller, Gloria Milliken, W. D. Milliken, Susan Morris, Gay H. Morton, Dolly Ann Myers.

Charlotte A. Neal, William F. Neal, Royce Neidert, Dove Neuman, Jane E. Newell, Richard Newton, Anne Nichols, Charles P. Nicholson, Chris Norris, N. Kay Norris, Linda Northrop.

Stuart Oakes, Elizabeth O'Connor, Holly Overton, J. B. Owen.

Brainard Palmer-Ball Jr., Gerald Papachristou, Paul S. Pardue, Johnny T. Parks, Patricia Parr, J. D. Parrish, Robert L. Parrish, David C. Patton, J. Thomas Patton, George R. Payne Jr., Jeanne B. Payne, Laurie S. Pearl, Chloe W. Peebles, Rob Peeples, Dan Petit, Lisa Petit, Janet A. Phillips, Rick A. Phillips, Robert L. Pierce, T. David Pitts, Dick Preston, Burline P. Pullin, C. Pyle.

John J. Quick.

William H. Redmond, Martha Lyle Reid, Virginia B. Reynolds, Sue E. Ridd, G. Dan Robbins, John C. Robinson, Erma L. Rogers, Ron Rogers, Tommie L. Rogers, James D. Rowell Jr., Robin A. Rudd, Martha A. Rudolph, John Rumancik, Michael G. Ryon.

Rebecca Satterfield, Steven Satterfield, Ann Schaefler, Ed Schell, Elizabeth M. Schilling, Richard Schrieber, Bill Schriver, Madeline Seefield, Donette C. Sellers, A. Boyd Sharp, Mary Jane Sharp, Clay Shelton, D. Silsbee, Damien J. Simbeck, Don Simbeck, Donna Simbeck, Regina Simbeck, Richard W. Simmers Jr., Mason Sinclair, Russ Skolgund, C. A. Sloan, James Slover, Arlo Smith, B. R. Smith, Donna J. Smith, Gerald R. Smith, Kim Smith, Noreen M. Smith, Ricky Smith, Tim and Angela Snow, David H. Snyder, John Spence, Debbie Spero, Veit Spero, Ann Stapp, D. Stark, Barbara H. Stedman, Stephen J. Stedman, Karen Stephens, John Stokes, Stan Strickland, Randy C. Stringer, J. Strom, L. M. Stubbs, Randy Suddarth, Betty Sumara, Glenn Swafford, Candy Swan.

James T. Tanner, Ann T. Tarbell, Richard G. Taylor, Norton and Annie Terry, Burney R. Tompkins, Sam Tucker, Bill Turner, Melissa Turrentine, Roy Turrentine.

David Vandegriff, David F. Vogt.

Joseph W. Wahl, James E. Waldron, Martha G. Waldron, Ellen J. Walker, Gary O. Wallace, B. Warden, J. Craig Watson, Susan L. Watson, Linda Watters, Ken Webster, Mike Wefer, Beth Weist, Richard L. West, Archie G. Whitehead, Eddie Wilbanks, Dan N. Williams Jr., Jack and Carol Williams, Barbara H. Wilson, Richard O. Wilson, Jeff R. Wilson, Terry J. Witt, Libby Wolfe, Pat Wood.

Harry C. Yeatman, Stanley W. York Jr.

Ralph Zaenglein, Mary A. Zimmerman.

And, finally, to my wife, Niki, for her help in so many ways.

Thanks to all of you.

The Atlas Project

The Atlas Project

The first breeding bird atlas based on a grid-based survey was *The Atlas of Breeding Birds in Britain and Ireland* (Sharrock 1976); its fieldwork was conducted from 1968 to 1972. The first similar North American project was a county bird atlas begun in Maryland in 1971; within the next few years statewide bird atlases were underway in several northeastern states (Laughlin, Kibbe, and Eagles 1982). The number of states and Canadian provinces with bird atlas projects grew rapidly during the 1980s, and by the end of the decade atlas projects were completed or underway in at least 35 states and provinces (Smith 1990).

As atlas projects were gaining popularity in the late 1970s and early 1980s, many Tennessee ornithologists, myself included, were reluctant to begin a state breeding bird atlas. Compared to other states and provinces with atlases projects, the pool of potential workers in Tennessee seemed small and poorly organized. Two organized surveys were also then underway: the annual county-based Breeding Bird Forays (see the next chapter, "The History of Tennessee Ornithology") and the Breeding Bird Survey. We soon realized, however, that these surveys were inadequate to provide the timely, detailed statewide breeding bird distribution and abundance information desired for natural resource planning and biogeographic studies. Serious discussion of a Tennessee Breeding Bird Atlas Project began in 1984. In the absence of a state biological survey or other similar organization, as well as a fully funded state nongame wildlife program, we decided to seek project sponsorship from the Tennessee Ornithological Society (TOS) and any assistance available elsewhere. The TOS Board of Directors, at its May 1985 meeting, approved the project proposal and sponsorship. The Tennessee Valley Authority, Tennessee Department of Conservation, and Tennessee Wildlife Resources Agency were soon recruited as cooperating agencies. On Memorial Day weekend, 1985, the annual TOS foray was held in Cumberland County and served as a trial of atlas techniques. The following winter, an atlas handbook was printed (Nicholson and Hamel 1986) and fieldworkers were recruited. Programs on the Atlas project and training sessions were presented to TOS and Audubon chapters across the state. Fieldwork began in the spring of 1986.

Organization

The state was divided into nine Atlas regions, and coordinators (listed in table 1) were recruited to manage Atlas work in each region. Because of their ready availability, U.S. Geological Survey 7.5-minute topographic maps ("topos" or "quads") were chosen as the base for the Atlas grid. Topographic maps were given code numbers based on the system used in the U.S.G.S. 1984 "Index to topographic and other map coverage" for Tennessee. Each topographic map was divided into six blocks, 3.75 minutes wide by 2.5 minutes tall or approximately 5.7 by 4.6 km, with an area of about 2,600 ha. Blocks, the basic units for recording bird observations, were numbered as shown in fig. 1. Excluding blocks along the state borders with less than half their area in Tennessee, the state contained about 4,200 Atlas blocks.

Although complete coverage of all 4,200 blocks would yield the most valuable results, project planners thought this was impossible with the limited number of potential workers in the state. Therefore, a systematic sampling scheme was used, where the east-central, or #5 block, on each topographic map was selected for complete coverage. These blocks are referred to as the #5 priority blocks. Part of the reason for choosing the #5 blocks was that three of the four surrounding blocks are on the same map sheet.

Atlas results from other blocks were also accepted, and many non-priority blocks, especially blocks in which atlasers lived, received thorough coverage. Several non-priority blocks containing unusual habitats were also examined carefully. These included marshes, Mississippi River bluff forests, unmined peaks in the Cumberland Mountains, and much of the Great Smoky Mountains. Many atlasers also surveyed the non-priority blocks surrounding their priority blocks, a type of coverage most common in the eastern part of the state. Results from non-priority blocks were not used in most statistical analyses.

In retrospect, the decision to cover a sample of the blocks instead of complete coverage was fortuitous, as the number of dedicated participants was lower than anticipated. Fieldwork was originally planned to be completed in five years. After reviewing results of the first three years, Atlas directors extended the fieldwork period to a sixth year, 1991.

Field Methods

Breeding Codes and Recording Observations

A standardized set of breeding criteria was adopted, modified from those in Laughlin, Kibbe, and Eagles (1982) and those used by bird atlas projects in other states. A brief description of these codes, listed in order of increasing certainty of breeding, follows:

Possible Breeding
  O—Species observed in block during breeding season but not in breeding habitat, and with no evidence of breeding. This code was primarily for species such as vultures or hawks soaring overhead and colonial nesting species away from the colony.
  H—Species observed in breeding habitat during the breeding season.

Probable Breeding
  F—Five or more singing males observed on single visit to block, widely distributed over block during breeding season.
  A—Agitated behavior or anxiety calls from adults.
  P—Pair observed in suitable breeding habitat during breeding season.
  T—Territorial behavior presumed by presence of singing male at same location on at least two occasions at least a week apart, or by chasing others of same species.

Table 1
Atlas Project Regional Coordinators

Region	Coordinator
1	Robert P. Ford
2	George R. Payne Jr.
3	Stephen J. Stedman (1986–88)
	Charles P. Nicholson (1988–91)
4	Morris D. Williams (1986–87)
	Charles P. Nicholson (1987–88)
	Damien J. Simbeck (1988–91)
5	David F. Vogt
6	Charles P. Nicholson
7	Linda Cartwright (1986–87)
	Robert L. Crawford (1987–91)
	Lillian H. & Kenneth H. Dubke (1990–91)
8	Peter I. Kalla (1986–87)
	Audrey R. & Ronald D. Hoff (1987–91)
9	Richard L. Knight, Fred J. Alsop III

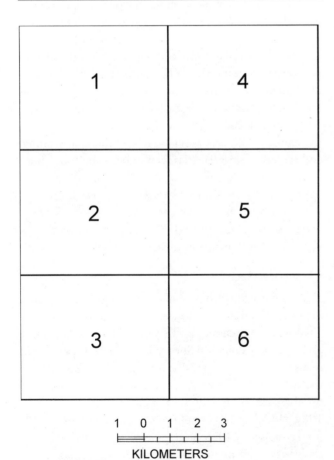

Fig. 1. Diagram of a 7.5-minute topographic map showing its division into 6 Atlas blocks.

C—Courtship behavior or copulation observed.
N—Visiting probable nest site. This code was primarily for hole-nesters.
B—Nest building by wrens, which may build dummy nests, or by woodpeckers, which also excavate roost holes.

Confirmed Breeding

CN—Carrying nest material.
NB—Nest building, where the actual unfinished nest is seen.
DD—Distraction display or injury feigning.
UN—Used nest or eggshell found and carefully identified.
FY—Adult carrying food for young.
FS—Adult carrying fecal sac.
FL—Recently fledged or downy young, still dependent on parents.
ON—Occupied nest observed. Usually applied to nests too high or enclosed, such as cavity nests, to see contents.
NE—Nest with eggs.
NY—Nest with young seen or heard.

During the first two years of the project, the codes CN and NB were combined into an NB code, and an additional confirmed code, TE, for the presence of 10 or more singing males observed on a single visit to the block, widely distributed over a block during the breeding season, was also used. After the 1987 field season, the Atlas directors dropped the TE code, and all TE records were changed to F unless there was evidence for a higher code. The NB code was split into CN and NB as described above, and all previous NB records were changed to CN unless retention as NB was warranted. The code changes, and reasoning behind them, were explained in project newsletters, and a redesigned field card (fig. 2) listed definitions of the new codes. At the same time, atlasers were requested to report the dates of all confirmed records.

When they first enrolled in the project, atlasers were given a copy of the *Tennessee Breeding Bird Atlas Handbook* (Nicholson and Hamel 1986), which contained explanations of project objectives, field methods, record keeping, detailed explanations of breeding codes with

SPECIES	Pb	Pr	Pv	PosDate
Cerulean Warbler		T	FL	6/18
Black-and-white Warbler		P	FL	6/18
American Redstart		F	FL	6/18
Prothonotary Warbler				
Worm-eating Warbler		F/A	FL	6/18
* Swainson's Warbler	H			
Ovenbird		C	FL	6/18
Louisiana Waterthrush	H		FL	6/18
Kentucky Warbler		P/C	FL	6/18
Common Yellowthroat		N/P/T		
Hooded Warbler			NE	5/13
Canada Warbler				
Yellow-breasted Chat	H		FL	6/18
Summer Tanager				
Scarlet Tanager		P	FY	6/18
Northern Cardinal			NB	5/13
Rose-breasted Grosbeak	H			
Blue Grosbeak none on 6/18				
Indigo Bunting			NE	5/13
** ~~Painted Bunting~~			NY	- 6/13
Dickcissel				
Rufous-sided Towhee			NB	5/13
** Bachman's Sparrow				
Chipping Sparrow	H			
Field Sparrow	H		NE	6/18
** Vesper Sparrow				
** Lark Sparrow				
* Grasshopper Sparrow				
Song Sparrow	H	T		
D.-e. (Sl.-col.) Junco				
Red-winged Blackbird				
Eastern Meadowlark				
** Western Meadowlark				
Common Grackle	O			
Brown-headed Cowbird			NE	5/13 — in Hooded W. nest
Orchard Oriole				
* N. (Baltimore) Oriole				
House Finch				
* Red Crossbill				
American Goldfinch H		C		
House Sparrow				

WR-0537

ADDITIONAL SPECIES	Pb	Pr	Pv	PosDate

COVERAGE

DATE	LOCATION	START	END	HRS
3/29/89	Gobey Rd.	1330	1430	1.0
47/8 5/13/89	to mine on Fork Mtn.; some on low road	0713	1500	7 3/4
6/18/89	mini-rt. along road + hike to mine on top	0545	1550	9.9
58/29	owling			17.6

NOTES

* Verification Form required for all observations in Positive category.
** Verification Form required for all observations within Safe Dates.

TENNESSEE BREEDING BIRD ATLAS

FIELD CARD

Quad Name: Gobey Quad Number: 368425
Block: 5 Region: 6 Year: 1989
Observer: C. P. Nicholson
Address:
Phone:

BREEDING CRITERIA & CODES

POSSIBLE - Pb column
O Species observed, not in breeding habitat
H Species heard or seen in breeding habitat

PROBABLE - Pr column
F Five or more singing males
A Agitated behavior or anxiety calls
P Pair seen
T Bird holding territory
C Courtship or copulation
N Visiting probable nest site
B Nest building-wrens or woodpeckers

POSITIVE - Pv column - record date in PosDate column
CN Carrying nest material
NB Nest building-unfinished nest seen
DD Distraction display
UN Used nest
FL Recently fledged young
FS Parent with fecal sac
FY Parent with food for young
ON Adult leaving/entering/on nest
NE Nest with eggs
NY Nest with young

Tennessee Wildlife Resources Agency

Fig. 2. An Atlas field card.

examples, and a list of references covering nest and egg identification, breeding biology, and local Tennessee avifaunas. The handbook also contained a list of species known to breed in the state, their abundance in different parts of the state, and their safe dates. Safe dates were the range of dates encompassing the species's breeding season, excluding unusually early and unusually late breeding and used with the codes O, H, F, and P. For migratory species, they started after most spring migrants had passed through and before fall migrants appeared. The fact that many migratory species begin nesting well before all members of the species have passed through was frequently emphasized.

Atlas workers recorded data for each block on a field card (fig. 2). If a miniroute (see below) was observed in the block, a summary sheet combining field card and miniroute results was also completed. An incidental observation card was used for blocks with only a few records. The names and addresses of the atlasers, dates, and number of days and hours spent working the block were also recorded on these forms. Written details on special verification forms were required for all observations of many rare species, and for reports in the confirmed category for some species whose range or breeding biology were not well known in Tennessee. Verification forms were also requested for some observations of species outside their normal breeding range and of unusually early or late breeding records.

Block Coverage

The basic objective of block coverage was for the Atlas worker to record all of the species breeding in a block and assign them to the highest breeding code. This required visiting all habitat types present in the block. In recognition of the fact that recording all species present in a block was unlikely, standards for adequate coverage were adopted.

Previous TOS county forays had shown that about 100 potentially breeding species could be found in most counties. The foray results also suggested that 60 to 80 species could be recorded in most blocks. Minimal adequate coverage standards were initially set at recording 60 species, with a quarter confirmed and half in the probable category. Following the change in breeding codes adopted in 1987, the target proportion of confirmed records was raised to 40%. Some atlasers had difficulty achieving this. They were encouraged to move on to another block after working 16–20 hours in the block and recording a reasonably complete species list. Species lists of "completed" blocks were reviewed by project directors, and lists of missed species likely present in the block were compiled. Many of these incomplete blocks were later revisited and completed.

Atlas workers were encouraged to make several visits to the block throughout the breeding season, both because of the different timing of breeding of various species and to increase records in the T category. They were also encouraged to visit the block at night to record owls and goatsuckers.

Abundance Estimates

To provide a measure of relative abundance across the state, birds were censused along a miniroute (Bystrak 1980) within each #5 priority block. The miniroute, an abbreviated version of a USFWS/National Biological Service Breeding Bird Survey (BBS) route (Robbins, Bystrak, and Geissler 1986), consisted of censusing birds at 15 points or stops spaced 800 m (0.5 miles) apart. At each stop, the observer recorded all species heard or seen within a 400 m (0.25 mile) radius during a 3-minute period.

In most blocks, miniroutes were censused along roads. Observers were instructed to begin the miniroute in the southeast corner of the block, to go toward the northwest corner, and then turn either south or east. These directions could be changed where lack of roads or heavy traffic made them unfeasible. In blocks with few roads, miniroutes were censused, at least in part, by hiking trails or, less commonly, from boats. In these cases, census points were also spaced at approximately 800 m intervals. Locations of miniroute stops were recorded on maps for future reference.

Miniroutes were censused between the last week of May and early July, with most done before the last week of June. To minimize the variability among stops caused by the predawn concentration of song, routes were begun at sunrise; most were completed in less than two hours.

The measure of abundance was the number of stops, out of 15, at which a species was recorded. The stops per route measure was chosen over the number of individuals of a species to simplify the censusing for the observers, many of whom were inexperienced with BBS routes. Recording the presence/absence of a species was easier than counting individuals, and inexperienced observers may thus record more different species. In addition, the variability between observers is probably less (Bart and Schoultz 1984, James, McCulloch, and Wolfe 1990). Both Bart and Klosiewski (1989) and James, McCulloch, and Wolfe (1990) showed a high positive correlation between use of the two abundance measures in detecting population trends. A disadvantage is that presence/absence is less effective in detecting the magnitude of population changes, especially in locally abundant species (Bart and Klosiewski 1989).

Many of the problems in using BBS routes for analyzing regional or temporal differences in species abundance (Droege 1990) also apply to miniroutes. They include undersampling of many species groups such as rails, owls, and goatsuckers, changes in a species's detectability during the census period, variability among observers, and the possibility of not sampling habitats in proportion to their area of a block. A relatively small number of skilled, experienced observers censused most of the miniroutes. Observer variability was also minimized by closely inspecting count results and having many questionable counts redone by more experienced observers.

Intensive Block Coverage Techniques

Forays and paid Atlas workers were used to ensure coverage of areas of the state with few resident observers. The forays were similar to those held prior to the Atlas project, except that priority blocks were assigned to participants. In 1986, the foray was held in Overton County on Memorial Day weekend. Beginning in 1987, 2 forays were annually held, on Memorial Day weekend and the third or fourth weekend in June. Locations were Paris and Pikeville in 1987, Rogersville and Selmer in 1988, Fayetteville and Jackson in 1989, Martin and Pulaski in 1990, and Columbia and Bolivar in 1991. During each foray, 5 to 10 parties of atlasers each worked 1 or 2 blocks. Most forays were attended by the same small group of atlasers, and attempts to increase participation by offering to pay travel expenses were unsuccessful.

Paid atlasers were hired during the last three years of the project to work blocks in remote areas and complete partially worked blocks. Most of the blocks worked by these "blockbusters" were in the western half of the state.

Blocks covered during the forays usually received one or two consecutive days of work. This was also true for some covered by blockbusters and other atlasers. The number of species and the percentage of records in the confirmed category was comparable to those of blocks that received more visits. The percentage of records in the probable category, however, was lower, as few species could be placed in the T category during a single visit. Many migratory species that began nesting in May were more difficult to confirm during the Memorial Day forays than later in the season when they were feeding young. Cedar Waxwings (scientific names of species are listed in Appendix 1) also posed a problem during the Memorial Day forays, as this was before most had begun nesting.

Other Sources of Atlas Data

In addition to the systematic and incidental block coverage by volunteer and paid atlasers, site-specific records were extracted from several other sources and incorporated into the Atlas files. These sources included USFWS Breeding Bird Survey routes, Breeding Bird Censuses published in the *Journal of Field Ornithology*, federal and state wildlife management studies, and special surveys of nesting Least Terns (courtesy of John Rumanicik, U.S. Army Corps of Engineers) and colonial herons (courtesy of Burline Pulline, TVA, and Gerald Smith, under contract with TWRA).

Data Management and Mapping

Regional coordinators collected results from atlasers, and, after editing the results, they typed the results into a microcomputer data management program. They then sent the results on floppy disk to the project chairman, who, after further editing, merged them into the central dBase-compatible databases. The two main databases were a species file, containing a record for each species observed in each block, with the species's breeding status, the year of observation, miniroute results, and date of the confirmed record, and a block coverage file, with information on the year, name of the atlaser, and number of hours and days spent in each block. Several ancillary files were also maintained. Following editing of the merged files, block reports were printed and given to the regional coordinators and atlasers to check for errors. Results for every block were again reviewed against the original data cards and summary sheets following the last year of fieldwork.

A committee reviewed the verification forms. Most were accepted as originally received. Some were accepted with a change in the breeding code, and a few were rejected because of unconvincing details. A few very unusual and questionable observations, for which the project directors were unable to get completed verification forms, were deleted from the results files.

Interim working maps of the distribution and breeding status of various species were made using Atlas*Graphics (Strategic Locations Planning, Inc.) software. These maps were useful in showing the progress of block coverage and in flagging occurrences outside of species' known ranges. The final breeding status maps were produced using Arc/Info (ESRI Inc.) software. Relative abundance maps were produced from miniroute results using Surfer and Surfer for Windows (Golden Software, Inc.) software.

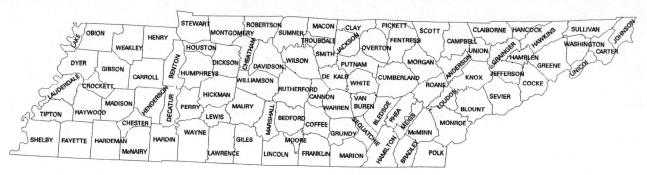

Map 1. Tennessee counties. The county boundaries are used as the background for most of the other maps in this book.

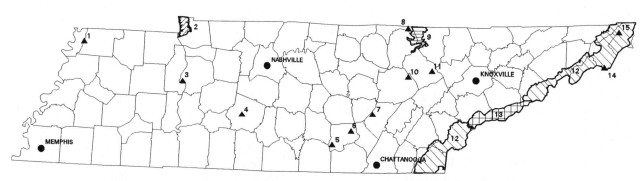

Map 2. The largest Tennessee cities, major federally owned lands, and other locations frequently mentioned in the text.
  Legend:
   1. Reelfoot Lake
   2. Land Between the Lakes
   3. Tennessee National Wildlife Refuge, Duck River Unit
   4. Monsanto Ponds
   5. Goose Pond
   6. Savage Gulf State Natural Area
   7. Fall Creek Falls State Park
   8. Pickett State Park
   9. Big South Fork National River and Recreation Area
   10. Catoosa Wildlife Management Area
   11. Frozen Head State Natural Area
   12. Cherokee National Forest
   13. Great Smoky Mountains National Park
   14. Roan Mountain
   15. Shady Valley

Landscape and Ornithology of Tennessee

The History of Tennessee Ornithology

Compared to most East Coast and Midwest states, Tennessee has a relatively short history of ornithological study. The first taxonomic and distributional studies began in the late nineteenth century, by which time there were already several publications on the flora, fishes, and mollusks of Tennessee. Until recently, little state government support for ornithology existed, and there is no comprehensive research collection of Tennessee birds housed within the state.

In spite of limited institutional support, the birds of Tennessee are now fairly well known, due to the efforts of a few resident and nonresident professional ornithologists and many dedicated amateurs. The development of this ornithology is described in this chapter. Shoup (1943) includes a few details on ornithology in his short history of zoology in Tennessee. Much of the history of the Tennessee Ornithological Society has been described by Davant (1965), Ganier (1916a, 1965), Ganier et al. (1935), and others.

The history of bird study in Tennessee falls roughly into three sections. The first section, the pioneer period, extends from the time of the first explorers to mention bird observations through 1870. The second period, characterized by the beginnings of systematic exploration, includes the first, mostly nonresident, collectors. Lastly, the period of the founding of the Tennessee Ornithological Society and modern bird study begins in 1915. Palmer's collection of biographies (1954) is a useful source of information on nonresident ornithologists. Information on resident amateurs and professionals was gleaned from my personal acquaintances with them, from their writings, and from discussions with others.

Pioneer Period: 1673–1870

While descending the Mississippi River with Father Marquette in the summer of 1673, Louis Joliet noted Carolina Parakeets along the river south of its junction with the Ohio River (Joliet 1673–77). In a review of Tennessee parakeet records, McKinley (1979) attributed this record to the portion of the river along the Tennessee border. If it had been made from the east side of the river, Joliet's observation of parakeets would be the first written bird record for Tennessee. Joliet's description of the location where he shot the parakeet, however, is vague, and the place could have been farther north along the Kentucky border. The next early record, 26 years later, was less ambiguously located along the Tennessee border. In December 1699, Father J. F. Buisson St. Cosmé, a French missionary, traveled down the Mississippi River. A few days' distance below the junction with the Ohio, and above the junction with the Wolf River, he noted seeing White Pelicans, which he described as being "as large as a swan, which has the bill about a foot long, and the throat of extraordinary size" (St. Cosmé 1699–1700:66).

During most of the next century, several explorers and pioneers briefly mentioned birds in their journals (many of these reports were reprinted in Williams 1928).

These were mostly ducks, geese, swans, and turkeys—edible and conspicuous species. These accounts are of greater historical than ornithological value.

Somewhat more detailed notes were left by the botanist André Michaux, who traveled through Tennessee four times between 1793 and 1796. On his second trip in 1795, while in Nashville, he recorded in his journal: "Sunday 21st of June 1795, killed and skinned some birds. Birds: Robin, Cardinal, *Tetrao* (grouse) *Lanius Tyrannus* rare, Quantities of the Genus *Muscicapa*; few species of the Genus *Picus*: Wild Turkeys" (Williams 1928, Thwaites 1904–7). After visiting Kentucky, Michaux returned to Tennessee in late December 1795, traveling up the Cumberland River by boat to Clarksville, then by horse to Nashville. While at Clarksville, he paused to pack his collections, which undoubtedly contained more Tennessee specimens than those mentioned above. Unfortunately, while returning to France, Michaux lost some of his bird collections and memoirs in a shipwreck off the coast of Holland (Savage and Savage 1986).

The first ornithologist to study birds in Tennessee was Alexander Wilson (1766–1813), who passed through the state in 1810 (Nicholson 1986). Wilson was traveling alone from Philadelphia to Pittsburgh to New Orleans to collect and illustrate birds and sell subscriptions to his *American Ornithology*, then in production. He was in Tennessee from about 22 April to 7 May 1810. Wilson was an astute observer, and his letters from Nashville and Natchez (reprinted in Hunter 1983) give interesting glimpses of pioneer Tennessee. Exactly how many bird species Wilson recorded in Tennessee is unclear; at least 18 species are attributed to the state in his letters and the species accounts in *American Ornithology* (Wilson 1811a, 1811b, 1812). Notable among his observations are the Nashville and Tennessee Warblers, which Wilson first described from specimens collected near Nashville, Greater Prairie Chicken, American Swallow-tailed Kite, Carolina Parakeet, and Common Raven.

John James Audubon (1785–1851) traveled overland across Tennessee at least once and sailed the Mississippi River several times between 1811 and 1830. There are, however, few specific references to Tennessee in his *Ornithological Biography*, and his only surviving journal detailing travel through Tennessee describes his 1820–21 flatboat trip from Cincinnati to New Orleans (Audubon 1929). These sources provide a few records of interest, including the Ivory-billed Woodpecker, Carolina Parakeet, and Bank Swallow. The parakeet and swallow observations are discussed below in the species accounts. Because there are no available records of the occurrence of the Ivory-billed Woodpecker in Tennessee, aside from its presence in archaeological sites, it is not discussed here. Deaderick (1940) lists the birds observed by Audubon on his 1820–21 trip.

Between the time of Audubon and the Civil War, there were few publications mentioning Tennessee birds. Gerald Troost served part of this period as natural history professor at the University of Nashville and as state geologist (Shoup 1943). Troost collected botanical and zoological specimens, including birds. He sent these to Europe before his death, and no other information on them is available. Both Troost and his successor as state geologist, James M. Safford, made brief, general mention of birds in various state geological survey reports. For the most part, these reports, as well as other mid-nineteenth-century scientific publications originating in Tennessee, contained little of ornithological value (Corgan 1977).

Beginnings of Systematic Exploration: 1870–1915

By the late nineteenth century, there were several publications on Tennessee fish and mollusks (cited in Shoup 1943), as well as a state flora (Gattinger 1887). Between 1879 and 1915, the scarcity of ornithological study ended as several people, including nationally known ornithologists, published information on Tennessee birds. At the start of this period, the descriptive phase of ornithology was coming to a close, and more attention was given to distributional and ecological studies. Several ornithological journals were also beginning publication, providing outlets for such works.

An 1870 article by Edward Drinker Cope (1840–1897) on the fauna of the "Southern Alleghanies" piqued zoological interest in the southern Appalachians (interestingly, many decades after botanists had discovered the region). Cope, best know for his work in paleontology and herpetology, listed birds observed during August 1867 in southwest Virginia and during September 1869 in western North Carolina. Although until this time the presence of typically northern birds nesting in the southern Appalachians was not known, Cope concluded that many of these species did breed there. In the same article, Cope mentions collecting salamanders in East Tennessee, but no account is available on birds he observed there.

In 1879, George Henry Ragsdale, of Gainesville, Texas, published a secondhand account of the Dark-eyed Junco nesting in East Tennessee. In the next few years, G. S. Smith of Boston published two notes on Red Crossbills he collected near the recently founded community of Rugby in August 1880 (Smith 1881, 1883). William Henry Fox (1857–1921) published a short note in 1882 on species observed during his

March and April 1882 visit to Lookout Mountain, giving details on six species. Four years later, Fox published a detailed list of 114 species observed during the spring of 1884 and 1885 in Roane County (Fox 1886). At the time of his Tennessee ornithological work, Fox was a student, first in Washington and then in New York City. He later practiced medicine in Washington, D.C., and was buried in Nashville. Many of the specimens he collected at Lookout Mountain and in Roane County are still in the U.S. National Museum of Natural History. Robert Ridgway, in a footnote to Fox's 1886 article, noted it was the first pertaining to the birds of Tennessee of which he had knowledge.

At about the same time, William Brewster (1851–1919), curator of ornithology at the Museum of Comparative Zoology, Harvard University, spent 12 days in May and June 1885 collecting birds in the mountains of western North Carolina (Simpson 1980). The publication describing his findings (Brewster 1886), the first detailed account of the Southern Appalachian avifauna, further stimulated interest in the area.

During the following year, Arthur LeMoyne published results of his studies in the Tennessee mountains (LeMoyne 1886). Citing his intention to visit East Tennessee and western North Carolina, and noting that Brewster (1886) had just published on western North Carolina birds, LeMoyne visited Greene, Cocke, Sevier, Blount, Roane, and Monroe Counties between March and early June 1886. He also refers to visits he made in 1881 and 1883. His findings, which he presented "with considerable diffidence" (LeMoyne 1886:116), must be interpreted with caution. LeMoyne correctly places Walden Ridge in Roane County (in part), but gives observations made at about 1310 m elevation there; the highest elevation in Roane County is less than 610 m. LeMoyne also incorrectly describes the county as bordering North Carolina. A few of his bird observations and his secondhand report of a Golden Eagle nest lack credibility. Moreover, he mentions leaving the state on 1 June but cites collections he made up to 6 June 1886. Stupka (1963) and Ganier (unpublished manuscript in Tennessee Ornithological Society Papers, Special Collections, University of Tennessee Library) also question the accuracy of LeMoyne's work. Little biographical information is available about LeMoyne. In his article, where the byline lists him as an M.D., he mentions growing up in East Tennessee, and, on completion of his visit, returning to Massachusetts. His species accounts suggest he was an experienced egg collector, and at least one egg set he collected in Tennessee is in the collection of the Western Foundation of Vertebrate Zoology.

The account by George Burritt Sennett (1840–1900), an Ohio businessman, of his observations during April and late summer 1886 (Sennett 1887) is more credible. Although dealing primarily with North Carolina, Sennett's numerous observations along the crest of Roan Mountain apply equally to Tennessee. This was the first report on the birdlife of Roan Mountain, an area which has since received much attention from ornithologists.

While Sennett was in the Roan Mountain area, Frank Warren Langdon (1852–1933), a physician and a leader of the Cincinnati Society of Natural History, led a party to Blount County. Their fieldwork during this August visit was primarily on Chilhowee Mountain and on Defeat Mountain in the Great Smokies. Langdon's list of 63 species (Langdon 1887) also included a few observations from Maryville and Knoxville.

Leon Otley Pindar (ca. 1870–1936), while in his teens, studied birds in Fulton County, Kentucky, and his publications (Pindar 1886, 1889, 1925) included observations from adjacent Lake and Obion Counties, Tennessee. With the exception of Pindar's limited work in the Reelfoot Lake area, however, ornithological exploration in West and Middle Tennessee lagged behind that in East Tennessee.

In 1895, Samuel Nicholson Rhoads (1862–1952) published the first comprehensive list of Tennessee birds (Rhoads 1895a), incorporating the studies mentioned above with the results of his collecting trip across the state. Rhoads, a well-rounded naturalist affiliated with the Academy of Natural Sciences in Philadelphia who collected throughout the Americas (Schorger 1953), was best known for his contributions to mammalogy. During his Tennessee trip, which began on 30 April 1895, he visited the Reelfoot Lake area, Raleigh in Shelby County, Bellevue in Davidson County, Chattanooga and Sawyer's Springs in Hamilton County, Harriman in Roane County, Allardt in Fentress County, Knoxville, Johnson City, and Roan Mountain, where his trip ended on 23 June (Rhoads 1895b). In addition to birds, he collected mammals, reptiles, amphibians, and mollusks, and published annotated lists of these groups (Rhoads 1895b, 1896a, Pilsbry and Rhoads 1896). In 1896, Rhoads published an addendum to his Tennessee bird list (Rhoads 1896b).

Shortly after its publication, Allen severely criticized Rhoads's bird list as a "bad example in the matter of a local list" (Allen 1896:245). This criticism was based on the inclusion of several species "of doubtful record or identity, although they all belong to the Tennessee fauna" (Rhoads 1895a:464). Including species thought to occur in the area, but for which no actual records existed, was common practice at the time. These questionable species were usually clearly noted as such. Rhoads, however, sometimes failed to do so, and his accounts of some species, such as the Trumpeter Swan, are ambiguous. Despite these shortcomings, Rhoads's list remains of great value to Tennessee ornithology, providing the only nineteenth-century information for many species and many parts of the state.

Bradford Torrey (1843–1912), a naturalist and writer from Boston, Massachusetts, visited Chattanooga in April and May 1894. He described his visit in *Spring Notes from Tennessee* (1896a). Torrey mentions birds throughout *Spring Notes* and concludes the book with an annotated list of the 93 species he observed in the Chattanooga area. Torrey also published a separate bird list, "Some Tennessee bird notes," in the *Atlantic Monthly* (Torrey 1896b), which adds little factual information to that in *Spring Notes*.

In 1907 two additional works appeared. James H. Fleming, of Toronto, Ontario, later to become president of the American Ornithologists' Union, published a list of birds observed in March and April 1907 near Surgoinsville, Hawkins County (Fleming 1907). A few of Fleming's records were based on specimens. Andrew Allison, a physician from Ellisville, Mississippi, published the results of his fieldwork during the spring of 1904 in Tishomingo County, Mississippi (Allison 1907). This work included several records from Grand Junction, Tennessee.

The following year, Arthur H. Howell (1872–1940), an ornithologist and mammalogist with the U.S. Bureau of Biological Survey—later to become the U.S. Fish and Wildlife Service (USFWS)—collected birds in Kentucky and Tennessee. After visiting Big Black Mountain and Barbourville, Kentucky, he traveled to Campbell and northern Hamilton Counties and Lawrenceburg, Tennessee, in August and September 1908. His account (Howell 1910) includes several important records, and many of his Tennessee specimens are in the U.S. National Museum of Natural History.

Of all of those listed above who studied birds in Tennessee for whom biographical information is available, none was apparently a Tennessee resident. The contributions by Tennesseans prior to 1915 were comparatively limited. A few oologists, some apparently in their teens, published short accounts of their egg-collecting experiences in the *Oologist* (Lyon 1893, McEven 1894, Wake 1893, 1897). McEven's collecting notes from Bell Buckle are the most informative of this series. Nothing is known about the egg collectors J. T. Overstreet and James Jackson, whose Common Merganser egg sets provide the only evidence of this species nesting in Tennessee (Kiff 1989). During this period, the first book on Tennessee's birds by a Tennessean, *Some Birds and Their Ways,* was written for schoolchildren by Alonzo C. Webb (1900). Webb later became a founder of the Tennessee Ornithological Society.

William R. Gettys (1876–1910), a resident of Athens, Tennessee, collected eggs there between 1897 and 1909. His notebook (transcribed copy in Tennessee Ornithological Society Papers, Special Collections, University of Tennessee Library, see also Ijams and Hofferbert 1934) contained records for approximately 585 nests of 70 species. Most of these species were represented in his egg collection. His records of nests of the Red-cockaded Woodpecker and Swainson's Warbler are discussed in the species accounts below. Gettys also contributed notes on migration dates to the U.S. Biological Survey series compiled by Cooke and Oberholser in *Bird-Lore*; these were later summarized by Ganier (1935a). Following his death, Gettys' collection was given to Tennessee Wesleyan College and is presently at the Louisiana State University Museum of Zoology. Gettys apparently did not publish any ornithological articles.

The first Christmas Bird Count in Tennessee was conducted in Knoxville by Mrs. Magnolia Woodward in 1902 (Woodward 1903). Counts in Tazewell were begun by Herman Y. Hughes in 1909 and begun in Nashville by Albert F. Ganier in 1914 (Trabue 1965). Both Hughes and Ganier later became founders of the Tennessee Ornithological Society.

Modern Bird Study: 1915–Present

With the founding of the Tennessee Ornithological Society in 1915, the study of Tennessee birds, particularly by Tennesseans, accelerated. The founding and early growth of the TOS, the first state bird club in the Southeast, has been covered in detail elsewhere (Ganier et al. 1935, Ganier 1965, Davant 1965, Goodpasture 1977a); major events and contributions of Tennessee bird students since 1915 are discussed in this section. Aside from some of the TOS founders, the persons described below generally produced a body of published articles of at least statewide importance. A few persons who have risen to prominence in the last decade are left for future evaluation by others.

The TOS was founded at a dinner meeting on 7 October 1915 at Faucon's Restaurant in Nashville. Its purposes included promoting ornithology in Tennessee, publishing results of its investigation, and promoting bird protection. The founders were Dixon Merritt (1879–1972), then editor of the Nashville *Tennessean,* who called the meeting (and later became famous for composing a limerick about pelicans); Albert F. Ganier (1883–1973), chief draftsman and later civil engineer for the Nashville, Chattanooga and St. Louis Railway; Alonzo C. Webb (1859–1939), art director of the Nashville public schools and author and illustrator of student readers on birds; George R. Mayfield (1877–1964), professor of German at Vanderbilt University and later the first editor of the *Journal of the Tennessee Academy of Sciences* and chairman of the Tennessee Conservation Commission; Judge Herman Y. Hughes (1863–1921), of the State Court of Appeals; and Dr. George M. Curtis (1890–1965), of the Vanderbilt School of Medicine. Except for Hughes, who lived in Tazewell but was frequently in Nashville for court sessions, all of the founders resided in Nashville.

Of the six TOS founders, Ganier made the greatest scientific contributions and was to dominate Tennessee ornithology for almost half a century. He not only held several offices within the TOS but was also the leading contributor to the *Migrant* (see below). Shortly after the founding of the TOS, Ganier, who became the society's curator, began collections of the skins of Tennessee birds and the eggs of the breeding species. These important collections are presently at the Louisiana State University Museum of Zoology. He served as secretary of the Wilson Ornithological Society from 1918 to 1923 and as president from 1924 to 1926. In 1934 Ganier became an elective member of the American Ornithologists' Union. Having a strong interest in conservation, during the 1930s he assisted in selecting areas to be purchased by the federal government and developed into state parks. More of Ganier's contributions are described below.

The first task of the newly founded TOS, as later described by Ganier et al. (1935), was to assemble a comprehensive list of the birds of Tennessee. A hypothetical list of about 200 species was drawn up, based on some of the published studies mentioned above and the experiences of the members. The group began holding regular local field trips, as well as expeditions throughout the state. During a trip to Reelfoot Lake that November, the TOS members listed 59 species of birds and collected 35 birds, which were prepared as study skins. The published results of this trip (Ganier 1916b) are the first in a long series of articles, mostly written by Ganier, describing results of these expeditions. The favored destinations of these expeditions, which continued into the 1940s, were West Tennessee swamps, wild areas of the Cumberland Plateau, and, as noted below, the higher elevations of the Blue Ridge.

In 1917, Ganier published "A preliminary list of the birds of Tennessee," which listed 270 species and subspecies (251 species) and their status in West, Middle, and East Tennessee. Rhoads's important list (1895a) was, curiously, not cited in the 1917 list, and Stone (1918), in a review of Ganier's list, noted its absence. Ganier had, however, briefly mentioned Rhoads's list in a 1916 announcement of the founding of the TOS. A lengthier, somewhat more detailed "Distributional list of the birds of Tennessee," similar in format to the 1917 list, was published in 1933 (Ganier 1933a). As he noted in the introduction to the 1933 list and elsewhere, Ganier intended to write a comprehensive book on the birds of Tennessee, a goal he never achieved.

The early years of the TOS were a time of rapid growth. Membership in the Nashville area grew, and chapters were founded in Knoxville (originally as the East Tennessee Ornithological Society) in 1923 and Memphis in 1930 (Davant 1965). Additional chapters have since been formed, with the number of chapters in recent years averaging about thirteen. The TOS became affiliated with the Wilson Ornithological Society in 1923 and for many years maintained a close affiliation with the Tennessee Academy of Sciences. When the TOS was incorporated in 1938, the constitution was revised to better deal with the increasing number of chapters. Most results of the early TOS studies were published in the *Wilson Bulletin* and the *Journal of the Tennessee Academy of Sciences*. With the increasing volume of studies of Tennessee birds and the need for better communication between TOS members, the society began a quarterly journal, the *Migrant*, in 1930. The goal of the *Migrant*, as stated on its masthead, has been to "record and encourage the study of birds in Tennessee." The *Migrant* has had 10 different editors since its founding.

A frequent destination for the early expeditions led by Ganier was the Great Smoky Mountains. His 1920 trip provided the first ornithological work in the area since the nineteenth-century visits of LeMoyne and Langdon. Ganier led eight other trips to the Smokies through 1933 and a week-long foray with 17 participants in 1938 (Ganier 1962). Results of the first few trips were described in "Summer birds of the Great Smoky Mountains" (Ganier 1926). Following the 1926 authorization of the National Park, Ganier began preparing a monograph on the birds of the Smokies (Ganier 1931a), which was never completed. Because of his difficulty in obtaining permits to collect specimens in the park, Ganier's later high elevation work was outside its boundaries. Others conducting bird work in the park included Thomas D. Burleigh, who collected many specimens and did important taxonomic work in the early 1930s; E. V. Komarek, who led trips for the Chicago Academy of Sciences in 1931, 1932, and 1933; Raymond J. Fleetwood, a wildlife technician with the Civilian Conservation Corps during 1934 and 1935;

James T. Tanner, zoology professor at the University of Tennessee, who studied chickadees, juncos, and altitudinal distribution in the late 1940s and 1950s (Tanner 1952, 1958); and Ben J. Fawver, who censused birds in different forest types in 1947 and 1948 (Fawver 1950, Kendeigh and Fawver 1981). Arthur Stupka, who became park naturalist in 1935, later compiled his observations with other studies into *Notes on the Birds of Great Smoky Mountains National Park* (Stupka 1963), which remains the most authoritative work on the park's birdlife. More recent published studies of note in the Tennessee portion of the park include ecological studies by Rabenold (1978), the census and nest predation studies of Wilcove (1985, 1988), and the census study of Alsop and Laughlin (1991).

The most important individual to contribute to West Tennessee ornithology was Ben B. Coffey Jr. (1904–1993). Coffey, a native Nashvillian, was a fire protection engineer for the Tennessee Inspection Bureau who began bird study as a Boy Scout. He moved to the Memphis area in 1928 and was instrumental in founding the TOS chapter there. Coffey and his wife and frequent co-worker, Lula, conducted important swift and heron banding studies, goatsucker surveys, and distributional studies in West Tennessee, Arkansas, and Mississippi. The Coffeys also conducted numerous USFWS Breeding Bird Surveys and Christmas Bird Counts throughout the same region. Over a four-year period during the 1950s, Coffey published 35 issues of *Mid-South Bird Notes,* a mimeographed newsletter containing items such as migration reports and important locality records. The results of many of the Coffeys' distributional studies await synthesis and publication. Ferguson (1992) gives a recent summary, including a bibliography and interesting anecdotes, of Coffey's ornithological accomplishments. In 1991, Ben Coffey was honored by being elected a Fellow of the American Ornithologists' Union.

Although the contributions of most others from West Tennessee pale in comparison to the Coffeys', several deserve mentioning. Wendell L. Whittemore (1916–1986) published a significant study on Reelfoot Lake (Whittemore 1937). John B. Calhoun, of Nashville, participated in a wildlife survey of Hardeman and McNairy Counties and published his findings in an important paper on the summer birds of that area (Calhoun 1941). Calhoun also published several notes dealing with the Nashville area. Among the more recent valuable contributors are Martha G. Waldron, longtime compiler of the West Tennessee "Season" report for the *Migrant* and of a Shelby County annotated checklist (Waldron 1987), and T. David Pitts, zoology professor at the University of Tennessee at Martin. Pitts has conducted important studies on such topics as bluebirds (e.g., Pitts 1981) and Reelfoot Lake (Pitts 1985); he recently served as editor of the *Migrant.*

In Middle Tennessee several persons have made valuable contributions. Henry S. Vaughan (1870–1958), a general contractor, joined the TOS in 1916 and was active in the Nashville area. He began collecting bird eggs while in his teens and eventually amassed a collection of several hundred sets, including most North American species. Vaughan was instrumental in founding the Children's Museum (later to become the Cumberland Museum and Science Center) in Nashville; part of his egg collection, along with that of his brother William, is presently in the Cumberland Museum. Vaughan also contributed observations to A. C. Bent's life histories. William M. Walker (1900–1947), a chemist, joined TOS in 1922 and made significant contributions from both the Nashville and Knoxville areas. Alfred Clebsch, a German immigrant and tobacco warehouse manager, was a founder of the Clarksville TOS chapter, a TOS secretary-treasurer, and frequently accompanied Ganier on expeditions across the state. Clebsch co-authored the reports on some of these expeditions (e.g., Ganier and Clebsch 1938, Ganier and Clebsch 1946). Clebsch's son Ed went on several of these expeditions while in his teens; he was later a professor of botany at the University of Tennessee.

Jesse M. Shaver (1888–1961), professor of biology at George Peabody College for Teachers, made many important, frequently overlooked contributions. His students completed several M.A. theses in the 1920s and 1930s, studying the influence of environmental factors on bird song and the nesting of several species. Several of these studies were published in the *Auk,* the *Wilson Bulletin,* and the *Journal of the Tennessee Academy of Sciences,* which Shaver edited for 25 years. Moreover, his "Bibliography of Tennessee Ornithology" (Shaver 1932) is a useful introduction to early publications on the state's birds. Shaver was also interested in botany and published a book on the ferns of Tennessee.

Harry C. Monk (1902–1982), joined the TOS in 1919 and soon served terms as president and curator. He conducted long-term, local studies in the Nashville area and compiled a voluminous file of published Tennessee bird records. Among his most significant publications are a study of the water birds of Radnor Lake (Monk 1932), based on 337 visits over a seven-year period, a compilation of migration dates for the Nashville area (Monk 1929), the first such article for the state, and a study of Mourning Doves (Monk 1949).

Amelia R. Laskey (1885–1973), a Nashville homemaker with no formal training in science, may nonetheless have gained the widest national reputation of all Tennessee ornithologists. After moving to Tennessee in 1921, she began banding and patiently observing birds around her house and in the Warner Parks. She published important studies on the life history of several species in the *Auk,* the *Wilson Bulletin,* and *Bird-banding,* as well as numerous articles in the *Migrant.* Her studies of the Brown-headed Cowbird (Laskey 1950) and the Northern Mockingbird (e.g., Laskey 1962) are still frequently cited. She began the long-term studies of television tower casualties in Nashville and, at the time of her death, was summarizing the results of 37 years of Eastern Bluebird nesting data she had collected at Warner Parks. One of the first women to become an elected member of the American Ornithologists' Union, she was honored with the title of Fellow of the AOU in 1966.

Henry O. Todd (1910–1974), a professional photographer from Murfreesboro, helped found the TOS chapter in that city. He is best known for his egg collection, probably the largest in Tennessee. One of his specialties was collecting vulture eggs, and he traded them with many other egg collectors. Todd also published several short papers (e.g., Todd 1935, 1938) describing nests he had found.

Henry Elmer Parmer (1914–1985), manager of a furnace and roofing business, was influential in the Nashville area. He accumulated records of his own and others for the Nashville Records Committee, eventually compiling them into *Birds of the Nashville Area,* now in its fourth edition (Parmer 1985), and among the best regional checklists for the state. Parmer also compiled Middle Tennessee "Season" reports for the *Migrant* for many years.

Katherine A. Goodpasture conducted important banding studies in the Nashville area and headed systematic monitoring of television tower casualties for many years. She served almost 20 years as state coordinator of the USFWS Breeding Bird Survey, helping to ensure that essentially all of the state's 42 routes were censused each year, a record matched by few other states. Others in Middle Tennessee making recent contributions of note include Michael L. Bierly, formerly a TOS leader in the Nashville area, and author of *Bird Finding in Tennessee* (Bierly 1980), the first book of its kind in the state. Morris D. Williams not only has published numerous articles from across the state in the *Migrant,* but also compiled a valuable index to that journal (Williams 1977c). John C. Robinson, formerly assistant refuge manager at Cross Creeks National Wildlife Refuge, conducted much fieldwork during the 1980s in the Stewart County area and co-authored a valuable annotated list of the county birds (Robinson and Blunk 1989). This list, unfortunately, was published in somewhat obscure symposium proceedings. In 1990, Robinson published *An Annotated Checklist of the Birds of Tennessee* (Robinson 1990). This very useful book, which gives the status, distribution, and dates of occurrence for each species recorded in the state, is the first of its kind for Tennessee and a major contribution to state ornithology.

A number of East Tennesseans have also made significant contributions to ornithological study in Tennessee. Henry Pearle Ijams (1876–1954), a commercial artist, helped found the Knoxville chapter of TOS, compiled Knoxville area bird counts for many years, and published several Knoxville area bird observations in the *Migrant.* Ijams is probably best known for drawing the cover long used on the *Migrant* and for donating his home, once a frequent meeting place for the Knoxville TOS chapter, to the city to become the Ijams Nature Center.

Bruce P. Tyler (1874–1957) and Robert B. Lyle, both of Johnson City, were influential in the growth of ornithology in upper East Tennessee. Tyler, a land agent for the Clinchfield Coal Corporation, and Lyle frequently worked together and collaborated on important papers on the winter birds (Tyler and Lyle 1933) and nesting birds (Lyle and Tyler 1934) of northeastern Tennessee. Lyle also accumulated an egg collection of several hundred sets, with good series of several high elevation species. Unfortunately, because Lyle did not individually mark many of his sets, their value is diminished. This collection is presently at East Tennessee State University.

Fred William Behrend (1896–1976) of Elizabethton was a translator-stenographer and later a journalist. He was a founder of the Elizabethton TOS chapter, a pioneer in hawk watching, and well-known for his winter surveys of Roan Mountain. Lee R. Herndon (1897–1979), a chemist by profession, was also a resident of Elizabethton and a founder of the TOS chapter there. Herndon authored several notes on locality records in northeastern Tennessee, as well as a longer article on the birds of Carter County (Herndon 1950b). He served as editor of the *Migrant* from 1956 to 1971 and operated a banding station for many years.

The University of Tennessee at Knoxville has been the major institution for ornithological research in Tennessee. Both James T. Tanner (1914–1991) and Joseph C. Howell were zoology professors there who made distinguished contributions to Tennessee ornithology through their own research and through their graduate students. Some of Tanner's work in the Great Smoky Mountains has been mentioned above; he also had a strong interest in population ecology and wrote important papers on population trends of Tennessee birds based on analyses of Christmas and spring counts (Tanner 1985, 1986) and a summary of range changes (Tanner 1988). He edited the *Migrant* from 1947 to 1955. Howell conducted roadside bird censuses in Knox County for many years prior to the establishment of the USFWS Breeding Bird Survey (Howell 1951a) and, with the collaboration of Muriel B. Monroe, wrote an annotated list of Knox County birds (Howell and Monroe 1957). Later, with James M. Campbell, a dedicated amateur, he began a long-term study of the birds of Campbell County, most results of which remain unpublished. Fred J. Alsop III, a student of Tanner's and later on the East Tennessee State University zoology faculty, edited the "Season" reports in the *Migrant* for several years, directed the TOS forays in the early 1970s, has published several notes on Tennessee birds, and published a pictorial guide to the birds of the Great Smoky Mountains (Alsop 1991).

Several bird studies have originated in the Forestry, Wildlife, and Fisheries department at the University of Tennessee, primarily under the direction of Ralph W. Dimmick. While these have primarily been of game species (e.g., Dimmick 1974, Kalla and Dimmick 1987), more recent studies (e.g., Ford 1990, Hardy 1991) have focused on other birds.

Another East Tennessee institution where important bird research has been carried out, at least during the 1970s and early 1980s, is Oak Ridge National Laboratory. Roger Kroodsma published several articles on bird use of forest edges along power-line corridors (e.g., Kroodsma 1984). H. H. Shugart Jr. best known for his work in modeling forest growth, co-authored studies of bird habitat selection (Anderson and Shugart 1974) and size variation in Ovenbird territories in relation to habitat structure (Smith and Shugart 1987).

Charles P. Nicholson, a biologist with the Tennessee Valley Authority, has made many contributions, including status reports on the Bachman's Sparrow and Red-cockaded Woodpecker (Nicholson 1976, 1977), coordinating the TOS forays in the late 1970s and early 1980s (described below), and serving as editor of the *Migrant* between 1981 and 1988. He directed the Tennessee Breeding Bird Atlas Project from 1987 through its completion.

An important, but poorly known, statewide survey was conducted in 1937 by the U.S. National Museum of Natural History under the guidance of Alexander Wetmore, then secretary of the Smithsonian Institution. The Tennessee survey was one of five that collected birds and mammals in east-central states during the late 1930s. The collecting team traveled across Tennessee from west to east between 8 April and 16 July, and returned to work from east to west, then to south Middle Tennessee, between 10 September and 10 November. William M. Perrygo led the field parties, assisted by Carleton Lingebach in the spring and Henry R. Schaefer in the fall. Wetmore and Herbert Friedmann, also of the U.S. National Museum, each joined the expedition for short periods. The resulting report (Wetmore 1939), is the only published detailed taxonomic treatment of Tennessee birds, with subspecific determinations based on the several hundred specimens collected in 1937, as well as earlier specimens collected by Fox and A. H. Howell. Several noteworthy locality records are included, and Perrygo's party was the first to study the summer birds of Polk County, in the extreme southeast corner of the state.

The annual TOS expeditions led by Ganier ceased in the 1940s. They were restarted as annual forays in 1971, with birders gathering to work a relatively little studied county, usually over Memorial Day weekend in late May. The first of these renewed forays, coordinated by Fred J. Alsop III, covered Campbell (Alsop 1971b), Lawrence (Alsop and Williams 1974), and Benton Counties (Alsop 1976). Later counties included Grundy in 1974 (Dubke and Dubke 1977, Nicholson 1980a), Lauderdale in 1975 (Coffey 1976), Johnson in 1976 (results unpublished), and Wilson and Smith in 1977, also unpublished. In 1978, Charles P. Nicholson began directing the forays, and counties studied were Decatur in 1978, (Nicholson 1980b), Fentress and Pickett in 1979 (Nicholson 1981a, 1982a), Monroe in 1980 (unpublished), McNairy in 1981 (Nicholson 1984b), White in 1982 (unpublished, codirected by Michael L. Bierly), and Polk in 1983 (unpublished). Reelfoot Lake was worked in 1984 (Pitts 1985). The 1985 foray to Cumberland County was used as a test of bird atlas procedures (results unpublished), and 1986–1991 forays concentrated on covering atlas blocks.

The Environment of Tennessee

Tennessee has a very diverse landscape, among the most diverse of any inland state in Eastern North America. With an area of approximately 120,200 sq km, Tennessee is not particularly large. However, it spans from about 81°40′ to 90°31′ W longitude, across several major physiographic provinces, and over an elevational difference of almost 2000 m. An understanding of this physiography, the climate, and the resultant vegetation types is necessary for interpreting the distribution of Tennessee birds.

Physiography

Physiographic regions are areas of similar land surfaces resulting from similar geologic history. The boundaries of the physiographic regions adopted here (map 3) are based on geologic formations mapped by Miller, Hardeman, and Fullerton (1966). They are slightly modified from the system of Fenneman (1938), who classified Tennessee into five major physiographic provinces. The main difference lies in the subdividing of the Gulf Coastal Plain province, which is further described below. Two other provinces, the Interior Low Plateaus and the Appalachian Plateaus provinces are also subdivided into sections. The eastern three provinces—the Blue Ridge, Ridge and Valley, and Appalachian Plateaus—together make up the Appalachian Highlands in Tennessee.

Following are descriptions of the physiographic areas, their geology, and their soils. The soil descriptions are mostly from Springer and Elder (1980), who recognized nine major soil areas in Tennessee. With one exception, the major soil areas conform closely with the physiographic regions. The exception is the Major Stream Bottoms soil area, which contains all of the Mississippi Alluvial Plain, as well as regions along streams in the Coastal Plain and Interior Low Plateaus provinces. Brief soil descriptions are presented here; further details are available in Springer and Elder (1980), as well as in the various county soil surveys.

Blue Ridge Province

The Blue Ridge province extends from north Georgia to Pennsylvania; its narrow Tennessee portion, separate from the main spine of the Blue Ridge in North Carolina, is often called the Unaka Mountains. Elevations in this region range from about 305 to 2024 m. The Unaka Mountains include several ranges: from north to south, these include the Iron, Holston, Stone, Unaka, Bald, Great Smoky, and Unicoi Mountains. Distinct mountains of lower elevation than those near the North Carolina border occur along the western edge of the province. These are, from north to south, English, Chilhowee, Starr, and Bean Mountains. Streams in the region generally flow toward the northwest or west into the Ridge and Valley province.

The Unaka Mountains are composed of the oldest rocks in the state, dating from the Precambrian era 600 million to a billion years ago (Miller 1974). In the southern and central Unakas, these rocks are of sedimentary and metamorphic origin, primarily sandstone,

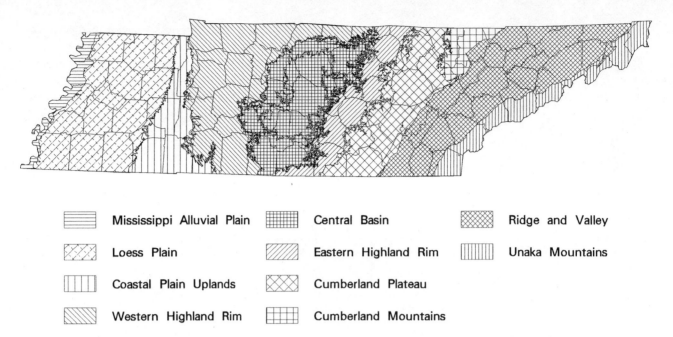

Map 3. *Physiographic regions of Tennessee. The boundary of the Inner Central Basin is shown within the Central Basin region.*

conglomerate, phyllite, and graywacke; granite and gneiss of igneous and metamorphic origin are more common in the north. Several mountains along the western edge are formed of quartzite, shale, and sandstone of slightly younger Cambrian age.

The Blue Ridge was formed in several cycles of mountain building resulting from the collision of continental and oceanic plates. The mountain building, known as the Appalachian orogeny, occurred as formerly flat rocks were folded, faulted, uplifted, and, along the western edge, thrust over younger rocks. It ended about 250 million years ago, at the end of the Paleozoic era. Once the area was above sea level, erosion began, reducing the heights of the mountains to their current elevations.

Springer and Elder (1980) grouped the province's soils into the Unaka Mountain soil area. They are high in weatherable minerals and usually shallow and loamy on upper slopes and deep and fertile on lower slopes. Soils in the valleys are deep, well-drained, loamy and fertile. In portions of the central Unakas, the overthrust sedimentary and metamorphic rocks have eroded down to the younger, underlying limestone. The best known of these "windows" is Cades Cove in the Great Smokies. Deep, well-drained loamy soils occur in these areas.

Ridge and Valley Province

Immediately west of the Blue Ridge is the Ridge and Valley province, which extends from the coastal plain of Alabama to the St. Lawrence Valley. Its Tennessee portion is often popularly known as the Great Valley. The province consists of parallel ridges and valleys, oriented from northeast to southwest. The ridges are higher at the northern end, where Clinch Mountain and Bays Mountain reach heights of 800 and 945 m, respectively. Valley floors slope gently toward the southwest, from an average elevation of about 300 m in the north to about 230 m in the south. Larger streams follow the parallel valley floors toward the southwest, forming a trellis drainage pattern.

The Ridge and Valley is underlain by dolomite, limestone, and shale formed from marine sediments during the Cambrian and Ordovician periods, about 600 to 430 million years ago (Miller 1974). More recent Pennsylvanian deposits, such as those that cap the Cumberland Plateau, have eroded away. Folding and faulting occurred during the Appalachian orogeny, and the present complex topography has resulted from weathering and erosion of the softer rocks.

The province's soils vary with the underlying rock formations and together make up the Ridge and Valley soil area (Springer and Elder 1980). Upland soils over shales and sandstones are generally strongly acid and of low fertility, while soils in the valleys tend to be deeper and more fertile.

Appalachian Plateaus Province

The Appalachian Plateaus province extends from Alabama into New York, and includes seven sections, two of which occur in Tennessee (Fenneman 1938). These sections, the Cumberland Plateau and Cumberland Mountains, are not differentiated by some authors (e.g.,

Miller 1974) because of their similar geology and surface rocks, and are often collectively referred to as the Cumberlands elsewhere in this book. They differ greatly, however, in their altitude, relief, and degree of dissection. Both are underlain by limestone deposited during the early Paleozoic era. During the Pennsylvanian period, about 300 million years ago, layers of sandstone and coal-bearing shales were deposited over the limestone (Miller 1974). Coal deposits occur throughout the Cumberlands, and large coal mines are most common in the Cumberland Mountains and in Bledsoe, Sequatchie, and Van Buren Counties.

Cumberland Plateau

The Cumberland Plateau is separated from the Ridge and Valley province on the east by a prominent escarpment up to 300 m in height, and on the west from the Highland Rim by a more irregular escarpment. The surface of the plateau, generally from 520 to 580 m elevation, is relatively flat, a result of the high resistance to erosion of the sandstone cap layers. Deep gorges, rimmed by sandstone bluffs, occur in several places where streams cut through the sandstone. A long archlike folding or anticline on the southern half of the plateau collapsed and eroded during the Mesozoic era to form the Sequatchie Valley.

Soils of the Cumberland Plateau and Cumberland Mountains make up the same soil area, and are formed either directly from the underlying sandstones and shales, or from materials weathered from them and moved downslope (Springer and Elder 1980). Most of the Plateau is covered by shallow, well-drained, acid soils low in natural fertility. Sandstone outcrops are common. Thicker, stony soils have accumulated on the lower slopes of the mountains and escarpments. Soils in the Sequatchie Valley are allied with those in the Ridge and Valley province.

Cumberland Mountains

The Cumberland Mountains are rugged, with elevations ranging from about 350 m along the lowest streams to 1075 m on Cross Mountain. The mountains were raised above the Cumberland Plateau by faulting during the Appalachian orogeny. The northeast corner of the region, from Pine Mountain to Cumberland Mountain, was pushed over a section of the plateau to form the Cumberland Thrust Block. Differential rates of erosion of the folded rock strata account for the high local relief.

Interior Low Plateau Province

This province extends from northern Alabama to Ohio, Indiana, and Illinois, encompassing most of Middle Tennessee. As its name implies, it is a low plateau, with a basal structure of limestone deposited during the Ordovician period, 430 to 500 million years ago (Miller 1974). Additional layers of limestone, chert, and shale were later deposited during the Silurian, Devonian, and Mississippian periods. Sandstones, similar to those on the Cumberland Plateau, were probably deposited across the area during the Mississippian period. As the Mesozoic era began, the center of the region was uplifted into a structure known as the Nashville Dome. The center of the dome fractured and eroded down to Ordovician limestones and outward, forming the oval Central Basin, surrounded by the less eroded Highland Rim. A distinct escarpment separates parts of the two regions. The southwest corner of the region identified by Fenneman (1938) as the Highland Rim is here, for geologic reasons, considered part of the Coastal Plain Province.

Highland Rim

The Highland Rim is often divided into the Eastern Highland Rim and the Western Highland Rim, separated by the Central Basin. The Eastern Highland Rim, with an average elevation of about 305 m, is narrower and generally flatter than the Western Highland Rim. Short Mountain, 632 m tall, is an erosional remnant, capped with Pennsylvanian sandstone, on the eastern rim. Several streams have cut narrow, steep valleys draining westward into the Central Basin. The topography of the Western Highland Rim is rolling to steep and dissected by several streams, most draining to the west. Elevations range from 110 m along the larger streams to 335 m in the southern portion of the region. Extensive, nearly flat areas occur in a karst plain at the northern edge of the Western Highland Rim, and in an area locally known as the Barrens in the western portion of the Eastern Highland Rim.

Soils of the Eastern and Western Highland Rims make up a single Highland Rim soil area, underlain by limestone or chert (Springer and Elder 1980). On hillsides soils have formed from limestone and have subsoils high in clay and chert. Soils on the flatter areas are formed from thin loess and limestone. Except for the extreme northern portion of the rim, most of the soils are highly leached, acid, and low in fertility.

Central Basin

The Central Basin, also known as the Nashville Basin, is an area of flat to hilly terrain and low gradient streams, surrounded by the Highland Rim. The basin is often divided into two sections, the Inner Basin and the Outer Basin (Miller 1974, Springer and Elder 1980). Compared to the Inner Basin, the Outer Basin is hillier and of higher elevation, averaging about 230 m. A few hills in the southern portion, geologically closely related to the Highland Rim, reach elevations of about 380 m. Most of the area is underlain by phosphatic limestone, and rock outcrops are common. The soils range from shallow to deep, are well drained, and, because of the high phosphate content, fertile (Springer and Elder 1980).

The relatively flat Inner Basin has an average elevation of about 180 m. Except where alluvium has been deposited near streams, the soils are shallow, and outcrops of the underlying limestone are common (Springer and Elder 1980). Karst features such as caves and sinkholes are also common, and surface drainage is limited in karst areas.

Coastal Plain Province

The Coastal Plain Province extends around the Atlantic and Gulf coasts and, in the "Mississippi embayment," up the Mississippi River to southern Illinois. Fenneman (1938) recognizes two distinct divisions in Tennessee: the Mississippi Alluvial Plain and the East Gulf Coastal Plain. The East Gulf Coastal Plain area, because of differing landforms and soils, is further divided into the Coastal Plain Uplands and the Loess Plain (the West Tennessee Plain of Miller (1974)) regions (map 3).

The Coastal Plain Province has been periodically inundated by the sea, most recently during the late Cretaceous and Tertiary periods. Both marine and nonmarine sediments, composed of sand with lesser amounts of clay and silt, were deposited in the embayment (Miller 1974).

Mississippi Alluvial Plain

This region, frequently referred to as the Mississippi River Valley and Mississippi River Floodplain, is a narrow strip between the western border of the state (parts of which are west of the current channel of the Mississippi River) and the Chickasaw Bluffs. Although underlain by alluvial sediments deposited over several million years, the near-surface and surface deposits are of recent origin and the region is the most geologically active area of the state. Its average elevation is about 77 m, and it slopes gently from the north to the south. Lateral migrations of the Mississippi River have left numerous oxbow lakes, meander scars, and natural levees, providing the only naturally occurring topographic relief. Much of the area periodically floods, although artificial levees have been constructed along Mississippi since the early 1800s. The northern end of the region is underlain by the deeply buried New Madrid Fault, whose movement in 1811 and 1812 formed Reelfoot Lake.

Springer and Elder (1980) combine this region with river bottoms farther east into a Major Streams Bottoms soil region. Two subunits of this soil region, one with loamy, silty and sandy soils and one with clayey soils, are mostly within the Mississippi Alluvial Plain. Most of the region is heavily farmed.

Loess Plain

The Loess Plain has a flat to rolling terrain with an average elevation of about 120 m. Its western edge is marked by the dissected Chickasaw Bluffs rising about 30 m above the Mississippi Alluvial Plain. Springer and Elder (1980) identified the Loess Plain as a distinct soil area, with soils formed from loess, a windblown rock powder deposited during Pleistocene glaciation. The loess varies in thickness from up to 27 m in the bluffs to about a meter at the eastern edge of the region. Loess-derived soils are silty and fairly fertile. Several low-gradient streams with wide floodplains flow from east to west across the region.

Coastal Plain Uplands

The Coastal Plain Uplands include the divide between the Mississippi and Tennessee River drainages, as well as an area east of the Tennessee River in Hardin and Wayne Counties. Compared to the rest of the province, it is hilly, with an average elevation of about 150 m, and some hills near the Tennessee River reach over 215 m. The area was covered by a thin layer of loess, much of which has since eroded away. A few rock outcrops of Silurian limestone occur near the Tennessee River. Springer and Elder (1980) included the region in their Coastal Plain soil area. The soils are mostly derived from marine sands and clays, and tend to be highly leached, strongly acid, low in fertility, and easily eroded.

Climate

Except for the high elevations of the eastern mountains, Tennessee has a humid, mesothermal climate as defined by Thornthwaite (1948). The four seasons are well-defined and of approximately equal length. The climate is strongly influenced by the flow into the Mississippi Valley region of the tropical airstream from the Gulf of Mexico, the Arctic airstream from Canada, and, to a lesser extent, airstreams from the west which have crossed the Rockies and Great Plains (Bryson and Hare 1974). The warmer, moister tropical airstream is prevalent in Tennessee during the spring and summer. During the late fall and winter, the southern anticyclonic airstream prevails, with winds radiating from high pressure cells over the southeastern United States. Much winter precipitation occurs along the frontal boundary between cold, dry Arctic airstreams and the warm, moist tropical airstream.

Because of Tennessee's relatively short north-south axis, latitudinal temperature differences are small. Topography, however, greatly affects local climates (map 4). In the Great Smokies, temperatures decrease an average of 1.2°C (2°F) with each 305 m increase in elevation (Shanks 1954). This effect results in the highest elevations of the eastern mountains having a climate similar to that 1600 km to the northeast.

Most of Tennessee receives about 114 to 140 cm (45 to 55 in) of annual precipitation (Dickson 1975) (map 5). As with temperature, precipitation differences are greater from east to west than from north to south. Precipitation increases with elevation, due to the cooling and condensation of moisture in ascending air masses. In the Great Smokies, precipitation increases by half between 460 and 1450 m elevation to a statewide maximum of about 230 cm/year (90 inches/year) (Shanks 1954). The lowest precipitation, in the northeast corner of the state, results from the rain shadow effect of the Great Smoky and Cumberland Mountains. Although occurring throughout the year, precipitation peaks in the winter and early spring. A secondary peak in midsummer is due to thunderstorm activity. Extended droughts are normally restricted to the summer and fall.

Past Climate Changes

At the peak of the most recent, Wisconsin glaciation, about 18,000 years ago, temperatures in Tennessee were several degrees cooler than at present during the summer, but similar during the winter (Bryson and Hare 1974); the global mean temperature was about 5°C cooler than at present. Following the glacial retreat about 10,000 years ago, the climate was relatively warm for about 5,000 years, and then somewhat cooler and wetter. As described in the next chapter, vegetation patterns shifted with these climate changes.

The period from the earliest European settlement to the late nineteenth century was unusually cool and often referred to as the Little Ice Age. From the 1890s to the present, the overall trend has been a temperature increase of about 1°C (Diaz and Quayle 1980). Much of this increase occurred from the 1890s through the 1950s; from the 1950s through the late 1970s, this warming trend was reversed and annual precipitation increased by about 5 cm. Since the late 1970s, average temperatures have increased and precipitation decreased.

Climate during the Atlas Period

The first three years of Atlas fieldwork was a period of extended drought, which peaked during the summer of 1988. During the late spring and early summer of 1986,

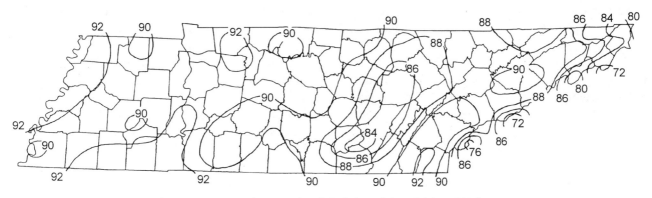

Map 4. Average maximum July temperatures in degrees Fahrenheit (adapted from Dickson 1975).

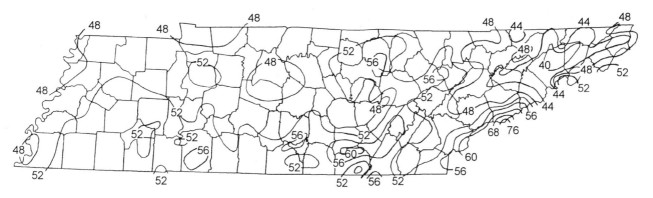

Map 5. Average annual precipitation in inches (adapted from Dickson 1975).

temperatures averaged from 0.6 to 1°C above normal in May to about 2.2°C above normal in July (National Oceanic and Atmospheric Administration weather data from monthly summaries in *Weatherwise*). May precipitation was near normal and June and July precipitation was 25–50% below normal. During 1987, average May temperatures were 2.2 to 3.3°C above normal while June and July temperatures 0 to 0.6°C above normal. May–July precipitation was about 75% of normal. In 1988, average monthly temperatures during May, June, and July were 0 to 1°C above normal; the June figures, however, are misleading. Record lows were set at the beginning of the month, and during the third and fourth weeks there were several days of record high temperatures. May precipitation was about half of normal and June and July precipitation about 70% of normal.

In contrast to the first three years, 1989 was wet and cool. Average temperatures were 1 to 2.2°C below normal in May, normal to 1°C below normal in June, and about normal in July. May precipitation ranged from about 75% of normal in the west to 125% of normal in the east. June precipitation was from 200 to 400% of normal, and July precipitation about 150% of normal. During 1990, average May temperatures across the state were 0 to 1°C below normal, June temperatures about 1°C above normal, and July temperatures near normal. May 1990 precipitation was near normal in the west and 150% of normal in the east; this pattern reversed in June, when the west had 150% of normal and the east 25–50% of normal. July 1990 precipitation was up to 150% of normal at each end of the state and about 75% of normal in Middle Tennessee. During 1991, May temperatures were about 3°C above normal, and June and July temperatures were both about 1°C above normal. May 1991 precipitation was near normal in most of East Tennessee and about 150% of normal in the rest of the state. Heavy rains in late May caused severe flooding in parts of Middle Tennessee. Precipitation in June 1991 varied from slightly above normal at each end of the state to about 25% of normal in Middle Tennessee. Precipitation in July 1991 was from 50 to 75% of normal across the state.

Future Climate Changes

Whether the climate changes in the last century have resulted in changes in the distribution of Tennessee birds is arguable, as there have been breeding range expansions from both the south, e.g., the Blue Grosbeak, and from the north, e.g., the Willow and Alder Flycatchers. The number of range expansions from the north is, in fact, probably greater than the number of expansions from the south as noted by Hall (1989).

Future climate changes may have a dramatic effect on the distribution of Tennessee birds. Climate models of the Intergovernmental Panel on Climate Change predict a 2°C increase in the average global surface temperature over the next century as a result of anthropogenic increases in atmospheric carbon dioxide and other greenhouse gases (Watson, Zinyowera, and Moss 1996) This is a greater rate of warming than any in the last 10,000 years. Such an increase would result in many changes in the distribution of plant and animal species in Tennessee and elsewhere (Peters and Lovejoy 1992, Watson, Zinyowera, and Moss 1996). Communities with primarily northern affinities, such as the spruce-fir and maple-beech-birch forests, would be eliminated from the state or further restricted to high elevations and new species assemblages would form from remnants of other communities.

Vegetation

Several broad-scale vegetation maps of Tennessee have been produced by plant ecologists and geographers in recent decades. A fine-scale vegetation map of the whole state does not yet exist, although one currently in production by the Tennessee Biodiversity Program (Reid 1992) should be available in 1997. The broad-scale maps include Braun's (1950) map of deciduous forest regions of eastern North America (map 6), Kuchler's (1964) map of the potential natural vegetation of North America (map 7), and Shanks's (1958) map of Tennessee floristic regions. Braun's classification was based, where possible, on analysis of virgin forests and climax communities. Kuchler defined potential natural vegetation as the vegetation which would exist in the absence of human influences. Shanks defined his floristic regions by analyzing ranges of both early- and late-successional species. Greller (1988) proposed a classification of deciduous forest regions, which, for Tennessee, is essentially the same as Braun's. The regions defined by all of these classifications, with the exception of Kuchler's, closely follow physiographic region boundaries.

Following is a description of the forest vegetation in each physiographic province, based primarily on the classifications of Braun, Kuchler, and Shanks. The chapter concludes with a brief description of Tennessee wetlands, which do not closely follow physiographic boundaries.

Blue Ridge Province

Braun (1950) included the Blue Ridge Province within the Southern Appalachian Section of the Oak-Chestnut Forest region. This region, similar in range to the Appalachian Oak Forest of Kuchler (1964), is characterized by the former dominance of oak-chestnut forest

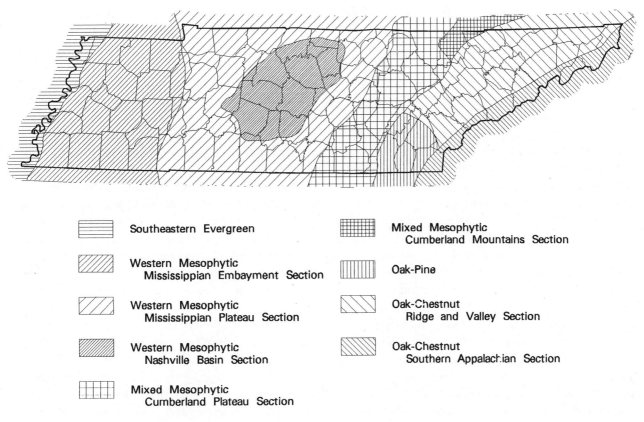

Map 6. Deciduous forest regions of Braun (1950).

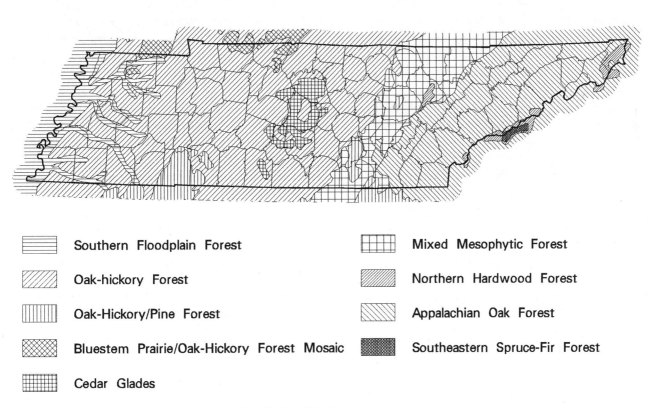

Map 7. Potential natural vegetation as mapped by Kuchler (1964).

The Environment of Tennessee

on most slopes. The descriptor "former" is applicable because American chestnut (scientific names of plants are listed in appendix 1), once a codominant or locally dominant species, has been virtually eliminated in the last 70 years by the fungus *Cryphonectria* (*Endothia*) *parasitica*, commonly known as chestnut blight. The change in forest composition following death of the chestnut has varied regionally; among the observed changes are the establishment of red maple, blackgum, and pines in the resulting canopy openings, increased dominance of oaks, and increased dominance of hickories (e.g., McCormick and Platt 1980).

Because of the great topographic variation, forest types vary with elevation. The former oak-chestnut forest type ranged from the lowest elevations up to about 1370 m. It includes several species of oaks as well as various other species according to elevation and available moisture. Dominant species typically include northern red oak, chestnut oak, white oak, and tulip-poplar. Ericaceous species, including azaleas and laurel, dominate the understory. Mixed mesophytic forests, usually referred to as cove hardwood forests in the Southern Appalachians, occur in coves and on lower north slopes. This forest type has the highest number of tree species of any in North America (Clebsch 1989). Dominant species vary locally, but frequently include buckeye, basswood, sugar maple, silverbell, tulip-poplar, beech, yellow birch, and hemlock. Several species of small trees and shrubs typically occur in the understory. White pine commonly occurs with hemlock and hardwoods in the northern Unaka Mountains, but is uncommon in the Great Smoky Mountains and farther south.

Oak and oak-pine forests occur on dry slopes and ridges at low to moderate elevations. Species present include scarlet, post, southern red, and black oaks, and shortleaf, Virginia, and pitch pines. On dry, higher elevation sites, a pine heath community may occur, composed of Table Mountain and pitch pines with an understory of ericaceous shrubs.

Between about 1370 and 1524 m, forest types typical of more northern latitudes occur. These include the northern hardwood and spruce-fir forests types. Dominant species in the northern hardwood forest type, which extend up to about 1768 m, are sugar maple, yellow birch, American beech, and buckeye. A nearly pure beech forest, apparently restricted to the Great Smoky Mountains and often referred to as a beech gap, occurs at the upper elevational limit of the northern hardwood forest (Whittaker 1956). Beech gaps typically have stunted trees with an open canopy and grassy understory.

The spruce-fir forest type occurs from about 1370 m to the highest elevations of the Unaka Mountains. Although it probably occupies the smallest area of any major forest type in Tennessee (Dull et al. 1988, May 1991), it is among the most ecologically distinctive. It is also, as further described in the next chapter, presently undergoing dramatic changes. At the lower elevational limit of the spruce-fir, red spruce occurs either as a component of the northern hardwood or cove hardwood types, or as a codominant with yellow birch. At higher elevations, spruce and Fraser fir, a Southern Appalachian endemic, occur as codominants or with one species dominating. At the highest elevations, above about 1900 m, Fraser fir occurs in pure stands. Mountain maple, mountain ash, and rhododendron often occur in the understory. *Rhododendron catawbiense* and other ericaceous species form a subclimax, heath community on exposed, rocky spurs. Treeless, mountaintop areas known as grassy balds, dominated by grasses and sedges, and heath balds, dominated by ericaceous shrubs, also occur. The origin of balds has long been debated; explanations include fire, aboriginal clearing, and local climatic influences, as well as grazing by large herbivores now extinct or extirpated from the region (Clebsch 1989, Weigl and Knowles 1995).

Kuchler (1964) mapped two potential natural vegetation types at high elevations in the Unaka Mountains. These are the northern hardwood and southeastern spruce-fir forests, comparable to the types described by Braun (1950). Although Kuchler's map (map 7) shows the spruce-fir forest restricted to the Great Smoky Mountains, both species occur to the northeast on Roan Mountain, and red spruce occurs on Unaka Mountain. Red spruce also formerly occurred at the unusually low elevation of about 853 m in boggy parts of Shady Valley, Johnson County (Ganier and Tyler 1934, Barclay 1957). The Great Smokies contain the most extensive tracts of spruce-fir in the Southern Appalachians (Dull et al. 1988).

Detailed studies of the vegetation communities of the Unaka Mountains are limited to Johnson County, the Great Smoky Mountains, and the Roan Mountain area. These include the seminal study by Whittaker (1956) of the vegetation of the Smokies and Brown's (1941) study of Roan Mountain. Clebsch (1989) gives a recent overview of vegetation research in the Unakas.

Ridge and Valley Province

Braun (1950) divided the Ridge and Valley Province between two forest regions, with the southern quarter in the Gulf Slope Section of the Oak-Pine Forest region and the remainder in the Ridge and Valley Section of the Oak-Chestnut Forest region. Shanks's (1958) division of the province is similar to that of Braun, while

Kuchler (1964) includes almost all of it in the Appalachian Oak Forest type. Martin (1989) gives an overview of the present forest communities in the province.

Within the Ridge and Valley Section of the Oak-Chestnut Forest region, Braun recognized different forest communities on the high ridges and on the low ridges and valley floors (Braun 1950). White oak dominates the valley floor forest community. Other frequently occurring species include tulip-poplar, hickories, and black, northern red, and southern red oaks; white pine occurs in the northern end of the region. The proportion of white oak is less on the low ridges, where a white oak-black oak-hickory type dominates, and scarlet and chestnut oaks and tulip-poplar are often present. Shortleaf and Virginia pines are common in young, second-growth stands. Red cedar is common on abandoned farmland and occurs as a subclimax, barrens community on thin soils over limestone.

The higher ridges north of Knoxville, because of their greater topographic diversity, contain a greater number of forest types. Chestnut formerly dominated mesic slopes; northern red oak is now dominant on many of these sites. White oak and red maple are also common on mesic slopes, and a white oak-beech community occurs on lower southern slopes. Mixed mesophytic communities more typical of the nearby Cumberland Mountains also occur on north slopes and in ravines.

The oak-pine forest region occurring in the southern Ridge and Valley Province is a transition belt between the central hardwood and southeastern evergreen forests. Although the climax forest type in this region is oak-hickory, pine is often dominant in early successional stands (Braun 1950). Shortleaf and Virginia pines are common and widespread; loblolly pine occurs naturally on some mesic sites and has been widely planted elsewhere. The transition between the Ridge and Valley oak communities and the oak-pine communities is gradual.

Appalachian Plateaus Province

Braun (1950) included this province in the Mixed Mesophytic Forest region, and recognized two sections in Tennessee, the Cumberland Mountains and the Cumberland Plateau. The boundaries of these sections correspond closely with those of Fenneman's physiographic regions. Shanks (1958) included the whole province in the Cumberland Plateau floristic region. Kuchler (1964) classified the potential natural vegetation of most of the province as Mixed Mesophytic Forest. The exceptions were an area of Northern Hardwoods centered over the southern end of the Cumberland Mountains and two areas of Oak-Hickory-Pine Forest on the Cumberland Plateau.

The Mixed Mesophytic Forest is the oldest and most complex of the deciduous forest types (Braun 1950). Widespread dominant trees in this type include beech, tulip-poplar, basswood, sugar maple, buckeye, northern red and white oaks, and hemlock. Dominance in a particular stand is usually shared by some of these species and other species of more local distribution. Several small trees and deciduous shrubs occur in the understory, and the herbaceous vegetation is usually rich and varied.

Within the Cumberland Mountains, mixed mesophytic forest occupies most of the slopes, with species composition varying with topography and microclimate. Hemlock is usually confined to ravines, and rhododendron and laurel often occur in its understory. Dry slopes and ridges are often occupied by subclimax oak-chestnut or oak-pine communities, with the dominant oak species usually chestnut oak. Yellow pines (shortleaf, Virginia, and uncommonly pitch pines) often occur in nearly pure stands on shallow sandy soils over sandstone.

Braun (1950) identified a "Cliff Section" subdivision of the Cumberland Plateau section of the Mixed Mesophytic Forest region. Most of the Cumberland Plateau in Tennessee is within this subdivision and forested with oak, oak-hickory, and oak-pine communities. The original forest vegetation of the plateau surface, although poorly known, was probably dominated by white and black oaks, hickory, tulip-poplar, and blackgum. A swamp forest dominated by red maple occurs on poorly drained areas. The mixed mesophytic forest occurs as the climax type on the slopes of the numerous ravines, often dominated by beech, tulip-poplar, basswood, shagbark hickory, and white and northern red oaks. Hemlock is abundant in the deeper gorges. A pine-heath subclimax, with an understory of ericaceous shrubs and grasses and a frequently open overstory of shortleaf and Virginia pines, occurs at the plateau margins above the gorges. Hinkle (1989) gives a recent analysis of Cumberland Plateau forest communities.

Interior Low Plateaus Province

Braun (1950) included the Interior Low Plateaus Province, as well as most of the Tennessee portion of the Coastal Plain Province, within the Western Mesophytic Forest region. She divided the Tennessee portion of this forest region into the Nashville Basin, Mississippian Plateau, and Mississippi Embayment sections. Her boundaries of the Nashville Basin section include the inner basin and part of the outer basin as shown on map 3. The remaining portion of the outer basin, as well as the Highland Rim, make up the Mississippian Plateau section. Shanks (1958) recognized a Highland Rim floristic region, with subdivisions for the Kentucky

The Environment of Tennessee

Prairie Barrens, Barrens of the Southwestern Rim, and Barrens of the Southeastern Rim. Kuchler (1964) considered the potential natural vegetation of most of the province to be Oak-Hickory Forest. He also recognized three additional types in the province: Cedar Glades, mostly in the inner basin; a mosaic of Bluestem Prairie and Oak-Hickory Forest, roughly corresponding to Shank's Kentucky Prairie Barrens; and Oak-Hickory-Pine in the extreme southwest corner of the region. Shanks included the southwest corner of the region in his Coastal Plain Upland floristic region.

The Western Mesophytic Forest region is a transition region, lacking a single climax type (Braun 1950). The forests are, in general, less luxuriant than those of the Mixed Mesophytic Forest region, and fewer species dominate in individual stands.

The most distinctive feature of the forests of the Central Basin is the Cedar Glade community, more properly referred to as "limestone cedar glades," as noted by Quarterman (1989). This forest type consists of open to dense stands of red cedar growing in shallow soil over limestone. Scattered deciduous trees, including post and chinquapin oaks, shagbark hickory, hackberry, redbud, and winged elm, also occur. These areas have distinctive herbaceous spring flora, and are very dry in the summer. Quarterman (1989), as well as numerous other authors, gives further information on the structure of glade communities. Rolling hills in the basin supported a forest of white oak, tulip-poplar, and sugar maple; only remnants of this community remain. Higher hills support forests similar to that on the surrounding Highland Rim.

Several forest types occur on the Highland Rim. The dissected portions of the eastern rim support a Mixed Mesophytic Forest type, although usually with a higher proportion of beech than occurs in the Cumberlands. Oak forest occupies much of the well-drained, moist, upland portions of the rim. Common species in this type include white, black, chinquapin, and black oaks, tulip-poplar, beech, hickory, and sugar maple. Swamp forests occur on poorly drained flats. Their species composition varies according to the moisture regime; common species include pin, overcup, willow, water, and swamp chestnut oaks, red maple, sweetgum, and blackgum (Ellis and Chester 1989). Upland, drought-prone flatwoods, often labeled barrens on maps of the region, are most prevalent on the southern part of the eastern rim, centered in Coffee County. Forests in these areas are dominated by southern red, post, blackjack, and scarlet oaks (McKinney 1989).

Portions of the Western Highland Rim along the Kentucky border, on the Pennyroyal Plateau, were once tallgrass prairie. This plant community, more extensive in Kentucky, was known as "the Barrens" to early settlers, and is the Kentucky Prairie Barrens floristic region described by Shanks (1958). The Tennessee and Kentucky prairies are floristically similar to but disjunct from the eastern "Prairie Peninsula" described by Transeau (1935) (Baskin and Baskin 1981). Kuchler (1964) identified the potential natural vegetation of this area as a mosaic of Bluestem Prairie and Oak-Hickory Forest. The vegetation is floristically similar to midwestern prairies; dominant species included big bluestem, little bluestem, Indian grass, switchgrass, and several forbs (Chester and Ellis 1989, DeSelm 1988, 1989). Barrens, small woodland openings dominated by prairie grasses, occur on the Highland Rim, as well as elsewhere in Tennessee (DeSelm 1989). Their maintenance is dependent on periodic disturbance.

Coastal Plain Province

Braun (1950) included the Loess Plain and Coastal Plain Uplands sections of this province in the Mississippi Embayment Section of the Western Mesophytic Forest Region and the Mississippi Alluvial Plain as a section of the Southeastern Evergreen Forest Region. She noted the similarity of the Mississippi Embayment Section to the Oak-Hickory Forest Region west of the Mississippi River, but included it in the Western Mesophytic Forest Region because of the mixed mesophytic forest on the loess bluffs. Shanks (1958) identified the Mississippi Alluvial Plain as a distinct floristic region, and included the rest of the Coastal Plain in the Mississippi Embayment floristic region, with distinct Coastal Plain Upland and Mississippi River Bluffs subdivisions. Kuchler (1964) identified the Oak-Hickory Forest, Oak-Hickory-Pine Forest, Bluestem Prairie–Oak-Hickory Forest Mosaic, and Southern Floodplain Forest potential natural vegetation types in the Coastal Plain Province.

The Mississippi Embayment Section of the Western Mesophytic Forest Region is composed of several vegetation types, including prairie, oak-hickory forest, swamp forest, and mixed mesophytic forest (Braun 1950). Swamp forest, restricted to the floodplains of the larger streams and more common in the Loess Plain than on the Coastal Plain Uplands, is described below with the Mississippi Alluvial Plain. Areas of prairie, contiguous with more extensive prairies in western Kentucky, formerly occurred in Henry, Weakley, and Obion Counties. The most common forest type is oak-hickory, with the species composition varying with topography and soil moisture. Southern red oak is dominant on dryer upland sites, and white oak, often in association with tulip-poplar and sweetgum, is dominant on more mesic sites. Hickories are common throughout the area. Shortleaf pine occurs on sandy upland soils of the Coastal Plain Uplands. Because of its prevalence at the southern end of the region, Kuchler (1964) identified the potential natural vegetation of this area as Oak-Hickory-Pine Forest.

Braun (1950) and Shanks (1958) identified distinctive floras on the loess bluffs at the western edge of the Loess Plain. Mixed mesophytic forest occurs on ravine slopes of the loess bluffs; common species include beech, tulip-poplar, northern red oak, sweetgum, sugar maple, and hickories (Braun 1950, Miller and Neiswender 1989). The proportion of oaks increases in forests on ridgetop sites.

The Mississippi Alluvial Plain was identified as a distinct section of the Southeastern Evergreen Forest by Braun (1950) and as distinct from the rest of the Tennessee coastal plain by Shanks (1958) and Kuchler (1964). Kuchler described the potential natural vegetation as Southern Floodplain Forest, dominated by tupelo, oaks, and bald cypress and extending eastward along the major Mississippi River tributary streams. The forest type varies with the geologic history and elevation of the site, and, consequently, its periodicity of flooding (Braun 1950, Christensen 1988, Guthrie 1989, Shankman 1993). Swamp forest occurs in areas with standing water present most of the year, typically in oxbow lakes resulting from stream channel migrations. Bald cypress and water tupelo are usually dominant, although tupelo is absent from the Reelfoot Lake area. Areas flooded during the winter and early spring support a diverse forest dominated by red maple, sweetgum, water hickory, and overcup and other oaks. Cane frequently occurs in the understory of this seasonally flooded forest. Dominant trees on the highest, rarely flooded sites include American beech, American elm, sweetgum, Nuttall, willow and cherrybark oaks, and shagbark hickory.

Wetlands

Wetlands are lands where the water table is usually near or at the surface or where lands are covered by shallow water, support vegetation typically found in wet habitats, have soils wet enough to produce anaerobic conditions periodically, and/or have a substrate saturated or covered with water at some time during the growing season (Cowardin et al. 1979). Wetlands are frequently prominent landscape features, especially in West Tennessee. Dahl (1990) estimated that wetlands covered about 784,000 ha of Tennessee (7.2%), in the late 1700s.

Tennessee wetlands can be broken down into three main types: riverine, lacustrine, and palustrine (Cowardin et al. 1979). Riverine wetland systems are usually contained within a channel of flowing water and lack trees, shrubs, and persistent emergent vegetation. Lacustrine systems are typically lakes, reservoirs or ponds, have less than 30% of their area covered by trees, shrubs, or persistent emergent vegetation, and are more than 8 ha in size. These wetlands were probably quite rare in prehistoric Tennessee outside of the Mississippi Alluvial Plain. Palustrine systems include wetlands dominated by trees, shrubs or persistent vegetation, or wetlands lacking this vegetation and less than 8 ha in size.

The most common types of palustrine wetlands are forested wetlands, scrub-shrub wetlands, and emergent wetlands. Forested wetlands include permanently flooded cypress and tupelo forests, periodically flooded bottomland hardwoods, and periodically flooded streamside (riparian) forests. This was historically the most common wetland type, and widespread in West Tennessee. Scrub-shrub wetlands are dominated by woody vegetation less than 6 m tall; common examples are alder swamps, buttonbush swamps, and young willow thickets. Emergent wetlands are dominated by herbaceous species such as cattail, sedges, rushes, giant cutgrass, maidencane, and smartweeds. This wetland type supports the most distinctive avifauna, including bitterns, gallinules, and rails.

The Landscape of Tennessee

The single most important factor controlling the local distribution and abundance of birds is the availability of suitable habitat. The previous chapter described the physiography, climate, and broad vegetation patterns of Tennessee, environmental factors that influence habitat availability. This chapter describes changes in the Tennessee landscape, particularly those attributable to human actions, that have influenced local habitat availability. It begins with a description of the prehistoric landscape, as it presumably appeared to the first European explorers.

The Prehistoric Landscape

In 1897, J. B. Killebrew, Tennessee commissioner of Agriculture, Statistics and Mines, conveyed what has long been a popular view of the prehistoric landscape of Tennessee: "The first settlers in the state found the whole country covered by a solemn, mysterious, and seemingly interminable forest. There was not a foot of prairie land in the state, unless the bald spots on the tops of the higher mountains may be so called. There were no breaks in the continuity of the woodlands, except on a few rocky points and in some narrow, winding openings where the streams flowed . . . the dark and melancholy woods that stretched from mountain to valley, from plain to river, from state line to state line . . ." (Sudworth and Killebrew 1897:3). A careful reading of accounts of the earliest explorers in Tennessee (compiled in Williams 1928), as well as recent archaeological and ecological studies, show that Killebrew, and many other writers, did not accurately describe the Tennessee landscape at the time of the first European contact in the sixteenth century.

The area that Tennessee occupies has never been glaciated. During the Quaternary glaciation cycles, however, glaciers approached within 300 km of the state and repeatedly altered its topography, climate, and vegetation. At the height of the most recent glaciation, the Wisconsin glaciation, about 18,000 years ago, most of Tennessee was covered by jack pine-spruce-fir forest, bordered on the south by mixed conifer-northern hardwood forest (Delcourt and Delcourt 1981). Mixed hardwood forest, precursor of the present mixed mesophytic forest, extended northward along the Mississippi River, and tundra capped the high peaks of the Blue Ridge. Remains of several boreal forest bird species, including the Hawk Owl *(Surnia ulula)*, Boreal Owl *(Aegolius funereus)*, Gray Jay *(Perisoreus canadensis)*, and Pine Grosbeak *(Pinicola enucleator)*, occur in Middle Tennessee cave deposits from this period (Parmalee and Klippel 1982). The presence of Sharp-tailed Grouse *(Pediocetes phasianellus)* remains in the same deposits indicates that short-grass prairie or savannah also occurred. None of these species has recently occurred in Tennessee.

By 10,000 years ago, following the glacial retreat, mixed hardwood forest had extended across the state (Delcourt and Delcourt 1981). Cypress-gum forest occurred along the Mississippi River, and spruce-fir was restricted to the Blue Ridge. During the mid-Holocene warming, 8000–4000 years ago, prairie and oak-hickory forest penetrated eastward as the range of the mixed hardwoods contracted. The extent of prairie probably peaked about 6000 years ago when summer temperatures were highest (COHMAP Members 1988). At the end of warming period, temperatures cooled, precipitation increased, and the current natural vegetation patterns stabilized, as mapped in map 7 and further described in the preceding chapter.

Humans first occupied Tennessee 10,000 to 12,000 years ago (Chapman et al. 1982). By the Late Archaic period, about 3500 years ago, semipermanent settlements and squash cultivation were established. The impact of the Native Americans on the landscape increased in the following centuries. Chapman et al. (1982), studying charcoal fragments recovered from Little Tennessee River Valley archaeological sites, found a shift from late-successional to early-successional tree and shrub species, which they interpreted as resulting from revegetation of abandoned croplands and fuelwood cutting areas. The impacted landscape was a mosaic of croplands near permanent settlements, early-successional forest, and remnant deciduous forest on higher ridges. Native Americans also impacted the forest environment by periodically burning large areas to encourage the growth of fruits and berries and improve the habitat for game animals; burning was also employed as a hunting technique (Pyne 1982, Williams 1989).

The Native American population and its impacts on the environment probably peaked about 1500 A.D. Following European contact, Native American populations collapsed due to epidemics of diseases carried by Europeans. In Tennessee, this decline may have been underway when Hernando DeSoto entered the state in 1540 (Smith 1987). Because of the limited archaeological data and incomplete historical accounts, estimates of the Native American population at the time of European contact are varied and controversial. Many authors (e.g., Dobyns 1983, Driver 1970) do agree, however, that the population density in parts of Tennessee was among the highest in inland North America. Permanent settlements were concentrated along river valleys, and at the time of European contact the highest population density was probably among the Cherokees in the southern half of the Ridge and Valley Province of East Tennessee (Satz 1979, Chapman 1985). Much of the Cumberland Plateau and Middle and West Tennessee apparently had a low resident population and was claimed as hunting territory by several tribes. Dobyns (1983) estimated a Native American population density at European contact of 2.53 persons per km^2 in the Mississippi River valley, including the Tennessee and Ohio River drainages. At this density, the pre-Columbian Tennessee population was approximately 270,000, more than the 1810 Tennessee population of 261,727 (Killebrew 1874). Based on estimates of 0.12 to 0.4 ha of cultivated cropland per person (Williams 1989), the Native American population cultivated from 32,400 to 108,000 ha in Tennessee. Except in very rich river bottom settlements, fields were periodically fallowed for several years; this process increased the amount of land disturbed by farming. The area periodically burned was probably several times that disturbed by cultivation, and often in hunting areas distant from villages (Williams 1989).

Repeated burning produced open, parklike forests with grassy understories and favored oak and pine forests (Pyne 1982). Burning on particular soil types probably maintained and spread prairie and barrens habitats. André Michaux, observing the prairies in southern Kentucky and north-central Tennessee in the late 1790s, wrote: "The spacious meadows in Kentucky and Tennessee owe their birth to some great conflagration that has consumed the forests, and they are kept up as meadows by the custom that is still practiced of annually setting them on fire" (Thwaites 1904–7). Alexander Wilson described Greater Prairie-Chickens in this area a few years later (Wilson 1811a).

Frequent burning in eastern North America has been attributed to creating habitat suitable for American bison, which probably spread east of the Mississippi River around 1000 A.D., and entered the Southeast around 1500 A.D. (Roe 1970). Bison were reported from many parts of Tennessee by early explorers (see accounts in Williams 1928), including along the Mississippi River, which they readily swam (Roe 1970). The "innumerable" and "vast" herds mentioned by settlers and explorers in the late eighteenth century, however, probably numbered no more than a few hundred. Bison affected the landscape by maintaining grasslands through grazing and trampling woody vegetation. They also influenced European settlement patterns, as Nashville was the site of French Lick, a salt spring frequented by bison and other large mammals. American bison were extirpated from Tennessee by 1810 (Roe 1970).

Even in areas not directly influenced by the activities of Native Americans, the landscape was in a continuous state of change. Natural disturbances, resulting in openings in the forest canopy, occurred as a result of hurricanes, tornadoes or other windstorms, wildfire, drought, icestorms, insect outbreaks, and other factors. The average rate of forest disturbance in many forest types is about 1% per year (Runkle 1985). The scale of the disturbance, however, varies greatly from about 30 sq m due to the death of individual trees to thousands of hectares affected by fire. Severe disturbances over large areas resulted in the establishment of relatively even-aged forests of early-successional, shade-intolerant species. Periodic fires, ignited by lightning or Native Americans, may have been typical of the oak-dominated forests of much of Tennessee, although the primeval disturbance regime for this forest type is poorly known (Runkle 1985). The fire response of the American chestnut, once a major component of Tennessee forests, is also poorly known.

The natural disturbance regime of the cove and mixed mesophytic forests of eastern Tennessee and elsewhere in the southern Appalachians is better known (Runkle 1982, 1985). Fires are uncommon and mostly restricted to south-facing slopes near ridgetops, and death of canopy trees is mostly due to small-scale disturbances such as ice damage, lightning strikes, or windthrow affecting one or a few trees at a time. The canopy openings or gaps created by these disturbances form and close at the rate of about 1% per year and make up about 7–10% of the forest area. Shade-intolerant species survive in the larger gaps, and the resultant forest is thus a mixture of shade tolerant and intolerant species of differing ages.

As a result of local geology, natural disturbances, and the activities of Native Americans, prehistoric Tennessee had a diverse landscape of uneven-aged and even-aged forests, open, parklike forests, limestone glades, fire-maintained grasslands, and cultivated and abandoned farmlands. The total area in cultivation at any time may have been over 100,000 ha, with a similar or larger area of abandoned farmlands reverting to forest. Forests probably occupied no more than 80–90% of the landscape of prehistoric Tennessee, and the high landscape diversity contributed to the high number of plant and animal species in modern Tennessee.

Historic Changes in the Landscape

Following the collapse of their populations after European contact, the Native Americans' effect on the landscape lessened. It continued, however, in the vicinity of the surviving villages, which were most numerous in southeast Tennessee (Chapman 1985, Smith 1987). By the time of the heavy immigration of European settlers in the late eighteenth and early nineteenth centuries, many areas that were formerly cultivated or burned by Native Americans had probably reverted to forest (cf. Rostlund 1957). Grasslands, parklike forests, and Native American "old fields," however, were still widespread and frequently noted by early travelers (see accounts in Williams 1928).

The first European settlements, aside from military and trade outposts, were established in northeast Tennessee around 1770 (Corlew 1981). Settlement of north Middle Tennessee began a decade later, and settlement of West Tennessee began after the 1818 Jackson Purchase. The state population grew relatively steadily throughout most of the nineteenth century and accelerated after 1920 (fig. 3). The urban population began rapid growth in the 1940s and exceeded the rural population in the 1960 and later censuses. Much of the population has recently been concentrated into a few urban areas (map 8); according to 1990 census results, 5.7% of the state was classified as urban land. Using a definition similar to that of urban land, the Soil Conservation Service (Soil Conservation Service 1989) classified 6.4% of the state as developed land.

The rest of this chapter describes major changes in the Tennessee landscape since the late eighteenth century. It is organized by major land-use categories; these categories, however, are not mutually exclusive.

Agriculture
Farming Practices

The greatest impact of the human population increase on the landscape has been in the clearing of forests for agriculture. The earliest agricultural clearing, as carried out by Native Americans and adopted by Europeans, involved girdling the trees, removing the understory vegetation, and then planting crops under the standing dead trees (Williams 1989). This method was still used in the late nineteenth century in the eastern mountains. The dead trees were removed or burned as they fell to the ground. Later clearing methods involved felling trees and grubbing out or burning the stumps.

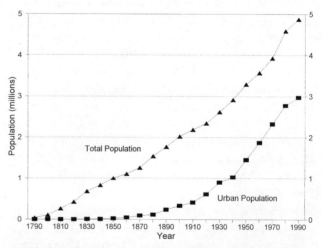

Fig. 3. Growth in the total human population and the urban population of Tennessee, 1790–1990. From U.S. Census Data in Vickers and Kirby (1991).

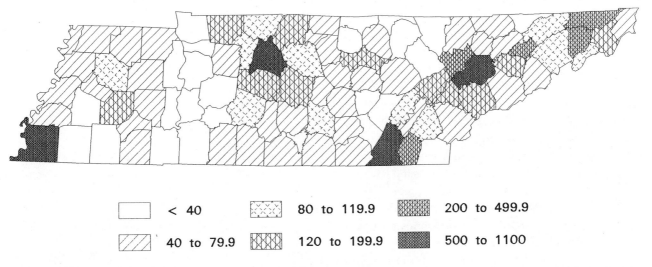

Map 8. Human population density per 2.59 sq km (1 sq mi) by county, 1990. From U.S. Census Data in Vickers and Kirby (1991).

In addition to clearing and keeping clear more of the forest than had Native Americans, European settlers introduced many nonnative grasses and herbs. Some of these introductions, such as clover, timothy, and orchard grass, were deliberate, as few native perennial grass species withstood continued grazing (Jones 1974). Other introduced species, however, rapidly spread as weeds well adapted to areas disturbed by agricultural practices. Several alien plants species, which spread into woodlands, were later introduced; some of these are further described below.

The introduction of livestock was also a great change from Native American agricultural practices. Except for dogs and possibly turkeys, Native Americans did not keep large, domestic animals (Chapman 1985). European settlers brought cows, horses, hogs, chickens, and domesticated turkeys. Cows and hogs were given free range of the unfenced forest to feed on acorns, chestnuts, beechnuts, grasses, and cane (Williams 1989). This practice was widespread until the latter half of the nineteenth century and locally practiced in the southern Appalachians into the early twentieth century.

Agricultural practices employed at the beginning of the nineteenth century were similar across the state. The first settlements were along streams and rivers. Floodplains were cleared and planted, primarily with corn, and wooded uplands were used for grazing livestock and supplying firewood and lumber. As the population grew, agricultural clearing moved into the less fertile uplands. As local areas became more heavily settled, fencing became more common, first to keep free-ranging livestock out of crops and later to contain the livestock. Fence construction, primarily from wood, contributed to the rate of forest clearing (Williams 1989).

Regional differences in agriculture soon developed in the early nineteenth century. Cotton became the dominant crop in West Tennessee and south Middle Tennessee (Bureau of Census 1925). Elsewhere in the state, farm operations were generally smaller, and corn remained the dominant crop; tobacco and livestock were also important, and increased area was devoted to raising forage crops for livestock. Productivity of these farms remained low into the early twentieth century.

During the 1930s and 1940s, agriculture changed greatly through New Deal programs, increased mechanization, and a reduced farm labor force (Fite 1979). Marginal croplands were converted to pasture or forest, use of chemical fertilizers increased, and the fewer remaining farms grew in size (Bureau of Census 1945, 1980). Livestock rose in importance, and its market value has recently surpassed that of other crops (Bureau of Census 1989). A further dramatic change occurred during and shortly after World War II, when soybeans replaced cotton as the dominant crop in West Tennessee (Fornari 1979). Concurrent with these changes has been a decrease in the varieties of crops grown by individual farmers, with most recently specializing in one or two crops (Census of Agriculture data). The area of the major row crops—corn, soybeans, and cotton—has recently fluctuated with crop prices. High soybean prices in the 1970s, for example, resulted in a great increase in soybean acreage and the drainage of West Tennessee bottomlands for soybean cultivation.

Area of Farmland

With the exception of the 1860s, the rate of agricultural clearing remained high throughout the nineteenth century (fig. 4). The total area of improved farmland, including both cropland and pasture, peaked about 1940 at slightly over half the area of the state (fig. 5).

Since 1940 farmland area has decreased by about one-third. The area of pasture (excluding grazed woodland) peaked in 1959; after holding fairly steady through the 1960s, it has since dropped by about 40% (Census of Agriculture data, Daugherty 1989). Grazed woodland decreased from 1,376,000 ha in 1950 to 566,600 ha in 1982 (Daugherty 1989). The proportion of cropland (excluding grazed woodland) by county from the 1987 Census of Agriculture is shown in map 9.

Decreases in the total amount of farmland have resulted from farm abandonment and conversion of farmland to other land uses, such as urban and suburban development. The rate of farm abandonment varies with the overall economy, crop prices, and government farm programs. Abandoned farmland, in the absence of urban or suburban development, usually reverts to forest, and reforestation has been greatest in areas of marginal farmland in Middle and East Tennessee (Hart 1968, May 1991).

One of the recent government farm programs, the Conservation Reserve Program, has resulted in about 185,000 ha of erodable farmland being removed from agricultural production to permanent grassland during the years 1986–91 (Osborn, Llacuna, and Linsenbigler 1992). An additional 12,000 ha were planted with trees. This program has had beneficial effects on several grassland bird species (Hays, Webb, and Farmer 1989).

The number of farms has steadily decreased with the total area farmed (Bureau of Census 1925, 1945, 1980, 1989). The average size of farms, however, increased from 27 ha in 1925 to 59 ha in 1987. Reasons for this change include a greater rate of farm abandonment among small, unproductive farms than among larger, more fertile farms and the consolidation of nearby farms. The average size

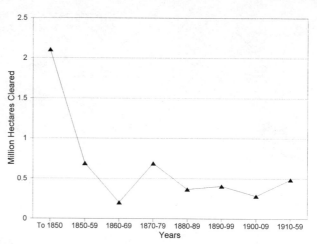

Fig. 4. Rate of clearing of forested land for agriculture in Tennessee through 1959. From U.S. Census of Agriculture data in Hart (1968), Williams (1989).

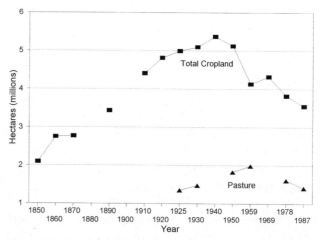

Fig. 5. Farmland area in Tennessee, 1850–1987. Gaps in trend lines are due to changes in census techniques, which made results of some censuses not directly comparable to others. From Bureau of Census (1925, 1945, 1980, 1989).

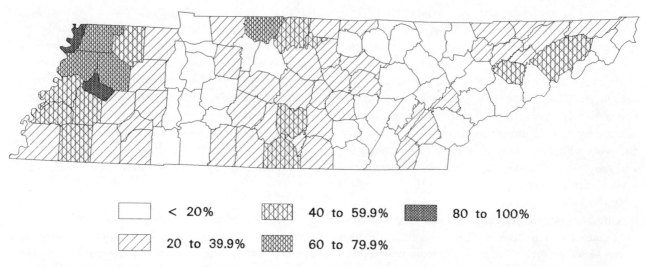

Map 9. Proportion of land area in cropland by counties in 1987. From 1987 U.S. Census of Agriculture.

of cropped fields has also increased, driven in part by increased mechanization and use of larger equipment. This change has resulted in the elimination of many kilometers of brushy and wooded fencerows.

Forests

Forest Area

The rate of agricultural clearing and, conversely, of reforestation of abandoned farms, had a major effect on the forested area of the state, at least until recent decades, when suburban and urban development became major influences. The U.S. Department of Agriculture estimated the forest area in 1870 to be 5.4 million ha, or 49% of the state (Killebrew 1874). This was probably an underestimate, as only about one-quarter of the state was then cleared farmland (fig. 5). Near the end of the century, Sudworth and Killebrew (1897) estimated that 6.4 million ha or 60% of the state were forested. Sudworth and Killebrew (1897) also noted that, by the mid-1800s, so little woodland remained in some of the best agricultural districts that landowners were concerned with firewood supplies and that parts of Middle Tennessee were almost totally cleared by the 1890s.

Accurate estimates of forest area became available in 1950; since then the forested area has fluctuated slightly (fig. 6). In 1989, 51.4% of the state was forested; the forested proportion was highest in the Blue Ridge, Cumberland Mountains, and Western Highland Rim (map 10). The changes in the proportion of forested land in recent decades have not been uniform across the state. Counties showing the greatest loss of timberland (forest in private or public ownership not reserved by statute from commercial harvest) between 1980 and 1989 clustered along the eastern border and in the Loess Plain and Coastal Plain Uplands regions, while counties with the greatest gain in timberland were in the Highland Rim and southern Central Basin regions and in the northeastern corner of the state (May 1991).

While the forested area has shown an overall increase, the size of contiguous forest blocks, those unbroken by human-made openings such as roads, utility corridors, and farmland, has probably decreased. Few accurate measurements of this forest fragmentation are available. According to the 1980 statewide forest survey (Birdsey 1983), 24% of timberland was in tracts of 40 ha or less, 55% in tracts of between 40 and 1012 ha, 14% in tracts of between 1012 and 2024 ha, and 7% in tracts of over 2024 ha. Because forest tract size effects many birds, as well as other species (see following chapter), trends in forest tract size deserve further study.

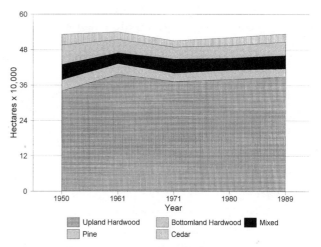

Fig. 6. Change in area of the major forest types, 1950–1989. From USDA Forest Service forest survey data in May (1991) and earlier reports.

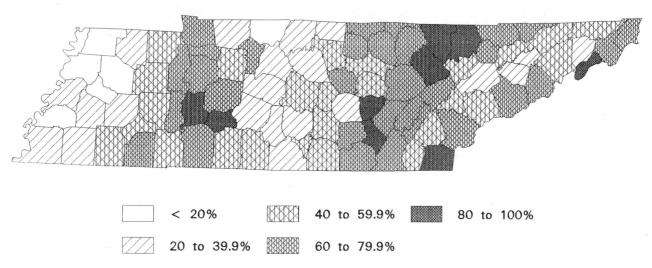

Map 10. Proportion of land area in forest by counties in 1989. From Vissage and Duncan (1990).

Forest Fires

During the period of open-range grazing, woodlands were burned to improve livestock forage; these fires frequently escaped control and burned large areas (Pyne 1982). Half of the 400,000 ha of forest (almost 4% of the state) that was reported burned in 1880 was burned to improve grazing (Sudworth and Killebrew 1897). This was probably an underestimate of the total area burned. Sudworth and Killebrew (1897) and other authors noted the damaging effects of free-ranging cattle and burning on the forest. The fires, however, helped maintain pine forests and barrens. The 1925 Census of Agriculture reported 96,000 ha burned. Effective forest fire control was then being organized, and burning to improve forest grazing soon died out (Pyne 1982).

Slash fires following timber harvesting were common in the southern Appalachians early in this century (Pyle 1984). As described below, these fires often changed the species composition of the regenerating forests. Slash fires were also common following nineteenth-century charcoal production on the Western Highland Rim, as further described later in this chapter.

Fire is presently used on a limited basis as a tool in forest management, especially in pine forests. Other human-caused and naturally ignited forest fires are aggressively fought, and the area burned annually is usually only a few thousand hectares.

Forestry

Until the late nineteenth century, most timber harvesting in Tennessee, as elsewhere in the South, was undertaken to clear land for agriculture, meet local lumber demands, and provide wood for fuel and fencing (Williams 1989). The demand for fuelwood decreased at the end of the century as the use of coal increased. Much early commercial logging was concentrated along the rivers, and the logs were rafted to sawmills in the larger towns, where the sawn lumber was sold locally or shipped by barge or train. Rafting of logs continued into the 1920s in East Tennessee (Allred, Atkins, and Fitzgerald 1939). Commercial logging rapidly increased in the 1870s and 1880s as lumber supplies were depleted in the northeastern United States and railroad transportation improved. Logging peaked in 1909, when 2.9 million cubic meters of lumber were cut (Allred, Atkins, and Fitzgerald 1939). By 1931, only 1.6% of the commercial forest area (excluding what was to become the Great Smoky Mountains National Park) in Tennessee was considered virgin or only lightly harvested (Allred and Fitzgerald 1939).

The high elevations of the eastern mountains were logged somewhat later than the rest of the state. Much of the high quality white pine, hemlock, and tulip-poplar in Johnson County was cut between 1897 and 1911 (Barclay 1957). In the Great Smokies, most logging between 1880 and 1900 was of high value, easily accessible species (Lambert 1961). Lumber and paper companies purchased large tracts of mountain land at the turn of the century; the Unaka Mountains made up the first region of the state where forest product companies owned large tracts of land. Mechanized logging, using cable skidders and railroads and cutting all trees larger than 0.3 m diameter, soon began. These operations continued in the Smokies until the 1920s, when land purchases for the Great Smoky Mountains National Park began.

The tree species sought by loggers have changed as the most valuable species were depleted. During the first half of the nineteenth century, white oak staves and red cedar crossties were the major lumber products exported from the state. By 1870, the supply of exportable cedar in many Middle Tennessee counties was exhausted (Killebrew 1874). Tulip-poplar and black walnut were the most widely sought species in the late nineteenth century (Allred, Atkins, and Fitzgerald 1939). Nashville had one of the largest hardwood markets in the country during this period (Sudworth and Killebrew 1897). Demand shifted to oaks during the twentieth century, and Tennessee led the nation in oak lumber production in 1930 (Allred, Atkins, and Fitzgerald 1939). The use of smaller pines and hardwoods for paper production has greatly increased since the 1950s with the construction of several pulp mills across the state (May 1991).

Timber harvesting has traditionally used selective and high-grading harvesting techniques, which remove only a portion of the trees from a tract; from 1980 to 1989, these techniques were used on 72% of the 1.1 million ha harvested (May 1991). Clearcutting, which removes virtually all of the trees, was used on 18% of the harvested area. The heavy reliance on selective harvesting and high grading, as well as effective fire control in recent decades, has resulted in a slight shift in the species composition of upland hardwoods from oaks toward more shade-tolerant species such as maples and beech. The rate of recent timber harvesting has been lower than the rate of growth of the forest. As a result, the proportion of trees in the sawtimber (hardwoods greater than 28 cm diameter and softwoods greater than 23 cm diameter) and poletimber (hardwoods 13 to 28 cm diameter and softwoods 13 to 23 cm diameter) size classes has increased, and the proportion in the seedling-sapling size classes (less than 13 cm diameter) has decreased (May 1991). The total volume of live trees increased threefold from 1950 to 1989.

Most major forest types have shown slight changes in area since 1950 (fig. 6). Both the pine category, which here includes the loblolly-shortleaf pine and white pine-hemlock types, and the mixed category, composed primarily of the oak-pine type, have shown an overall decrease. Much of this change is due to the natural succession from pines to hardwoods. This loss has recently been reversed by the increase in pine plantations (some of which are classed as mixed because of the number of hardwood trees present). Over 200,000 ha of pine plantations presently exist, mostly of loblolly pine and mainly on the Cumberland Plateau, Western Highland Rim, and Coastal Plain Uplands (May 1991). Between 1980 and 1989, 61,000 ha were planted to pine. In addition to the commercial pine plantations, numerous small groves of pine, primarily loblolly, have been planted since the 1930s for erosion control and landscaping throughout the state, including areas where pine did not naturally occur.

Bottomland hardwoods, including the oak-gum-cypress and elm-ash-cottonwood forest types, have also decreased since 1950. Most of this loss is due to agricultural clearing in West Tennessee. The loss of bottomland hardwoods since 1950, however, has probably been relatively minor compared to earlier losses from agricultural clearing and reservoir construction.

Other Forest Changes

Introduced species, including insects, fungi, and plants, have caused several changes in Tennessee forests and will likely continue to do so in the future. The chestnut blight fungus virtually eliminated mature American chestnut trees during the first half of this century; only stump sprouts, which rarely live to reach a diameter of 5 cm, survive. American chestnut formerly occurred in the eastern two-thirds of the state and was once a dominant species in much of East Tennessee and parts of the Highland Rim. Its rapid death resulted in numerous forest openings, which probably affected several bird species, as Brooks (1947) noted was the case farther north. As previously described, changes in tree species composition also occurred following the death of the chestnut. The loss of the chestnut has probably also resulted in an overall drop in the nut crop available to Wild Turkeys and other animals.

The balsam woolly adelgid, an insect native to Europe, was first detected in the southern Appalachians in 1957, several decades after its probable arrival (Eagar 1984). It feeds on the sap of firs and has killed most of the mature Fraser firs in the southern Appalachians. Following the death of the mature firs, a thick layer of fir seedlings, blackberry, and other species develops; standing dead trees are numerous for a decade or more (Nicholas et al. 1992, Pauley and Clebsch 1990, Witter and Ragenovich 1986). The effects of these changes on birds inhabiting the spruce-fir forest are mentioned elsewhere in the species accounts.

Dutch elm disease, caused by a fungus *(Ceratoctstis ulmi)* transmitted by introduced elm bark beetles, has killed many American elms across the state in recent decades. Because American elm is a minor component of Tennessee forests (May 1991), its partial loss has not resulted in a widespread, easily noticeable change in Tennessee forests. The death of American elms has been more noticeable in urban areas, where American elms were once widely planted for landscaping.

A more noticeable change in Tennessee forests will probably result from the predicted spread of the gypsy moth into the state. This alien species, first introduced in Massachusetts in 1869, has slowly spread throughout much of the Northeast and into Virginia, West Virginia, and North Carolina (Gottschalk, Twery, and Smith 1990). Gypsy moths have been found at several locations in Tennessee since the mid-1980s, although sustained outbreaks have not yet occurred. Gypsy moths feed on the foliage of numerous tree species. Oaks, particularly black and chestnut oaks, are preferred foods, and dry, ridgetop sites are most susceptible (Gansner and Herrick 1985). These conditions are widespread in Tennessee. Although gypsy moth infestations rarely kill trees directly, tree death often occurs from the defoliated tree's increased susceptibility to secondary insect and fungal attacks. Resultant changes in the forest include an increase in the number of dead snags, a more open canopy, an increase in understory vegetation, and a decrease in nut crops (Twery 1990). All of these changes can affect local bird populations.

In addition to the balsam woolly adelgid, other anthropogenic factors have greatly altered the spruce-fir forest. The low-elevation spruce forest in Shady Valley, Johnson County, has been largely cleared (Barclay 1957). Logging of the spruce-fir on Roan Mountain began in 1929, and the structure of the regenerating, secondary stands differs from the virgin forest (Warden 1989). The density and basal area of both red spruce and Fraser fir is greatly reduced, and fir is no longer present in lower-elevation (1646 m) stands. Hardwoods have greatly increased in both the lower and higher elevation (1768 m) stands. About one-quarter of the extensive spruce-fir forest in the Great Smoky Mountains was disturbed by logging and slash fires prior to formation of the national park (Pyle 1984, Pyle and Schafale 1988). Most of this disturbed area reforested to types other than spruce-fir. Furthermore, some evidence shows a decline in red spruce growth in the past 25 years as well as canopy crown deterioration during the mid-1980s (LeBlanc, Nicholas, and Zedaker 1992, McLaughlin et al. 1987, Peart et al. 1992).

Cane was formerly a common understory species on bottomlands and slopes adjacent to rivers, where it formed extensive canebrakes (e.g., accounts in Williams 1928). The presence of cane was considered an indicator of fertile soils by early settlers, and cane is palatable to livestock (Williams 1989). Because of agricultural clearing and grazing, canebrakes have declined to a fraction of their former area. Fire control may have also been a factor, as canebrakes can deteriorate in the absence of periodic burning (Hughes 1966).

Numerous nonnative plant species, most originally introduced for landscaping, erosion control, and improving wildlife habitat, have escaped control and spread over large areas. Among the species that altered large areas of the landscape are Japanese honeysuckle, bush honeysuckle, multiflora rose, and kudzu; the first three have fruits readily eaten by birds, who have contributed to their wide distribution. The honeysuckles and multiflora rose spread readily into disturbed woodland and woodland edges, where they have greatly changed the understory in many areas. Kudzu was once widely planted on bare slopes for erosion control; from these openings, it slowly spread into adjacent woodlands, killing trees in the process.

Mining and Mineral Processing
Iron

Tennessee had a thriving iron-mining, smelting, and forging industry during the nineteenth century. Hematite and limonite ores were mined from open pits, of which at least 279 have been identified on the Western Highland Rim (Smith, Stripling, and Brannon 1988). The area disturbed by these iron ore mines was relatively small, rarely more than 5 ha in area or more than 18 m deep. The production of the charcoal used to fuel the iron smelters and forges, however, had a much more widespread impact on the landscape.

Prior to the Civil War, charcoal-fueled furnaces were located in the northern half of the Ridge and Valley and the Western Highland Rim (Schallenberg and Ault 1977). The East Tennessee furnaces either closed or switched to burning coke or coal by the 1870s. Many of the Western Highland Rim furnaces also closed; the few remaining in operation became larger, and some continued operating into the early twentieth century. Those operating in the early 1870s required the annual harvest of 2428 ha of timberland to provide charcoal; all of the forest was completely cut for up to 6 km around some older furnaces (Killebrew 1874). Later furnaces and charcoal-production techniques used wood more efficiently. Because of their much greater capacity, however, large areas of forest were still felled. Based on an average annual production of 27,885 metric tons of iron per year between 1872 and 1899 and an average requirement of 67 ha of forest to produce the charcoal to smelt 1000 metric tons of iron, 1870 ha of forest were annually felled to fuel the late-nineteenth-century furnaces (Schallenberg and Ault 1977, Williams 1989).

Many areas timbered for charcoal were slow to reforest, as wildfires were frequent in the cutover areas (Killebrew 1874). Areas protected from fire were sufficiently reforested in 25 to 30 years to provide a second cutting for charcoal. These areas, however, were uncommon. Because of the fires, a much larger area of timberland was required to fuel the furnaces than would have been the case with widespread fire protection. Sudworth and Killebrew (1897) noted this problem as the charcoal-fueled iron industry was dying out. Current topographic maps identify much of this Western Highland Rim area as "Coalings Land" or "The Coalings." The long-term effect of charcoaling in much of this area remains poorly known; it has probably resulted in lower soil fertility and a higher proportion of oaks and other shade-intolerant species than would have otherwise occurred.

Copper

Copper mining and smelting began in the Copper Basin of Polk County, in the extreme southeast corner of the state, in the 1850s (Quinn 1992). Although the ore was extracted from deep mines, the smelting process had severe impacts on the landscape from the intensive logging to fuel the process and the sulfur dioxide emitted by smelters and ore roasters. Production remained low until 1891, when a railroad into the basin was completed. By the time sulfur dioxide emissions and open-air ore roasting were halted about two decades later, about 3000 ha were completely denuded of all vegetation and topsoil; a surrounding area of about 7000 ha was heavily gullied and vegetated with scattered patches of grasses (Hursh 1948). Efforts to revegetate the area began in 1930, but made little noticeable progress until the 1970s. Accelerated efforts during the 1970s and 1980s resulted in revegetation by 1991 of most of the area, predominantly with loblolly pine (Quinn 1992). The bird population of this area remains low and is dominated by early successional species (Atlas results).

Coal

About 11,900 square km of the Cumberland Plateau and Cumberland Mountains are underlain by coal. Commercial mining of this coal began in the mid-1800s and increased greatly in the late 1800s (Killebrew 1874). Coal was extracted from deep mines, with relatively little direct impact on the landscape. Although surface or strip mining was first used in the early 1900s, it remained a minor method of coal extraction until the 1950s (Tennessee Department of Conservation and Tennessee Valley Authority 1960). During much of the 1970s and 1980s, more coal was mined from surface mines than from deep mines. Early surface mines rarely covered more than a few hectares, and they were usually abandoned with little reclamation. The size of surface mines increased in the 1960s; in mountainous areas, rock cliffs ("highwalls") and partially vegetated rock and soil benches and outslopes remained following mining. Streams draining mined areas were heavily polluted. Since 1977, reclamation of mined areas by the restoration of disturbed areas to their approximate original contours and by revegetation has been required by law. Recent surface mining has resulted in relatively long-lasting grasslands and shrublands, often 50 ha or more in size, in otherwise heavily forested areas. Approximately 22,613 ha have been disturbed by surface mining, 19,993 ha before 1977, and 2620 ha since then (Roger W. Bollinger, TVA, pers. comm.).

Annual coal production peaked in 1981 at 9.6 million metric tons and has recently declined to half that amount; most recent mining has been in the Cumberland Mountains and the southern Cumberland Plateau. Coal mining will continue to alter the landscape in the future, although the rate of alteration is difficult to predict. The remaining coal reserves in Tennessee, recoverable with current technology, were about 101 million metric tons in 1988 (Vickers and Kirby 1991). About one-third of these reserves are recoverable by surface mining.

Other Mining

Numerous other former and active mining operations occur across the state; compared to the previously described iron, copper, and coal operations, however, their local impacts are usually limited to only a few hectares and the cumulative area disturbed is probably only a few thousand hectares. The major nonfuel minerals mined include clays, phosphate rock, sand and gravel, building stone, limestone, and zinc ore (Vickers and Kirby 1991). By tonnage, zinc ore is mined in the highest volume among these minerals. It is extracted from deep mines, and the major surface disturbances are tailings piles, an unusual habitat used for nesting by Bank Swallows in Knox and Grainger Counties. Phosphate is extracted from surface mines in the southern Central Basin, where recent mines have been reclaimed to resemble the pre-mining landscape. Some earlier mines, and especially settling ponds used in phosphate processing, have become unusual lake and wetland habitats, such as the Monsanto Ponds complex in Maury County. The other important minerals are primarily extracted from quarries and sand and gravel pits concentrated in the eastern two-thirds of the state.

Wetlands, Streams, and Reservoirs

From an estimated prehistoric area of about 784,000 ha, Tennessee wetlands were reduced by 59% to 319,000 ha (2.9% of the state) by the early 1980s (Dahl 1990). This loss is greater than both the 53% average loss for the lower 48 United States, and the 50% average loss for the 11 southeastern states. The greatest cause of wetlands loss has been from drainage for agricultural use, primarily through stream channelization. In terms of the area affected, the loss has been greatest in West Tennessee forested wetlands; every major direct tributary stream to the Mississippi River, except for the main stem of the Hatchie River, has been altered, some repeatedly since the early 1900s (Hupp 1992). Other wetland types across the state have also been greatly altered, such as the once extensive bogs in Johnson County (Ganier and Tyler 1934, Barclay 1957). About 90% of the remaining wetland area is in West Tennessee, where forested wetlands make up the most extensive type. Forested wetlands account for about 40% of the wetland area in Middle and East Tennessee (Technical Working Group 1993).

Although few quantitative data are available on recent wetland trends, the rate of wetland loss through agricultural conversion has apparently slowed, in part due to federal regulations and farm subsidy programs enacted in the 1980s (Technical Working Group 1993). Some marginal cropland has also been abandoned and reverted to wetlands, and, in the absence of regular channel maintenance, woody wetland vegetation is recovering along some channelized streams (Hupp 1992). Extensive emergent wetlands have also recently developed along a few West Tennessee streams. Wetland loss due to urbanization and highway construction may be increasing (Technical Working Group 1993); this loss is sometimes "mitigated" by the construction of artificial wetlands. However, the long-term value of these constructed wetlands for wildlife is poorly known.

The TWRA, U.S. Fish and Wildlife Service, and U.S. Army Corps of Engineers have recently purchased large areas of wetlands, primarily in West Tennessee.

These purchases have been aided by a state real estate transfer tax dedicated to wetlands acquisition and by the North American Waterfowl Management Plan (Tennessee Wildlife Resources Agency 1990a). The state Wetlands Conservation Strategy sets a goal of restoring 28,340 ha of marginal cropland in West Tennessee to wetlands by the year 2000 (Technical Working Group 1993).

Beavers are also an important factor in wetland dynamics. They were presumably widespread in prehistoric Tennessee, and their ponds and the early successional woodlands that followed pond abandonment increased the diversity of the landscape. Because of unregulated fur trapping, beavers were eliminated from much of the state by the early twentieth century (Rhoads 1896a, Kellogg 1939). The beaver population has greatly increased in recent decades as a result of deliberate reintroductions for erosion control and wildlife habitat improvement, through the natural increase of remnant populations, and by immigration from adjacent states. Beavers are now found at lower elevations across the state, and local control of nuisance beavers is often necessary to prevent damage to roads and commercial forests.

The construction of reservoirs and ponds has also greatly altered the Tennessee landscape. The first hydroelectric and navigation improvement dams were built in the eastern half of the state around 1900. The reservoirs created by these dams had surface areas of a few hundred hectares. Following the establishment of the Tennessee Valley Authority (TVA) in 1933, numerous large reservoirs with surface areas of thousands of hectares have been built. The TVA and the U.S. Army Corps of Engineers have built dams on the main channels and most major tributaries of both the Tennessee and Cumberland Rivers. These reservoirs have a total surface area of about 234,000 ha, 2.2% of the state. Of the major river systems in Tennessee, only the Hatchie in West Tennessee, the Buffalo, lower Duck, and Harpeth in Middle Tennessee, and the Sequatchie in East Tennessee remain relatively free of the effects of impoundments and channelization. When the natural lakes, unimpounded rivers, and the thousands of small farmponds and recreational lakes built across the state are added to the area of the larger reservoirs, the total area of the state covered by water is about 3% (Soil Conservation Service 1989).

Historic Changes in the Tennessee Avifauna

The avifauna of late prehistoric Tennessee, as partially reconstructed from archaeological material (P. W. Parmalee pers. comm.), was similar to the present avifauna. Obvious differences include the presence of remains of species that have become extinct or extirpated, such as the Passenger Pigeon and Ivory-billed Woodpecker, and the absence of recently introduced species, such as the Rock Dove and European Starling. The distribution and abundance of many extant, native species have also changed during this period, primarily as a result of human manipulation of their habitats. Most species occupying grassland habitats, for example, are undoubtedly much more numerous and widely distributed than they were in prehistoric Tennessee. Conversely, most forest-dwelling species are less common, and two, the Passenger Pigeon and Carolina Parakeet, are extinct. Many waterbirds are probably more numerous and widespread as a result of the numerous ponds and reservoirs constructed across the state.

The sketchy historical information available for many species makes it difficult to reconstruct their pre-European settlement ranges and describe their postsettlement changes in distribution and abundance. Although the accounts of the early explorers, and particularly those of the naturalists André Michaux, Alexander Wilson, and John James Audubon (see the second chapter, "The History of Tennessee Ornithology") contain valuable records, they deal with few species and limited areas of Tennessee. The first reasonably comprehensive local list was that of Fox (1886). It was followed a decade later by Rhoads's statewide survey (Rhoads 1895a). By the time of Rhoads's survey, the human population of Tennessee was approaching 2 million and 40% of the forest had been cleared (see previous chapter, "The Landscape of Tennessee").

The remainder of this chapter summarizes changes in the distribution and abundance of Tennessee birds. It concludes with a description of past and current conservation efforts.

Changes in Distribution

Tanner (1988) summarized the major historic range changes of many Tennessee birds. Of the 26 species he identified as having expanded ranges, I delete the Brown Creeper, which has probably only reclaimed its historic West Tennessee range, and add six more species (table 2). To his list of nine species with decreased ranges, I add three more species (table 2), and remove the Wild Turkey, which has now reoccupied much of its former range (Tennessee Wildlife Resources Agency 1994c), and the Anhinga, which does not appear to have established nesting populations outside the Reelfoot Lake area. This analysis identifies 28 species with expanded ranges and 10 species with decreased ranges. Most of these range changes are the result of changes in the availability of the species's preferred habitats (table 2). Each of these range changes is further described in the species accounts elsewhere in this book.

A problem in conducting an analysis of changes in distribution is determining, from the sometimes sketchy

Table 2
Permanent Resident and Migrant Species Breeding in Tennessee Showing Major Range Changes and Primary Causes of Those Changes.

Category	Primary Causes
SPECIES WITH EXPANDED RANGES	
Great Blue Heron*	Increase in habitat
Cattle Egret*	Natural range expansion
Black-crowned Night-Heron*	Increase in habitat
Yellow-crowned Night-Heron*	Increase in habitat
Canada Goose*	Introduction into newly created habitat
Osprey*	Increase in habitat
Bald Eagle*	Increase in habitat
Black-necked Stilt*	Natural expansion into newly created habitat
Rock Dove*	Human introduction
Alder Flycatcher*	Natural range expansion
Willow Flycatcher*	Natural range expansion
Eastern Phoebe*	Increase in habitat
Tree Swallow* (breeding range)	Natural Range expansion
Bank Swallow (breeding range)	Increase in habitat
Fish Crow	Natural range expansion
Brown-headed Nuthatch*	Natural range expansion
House Wren* (breeding range)	Increase in habitat
European Starling*	Human introduction
Pine Warbler	Increase in habitat
Blue Grosbeak*	Increase in habitat
Eastern Towhee	Natural range expansion
Savannah Sparrow (breeding range)	Increase in habitat
Song Sparrow* (breeding range)	Increase in habitat
Brown-headed Cowbird*	Increase in habitat
House Finch*	Human introduction
House Sparrow*	Human introduction
SPECIES WITH DECREASED RANGES	
Peregrine Falcon*	Pesticide poisoning
Ruffed Grouse	Habitat loss, human persecution
Greater Prairie-Chicken*	Human persecution, habitat loss
Passenger Pigeon*	Human persecution, habitat loss
Carolina Parakeet*	Human persecution, habitat loss
Red-cockaded Woodpecker	Habitat loss
Common Raven*	Human persecution, habitat loss
Bewick's Wren*	?
Golden-winged Warbler	Habitat loss, interspecific competition (?)
Bachman's Sparrow*	Habitat loss

NOTE: *Discussed in Tanner (1988).

historical data, what constitutes a range change, as well as what constitutes an established population. Also, the recent records of several species are too few to determine confidently their present breeding range. In some cases, the "first" breeding location of a species in a part of the state was not revisited in later years to determine if the species was truly established there. The problem of poor historic and recent information is especially acute for cryptic species and species in habitats difficult to survey, such as American and Least Bitterns, King and Virginia Rails, Common Moorhen, and Purple Gallinule. In these cases, while there is little doubt that their preferred marsh habitats have decreased, it is difficult to state that their overall ranges in the state have changed.

Changes in Abundance

The best information on long-term population trends of birds breeding in Tennessee is provided by the Breeding Bird Survey, which dates from 1966. Of 123 species censused by Tennessee BBS routes (about three-quarters of the breeding avifauna), 93 occurred on enough routes (14) to analyze their population trends by route regression techniques (Geissler and Sauer 1990, Link and Sauer 1994). Fifty-six of these species showed either significant ($p < 0.10$) increasing or decreasing overall trends from 1966 to 1994 (table 3). Thirty-nine (70%) of the species with significant population trends showed decreases, a greater proportion than expected by chance (chi square = 8.64, $p < 0.005$).

Table 3
Breeding Species Showing Significant (p < 0.10) Increasing or Decreasing Population Trends as Measured by Breeding Bird Surveys, 1966–94, and Analyzed by the BBS Staff of the National Biological Service

Species	Annual Trend, % 1966–94	1966–79	1980–94	Migratory Status	Nest Site	Habitat
NONPASSERINES						
Great Blue Heron	+23.8‡		+13.2‡	S	OH	Wetland/Lake
Green-backed Heron	−2.2‡		−5.0†	M	OH	Wetland/Lake
Wood Duck	+6.9*			S	C	Wetland/Lake
Turkey Vulture	+2.2*			S	OL,C	Unclassified
Broad-winged Hawk			−10.8†	M	OH	Large, Small forest
American Kestrel	+2.3*	+12.0‡		P	C	Grassland/Agricultural, Scrub
Northern Bobwhite	−3.0‡	−1.6‡	−4.3‡	P	OL	Grassland/Agricultural
Killdeer	+1.7†	+6.7‡		P	OL	Grassland/Agricultural
Yellow-billed Cuckoo	−1.8†			M	OL	Small forest
Great Horned Owl			+21.7†	P	OH	Small forest
Common Nighthawk			−8.3*	M	OL	Suburban/Urban, Grassland/Agricultural
Chuck-will's-widow			−7.1‡	M	OL	Small forest
Whip-poor-will	−5.8†			M	OL	Large forest
Chimney Swift	−1.5†		−2.3†	M	C	Unclassified
Belted Kingfisher	−2.0*			P	C	Wetland/Lake
Red-headed Woodpecker	+6.5†			P	C	Small forest
Yellow-shafted Flicker	−2.8‡	−2.5*	−4.1†	S	C	Small forest
PASSERINES						
Acadian Flycatcher		−2.0‡		M	OH	Large forest, Riparian
Eastern Phoebe			+4.2‡	S	OL	Unclassified
Eastern Kingbird	−1.4*			M	OH	Grassland/Agricultural, Scrub
Horned Lark	−7.9*			P	OL	Grassland/Agricultural
Purple Martin	+3.2†			M	C	Suburban/Urban
Barn Swallow	−2.5‡		−2.8‡	M	OL	Unclassified
Blue Jay	−2.7‡	−2.2‡		P	OH	Small forest
American Crow	−0.6*	−1.3*		P	OH	Small forest
Carolina Chickadee			+2.2†	P	C	Small forest
Tufted Titmouse			+1.9†	P	C	Small forest
White-breasted Nuthatch	+8.3‡	+8.8‡	+3.7*	P	C	Small forest
Carolina Wren			+6.2‡	P	C	Small forest, Suburban/Urban
Bewick's Wren	−22.0‡	−16.0‡		S	C	Scrub, Suburban/Urban
Eastern Bluebird		−4.2‡	+7.3‡	P	C	Grassland/Agricultural, Scrub
Wood Thrush	−2.3‡		−3.2	M	OH	Small forest
American Robin	+2.9‡	−3.1‡	+2.1‡	P	OH	Suburban/Urban
Gray Catbird	−5.0‡	−3.3‡	−10.2‡	M	OL	Scrub, Riparian
Northern Mockingbird	−1.1‡	−2.0‡	+0.8	P	OL	Suburban/Urban, Grassland/Agricultural
Brown Thrasher	−1.0*		−2.7†	S	OL	Scrub, Small forest, Suburban/Urban
Cedar Waxwing	+14.6‡		+25.2‡	S	OH	Small forest, Suburban/Urban
Loggerhead Shrike	−7.0‡	−4.7‡		P	OL	Grassland/Agricultural, Scrub
European Starling		−3.1‡	+3.8‡	P	C	Suburban/Urban, Small forest, Grassland
White-eyed Vireo	−1.2*		−3.3‡	M	OL	Scrub
Red-eyed Vireo			+2.1‡	M	OH	Small forest, Large forest
Northern Parula			+4.2*	M	OH	Small forest, Large forest, Riparian
Yellow Warbler	−3.0†		−7.2‡	M	OL	Scrub, Riparian
Yellow-throated Warbler	+2.0*	+4.8‡		M	OH	Small forest, Riparian
Pine Warbler	+9.8‡	+12.9‡	+7.3	S	OH	Small forest
Prairie Warbler	−3.6‡	−2.9*	−2.2*	M	OL	Scrub
Cerulean Warbler		−5.1‡		M	OH	Large forest
Black-and-White Warbler	−5.5‡		−5.6‡	M	OL	Large forest
American Redstart	−6.4‡	−6.4†		M	OH	Large forest
Worm-eating Warbler	−2.6*			M	OL	Large forest
Ovenbird	−2.3‡		−3.1‡	M	OL	Large forest
Louisiana Waterthrush	−2.3*			M	OL	Riparian, Large forest
Kentucky Warbler	−3.0†	−3.3*		M	OL	Small forest
Common Yellowthroat	−1.2†		−2.7‡	M	OL	Scrub, Grassland/Agricultural

Continued on page 44

Table 3—Continued

Species	Annual Trend, % 1966–94	1966–79	1980–94	Migratory Status	Nest Site	Habitat
Hooded Warbler	-1.5‡			M	OL	Large forest
Yellow-breasted Chat	-4.2‡	-5.3‡		M	OL	Scrub
Scarlet Tanager	+2.2*	+3.8‡		M	OH	Large forest
Northern Cardinal	-1.2‡	-2.1‡		P	OL	Scrub, Small forest
Blue Grosbeak	+3.6‡	+8.1‡	+2.8*	M	OL	Scrub, Grassland/Agricultural
Indigo Bunting	-1.8‡	-2.5‡		M	OL	Scrub
Dickcissel		-8.9‡	+3.2*	M	OL	Grassland/Agricultural, Scrub
Eastern Towhee	-2.1‡			P	OL	Scrub
Chipping Sparrow	-1.9†	-6.7‡		S	OL	Suburban/Urban, Small forest
Field Sparrow	-3.0‡	-6.6‡		P	OL	Scrub
Grasshopper Sparrow		-10.2‡		M	OL	Grassland/Agricultural
Song Sparrow	+2.6‡	+5.6‡	-2.1‡	P	OL	Suburban/Urban, Scrub, Riparian
Red-winged Blackbird		+4.5‡	-1.6†	P	OL	Grassland/Agricultural, Wetland
Eastern Meadowlark	-2.7‡	-2.0‡	-2.3‡	P	OL	Grassland/Agricultural
Common Grackle	-3.4‡	-2.1‡	-3.3‡	P	OH	Small forest, Riparian
Brown-headed Cowbird	-1.6*			P	—	Unclassified
Orchard Oriole	-1.5‡		-2.5‡	M	OH	Scrub, Grassland/Agricultural
Baltimore Oriole	+5.1†			M	OH	Small forest, Riparian
House Finch	+60.9‡		+63.3‡	P	OL	Suburban/Urban
American Goldfinch		-4.9‡	+3.8‡	P	OH	Scrub
House Sparrow		-4.2‡	-2.7†	P	C	Suburban/Urban, Grassland/Agricultural

NOTES: The absence of an entry in a trend column signifies that the trend during that period was not significant. This list excludes species found on fewer than 14 BBS routes. Migratory status codes are P = permanent resident; S = short-distance migrant; M = long-distance, neotropical migrant. Nest site codes are C = cavity; OH = other sites (primarily open cup) in trees; OL = other sites (primarily open cup) in low vegetation or on ground.
*$p < 0.10$.
†$p < 0.05$.
‡$p < 0.01$; chi-square tests.

The species with significant population trends were categorized by life history traits (tables 3, 4) to investigate possible explanations for the trends, as has been done by several other authors (e.g., Robbins et al. 1989, Sauer and Droege 1992, Böhning-Gaese, Taper, and Brown 1993) for larger geographic areas. Species primarily nesting in forested habitats were classified into small forest species, whose probability of occurrence shows little increase with the size of the forested area (e.g., Robbins, Dawson, and Dowell 1989) and frequently occur in areas of less than 5 ha, and large forest species, whose probability of occurrence increases with the forested area and which infrequently occur in areas of less than 5 ha. Nest site was used as an indicator of a species's susceptibility to both nest predation and Brown-headed Cowbird parasitism. Cavity-nesting species generally have higher nest success than species with other nest types (Nice 1957, Martin and Li 1992), and, for potential cowbird hosts (many of the passerines in table 3, see also appendix 3), usually lower parasitism rates.

The results of this analysis by life history traits (table 4) are generally consistent with the analyses for larger areas cited above. Among long-distance neotropical migrant species, species occupying scrub habitats, and species building open nests low to the ground, the proportion of decreasing species is greater than expected by chance. Among species requiring large forested areas, only the Scarlet Tanager, a neotropical migrant, shows an increase. The only increasing species primarily using scrub (shrubby old fields, early successional forests)

Table 4
Numbers of Bird Species Showing Significant Decreasing or Increasing Population Trends on 1966–94 Tennessee BBS Routes in Relation to Life History Traits

Life History Trait	Population Trend	
	Decreasing	Increasing
Migratory status		
Permanent resident	13	7
Short-distance migrant	4	5
Long-distance neotropical migrant*	22	5
Principal breeding habitats		
Wetland/Lake	2	2
Riparian	1	0
Large forest (> ca. 5 ha)	6	1
Small forest (< ca. 5 ha)	7	6
Scrub*	14	1
Suburban/Urban	1	4
Grassland/Agricultural	4	2
Unclassified	3	1
Nest site—all species		
Cavity	4	5
Other, in tree	10	6
Other, in low vegetation or on ground*	26	5

NOTE: *$p < 0.01$, chi-square test. Significant difference in the proportion of decreasing and increasing species.

habitat is the Blue Grosbeak, which has greatly expanded its Tennessee breeding range in recent decades.

A few researchers, among them Robbins et al. (1989) and Sauer and Droege (1992), have noted an increase since the late 1970s in the proportion of species in eastern North America, especially neotropical migrant species, with decreasing population trends. This was not as evident in Tennessee BBS results, which show 22 neotropical migrants with long-term (1966–94) decreases and 15 neotropical migrant species with recent (1980–94) decreasing trends (table 3). An additional three neotropical migrants that did not have a significant long-term trends showed significant recent decreasing trends. Four neotropical migrants—the Red-eyed Vireo, Northern Parula, Blue Grosbeak, and Dickcissel—showed recent increasing trends, compared to five with long-term increases.

The primary cause of these changes in abundance, as with the changes in distribution described earlier, is habitat change, on the breeding grounds as well as, for migrant species, on their wintering grounds. The relative contribution of habitat changes on breeding versus wintering grounds is controversial (e.g., Robbins et al. 1989, Böhning-Gaese, Taper, and Brown 1993); both are undoubtedly important. Habitat changes can be either direct, through changes in the area of a particular habitat type, or indirect. The decreases among scrub species are probably directly related to the decreased area of early successional forests (see previous chapter), and changes in agricultural practices which have reduced the area in fencerows and shrubby ("unimproved") pastures. Although the overall area of forest in Tennessee is increasing (as outlined in the previous chapter), the quality of this forest for species requiring extensive tracts of mature forest (forest interior species) may be decreasing as road building and suburban/urban development fragment contiguous tracts of forest. Unfortunately, no long-term analysis of this change presently exists.

The decrease in the size of forest tracts increases the amount of edge, the boundary between the forested and unforested area. Several environmental changes occur at edges; among the most important to birds are higher rates of both nest parasitism by the Brown-headed Cowbird and nest predation near forest edges than in the forest interior (Askins, Lynch, and Greenberg 1990). As the size of the forest decreases, the proportion of the area not exposed to edge effects (the forest interior) decreases, and forested tracts of only a few hectares have no forest interior habitat. Although less studied, these effects may also occur when extensive areas of other habitats, such as scrub, are fragmented.

Habitat fragmentation is probably a reason why,

among species using open (not cavity) nests, a greater proportion are declining rather than increasing, a comparison also true when the analysis is restricted to passerines. It may be even greater when the fragmentation is caused by suburban/urban development because of increased densities of the domestic cat, a frequent predator on adult and fledgling birds, and of the raccoon, a nest predator frequently attracted by household garbage. Bird feeders may also increase local densities of the Blue Jay, a nest predator, and the Brown-headed Cowbird.

The population trends of at least the neotropical migrant species are apparently not uniform across the state. Although the number of BBS routes is too low to allow separate analyses by Tennessee physiographic provinces, multistate analyses by physiographic provinces show regional differences in trends for individual warbler species and for migrant species as a whole (James, Wiedenfeld, and McCulloch 1992, Sauer and Droege 1992). Several individual warbler species, as well as migrants as a whole, show more decreasing trends in the Cumberlands, Ridge and Valley, and Blue Ridge than in provinces farther west. This is probably due in part to regional differences in trends in habitat availability. Some of it may also be due to sampling artifacts, as the few routes in the Cumberlands and Blue Ridge are mostly along low-elevation public roads, where the rate of development is highest and perhaps not representative of the regions as a whole.

Conservation Efforts

During the first half of this century, specific efforts to reverse range reductions and population declines of Tennessee birds were directed toward game species, notably waterfowl and the Wild Turkey. Populations of wintering waterfowl increased through restrictive hunting regulations and habitat development, particularly in wildlife refuges and management areas. Since peaking throughout much of the United States in the 1950s, waterfowl have slowly declined through loss of breeding and wintering habitat. The North American Waterfowl Management Plan was established in 1986 to halt this decline (Tennessee Wildlife Resources Agency 1990a) and has contributed to the establishment of several new wildlife management areas in Tennessee. Wood Duck numbers have greatly increased as a result of restrictive hunting regulations, riparian habitat protection, and, at least locally, by providing nest boxes. Early attempts to restore Wild Turkey populations, once extirpated from most of the state, consisted of closing hunting seasons and stocking farm-reared birds (Holbrook and Lewis 1967, Schultz 1955). More recent efforts in transplanting wild birds have succeeded in restoring turkeys to much of their former range (Tennessee Wildlife Resources Agency 1994c). Throughout this period, management areas were established for several game species besides waterfowl. The land in these areas, as well as the waterfowl management areas, has benefited many nongame species (e.g., species not hunted).

Specific efforts toward restoring nongame birds began in the 1970s with passage of the U.S. Endangered Species Act and state legislation protecting species classified as Endangered, Threatened, and In Need of Management (table 5) (Tennessee Wildlife Resources Agency 1975, 1976 and subsequent proclamations, U.S. Fish and Wildlife Service 1991, 1992, 1994). These efforts have been very successful for a few species. The Bald Eagle nesting population presently exceeds the target set in its southeastern recovery plan (U.S. Fish and Wildlife Service 1989), and the Peregrine Falcon will probably begin nesting in the state in the near future. The East Tennessee Osprey population has grown rapidly as the lingering effects of pesticides have waned and secure nest sites were provided. New Osprey populations have become established elsewhere in the state by pioneering birds and deliberate releases of young birds. The Black-crowned Night-Heron, as well as other wading birds, has benefited from cooperative efforts to protect nesting areas, and populations of most wading birds are increasing. Because of the lack of a means of measuring their success, the long-term results of programs releasing young Mississippi Kites and Barn Owls are difficult to evaluate.

Table 5
Species Listed as Endangered, Threatened, or In Need of Management in Tennessee, and Recommended State Listings

Species	Federal Listing[a]	State Listing[b]	Recommended State Listing
Double-crested Cormorant	—	In Need of Management	In Need of Management
Anhinga	—	In Need of Management	In Need of Management
Least Bittern	—	In Need of Management	In Need of Management
Great Egret	—	In Need of Management	In Need of Management
Snowy Egret	—	In Need of Management	In Need of Management
Little Blue Heron	—	In Need of Management	In Need of Management
Black-crowned Night-Heron	—	—	In Need of Management
Osprey	—	Threatened	In Need of Management
Mississippi Kite	—	In Need of Management	In Need of Management
Bald Eagle	Threatened	Threatened	Endangered
Northern Harrier[c]	—	In Need of Management	In Need of Management
Sharp-shinned Hawk	—	In Need of Management	In Need of Management
Cooper's Hawk	—	In Need of Management	In Need of Management
Golden Eagle	—	Threatened	In Need of Management
Peregrine Falcon	Endangered	Endangered	Endangered
Black Rail	Former Candidate[d]	—	—
King Rail	—	In Need of Management	In Need of Management
Virginia Rail[e]	—	—	Threatened
Purple Gallinule	—	—	In Need of Management
Common Moorhen	—	—	In Need of Management
Sandhill Crane[c]	—	In Need of Management	In Need of Management
Least Tern	Endangered	Endangered	Endangered
Barn Owl	—	In Need of Management	In Need of Management
Northern Saw-whet Owl	—	In Need of Management	In Need of Management
Yellow-bellied Sapsucker[e]	—	In Need of Management	Threatened
Red-cockaded Woodpecker	Endangered	Endangered	Endangered
Olive-sided Flycatcher	—	In Need of Management	Threatened
Common Raven	—	Threatened	In Need of Management
Bewick's Wren[f]	Former Candidate	Threatened	Threatened
Loggerhead Shrike[g]	Former Candidate	—	—
Cerulean Warbler	Former Candidate	—	—
Swainson's Warbler	—	In Need of Management	In Need of Management
Bachman's Sparrow	Former Candidate	Endangered	Endangered
Vesper Sparrow	—	In Need of Management	In Need of Management
Lark Sparrow	—	Threatened	Endangered
Grasshopper Sparrow	—	In Need of Management	In Need of Management

NOTES: [a]USFWS 1992, 1994a, 1996.
[b]TWRA 1994a, 1994b.
[c]Migrant/wintering populations.
[d]A candidate for federal listing as Endangered or Threatened, for which insufficient information to justify listing is presently available.
[e]Breeding population only.
[f]Federal candidate status applies to the subspecies altus nesting in the eastern third of the state.
[g]Federal candidate status applies to the subspecies migrans nesting in the eastern third of the state.

Not all species, however, have fared well, despite the added attention given to them by their endangered or threatened status. The Red-cockaded Woodpecker has recently disappeared from the state, and the range of the Bachman's Sparrow has continued to contract. Recommended changes and additions to the state list, based in part on the Atlas results, are given in table 5; the justifications for these are described in the individual species accounts.

More broad-based efforts to conserve Tennessee's birds, as well as other animals and plants, began as the Atlas project neared completion. They are cooperative efforts of several state, federal, and private organizations, including the Tennessee Ornithological Society. One of these efforts, the Tennessee Biodiversity Program, is identifying unprotected areas with high species diversity or populations of rare species. This "gap analysis" (Scott et al. 1993) combines a computer-based vegetation map of the state with maps of Breeding Bird Atlas results, range maps of amphibians, reptiles, wintering birds, and mammals, and mapped locations of rare plants and animals. The results are then used by land planners to better consider conservation needs when making land use decisions.

The second effort is the Tennessee part of the Neotropical Migratory Bird Conservation Program. This international program, also known as Partners in Flight, is an international effort begun in 1990 to reverse the decline of neotropical migrant birds. The Tennessee portion is adding to the Biodiversity Program described above to promote management of migratory songbirds on public and private lands (Ford and Cooper 1993). It is strongly supported by the Tennessee Wildlife Resources Agency and has already resulted in an increase in bird surveying and monitoring, as well as more specific research on nongame birds.

An Overview and Analysis of Atlas Results

During the six years of Atlas fieldwork, observers reported at least one bird species from each of 2736 Atlas blocks (map 11). These blocks comprised 65% of the 4200 blocks in the state. The Atlas project came very close to its goal of a complete species list and a completed miniroute in all 700 priority blocks. All priority blocks received some coverage, and the species list is considered complete in 655 (93%) of them (map 12). Coverage for nocturnal species, primarily goatsuckers and owls, was uneven, and the absence of these species was not used as a criterion in determining completed species lists. Acceptable miniroutes were censused in 692 (99%) of the priority blocks (map 13). The observers, mostly volunteers, reported spending 35,540 hours working the Atlas blocks and almost one-third of that time (11,387 hours) was spent completing the priority blocks.

The total number of Atlas records was 94,097, and records from completed priority blocks totaled 43,520. Confirmed breeding records made up 37% of the records from completed priority blocks (table 6), close to the goal of 40% adopted after the second year of fieldwork. The proportion of probable records, 26%, was lower than desired, but, as explained in the first chapter, was partly due to the "blockbusting" method used in working many priority blocks. Of the 170 total species reported by Atlas workers, 164 occurred in completed priority blocks. Breeding was confirmed for all but 8 of the 170 species (table 7). These numbers do not include a White-tailed Kite found in a Humphreys County block in June 1991 (Purrington 1991), the second state record of a species whose eastern population nests as far north as Louisiana (American Orni-

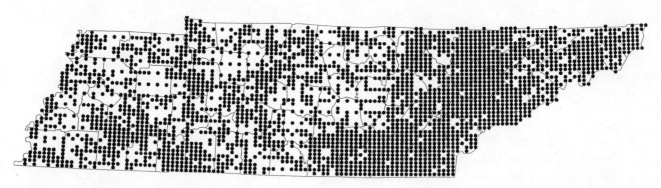

Map 11. Atlas blocks from which at least one species was reported.

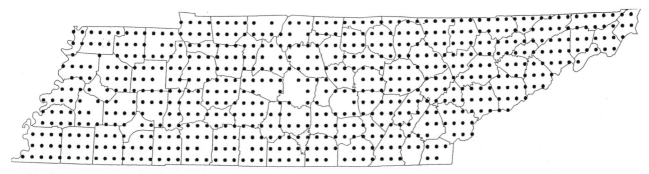

Map 12. Atlas priority blocks with complete species lists.

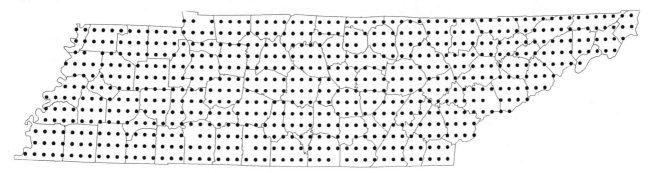

Map 13. Atlas priority blocks with completed miniroutes.

Table 6
Number and Proportion of Atlas Records by Breeding Status and Block Type

	Breeding Status		
Block Type	Possible	Probable	Confirmed
All blocks	47,846 (51%)	20,111 (21%)	26,140 (28%)
Completed priority blocks	16,045 (37%)	11,508 (26%)	15,967 (37%)

Table 7
Species Reported during Atlas Fieldwork with Only Probable or Possible Evidence of Breeding

Breeding Evidence	
Probable	Possible
Double-crested Cormorant	Purple Gallinule
Peregrine Falcon	Olive-sided Flycatcher
American Coot	
Hermit Thrush	
Magnolia Warbler	
Painted Bunting	

Table 8
The Twenty Most Frequently Reported Breeding Birds

Species	Number of Blocks	% Completed Priority Blocks
Indigo Bunting	1980	99.8
Northern Cardinal	1898	99.1
American Robin	1776	95.4
American Crow	1766	100.0
Mourning Dove	1766	97.6
Carolina Wren	1727	99.2
Common Grackle	1699	96.0
Eastern Bluebird	1660	94.5
Barn Swallow	1623	94.8
Tufted Titmouse	1605	99.7
Blue Jay	1605	99.4
Eastern Meadowlark	1582	91.9
European Starling	1580	90.4
Northern Mockingbird	1566	87.8
Eastern Towhee	1555	99.4
Carolina Chickadee	1554	99.4
Red-winged Blackbird	1535	94.2
Chimney Swift	1487	97.7
Common Yellowthroat	1472	97.1
Eastern Phoebe	1406	95.9

thologists' Union 1983). They also do not include the Ring-necked Pheasant, which was reported in two blocks. Because no self-sustaining populations of this formerly introduced species are known, these records were probably of escaped birds, as described in the pheasant species account.

The 20 most frequently reported species were all easily detected from roadsides (table 8), and 14 of these 20 were found in 95% or more of the completed priority blocks. At the opposite extreme, five species were found in only one or two blocks (table 9). All of these rare species occupied habitats of very limited area

An Overview and Analysis of Atlas Results

Table 9
The Least Frequently Reported Breeding Birds

Species	Number of Blocks
Virginia Rail	2
Sora	1
Red-cockaded Woodpecker	1
Yellow-bellied Sapsucker	1
Magnolia Warbler	1

in Tennessee and, in the case of the rails and gallinule, were relatively difficult for Atlas workers to observe.

The number of species in completed priority blocks ranged from 20 to 91, and averaged 66.5 (s.d. = 7.9). Thirty blocks had 80 or more species and 22 had fewer than 50 species. All but one of the blocks with fewer than 50 species were in the Unaka Mountains; the exception was the Stockton block on the northern Cumberland Plateau in Fentress County.

The only breeding species found in the state for the first time during the Atlas period, and one which I did not expect we would find, was the Sora, confirmed by the presence of a brood of young at Memphis. The confirmed breeding record of the Northern Saw-whet Owl in Claiborne County was both outside of its previously known range and, although juveniles had previously been found in the Smokies, the first unequivocal confirmed breeding record for the state. Other confirmed breeding records that were very unusual because of their location were the Red Crossbills at Memphis and the Brown Creepers near Kingsport. The Willow Flycatcher, Tree Swallow, and House Finch, as expected, continued their spread across the state. The Cedar Waxwing, however, had what was probably the most explosive increase in breeding distribution, as its breeding was confirmed in 23 new counties and for the first time in West Tennessee. Because the Cedar Waxwing population on Breeding Bird Survey routes also increased during the Atlas project, its breeding range expansion was more than an artifact of the increased fieldwork.

The systematic fieldwork throughout the state produced, as expected, much valuable information about the distribution of Tennessee's Endangered and Threatened bird species. The wide distribution and number of records of the Sharp-shinned and Cooper's Hawks and Grasshopper Sparrows justify downgrading their status from Threatened to In Need of Management. The Red-shouldered Hawk is common enough to justify removal from the list of species In Need of Management. Much of the Atlas information on listed species, however, is not good. The Red-cockaded Woodpecker was found in only one block in the southeastern corner of the state, and, since the Atlas period, it has disappeared from the state. The Yellow-bellied Sapsucker was also found in only one block, although its status is not as critical as was the Red-cockaded's in 1991. The number of records of several marsh birds was also disappointingly low, and several of them deserve a more protective status in Tennessee.

Trends in Species Richness

To explore trends in species richness (the number of species occurring in a defined area) of birds breeding in completed Atlas priority blocks, the species lists of these blocks were standardized by eliminating the following 10 poorly sampled species:

Wild Turkey
American Woodcock
Eastern Screech-Owl
Great Horned Owl
Barn Owl
Barred Owl
Northern Saw-whet Owl
Common Nighthawk
Whip-poor-will
Chuck-will's-Widow

After this compensation for differences in block coverage, the maximum possible richness was 154 species. The adjusted total number of species per completed priority block (species/block) ranged from 20 to 84 and averaged 64.4 (s.d. = 7.3, n = 655) (map 14). The blocks with the lowest number of species were located in the Cumberlands and the Unaka Mountains and were almost entirely forested lands. The blocks with the highest number of species were, in contrast, scattered across the state and composed of a combination of extensive forested tracts, farmland, and water bodies. The proportion of neotropical migrant species/block ranged from 30.4 to 58.9% and averaged 44.4% (s.d. = 5.1) (map 15); it was generally highest in heavily forested areas.

The total number of species/block increased slightly from west to east and from south to north. When analyzed by simple linear regression, both increases are significant ($p < 0.01$), although the proportion of the variation in species numbers explained is very low for both longitude ($r^2 = 0.011$) and latitude ($r^2 = 0.028$). The overall geographic trends in total species numbers are similar to the trends mapped by Cook (1969) for North America, except that the increase found by Cook in the southern Appalachians was not as pronounced in the Atlas data. This difference is due to the more limited range of habitats within the Atlas blocks, which are many times smaller than the mapping units used by Cook.

The proportion of neotropical migrant species/block significantly increased with latitude ($p < 0.05$) but not

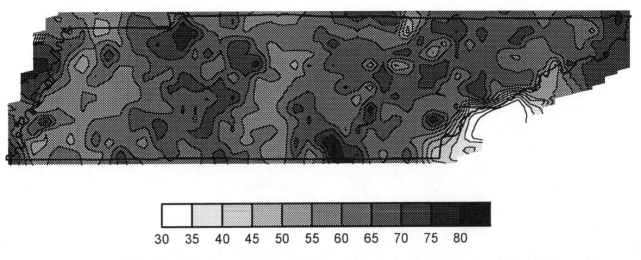

Map 14. *Contour map of the adjusted total number of species per completed priority block. The contour interval is five species.*

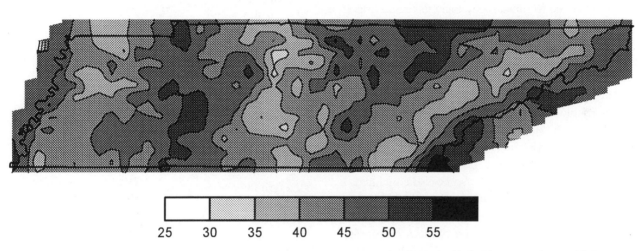

Map 15. *Contour map of the proportion of neotropical migrant species per completed priority block. The contour interval is 3 percent.*

with longitude. The northward increase is due in large part to the high proportion of neotropical migrants in the Cumberland Mountains, the northern Cumberland Plateau, and the northeastern Highland Rim.

Both the total number and proportion of neotropical migrant species/block were significantly related to several gross environmental measures, although the predictive ability of these measures was very low (table 10). As described in the third chapter, "The Environment of Tennessee," rainfall, temperature, and elevation are interrelated. The blocks with the lowest number of species were at the highest elevations of the Unakas. Their low species numbers were due both to the limited variety of habitats they contained and, for the few dominated by spruce-fir forest, the low number of species in that habitat, as noted by Rabenold (1993) and others.

Both the total number and the proportion of neotropical migrant species/block also significantly varied (analysis of variance, $p < 0.01$) between landscape units classified by physiographic regions (map 3) and by potential natural vegetation (map 7). Of the two landscape classifications, physiographic regions had the greatest predictive value for both species richness measures, and the differences between regions were greater for the proportion of neotropical migrant species/block than for the total number of species/block (table 11). The proportion of neotropical migrant species/block in

Table 10

Results of Regression Analysis of Relationships between Meausures of Species Richness and Gross Environmental Attributes of Atlas Blocks

	Annual Rainfall	Maximum July Temp.	Average Elevation	Relief
Total species	-0.442	—	-0.007	-0.005
Neotropical migrants	—	—	—	+0.003
% neotropical migrants	+0.238	—	+0.005	+0.009

NOTE: The slope of the regression line is given for significant correlations ($p < 0.05$); dash (—) indicates nonsignificant correlations.

An Overview and Analysis of Atlas Results

Table 11
Average Number of Total Species and Proportion of Neotropical Migrant Species per Block in Physiographic Region

Physiographic Region	Total Species	% Neotropical Migrant Species
Mississippi Alluvial Plain	66.2	46.8
Loess Plain	60.4	42.6
Coastal Plain Uplands	66.1	47.0
Western Highland Rim	67.3	46.8
Central Basin	62.4	40.9
Eastern Highland Rim	65.7	43.4
Cumberland Plateau	66.2	47.2
Cumberland Mountains	66.0	51.9
Ridge and Valley	66.2	41.1
Unaka Mountains	59.8	47.7

NOTE: The differences between regions are significant (analysis of variance, $p < 0.01$) for both species measures.

the Cumberland Mountains physiographic region significantly differed (Student-Newman-Keuls test, $p < 0.05$) from the other physiographic regions. The spruce-fir potential natural vegetation area was the only landscape unit among the two classifications in which the total number of species/block differed from all other units.

The 14 blocks with the least human disturbance were located in the Cumberland Plateau and Unaka Mountains regions. These blocks contained no sizable human-maintained openings except for grassy balds in some Unaka Mountain blocks; some also contained early-successional forests that probably would have been present under a completely natural management regime. In comparison to the rest of the Atlas blocks, these low-disturbance blocks had a relatively low richness, averaging 40.8 species (s.d. = 9.2, range 20–54). Their proportion of neotropical migrants was a relatively high 50.1% (s.d. = 6.3, range 35.0–55.8).

Breeding Bird Communities

To identify areas with homogeneous bird communities, based on their species composition, the presence/absence species lists of the 655 completed priority blocks were analyzed by cluster analysis. A similar analysis using the results of the first five years of Atlas fieldwork (472 completed priority blocks) was described by Nicholson (1991). The block species lists were adjusted by eliminating the poorly sampled species as described above, as well as very rare (in less than 1% of blocks) and ubiquitous (in more than 95% of blocks) species, listed below:

Very rare species
  Anhinga
  Double-crested Cormorant
  Least Bittern
  American Black Duck
  Blue-winged Teal
  Hooded Merganser
  American Coot
  Common Moorhen
  Sora
  Spotted Sandpiper
  Black-necked Stilt
  Black-billed Cuckoo
  Northern Saw-whet Owl
  Red-cockaded Woodpecker
  Yellow-bellied Sapsucker
  Alder Flycatcher
  Least Flycatcher
  Scissor-tailed Flycatcher
  Golden-crowned Kinglet
  Black-capped Chickadee
  Blackburnian Warbler
  Painted Bunting
  Savannah Sparrow
  Bachman's Sparrow
  Red Crossbill

Ubiquitous species
  Mourning Dove
  Yellow-billed Cuckoo
  Chimney Swift
  Red-bellied Woodpecker
  Downy Woodpecker
  Eastern Wood-Pewee
  Eastern Phoebe
  Blue Jay
  American Crow
  Carolina Chickadee
  Tufted Titmouse
  Carolina Wren
  American Robin
  Common Yellowthroat
  Yellow-breasted Chat
  Northern Cardinal
  Indigo Bunting
  Eastern Towhee
  Brown-headed Cowbird
  Common Grackle

Some of the very rare species may have been inadequately sampled, and ubiquitous species contribute little to the statistical analyses used (Gauch 1982). The adjusted data matrix contained 108 species.

The bird data matrix, consisting of presence/absence values for each species in each block, was first analyzed with COMPCLUS, a nonhierarchical clustering program (Gauch 1979, 1982) that joins similar blocks into composite samples. The distance measure used was percentage dissimilarity and the maximum within-cluster distance was 60. The composite samples were then aggregated with the SAS routine CLUSTER (SAS Institute 1987) using the UPGMA distance measure. Lastly, detrended correspondence analysis (DECORANA) (Hill and Gauch 1980) was used to form an ordination of the composite samples.

The initial clustering produced 33 composite samples, 22 of which were each composed of one block. These single-block samples (map 16a), although scattered across the state, were mostly grouped along the eastern and western borders; they were considered outliers and excluded from the later analyses. The eastern outliers included blocks containing spruce-fir forest, and the western outliers included extensive Mississippi Alluvial Plain wetlands. Each of these habitats supports several birds of very limited distribution in the state, which contributed to the distinctive species lists of these outliers.

The remaining 11 composite samples ranged in size from 3 to 158 blocks. Clustering joined the 11 composite samples into 6 clusters at an average normalized between-cluster distance of 1.80 (maps 16a, 16b, and 16c; fig. 7). The most distinctive cluster, Cluster A, contained a single composite sample of three high-elevation hardwood forest blocks in the Unaka Mountains. Frequently occurring species in these blocks included the Ruffed Grouse, Winter Wren, Veery, Solitary Vireo, and Scarlet Tanager. The second most

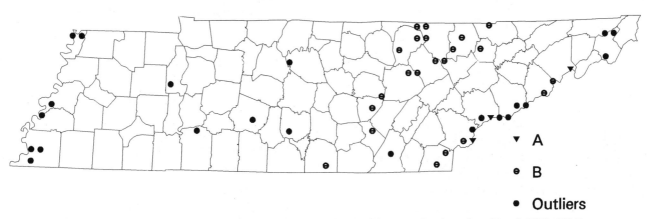

Map 16a. The 22 outliers and six bird community clusters (A–F) identified by composite clustering (Gauch 1979, 1982) of individual blocks and UPGMA clustering (SAS Institute 1987) of composite block samples.

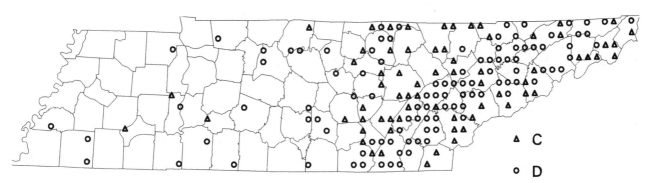

Map 16b. Bird community clusters C and D.

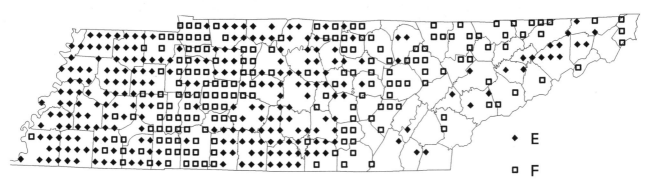

Map 16c. Bird community clusters E and F.

An Overview and Analysis of Atlas Results

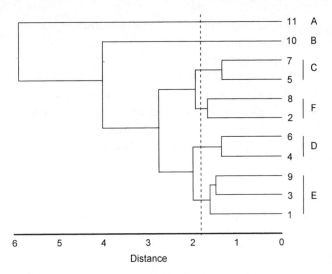

Fig. 7. Dendrogram showing results of UPGMA clustering (SAS Institute 1987) of composite atlas block samples. Six bird community clusters were identified at an average between-cluster UPGMA distance of 1.80.

distinctive cluster, Cluster B, contained a single composite sample of 22 heavily wooded blocks in the Cumberlands and at low elevations in the Unaka Mountains. Its single member outside these regions was in the northwest Ridge and Valley, in the heavily wooded Chuck Swan Wildlife Management Area in Union County. Frequently occurring species in Cluster B blocks included the Pileated and Hairy Woodpeckers, Acadian Flycatcher, Wood Thrush, Worm-eating and Black-and-white Warblers, and Ovenbird.

The remaining clusters were less geographically cohesive. Cluster C was made up of two composite samples and 68 blocks, mostly in the Cumberlands and at low elevations in the Unaka Mountains. Although for the most part predominantly wooded, they also contained some farmed or urban areas. Cluster D contained two composite samples and 104 blocks. Most of its members were located in the Ridge and Valley, the Sequatchie Valley, and the Eastern Highland Rim, and composed of a mix of farmland and forest. With three composite samples and 256 blocks, Cluster E was the largest. Most of its members were located in the western half of Tennessee, in the Loess Plain, Western Highland Rim, and Central Basin regions. Its member blocks located in the Western Highland Rim, as well as those in the Ridge and Valley, were mostly dominated by farmland, as is true of the Loess Plain and Central Basin. The final cluster, Cluster F, contained two composite samples and 180 blocks. They were almost evenly divided between heavily forested parts of the Coastal Plain Uplands and Western Highland Rim in the west and the Eastern Highland Rim and Cumberland Plateau and Mountains in the east. Several members also occurred in the northwestern Ridge and Valley, which is more heavily forested than the region as a whole, and in farmed portions of the Unaka Mountains.

Fig. 8 shows the relationships between the composite samples and clusters, based on ordination using DECORANA, of the composite samples' species lists. The composite samples are arranged by their standardized, weighted average species scores shown along each axis. Samples farther apart have fewer species in common. Axis 1 represents a west to east longitudinal gradient, along which most range limits in the state occur; west is on the left. The clusters greatly overlap along Axis 2, which has a less obvious geographic or environmental explanation.

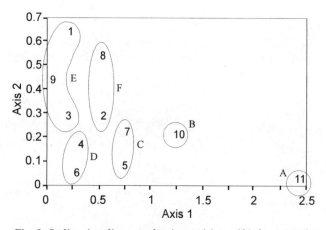

Fig. 8. Ordination diagram showing positions of bird community clusters, identified by letters, and their composite sample members, marked by triangles. The ordination was formed by detrended correspondence analysis (DECORANA) (Hill and Gauch 1980), and the eigenvalues of Axis 1 and Axis 2 were 0.33 and 0.20, respectively.

Species Accounts

Introduction to the Species Accounts

The following species accounts describe the Atlas results for each of the 170 species recorded during the Atlas survey and confirmed breeding in Tennessee prior to or during the Atlas survey. A separate section following the main species accounts contains short accounts of 1 hybrid and 19 species in the following categories: 1) extirpated or extinct breeding populations; 2) widely suspected but not confirmed breeding in Tennessee; 3) extant with breeding confirmed in Tennessee before 1986 and not found during the Atlas survey; and 4) unsuccessfully introduced.

Maps

A main feature of each species account is the thematic map showing the category of breeding evidence (Possible, Probable, Confirmed) for each Atlas block in which the species was found. The accompanying table gives the number and proportion of records in each breeding category for all blocks and for the completed priority blocks. With one exception, the map symbols are centered within the block boundaries. The exception is the Bald Eagle map, where confirmed records shade the entire quad map to protect nest locations.

Many of the Atlas distribution maps also highlight counties where a subject species historically bred (i.e., met Atlas-confirmed breeding criteria before 1986) and was not found during 1986–91 Atlas fieldwork. For a few species of limited distribution with nonbreeding individuals often present in early summer, such as some herons, all historically confirmed counties are highlighted except for those in which Atlas workers reported confirmed breeding. These historical breeding records were extracted from the same published and unpublished sources listed below in the description of the text of the species accounts.

While the Atlas distribution maps show the frequency of occurrence or proportion of blocks in a particular area containing a species, they are not an accurate measure of species abundance or population size. This abundance information is portrayed by mapped miniroute results for species adequately sampled by these primarily roadside counts. Miniroute results are displayed as contour maps, showing the proportion of 15 miniroute stops at which a species was found. The miniroute maps were generated with the program Surfer for Windows (Golden Software Inc., Golden, CO). This program creates a smoothed, three-dimensional grid from the latitude and longitude coordinates of the block centerpoints and their miniroute results. The miniroute result at each grid intersection is interpolated from the results of miniroutes within a search radius of two quad maps using the quadrant search and Kriging options.

The three-dimensional surface grid is then displayed in the form of a contour map, with contour lines connecting grid intersections of equal relative abundance (stops/route), similar to the elevation contour lines on a topographic map or the isotherms on a weather map. To make interpreting the maps easier, areas of equal abundance are also shaded. Because of the methods used in constructing the maps, the shaded contours are often projected beyond the state boundaries. This effect is most pronounced for species that are abundant along a border of the state.

Because of the differences in abundance between species, the contour intervals vary. For species whose abundance ranges from rare to relatively common across the state, the minimum contour was usually half the interval between the other contours (e.g., intervals of 1.5, 3, 6, 9, and 12 stops/route, as shown for the Song Sparrow in map 17). The interpolation procedure often produces zero and negative abundance contours; to simplify the maps, these are not shown. The interpolation procedure also frequently masks single blocks recording a species at a few miniroute stops and surrounded by blocks not recording the species. The Atlas distribution map is therefore better than the miniroute map at portraying a species's local presence or absence.

The number or proportion of miniroutes, out of 692, on which a species was recorded and the average miniroute abundance for those routes recording the species are given for each species. The species accounts also frequently compare average abundances among physiographic regions.

Text

The species accounts accompanying the maps of Atlas results have been written to provide a concise description of the seasonal occurrence, habitat, present and historical distribution and abundance, breeding density, accuracy of Atlas results, and breeding biology of each species. The information available on these topics varies among species. The breeding biology discussion concentrates on the chronology, nest placement, clutch size and, where applicable, rate of Brown-headed Cowbird parasitism in Tennessee. The rate of cowbird parasitism is usually expressed as the proportion of nests containing cowbird eggs or nestlings. Because the nest observations occurred at all stages of the nest cycle, they are probably a conservative estimate of the true parasitism rate. As often noted in the accounts, the breeding biology of many species in Tennessee is poorly known.

The gross distribution and abundance of most birds breeding in Tennessee has been fairly well documented since the mid-1930s. Prior to that, the main sources are Rhoads's (1895a) statewide survey and annotated list and Ganier's 1917 and 1933 checklists. Other early articles, many cited in the preceding chapter on the history of Tennessee ornithology, deal with small areas of the state or a limited number of species. Historical changes in distribution are described in the text of the accounts. Range changes for species that have undergone a gross range contraction are mapped by shading counties where the species formerly bred (i.e., met Atlas-confirmed criteria before 1986) on the Atlas distribution maps.

The best source of recent population trend information is the U.S. Fish and Wildlife Service/National Biological Service Breeding Bird Survey (BBS), established in Tennessee in 1966 (Robbins, Bystrak, and Geissler 1986, Droege 1990). Forty-two BBS routes occur in Tennessee (map 18); a majority of them have been censused each year since 1966. Population trends for 1966–94, 1966–79, and 1980–94 were determined from BBS results using the route-regression method (Geissler and Sauer 1990, Link and Sauer 1994). These analyses were performed by the National Biological Service, and Tennessee BBS results presented in the species accounts without a citation are from this data set. Comparable rangewide trend information is mostly from Peterjohn, Sauer, and Link (1994). For several

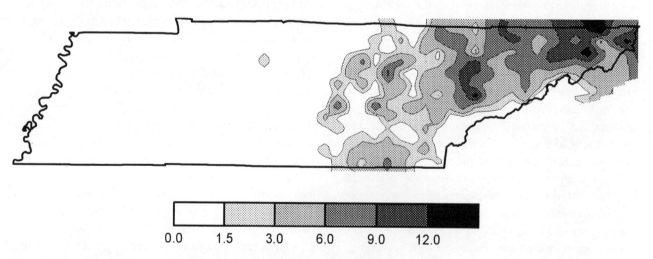

Map 17. Song Sparrow relative abundance mapped from Atlas miniroute results.

permanent residents, Tanner's (1985) analysis of Tennessee Christmas Bird Counts also provides useful general trend information.

Information on breeding densities in different habitats comes primarily from the Breeding Bird Census studies published in *American Birds* and the *Journal of Field Ornithology*. These censuses, conducted with the spot-mapping method, have all been in the eastern half of the state and primarily in forested and mixed habitats (map 19, table 12). Few plots have been censused for more than two years. Plot sizes vary, and the densities given in the species accounts have been standardized to the number of pairs/100 ha. The only recent census study in West Tennessee was by Ford (1990), who used line transect counts to census many forested wetland sites. His density estimates are not directly comparable to spot map results.

For brevity, records from the "season" reports in *Audubon Field Notes, American Birds,* and the *Migrant* are cited with the journal name, volume number, and page number.

Information on breeding biology comes from numerous published and unpublished sources. The main unpublished sources are egg collections and Nest Record Cards. Major extant egg collections are those of W. R. Gettys, from McMinn County between about 1897 and 1920; A. F. Ganier, mostly from around Nashville between about 1917 and 1950; H. C. Monk, mostly from Rutherford County between about 1930 and 1955; and R. C. Lyle, from northeast Tennessee between about 1925 and 1950. Except for waterbirds at Reelfoot Lake, West Tennessee is poorly represented in egg collections. Nest Record Cards are maintained by the Cornell Laboratory of Ornithology and, for Tennessee, date primarily from the mid-1960s through 1991; although the information they contain varies greatly in quality, they are one of the best sources on the rate of Brown-headed Cowbird parasitism. Most Nest Record Cards are from the eastern two-thirds of Tennessee. The unpublished notes of several ornithologists, such as those of A. R. Laskey housed at the Cumberland Museum and Science Center, Nashville, were also important sources.

The summary statistics describing clutch size and nest height are generally presented without citing their sources. Exceptions are very unusual records, for which the sources are cited, and published studies with large sample sizes for individual species, which are presented separately from the combined information from other sources. Appendix 2 contains the early and late dates of nests with eggs, nests with young and fledglings. For both altricial and precocial species, the fledgling dates are those of young which have left the nest and are still dependent on their parents. Many of the dates in this appendix are from the dates of confirmed breeding observations recorded by Atlas workers.

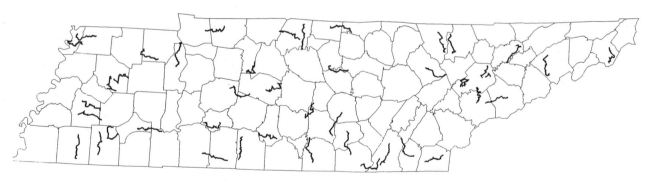

Map 18. Location of Tennessee Breeding Bird Survey routes.

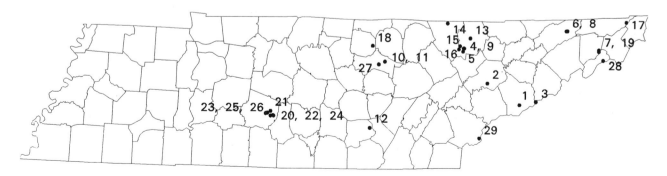

Map 19. Locations of spot-map breeding bird census plots.

Introduction to the Species Accounts

Table 12
Summary of Tennesssee Breeding Bird Census Results through 1992

Location*	Habitat	Year of Census	Area (ha)	Species	Pairs 100 ha	Citation[†]
1	Mature pine-oak forest[‡]	1947–48	12	23	351	Fawver 1950, Kendeigh and Fawver 1981
	Seral pine-oak forest[‡]**	1947–48	6, 10.2	23	431	Fawver 1950, Kendeigh and Fawver 1981
	Pine heath[‡]	1947–48	12	10–12	183–363	Fawver 1950, Kendeigh and Fawver 1981
	Pine heath[‡]	1948	10	10	270	Fawver 1950, Kendeigh and Fawver 1981
	Virgin cove hardwoods forest[‡]	1946	23	15	820	Aldrich and Goodrum 1946
	Virgin cove hardwoods forest[‡]	1947	9.2	22	725	Fawver 1950, Kendeigh and Fawver 1981
	Virgin cove hardwoods forest[‡]	1948	7.2	22	740	Fawver 1950, Kendeigh and Fawver 1981
	Virgin cove hardwoods forest[‡]	1982	6.8	15	744	Wilcove 1988
	Virgin cove hardwoods forest[‡]	1947–48	6.9	10–16	450–63	Fawver 1950, Kendeigh and Fawver 1981
	Virgin cove hardwoods forest[‡]	1982–83	5.6	10–11	447–679	Wilcove 1988
	Cove hardwoods forest[‡]	1948	6	13	925	Fawver 1950, Kendeigh and Fawver 1981
	Cove hardwoods forest[‡]	1948	12.8	22	488	Fawver 1950, Kendeigh and Fawver 1981
		1982	12.8	21	546	Wilcove 1988
	Hemlock-deciduous forest[‡]	1948	10	19	1075	Fawver 1950, Kendeigh and Fawver 1981
		1982–83	4.2	12–15	881–989	Wilcove 1988
	Hemlock-deciduous forest[‡]	1948	8	17	1013	Fawver 1950, Kendeigh and Fawver 1981
		1982–83	7.8	20–22	848–1182	Wilcove 1988
	Hemlock-deciduous forest[‡]	1947–48	11.6	20–21	813–73	Fawver 1950, Kendeigh and Fawver 1981
		1982	9	16	994	Wilcove 1988
	Chestnut oak (chestnut) forest[‡]	1947–48	7.6	24–25	650–710	Fawver 1950, Kendeigh and Fawver 1981
		1982	4.8	18	958	Wilcove 1988
	Chestnut oak (chestnut) forest[‡]	1948	9	19	505	Fawver 1950, Kendeigh and Fawver 1981
		1982–83	11	14–15	345–66	Wilcove 1988
	Northern red oak (chestnut) forest[‡]	1948	7.5	14	515	Fawver 1950, Kendeigh and Fawver 1981
		1982	11	14	605	Wilcove 1988
	Beach Gap Forest	1947	7.6	11	418	Fawver 1950, Kendeigh and Fawver 1981
		1982	7	7	378	Wilcove 1988
	Beach Gap Forest	1948	8	13	463	Fawver 1950, Kendeigh and Fawver 1981
	Heath bald[‡]	1947	10	2	50	Fawver 1950, Kendeigh and Fawver 1981
	Heath bald[‡]	1948	4	3	113	Fawver 1950, Kendeigh and Fawver 1981
	Virgin spruce-fir[‡]	1947–48	11.6	16–17	645–915	Fawver 1950, Kendeigh and Fawver 1981
	Virgin stunted Fraser fir[‡]	1948	6.8	6	423	Fawver 1950, Kendeigh and Fawver 1981
	Early successional spruce-fir[‡]	1947–48	6.6	7–12	915–30	Fawver 1950, Kendeigh and Fawver 1981
2	Ridge and Valley Hardwood forest[‡]	1965–75	24.3	16–28	437–823	Howell 1965–74

Continued on page 61

Table 12—Continued

Location*	Habitat	Year of Census	Area (ha)	Species	Pairs 100 ha	Citation†
3	Virgin spruce-fir forest‡	1967	24.3	13	700	Alsop 1969
		1985	24.3	14	455	Alsop and Laughlin 1991
4	Mixed deciduous forest—strip mine	1972–73	25.9	36–37	734–834	Yahner 1972, 1973
5	Upland mixed deciduous forest with strip mines	1973	22.7	40	693	Garton 1973
6	Cutover mixed deciduous forest‡	1974	20	27	529	Smith 1975
7	Deciduous clearcut‡	1975–84	20.2	13–21	158–450	Lewis and Smith, 1975, Lewis 1977–84b
		1991	20.2	27	330	Lewis 1991
8	Pasture with hedgerows	1974	36.5	17	305	Lewis 1975a
	Pasture	1974	20.6	15	242	Lewis 1975b
9	Maple-gum-hickory forest‡	1976	20	32	953	Smith 1977
10	Pasture with brush, wooded strips	1977	8.7	30	517	Simmers 1978
11	Mixed mesophytic woods, field sand brush	1978–88	27	35–44	300–474	Simmers 1979–89
		1989–90	22.9	35	300–323	Simmers 1990, 1991
12	Virgin mixed-mesophytic forest‡	1977	11.8	35	811	Robertson 1979
13	Deciduous forest and contour strip mine	1978–79	23.4	29–35	312–95	Nicholson 1979a, 1980c
14	Strip mine and deciduous woodlot	1978–79	20.1	21–22	234–49	Nicholson 1979b, 1980d
15	Oak-maple forest‡	1980	8.1	19	371	Turner and Fowler 1981a
16	Oak-maple forest‡	1980	8.1	16	284	Turner and Fowler 1981b
17	Deciduous clearcut (2 yr.)‡	1980	8.1	4	210	Lewis 1981b
18	Disturbed mixed-mesophytic woodland ravine‡	1981	7.9	28	880	Simmers 1982a
19	Mixed deciduous forest‡	1982	25.4	22–26	286–90	Lewis 1983a, 1984a
20	Cedar forest I‡	1982	6.1	22	264	Fowler and Fowler 1983a
21	Cedar forest II‡	1982	6.1	16	222	Fowler and Fowler 1983b
22	Cedar forest III‡	1982	6.1	17	222	Fowler and Fowler 1983c
23	Upland hardwood forest I‡	1983	6.1	17	362	Fowler and Fowler 1984a
24	Upland hardwood forest II‡	1983	6.1	14	231	Fowler and Fowler 1984b
25	Abandoned agricultural lands I‡	1983	6.1	14	404	Fowler and Fowler 1984c
26	Abandoned agricultural lands II‡	1983	6.1	14	428	Fowler and Fowler 1984d
27	Mature deciduous-coniferous forest with stream	1991–92	10.2	47	535	Stedman and Stedman 1992, 1993
28	Red spruce forest‡	1992	9.9	13	405	Mayfield 1993
29	Mature maple-beech-birch forest‡	1992	10.2	21	863	Mitchell and Stedman 1993

NOTES: *The numbers in the column labeled "Location" refer to points on the map in map 19. All of the censuses were conducted with the spot-map method, except for some of the Fawver (1950) and Wilcove (1988) censuses, which used a somewhat similar "cruising count" technique. For censuses conducted on the same plot in consecutive years, the minimum and maximum number of species and pairs/100 ha are given. Separate results are given for censuses replicated on the same plot after an interval of several years.

†Only those citations referenced in the species accounts are listed in the literature cited section. The other censuses dated earlier than 1971 are published in *Audubon Field Notes* and 1971–83 census are published in *American Birds*. Censuses conducted form 1984 to 1987 are not published and are available by mail from the Cornell Laboratory of Ornithology, Ithaca, New York. More recent censuses are published in a *Journal of Field Ornithology* annual supplement.

‡Relatively homogenous plot.

**Average of 2 plots.

Scientific names of plants mentioned in the text are listed in Appendix 1. The following abbreviations are commonly used in the species accounts:

AOU—American Ornithologists' Union
BBS—Breeding Bird Survey
CBC—Christmas Bird Count
cm—centimeter
ha—hectare
km—kilometer
LBL—Land Between the Lakes
m—meter
ms.—manuscript
n—number of observations in the sample
NRC—Cornell Nest Record Cards
NWR—National Wildlife Refuge
pers. comm.—personal communication from the person referenced
pers. obs.—original observation by the person referenced, usually the account author
s.d.—standard deviation
SP—State Park
TVA—Tennessee Valley Authority
TWRA—Tennessee Wildlife Resources Agency
unpubl.—unpublished
USFWS—U.S. Fish and Wildlife Service
WFVZ—Western Foundation of Vertebrate Zoology
WMA—Wildlife Management Area
$\bar{x}$—average or mean

Confirmed Breeding Species

Pied-billed Grebe
Podilymbus podiceps

The Pied-billed Grebe is a small diving bird that feeds on aquatic insects, crustaceans, and fish and nests throughout much of North and South America. In Tennessee it is a fairly common migrant and winter resident and a rare summer resident (Robinson 1990). Migrant and wintering birds occur on lakes and large ponds, more abundantly when submerged aquatic vegetation is present. Most breeding records have been from shallow ponds and lakes with areas of open water and emergent aquatic vegetation such as rushes, cattails, and grasses. The first fall migrants arrive in late summer, frequently by early August. Spring migration extends into early May.

The Pied-billed Grebe was first reported in the state by Rhoads (1895a:465) at Reelfoot Lake between 30 April and 6 May 1895. He felt it "not unlikely" that the grebe nested there. Grebes nested within a few km of the Tennessee border in western Kentucky in 1919 (Ganier 1933b) and in northwest Mississippi beginning in 1932 (*Migrant* 3:28). The first Tennessee nest record, however, was not until 20 May 1934, when Ganier found a nest with 5 eggs at Long Pond near Belvidere, Franklin County (*Migrant* 5:29).

Within the next few decades, confirmed evidence of nesting was reported from Obion County at Reelfoot Lake (Whittemore 1937), Warren County (Todd 1944, erroneously placed in Rutherford County by DeVore (1975) and Robinson (1990)), a small lake in Knox County (*Migrant* 34:52), Buena Vista Marsh and Radnor Lake, Davidson County (Parmer 1985), Amnicola Marsh, Hamilton County (*Migrant* 39:65), Monsanto Ponds, Maury County (*Migrant* 46:66), and Shelby County (*Migrant* 54:59). Herndon (1950b) reported probable breeding and young at Lake Phillip Nelson, Carter County, although it is unclear whether the young

Pied-billed Grebe. Chris Myers

were a family group. Other grebes reported before 1986 from late May through mid-July in Benton, Humphreys, Sumner, Grundy, Campbell, Sullivan, and Washington Counties (*Migrant*) may also have been breeding birds.

During Atlas fieldwork, the Pied-billed Grebe was reported from 15 blocks. Because of the grebe's often secretive nature and the occasional difficulty of inspecting its habitat, it may have been missed in some blocks. All of 6 confirmed records were of adults with fledglings. The confirmed records in the Mississippi Alluvial Plain and Loess Plain regions were in shallow, marshy ponds. The Shelby County record was from a 13.3 ha shallow lake fringed with emergent vegetation. In Hardin County, a family of grebes was on Pickwick Reservoir near the dam. The Davidson County brood was at Radnor Lake and the Cumberland County brood at the lake in Cumberland Mountain State Park; both of these lakes are surrounded by woods and have sheltered coves with scattered emergent vegetation. The Meigs County brood was in a cove of Chickamauga Reservoir with dense emergent and submerged vegetation. The possible and probable records were from Reelfoot Lake and both small and large artificial lakes. An additional possible breeding record was of an adult in an unspecified block in Putnam County in June 1991 (*Am. Birds* 45:1124).

Too few records of the Pied-billed Grebe exist to show trends in the breeding population. The future of the Tennessee nesting population, however, is probably secure. Some wetlands where nesting formerly occurred, such as Buena Vista Marsh, have been altered to the point that they are no longer suitable nesting habitat. In other areas, such as Reelfoot Lake and the Monsanto Ponds, nesting has occurred at infrequent intervals despite little apparent change in the quality of nesting habitat. The recent increase in aquatic vegetation on many of the large reservoirs, in addition to being a factor in the recent increase in the wintering grebe population (Tanner 1985), has probably improved nest habitat in sheltered coves; elsewhere on these lakes there is probably too much disturbance from waves. Other seemingly suitable nest habitat has been recently created through construction of settling ponds and borrow pits. Listing as threatened or endangered is not presently warranted.

Breeding Biology: The only Tennessee egg records of the Pied-billed Grebe are from mid-April and May. Observations of young, however, show that nesting begins as early as late March. Pied-billed Grebe pairs are strongly territorial, and they breed on ponds as small as a quarter hectare. The nest is a floating mass of vegetation 0.3 m or more in diameter, built by both adults and anchored to dead or live emergent plants, usually adjacent to open water, and often fairly conspicuous (Palmer 1962). The adults continue to add vegetation during incubation.

Clutches of 5 fresh eggs and 8 incubated eggs have been reported in Tennessee. Clutches elsewhere are most frequently of 4 to 8 eggs, with an average size around 7 (Palmer 1962, Sealy 1978b). Two groups of 14 and 16 young and at least 3 adults observed in Shelby County (Atlas results, *Migrant* 62:22) were likely the product of 4 or more pairs of adults. Grebe eggs are pale bluish white or greenish when first laid and soon turn buffy or brown. Both adults incubate for about 23 days, beginning before the clutch is complete. The eggs are usually covered with vegetation when the birds are off the nest. The young are down-covered at hatching and soon leave the nest, often by riding on a parent's back (Palmer 1962). They are fed by the parents for at least several days after hatching; their age at independence is poorly known. Studies elsewhere have found that the grebe is often double-brooded (Palmer 1962). The young observed in Tennessee in August and September (*Migrant*, Atlas results) may have been from second broods.—*Charles P. Nicholson.*

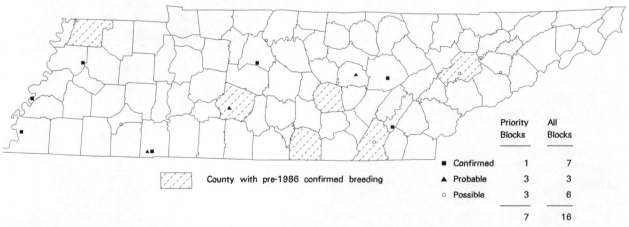

Distribution of the Pied-billed Grebe.

Double-crested Cormorant
Phalacrocorax auritus

The Double-crested Cormorant is an uncommon to locally common migrant, uncommon winter resident, and rare summer resident of Tennessee. Formerly a locally common nesting species at Reelfoot Lake and on Kentucky Reservoir, its population in Tennessee and elsewhere dropped precipitously during the 1960s and 1970s. Its population is presently increasing, although it has yet to return to its former status in Tennessee.

Rhoads (1895a) described the Double-crested Cormorant as abundant at Reelfoot Lake and as also occurring in Haywood and Lauderdale Counties. Whether it nested at the latter two sites is unknown. It was probably nesting then at Reelfoot, as an area there was called the "Turkey Roost" after Water Turkeys, a name applied to both cormorants and Anhingas. Ganier (1933b) observed 75 cormorant nests in 1919 at Reelfoot in a colony near the Kentucky border. Later nests were in the large mixed-species Cranetown colony, where the number of cormorant nests was estimated at 50 in 1921, 200 in 1932 and 1934, 400 in 1936, and 185 in 1938 (Gersbacher 1939). Most birds had abandoned this site by 1963, when only 1 pair of cormorants was present (Gersbacher 1963). A colony at the northern end of the lake in Kentucky, possibly the same one active in 1919, contained nesting cormorants until at least 1949 (Mengel 1965, Palmer-Ball 1991).

The only other site where cormorants have nested is on Kentucky Reservoir. Soon after impoundment of the reservoir in 1943, large numbers of cormorants began roosting in dead trees in shallow water at the mouth of the Duck River, within the Tennessee National Wildlife Refuge. About 100 pairs nested there in 1949. Nesting at this site continued until at least 1955 (*Migrant* 26:27); the date nesting ceased is not known but was probably soon after 1955.

The decline in cormorant numbers was due to several causes. They were frequently shot because of their perceived competition with fishermen, and nestlings were killed for use as fishbait (Rhoads 1895a, Ganier 1933b); this persecution continued at Reelfoot until at least 1936 (Ganier 1937b). Other colonies in the Mississippi Valley were lost by wetland drainage (James and Neal 1986). Populations breeding around the Great Lakes, source of many of the cormorants that migrated through and wintered in Tennessee (Dolbeer 1991), were also affected by control programs, sometimes officially sanctioned, to reduce their perceived impact on commercial fisheries (Ludwig 1984). The most important cause, however, was probably nesting failure due to pesticide contamination, especially by DDT. Although no information is available on pesticide levels in cormorants nesting in Tennessee, pesticides affected Great Lakes populations (Ludwig 1984), and pesticide contamination occurred in Red-winged Blackbirds and herons nesting in Tennessee (Alsop 1972, Fleming, Pullin, and Swineford 1984).

Following the banning of the use of DDT in 1973, Great Lakes cormorant populations increased exponentially between 1976 and 1981 (Ludwig 1984). From 1966 to 1993, the continental population increased at an average annual rate of 5.5%/year ($p < 0.01$) on BBS routes (Peterjohn, Sauer, and Link 1994). The number of transient and wintering cormorants in Tennessee has also increased (Pitts 1985, Christmas Bird Count results, *Migrant*). Several cormorants have recently been present across the state during the late spring and early summer (Atlas results, *Migrant*). On 3 June 1987, a pair of cormorants was found perched within the Great Blue Heron colony at the mouth of the Duck River on Kentucky Reservoir (Atlas results). One of them flew several times to the water and returned with pieces of aquatic plants, which it passed to the other bird, who placed it on a limb. Neither the cormorants nor evidence of a nest were observed on later visits. All of the other Atlas records were of single or small groups of birds without evidence of nesting. In 1992, after Atlas fieldwork, 5 cormorants and at least 1 nest were discovered along the Holston River in eastern Hawkins County (R. Caldwell pers. comm.).

The Double-crested Cormorant was listed as In Need of Management in 1976 (TWRA 1976), when none were known to nest in the state and the wintering and non-breeding summer population was much lower than

Double-crested Cormorant. Chris Myers

Confirmed Breeding Species

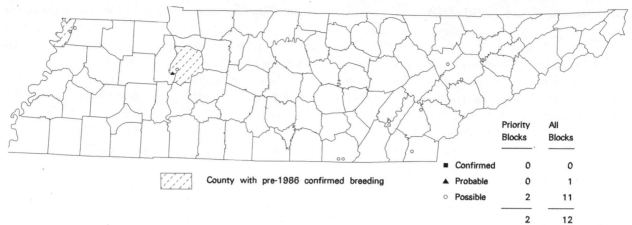

Distribution of the Double-crested Cormorant.

at present. Since then the non-breeding population has greatly increased, and a few cormorants have begun nesting at Reelfoot and elsewhere. Because of its small size, the population of cormorants breeding in Tennessee should continue to be listed as In Need of Management.

Breeding Biology: Cormorants usually first nest when three years old (Palmer 1962). Some immature birds return to nesting colonies and engage in courtship and nest building, while others remain south of the nesting areas. Many of the cormorants reported by Atlas workers were probably summering immature birds. Cormorant nests at the Reelfoot Cranetown colony were at heights of 24.4 to over 30.5 m in the tops of cypress trees (Ganier 1933b, Gersbacher 1939). Herons and egrets frequently nested in the same trees, although usually at a lower height than the cormorants. At the Duck River site, cormorants nested at heights of 3 to 9.1 m in dead oaks standing in water (Cypert 1949). Great Blue Herons nested in a separate colony about 800 m from the cormorants.

Double-crested Cormorants begin nesting activities as early as late March, as Cypert (1949) reported egg-laying in progress at the Duck River colony on 3 April. Egg laying at Reelfoot apparently occurred over the period of a few weeks. On 26 April 1919, Ganier (1933b) reported egg-laying in progress and collected several sets of eggs which had been incubated up to 8 days (egg coll. data). On 29 May 1932, some nests contained large young while others contained eggs. Clutch size ranged from 3 to 5 eggs and averaged 4.0 (s.d. = 0.77, n = 11). Both adults incubate for 25 to 29 days; the young first fly in 5 to 6 weeks and are independent in 10 weeks (Palmer 1962). The cormorant is single-brooded.—*Charles P. Nicholson.*

Anhinga
Anhinga anhinga

The Anhinga is a long-necked, long-tailed, fish-eating resident of wooded sloughs and shallow lakes. It is a poorly known, rare summer resident of West Tennessee and a very rare vagrant elsewhere in the state. The largest breeding population formerly occurred at Reelfoot Lake, the northern limit of this primarily tropical species's breeding range (Palmer 1962). Confirmed or probable nesting has been reported at four different sites, two of which were extant during the Atlas project.

Rhoads (1895a) first mentioned the occurrence of the Anhinga in Tennessee, noting its presence near Memphis and a probable nesting at Reelfoot Lake. Based on the reports of Benjamin Miles, Rhoads also described it as nesting in the river bottoms of Haywood and Lauderdale Counties. Pindar (1889) reported it from Fulton County, Kentucky, which encompasses the north end of Reelfoot Lake. According to Mengel (1965), however, Pindar con-

Anhinga. Chris Myers

fused it with the Double-crested Cormorant. It was observed with certainty at Reelfoot by Ganier in 1919, when he collected a set of its eggs there. Its numbers at Reelfoot apparently peaked during the 1930s. In 1932, at least 50 pairs were nesting at the large mixed-species Cranetown heronry (Ganier 1933b). Fifty nests were again present in 1934, 100 in 1936, and 40 in 1938 (Gersbacher 1939). This was apparently the last observation from the Tennessee portion of Reelfoot Lake for several decades, although the Anhinga nested at the Kentucky Cranetown heronry until 1950 (Mengel 1965, Palmer-Ball 1991). It also ceased nesting in the northeast corner of Arkansas during this period (James and Neal 1986).

In 1977, an Anhinga was observed on a nest in the recently reestablished mixed-species heronry near Little Ronaldson Slough at Reelfoot Lake (Pitts 1982b). Pitts also reported a pair circling over the colony on 9 June 1981. One Anhinga was reported from Reelfoot during late June 1984 (*Migrant* 55:87). During the Atlas period, single nests, and up to three adult birds, were reported at Reelfoot during 1986, 1987, and 1988. Nestlings were present in June 1987 (M. Waldron pers. comm.).

Single Anhingas were observed in late May 1948 and 1949 in a mixed-species heronry in a large flooded bottom at the mouth of the Duck River in Humphreys County (Cypert 1949). Nesting of this species at the Duck River site was confirmed in 1953, when 5 Anhinga nests were present (Cypert 1955). Fifty birds were present in early May 1955 (*Migrant* 26:27). Anhingas apparently quit using this site a short time later, although herons continue to nest at this site.

Twelve Anhingas were present in late March 1980 at the recently constructed lake at Big Hill Pond State Park, McNairy County (Waldron 1980). One pair remained to nest; the outcome of this effort is unknown. No other Anhingas have been reported during several more recent searches of this site.

During the Atlas period, a pair of Anhingas probably nested at a mixed-species heronry near a pond in the Anderson-Tully Wildlife Management Area, Lauderdale County. The pair was observed perching near a nest that differed from those of nearby herons and egrets in being more compact and constructed of heavier sticks (P. B. Hamel pers. comm.). The Anhingas, however, were not seen on the nest. This colony was apparently not active in 1987, and only Great Blue Herons were observed there in 1989 (M. Waldron pers. comm.).

Breeding Biology: All of the areas where Anhingas have nested in Tennessee have been relatively shallow lakes with numerous trees, frequently cypress, emerging from the water. This is typical of Anhinga habitat elsewhere (Palmer 1962). With the exception of Big Hill Pond, all of the Anhinga nests have been in colonies where the most numerous species were usually Great Blue Herons and Great Egrets. In the Cranetown colony at Reelfoot, Anhingas nested in the tops of the tallest dead trees, at heights of 30.5 m or more, at the outer edge of the colony in deep water (Gersbacher 1939). The single nest at Big Hill Pond was 4.5 m above water in a dead tree 45 m from shore (Waldron 1980). South of Tennessee, Anhinga nests are frequently less than 3 m above water (Palmer 1962). They typically differ from heron and egret nests by being more compact and lined with leaves (Allen 1961, Palmer 1962).

Ganier (1933b) reported clutches of 4 and 5 eggs in the Reelfoot colony. Single clutches of 2, 3, 4, and 5 eggs are present in Tennessee egg collections; whether the smaller clutches were complete is unknown. Clutches of 4 and 5 eggs are most frequent elsewhere (Palmer 1962). Both adults incubate and feed the nestlings; incubation takes about 25 to 28 days (Palmer 1962).
—*Charles P. Nicholson.*

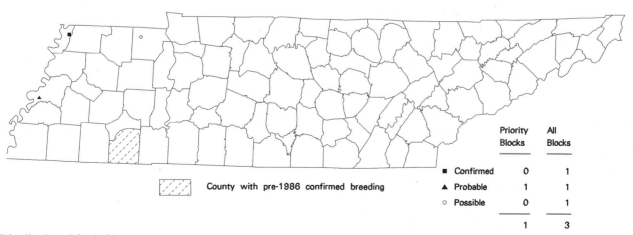

Distribution of the Anhinga.

Least Bittern
Ixobrychus exilis

The Least Bittern, smallest of the herons, is a rare to locally common summer resident. It is a fairly late migrant, usually arriving in late April. Its fall migration is poorly known, and most birds probably depart by late September. The Least Bittern inhabits marshes with tall, emergent vegetation, such as rushes and cattails, bordering open water up to a meter or more deep (Weller 1961).

The first reported nesting of the Least Bittern in Tennessee was probably about 1919 at Reelfoot Lake (Ganier egg coll. data). The Least Bittern has been regularly reported from Reelfoot Lake since then, and Mengel (1965) noted that its population there was probably the largest in the southern United States. Several pairs of Least Bitterns were regularly reported from the shallow lakes and borrow pits along the state line south of Memphis during the 1930s (e.g., *Migrant* 9:28). Other sites across the state that have supported populations of several pairs include a marsh near Tatumville, Dyer County (*Migrant* 49:90), a marsh near Trenton, Gibson County (*Migrant* 47:99), Goose Pond in Grundy County (*Migrant* 43:75), Hoover and Mason Lake in Maury County (Todd 1944), Amnicola Marsh in Hamilton County (DeVore 1968a), and the Alcoa Marshes in Blount County (Williams 1975b). Nesting bitterns have also been historically reported from at least eight other counties across the state, and breeding season records exist for an additional eight counties.

Marshes inhabited by Least Bitterns in Tennessee, for which habitat information is available, have mostly been dominated by giant cutgrass (often referred to as sawgrass), great bulrush, or cattail and have areas of open water. A few records from marshes dominated by buttonbush and willow have also been reported. The minimum marsh size capable of supporting Least Bitterns is not well known, and probably varies with the productivity of the marsh. Nesting bitterns have been reported from 0.8 ha marshes (*Migrant* 5:29, Ganier 1935b), and bitterns have been reported during the breeding season from marshes of less than 0.4 ha (Nicholson pers. obs.). Least Bitterns typically feed by clinging to vegetation and spearing prey in adjacent open water; small fish, aquatic insects, and crayfish make up most of their diet (Weller 1961). The diet of seven birds from Reelfoot Lake examined by Simpson (1939) was dominated by small fish and also included a small frog *(Acris crepitans)*, aquatic insects, and the remains of a small mammal.

Atlasers reported Least Bitterns from only eight blocks, all but two of them in northwest Tennessee. This is probably an underestimate of the species's present range, as some atlasers were reluctant to enter marshes, few used tape-recordings to elicit bittern calls, and several were probably unfamiliar with bittern calls. Some historical nesting sites, including Goose Pond in Grundy County, Amnicola Marsh, and the Alcoa Marshes, were not sufficiently worked by atlasers to conclude that they no longer support Least Bitterns. There is little doubt, however, that some historical nesting sites, among them the borrow pits south of Memphis, Buena Vista Marsh at Nashville, and some marshes on the Eastern Highland Rim, no longer provide Least Bittern habitat. Although the bittern remains fairly common at Reelfoot Lake (Pitts 1985), its population there may also have decreased due to the replacement of giant cutgrass with less suitable woody vegetation since water levels were stabilized in the 1940s (Henson 1990).

Because of concern over apparently declining numbers and the loss of suitable wetland habitat, the Least Bittern was listed as In Need of Management in 1976 (TWRA 1976); the continuation of this listing is presently justified. Few specific management programs for Least Bitterns have been initiated, although recent wetland acquisition and protection programs of the TWRA and the Nature Conservancy are preserving bittern habitat.

Breeding Biology: Least Bitterns nest in dense stands of emergent marsh vegetation, such as giant cutgrass, bulrush or cattail, or less commonly in wetland shrubs or small trees (Palmer 1962). Pairs may nest solitarily or in loose colonies, as at Amnicola Marsh (DeVore 1968a). The nest is usually placed close to open water, and the male apparently does most of the nest building (Weller 1961). The platform nest is formed in a clump of

Least Bittern. Chris Myers

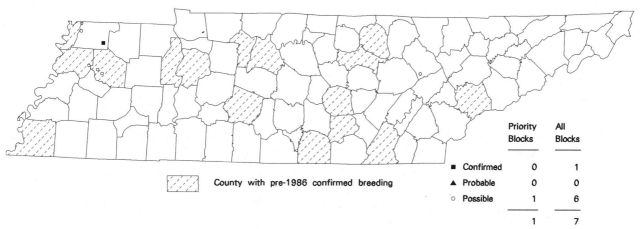

Distribution of the Least Bittern.

vegetation by bending down vegetation and then adding additional plant stalks (Weller 1961, DeVore 1968a, Williams 1975b). Of 18 nests at Amnicola Marsh, 17 were built in great bulrush and 1 in bulrush and cattail (DeVore 1968a). Ten of 11 nests at Alcoa were in either cattail of bulrush; the other nest was in a willow (Williams 1975b). Nests at Reelfoot Lake have been reported in giant cutgrass. Nests heights in Tennessee range from 0.1 to 1.4 m above the water. DeVore (1968a) found an average height of 0.46 m (n = 18) and Williams (1975b) found an average height of 0.9 m (n = 11).

Egg laying occurs throughout May and June, with a poorly defined peak in mid-May. Clutch sizes in Tennessee are 3–6 eggs, with 4 and 5 most common, and an average of 4.43 (s.d. = 0.69, n = 46), very close to the 4.48 average of several studies given by Weller (1961). Incubation, by both adults, begins after the first egg is laid, and lasts 17–18 days (Weller 1961). Nest material may be added until after the young hatch. Both adults feed the young, which begin leaving the nest when 6–9 days old. The age at which the young are independent is not well known (Palmer 1962). DeVore (1968a) found a low hatching success rate of 36% and attributed nest losses to predation by minks. Renesting is common, and Least Bitterns are probably double-brooded (Weller 1961, Palmer 1962).—*Charles P. Nicholson.*

Great Blue Heron
Ardea herodias

The Great Blue Heron, largest and hardiest member of its family in Tennessee, is a common winter resident and locally common summer resident. It occurs along lakeshores, rivers, and ponds, where it feeds on fish as well as a variety of other primarily aquatic animals including crustaceans, amphibians, and insects. It typically nests in colonies, sometimes numbering several hundred pairs, although a few solitarily nesting pairs have been reported (Pullin 1983, 1985–88, 1990).

The earliest historic report of the Great Blue Heron in Tennessee was by Merriam (in Rhoads 1895a), who observed one in the Little Tennessee River valley on 29 July 1887, by which time dispersal from nest colonies has normally occurred. Rhoads (1895a) observed it at Reelfoot Lake and in Shelby and Roane Counties during the species's normal nesting period. At that time a wading bird colony existed at Reelfoot, although it is unclear whether Great Blue Herons nested there. The first confirmed nesting was reported in 1921 at the Cranetown colony at Reelfoot Lake (Ganier 1933b). About 250 Great Blue Heron nests were then present. Great Blues continued to nest at this site, usually outnumbered by other species, until about 1960 (Gersbacher 1939, 1964). Kentucky Cranetown, at Reelfoot a short distance north of the Tennessee line, was occupied by Great Blues and other species from at least 1909 to 1963 (Gersbacher 1963).

Great Blue Heron. Chris Myers

Until the completion of Kentucky Reservoir in 1943, the only other colonies were near the western arm of the Tennessee River at Sulphur Well in Henry County, near the mouth of the Duck River in Humphreys County, and near Pittsburg Landing in Hardin County (Ganier 1951a). The first two of these colonies were destroyed during reservoir construction. Each relocated to a nearby site; the Henry County colony apparently lasted only a few more years, while the Duck River colony still exists. The Hardin County colony persisted until about 1970, when the site was cleared for agriculture (Pitts 1973). Another colony was active during this period near Hop-In in Weakley and Obion Counties (Pitts 1973).

The low point in the Great Blue Heron population was probably reached in the 1960s or early 1970s. Part of this decline was due to loss of colonies from direct human disturbance and habitat destruction (Gersbacher 1964, Pitts 1973). Reduced reproductive success from pesticide contamination may also have been a factor. Fleming, Pullin, and Swineford (1984) found significantly thinner eggshells and higher pesticide concentrations in heron eggs collected from Tennessee colonies in 1980 than in pre-1947 samples. Although the pesticide levels and eggshell thinning in the 1980 samples were probably not high enough to affect reproduction, they were taken almost a decade after use of some of the most detrimental pesticides ceased. During the period of the lowest population, however, new colonies were established at Sinking Pond in Coffee County and Armstrong Bend adjacent to Chickamauga Lake in Meigs County (Pitts 1977). These were the first colonies east of the western arm of the Tennessee River and the first to exploit the habitat created by Middle and East Tennessee reservoirs.

Systematic aerial censuses of heron colonies began in 1977 and were carried out through 1988 in Middle and East Tennessee and 1989 in West Tennessee (summarized in Pullin 1990, Smith 1989–90). These censuses showed a large and fairly steady growth in the number of both colonies and Great Blue Herons. Pullin (1990) found an average compound annual growth rate of 39.6% (s.d. = 33.9, range 14.4–105.2) for nine of the largest colonies from 1977 to 1988. This rapid growth is also evident in BBS results, which show a significant ($p < 0.01$) increasing trend of 23.8%/year from 1966 to 1994.

During the Atlas period, Great Blue Herons were reported from 40% of the completed priority blocks, and breeding was confirmed in 66 priority and nonpriority blocks. Nesting colonies were concentrated at Reelfoot Lake and other Mississippi Alluvial Plain swamps, along the main stems and branches of the Obion, Forked Deer, Hatchie, and Big Sandy Rivers, near Kentucky Lake at the mouth of the Duck River, and along upper Tennessee River reservoirs upstream to Knoxville. The absence of colonies in the Cumberland River drainage is difficult to explain. Most of the possible records in West Tennessee and in the vicinity of colonies in Middle and East Tennessee were probably of foraging birds nesting at nearby colonies. The possible records elsewhere are either birds nesting in undiscovered colonies or non-breeding birds; the Great Blue Heron usually does not begin nesting until at least two years old (Palmer 1962).

The Great Blue Heron was listed as In Need of Management in Tennessee from 1976 through 1986, when it was taken off the list because of its increasing population (TWRA 1976, 1986). Its future in the state is probably secure, and its breeding range will probably spread to presently unoccupied reservoirs. Individual colonies, however, remain vulnerable to human disturbance and will occasionally require protective measures.

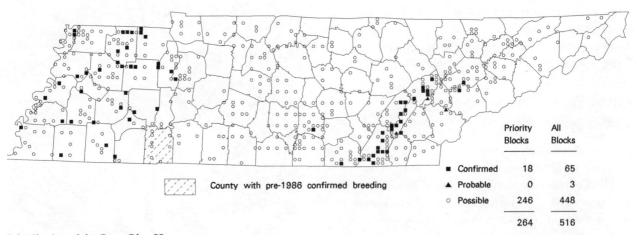

Distribution of the Great Blue Heron.

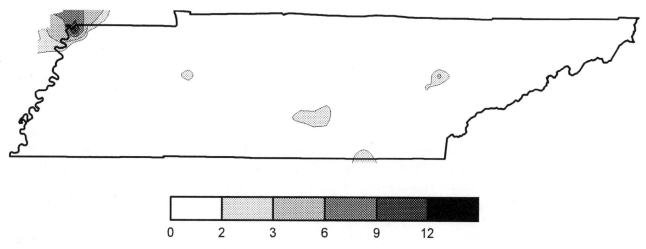

Abundance of the Great Blue Heron.

Breeding Biology: Great Blue Herons usually nest in trees near water. Nests at Reelfoot Lake have been in large cypress trees, typically near the top of the canopy at heights of 27.4 to 30.5 m (Gersbacher 1939, Pitts 1982b). Most other West Tennessee colonies were also in cypress trees. Colonies farther east have been in pines, both planted and natural, and both upland and bottomland hardwoods. Some colonies have been short-lived because of the deterioration and death of nest trees from guano accumulations; this has occurred more frequently at colonies not subject to periodic flooding. Herons have also nested several times on tall, steel transmission line towers (Pullin 1983, pers. comm.). The average number of Great Blue pairs in colonies in the eastern half of the state in 1988 was 84.7 pairs (s.d. = 129.4, n = 18). In the western half of the state, the 1990 average, excluding the Reelfoot colony, was 47.9 pairs (s.d. = 73.9, n = 22). The Reelfoot colony contained about 2200 nests in 1990–91, about 90% of them Great Blues (Greene, Knox, and Pitts 1991). The next largest colony was at Armstrong Bend, which contained 529 Great Blue pairs in 1988 (Pullin 1990).

Great Blue Herons begin reoccupying nest sites on warm winter days in January (Nicholson pers. obs.). Both adults build the nest, a platform of sticks with a shallow bowl lined with twigs and green leaves or pine needles (Palmer 1962). Nests from previous years are usually rebuilt and may be a meter in diameter. Egg laying probably peaks in mid-March. The few available records suggest that clutches of 3 and 4 eggs are most common. Both adults incubate for about 28 days, and both feed the nestlings, which fledge in slightly under 2 months (Palmer 1962, Hancock and Kushlan 1984).
—*Charles P. Nicholson.*

Great Egret
Ardea alba

The Great Egret, largest of the white herons to regularly occur in Tennessee, is an uncommon summer visitor and rare winter visitor (Robinson 1990). It is most numerous in West Tennessee, where, as a nesting species, it is locally common. It is usually present by late March and departs by late November. Except for near nesting colonies, the largest number of birds are present during late summer; these birds are probably post-breeding migrants from colonies outside the state.

During the Atlas period, Great Egrets nested in at least 6 Tennessee colonies, all but 1 in or near the Mississippi Alluvial Plain. The egret population increased from at least 32 pairs in 2 colonies in 1986 to at least 270 pairs in 5 colonies in 1990–91 (Smith 1989–90, Greene,

Great Egret. Chris Myers

Confirmed Breeding Species

Knox, and Pitts 1991, Atlas results). The largest concentration was at Reelfoot, where about 220 pairs nested in 1990–91. The only nest record distant from the Mississippi River was on a Chickamauga Reservoir island, Hamilton County, where a pair nested with Great Blue Herons in 1990. These birds may have originated from an egret restoration project in northeast Alabama (Pullin 1987); an egret pair also nested with Great Blue Herons along the Tennessee River less than a kilometer south of the Tennessee–Alabama line in 1990 (B. Pullin pers. comm.). All of the Atlas records of possible breeding status, except those near colonies, are probably non-breeding birds or early dispersers from colonies outside Tennessee. Great Egrets were recorded on 6 Atlas miniroutes at an average of 2.2 stops/route and all near the Mississippi River.

The earliest systematic survey of Tennessee birds, by Rhoads (1895a), did not mention Great Egrets. At that time, Great Egret populations in North America were greatly depressed by hunting to supply their long plumes, or "aigrettes," to the millinery industry. Before this population reduction, Great Egrets regularly occurred along the Mississippi River far north of Tennessee (Widmann 1907, Graber, Graber, and Kirk 1978), and it is likely they nested in Tennessee in the mid-nineteenth century. Their North American population probably reached its lowest level around 1902–3, and began increasing following legal protection a few years later (Palmer 1962). The exact year they again began nesting in Tennessee is unknown. None were nesting at Reelfoot Lake in 1919 or 1921, but by 1932, 450 nests were present in the Cranetown colony (Ganier 1933b). By 1938, the Cranetown egret population had increased to 655 nests and 3500 birds, outnumbering all other species in the colony (Gersbacher 1939). Another colony at the north end of Reelfoot Lake in Kentucky, known as Crane Roost or Kentucky Cranetown, grew from 2 egret pairs in 1932 to 200 pairs in 1949 (Gersbacher 1964, Mengel 1965).

A colony established in the 1940s at the mouth of the Duck River in Humphreys County contained 150 egret nests in 1949 (Ganier 1951a). A colony at Dyersburg, Dyer County, probably established in the late 1950s, contained about 60 Great Egret nests in 1960 (Ganier 1960).

The Great Egret population in Tennessee collapsed after this rapid growth. The decline of Cranetown began in the 1950s, and it was inactive in 1963 (Gersbacher 1964). The Dyersburg heronry was last active in 1969, Kentucky Cranetown was inactive by the early 1970s, and Great Egrets last nested at the Duck River heronry in 1972 (Pitts 1973, 1977). After a few years when none were known to nest in Tennessee, Great Egrets returned to Reelfoot Lake in the late 1970s (Pitts 1982b). All of the other sites where egrets nested during the Atlas period were first discovered during the Atlas period.

Because of declining numbers and threats to wetland habitats, the Great Egret was listed as In Need of Management in Tennessee 1976 (TWRA 1976). Although the recent increase in its numbers and current wetland conservation programs in West Tennessee bode well for its future, the In Need of Management designation should be retained for the time being.

Breeding Biology: At all current and historical Tennessee colonies, Great Egrets have nested in association with other heron species. Great Blue Herons have usually been present and, except at Cranetown, outnumbered the egrets. The Dyersburg colony was unusual in that no Great Blues were present, and Little Blue Herons were dominant. Most nesting colonies have been in seasonally or permanently flooded forested wetlands. Bald cypress has been the dominant tree species at past and present Reelfoot colonies (Gersbacher 1939, Pitts 1982b), as well as at the Duck River (Ganier 1951a) and Shelby County (M. Greene pers. comm.) colonies. Nests at Dyersburg were in boxelder, "water" (silver?) maple, overcup oak,

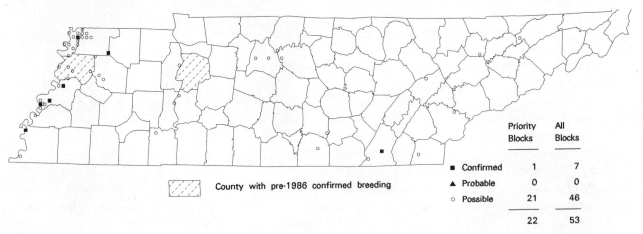

Distribution of the Great Egret.

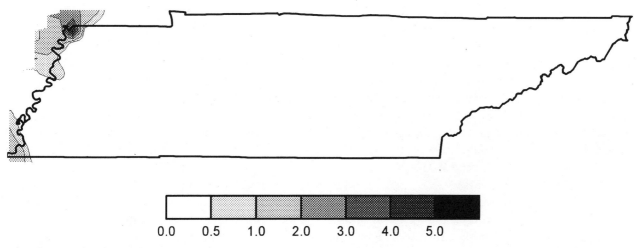

Abundance of the Great Egret.

and elm (Ganier 1960), and at a Lauderdale County colony in a cypress-tupelo swamp (P. B. Hamel pers. comm.).

Great Egret nests at Cranetown were built about 23–27 m above the water, usually lower in the trees than Double-crested Cormorant and Great Blue Heron nests and higher than Black-crowned Night-Heron nests (Gersbacher 1939). Individual trees often contained several nests. At the Dyersburg colony, where the average tree height was much less than at Cranetown, most nests were between 5 and 8 m high (Ganier 1960).

Great Egrets begin arriving in nesting colonies in March. The male defends a small territory around the nest site, and nests from previous years may be reused (Hancock and Kushlan 1984). Courtship includes display flights as well as perched displays when the plumes are prominently erected. The nest, begun by the male during courtship, is a platform of sticks placed on the fork of a tree limb (Gersbacher 1939, Palmer 1962).

Egg laying occurs during April and May. The peak is difficult to determine from the relatively few available detailed observations, although Ganier (1951a) stated that most egrets began to incubate during late April. Clutches of 2 to 5 eggs have been found in Tennessee, and the most common clutch size is 4 eggs (Gersbacher 1939, egg coll. data). The eggs are pale bluish green. Both adults incubate for about 25 days (Hancock and Kushlan 1984). Both adults also feed the nestlings by regurgitating fish, frogs, and crayfish into the mouths of the young. The young begin leaving the nest and climbing in the tree limbs when 2 to 3 weeks old, and at about 6 weeks old fly with their parents to feed (Gersbacher 1939). Most of the nests at Cranetown were empty by the middle of July. Nesting success in Tennessee is poorly known. Gersbacher (1939) noted that while 4-egg clutches at Cranetown were most common, most nests contained 3 young.—*Charles P. Nicholson.*

Snowy Egret
Egretta thula

The Snowy Egret is among the most beautiful and elegant Tennessee birds. Adults are most easily recognized by their snow-white plumage, "golden slippers," and active, yet graceful, foraging behavior. Their plumes are long and attractive, and are often displayed during courtship and when approaching the nest. The Snowy Egret is a rare migrant and summer visitor in most of Tennessee (Robinson 1990). It is most numerous in West Tennessee, especially along the Mississippi River, where it is an uncommon migrant and local breeding species. Most are present between mid-April and October (Waldron 1987, Robinson 1990).

The first Tennessee report of the Snowy Egret was by Rhoads (1895a), who gave secondhand accounts of small numbers of migrating birds in West Tennessee.

Snowy Egret. Chris Myers

At that time, Snowy Egrets were in great demand for their long plumes by the millinery trade, and many local populations were extirpated. This situation affected populations for decades after the plume trade was halted in the early twentieth century (Palmer 1962). Coffey (1952) speculated that Snowys probably nested close to Memphis and farther north prior to the intensive harvests of the early 1900s. Until 1950, the colonies closest to Tennessee were in Phillips County, Arkansas, in 1913 (James and Neal 1986) and at Moon Lake, Mississippi, in 1943 (Coffey 1943a).

The first documented nestings in Tennessee occurred in 1950 near Ridgely, a short distance south of Reelfoot Lake, and on Redman Bar in the Mississippi River near Memphis (Ganier 1951a, Coffey 1952). The Ridgely heronry contained 70–100 Snowy Egret nests, and 150 nests were present in the Redman Bar heronry. Local residents stated the Ridgely heronry had been present for years; it disbanded in the late 1950s (Ganier 1960). The Redman Bar heronry apparently only lasted a few years. Another heronry near Dyersburg, Dyer County, established in the late 1950s, contained at least 20 Snowy Egrets in 1962 (Ganier 1960, *Migrant* 33:47). Snowy Egrets were apparently not present each year in this heronry, which disbanded in 1969 (Leggett 1970). These 3 sites are the only known nesting locations of Snowy Egrets in Tennessee prior to the Atlas project.

The Snowy Egret was reported from 13 blocks during the Atlas period. Most of these reports were in the "observed" category, birds foraging during early summer in shallow sloughs and ponds away from nesting habitats. It was found on 4 miniroutes at an average relative abundance of 1.8 stops/route. Breeding was confirmed in 2 blocks, each at a colony along the Mississippi River, consisting of Snowy Egrets, Little Blue Herons, Cattle Egrets, and Black-crowned Night Herons, and each first discovered in 1990 (Ford 1992). One of these heronries was at Plum Point, Mississippi River mile 785, Lauderdale County, and the other at Cypress Bend, Mississippi River mile 863, Lake County.

During the period from the 1960s through 1980s when Snowy Egrets were not known to nest in Tennessee, they regularly occurred as late summer visitors, probably in part from colonies in adjacent states (see Pitts 1973). Snowy Egrets nested near Sikeston, Missouri, about 56 km north of Reelfoot Lake, at Burdette, Arkansas, west of Lauderdale County, Tennessee, and possibly at Cayce, Kentucky, 16 km north of Union City, Tennessee (Pitts 1973, James and Neal 1986). A large heronry containing a small proportion of Snowy Egrets throughout the Atlas period was at Caruthersville, Missouri, about 20 km from the Cypress Bend heronry (Robbins and Easterla 1992).

Presumably because of the absence of nesting birds until very recently, the Snowy Egret had not received any special protective status in Tennessee. Following the establishment of nesting colonies discovered during Atlas fieldwork, it was listed as In Need of Management in 1994 (TWRA 1994a).

Breeding Biology: All of the Tennessee Snowy Egret colonies have been in or very close to the Mississippi Alluvial Plain on islands, sandbars, or other periodically flooded sites. Snowy Egrets are highly colonial and typically nest in dense, mixed-species colonies (Eckert 1981). In Tennessee, heronries containing Snowy Egrets have been about 50 by 100 m in size (Ridgely, Ganier 1951a), 100 by 150 m (Redman Bar, Coffey 1952), and 75 by 120 m (Plum Point, Ford 1992). The proportion of Snowy Egrets in Tennessee heronries has ranged from less than 1% to about 20% (Coffey 1952, Ganier 1951a, Ford 1992, *Migrant* 33:47). Habitats and nest trees for Snowy Egrets are typically thick growths of willow from 3 to 4 m tall (Bent 1926). The Plum Point heronry was in thick willows

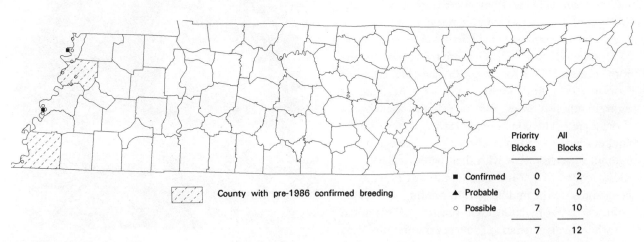

Distribution of the Snowy Egret.

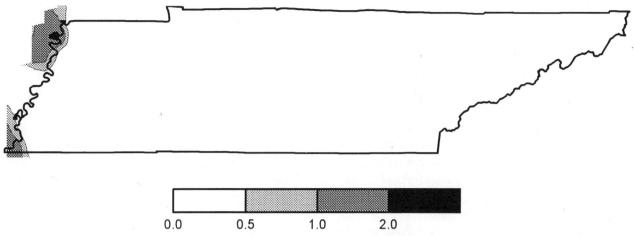

Abundance of the Snowy Egret.

up to about 28 cm diameter and mostly 5–6 m tall on sandy soil with much recent flood debris (Ford pers. obs.).

The male Snowy Egret selects the nest site, which, after mating, is vigorously defended by both the male and female (Palmer 1962). Nests are typically built by both sexes and consist of a flimsy, flat platform (Eckert 1981), often indistinguishable from the nests of Little Blue Herons or Cattle Egrets. The nests are rarely over 30 cm in diameter, and generally are made up of small twigs and reeds less than a meter long (Palmer 1962, Eckert 1981). Nests in the Ridgely heronry were 2–3 m high (Ganier 1951a) and in both the Redman Bar and Plum Point heronries, from 2 to 5 meters high (Coffey 1952, Ford pers. obs.).

The normal initiation of nesting in Tennessee probably occurs in May; at the 1990 Cypress Bend heronry, nesting was delayed at least a month, probably because of flood conditions in May and June (J. Wilson pers. comm.). Little clutch size information is available from Tennessee heronries. Clutches of 3–5 eggs are common elsewhere (Hancock and Kushlan 1984). The pale bluish green eggs are laid at 2-day intervals, and incubation begins before the clutch is completed (Eckert 1981). Both sexes incubate and feed the nestlings. The length of the incubation period is poorly known; the young climb from the nest in about 3 weeks and fly in about 4 weeks (Hancock and Kushlan 1984). Because of the asynchronous hatching of the young, the youngest chicks frequently do not survive.—*Robert P. Ford.*

Little Blue Heron
Egretta caerulea

The Little Blue Heron is an uncommon migrant, an uncommon to fairly common late summer visitor, and a rare nesting bird in Tennessee. It is usually present from early April to October, although a few early winter records exist (Robinson 1990). The Little Blue Heron is unusual among members of its family in having a mostly white immature plumage and dark adult plumage acquired by two years of age (Palmer 1962). Year-old birds in white and mixed white and dark ("calico") plumage have been found nesting in Tennessee and elsewhere (Ganier 1951a, Palmer 1962).

The Little Blue Heron is most common during the summer in West Tennessee, and all nesting colonies have been along the western arm of the Tennessee River or near the Mississippi River. Its nesting population has fluctuated greatly in the last 90 years, and during the Atlas period it nested at 2 sites along the Mississippi River, each first discovered in 1990. The Atlas observations away from the colonies were of adult, calico, and white-plumaged birds. Those near the Mississippi River were either of birds from the Tennessee colonies, colonies

Little Blue Heron. Chris Myers

Confirmed Breeding Species

in Arkansas or Missouri, or of non-breeding birds. Records farther east were probably of non-breeding birds.

The nineteenth-century distribution of the Little Blue Heron in Tennessee is poorly known. It nested as far north as southwest Indiana in the late 1800s (Graber, Graber, and Kirk 1978), suggesting that this typically southern species probably nested in West Tennessee. It was likely not a target of plume collectors at the turn of the century, although colonies may have been disrupted by the collection of associated nesting species (Hancock and Kushlan 1984). The first state record was in May 1919, when Ganier (1933b) observed 5 near the north end of Reelfoot Lake. Two years later Ganier (1933b) found a colony of about 12 pairs along the canal draining the southern end of Reelfoot Lake. This colony apparently did not persist, although the Little Blue Heron continued to be common in late summer (Whittemore 1937).

The next reports of nesting Little Blue Herons occurred in about 1950 @in Lake (Ganier 1951a) and Shelby (Coffey 1952) Counties. The Lake County colony, near Ridgely, in 1950 contained between 700 and 1000 nests, 90% of them Little Blue Heron nests. The colony had apparently been occupied for several years prior to 1950 and disbanded by 1960. The Shelby County colony, on Redman Bar in the Mississippi River at Memphis, contained about 750 nests, 80% of them Little Blue Heron nests. It was only active for a few years. Another large heronry was established in Dyer County in the 1940s; in 1960 it contained about 540 Little Blue Heron nests (Ganier 1960, Leggett 1968). By 1964, it had grown to about 2000 nests (Coffey 1964) and apparently peaked in size in 1965 before disbanding after the 1969 nesting season (Leggett 1968, 1970). Up to 15 pairs also nested on the Tennessee National Wildlife Refuge between 1962 and 1970 (Pitts 1973).

Between 1970 and 1985, Little Blue Herons nested in small numbers in at least 4 different sites, each apparently short-lived. These sites were in eastern Obion County, active in 1971 (Pitts 1973); Benton County, 8 nests active only in 1982; Houston County, 1 nest with 3 young in 1984; and Lauderdale County, active from 1983 to 1985 with 20 nests in 1985 (Pullin 1990). A few pairs also nested from 1981 to 1983 on an island in Lake Barkley, a short distance north of the Tennessee line (Palmer-Ball 1991).

Despite its tendency to nest at very few, potentially vulnerable sites, often in large numbers, the Little Blue Heron was, until recently, given no special protective status in Tennessee. The Ridgely colony was probably destroyed by conversion to agriculture, and the Dyersburg colony declined following residential and industrial development. Reasons for the short life of some other colonies are unknown. The Little Blue Heron was listed as In Need of Management in 1994 (TWRA 1994a), although an official status of Threatened is probably warranted.

Breeding Biology: All Little Blue Heron nest sites, for which detailed descriptions are available, have been in thickets of fairly small hardwood trees, less than 11 m tall. Dominant tree species have been black willow, boxelder, silver maple, overcup oak, and elm. Most nests have been between 1.5 and 7.6 m above ground or water. Even in the large colonies, the herons nested at high density in an area of 1.5 ha or less. Shallow water covered most sites early in the nesting season. These colony site characteristics are typical of other inland Little Blue colonies (Palmer 1962, Hancock and Kushlan 1984). Little Blues were the only species present in some of the smaller colonies. At the large colonies, other species nesting with the Little Blue Herons have been Great Blue Herons, Great, Snowy, and Cattle Egrets, and Black-crowned Night-Herons. Except for the 1983–85 Lauderdale County colony and both 1990 colonies, where Cattle Egrets were most numerous, Little Blue

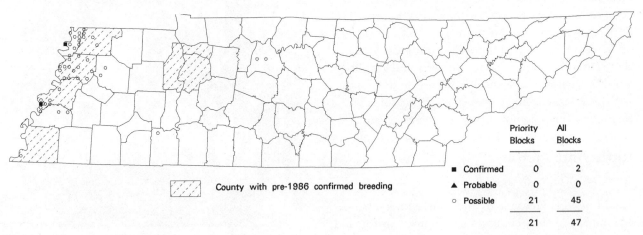

Distribution of the Little Blue Heron.

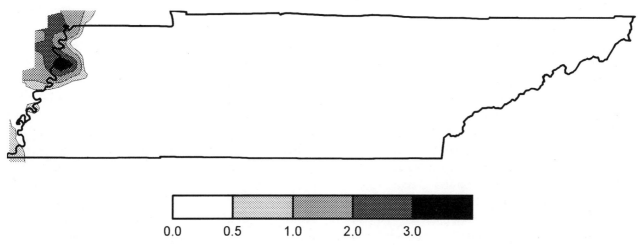

Abundance of the Little Blue Heron.

Herons made up 80 to 90% of the herons present. At the 1990 colonies, about 10% of the birds were Little Blue Herons (Ford 1992). Little Blue Heron nests are built of sticks and vary from a flimsy platform to substantially built with a well-defined bowl (Palmer 1962).

The chronology of Little Blue Heron nesting varies greatly both between different colonies and within colonies. Ganier (1951) and Coffey (1952) both observed nests under construction as well as nests containing eggs and small young in mid-May. Egg laying probably begins as early as mid-April. Ganier (egg coll. data) collected 2 clutches of 4 eggs at the Ridgely colony, and Coffey (1952) reported 1–3 eggs in nests at the Shelby County colony. Whether the Shelby County clutches were complete is not known; normal clutch size is 2–5 eggs, with 4 and 5 most common (Palmer 1962, Hancock and Kushlan 1984). Both adults incubate for 21–23 days, beginning after the second egg is laid. The young begin climbing out of the nest onto tree limbs when about 13 days old and are independent of their parents when about 50 days old (Hancock and Kushlan 1984).—*Charles P. Nicholson.*

Cattle Egret
Bubulcus ibis

The Cattle Egret, smallest of the white herons and egrets, is a rare summer resident of East and Middle Tennessee, and a fairly common summer resident of West Tennessee. It is usually present in the state from early April into November (Robinson 1990). The Cattle Egret is much less dependent on wetlands for foraging than other herons and egrets, and it frequently forages in freshly plowed fields and pastures, where it often catches insects flushed by livestock. Its nesting habitat requirements, however, overlap with those of other colonial herons and egrets.

The Cattle Egret is a native of Africa that immigrated to South America in the late nineteenth century, and then to North America by 1942 (Palmer 1962). The first Tennessee record was in Anderson County in 1961 (Tanner 1988). The first breeding record was of 8 nests in the large mixed-species Dyersburg colony in 1964 (Coffey 1964). The number of Cattle Egrets nesting at this site increased until the site was abandoned after the 1969 nesting season (Leggett 1970). The next nest record was in 1975, when 4 pairs nested in a Black-crowned Night-Heron colony on Cherokee Lake, Grainger County (Pitts 1977). They continued to nest there into the 1980s (Williams 1977b, Tanner 1988).

No other nests were reported until the mid-1980s, when large numbers nested at 2 sites along the Mississippi River. The first of these, at Hales Point, Lauderdale County, was established in 1983, not far from colonies in adjacent Mississippi County, Arkansas, where

Cattle Egret. Chris Myers

Confirmed Breeding Species

Cattle Egrets had nested since 1968 (James and Neal 1986). This colony contained 357 Cattle Egret nests in 1985, and smaller numbers of 4 other heron species (Pullin 1985–88). It was not used in later years. The second was on an island a short distance north of the Mississippi state line. Exact nest counts are not available, but between 1985 and 1987 the population grew to about 1000 birds and 300 to 500 nests (J. Rumancik pers. comm.). No nesting occurred there in 1988, although large numbers of several heron species continued to roost on the island (Robinson 1988).

During Atlas fieldwork, Cattle Egrets were discovered nesting at 8 additional colonies scattered across the state. The majority of the Atlas records, however, were of birds observed feeding in fields and wetlands, a fairly common sight in far western Tennessee. Many of these birds were in breeding plumage and probably from known colonies in Tennessee, Arkansas, and Missouri. Others, such as those observed at 3 miniroute stops in a Henry County block, may have been from undiscovered colonies. Still others were probably non-breeding birds. From east to west, the newly discovered nest sites were at a small lake near the Holston River, Jefferson County, in 1991; at least 1 nest, 1987 through 1991 near Fort Loudoun Dam, Loudon County; at least 2 nests in 1991 on an island in Chickamauga Lake, Meigs County; at least 2 nests in 1991 on an island in lower Chickamauga Lake, Hamilton County; at least 1 nest in 1987 on an island in Old Hickory Lake near Gallatin, Sumner County (Stedman 1988a); on an island in lower Old Hickory Lake, several nests in 1991; and along the Mississippi River in 1990, about 600 nests in Lake County and about 950 nests in Lauderdale County. The Lake County colony was about 20 km from a colony in Missouri that contained over 1000 Cattle Egret nests during the Atlas period (Robbins and Easterla 1992).

The Cattle Egrets in Loudon, Sumner, and Davidson Counties nested with larger numbers of Black-crowned Night-Herons. At the Meigs County site, most of the birds present were Great Blue Herons. At the 1990 Lauderdale and Lake County sites, about 80% of the birds present were Cattle Egrets. Other nesting species there were Little Blue Herons, Snowy Egrets, and Black-crowned Night-Herons. Only Cattle Egrets were present at the Jefferson and Hamilton County sites.

Since first nesting in Tennessee in 1964, the Cattle Egret population has greatly increased, reaching a peak in the number of colonies and number of nests in 1990 and 1991. This growth, however, has not been continuous, as no nests were known during part of this period, and some large colonies were active for only a few years. Throughout North America, Cattle Egrets increased from 1966 to 1993 at the annual rate of 1.7%/year ($p = 0.02$) on BBS routes (Peterjohn, Sauer, and Link 1994). The Cattle Egret is adaptable in its foraging and nesting habitat requirements, and the future of this non-native species in Tennessee is probably secure. Competition between nesting Cattle Egrets and native herons has been reported elsewhere in North America (Burger 1978), although it is not always apparent (Weber 1975). Whether it will affect native heron and egret populations in Tennessee remains to be seen.

Breeding Biology: Both adults participate in nest construction, with the male gathering nest material and the female placing it on the nest (Hancock and Kushlan 1984). Nests at the Dyersburg site were bulky and constructed of sticks and leaves (Coffey 1964). Except for the colonies in Grainger and Loudon Counties, which were in pines mixed with a few deciduous trees, most Cattle Egret nests have been in deciduous trees. Nests at the Shelby, 1990 Lauderdale, Jefferson, and Hamilton County sites, were in willows, either over water or seasonally flooded. Nests are usually built below the canopy, and

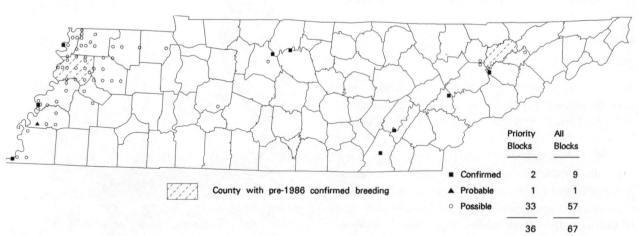

Distribution of the Cattle Egret.

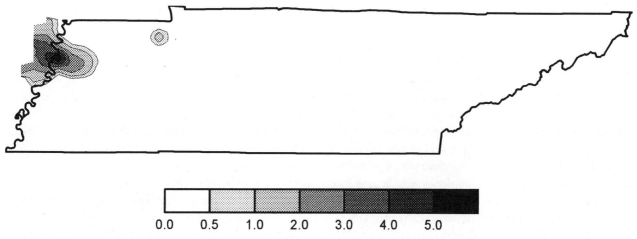

Abundance of the Cattle Egret.

their heights vary between colonies and with the tree species used. Average heights within colonies range from about 2.1 m in Jefferson County to 11 m at the Dyersburg colony. When other heron or egret species have been present, Cattle Egrets have frequently nested in the same trees as the other species, with no apparent difference in nest site selection. Studies elsewhere have shown much variation in nest placement (Hancock and Kushlan 1984). The Cattle Egret is normally highly colonial, nesting in very dense concentrations. Nesting with other species may be typical of the Cattle Egret when it is colonizing new areas (Palmer 1962).

Clutch initiation, determined primarily from the estimated age of nestlings, has been between early May and late July. It varies greatly between colonies; too few detailed records are available to describe variation within colonies, although it is high elsewhere (Weber 1975). Most clutches at the Shelby County colony contained 3 eggs (J. Rumancik pers. comm.); little other information on clutch size is available. Both adults incubate for about 24 days, starting before egg laying is complete (Weber 1975). The nestlings remain in the nest about 20 days, then climb into surrounding tree branches. They begin to fly when about 30 days old and are independent at about 45 days old (Hancock and Kushlan 1984).—*Charles P. Nicholson.*

Green-backed Heron
Butorides virescens

The Green-backed Heron is a fairly common summer resident, found at low elevations across Tennessee. It usually arrives in the state in late March or early April and departs by early November (Robinson 1990). Many winter records exist, mostly from late December Christmas Bird Counts.

The Green-backed Heron is the most widely distributed and probably the most abundant member of its family in Tennessee. It occurs along streams and the shorelines of ponds, small lakes, and large reservoirs, where it feeds primarily on small fish and aquatic invertebrates. Although usually inconspicuous in the vicinity of its nest, the Green-backed Heron is easy to observe around farm ponds and in flight; it was observed in about 75% of the completed Atlas priority blocks. In the Unaka Mountains, probable breeding birds were recorded in agricultural areas as high as 853 m. It was most noticeably absent from heavily wooded portions of the Cumberland Plateau and Cumberland and Unaka Mountains, where habitat is limiting, as it rarely occurs along high-gradient woodland streams (Nicholson pers. obs., Stupka 1963). Green-backed Herons were probably present in those priority blocks elsewhere in the state where atlasers failed to record them.

The overall range of the Green-backed Heron has probably not historically changed in Tennessee. Its

Green-backed Heron. Chris Myers

Confirmed Breeding Species

numbers in parts of the state, however, have probably changed due to drainage of wetlands and construction of farm ponds and reservoirs. At the time of European settlement, it was probably most common in West Tennessee because of the large area of wetlands there. Rhoads (1895a) observed the heron throughout the state, but gave no indication of regional differences in abundance. Clearing of forests for agriculture probably had little effect, as the heron is not dependent on large areas of forest. Construction of reservoirs and farm ponds probably increased suitable habitat, especially in Middle and East Tennessee. In West Tennessee, this increase was probably offset by wetlands drainage and stream channelization.

Atlasers recorded Green-backed Herons on slightly over a quarter of the miniroutes at an average relative abundance of 1.2 stops/route. The herons were somewhat more common in the Central Basin and Loess Plain than elsewhere in the state. This may be more an artifact of their increased visibility in these areas of little forest cover than an actual greater abundance. The Atlas miniroutes also recorded more Green-backed Herons than the BBS routes, where the average abundance of 1.8 stops/50-stop route during the Atlas period was about half the miniroute results. In a survey by boat of a portion of the Duck River in Maury County, Fowler and Fowler (1985) reported an average of 0.22 herons/km. Comparable information from other parts of the state is unavailable.

Information on population trends has only become available in recent decades. Tanner (1986) noted an increase in Green-backed Heron numbers on Tennessee spring bird counts after about 1970, which he noted may have been due in part to the ban on use of DDT. Results of Tennessee BBS routes, however, show a significant decline ($p < 0.01$) of 2.2%/year from 1966 to 1994. BBS results from throughout North America showed no significant trend from 1966 to 1993 (Peterjohn, Sauer, and Link 1994).

Breeding Biology: Unlike most of the other herons nesting in Tennessee, the Green-backed is usually a dispersed, solitary nester. Small colonies of up to 10 nests have been reported in Shelby, Williamson, Sullivan, Franklin, Hawkins, and Blount Counties (Coffey 1981, W. Coffey 1966, Williams 1975b). Pairs defend feeding areas as well as the nest area (Palmer 1962). Confirming breeding was difficult for atlasers, and less than 10% of the Atlas records were of confirmed breeding.

Green-backed Heron nests are often built near or over water, although they may be one km or more from water. The nests are built in shrubs, saplings, or trees and consist of an unlined platform of sticks, sometimes so lightly built that the eggs are visible from below the nest. The male begins nest construction prior to mating, and, after mating, both birds work on the nest (Palmer 1962). They continue to add material during incubation. Nests may be reused in subsequent years and these reused nests are more substantially built (Palmer 1962, Williams 1975b). Nest heights reported in Tennessee range from 1 to 12.2 m. W. Coffey (1966) gave an average height of 7.9 m for a colony of 9 nests in Virginia pines in Sullivan County, and Williams (1975b) found an average height of 3.1 m for 19 nests in the Alcoa Marshes, Blount County, most of which were in willows. The average height of 34 other nests is 4.7 m (s.d. = 2.3 m). Commonly used tree or shrub species include eastern red cedar, willow, buttonbush, hackberry, pines, and oaks.

Egg laying in Tennessee, as determined primarily from egg collections, appears to peak during the last third of April. There may be much variation in timing of egg laying, however, within colonies during the season and between seasons (Coffey 1981). Several records of half-grown young in mid-May (e.g., W. Coffey 1966) suggest egg laying during the first half of April is not uncommon. Clutches of 3 to 6 eggs have been reported

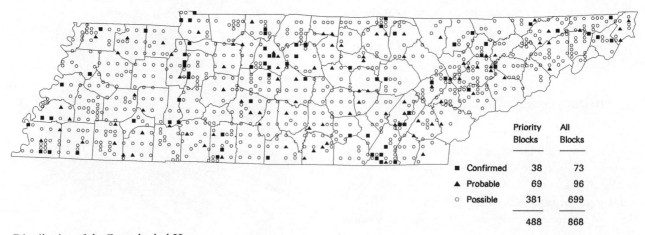

Distribution of the Green-backed Heron.

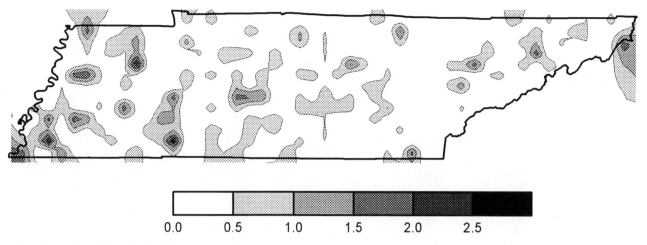

Abundance of the Green-backed Heron.

in Tennessee, with 4 and 5 eggs most common and an average size of 4.4 eggs (s.d. = 0.6, n = 45), similar to that given by Palmer (1962). The eggs are pale greenish or bluish green. Incubation, performed by both sexes, begins after the laying of the third egg, and lasts about 20 days (Palmer 1962, Douglass et al. 1965).

Both adults brood the nestlings and feed them by regurgitating food into the nestlings' mouths or, as the young mature, onto the nest (Palmer 1962). The young begin climbing from the nest into the nest tree when about a week old, fly when about 3 weeks old, and are independent by about 5 weeks after hatching (Palmer 1962). Replacement clutches are laid following loss of a nest, and second broods uncommonly occur (Palmer 1962). Second broods have not been reported in Tennessee.—*Charles P. Nicholson.*

Black-crowned Night-Heron
Nycticorax nycticorax

The Black-crowned Night-Heron is a permanent resident, fairly common in East and Middle Tennessee during the summer, rare in West Tennessee during the summer, and rare statewide during the winter. Most wintering birds occur near nesting colonies. It is probably the second most numerous colonial nesting heron in the state. Most recent colonies have been in the eastern two-thirds of the state, a reversal of the situation during the first half of this century.

The Black-crowned Night-Heron was first reported from the state in Roane County on 24 April 1885 by Fox (1886). The first published nest record was from the Reelfoot Lake Cranetown colony in 1933 (Vaughn 1933); Ganier (1933b) believed the night-heron nested there earlier. This colony grew to at least 45 night-heron nests in 1938 (Gersbacher 1939). It persisted into the late 1950s, when the number of night-herons rapidly declined, and was gone by 1963 (Gersbacher 1964). Two night-heron nests were present in the Ridgely heronry, a few kilometers south of Reelfoot, in 1950 (Ganier 1951a). This site was abandoned by the late 1950s. During the 1960s, at least 2 night-herons were reported in the Dyersburg heronry (e.g., Ganier 1960), but their nesting was apparently not confirmed. Black-crowneds nested in two large mixed-species colonies along the Mississippi River discovered in 1990. Several hundred birds, predominantly Cattle Egrets, nested at each of these sites; Black-crowneds made up less than 5% of the birds present (Ford 1992, Atlas results).

The first Middle Tennessee colony was reported by Ganier (1951a) at Bordeaux in suburban Nashville. Black-crowned Night-Herons apparently first nested at this site in 1908 (Pitts 1973). Their population was fairly stable at about 50 pairs into the mid-1970s (Pitts 1977).

Black-crowned Night-Heron. Chris Myers

Confirmed Breeding Species

The site came under state protection in 1978, and the night-heron population grew to a peak of 370 nests in 1984 (Pullin 1990). It then declined, perhaps because of continued nearby urban development, and was abandoned in 1991 (Atlas results). Other nearby small colonies occurred during the mid-1970s on islands in lower Old Hickory Lake (Pitts 1977) and from 1979 to 1982 along Mill Creek in south Nashville (Pullin 1990). A colony was established on an island in Old Hickory Lake near Gallatin, Sumner County, in 1985 and grew to 200 pairs by 1987 (Pullin 1990). Another colony was again present on a lower Old Hickory Lake island in 1991 (Atlas results).

The first East Tennessee nesting was in 1952, when about 15 pairs nested in a colony on the Blount County side of Fort Loudoun Lake (*Migrant* 23:53). Another small colony was active in 1966 at Concord, Knox County, also near Fort Loudoun Lake (Pitts 1977). Then three large East Tennessee colonies were active during the 1970s and 1980s. In 1974 a colony was initiated near the 1966 site (Pitts 1977). This population grew to a peak of 206 nesting pairs in 1982 and, because of conflicts with private landowners, moved 7 times between 1977 and 1986 (Pullin 1990). A few nested with Great Blue Herons on an island near Kingston, Roane County, in the mid-1980s (Pullin 1990); these birds may have dispersed from the Fort Loudoun population. In recent years the Fort Loudoun population has nested on public property near the dam; a few also nested with Great Blue Herons near Louisville, Blount County, in 1990 (Atlas results). The second large East Tennessee colony was discovered on Cherokee Lake, Grainger County, in the early 1970s (Pitts 1977). This population grew to a maximum of 764 nesting pairs in 1984, despite 5 relocations (Pullin 1990). Its location and size after 1985 is unknown, although the continued presence of the birds (Atlas results) suggests they still nest in the Cherokee Lake area.

The third large East Tennessee colony, near Sevierville, was first active about 1972 or 1973 (Nunnally and Williams 1977). This colony grew to a peak of 493 nesting pairs in 1985, despite moving at least once (Pullin 1990). In 1986 it settled near a Pigeon Forge subdivision; wildlife agents scared the birds away in early 1987. The present status of this population is unknown. A small colony of about 10 pairs was discovered near Kingsport in 1990 (Atlas results). Additional colonies probably exist in northeast Tennessee, as atlasers observed adults with fledged young below Boone Dam, Washington County, and below South Holston Dam, Sullivan County. Some of the scattered observations elsewhere in the state may have been birds from nearby, unknown colonies.

The Black-crowned Night-Heron was listed as In Need of Management by the TWRA in 1976, and elevated to Threatened in 1978. Following protection of the Bordeaux colony and increases in numbers at several colonies, its status was downgraded to In Need of Management in 1986, and it was taken off the list in 1994 (TWRA 1986, 1994a). Although some of the present colonies are on public lands, they are still vulnerable to human impact (Pullin 1990), and the current status of two of the large East Tennessee populations is unknown. Some colonies have been in pines, which remain suitable colony sites for only a few years, and, as in the case of the present Fort Loudoun colony, few secure alternative sites may exist. Thus the future of some night-heron populations is questionable and continuation of the In Need of Management status is warranted. Other reservoirs across the state, however, offer areas suitable for eventual colonization.

Breeding Biology: In West Tennessee, all Black-crowned Night-Heron nests have been in large, mixed-species colonies with other herons greatly outnumbering the Black-crowneds. In Middle Tennessee, the colonies have been

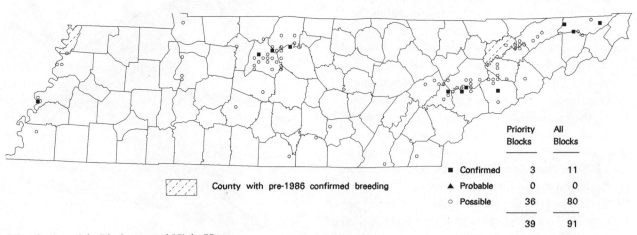

Distribution of the Black-crowned Night-Heron.

made up of only Black-crowneds or mixed with a few Yellow-crowned Night-Herons or Cattle Egrets. The large East Tennessee colonies have been of all Black-crowneds or Black-crowneds with a few Cattle Egrets; at least 2 small colonies were dominated by Great Blue Herons.

All Black-crowned Night-Heron colonies have been in wooded swamps or in upland woodlands within 3 km of a lake or river. Nests at Cranetown were in the tops of ash, red maple, and young cypress trees at 4.6 to 7.6 m (Vaughn 1933, egg coll. data) and 18.3 m above water (Gersbacher 1939), below large cypress trees holding the nests of other species. Nests at the Bordeaux colony were in large cedars and locust trees. Most East Tennessee colonies have been in Virginia, shortleaf, or loblolly pines, sometimes with cedars or deciduous trees also present. Nests at different East Tennessee colonies averaged about 12 m (Nunnally and Williams 1977), 13 m (Williams and Nicholson 1977), and 7 m (Nicholson pers. obs.) above ground.

Black-crowned Night-Herons begin returning to colony sites in late winter and initiate nesting activities in late March (Pullin 1990). Their nests are a platform of sticks placed among tree branches; nests built in previous years are frequently refurbished. The timing of egg laying varies greatly within colonies (e.g., Williams and Nicholson 1977) and probably peaks during the second half of April. Clutches of 1 to 5 eggs have been reported, with 3 to 5 egg clutches most common (Williams and Nicholson 1977); the smaller clutches reported may have been incomplete. Both adults incubate the eggs for 24 to 26 days, beginning after the first egg is laid (Palmer 1962). The young climb out of the nest in about 3 weeks and fly in about 6 weeks (Hancock and Kushlan 1984).—*Charles P. Nicholson.*

Yellow-crowned Night-Heron
Nycticorax violacea

The Yellow-crowned Night-Heron is a rare to uncommon summer resident of Tennessee. It arrives by late March and usually departs by late October, although a few winter records exist (Robinson 1990). As its name suggests, the Yellow-crowned Night-Heron is most active at dusk, during the night, and at dawn. It occurs in swamps, marshes, woodland ponds, and along sluggish woodland streams and feeds largely on crustaceans, particularly crayfish (Hancock and Kushlan 1984).

The early history of the Yellow-crowned Night-Heron in Tennessee is poorly known. Tanner (1988) described a twentieth-century expansion of its breeding range in the state. Whether this was a real change in distribution or a reflection of our poor knowledge of this cryptic

Yellow-crowned Night-Heron. Chris Myers

species is debatable. The species has, however, expanded its breeding range in the northeastern United States, as well as into northern Kentucky (Palmer 1962, Mengel 1965). The first Tennessee nest record was in 1926 near the Tennessee River in Hardin County (Ganier 1951a). In the next 15 years, Yellow-crowneds were reported nesting at Reelfoot Lake (Ganier egg coll.), Henry County (Pickering 1937), and Hardeman County (Calhoun 1941). The first Middle Tennessee nest record was in 1949 (Ganier 1951a) and the first East Tennessee nest record in 1954 (Nevius 1955). At several locations, Yellow-crowneds were reported on late April or early May spring counts, when nesting was probably underway, for several years before nesting was locally confirmed.

Atlas workers found the Yellow-crowned Night-Heron in 58 blocks and confirmed its nesting in 8 counties scattered across the state. Most of the sites where nesting was confirmed were first used before 1986, and it is unclear whether some other sites used in the 1970s and early 1980s, such as in Rutherford County (Henderson 1985), were searched during Atlas fieldwork. Because of this factor and, perhaps more importantly, because of the Yellow-crowned's inconspicuous behavior, the Atlas results probably fall short of portraying its true Tennessee distribution. Ford (1990) found Yellow-crowneds at 17 of 59 forested wetland sites in West Tennessee, indicating they are fairly common and widespread in that habitat. Yellow-crowneds were reported on 7 Atlas miniroutes at an average abundance of 1.1 stops/route and were concentrated in the Mississippi Alluvial Plain.

The Yellow-crowned Night-Heron has never received any special protective listing by the TWRA, presumably because of the lack of any noticeable population decline. It is very adaptable in its choice of nest sites, and there

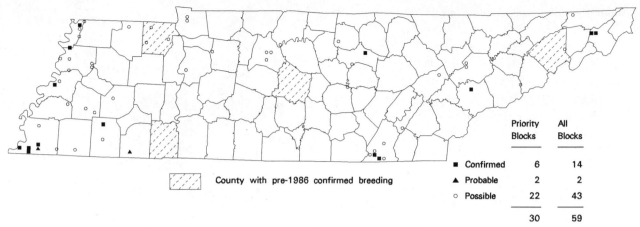

Distribution of the Yellow-crowned Night-Heron.

is only limited evidence of abandonment of nest sites due to human disturbance (e.g., Ganier 1951a). The greatest threat to the Tennessee population, which probably numbers a few hundred birds, is the destruction of forested wetlands, a loss that has slowed in recent years.

Breeding Biology: Throughout their range, Yellow-crowned Night-Herons nest in a variety of sites, either singly or in small colonies. With the exception of a few larger colonies in Shelby, Hardeman, and Blount Counties, colonies have generally contained four or fewer nesting pairs. One or two Yellow-crowned nests have been occasionally reported in colonies dominated by Black-crowned Night-Herons (e.g., *Migrant* 40:45, 49:73). As noted by Watts (1989) in coastal Virginia, Yellow-crowneds are often very tolerant of humans; the largest Tennessee colony, of 67 nests, was in the urban Riverside Park at Memphis (*Migrant* 48:75), and smaller numbers have more recently nested in suburban areas of Memphis and Chattanooga (Atlas results).

Yellow-crowned Night-Heron nests are usually substantial platforms of sticks, often lined with grass or leaves. Both members of the pair build, with the male delivering sticks to the nest (Hancock and Kushlan 1984). Most Tennessee nests have been built on the lower limbs of large, broad-canopied deciduous trees. Pines have been used at a few East Tennessee sites, as were large cedars in Rutherford County (Henderson 1985) and cypress at Reelfoot Lake (Spofford 1942b). Nest heights range from 4.6 to 23 m. Some individual nests have been used for 4 or more consecutive years (*Migrant* 48:104–5, Atlas results), although the longevity of larger colony sites is apparently less than that of more colonial herons, such as the Great Blue.

Egg laying begins by the end of March. Apparently complete Tennessee clutches contained from 4 to 8 eggs and averaged 5.1 (s.d. = 1.3, n = 9) within the range reported elsewhere (Hancock and Kushlan 1984). Many Tennessee nests have also contained 4–5 young. Both adults incubate the pale bluish green eggs for 21–25 days, and both adults feed the nestlings (Palmer 1962, Hancock and Kushlan 1984). The nestlings climb into nearby tree branches several days before flying at about 25 days old.—*Charles P. Nicholson.*

Canada Goose
Branta canadensis

The Canada Goose is presently a fairly common breeding bird in much of East and Middle Tennessee and rare in West Tennessee. During the summer, it is usually observed around lakes and large ponds, either swimming or grazing in areas of short grass. It has only become established as a widespread breeding bird in the last 3 decades, and its breeding range and numbers continue to increase.

Canada Geese breeding in Tennessee belong to the subspecies *B. canadensis maxima,* the Giant Canada Goose (Gore and Barstow 1970). This subspecies is differentiated by several characters, including its large size, and is nonmigratory throughout much of its range (Hanson 1965). During the 19th century, this subspecies bred through most of the upper Midwest and as far southeast as northeast Arkansas, northwest Tennessee, and southwest Kentucky (Hanson 1965). Evidence of their breeding at Reelfoot Lake is given by Kliph (1881, in Hanson 1965), who stated, "large numbers remain here during the entire year, rearing broods of young. The native goose is much heavier than his migratory brother from Canada. . . ." Pindar (1886) observed tame geese at Reelfoot raised from eggs taken from nests of wild birds. Soon after these reports, the Giant Canada Goose was extirpated from Tennessee and much of the rest of its range (Hanson 1965).

Canada Goose. Chris Myers

Several game farms and aviculturalists, however, maintained flocks of Giant Canada Geese. In the late 1950s, a private individual in Sumner County, near Old Hickory Reservoir, started a flock with three or four pairs obtained from a North Carolina game farm (Gore and Barstow 1970). The TWRA joined the effort in 1966, with the goal of producing a free-flying resident, huntable population to offset decreasing numbers of wintering birds (Gore and Barstow 1970, Yates and Whitehead 1978). The local area was closed to goose hunting, nest structures were built, adults and young banded, and by 1969 the Old Hickory flock numbered over 400 birds (Gore and Barstow 1970). A resident flock was also established at Cross Creeks NWR, Stewart County, by importing 21 geese from Missouri, Illinois, and Arkansas between 1967 and 1970.

By 1977, the Old Hickory flock numbered approximately 3000 birds (Cromer 1978) and was estimated to be 10,000 birds in 1980 (TWRA 1981). The statewide population at this time was estimated to be 12,000 birds. Evidence of its nonmigratory status was given by Cromer (1978), who reported only 3 recoveries of banded geese outside of Tennessee out of 4568 banded between 1967 and 1977.

In 1970 the TVA joined TWRA in the introduction effort, and artificial propagation of geese at the Buffalo Springs Game Farm in Grainger County was soon begun. Farm-raised geese were first stocked on Melton Hill Reservoir in 1972. By 1977, 19 free-flying flocks had been established by releasing game farm birds, adults obtained out of state, and birds captured on Old Hickory (Yates and Whitehead 1978). Artificial propagation was halted in 1978 (James 1979), although establishment of additional flocks by transplanting birds continued for a few more years. Hunting of the resident birds began at Old Hickory in 1979 and expanded to much of Middle and East Tennessee by 1987; the fall 1992 population was estimated to be about 38,000 birds (E. Warr pers. comm.).

The rapid growth of the resident goose flock is reflected in Tennessee BBS results. The first goose was recorded on a BBS route in 1974, and by 1991 geese had been recorded on 15 routes, with an average of 9.9 birds/route recording the species. Because the Canada Goose regularly occurs on few routes, however, a reliable estimate of its population trend is not available. North American BBS routes, which better sample the widespread resident populations than the northern migratory populations, show an increasing annual trend ($p < 0.01$) of 11.3%/year (Peterjohn, Sauer, and Link 1994).

Because geese are large, conspicuous, and often relatively tame, they were probably rarely missed by atlasers; consequently, the Atlas results should give an accurate outline of the species's 1991 breeding range in Tennessee. Additional work in nonpriority blocks along the reservoirs in East Tennessee and Old Hickory and Percy Priest Reservoirs in Middle Tennessee would probably

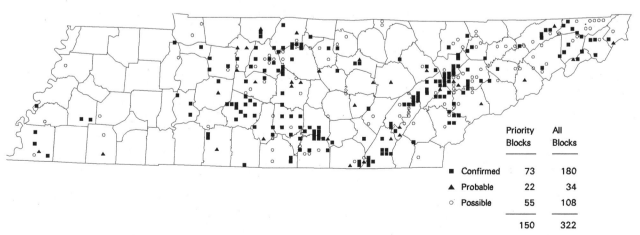

Distribution of the Canada Goose.

Confirmed Breeding Species

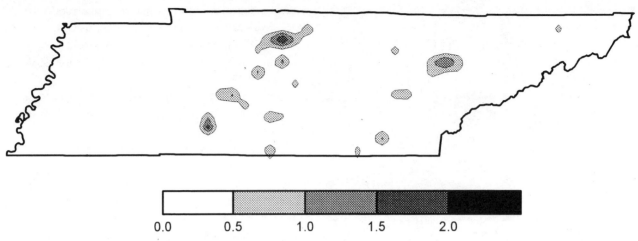

Abundance of the Canada Goose.

have shown its range to be continuous in these areas. Geese were recorded on 19 miniroutes, at an average of 1.47 stops/route. Over half of all Atlas records were confirmed, and 80% of these were of broods of goslings. Some of the possible and probable records were likely of non-breeding, immature birds; Giant Canada Geese do not breed until 2 or 3 years old (Hanson 1965).

Breeding Biology: The breeding season of geese in Tennessee begins in February, when unmated birds form pairs (Hubbard 1976). Nest sites, usually selected by the end of March, are near water, on either islands or the mainland. The nest is usually built on the ground in an open area, although elevated sites, such as duck blinds and large horizontal branches of trees, are occasionally used (James 1979, Poly 1979, Nicholson pers. obs.). The female builds the nest of grasses and weed stems, and after egg laying has begun, lines the nest with down (Hanson 1965). Clutch sizes of 1–10 eggs have been reported in Tennessee, and average 5.36 (s.d. = 1.50, n = 64, mode = 6) in an Old Hickory Reservoir sample (Poly 1979), and 5.39 (s.d. = 1.51, n = 111, mode = 5) in East Tennessee (James 1979, TVA unpubl. data, NRC). Goslings begin to fly in about 10 weeks and remain with their parents into the winter (Bellrose 1976).

Adult geese are strongly, often aggressively, protective of nests and goslings, and the survival rate of both is high. This factor, along with their abilities to tolerate relatively high levels of human disturbance and to disperse into unoccupied habitats insure the continued growth of both their numbers and range. Because of their attraction to mowed lawns, goose flocks have become a nuisance at some recreation areas and lakefront subdivisions, a problem that is also likely to increase.
—Charles P. Nicholson.

Wood Duck
Aix sponsa

The Wood Duck is an uncommon to locally common summer resident. Many migrate to winter a short distance south of Tennessee, and it is an uncommon winter resident (Robinson 1990). It is the most common duck species nesting in Tennessee and is found in swamps and woodland-bordered lakes and streams across the state. It nests in naturally occurring tree cavities and artificial nest boxes.

The Wood Duck was probably among the waterfowl noted by early explorers, and was probably a common species in prehistoric Tennessee. The earliest definite record is by Fox (1886), who observed a pair in Roane County on 4 April 1885. Rhoads (1895a:468) described the Wood Duck as "noted all across the state, where it is a summer resident." Its numbers probably declined

Wood Duck. Chris Myers

throughout the nineteenth century as river bottoms were cleared for agriculture and uncontrolled hunting occurred throughout the year. This decline, most severe late in the century, occurred throughout much of the range of the Wood Duck and was severe enough that in the early twentieth century many feared it was in danger of extinction (Palmer 1976).

The Wood Duck population in Tennessee, however, was probably never in danger of extirpation. It received protection from hunting by federal regulations in 1913 and the 1918 Migratory Bird Treaty Act (Bellrose 1990). Following this protection, Wood Duck numbers increased. Ganier (1933a) described it as fairly common in West Tennessee, rare in Middle Tennessee, and very rare in East Tennessee. The hunting season was reopened in 1942, and the Mississippi flyway population, including Tennessee, generally increased through the 1940s. Habitat continued to be lost during this period from the impoundment of large reservoirs and, especially in West Tennessee, stream channelization and wetland drainage. In the 1950s, a decline in the flyway population was again noted. In Tennessee, it was stable to decreasing in 1952–53 and decreasing in 1953–54 (Bellrose 1990); hunting was again prohibited during the mid-1950s.

This decline was reversed by the early 1960s and from 1966 to 1994, the Wood Duck population, as measured by Tennessee BBS routes, significantly ($p < 0.10$) increased at the rate of 6.9%/year. Most of this increase has occurred since 1979. The reasons for this increase include conservative hunting regulations and increased nest site availability, resulting from both maturing forests along streams and reservoirs and the initiation of nest box programs. Nest boxes were first introduced on the National Wildlife Refuges in the late 1940s. Fifty boxes were provided between 1949 and 1952; by 1984, at least 3300 boxes were present across Tennessee (Bellrose 1990).

During the Atlas period, the Wood Duck was reported in 39% of the completed priority blocks. Because its habitat was sometimes difficult for atlasers to investigate, the Atlas results are probably an underestimate of the duck's occurrence. It was absent from most of the Unaka and Cumberland Mountains. Many of the records on the Cumberland Plateau—where, as in the mountains, naturally occurring habitat was limited—were from artificial lakes and ponds. The Wood Duck was recorded on 42 miniroutes at an average relative abundance of 1.1 stops/route; these observations were too widely scattered to show regional differences in abundance. Almost 90% of the confirmed breeding records were of broods of ducklings.

Although it is unlikely to return to its prehistoric abundance, the Tennessee Wood Duck population is secure and may increase further. With its relatively large clutch size, persistent renesting, and high nest success rate, Wood Ducks are capable of rapid population increases (Nichols and Johnson 1990). With the recent decrease in the rate of wetland drainage and the increasing beaver population, habitat trends are encouraging. As hunting of locally raised Wood Ducks is increasingly being promoted as a replacement for decreasing populations of other ducks (Nichols and Johnson 1990), intensive management of Wood Ducks by state and federal agencies will likely continue in the future.

Breeding Biology: Pair formation occurs during the winter, and the male Wood Duck follows the hen to the nesting area. Females show a high rate of return to the area where they previously nested or, if yearlings, to where they were reared (Kirby 1990). Pairs begin investigating nest cavities in late February and early March. Trees containing suitable nest cavities tend to be older and larger than the average in current second-growth forests (Haramis 1990). Because of their susceptibility to heartrot decay and frequent occurrence in riparian

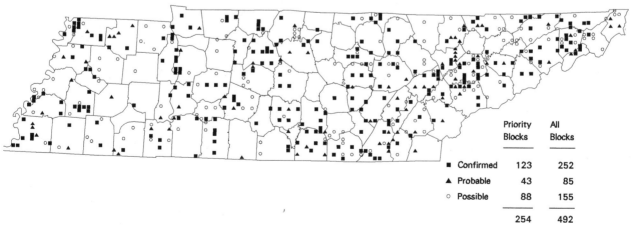

	Priority Blocks	All Blocks
■ Confirmed	123	252
▲ Probable	43	85
○ Possible	88	155
	254	492

Distribution of the Wood Duck.

Confirmed Breeding Species

areas, sycamore and beech trees often provide suitable nest sites. Cavities excavated by Pileated Woodpeckers are also frequently used (Haramis 1990). Nest sites over water are preferred, although sites up to 2 km from wetland habitats are used.

Egg laying begins as early as the first week of March, peaks from late March through mid-April, and extends at least into early June (Goetz and Sharp 1980, Spero et al. 1983). Normal clutches commonly contain 10 to 13 eggs, with 12-egg clutches most frequent (Haramis 1990). Spero et al. (1983) found an average clutch size, excluding dump nests, of 10.7 eggs in a Kentucky Reservoir population, and Cottrell, Prince, and Padding (1990) found average annual clutch sizes of 12.1–14.7 eggs along the Holston River in northeast Tennessee. Dump nests, produced by more than 1 female, contain as many as 34 eggs (Spero et al. 1983). Dump nests are most common early in the nesting season and are apparently an adaptation to high breeding densities in areas of limited nest sites (Spero et al. 1983, Haramis 1990). Dump nests constituted 25% of all nests in a dense Kentucky Lake population (Spero et al. 1983).

The female Wood Duck incubates for an average of 30 days (Haramis 1990). The ducklings leave the nest soon after hatching and are led by the female to the nearest wetland habitat. Typical brood-rearing habitat consists of wetlands with a combination of emergent vegetation, shrubs, fallen trees, and open water. Along the unimpounded Holston River, Wood Duck broods preferred wetlands with submerged or emergent vegetation and grazed and ungrazed lowland hardwood, nonforested herbaceous, and shrubby shoreline habitats (Cottrell, Prince, and Padding 1990). The female tends the brood for about 5 weeks, and the young fly in 8 to 9 weeks (Palmer 1976). Renesting by a small proportion of females has been reported from several southern states (Haramis 1990), but apparently not yet from Tennessee.—*Charles P. Nicholson*.

American Black Duck
Anas rubripes

The American Black Duck is a common migrant and winter resident and very rare summer resident of Tennessee. Migrant birds begin arriving by late August and the last depart by mid-May (Robinson 1990). The greatest numbers of wintering birds, however, are present from November into March; they occur on reservoirs, lakes and wildlife refuges across the state. It is a popular game species, although the harvest has been greatly restricted in recent years. The few breeding records are from wetlands scattered across Tennessee.

American Black Duck. Chris Myers

The main breeding range of the American Black Duck is in the forested wetlands of northeastern North America (AOU 1983), where it has recently declined in numbers (Rusch et al. 1989). Inland, it extends south to central Indiana, Ohio, and West Virginia; along the coast it extends south to North Carolina. Isolated nesting has occurred in northern Alabama, coastal South Carolina, and Georgia (Palmer 1976) as well as Tennessee.

Information on the breeding of the American Black Duck in West Tennessee is sketchy. Palmer (1976) maps a breeding record in the Reelfoot Lake area, although the source of this record is not given. It was not considered to breed there by either Ganier (1933b) or Pitts (1985). Calhoun (1941) reported it, according to local residents, occasionally nested in the creek bottoms of McNairy and Hardeman Counties.

Prior to the Atlas project, the only Middle Tennessee breeding record was of 13 young hatched from 2 nests at Cross Creeks NWR, Stewart County, in June 1967 (Robinson and Blunk 1989). The majority of nest records, however, have been from the East Tennessee at Alcoa Marsh, Blount County, where Williams (1975b) noted at least 8 nesting attempts from 1972 through 1975. A pair also nested on Fort Loudoun Lake, Knox County, in 1974 (*Migrant* 45:77). Adult ducks, without further evidence of breeding, have also been reported during the summer from several locations across the state (Robinson 1990).

During the Atlas period, breeding was confirmed in one block in Tennessee NWR in Humphreys County and a pair was observed in late May in an adjacent block. Single adults considered possible breeders were observed in late May in Sumner County and in early July in Stewart

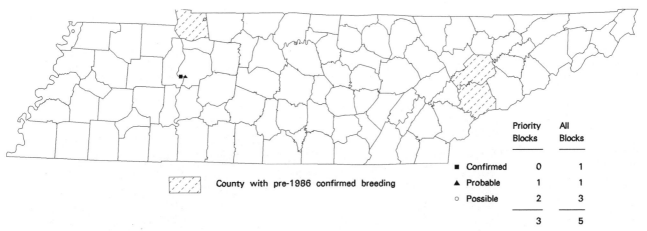

Distribution of the American Black Duck.

County. The confirmed record was of an adult with a brood of 8 young on 27 May 1986 in a partially wooded, shallow, permanently flooded impoundment. In the same block, a hen sat in incubating posture on an abandoned Osprey nest for about 2 weeks in early June 1986 (Atlas results, *Migrant* 57:106). No evidence of eggs was found when the nest was later inspected.

Breeding Biology: Egg records are limited to the 2 East Tennessee sites, where egg-laying began as early as the second week of February (Williams 1975b). Presumably completed clutches ranged from 8 to 22 eggs, and averaged 12.86 eggs (s.d. = 5.49, n = 7). The largest clutch may have been the product of more than 1 female. Clutches elsewhere average 9.3 eggs, and clutches of 7 to 12 eggs are most common (Bellrose 1976). Most nests were on mounds of earth at the base of willows, surrounded by water. Exceptions were nests in a clump of cattails and in a clump of St. John's wort. Much variation in nest site selection occurs elsewhere in the ducks' range. A clutch of 6 eggs was found at a nest site 13 days after the disappearance of an earlier clutch. The male American Black Duck abandons the hen soon after she begins incubation, which lasts about 26–29 days (Bellrose 1976). The ducklings leave the nest within hours of hatching and remain with the hen until they begin flying in about 60 days.—*Charles P. Nicholson.*

Mallard
Anas platyrhynchos

The Mallard is a common migrant, common winter resident, and uncommon summer resident. Its breeding population, which has received little attention in the ornithological literature, has greatly increased in the last 60 years, and presently occupies ponds and lakes across the state. Most of this population is probably descended from domestic ducks, crippled wild ducks, live decoys, and ducks stocked by waterfowl enthusiasts and conservation agencies. Over half of all ducks harvested by Tennessee hunters are Mallards (E. Warr pers. comm.).

Tennessee is south of the main, historic breeding range of the Mallard (Palmer 1976), and the typical breeding habitat of shallow ponds surrounded by grasslands was, at best, very rare in prehistoric Tennessee. The early checklists of Rhoads (1895a) and Ganier (1917, 1933a) did not list it as a summer resident. An early mention of nesting in Tennessee was by Ganier (1933b), who noted crippled ducks nesting at Reelfoot Lake, and doubted that a wild population would become established because of poaching. Calhoun (1941) reported occasional nesting in creek bottoms of Hardeman and McNairy Counties.

The potential for a sizable breeding population has steadily increased since the 1930s with the construction of numerous ponds and reservoirs. Attendant with this

Mallard. Chris Myers

Confirmed Breeding Species

habitat increase has been an increase in the number of semi-domesticated Mallard flocks kept by lakefront homeowners, marina owners, and in urban parks. Active efforts were also undertaken to establish wild breeding populations. In 1974, for example, the TVA, cooperating with TWRA in an attempt to establish a huntable local population, released almost 300 "dusky" Mallards in East Tennessee (TVA 1975). This effort was abandoned because of the low survival rate and the tendencies of the ducks to domesticate (TVA 1977). Some local populations, regardless of their origin, have become quite large. An estimated 400 ducks were present on Old Hickory Lake in the summer on 1972 (*Migrant* 43:76).

The Mallard was reported in about 15% of the completed Atlas priority blocks and several other blocks across Tennessee. The proportion of these ducks that were truly wild, however, is unknown. Atlas workers were instructed to report only Mallards showing no signs of domestication such as atypical plumage or tame behavior. Some atlasers were more liberal in interpreting these guidelines than others, and many of the records in urban areas and on reservoirs with heavy recreational use are probably of relatively domesticated ducks. These populations will probably continue to grow. Observations were concentrated around the East Tennessee reservoirs and Old Hickory and Percy Priest Lakes in Middle Tennessee. West Tennessee observations were concentrated at Reelfoot Lake and in Shelby County, where Mallards occurred in a variety of aquatic habitats. Mallards were recorded on 22 miniroutes at an average relative abundance of 1.1 stops/route, too few to map regional differences.

Breeding Biology: Most (85%) of the confirmed Atlas observations were of broods of ducklings. The breeding biology of the Mallard in Tennessee, however, is relatively poorly known. Most of the detailed nest records have been from marshes such as Alcoa Marsh and Amnicola Marsh. Egg laying begins as early as late February and extends at least into May. Mallard nests are typically a bowl built of grasses, reeds, or other plant material gathered by the hen at the nest site. The nest is lined with down. Nest sites include fallen logs and small islands in marshes and low-lying brushy areas adjacent to reservoirs. Seemingly complete clutches range from 9 to 14 eggs. Seven to 10 eggs are most common elsewhere (Palmer 1976). A brood of 16 ducklings (*Migrant* 58:101), if all from eggs laid by the single hen observed tending them, was unusually large. The drake abandons the hen soon after incubation begins, and the hen incubates and cares for the ducklings. Incubation lasts 27–28 days (Palmer 1976), and the downy young leave the nest soon after hatching. They begin to fly in about 8 weeks. Replacement clutches are laid following loss of a nest. Wild Mallards are normally single-brooded, and semi-domesticated Mallards are occasionally double-brooded (Palmer 1976).—*Charles P. Nicholson.*

Blue-winged Teal
Anas discors

This small, handsome duck is a common migrant through Tennessee and a rare winter and summer resident, occurring in marshes, ponds, and shallow lakes. Its main breeding range is in the northern prairies and central Canadian parklands (Palmer 1976); sporadic Tennessee breeding records date back to 1935. The Blue-winged Teal is a late spring migrant, frequently present into late May, and an early fall migrant, often present by early August.

Rhoads (1895a) observed pairs of Blue-winged Teal on Reelfoot Lake between 30 April and 6 May 1895 and speculated that they may have been summer residents. There was until recently, however, no confirmed

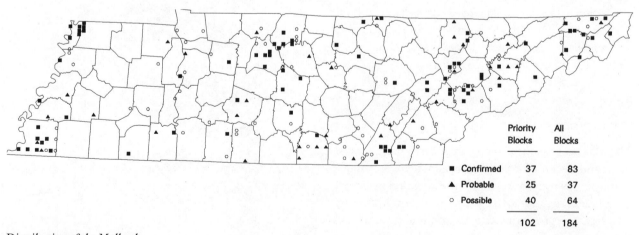

Distribution of the Mallard.

Blue-winged Teal. Chris Myers

evidence of teal breeding at Reelfoot (Pitts 1985). The first confirmed breeding was the observation by Ganier and others of a duckling about a month old at Goose Pond, Grundy County, on 26 May 1935 (Ganier 1935b). On 2 May 1936 Todd (1944) collected a clutch of 8 teal eggs, incubation advanced, at Hoover and Mason Lake, Maury County; teal also nested there in 1935. On 1 May 1938 Todd observed 2 broods of young at Goose Pond (DeVore 1975), and on 22 June 1937 he observed another brood at Walterhill, Rutherford County (Todd egg coll. data, DeVore 1975). Another report during this period, of a sick "young" teal in Montgomery County on 5 or 6 September 1936 (*Migrant* 7:70) is insufficient to be evidence of breeding.

The next breeding records were broods in Davidson County, at Buena Vista Marsh in 1961 and at Radnor Lake in 1964 (Parmer 1985). A pair nested in Anderson County in 1978 (*Migrant* 49:93) and broods were observed in Coffee County in 1979 (*Migrant* 50:87) and in Sumner County in 1980 (*Migrant* 57:81). Some of the many reports of adult teal from late May through July (Robinson 1990) may also have been of nesting birds.

During Atlas fieldwork, Blue-winged Teal were observed in 7 blocks; breeding was confirmed in 4 blocks, each in a different year and each by the presence of ducklings. All of the confirmed records were in the Mississippi Alluvial Plain, on shallow ponds with abundant vegetation. Observations of ducklings of known age and the 1936 egg record suggest that nesting begins as early as late March, when migrating teal are present. Because of this overlap, possible and probable records earlier than the last week of May were not accepted. This conservative approach may have resulted in an underestimate of the teal's breeding distribution.

In Louisiana, Texas, and southern Illinois, south of its main breeding range, the Blue-winged Teal has nested in large numbers when suitable habitat resulted from unusually heavy rainfall (Bellrose 1976). No similar response to high water levels has been reported in Tennessee, where the greatest concentration of breeding records was in the years 1935 to 1938. Five of these 6 breeding records, in Goose Pond and Hoover and Mason Lake, were in marshy areas not greatly affected by high rainfall. No more recent breeding records exist from either of these areas, despite fairly regular fieldwork, especially in the 1970s and 1980s. Compared to other ducks, the return rate of Blue-winged Teal to their place of birth is relatively low (Bellrose 1976).

An effort to increase the breeding teal population by releasing farm-reared young teal was initiated by the TVA and TWRA in the 1970s. The potential release sites, ponds surrounded by pasture, were chosen based on their apparent similarity to teal nesting habitat in Nebraska and the western Dakotas (TVA 1975). Because of captive breeding problems, however, the project was eventually discontinued (James 1979).

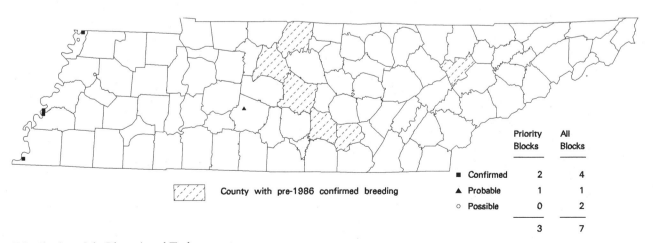
Distribution of the Blue-winged Teal.

Breeding Biology: The only clutch of Blue-winged Teal eggs reported in Tennessee is the set of 8 collected by Todd (1944) on 2 May 1936. The nest was in grass 3 m from the edge of the water. The broods of young teal reported in Tennessee had from 4 to 10 ducklings. Within their main breeding range, Blue-winged Teal usually nest in grassy upland areas near water; the nest consists of a bowl scraped into the ground and lined with vegetation (Bellrose 1976). The female adds down to the nest after egg-laying has begun. Clutches average about 10 eggs and are incubated by the female an average of 24 days (Bellrose 1976). The young are usually raised in ponds or sloughs with abundant emergent vegetation, and they begin flying in 5 to 6 weeks.—*Charles P. Nicholson.*

Hooded Merganser
Lophodytes cucullatus

The Hooded Merganser is a permanent resident, rare during the summer, uncommon during the winter, and fairly common during the spring and fall (Robinson 1990). Migrant or wintering birds usually occur in small flocks and are present from late October through early May. The Hooded Merganser feeds on small fish and aquatic invertebrates caught during short dives, and occupies swamps, ponds, and shallow, sheltered coves of lakes. It nests in hollow trees or nest boxes close to or over water.

Phillips (1926) showed the historic breeding range of the Hooded Merganser extending along the Mississippi River to south of Tennessee. The first Tennessee records were by Rhoads (1895a) who observed one on the Tennessee River near Chattanooga in late May, and noted it probably bred in Haywood County. Ganier (1933b) stated it nested regularly at Reelfoot Lake, but presented no further details. The first verified breeding record was the observation of a hen with 8 nestlings on 6 May 1934 at Horn Lake, along the Mississippi state line south of Memphis (*Migrant* 5:26). The lack of other early breeding records may have been due to the widespread decline which, as with the Wood Duck, occurred in the late nineteenth century throughout much of its range due to deforestation, the loss of nest trees, and unrestricted hunting (Palmer 1976). A reversal of this decline occurred by the 1930s. Since then the Hooded Merganser has probably nested regularly in small numbers in West Tennessee swamps and sporadically further east.

The largest local breeding population has been at Hatchie National Wildlife Refuge in Haywood and Lauderdale Counties, where nest boxes have been maintained, primarily for Wood Ducks, since the 1960s. Between 1967 and 1972, the proportion of nest boxes occupied by Hooded Mergansers varied from none in 1967 to 43% in 1970, and averaged 28% (Waldron 1982). More recent, less intensive nest box inspections carried out since then found about 20% of about 400 boxes were used by mergansers from 1977 through 1981 (Waldron 1982), 12 nests present in 1986, and 28 nests present in 1987 (Robinson 1990). These figures suggest a decline in the nesting Hooded Merganser population since the early 1970s. Even with this decline, the proportion of nest boxes used by Hooded Mergansers is much higher that reported elsewhere in the southeast (Kennamer et al. 1988).

Pre-1986 breeding records outside the Hatchie area are limited. Small numbers have regularly nested at Reelfoot Lake (Pitts 1985), and a brood of ducklings was observed there in early May 1983 (*Migrant* 54:50). Morse (in Mengel 1965) reported nesting on Kentucky Lake in 1950 or earlier. Bellrose (1976) mentioned Hooded Mergansers nesting in Wood Duck boxes at Cross Creeks NWR, Stewart County; this statement, however, is not supported by data in refuge files (Robinson and Blunk 1989). A road-killed duckling was observed at Chattanooga on 16 June 1974 and a brood of 5 ducklings found there in a flooded woodland, probably dammed by beavers, on 31 May 1982 (*Migrant* 45:102, 53:67–68; K. H. Dubke pers. comm.). Hooded Mergansers, usually 1 or 2 birds without evidence of breeding, have been reported between late May and early July from a few locations across the state (e.g., Robinson 1990). The sex and age of these birds have not always been given, and they could be non-breeding juveniles or males which have completed breeding. Hooded Mergansers probably begin nesting when 2 years old, and the male abandons the female after incubation has begun (Palmer 1976).

Hooded Merganser. Chris Myers

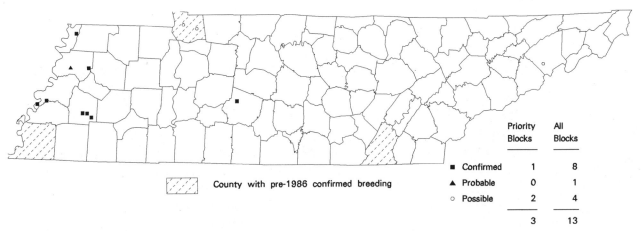

Distribution of the Hooded Merganser.

During the Atlas period, Hooded Mergansers were reported from 14 blocks, and broods of young observed in 8 of these blocks. The greatest concentration of records was along the Hatchie River. Further east, a hen with young was observed in 1987 on a Maury County phosphate settling pond with numerous Wood Duck nest boxes.

The Hooded Merganser probably nests more widely in Tennessee than the Atlas records indicate. Its eggs superficially resemble those of the Wood Duck; Soulliere (1987) describes a technique useful in distinguishing hatched eggs in the field. The systematic inspection of Wood Duck boxes in late summer would probably yield more Hooded Merganser nest records.

Breeding Biology: Hooded Mergansers begin nesting by mid-March, as suggested by observations of young in early May. The only detailed record of a nest in a natural cavity was in a hollow tree trunk about 3 m above water (Sights 1943). This nest contained 14 eggs. The average number of hatched eggs in boxes at Hatchie Refuge was 13.1 (Waldron 1982). Normal clutch size elsewhere is 8 to 12 eggs and averages about 10 eggs (Palmer 1976, Kennamer et al. 1988). The female incubates for an average of about 33 days. The young leave the nest cavity soon after hatching, and are abandoned by the female when about 5 weeks old. They begin to fly in about 10 weeks (Palmer 1976).—*Charles P. Nicholson.*

Black Vulture
Coragyps atratus

The Black Vulture is an uncommon to fairly common permanent resident of Tennessee. During the summer it occurs at low elevations throughout the state. During winter, it is least common in East Tennessee and most common in Middle Tennessee (Tanner 1985). Although the Black Vulture is seen in flight over a variety of habitats, its distribution is apparently determined by its food supply and the availability of suitable nesting sites. It feeds primarily on carrion, preferring large carcasses over small ones, usually in open areas, and infrequently kills live animals (Coleman and Fraser 1987, Jackson 1988a). The nest is typically in a dark recess such as a cave or hollow tree.

Atlas workers found the Black Vulture in 41% of the completed priority blocks. Most of the observations were of soaring birds, and coded as possible breeders. The proportion of these birds which actually nested in the blocks where observed is unknown. An unknown portion of the birds were also juveniles, as the Black Vulture probably first nests when 3 years old (Jackson 1988a). The Atlas observations suggest the Black Vulture most often occurs in rural areas with a high percentage of cleared land. It was found in 90% of the priority blocks in the Central Basin, followed by 59% in the Ridge and Valley and 50% in the Eastern Highland

Black Vulture. Chris Myers

Confirmed Breeding Species

Rim. In all of these regions, a large proportion of the farmland is pasture and hayfields, and nest sites are present on cliffs along rivers, on wooded ridges, and in the extensive areas of karst terrain. Its absence from much of heavily agricultural West Tennessee may be due both to poor foraging in the extensive areas of row crops and few nest sites in the remaining woodlands, some of which are frequently flooded. The Black Vulture was recorded on too few miniroutes to show regional differences in abundance.

The Black Vulture is probably more common in Tennessee now than it was prior to widespread agricultural clearing. Audubon (1834, in Mengel 1965) described its year-round occurrence in Kentucky, which suggests this southern species then occurred in Tennessee. It was first reported in Tennessee by Fox (1882), who observed it in early spring on Lookout Mountain, Hamilton County. A few years later he reported several small flocks in Roane County during March and April (Fox 1886), when nesting is usually underway. Lyon (1893) found many nests in Montgomery County. Rhoads (1895a) observed it from the Mississippi River eastward to Chattanooga and Knoxville; he was surprised at its abundance and observed 1 Black Vulture for each 3 Turkey Vultures. It was apparently absent or rare in McMinn County around 1900, as Gettys did not collect its eggs. In northeastern Tennessee, it apparently occurred only in a small area along the South Holston and lower Watauga Rivers (Lyle and Tyler 1934, egg coll. data). Ganier (1933a) described it as common in West and Middle Tennessee and fairly common in East Tennessee.

The Black Vulture has apparently increased in numbers in East Tennessee in recent decades, as has occurred elsewhere in the Appalachians (Coleman and Fraser 1990). Williams (1977a) reported an increase in the Great Smoky Mountains since Stupka (1963) described it as rather rare. It is now regularly observed in Carter County, where it was very rare a few decades ago (Herndon 1950b). A change in the Black Vultures' numbers is less apparent in the western two-thirds of Tennessee. Tennessee BBS routes show no significant trend from 1966 to 1994, although the proportion of routes showing increases is somewhat greater than the proportion showing decreases. Black Vultures increased significantly ($p < 0.05$) on Tennessee Christmas Bird Counts between 1963 and 1987; most of this increase was between 1963 and 1970 (Coleman and Fraser 1990). Throughout North America, the Black Vulture population showed a nonsignificant increasing trend from 1966 to 1993 (Peterjohn, Sauer, and Link 1994).

The Black Vulture has been listed as In Need of Management in Tennessee since 1976 (TWRA 1976), although there is little evidence that this listing has had any effect on its population. Its population, which appears rather healthy, would be helped by preservation of large, hollow trees suitable as nest sites, control of dog populations, and relaxation of laws requiring burial of domestic animal carcasses.

Breeding Biology: Most available information on Black Vulture nesting has been gathered by Middle Tennessee egg collectors. One collector in particular, H. O. Todd Jr., kept notes on over 300 nests found between 1936 and 1966, mostly in Rutherford County. He collected eggs from a large proportion of the nests. Todd and other collectors returned yearly to nest sites, and Todd noted Black Vulture use of a nest site for 40 years (egg coll. data).

The most frequently reported nest sites have been on the ground in crevices among rocks, under rock ledges, in standing hollow trees, and in caves. In an analysis of nest sites from throughout the species's range but including few Tennessee records, Jackson (1983) found an increased proportion of nests since 1920 in caves and

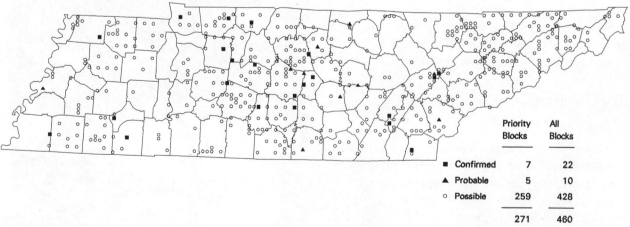

Distribution of the Black Vulture.

thickets and fewer in tree cavities or on bare ground. This change was attributed in part to the lower numbers of large hollow trees in modern forests. The proportion of Tennessee nests, most in Middle Tennessee, in different sites through 1940 (n = 172) and since 1940 (n = 158) is as follows: on ground in rock crevice, 27% vs. 31%; under rock ledge, 21% vs. 29%; standing hollow tree, 17% vs. 16%; cave, 15% vs. 8%; bare ground, usually near tree or rock, 8% vs. 4%; sink hole, 6% vs. 4%; building, 0% vs. 3%; hollow log or stump, both 2%; and thick brush, both 1%. Of 13 nests reported by atlasers, 3 were in old buildings, 3 in caves along rivers, 2 in hollow trees, and 1 each in a hollow stump, a sink hole, a rock crevice, a duck blind, and on the ground among branches of a fallen tree.

Black Vulture pairs begin perching near nest sites in late January and early February. Egg laying begins as early as mid-February and extends into late May. Most clutches are laid in March, with a slight peak during the last 10 days of the month, somewhat earlier than given by Jackson (1983) for the latitude of Tennessee. The normal clutch size is 2 eggs. Following loss of a clutch, a replacement clutch is laid in 3 to 4 weeks, frequently in the same nest site (Todd 1938, egg coll. data). Todd (1938) described a pair laying 2 replacement clutches, and a large portion of the clutches laid after early April are probably replacements.

Both adults incubate the eggs for 35–41 days (Jackson 1988a, Todd 1938). The young may first wander from the nest site in about 2 months, and begin to fly in 75–80 days. Young in deep nest cavities may be unable to leave until they are fully grown at 81 days or older (Todd 1938). The fledglings may be fed by their parents for several months (Jackson 1988a), although this phase of their life is poorly known in Tennessee.—*Charles P. Nicholson.*

Turkey Vulture
Cathartes aura

Turkey Vultures are master gliders. Even at a distance, their characteristic flight style, with wings held in a shallow "vee" and only occasional flapping, is easily recognized. They are a fairly common sight in rural Tennessee throughout the year.

Atlas workers found Turkey Vultures throughout the state, in almost 90% of the completed priority blocks. Because of their conspicuous flight, it is unlikely that vultures were missed in many blocks. They were found in at least two-thirds of the blocks in each physiographic region except the Mississippi Alluvial Plain, where they were found in 40% of the priority blocks. Their infrequent occurrence here and in parts of the adjacent Loess

Turkey Vulture. Chris Myers

Plain may be due both to a lack of nest sites, further described below, and the low suitability of the rowcrop-dominated landscape for foraging on carrion. They were also infrequently reported from some large urban areas and parts of the Unaka and Cumberland Mountains. While they may have been overlooked in the heavily forested mountain blocks, their absence from other areas, such as parts of the southern Ridge and Valley in Polk and Bradley Counties, is hard to explain. Turkey Vultures were found on about 15% of the miniroutes at an average relative abundance of 1.5 stops/route; because many miniroutes were completed before vultures began soaring, they are probably not an accurate measure of this vulture's regional abundance.

Most Turkey Vulture records were of birds observed soaring and coded in the "Observed" category. Vultures forage over a large area (Coleman and Fraser 1987), and these observations of possible breeding may not accurately represent the species's present breeding range. A few observations were of vultures visiting probable cliff nest sites, predominantly on the Cumberland Plateau.

Jackson (1983) described Turkey Vultures as widespread and numerous in North America in the 1800s and declining in numbers as populations of bison and other large mammals declined. The magnitude of this decline in Tennessee is unknown; open-range herding of cattle and hogs in woodlands, common until early this century, probably largely compensated for decreased numbers of large native mammals, particularly deer. Rhoads (1895a) recorded the Turkey Vulture statewide and described it as breeding in great numbers along the eastern escarpment of the Cumberland Plateau. The large number of egg sets (at least 30) collected by Gettys in McMinn County at the turn of the century suggests the

Confirmed Breeding Species

species was common there. Ganier (1933a) described the Turkey Vulture as a common summer resident statewide. More recently they probably declined again because of the loss of forest breeding sites, pesticide contamination, and laws requiring the burial of large animal carcasses (Jackson 1983). They probably also declined in the Great Smokies after cattle were removed from most of the area following establishment of the national park; more recently they have been concentrated in the Cades Cove area, where grazing still occurs (Stupka 1963). Although BBS routes do not sample vultures well, Tennessee routes show an increasing trend of 2.2%/year ($p < 0.10$) from 1966 to 1994. Rangewide BBS routes show an increasing trend of 1.1%/year ($p = 0.02$) from 1966 to 1993 (Peterjohn, Sauer, and Link 1994). Tanner (1985) did not note a significant change in numbers recorded on Tennessee Christmas Bird Counts. However, because of concerns over a perceived population decline, the Turkey Vulture was listed as In Need of Management in Tennessee from 1976 to 1986 (TWRA 1976, 1986).

Breeding Biology: Turkey Vultures nest in a variety of secluded sites. Of the few Turkey Vulture nests reported by Atlas workers, the East Tennessee nest and 1 in Middle Tennessee were in shallow caves on bluffs along rivers. The other Middle Tennessee nest was under limestone boulders, and the 2 West Tennessee nests were in abandoned buildings. The much larger sample from egg collection data shows regional differences in nest sites. All of 30 egg sets collected by Gettys between 1898 and 1908 in McMinn County, an area with few caves or bluffs, were from nests in hollow trees, logs, or stumps. Stupka (1963) described one nest on a cliff in 1925 and another in a stump about 1932, both in the Smokies. Out of 13 sets collected in northeast Tennessee by Lyle between 1913 and 1938, 11 were in caves in bluffs along rivers, 1 in a hollow tree, and 1 in a stump. Lyle and Tyler (1934) reported a nest in a tree cavity 12.2 m above ground. Seven sets collected in Middle Tennessee between 1918 and 1940 were in caves (6 on a river bluff), 1 in a sinkhole, 2 in a stump on a river bluff, and the rest under rock ledges in wooded or brushy areas. Jackson (1988b) described decreased use of tree cavities and increased use of thickets and buildings since 1920 and attributed it to the clearing of old-growth forests. Although too few Tennessee records are available to show a pronounced shift in nest sites, hollow trees, stumps, and logs large enough for nesting vultures are undoubtedly much rarer than in Gettys's time. In far western Tennessee, nest sites may be limited because of few large hollow trees and frequent flooding of much of the remaining forested areas, reducing opportunities for ground nesting.

Turkey Vultures are monogamous and pair bonds may last many years (Jackson 1988b), probably in part due to their tendency to reuse nest sites. Nesting activity in Tennessee probably begins in late February or March with courtship and selection of the nest site. The earliest egg date is 3 April, although a set collected in Washington County on 10 April, with incubation far advanced, would have been laid in early March. Egg laying peaks in late April. Although Jackson (1988b) noted the lack of documented replacement clutches, some Tennessee egg collectors believed relaying occurred. Ganier collected sets of 2 fresh eggs from the same site in Davidson County on 10 April and 8 May 1920. Many of the fresh sets collected in May are probably replacement clutches. Of 58 egg sets collected in Tennessee, 55 contained 2 eggs and the others 1 egg. Sets of 3 eggs have been reported elsewhere (Jackson 1988b). The eggs are white or creamy white, sparsely to heavily marked with spots and/or blotches of brown (Harrison 1978). The first egg laid is frequently more heavily blotched than the second (Jackson 1988b). Both adults incubate the eggs

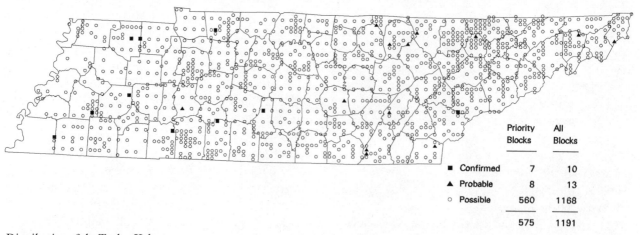

Distribution of the Turkey Vulture.

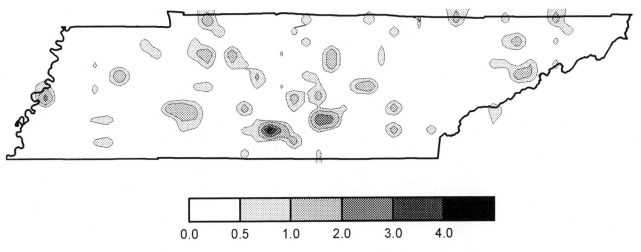

Abundance of the Turkey Vulture.

for 28–41 days (Jackson 1988b). The young are fed by regurgitation, and they first fly at about 9 weeks of age. The family group may stay together into the following year (Jackson 1988b).—*Ron D. Hoff.*

Osprey
Pandion haliaetus

The Osprey is an uncommon migrant, a rare winter visitor, and a local summer resident of Tennessee. Most of the spring migrants pass through during April and early May, and most fall migrants pass through during September (Robinson 1990). Most winter records are from Christmas Bird Counts; the proportion of these birds that remain throughout the winter is unknown. As a breeding species, the Osprey is common on Watts Bar Reservoir in East Tennessee and rare to uncommon elsewhere in the state.

The Osprey feeds almost exclusively on fish (Poole 1989), and, except while migrating, occurs in the vicinity of larger rivers, lakes, and reservoirs. Prior to the construction of large reservoirs during this century, the Osprey was a rare breeding bird in Tennessee. Rhoads (1895a) gave a secondhand account of its occurrence at Reelfoot Lake, and it has nested there since at least the early 1930s (Pitts 1985). The only other natural lake where nesting has been reported is Andrew Jackson Lake in Knox County, where a pair unsuccessfully attempted nesting in 1937 (Beddow 1990). Ospreys may have once nested along the lower Tennessee and Cumberland Rivers prior to their impoundment, and at natural Mississippi Alluvial Plain lakes besides Reelfoot.

The amount of suitable Osprey habitat greatly increased with the construction of twentieth-century reservoirs. Ospreys, however, were slow to take advantage of this newly created habitat. Nest attempts were reported on Hales Bar Reservoir (presently part of Nickajack Reservoir) in Marion County in 1933 and at Davy Crockett Reservoir in Greene County in 1940 (Alsop 1979b). Large impoundments along the Tennessee and Cumberland River systems were completed from the late 1930s through 1970s. Throughout much of this period the only known nest was on the Euchee navigation light on Watts Bar Reservoir; it has been active since at least the late 1950s (Beddow 1990). Other nests were reported in the late 1960s and early 1970s from two sites on Chickamauga Reservoir and a another site on Watts Bar. None of these early nests are known to have fledged young; this reproductive failure was probably because of the effects of the insecticide DDT (Poole 1989). Because of its low numbers and poor breeding success, the Osprey was listed as Endangered by the Tennessee Wildlife Resources Commission in 1975 (TWRA 1975).

The use of DDT and related pesticides was halted in North America by the early 1970s, and many North

Osprey. Chris Myers

Confirmed Breeding Species

American Osprey populations, including the Tennessee population, soon increased rapidly (Poole 1989). From 1966 to 1993, Osprey numbers, as measured by BBS routes, increased significantly (p < 0.01) throughout North America by 5.6%/year (Peterjohn, Sauer, and Link 1994). The Euchee nest successfully fledged young in 1977. Following the construction of secure artificial nest structures, which continues to the present, the Watts Bar population rapidly increased (Beddow 1990). This population expanded to nest on nearby Melton Hill Reservoir in 1988, Tellico Reservoir in 1989, and Fort Loudoun Reservoir in 1991; by 1991, it consisted of at least 29 nesting pairs (Beddow 1990, T. E. Beddow pers. comm.). The population increase has continued since 1991.

Another phase of Osprey management began in 1979 with the hacking of young Ospreys (Hammer and Hatcher 1983). Hacking is the process of releasing young birds into suitable but unoccupied habitat. The young birds are either hatched in captivity or removed from nests in areas with healthy populations. Four young at two East Tennessee sites were hacked in 1979, 27 at 14 sites in 1980, and 20 at 10 sites in 1981. Ospreys were also hacked in Middle and West Tennessee during the early 1980s. Little information is available on the success of the hacking. One member of a pair that built a nest at Percy Priest Reservoir in 1986 was earlier hacked there (*Migrant* 57:106), and other nests during the late 1980s on Old Hickory Reservoir and near the Holston River in Hawkins County may have built by hacked birds. There is little evidence that hacking elsewhere in Tennessee resulted in the establishment of nesting populations.

Breeding Biology: Ospreys do not begin nesting until they are at least three years old, and the age at first nesting is greater in dense populations where nest sites are limiting than in rapidly growing populations (Poole 1989). Yearling Ospreys usually remain in the southern Central America and northern South America wintering areas, and first return to breeding grounds when two years old (Poole 1989). Immature birds probably account for most of the numerous Tennessee summer records of non-breeding birds (Robinson 1990, Atlas results). Ospreys are normally monogamous, and pairs usually return to nest sites used in previous years (Poole 1989). Most breeding pairs in Tennessee return to nest sites by late March, and older, experienced pairs often return by the second week of March (Beddow pers. comm.).

Important features of Osprey nest sites are the closeness of water, good visibility, and easy access to the nest unimpeded by tree branches. Almost all East and Middle Tennessee Osprey nests have been on human-made structures over water, including navigation markers, electrical transmission towers, silos, and especially pole-mounted nest platforms. The Reelfoot Lake Osprey nests and a few East Tennessee nests have been in dead trees. Both adults build the nest, which is constructed of large and small sticks (Poole 1989). The bowl of the nest is lined with grass and a wide variety of other materials. Additional material is added in later years, and nests may be a meter in width and over 0.3 m tall.

Most Tennessee Osprey pairs lay eggs during the second half of March. Little information on clutch size is available; counts of nestlings suggest that clutches of 2 and 3 eggs are most common. Clutch size elsewhere is usually from 2 to 4 eggs, with 3 most common (Poole 1989). The eggs are white to pale cinnamon, heavily marked with brown. They are laid at intervals of 1–2 days, and incubation begins after the first egg is laid. Both adults incubate for 5 to 6 weeks. The young, which are fed by both parents, fledge in about 7 to 8 weeks; they are dependent on their parents for up to 3 weeks after fledging (Poole 1989). Most Tennessee birds fledge during late June and early July (Beddow pers. comm.). Active East Tennessee nests in 1991 fledged an average

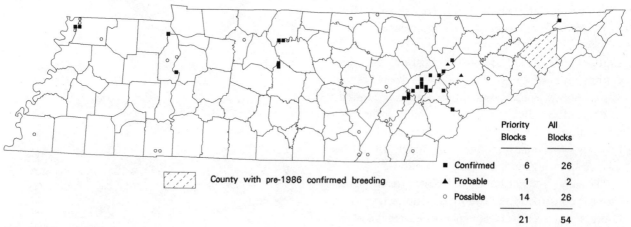

Distribution of the Osprey.

of 2.1 young (Beddow 1990), well above the 0.8–1.4 young/active nest needed to maintain a stable population (Poole 1989). This high reproductive rate and continued efforts to provide nesting structures bode well for the future of the Osprey in Tennessee. In 1994, its status was lowered to Threatened (TWRA 1994b); an even lower status of In Need of Management is currently justified.—*Charles P. Nicholson.*

Mississippi Kite
Ictinia mississippiensis

The Mississippi Kite, whose flight is among the most graceful of all raptors, is an uncommon summer resident of extreme western Tennessee. It is easy to recognize due to the distinctive shape of its long, pointed wings and long, notched tail, and is most often observed gliding gracefully above woodlands. Its diet is unusual among Tennessee raptors in that it feeds heavily on insects, usually caught and consumed in flight (Parker 1988). Flocks of 15 of more kites sometimes gather to forage on locally abundant food sources, such as large groups of dragonflies (Martin pers. obs.). The Mississippi Kite usually arrives in the state during the second half of April and departs by early September (Robinson 1990).

The original breeding range of the Mississippi Kite extended up the Mississippi River valley to southern Illinois (Parker and Ogden 1979). By the early twentieth century, however, it had disappeared from most of its range north of Memphis. The first mention of its occurrence in Tennessee was by Rhoads (1895a), who observed a bird he felt to be a Mississippi Kite soaring over a Shelby County meadow in May 1895 and quoted Miles's report of its probable occurrence in Haywood County. The first confirmed state record, however, was not until 11 August 1926, when Ganier observed birds in Henderson and Madison Counties (Coffey 1979). Ganier reportedly observed two old kite nests at the Madison County site the following winter. Mississippi Kites were regularly reported in Shelby County during the 1930s (Coffey 1940). First records for the other Tennessee counties adjoining the Mississippi River and Obion County were in the 1950s and 1960s (Coffey 1979); kites have been reported regularly from these counties since then. This apparent recovery has also occurred elsewhere in the eastern portion of its breeding range (Parker and Ogden 1979).

Because of concern over its small population and the loss of its apparently preferred bottomland forest habitat, the Mississippi Kite was listed as Endangered in Tennessee in 1975 (TWRA 1975). This listing stimulated a survey of its distribution by Kalla in 1978 (Kalla 1979, Kalla and Alsop 1983). Kalla's survey was mostly confined to the Mississippi Alluvial Plain, where he observed kites in Shelby, Lauderdale, Dyer, Lake, and Obion Counties. He estimated a minimum population of 162 birds concentrated in the metropolitan Memphis area and to the north in the larger areas of mature bottomland forest. Vegetation analysis of areas where large numbers of kites were observed suggested a preference for mature forests.

During the Atlas project, Mississippi Kites were reported in 55 blocks, 21 of them completed priority blocks. As previous reports of its distribution predicted, it was most frequently observed in the Mississippi Alluvial Plain, where it occurred in 10 of the 14 completed priority blocks. The Atlas records also show a significant eastward expansion along the rivers into the Loess Plain. This expansion probably began in the 1970s; kites were reported in Haywood County in 1974 and Hardeman County in 1976 (Coffey 1979). Ganier's 1926 records were also in the Loess Plain. The only breeding season report outside of West Tennessee was of an immature bird on 14 June 1987 in Robertson County (*Migrant* 58:140), probably a vagrant.

While often described as preferring extensive, mature wooded areas (e.g., Kalla 1979), many recent Mississippi Kite observations have been in urban areas. Kites have nested in a back yard in Dyersburg (Kalla 1979) and on the grounds of the zoo and Dixon Art Museum in Memphis, where they are frequently observed over residential areas, parks, and golf courses (Martin pers. obs.). In the western portion of its range, Mississippi Kites nest in large numbers in urban areas (Parker and Ogden 1979).

In an effort to increase the breeding population, young kites taken from nests of nuisance urban pairs in Kansas have been released in West Tennessee since 1983 (Martin

Mississippi Kite. Chris Myers

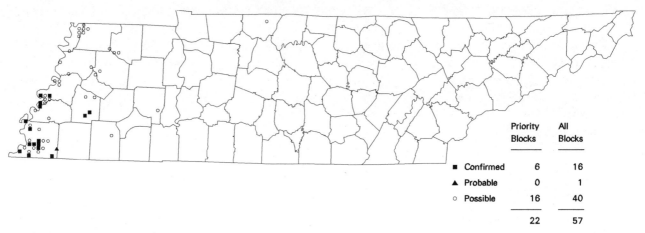

Distribution of the Mississippi Kite.

and Parker 1991). From 1983 to 1990, 117 birds were released at sites in the Memphis area, at Hatchie NWR, and at Paris Landing SP, the latter site probably outside the kite's original range. Most of these releases have been in areas where kites already occurred, and there is no evidence that the releases have resulted in the establishment of new breeding populations.

Regardless of the success of the release program, the numbers and range of the Mississippi Kite have increased since it was first listed as Endangered in Tennessee. Deforestation in West Tennessee, although still occurring faster than elsewhere in the state, has slowed since its peak in the 1970s (May 1991). New wildlife refuges and management areas protect much kite habitat. The kite has also proven to be less restricted to mature bottomland forests than was earlier thought, and its Tennessee status was recently downgraded to In Need of Management (TWRA 1994a).

Breeding Biology: The breeding biology of the Mississippi Kite in Tennessee is poorly known. Its nests, located high in trees, are difficult to find and observe and accounted for only 3 of the confirmed Atlas observations. Nest building begins soon after the kite's spring arrival. The nest is a rather substantial structure, built of sticks and about the size of a crow nest. It is placed near the top of the tree either in a central crotch or at the end of a sturdy branch. Nests have been at heights of from 12 to over 30 m; few precise measurements are available. Species of trees used for nesting have been bald cypress, tulip-poplar, sweetgum, sycamore, and white and southern red oaks. Colonial nesting, as occurs in the western portion of the kite's range (Parker 1988), has not been reported in Tennessee. Although reuse of old nests has been reported elsewhere (Parker 1988), a pair nesting on the grounds of the Memphis Zoo and Aquarium used a different tree each year for several consecutive years (Martin pers. obs.).

No records of Mississippi Kite eggs are available from Tennessee. Observations of nestlings and fledglings, however, suggest that egg laying occurs during the first half of May. The normal clutch size reported elsewhere is 2 eggs (Parker 1988), and observations of nestlings and fledglings suggest that 2 eggs is the normal clutch size in Tennessee. Both adults incubate the eggs for 29 to 31 days and feed the nestlings, which fledge in about 34 days (Parker 1988). The young fly well at about 50 days old, and are fed by the adults until at least 60 days old. Family groups probably form the nucleus of the sometimes large, premigratory flocks apparent in August.—*Knox Martin.*

Bald Eagle
Haliaeetus leucocephalus

Bald Eagles are permanent residents in Tennessee, rare during the summer, and uncommon in winter. Wintering birds begin arriving in October, peak in numbers in early February, and depart by early April. Most wintering birds concentrate around a few lakes and reservoirs (Robinson 1990).

The size and distribution of the historic breeding population of eagles in Tennessee prior to its temporary extirpation in the 1960s and 1970s is poorly known. Nesting Bald Eagles were first reported by Audubon (1929), who noted 2 nests in November 1820 along the Mississippi River in what is now Lake or Dyer County and saw a pair at Memphis. Rhoads (1895a) gave a secondhand report of their nesting at Reelfoot Lake. Ganier (1931c) described 3–4 pairs nesting at Reelfoot Lake, "perhaps" 1–2 pairs along the Mississippi River, and 3 pairs a short distance south of Memphis in Arkansas. Ganier (1933b) reported 5–6 pairs nesting at Reelfoot, and a similar number of pairs probably nested there into the 1950s. The count of 14 active nests in 1954

Bald Eagle. Chris Myers

tion from pesticides, especially DDT (USFWS 1989). Southern eagle populations were listed as Endangered in 1967, and given additional protection by the federal Endangered Species Act of 1973 (USFWS 1992). The use of DDT was prohibited in the United States in 1972 and in Canada in 1973, and northern and Florida eagle populations slowly increased (Stalmaster 1987).

Active recovery efforts began in Tennessee in 1980 with a hacking program at LBL in Stewart County. Because eagles tend to return to the area where they were raised when ready to breed (Simons et al. 1988), eagle hacking should result in the establishment of new breeding populations. A pair of unknown origin fledged 1 young from a nest near Cross Creeks NWR in 1983 and a bird hacked at LBL successfully nested at a second site near Cross Creeks in 1984. Hacking was begun at Reelfoot Lake in 1981. This project, and the LBL hacking project, continued until 1988, releasing 43 and 44 birds, respectively. Other hacking projects were later established on upper Cheatham, Dale Hollow, Chickamauga, Douglas, and South Holston Reservoirs. By the time of the projected termination of these projects in 1996, 248 eagles will have been released. About a quarter of the 183 birds released through 1991 were captive reared; most of the rest were taken from wild nests in Alaska and Wisconsin (Hatcher unpubl. data).

(W. Crews pers. comm.) probably included the nesting pairs' alternate nests; "active" was not defined. Nesting there continued until 1964 (Ganier 1964b, Crews pers. comm.). A nest was present near Kentucky Lake in Stewart County in 1948 and 1949 (Ganier 1951b). Other reports of nests in Clay, Pickett, and Montgomery Counties (Hassler and Hassler 1972, Pickering 1938) are unconvincing. The place name of Eagle Nest Island, in the Tennessee River in Decatur County, is suggestive of eagle nesting.

Bald Eagles feed primarily on fish and usually nest near water bodies (Stalmaster 1987). Following the construction of over 200,000 ha of new reservoirs in Tennessee since the 1930s, more potential eagle habitat existed than ever before. Despite this large increase in habitat, the nesting population in Tennessee, as in much of the rest of the United States, declined due to a combination of human encroachment, shooting, and, most importantly, reproductive failure caused by contamina-

The present nesting population grew from the 1 nest in 1983 to 17 occupied nests, 11 of them successful, in 1991. This exceeded the goal of 15 set in the U.S. Fish and Wildlife Service's Recovery Plan (USFWS 1989). Based on the results of the ongoing hacking projects and current nesting success, Hatcher (1991) projected an increase to 42 successful nests in Tennessee by the year 2000. Because of its increasing populations, both rangewide and in Tennessee, its federal and state status were recently changed from Endangered to Threatened (TWRA 1994b, USFWS 1995).

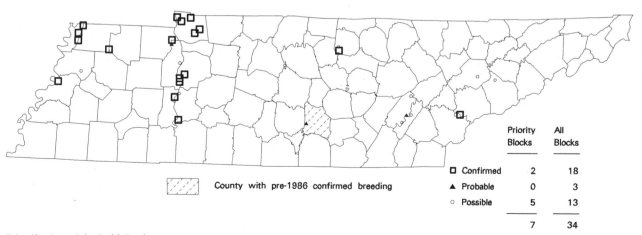
Distribution of the Bald Eagle.

Confirmed Breeding Species

Breeding Biology:

Bald Eagles are monogamous, and the pair bond often lasts the life of the birds (Stalmaster 1987). They normally produce their first young at 4 or 5 years of age, shortly after molting into adult plumage. Nesting by younger birds is rare (Stalmaster 1987). A 3-year old, hacked bird in subadult plumage successfully nested in Tennessee, and another, from a natural nest in Stewart County, successfully nested in Kentucky when 3 years old. In at least 2 cases, 1 or 2 adults were present for a few years in the nest area before nesting occurred.

Floyd (1990) observed Bald Eagles in the vicinity of a nest in Stewart County from September through July. Courtship and repair of the previously used nest began in October, and lasted for several weeks. Nests are typically in large, isolated tracts of mature forest, often on the edge of an opening. They are built near the top of a tall tree, often above the canopy of the adjacent forest (Stalmaster 1987). Tennessee nests active in 1991 averaged 547 m (s.d. = 744, n = 17, range 0–2100 m, Hatcher unpubl. data) from water. Most nests in the Reelfoot area have been in cypress trees; nests have also been reported in oak, hickory, cottonwood, tulip-poplar, and sycamore trees. The pair at Tellico Reservoir used a nest previously built by Ospreys. The nest is used for several years with nest material added each year; Ganier (1931c) described a nest at Reelfoot Lake as 1.2 m high and over 1.5 m wide. Bald Eagles often have more than 1 nest in their territory (Stalmaster 1987); at least 3 territories in Tennessee contain alternate nests (Floyd 1990, Hatcher unpubl. data).

Recent egg-laying dates, based on observations of behavior of nesting birds and calculated ages of nestlings, extend from early February through late April and peak about 20 February (Floyd 1990, Hatcher unpubl. data). This is later than the few historical records indicate, as the observations of Ganier (1931c) and Spofford (1948b) describe egg laying in November and December. No direct clutch observations are available; clutch size probably averages somewhat larger than the average of 1.8 young fledged per successful nest (s.d. = 0.7, range 1–3, n = 45, Hatcher unpubl. data). Stalmaster (1987) lists an average clutch size of 1.95 eggs elsewhere in the species's range. Both adults incubate the eggs for 34–36 days, and both feed the nestlings, which fledge in 70–84 days (Stalmaster 1987). Young leave the nest in late May or early June. Fledglings remain close to the nest for about a month (Floyd 1990). Recoveries and observations of tagged, hacked young show a pronounced northward dispersal (Hatcher unpubl. data), similar to the movements described by Broley (1947) in Florida. No information is available on the dispersal of young from natural nests in Tennessee. Summering, non-breeding eagles have been present in Tennessee for many years *(Migrant)*. The Atlas observations of possible breeding birds distant from known nests may have been of birds from nest sites south of Tennessee; none of these birds were apparently wing-tagged, as hacked birds have been. —*Robert M. Hatcher.*

Sharp-shinned Hawk
Accipiter striatus

The Sharp-shinned Hawk is an uncommon permanent resident of Tennessee, most numerous in September and October, when migrants move into and through the state. Wintering birds occur across the state and depart by early May. During the summer, the Sharp-shinned Hawk usually occurs in coniferous and mixed woodlands and decreases in abundance from east to west.

Based on their preference for coniferous trees, especially pines, breeding Sharp-shinned Hawks were probably fairly common in prehistoric East Tennessee, and locally distributed in other parts of the state where pine occurred. The few historic and recent records from the Central Basin suggest it rarely occurs in cedar woodlands. The scarcity of early records, however, makes reconstruction of its range speculative. The Sharp-shinned Hawk was first reported in Tennessee by LeMoyne (1886), who described it as uncommon in the Great Smoky Mountains. Rhoads (1895a) did not observe it on his statewide survey.

The first nest record was in 1919, from a naturally occurring grove of pines on a bluff at Craggie Hope, Cheatham County (Ganier 1923b). Ganier and others continued to collect Sharp-shinned eggs from this area almost annually until at least 1946. Ganier (1933a) de-

Sharp-shinned Hawk. Chris Myers

scribed it as a rare, permanent resident in Middle and East Tennessee. Local East Tennessee annotated lists published from the 1930s through 1950s (e.g., Pickett Forest, Ganier 1937a; Fall Creek Falls, Ganier and Clebsch 1940; Unicoi Mountains, Ganier and Clebsch 1946; Carter County, Herndon 1950b; Knox County, Howell and Monroe 1957), listed the Sharp-shinned Hawk as either rare or absent during the summer. The first summer record from West Tennessee was from Hardeman County in 1939 (Calhoun 1941).

Sharp-shinned Hawk populations in northeastern North America declined into the 1920s due to heavy shooting during fall migration (Palmer 1988). They also declined throughout much of their range in the 1950s and 1960s as reproductive success was affected by pesticides; according to migration counts, this trend was reversed in the 1970s (Palmer 1988). Because of the few historic records, the trend of the Tennessee breeding population is difficult to determine. Few Sharp-shinned Hawks have been recorded on BBS routes, which show no significant trend in Tennessee or North America (Peterjohn, Sauer, and Link 1994). The wintering population, most of which nests north of Tennessee, has shown no overall, long-term trend on Christmas Bird Counts, although it increased between 1975 and 1986 (Tanner 1985, Adkisson 1990).

Atlasers found the Sharp-shinned Hawk in a total of 92 Atlas blocks and in 8% of the completed priority blocks. Because the Sharp-shinned Hawk is inconspicuous and seldom occurs above the forest canopy during the breeding season except in display flights (Palmer 1988), it was likely missed in some priority blocks. By physiographic regions, it was most frequently reported in the Cumberland Mountains, where it occurred in 19% of the completed priority blocks, followed by the Ridge and Valley (15%) and the Unaka Mountains (11%). Within the Unakas, the Sharp-shinned occurs throughout most of the elevational range and probably breeds as high as 1524 m (Stupka 1963, McNair 1987c). It was found in less than 10% of the completed priority blocks in the other physiographic regions. Most of the observations for which habitat information is available were in the vicinity of pines and, except for the observations from Rutherford and Shelby Counties, within the natural range of pine. With the exception of the Shelby County observation, the range depicted by the atlas map agrees with previous descriptions of the hawk's Tennessee range. The Sharp-shinned Hawk was observed at 1 stop on 5 Atlas miniroutes, too few to show regional differences in abundance.

Because of concern over apparent population declines, the Sharp-shinned Hawk was listed as Threatened by the Tennessee Wildlife Resources Agency in 1975 (TWRA 1975). Based on the Atlas results, which include new county records of possible, probable, and confirmed breeding, and the hawk's ability to nest in pine plantations (see below), it appears to be in no danger of extirpation. Its status was lowered to In Need of Management in 1994 (TWRA 1994a).

Breeding Biology: Most of the Sharp-shinned Hawk nest records have been from two locations on the Western Highland Rim in Cheatham and Williamson Counties, where the species nested for several consecutive years (e.g., Ganier 1923b, Goodpasture 1956). All nests, except for one, have been in pines; the exception, in a red cedar, was near Virginia pines used in previous years (Goodpasture 1956). This preference for nesting in conifers is typical of the species elsewhere (Palmer 1988, Wiggers and Kritz 1991). All of the pines holding nests and identified to species were Virginia pines, except for one white pine in a white pine plantation (Yeatman 1974). Some of the unidentified pines holding nests were probably shortleaf pines, a common native species in much

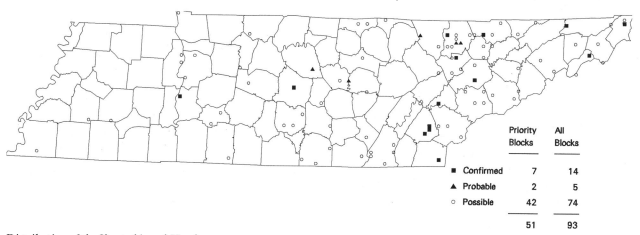

Distribution of the Sharp-shinned Hawk.

of the hawk's Tennessee range. Nests have been in pure groves of pine as well as predominantly deciduous woods with scattered pines. The hawk shows some preference for nesting on ridgetops (egg coll. data, Atlas data, Wiggers and Kritz 1991); in parts of Tennessee, natural pine most frequently occurs in such sites. Although only 1 nest was specifically described as occurring in a pine plantation, Sharp-shinned Hawks readily nest in plantations elsewhere (Wiggers and Kritz 1991), and plantations may be providing habitat suitable for an expansion of the species's range in Tennessee.

Sharp-shinned Hawk nests averaged 9.8 m above ground (s.d. = 3.8, n = 23, range 5.5–18.3 m). The nest is a relatively large platform of sticks, built primarily by the female; nests built during the previous year are sometimes reused. It is often lined with pieces of bark, and usually within the tree canopy next to the trunk (Palmer 1988). Egg laying peaks about 5 May through 15 May. Clutch size ranges from 3 to 6 eggs, with 5 eggs most common and an average of 4.82 (s.d. = 0.67, n = 28). Incubation begins before the last egg is laid, and lasted 24 days for a Tennessee clutch (Yeatman 1974). The young leave the nest in 3–4 weeks, although their plumage is not fully developed until about 40 days old; they are independent in about 7 weeks (Palmer 1988).
—*Charles P. Nicholson.*

Cooper's Hawk
Accipiter cooperii

The Cooper's Hawk is an uncommon permanent resident of woodlands and woodland edges. It is somewhat more numerous during its September–October and April–May migration periods, and its winter population is at least partially made up of birds nesting north of Tennessee. The Cooper's Hawk was historically a common species, especially in Middle and East Tennessee. Between the late 1940s and 1970s, it became very rare; this trend appears to have recently reversed.

The earliest record of the Cooper's Hawk in Tennessee is by Gettys, who collected its eggs in McMinn County in 1899 and 1905 (egg coll. data). In his preliminary checklist, Ganier (1917) described it as fairly common across the state; in the next edition it was listed as fairly common in West Tennessee and common in Middle and East Tennessee (Ganier 1933a). Todd (1935) described it as the commonest hawk in Rutherford County and Wetmore (1939) also described it as common. Given its abundance in the 1930s, it is curious that it was not mentioned by Rhoads or other late-nineteenth-century ornithologists. Based on its abundance in the 1930s, when

Cooper's Hawk. Chris Myers

the amount of forest land was at a minimum, it is possible that the Cooper's Hawk population increased with the openings and forest edge created by nineteenth-century European settlers.

Throughout much of this century, the Cooper's Hawk, colloquially known as the chicken hawk, was frequently shot by sportsmen, farmers, and others for preying on game birds and poultry. This killing was often condoned by early Tennessee naturalists (e.g., Ijams 1931); it is unlikely that it had a widespread effect on the hawk's population. The Cooper's Hawk population, however, did greatly decline from the late 1940s into the 1970s in much of its range (Meng and Rosenfield 1988), mostly due to poor reproductive success resulting from pesticide (primarily DDT) poisoning. While there is little quantitative data documenting this decline in the Tennessee breeding population, many recent annotated lists, such as the 1971 through 1984 TOS foray reports *(Migrant)*, list either very few or no Cooper's Hawks. These reports suggest the Cooper's Hawk was much less common than earlier in the century. BBS results show no significant trend since 1966, and Christmas Bird Counts likewise show no significant long-term trend (Tanner 1985). This decline, however, has probably reversed. The number of birds/party hour on Tennessee Christmas Bird Counts has increased in recent years (Adkisson 1990). An increase is also apparent elsewhere in the hawk's range (Meng and Rosenfield 1988); Pennsylvania migration counts show the post-DDT recovery of the Cooper's Hawk started later and occurred more gradually that did the Sharp-shinned Hawk's recovery (Bednarz et al. 1990). Rangewide BBS results show an average increase of 7.2%/year from 1966 to 1993 (Peterjohn, Sauer, and Link 1994).

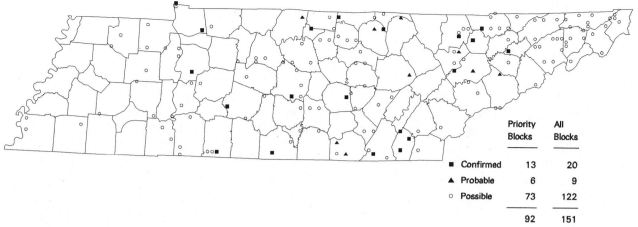

Distribution of the Cooper's Hawk.

During Atlas fieldwork, the Cooper's Hawk was found in a total of 150 blocks and 14% of the completed priority blocks. Because of its low numbers and frequently inconspicuous behavior, the Cooper's Hawk was likely missed in some blocks. It was most frequently observed in the Eastern Highland Rim physiographic region (25% of priority blocks), followed by the Ridge and Valley (22%), and the Central Basin (17%). The reasons for the concentration of records in the northern end of the Ridge and Valley are not clear. The numerous records in populous areas such as Davidson, Knox, and Washington Counties suggest both the higher probability of it being observed in areas with resident observers and some adaptability to suburban habitats. Although there have been few previous records from heavily wooded, high-elevation areas in the Blue Ridge, the hawk was reported from 2 such blocks in the Smokies. The highest elevation nest record is from Shady Valley, about 860 m, in 1939 (egg coll. data). The Cooper's Hawk was recorded at 1 stop on 7 Atlas miniroutes, too few to show regional differences in abundance. A nest with 3 young on 7 July 1986 in Henderson County (*Migrant* 57:103) was in an unknown block and is not shown on the Atlas map. This is one of very few recent or historic West Tennessee confirmed breeding records.

Because of concern over its population decline, the Cooper's Hawk has been listed on most editions of the Audubon Blue List since 1972 (Tate 1986). It was also listed as Threatened in Tennessee in 1975 (TWRA 1974). Although it is presently much less common than it was in the 1930s, its numbers are increasing on Christmas Bird Counts, and Atlas results show that it is distributed throughout much of the state. With this recent increase and the hawk's adaptable habitat requirements, its status was changed to In Need of Management in 1994 (TWRA 1994a).

Breeding Biology: The Cooper's Hawk is fairly versatile in its nesting habitat requirements. Nests have been reported in rural woodlots, urban cemeteries, wooded urban parks, suburban areas, and near the edge of more extensive forests. A variety of tree species have been used for nesting, with oaks the most frequent; nests have also been built in hickory, beech, elm, sugar maple, hackberry, tulip-poplar, red cedar, and Virginia and white pines. Pines and cedar were used for 6 of 30 nests. As observed elsewhere (Meng and Rosenfield 1988), the same nest is often reused for several years; Todd noted a nest used at least 6 years (egg coll. data). Nests averaged 11.1 m above ground (s.d. = 5.4 m, n = 29, range 6.1–34 m). Excluding an unusually high nest, 34 m high in a hickory, they averaged 10.3 m (s.d. = 3.1 m, n = 28, range 6.1–15.2 m). The nest is usually built of twigs, lined with bark, and placed in a main crotch of the tree below the canopy (Meng and Rosenfield 1988, Wiggers and Kritz 1991). A few Tennessee nests were built on old squirrel or crow nests, a practice also reported elsewhere (Meng and Rosenfield 1988).

Egg laying begins in early April and peaks during the last third of April. Clutch size averages 4.00 eggs (s.d. = 0.73, n = 35, range 3–5), and clutches of 4 are most common. Clutches of 3 to 6 eggs are normal elsewhere, with 4 and 5 most common (Meng and Rosenfield 1988). The eggs are usually whitish without markings, although the Todd collection contains a well-marked set from Rutherford County. The female incubates for 30 to 34 days, beginning before the clutch is complete (Meng and Rosenfield 1988). The young begin to fly when about 5 weeks old and leave the nest soon thereafter. They are fully developed in about 8 weeks.—*Charles P. Nicholson.*

Red-shouldered Hawk
Buteo lineatus

The Red-shouldered Hawk is an uncommon to fairly common permanent resident found throughout much of the state. Although it inhabits forest and has been described by several authors as shy, its frequent loud calls make it relatively conspicuous, and it is usually easy to identify.

Atlas workers found Red-shouldered Hawks in slightly over one-third of the completed priority blocks, with the highest frequency of occurrence in the Cumberland Plateau and Mountains, the Western Highland Rim, and the Coastal Plain Uplands. They were locally common in forested bottomlands in West Tennessee and along Chickamauga and Watts Bar Reservoirs in East Tennessee, as well as in the northwest corner of the Ridge and Valley. Within the Cumberland Mountains, the Red-shouldered was the most frequently reported hawk.

Ganier (1933a) described the Red-shouldered Hawk as fairly common in West Tennessee, rare in Middle Tennessee and very rare in East Tennessee, and it has traditionally been considered to inhabit bottomland and swamp forests (Robinson 1990). Calhoun (1941) found it to be the only common hawk in portions of Hardeman and McNairy Counties. The few published annotated lists from the Western Highland Rim and Cumberlands make it difficult to determine whether the Red-shouldered Hawk has increased in abundance in these areas. There is little doubt, however, that its numbers have decreased in West Tennessee with the destruction of bottomland forests. This decline was noted by Butler (1948) in the mid-1940s. Red-shouldered Hawks have apparently never been common in the Central Basin, Eastern Highland Rim, Unaka Mountains, and northeast Tennessee.

Red-shouldered Hawk. Chris Myers

Red-shouldered Hawks usually occur in mature deciduous forest near open water and clearings (Palmer 1988). No quantitative studies of their habitat use have been conducted in Tennessee. Studies elsewhere have shown they require forest tracts of at least about 250 ha (Bednarz and Dinsmore 1981, Preston et al. 1989). Bottomland hardwood and mesic upland forests are used more than dry upland forests, and territories include marshes or wet meadows used for foraging. In Tennessee, areas of the Western Highland Rim, Cumberland Mountains, and adjacent portion of the Ridge and Valley where Red-shouldered Hawks are relatively common are predominantly upland forest with numerous streams and cleared fields in the valleys. In similar habitat in western Maryland, Titus and Mosher (1981) found most nests within 100 m of water.

Concern over declining numbers on migration and Christmas counts (Palmer 1988) has led to the inclusion of the Red-shouldered Hawk on the Blue List since its inception (Tate 1986) and its listing as In Need of Management in Tennessee (TWRA 1976). Any statewide decline probably occurred prior to the late 1960s, as Red-shouldered Hawk numbers on BBS routes suggest an increase from 1966 to 1994, although the trend is not significant. Tanner (1985) found no significant change in numbers on Christmas counts, although a separate analysis found the hawk's reporting increased from 1969 to 1985 (Mitchell and Millsap 1990). The wintering population is composed of the locally breeding, nonmigratory birds as well as birds from north of Tennessee (Henny 1972). Rangewide BBS routes also show an increasing trend of 2.4%/year (p = 0.02) from 1966 to 1993 (Peterjohn, Sauer, and Link 1994).

With its wide distribution, local abundance, and increasing population, the special protective status of In Need of Management is no longer justified and was removed in 1994 (TWRA 1994a). Its numbers, however, should continue to be monitored, as further loss of wetlands and riparian habitat would be detrimental, as would a widespread change in the relatively mature age composition of much of Tennessee's forest.

Breeding Biology: Red-shouldered Hawks begin occupying their breeding territories around the first of the year, and courtship behavior occurs during warm days in January and February (Nicholson unpubl. data). Both birds soar in circles high over their territory, calling loudly; one bird frequently dives toward the other. The nest is usually built in a large tree in mature forest with a dense understory and close to water, and these factors may be more important than the tree species composition (Titus and Mosher 1987, Palmer 1988,

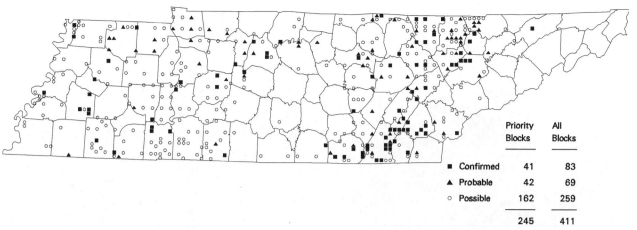

Distribution of the Red-shouldered Hawk.

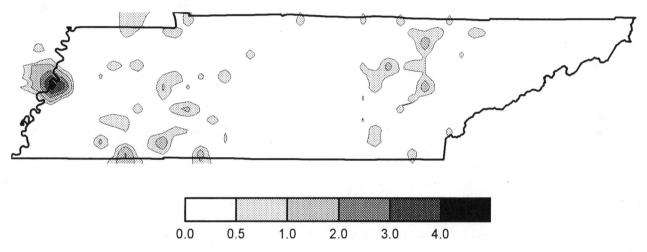

Abundance of the Red-shouldered Hawk.

Preston, Harger, and Harger 1989). It is usually placed in a fork of the main tree trunk, below the canopy. The large nest is made of sticks and lined with twigs, bark, or leaves. The nest may be used for several years and relined each year. Twigs with fresh, deciduous, or coniferous leaves are brought to the nest during the late incubation and nestling periods (Palmer 1988).

Forty-five Tennessee nests, most from Middle and East Tennessee, ranged from 7.6 to 19.8 m above ground, with an average of 13.6 m (s.d. = 3.3 m), close to the rangewide average of 14.3 m (Palmer 1988). Several tree species are used for nesting, with oaks, sweetgum, beech, and tulip-poplar most frequently reported. Two nests were reported in pines, and 1 in a cypress.

Egg laying begins in late February or early March, and most records of completed clutches are from mid- to late March. Clutch size varies geographically; clutches in Tennessee ranged from 2 to 4 eggs, with an average of 3.0 (s.d. = 0.56, n = 23), somewhat higher than that suggested by Henny (1972). The eggs are white with blotches and spots of brown. Both adults incubate, beginning before the clutch is complete. The eggs hatch in 33 days and the young leave the nest in 32–45 days (Palmer 1988, Portnoy and Dodge 1979). Several Tennessee nests with young contained infertile eggs, as noted in studies elsewhere (Palmer 1988). The fledglings begin hunting about 2 weeks after first flight and may be fed by the parents for 8–10 weeks.—*Charles P. Nicholson.*

Broad-winged Hawk
Buteo platypterus

The Broad-winged Hawk is a fairly common to common migrant and an uncommon summer resident found in forests throughout Tennessee. Migrants usually arrive in early April and depart by early October. During migration, particularly in the fall, it occurs in large flocks often numbering hundreds of birds. These migratory flocks are most prominent over the ridges of East Tennessee.

Broad-winged Hawk. Chris Myers

During the summer, this small, tame hawk spends much of its time within the forest canopy and is as likely to be heard as seen. Its call is a distinctive, clear, monotone whistle.

The Broad-winged Hawk occupies forests, both deciduous and mixed. It occurs at all elevations of the eastern mountains, including the high-elevation spruce-fir forest, although there are no nest records from this habitat (Stupka 1963). It usually nests within 150 m of water and a forest opening, and nests occur in woodlands as small as a few hectares (Titus and Mosher 1981, Mosher and Palmer 1988). Much of its foraging occurs at forest edges. The Broad-winged Hawk has been recorded on Tennessee breeding bird census plots in Ridge and Valley hardwood forest (Howell 1973), Cumberland Mountain mixed mesophytic forest with and without strip mines (Yahner 1972, 1973, Smith 1977), Highland Rim mixed mesophytic forest and old fields (Simmers 1978–84), low-elevation Unaka Mountain deciduous forest (Lewis 1983a), and Central Basin deciduous forest (Fowler and Fowler 1984b). Stedman and Stedman (1992) found fledgling Broad-wingeds in an 11 ha Highland Rim, primarily deciduous woodland.

The Broad-winged Hawk was probably fairly common to common in prehistoric Tennessee, and it was likely the most numerous nesting hawk. It probably bred throughout the state except in areas of extensive grasslands or cedar glades. Suitable habitat declined with the agricultural clearing by European settlers. The Broad-winged Hawk's early history in Tennessee is poorly known. It was first mentioned by Rhoads (1895a), who described it as the most abundant hawk in East Tennessee, including the mountains. He did not describe its status in the rest of the state. Ganier (1917, 1933a) described it as fairly common in West and East Tennessee and rare in Middle Tennessee. His description of it as rare in Middle Tennessee was probably true for the Central Basin, but an underestimate of its abundance on the Highland Rim.

Its statewide numbers were probably lowest in the 1930s, when forest area was at a minimum. Since then its numbers have likely increased with reforestation on the Coastal Plain Uplands, Highland Rim, Cumberlands, and Unaka Mountains. Continued declines have probably occurred with clearing and development in the Loess Plain, Central Basin, and Ridge and Valley. Tennessee BBS routes show no significant population trend from 1966 to 1994, although it decreased by 10.8%/year ($p < 0.05$) from 1980 to 1994. From 1966 to 1993, it increased by 2.0%/year ($p = 0.05$) on BBS routes throughout its range (Peterjohn, Sauer, and Link 1994).

Atlas workers found the Broad-winged Hawk in 47% of the completed priority blocks; most records were in areas with a high proportion of forest cover. Because of the hawk's often inconspicuous behavior, it was probably missed in some blocks. In extreme western Tennessee, many records were in bluff forests, which are among the only extensive upland forests in the area. The few

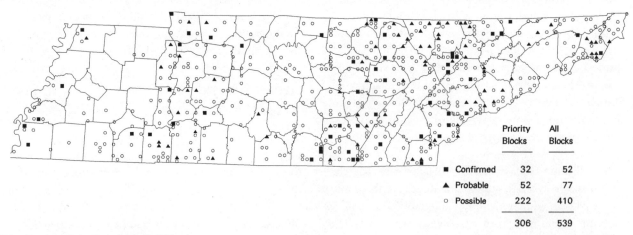

Distribution of the Broad-winged Hawk.

	Priority Blocks	All Blocks
■ Confirmed	32	52
▲ Probable	52	77
○ Possible	222	410
	306	539

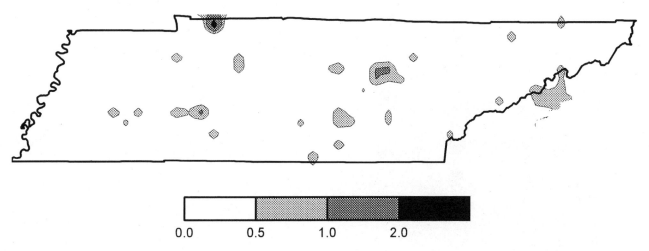

Abundance of the Broad-winged Hawk.

records elsewhere in West Tennessee suggest it requires some minimum area of forest within its territory and avoids bottomland forests. The Broad-winged Hawk was recorded on 48 miniroutes, at an average abundance of 1.1 stops/route; no regional differences in abundance are evident in these results.

Breeding Biology: Confirming breeding of the Broad-winged Hawk was difficult for Atlas workers, and detailed descriptions of only a few Tennessee nests are available. The Broad-winged Hawk is monogamous, and most do not breed until at least 2 years old (Mosher and Palmer 1988). A single brood is raised. Tennessee nests were between 7.6 and 19.8 m above ground, with an average height of 12.8 m (s.d. = 4.5, n = 8). Tree species used include shortleaf pine, black and white oaks, hickory, tulip-poplar, and sweetgum, with no obvious preference. Sharp (1931) noted that Middle Tennessee nests were usually in oaks or chestnuts. Elsewhere in its range, Broad-winged Hawk nests are usually from 7.6 to 12.2 m above ground, in the first main crotch of the nest tree, and in the lower third of the canopy (Mosher and Palmer 1988). The nest tree is usually at least 25 cm in diameter, and among the largest trees in the nesting area (Titus and Mosher 1987). Although both adults carry sticks to the nest, the female does most of the construction (Mosher and Palmer 1988). The nest is small relative to the bird's size and rather poorly built. It is seldom used 2 consecutive years. The nest is made from sticks and twigs, and the bowl is lined with bark chips. Sprigs of fresh green leaves are continually added during incubation and the early nestling period.

Egg laying peaks in the late April and early May. Clutch size in Tennessee is from 2 to 4 eggs, with an average of 2.71 (s.d. = 0.69, n = 16). Elsewhere clutches of 2 and 3 eggs are most common, and clutches of 1 and 4 eggs rare (Mosher and Palmer 1988). The female incubates the eggs for about 31 days, beginning when the first egg is laid. At hatching, the young are covered with down and have their eyes open. The female broods them for most of the first week, while the male delivers food to the nest. Later, both parents feed the young. The young first leave the nest, by climbing onto limbs of the nest tree, when 29–31 days old (Mosher and Palmer 1988). They begin flying when about 6 weeks old and are not independent until about 8 weeks old.
—*Charles P. Nicholson.*

Red-tailed Hawk
Buteo jamaicensis

The Red-tailed Hawk is a common, statewide, permanent resident. Compared to other hawks, it is relatively conspicuous because of its habit of hunting from high, exposed perches and frequent soaring. It is probably the most frequently seen hawk across Tennessee and is a common sight along rural highways. Although present throughout the year, a noticeable influx of migrant birds occurs in late fall; these wintering birds have usually departed by early April. Locally breeding adults are nonmigratory, and young birds disperse from natal territories their first fall, sometimes up to several hundred km (Palmer 1988, Williams 1980a).

The Red-tailed Hawk is adaptable in its habitat use, requiring a suitable nest site—usually a large tree in an open area or emerging above the canopy of surrounding woodland—and fields with exposed perches or open woodland for foraging (Bednarz and Dinsmore 1982, Palmer 1988). Because of their foraging habitat requirements, Red-tails were probably uncommon in prehistoric Tennessee and most numerous in the prairies of north-central and northwest Tennessee, in cedar glades,

Red-tailed Hawk. Chris Myers

and in frequently burned, open-canopied woodlands. European settlement, and the resulting landscape of croplands and unimproved pastures in the valleys and wooded uplands, probably resulted in an increase in Red-tail numbers.

Rhoads (1895a) described the Red-tailed Hawk as rare, although he admitted that he had trouble identifying it. Ganier's 1917 and 1933 lists described the Red-tail's summer status as rare in West Tennessee, rare (1917) and then fairly common (1933a) in Middle Tennessee, and fairly common in East Tennessee. Compared to its present status, these early compilations suggest that the Red-tail was much less common in the late nineteenth and early twentieth centuries than it is at present. Shooting of hawks, legal during this early period and still widespread into the 1960s, probably had some effect on Red-tail populations (Henny and Wight 1972); suitable nest sites were, and probably still are, limiting in some heavily agricultural areas. Vaughn's (1932:4) statement that Red-tails nested "only in the larger tracts of big timber and are rarely found within five or six miles of cities" is not true today, and if true then, would imply that Red-tails have adapted to habitat changes. They presently nest near cities and in small woodlots surrounded by cultivated or fallow fields.

The Red-tailed Hawk was found in 74% of the completed priority blocks; both this proportion and the total number of Atlas blocks from which it was reported were considerably higher than for any other hawk. The proportion of completed priority blocks in which it occurred was lowest in the Unaka Mountains (38%), the Cumberland Plateau (50%), and the Cumberland Mountains (62%); all of these areas are relatively heavily wooded. In all of the other regions its frequency of occurrence equaled or exceeded the statewide average. Within the Unaka Mountains, Red-tails occasionally soar over the highest elevations (Stupka 1963); the highest elevation at which nesting occurs is unknown.

The Red-tailed Hawk was reported on 107 miniroutes at an average relative abundance of 1.1 stops/route. The mapped miniroute results show little regional difference in abundance. The miniroute abundance was over twice as high as its abundance of 1.65 stops/route and 1.86 birds/route on the 50-stop Tennessee BBS routes that recorded Red-tails during the Atlas period. BBS results show no significant trend in the Tennessee breeding population from 1966 to 1994; rangewide, its population increased significantly ($p < 0.01$) by 2.9%/year from 1966 to 1993 (Peterjohn, Sauer, and Link 1994). The Tennessee wintering population, composed of both resident and migrant birds, has also recently increased (Tanner 1985).

During the Atlas period, breeding of the Red-tailed Hawk was confirmed in about two-thirds of the counties, although Red-tails probably nest in every county. Of the confirmed observations, about half were of fledged

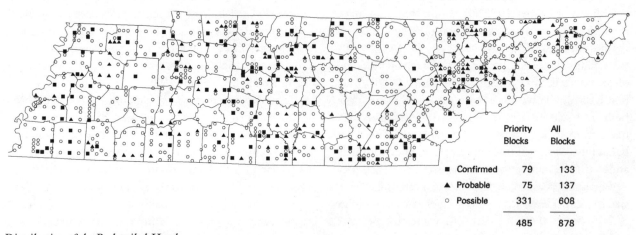

	Priority Blocks	All Blocks
■ Confirmed	79	133
▲ Probable	75	137
○ Possible	331	608
	485	878

Distribution of the Red-tailed Hawk.

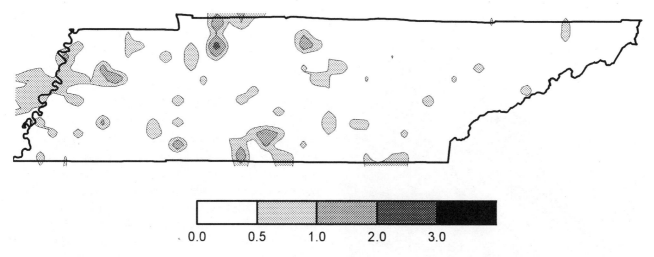

Abundance of the Red-tailed Hawk.

young and 35% were of nests. The nests are often easy to find during the winter and early spring before trees leaf out, and the adults are noisy and conspicuous during this period. Little Atlas work, however, was done during this period, which in part accounts for the low number of nests reported.

Breeding Biology: Red-tailed Hawks rarely begin breeding until at least 2 years old (Palmer 1988). Once mated, the monogamous pair bond usually lasts until death of a mate. Studies of nonmigratory birds elsewhere have shown that the pair remains on its territory throughout the year (Palmer 1988); this is probably the case in Tennessee. A conspicuous courtship display, given as early as January and frequently in late winter, is high-circling, the pair soaring together at great heights, sometimes touching talons or wings.

Both adults build the nest, which is usually placed high in a large tree. Nests heights range from 7.6 to 30.5 m and average 19.9 m above ground (s.d. = 3.8, n = 53). Oaks and hickories are the most commonly used tree species; others include sugar maple, American beech, elm, chestnut, blackgum, tulip-poplar, and, on at least 3 occasions, pines. The species of nest tree is apparently less important than its size, relative isolation from disturbance, and unobstructed access (Palmer 1988). A few nests have been reported on rock bluffs (Ganier 1931b, NRC). The nest is large and well-constructed, mostly of sticks. Other materials used are cornstalks and husks, grape vines and bark; fresh conifer springs and deciduous leaves are often placed in the nest during incubation. Vaughn (1943) found that territories in Middle Tennessee usually contained 2 old nests at the beginning of the nesting season, 1 of which was reused. With reuse the nests become quite large, occasionally over a meter in diameter.

Egg laying peaks in the middle of March, although it begins as early as late February. Clutches average 2.18 eggs (s.d. = 0.56, n = 51, range 1–4). The most common clutch size is 2 eggs, and clutches of 1 and 4 eggs are rare. Clutches of 2–3 eggs are most common elsewhere (Palmer 1988). The eggs are bluish white to white, marked with variable amounts of brown, red, or purple spots and blotches.

Both adults incubate the eggs for about 34 days, beginning before the clutch is complete (Palmer 1988). The female broods the young nestlings closely while the male delivers food to the nest. After 42–46 days, the young birds become "branchers," perching on limbs or the ground near the nest, while still being fed by the parents and sometimes returning to the nest to roost. The young often noisily chase and beg food from the parents during this period. Fledglings fly well at about 9 weeks, and remain with their parents for up to 10 weeks after their first flight (Palmer 1988). Only a single brood is raised, and a replacement clutch is laid 3–4 weeks after loss of the first clutch.—*Candy L. Swan.*

American Kestrel
Falco sparverius

The American Kestrel is the smallest and most colorful hawk found in Tennessee. It is present year-round and throughout the state in open areas such as croplands, pastures, and along roadsides. Kestrels are probably more common in urban areas than any other hawk. Migrant Kestrels from north of Tennessee are also present during the winter, arriving in October or November and departing by April. Limited information from recovery of banded birds suggests that Kestrels of breeding age are resident in Tennessee, while young birds may migrate a short distance (Bird 1988, Williams 1980a).

Confirmed Breeding Species

American Kestrel. Chris Myers

American Kestrels were probably uncommon and local in Tennessee until the widespread forest clearing that accompanied nineteenth-century settlement, when their numbers probably greatly increased with the availability of open habitats. At the end of the century, Rhoads (1895a:475) noted it at several points across the state, although "not as common as in the Middle States." Ganier (1933a) described it as common statewide. Its population, as measured by Tennessee BBS routes, increased by 2.3%/year (p < 0.10) from 1966 to 1994. This increase was greatest, 12.0%/year (p < 0.01), from 1966 to 1979. Throughout the North American portion of its breeding range, the American Kestrel population showed no significant long-term trend from 1966 to 1993 (Peterjohn, Sauer, and Link 1994).

Kestrels are relatively conspicuous during the breeding season and often sit on conspicuous perches in open areas and at woodland edges, or hover over fields as they hunt insects, small mammals, and birds. Atlas workers found them in 43% of the completed priority blocks, most frequently in the Central Basin. Other concentrations occurred in the northern portion of the Western Highland Rim and northeast Tennessee, where Lyle and Tyler (1934) reported it to be fairly common. They were also found in all major urban areas. Kestrels were not found in a few areas where they previously occurred, such as Pickett Forest (Ganier 1937a), where forests have matured to the point that little suitable habitat exists. The lack of records in parts of the Ridge and Valley, Cumberland Plateau, and West Tennessee with apparently suitable habitat is difficult to explain.

Atlas workers recorded American Kestrels on 11% of the miniroutes, at an average relative abundance of 1.22 stops/route. Their highest abundance was in the Central Basin. Kestrels have also been recorded on all but 2 of the state BBS routes, at an average density of 0.5 birds/route.

Breeding Biology: American Kestrels nest in cavities, which they do not excavate themselves. Little or no nest material is used. Hamerstrom, Hamerstrom, and Hart (1973) showed that availability of nesting cavities can be a limiting factor to Kestrel numbers, and providing suitable nest boxes can result in a population increase. Kestrels use a wide variety of cavities. Sites used in Tennessee include woodpecker holes, natural cavities in tree snags, nest boxes specifically built for Kestrels as well as Wood Duck and Purple Martin boxes, and holes in buildings. Nests have also been found in old European Starling and House Sparrow nests in cavities. Ganier (1946a) and Spofford (1948a) reported Kestrel nests on cliffs on the Cumberland Plateau, in the latter case on a site

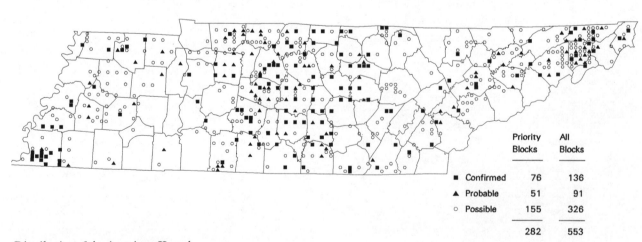

	Priority Blocks	All Blocks
■ Confirmed	76	136
▲ Probable	51	91
○ Possible	155	326
	282	553

Distribution of the American Kestrel.

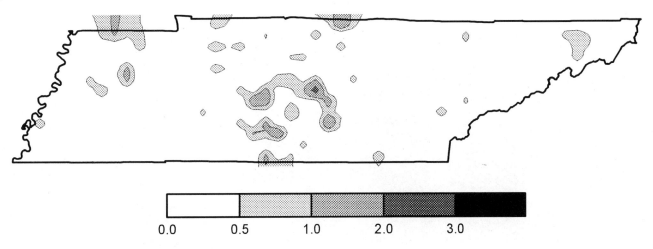

Abundance of the American Kestrel.

previously used by Peregrine Falcons. No nests have recently been reported from "natural" cliffs, although recent nests reported on quarry walls and buildings were on sites resembling cliffs. Nest heights range from 2 to at least 18.3 m above ground; the highest nests have been on tall buildings (e.g., Dubke and Dubke 1975).

Kestrel breeding activity begins in March, when the male establishes a territory containing the nest site and foraging area. No information is available on territory size in Tennessee. American Kestrels are normally monogamous and may be paired to the same mate over several breeding seasons (Bird 1988). Upon arrival of the female, the male begins courtship, consisting of aerial displays, calling, courtship feeding, nest site inspection, and copulation (Willoughby and Cade 1964). Once a nest site is accepted by the female, she remains in the territory and becomes increasingly dependent on the male for food.

Egg laying begins as early as late March and peaks in early to mid-April. Clutch size is 4 to 6 eggs, with an average of 4.74 (s.d. = 0.65, n = 19), similar to that reported elsewhere in North America (Bird 1988). The eggs are white to buff, with varying amounts of brown mottling. Both sexes incubate the eggs for 28–29 days (Bird 1988). The female broods the nestlings for several days after hatching, while the male provides food for both her and the young. She also later hunts for the young, which usually leave the nest at 29–31 days old (Bird 1988). Because incubation begins before the clutch is complete, the nestlings are of different ages and may fledge over a period of days. The young begin hunting shortly after leaving the nest, although the parents continue to feed them for about 2 weeks. The family unit remains together for a month or more. These family groups are conspicuous, and fledglings accounted for two-thirds of the confirmed Atlas records.

After losing a clutch, Kestrels readily renest, either in the same nest or, if available, in an alternate site (Bird 1988). Second broods have been reported (based on reuse of the same nest site) in Tennessee from Memphis (Irwin 1959), Nashville (A. R. Laskey unpubl. notes), and Chattanooga (Dubke and Dubke 1975). The several reports of fledglings in late July and August may also have been young of second broods. Evidence of second broods elsewhere is given by Bird (1988), who noted renesting may begin before the first brood is independent.—*Charles P. Nicholson and Audrey R. Hoff.*

Peregrine Falcon
Falco peregrinus

Great speed, power, and agility of flight make the Peregrine Falcon one of the most thrilling birds to observe. Just a few decades ago it was feared that they might become a faded memory. The Peregrine had been extirpated as a breeding bird from Tennessee and the rest of its eastern U.S. range by the middle of this century; it is now, however, making a comeback as a result of a captive-breeding and release program. This falcon's worldwide range, perhaps the broadest of any bird, served as insurance, and now the Peregrine is a symbol of conservation (Cade 1982). The Peregrine is presently a very rare summer visitor, a rare winter resident, and an uncommon migrant in Tennessee (Robinson 1990).

Once called the Duck Hawk in American ornithological literature, the Peregrine Falcon had a historic population of over 350 pairs in the eastern United States (Hickey 1942). At least 25 eyries had been located in Tennessee by the 1940s (Alsop 1979b) in 15 counties scattered across the eastern one-third of the state and in the northwestern corner. This range was limited by the availability of nesting sites, predominately cliffs. The eastern sites were remote mountain cliff eyries, canyon

Peregrine Falcon. Chris Myers

eyries on Cumberland Plateau escarpments, and river bluff eyries on the upper Tennessee River system (Spofford 1942a). A small, disjunct population nested in tree cavities in some West Tennessee swamps until 1947, mainly at Reelfoot Lake, but also at a site near the Dyer–Lauderdale county line (Ganier 1932a, Spofford 1943, 1947b). This apparently was the last remnant of a more widespread small population occupying the middle Mississippi River valley (Hickey and Anderson 1969). The earliest Tennessee nest was found 4 April 1893 on a river bluff near Knoxville (Ganier 1931b), although Stupka (1963) mentions the probable occupancy of a Smoky Mountain site in 1848. Most of the state's former breeding Peregrines appeared to have been permanent residents (Ganier 1934c, Spofford 1950).

Nest site abandonment in Tennessee was noted as early as the 1930s (Ganier 1940a), and the decline escalated dramatically in the 1940s. Representative last known nestings were in 1943 in the Great Smoky Mountains (Stupka 1963), 1944 in Fall Creek Falls SP (Spofford 1944), and 1947 at Reelfoot Lake (Spofford 1947a). Single birds were present at many of these sites for a few additional years. A pair in the Doe River gorge in Carter County in 1946 may have nested (Herndon 1950b). Two immature peregrines near Jellico in Campbell County in 1952 (Mengel 1965) suggested local nesting; there are suitable cliffs on both sides of the Tennessee–Kentucky border. A 1964 survey of many former eyries failed to locate any active sites in Tennessee or elsewhere in the eastern United States (Berger, Sindelar, and Gamble 1969); many of the historic cliff sites in Tennessee, however, are still suitable today. The Peregrine Falcon was among the first organisms to become federally listed as an Endangered Species, and it remains so listed today.

Reproductive failure due to the effects of accumulated DDT and other pesticides proved to be the main cause of decline in North American Peregrines (Hickey 1969, Ratcliffe 1980, Cade et al. 1988). Tennessee's population, however, seems to have largely disappeared before the widespread use of DDT, which came shortly after World War II, could have taken effect. As it is possible that some of the last nesting records were kept secret to try to protect the birds, our knowledge of the last nestings may be inaccurate. Shooting has always taken a toll on raptors. Egg collecting also had an impact at some Tennessee sites (Jones 1933), but it did provide a valuable source of pre-DDT eggs for analysis.

Captive breeding of Peregrines and release (called hacking) techniques were worked out by the early 1970s (Cade et al. 1988). The hacking of fledgling-aged falcons was begun in 1974 in the northeastern United States and, with success there, was expanded to the southern Appalachians in 1984. Thirty-four young Peregrines were hacked at 2 sites in the Unaka Mountains from 1984 to 1989 (TWRA unpubl. data), in addition to over 100 released in adjacent states. An active nest in western North Carolina in 1987 was the first by released birds (A. Boynton pers. comm.); as of 1996, however, no breeding had been confirmed in Tennessee. Atlas records include a territorial pair at a historic eyrie in the Smokies, a single bird on territory in downtown Knoxville, and a single sighting at Savage Gulf in Grundy County. Special surveys have been made for Tennessee nests, but not all suitable sites have been inspected. Continued surveys should soon find an active nest in Tennessee.

Breeding Biology: Most Peregrine Falcon nests are simple scrapes in the loose substrate of inaccessible cliff ledges or potholes, often with a protective overhang. Nest ledges are "well marked by white excreta" (Ganier 1931b:8), which aids in their discovery. Nesting cliffs are usually from 30 to 60 m high or higher, with a commanding view, and often overlook waterways. Favored sites may be used for many years. An unusual Tennessee eyrie was an old Red-tailed Hawk nest built in a niche on a cliff in Grundy County and used by Peregrines for 9 years (Ganier 1931b, 1933d). The small population that nested in West Tennessee swamps used natural cavities 20–25 m up in giant bald cypress or sycamore trees from which the tops had broken off or large limbs fallen (Spofford 1942a, 1947a, 1947b). These giant, supercanopy trees also had commanding views (photos in Hickey 1969). Some of the restored birds in other states have nested on bridges and tall urban buildings.

Peregrines are monogamous and generally mate for life. Spofford (1947a) reported a most unusual occurrence of a successful nest at Reelfoot Lake with 3 adults present, the 2 females cooperating in incubation. Court-

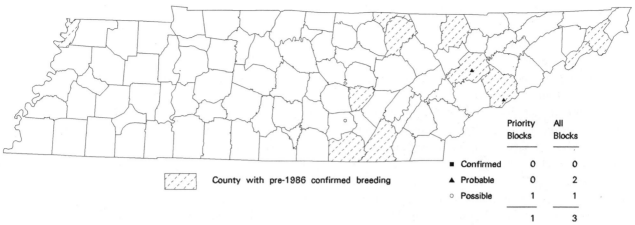

Distribution of the Peregrine Falcon.

ship may occur at any season at permanently occupied sites, but intensifies in late winter through early spring. Courtship displays include spectacular aerial maneuvers (Cade 1982). Although Tennessee egg dates were mostly from March to May, a Reelfoot eyrie with fledglings present on 30 April (Spofford 1944) and a nest near Johnson City with 3 young and 2 eggs on 29 March (Lyle and Tyler 1934) both suggest February egg laying. Tennessee clutches ranged from 3 to 5 eggs, typical of the species. Eggs are laid at 1–2 day intervals and incubated mostly by the female for about 4 weeks (Harrison 1975). Both parents feed the young, which fledge in 6–8 weeks. Because of the species's specialized hunting technique, the young may be dependent for 2 months or more after fledging (Sherrod 1983).—*Richard L. Knight.*

Ruffed Grouse
Bonasa umbellus

This mid-sized, cryptically colored bird of forested habitats often escapes detection due to its secretive habits and sporadic territorial behaviors. It is most commonly detected by the dull, muffled drumming sound of its territorial announcement, a sound created by the rapidly accelerating beating of its wings. Ruffed grouse are fairly common permanent residents of the eastern one-third of Tennessee, but within this range its distribution is spotty, correlated with the general abundance of hardwood forests of suitable ages and types.

Ruffed Grouse habitats are predominantly deciduous forests with a mix of age classes and understories. Patches of young forests (5–15 years old) interspersed with older forests are prime habitat. Mature forests with patchy understories of laurel, rhododendron, or blueberries are suitable. Moist sites with Christmas ferns, Japanese honeysuckle, and various other types of succulent green vegetation provide food and cover for mature birds during winter (Stafford and Dimmick 1979). Patches of open areas covered with lush green vegetation produce high insect populations required by chicks during June and early July. Older juveniles and adults feed heavily on blueberries, huckleberries, dogwood berries, and grapes during mid-summer to late autumn.

The aboriginal range of Ruffed Grouse in Tennessee extended diagonally northwest from Lincoln County on the southern border to Henry County on the northern border of the state (Bump et al. 1947). By 1950, the primary range had declined to include the Cumberland Plateau and Mountains, the northern Ridge and Valley, and the Unaka Mountains (Schultz 1953). West of this range, Ruffed Grouse populations were small, isolated, and concentrated in the extreme northern Highland Rim and Central Basin. For the late 1970s, White and Dimmick (1979a) redefined a more restricted range than Shultz (1953), now lacking the westward extension across the northern portion of the Central Basin. Cole and Dimmick

Ruffed Grouse. Chris Myers

(1991) estimated that by 1990 Ruffed Grouse occupied nearly 40,000 sq km in Tennessee (37.4% of the state).

Too few Ruffed Grouse have been found on Tennessee BBS routes to show a trend in its abundance. The Ruffed Grouse is a popular game species in the eastern one-third of Tennessee. More than 21,000 hunters harvested approximately 50,000 birds during the 1989–90 hunting season (Whitehead 1991). Population density, hunter effort, and harvest vary between years, but the trends appear relatively stable.

Beginning in 1975 and continuing sporadically through 1986, wild-trapped Ruffed Grouse were released by Tennessee Wildlife Resources Agency and the University of Tennessee at 4 locations in the northwestern portion of their primordial range on the Western Highland Rim. Birds were released in Benton, Cheatham, and Humphreys Counties (Jones 1979, White and Dimmick 1979b, Gudlin and Dimmick 1984, Kalla and Dimmick 1987). Periodically for several years after their release, Ruffed Grouse, including mature individuals and broods, have been reported to TWRA. Follow-up field reconnaissances to locate drumming males were successful 2 to 4 years post-release in Humphreys and Benton Counties, but self-sustaining populations were not apparent. Additionally, Ruffed Grouse were released in the Kentucky portion of Land Between the Lakes by the Kentucky Department of Fish and Wildlife; subsequently, broods and adults were observed in Stewart County, Tennessee. The Atlas observations on the Western Highland Rim were the result of these restocking efforts.

Ruffed grouse were reported in 82 of the 655 completed priority blocks (13%), substantially less than would be expected on the basis of its occupied range in Tennessee. The outline of its range in the eastern half of the state, as portrayed by Atlas results, corresponds well with the range described by Cole and Dimmick (1991), except that Atlas workers probably missed it in several Ridge and Valley counties. Only 3 miniroutes recorded this species, each at only 1 stop.

Breeding Biology: The low reporting rates for this species result from the timing of its breeding season, the sporadic nature of its territorial display, and the difficulty of access to prime breeding habitat. Territorial display (referred to as "drumming") peaks in mid-April and ceases almost entirely by mid- to late May. Thus, grouse were silent during the period when most Atlas fieldwork and BBS censuses were conducted. Its breeding display usually occurs on high ridges or is associated with dense patches of laurel or rhododendron. Most of the grouse observed during Atlas fieldwork were adults (53% of Atlas records) or broods (29%) along road edges and trails. Only 2 nests were reported.

Female Ruffed Grouse visit the promiscuous, drumming males to be fertilized, and the male takes no further role in nesting. Ruffed Grouse nest on the ground, usually at the base of a tree, near a log, or in a brushpile in mature forests with semi-open understory (Bump et al. 1947). The nest is constructed of grasses and dead tree leaves and lacks a canopy. Egg laying probably peaks about the middle of April. The clutch contains 10–12 eggs, laid at a rate of 2 eggs every 3 days (Johnsgard 1975). The eggs may be milky white to buff and may have small spots or blotches. Incubation lasts 23–24 days. The precocial young leave the nest within a few hours after hatching, and are often aggressively defended by the hen. The chicks begin to fly at about 2 weeks. The brood breaks up at about 12 weeks, and the chicks disperse into new individual home ranges during early autumn. Renesting may occur if the first nest is destroyed (Johnsgard 1973), but there is no evidence that Ruffed Grouse raise 2 broods in a single year. By September, distinguishing between adults and juveniles in the field is not reliable.—*Ralph W. Dimmick.*

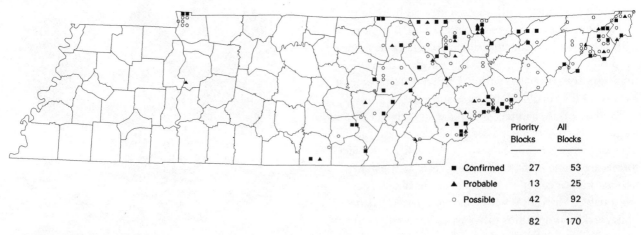

Distribution of the Ruffed Grouse.

Wild Turkey
Meleagris gallapavo

The Wild Turkey, the largest bird nesting in the state, is an uncommon, permanent resident found throughout the state in mature woodlands with scattered openings or adjacent fields. This status represents a major increase in its population and distribution in the last 40 years.

The Wild Turkey was probably common in prehistoric Tennessee, and its remains frequently occur in Indian middens. Schorger (1966) estimated a pre-Columbian population of 334,376 turkeys in what is now Tennessee. It was mentioned in the journals of several early explorers, among them the naturalist Michaux, who observed and possibly collected them near Nashville on 21 June 1795 (Williams 1928). Additional early records are presented by Wright (1915); they suggest the turkey was common and often encountered in flocks of up to 50 birds. The turkey population began declining with the spread of European settlers due to several causes, among them clearing of the forests, unrestricted hunting, diseases carried by domestic birds, and competition with free-ranging livestock for nuts and fruits (Aldrich 1967).

Several late-nineteenth-century ornithologists reported turkeys in Tennessee, although few of them personally observed the birds. Fox (1886) described the turkey as still quite common on the Cumberland Plateau in Roane County. Langdon (1887) described the occurrence, a few days before his visit, of a flock and brood of young in the Chilhowee Mountains of Blount County. Rhoads (1895a:475) stated that the Wild Turkey existed in "considerable numbers" in secluded parts of the Cumberland Plateau and Unaka Mountains, and was reported to him from Walden Ridge in Hamilton County, Fentress County, Roan Mountain, and Haywood County, where it was a "rather scarce bird." Over a decade later Howell (1910:301) described the turkey as occurring in moderate numbers on Walden Ridge near Chattanooga, but "is fast disappearing, as it is shot at all seasons by the residents. Two men told of killing all but one from a bunch of six or seven the day before I arrived [24 August 1908]." Ganier (1933a) described the turkey as very rare in West Tennessee, very rare on the Highland Rim of Middle Tennessee, and rare in East Tennessee, chiefly occurring in the mountains. The die-off of the American chestnut during this period probably hurt turkey populations by eliminating an important fall and winter food source.

Turkey hunting was closed from 1923 to 1929 and from 1941 to 1951 (Holbrook and Lewis 1967). Efforts by the Tennessee Game and Fish Commission and later TWRA to restore the Wild Turkey to its original range began with the release of 3717 game farm birds from 1941 through 1950. This effort was generally unsuccessful. In the early 1950s, the range of the Wild Turkey was restricted to the Unaka Mountains, the Cumberland Mountains and Plateau, portions of the Western Highland Rim, and portions of the Mississippi Alluvial Plain (Schultz 1955). In 1951, a restrictive hunting season on males during the spring was opened, and in 1954 restoration efforts concentrated on transplanting wild birds (TWRA 1990b). Various statewide population estimates were 6000 in 1952 (Mosby 1973), 3200 in 1959, 3700 in 1962 (Lewis 1962), and 5000 in 1968 (Mosby 1973). The drop in the 1950s was due to the loss of several small populations across the state. The modern restoration efforts have been very successful, with turkey populations now present in all 95 counties (TWRA 1994c). The number of turkeys annually harvested by hunters has grown dramatically from 1136 in 1986 to 3421 in 1991 and to 6690, from 86 counties, in 1994 (TWRA 1994c).

Within the Unaka Mountains, broods of young turkeys have been reported as high as 1768 m, and the highest nest was at 1585 m (Stupka 1963). At the time of the formation of the Great Smoky Mountain National Park in 1934, turkeys there were more common above 1067 m than lower, presumably because of greater human predation at lower elevations (Stupka 1963). This situation has since reversed, as turkeys have become concentrated around the few maintained low-elevation open areas.

Until relatively recently, the Wild Turkey was assumed to require forested tracts of several thousand hectares with scattered small openings (Mosby 1973). As shown by its current range, the turkey presently thrives in areas of mixed, smaller forests and farmland.

Wild Turkey. Chris Myers

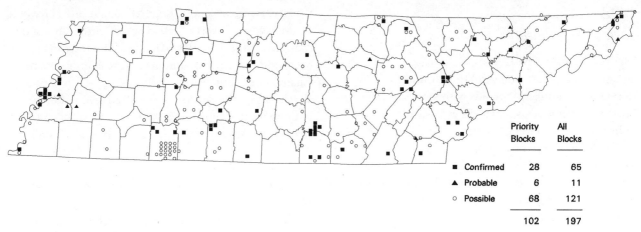

Distribution of the Wild Turkey.

The present population, which includes offspring of birds from other states, may be more adaptable than the original stock. A more likely explanation for this difference, however, is that earlier conceptions of its habitat requirements were based on habitats it occupied early in this century, when, because of human persecution, it only survived in areas of extensive forest.

The Wild Turkey was often difficult for atlasers to observe, and they found it in about 15% of the priority blocks. This low proportion is an underestimate of the bird's occurrence, and the mapped results include several observations by wildlife officers and other sources. The locations of many other observations published in TWRA's annual Wild Turkey reports (e.g., TWRA 1990b) were not accurate enough to include in the Atlas map. From 1986 through 1991, turkeys were legally harvested in at least 67 counties (TWRA 1990b).

Breeding Biology: Breeding activities begin in late winter when winter flocks begin breaking up and males establish territories. The dominant males hold these territories throughout much of the spring and attract hens with their gobbling calls and fan-tailed, droop-winged strutting displays. The dominant males mate with several hens, and the hens incubate the eggs and care for the poults. In East Tennessee, gobbling, and presumably the initiation of egg laying, peak during the last 2 weeks of April and first week of May (Holbrook and Lewis 1967). The peak of hatching is around the first of June (TWRA 1990b).

Turkey hens lay their eggs in a leaf- or grass-lined depression on the ground, usually near the base of a tree or bush. Nests are in brushy open areas or within the forest and usually in areas of thick, low vegetation (McGuiness 1989). Clutches contain from 7 to 14 eggs and average 10.79 eggs (s.d. = 2.15, n = 14). Following loss of a clutch, the hen usually lays a smaller replacement set; McGuiness (1989) found an average of 11.3 eggs in first clutches (n = 9) and 9.0 in second clutches (n = 2). Incubation lasts about 28 days, and the precocial young leave the nest soon after hatching. The young begin to fly when 6 to 10 days old and fly strongly in 3 weeks (Bailey and Rinell 1967). Family groups remain together into the fall.—*Charles P. Nicholson.*

Northern Bobwhite
Colinus virginianus

This small, gallinaceous bird is often identified by its clear 2- or 3-noted "bob-white" or "ah-bob-white" territorial call during spring and summer. It is popularly known as the bobwhite or bobwhite quail and is an abundant, statewide, permanent resident. It occurs at almost all elevations in Tennessee, though it is not common in forested habitats at the highest elevations of East Tennessee.

Northern Bobwhites have relatively unspecialized habitat requirements. They populate farmlands, particularly where grain crops such as soybeans, corn, and wheat are grown. They also occupy pinelands and deciduous forests. In Tennessee, they reach highest abundance where croplands, wooded fencerows, and idle lands dominated by broom sedge are interspersed in a mosaic fashion. This type of habitat is most common in West Tennessee and scattered parts of Middle Tennessee. In the Unaka Mountains, they occur at low densities on high-elevation balds (Stupka 1963, Atlas results); these populations probably migrate downslope to winter at lower elevations. Stupka (1963) reported a brood at 1768 m on Andrews Bald, North Carolina.

In pre-Columbian Tennessee, Northern Bobwhites probably were scattered and relatively low in abundance. The bird is typically associated with early successional plant communities. Consequently, when the Tennessee landscape was largely mature deciduous forest land, bobwhites likely populated such habitats as remnant prairies

Northern Bobwhite. Chris Myers

and patches of forest land recently ravaged by fires and/or severe windstorms. American Indians frequently burned forests and fields to improve habitat for wild grazing mammals; those fires benefited bobwhites as well. With the advent of settlement and small patch farms, bobwhites reached their peak of abundance. Rhoads (1895a) described the bobwhite as more abundant in West and Middle Tennessee than he had observed elsewhere in the United States. Bobwhite populations probably reached their peak shortly after Rhoads's report.

Northern Bobwhite populations have declined throughout much of this century. Reforestation has eliminated some habitat as have modernized agricultural practices, which have resulted in larger, cleaner fields with reduced available food and cover for bobwhites. Also, the transition from grain production to livestock production in many areas produced landscapes dominated by fescue pastures and hayfields. This particular type of grass yields little food and unsuitable nesting habitat for bobwhites, and is believed to be partly responsible for the long-term downward slide of the species's abundance in Tennessee. BBS route results show a significant annual decline of 3.0% ($p < 0.01$) since 1966. Between 1980 and 1994, the rate of annual decline increased to 4.3% ($p < 0.01$). Bobwhites have been found on all 42 BBS routes in Tennessee, at average densities of 22.4 birds/route and 16.9 stops/route during the Atlas period.

Bobwhites are a popular game species in Tennessee and are pursued by more than 80,000 hunters every year. During 1970, hunters harvested an estimated 1.7 million birds in Tennessee (Johnsgard 1973), but the long-term downward trend in the statewide population has been paralleled by a similar decline in harvest. The harvest in 1989–90 was estimated to be about 825,000 birds (Whitehead 1991).

Northern Bobwhite density during December ranged from 1.1 to 3.7 birds/ha over a span of 22 years on 850 ha of farmland habitat in southwest Tennessee (Dimmick 1992). At the initiation of the breeding season, population densities on this same area ranged from 1.6 to 3.0 birds/ha during 1967–74 (Dimmick 1974). However, winter densities as low as 1 bird/40 or 50 ha predominate in many landscapes in Tennessee.

Atlasers reported Northern Bobwhites in 95% of the completed priority blocks. This high reporting rate accurately reflects its abundance and widespread distribution; its persistent, highly audible calling during May through July insures its detection by atlasers wherever it occurs. Northern Bobwhites were recorded on 88% of the mini-routes at an average of 5.7 stops/route, slightly higher than their abundance on BBS routes during the same period. Regions of highest abundance were the Loess Plain, Coastal Plain Uplands, Mississippi Alluvial Plain, Central Basin, and Eastern Highland Rim. Lowest numbers were in the Cumberland and Unaka Mountains.

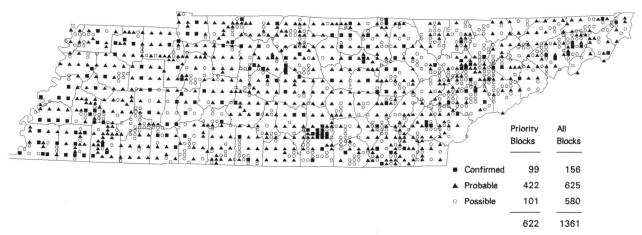

	Priority Blocks	All Blocks
■ Confirmed	99	156
▲ Probable	422	625
○ Possible	101	580
	622	1361

Distribution of the Northern Bobwhite.

Confirmed Breeding Species

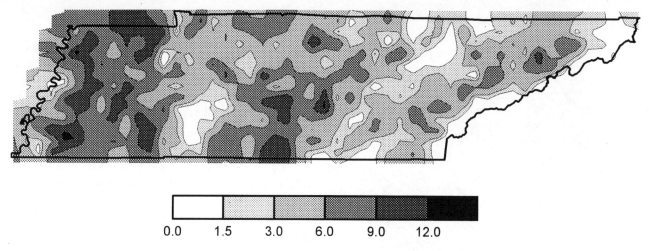

Abundance of the Northern Bobwhite.

Breeding Biology: Northern Bobwhites characteristically are monogamous and breed during their first year of life. Nests are constructed on the ground in shallow, saucer-shaped depressions that are formed with the feet and beak (Stoddard 1931). Either or both sexes may participate in nest construction. Stems and leaves of dead grass are woven into a hollow, covered structure about the size of a softball. Usually the canopy is complete, with one end open for access. Broom sedge typifies nesting habitat in Tennessee.

Clutch sizes ranging from 5 to 24 were observed during a 7-year nesting study in Fayette County, Tennessee. Mean clutch size for 233 clutches was 11.9 eggs (Dimmick 1974). The eggs are white, blunt at one end, and pointed at the other. The female typically incubates the eggs, but Klimstra and Roseberry (1975) observed males incubating eggs at 28 (26.4%) of 106 nests. In West Tennessee, about 15% of clutches were tended partially or entirely by males (Dimmick, unpublished data). Incubation lasts 23 days.

Northern Bobwhites typically rear only a single brood in a nesting season. Second or third nesting attempts may occur following destruction of prior nests, but the extent of renesting is not well quantified. Stanford (1972) documented cases of females laying, incubating, and hatching 2 sequential clutches in a single nesting season, but this is probably not common and has not been verified in Tennessee. Nest success rates are low; Dimmick (1974) reported that 23% of 761 nests in West Tennessee produced chicks. Similar rates were observed elsewhere (Simpson 1972, Klimstra and Roseberry 1975). Both parents participate in rearing the precocial chicks, and are very protective of them until they are several weeks old.

In Tennessee, Northern Bobwhites are mostly paired by late April, but territorial whistling of males continues well into summer, peaking in intensity as late as the latter part of June through July (Saunders 1973). Nest building, egg laying, and incubation are most intensive in May, June, July, and August (Dimmick 1971), and both mated and unmated males continue to whistle. In most years, the proportion of bobwhite chicks hatched after the end of August is insignificant. Bobwhites are capable of limited flight by 2 or more weeks of age. They remain with the parents after fledging. The fledged brood typically forms the nucleus of a covey of 12–15 birds, which is the social unit for bobwhites during the non-breeding season.—*Ralph W. Dimmick.*

King Rail
Rallus elegans

The King Rail is a rare summer resident of Tennessee, normally present between early April and early October (Robinson 1990). A few winter records also exist. The King Rail inhabits marshes, where it stalks among the vegetation, feeding on crayfish and aquatic insects. It is most easily observed in the spring, when advertising and courtship calls are frequently given during both day and night.

The King Rail was first reported in Tennessee by Rhoads (1895a), who observed one in Chattanooga in late May or early June, 1895, at which time it was probably nesting in the area. The first dated nest records, both in West Tennessee, were in 1921, from Reelfoot Lake and Carroll County (Ganier 1933b, Ganier egg coll.). It was first reported nesting in Middle Tennessee in Robertson County in 1934 (*Migrant* 5:28–29), and in East Tennessee in Cocke County prior to 1935 (Walker 1935). During the late 1930s and 1940s, confirmed and probable nesting was reported from several additional counties in Middle and West Tennessee. The number of known extant nesting localities peaked during this period.

During the Atlas period, the King Rail was reported from 3 sites in 3 counties in the western half of the state. It was also reported as possibly breeding at Goose Pond

King Rail. Elizabeth S. Chastain

in Grundy County, where nests were frequently reported in the past. It was not reported from 2 sites in the eastern half of the state where it was most consistently reported during the 1960s and 1970s: Alcoa Marsh in Blount County and Amnicola Marsh in Hamilton County. Each of these sites received some Atlas fieldwork, although not enough to determine definitively the status of the King Rail. Its present status at Reelfoot Lake is also poorly known (Pitts 1985). Some other sites where it has nested in the past, such as Buena Vista Marsh in Davidson County (Parmer 1985), have been drained or otherwise modified to the point where they no longer provide suitable rail nesting habitat. Other than the anecdotal sightings and nest records of the King Rail published primarily in the *Migrant,* no information is available on population trends of the King Rail in Tennessee.

The lack of population trend information is also true elsewhere in the King Rail's range (Eddleman et al. 1988). Concern over its numbers has resulted in it being listed as a species of special concern on the Audubon Blue List since 1976 (Tate 1986). It was, until recently, legally hunted in Tennessee; the very small harvest, however, probably had little effect on the size of the breeding population. Drainage and other alterations of Tennessee marshes has probably caused a population decline. Current wetland protection programs in Tennessee should slow the loss of rail habitat, although to date these programs have focused more on forested wetlands than more ephemeral marshes. The King Rail was justifiably listed as In Need of Management in 1994 (TWRA 1994a).

Breeding Biology:
The King Rail nests in marshes with shallow water, typically less than 25 cm, and grasslike emergent vegetation such as rushes, sedges, cattails, and grasses (Eddleman et al. 1988, Meanley 1992). DeVore (1968a) found nests at Amnicola Marsh in areas of dense vegetation dominated by great bulrush as well as in areas of thin vegetation dominated by pondweed. The minimum marsh area in which the King Rail can successfully reproduce is poorly known. Descriptions of Tennessee nests occasionally mention their occurrence in small marshes (e.g., Howell and Monroe 1957), and Williams (1975b) found single nesting pairs in 0.8 and 1.2 ha marshes, not all of which was optimal rail habitat. Most foraging is in water less than 10 cm deep, and a decreasing water level, such as from natural drying, improves brood foraging (Eddleman et al. 1988, Meanley 1992). The recent shift in waterfowl management on National Wildlife Refuges from rowcrops to natural vegetation in shallow impoundments is probably beneficial for the King Rail, particularly when the impoundments hold water during the summer. Half the Atlas observations were in shallow, managed impoundments on refuges.

The King Rail is monogamous, and pairs often return to the nest site used the previous year (Williams

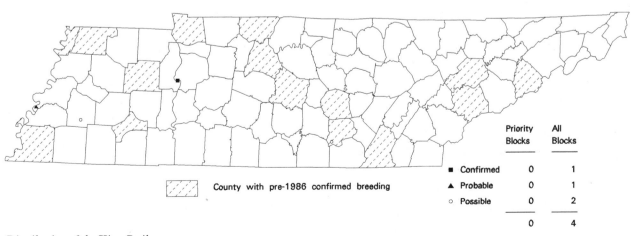

Distribution of the King Rail.

1975b, Meanley 1992). The male's courtship displays include walking with its tail uplifted exposing the white undertail feathers, flicking its tail up and down, and following the female while stretching his head and neck upward with the bill open. The male also builds incomplete nests and feeds the female (Meanley 1992). Further, the male constructs most of the nest in which eggs are laid. It is placed in a clump of vegetation or between adjacent clumps and is constructed with a base of wet, decaying vegetation and a rounded platform or bowl of dry, dead vegetation, gathered near the nest. A canopy is formed over the nest by bending stems of adjacent plants over the nest, and a ramp is added below the front of the nest. All 5 nests DeVore (1968a) found at Amnicola Marsh were in clumps of giant bulrush and were built mostly with bulrush. Williams (1975b) reported nests at Alcoa Marsh were most often in bulrush, cattails, and sweet flag; a single nest was in a buttonbush. The heights of these nests above water varied from 6 to 51 cm, and DeVore (1968a) found an average height of 9 cm. The lowest nests in DeVore's study were flooded by a rise in the water level.

Egg laying begins as early as the second half of March (Williams 1975b), although most Tennessee clutches were begun between mid-April and mid-May. The King Rail has an unusually large clutch compared to other rails and to other semiprecocial species (Meanley 1992). Apparently completed clutches in Tennessee contained 7 to 13 eggs and averaged 10.1 eggs (s.d. = 2.02, n = 25). Meanley (1992) gave average clutch sizes of 10.5 to 11.2 eggs in studies elsewhere and noted that early clutches are larger than late clutches. The eggs are pale buff colored, with irregular small spots of various shades of brown.

Both adults incubate the eggs for 21 to 23 days (Meanley 1992). The newly hatched semiprecocial young are covered with black down and quickly leave the nest. For the first month, they get most of their food from the adults and, by 2 months of age, are mostly feeding themselves. The brood stays together for 9 weeks or more (Meanley 1992).—*Charles P. Nicholson.*

Virginia Rail
Rallus limicola

The Virginia Rail is a permanent resident of Tennessee, uncommon during migration and rare during the summer and winter (Robinson 1990). It inhabits marshes where it inconspicuously stalks among the vegetation and, particularly during the spring, is as likely to be heard as seen. Its vocalizations, unfamiliar to most Tennessee bird-watchers,

Virginia Rail. Elizabeth S. Chastain

include a distinctive "kid-kid-kidic-kidic" song. Once seen, it is easily identified by its small size and long bill.

The first Tennessee nest record of the Virginia Rail was at Amnicola Marsh, Hamilton County, in 1963 (West 1963). In 1974, a single nest was reported at Alcoa Marsh, Blount County (Williams 1975b), and 3 nests were reported at a small Hawkins County marsh (Smith et al. 1975). Nesting was reported at Meadowview Marsh in Sullivan County in 1976 (*Migrant* 47:77) and 1987 and 1988 (Phillips 1989). Single pairs with young were observed at Monsanto Ponds, Maury County, in 1984 (Lochridge and Lochridge 1984) and 1985 (*Migrant* 56:110). During Atlas fieldwork, the only Virginia Rail observation, other than at Meadowview Marsh, was of a single bird on 25 May 1990 in a marshy area along the Watauga River in Washington County.

Prior to the first Tennessee nesting, the only nest record of the Virginia Rail south of Tennessee was from northern Alabama in 1945 (Imhof 1976). Nesting was reported in Louisiana in 1969 (*Aud. Field Notes* 23:668) and in Georgia in 1970 (*Aud. Field Notes* 24:672). These nest records, along with the Tennessee records, may represent an expansion of the Virginia Rail's nesting range to the southeast. Nesting in Tennessee, and elsewhere in the Southeast, could easily have been overlooked and may be more common than the few records indicate. The dates of the few Tennessee nest records indicate nesting activity begins in March, when migrating rails are present. The Tennessee literature contains many records of Virginia Rails present during May, and at least two July records, reported without evidence of nesting.

No information is available on population trends of either nesting, migrant, or wintering Virginia Rails in

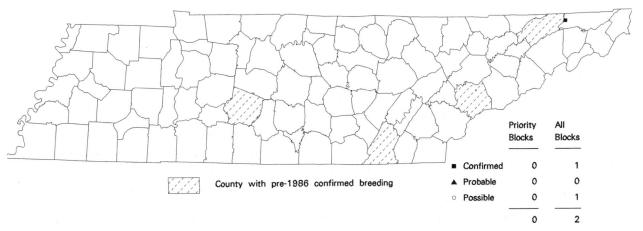

Distribution of the Virginia Rail.

Tennessee, as is the case elsewhere in the species's range (Zimmerman 1977). Much rail habitat, however, has been lost through destruction of wetlands in Tennessee and elsewhere (Eddleman et al. 1988). The Virginia Rail is legally hunted in Tennessee during the waterfowl season, and the daily bag limit is large. The size of the harvest, although not well known, is probably very small and unlikely to have greatly affected its local breeding population. Systematic surveys of potential breeding habitat by observers familiar with its calls and using broadcast tape recordings are necessary to better define its nesting distribution and population trends. Until such surveys show the Virginia Rail to be common and until the loss of marshland is reversed, its Tennessee breeding population should be listed as Threatened.

Breeding Biology: The Virginia Rail nests in marshes with emergent vegetation such as cattails, sedges, and rushes. A mixture of robust vegetation, such as cattails, with finer sedges or rushes is probably an important habitat component, as is a high proportion of vegetation to water edge (Johnson and Dinsmore 1986, Kaufmann 1989, Zimmerman 1977). It often occurs in relatively small marshes, with marshes as small as 0.2 ha supporting several pairs (Zimmerman 1977).

Tennessee marshes where the Virginia Rail has nested have been from 3.2 ha to 8.1 ha in size (Smith, Lewis, and Alsop 1975, Phillips 1989). The dominant vegetation was rushes at Amnicola and sedges and grasses at the Hawkins County marsh (West 1963, Smith, Lewis, and Alsop 1975). The Monsanto site and the Alcoa and Meadowview Marshes contained a variety of vegetation types, including rushes, cattails, sedges, grasses, and shrubs (Williams 1975b, Phillips 1989, Nicholson pers. obs.). The nest at Amnicola was built in a clump of common rush, and the nest at Alcoa in a clump of cattails (West 1963, Williams 1975b). A nest at the Hawkins County site was in a clump of bulrush and sedge, and 3 of the Meadowview nests were in rushes and 1 in cattails (Smith, Lewis, and Alsop 1975, Phillips 1989).

The Virginia Rail begins breeding activities in Tennessee in March, when migrating rails are present. Pairs defend their territories by calling and chasing other rails (Kaufmann 1989). Courtship activities include the birds preening and chasing each other and the male feeding the female. Nest construction begins about the time the female lays the first egg and is continued during egg laying (Kaufmann 1989). Both birds construct the bowl-shaped nest from vegetation gathered nearby. Tennessee nests have been built of cattails, rushes, grasses, and sedges. A canopy is often built over the nest, although this has not been mentioned in the descriptions of Tennessee nests. Tennessee nests range in height from 9 to 25 cm above water.

Egg laying begins as early as late March. Two clutches of 6 eggs, 3 of 7 eggs, and 1 of 9 eggs have been found in Tennessee. Because at least 1 of these clutches was probably abandoned when found, some may have been incomplete. Clutch sizes of 4 to 13 eggs, with an average of 8.5, have been reported elsewhere in the rail's range (Kaufmann 1989). The eggs are pale buff to white with sparse, irregular brown spots (Harrison 1975). Both adults incubate the eggs, beginning before the clutch is complete. Incubation lasts 18 to 20 days, and the eggs hatch over a 2-day period (Kaufmann 1989). The young are covered with down at hatching, and first leave the nest when 2 to 3 days old. The adults brood the young on specially built brood nests, constructed by the male near the original nest. Although the young begin to feed themselves when a few days old, they are fed by both adults until at least 3 weeks old (Kaufmann 1989). The Virginia Rail is single-brooded.—*Charles P. Nicholson.*

Sora
Porzana carolina

The Sora is a fairly common migrant, a rare winter resident, and a very rare summer resident of Tennessee (Robinson 1990). During its spring and fall migration, it is probably the most numerous rail in the state. Most spring migrants are present from early April through mid-May, and fall migration is from late August through early November. The Sora is usually found in marshes, although during migration it occasionally occurs along lakeshores and in wet fields, including winter wheatfields. Like other rails, it is usually difficult to see as it stalks through the vegetation.

The breeding range of the Sora extends southward to southern Missouri, central Illinois, central Indiana, central Ohio, and West Virginia (AOU 1983). The first and only report of its nesting in Tennessee was in 1990, when an adult and 2 young were observed from 17 April to 8 June in southwest Shelby County (Waldron 1990). The brood occupied a shallow emergent wetland dominated by grasses and sedges.

The Sora nests in marshes dominated by emergent vegetation such as cattails, sedges, and rushes. Its distribution and habitat use overlap greatly with those of the Virginia Rail (Johnson and Dinsmore 1986, Kaufmann 1989). The Sora makes much use of edges between water and vegetation, and between different vegetation types. It readily swims, and it feeds mostly on seeds that it picks from vegetation mats. During both the nesting and brood-rearing periods, the Sora shows little preference for particular species of emergent vegetation (Johnson and Dinsmore 1985).

Very little information on population trends of the Sora in Tennessee is available. Like other marsh-dwelling species, it has lost habitat through drainage of wetlands. It is legally hunted in Tennessee, and the bag limit is generous. Although no good measure of the size of the harvest is available, it is probably very small and much less of a threat to the species's population than is the loss of suitable habitat.

Tennessee ornithologists should be aware of the possibility of the Sora nesting elsewhere in the state and report in detail any suspected nesting activity. Only through such observations will we determine whether the Soras nesting in Shelby County were vagrants or part of a population regularly nesting in Tennessee.

Breeding Biology: Soras are monogamous, and they vigorously defend their territories against other rails (Kaufmann 1989). The female begins nest building about the time the first egg is laid, and both birds add material to the nest during egg laying. The nest, built of cattail, sedge, or other vegetation, is a well-made, inconspicuous, cup-shaped basket placed low in a clump

Sora. David Vogt

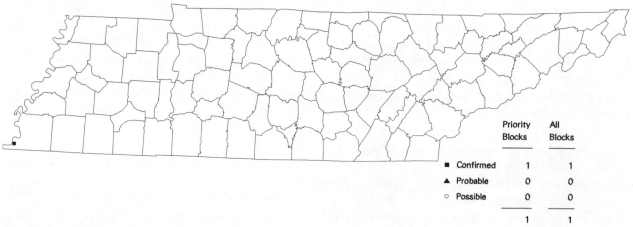

Distribution of the Sora.

of vegetation. A canopy is usually built over the nest from the surrounding vegetation. Clutch size ranges from 6 to 16 eggs, with an average of 10.5 eggs (Kaufmann 1989). The clutch of the sole Tennessee nesting was probably completed during the second half of March, when numerous migrating Soras are present in the state. The eggs are a rich buff color with brown spotting, and are incubated for about 18 days (Harrison 1975). Incubation, carried out by both adults, begins before the clutch is complete, resulting in the eggs hatching over a period of several days. The young are feathered at hatching and leave the nest 2 or 3 days later. They are frequently brooded in special brood nests, which resemble the original nest. They are fed by their parents for 2 to 3 weeks (Kaufmann 1989).—*Charles P. Nicholson.*

Purple Gallinule
Porphyrula martinica

The Purple Gallinule is a southern and tropical species whose breeding range extends northward to northwest Tennessee and isolated locations in Illinois and Ohio (Ripley 1977, AOU 1983); in Tennessee it is a rare migrant and a very rare summer resident (Robinson 1990). Like the closely related Common Moorhen, the Purple Gallinule typically occurs in marshes. It swims readily, often walks on floating plants, and climbs about in shrubs and low trees over water. It is, however, less frequently reported in Tennessee than the moorhen; most records are from the spring and early summer (Robinson 1990).

The Purple Gallinule was first reported nesting in Tennessee by Ganier (1933b). On 17 June 1923 he found a nest with 1 egg as well as incomplete nests in a small colony in the Brewer's Bar area of Reelfoot Lake.

Purple Gallinule. Elizabeth S. Chastain

Least Bitterns, Common Moorhens, and American Coots were nesting in the same area. Although Purple Gallinules were periodically reported from Reelfoot Lake during the next several decades, mostly during early May, the next confirmed nesting there was apparently not until 21–23 June 1984, when at least 7 adults and a nest with 5 eggs were observed (Pitts 1985, Nicholson pers. obs.). Most of the gallinules and the nest were near the south end of Brewer's Bar, in the same general area as Ganier's earlier observation.

The first breeding record east of Reelfoot was at Goose Pond, Grundy County, where Ganier and others found a pair and a nest with 6 slightly incubated eggs in 1935 (Ganier 1935b, 1952). Purple Gallinules were again nesting at Goose Pond on 1 June 1964, when an adult was flushed from 7 eggs (Dubke 1974). The only other location where breeding has been confirmed is at a marsh-bordered small lake in northeast Warren County, where Ganier (1952) found a pair and empty nest on 30 July 1952. Purple Gallinules have also been reported between late May and late July, without confirmed evidence of breeding, from Maury (*Migrant* 8:21), Hamilton (*Migrant* 38:67), Hamblen (*Migrant* 39:65), and Washington (*Migrant* 44:86) Counties.

During Atlas fieldwork, Purple Gallinules were reported from 3 blocks; none of the records were of confirmed breeding. One of the records was from the Brewer's Bar area of Reelfoot Lake. Another record, from Bear Creek WMA, Stewart County, was of a bird present from 18 May to 1 June 1986 (*Migrant* 57:78). The third record was of a bird observed on 4 June 1988 in a shallow lake at Hatchie NWR, Haywood County. Several later attempts to find this bird were unsuccessful (R. P. Ford pers. comm.).

The Purple Gallinule has apparently always been a rare bird in Tennessee, and the only site where nesting probably occurs regularly is Reelfoot Lake. The decreased area of emergent vegetation there (Henson 1990) has probably resulted in a loss of suitable gallinule habitat. Purple Gallinules are legally hunted during the Tennessee waterfowl season; although no information is available on the size of the harvest, it is probably insignificant as most have probably migrated south by the time the season opens. The greatest threat to the Purple Gallinule is the loss of marshlands, and because of this threat it should be listed as In Need of Management.

Breeding Biology: The Purple Gallinule nests in marshes with abundant emergent vegetation. Its nesting habitat requirements are similar to those of the Common Moorhen, and the 2 species are often sympatric (Helm, Pashley, and Zwank 1987), as at Reelfoot Lake (Ganier 1933b,

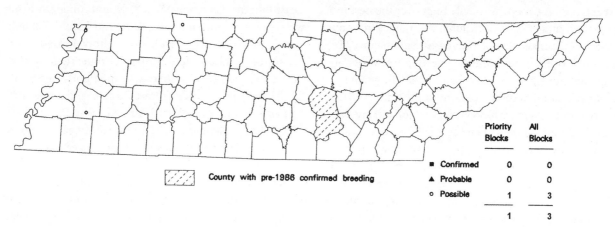

Distribution of the Purple Gallinule.

Nicholson pers. obs.). Both adults build the nests, as well as trial nests and nestlike brood platforms. Nests are either floating structures placed on low marsh vegetation or suspended from upright vegetation some distance above the water (Helm, Pashley, and Zwank 1987); both types have been found in Tennessee. The nests described by Ganier (1933b) at Reelfoot were suspended from tall grass (probably giant cutgrass) about 0.3 m above meter-deep water and were formed of grass blades bent over and woven together. The 1935 Goose Pond nest was similarly built but about 0.4 m above water, and the 1952 Warren County nest was in cattails 0.75 m above water (Ganier 1935b, 1952). The 1984 Reelfoot nest, built of dead cattail leaves and green knotweed leaves, was floating in a bed of knotweed about 10 m wide between an open boat channel and a thicker area of cattail and shrubs (Nicholson pers. obs.). Its shallow bowl was about 5 cm above the water.

The few nest records show that egg laying occurs from mid-May through mid-June. Apparently complete clutches have been of 5 and 6 eggs. Clutches reported elsewhere averaged 6.4 eggs and range in size from 3 to 13 eggs (Helm, Pashley, and Zwank 1987). Incubation, performed by both adults, begins before the clutch is completed and lasts about 22 days. The young are covered with down at hatching and fed by the adults for a few weeks. They return to the original nest or a similarly constructed brood nest to be brooded until they are about 3 weeks old (Ripley 1977). Second broods have been reported elsewhere.—*Charles P. Nicholson.*

Common Moorhen
Gallinula chloropus

The Common Moorhen, formerly known as the Florida or Common Gallinule, is a rare migrant, a rare summer resident, and a very rare winter visitor (Robinson 1990).

Its main migration periods are from late April to mid-May and from late September to early October. The Common Moorhen typically occurs in marshes and ponds with areas of open water and abundant aquatic vegetation. It swims readily and is easily recognized by the red shield on its forehead, its blackish plumage, and white-striped flanks.

The earliest mention of the Common Moorhen in Tennessee was by Miles (in Rhoads 1895a), who considered it to breed in the Haywood County area and noted that it was less common than it had been in the 1870s. The first confirmed record of its nesting was on 17 June 1923 (contra Robinson 1990) at Reelfoot Lake, where Ganier (1933b) found several occupied nests, including 1 from which he collected a set of 8 fresh eggs. The moorhens were nesting in a colony with Least Bitterns, Purple Gallinules, and American Coots in tall grass (probably giant cutgrass) in water about 1 m deep near where Bayou du Chien crosses Brewer's

Common Moorhen. Elizabeth S. Chastain

Bar, Obion County. Ganier (1933b) also found about 25 moorhens and several incomplete nests on 30 May 1932 in an extensive grassy flat along the Lake/Obion County line northwest of Samburg. Breeding at Reelfoot Lake was later observed in 1936 by Whittemore (1937) and in 1941 by Pickering (1941) and Spofford (1941). Mengel (1965) collected a female containing enlarged ova at Reelfoot on 27 May 1949.

Although a few Common Moorhens were reported from Reelfoot Lake during late spring and early summer over the next several decades (*Migrant*), the next confirmed breeding record there was apparently not until 1984 (Pitts 1985, Nicholson pers. obs.). This gap in breeding records is probably due to infrequent visits by ornithologists to the few areas of suitable nest habitat, which are only accessible by boat (Ganier 1933b, Nicholson pers. obs.). On 21–22 June 1984, about 13 moorhens, including a brood and a pair building either a nest or brood platform, were observed at Reelfoot; most were in the Brewer's Bar area where Ganier (1933b) reported them. Another nest with eggs was found there during Atlas fieldwork on 31 May 1988.

The first breeding record outside of Reelfoot Lake was of a nest with 7 eggs and 1 downy young in a small marsh at Powell, Knox County, on 26 May 1970 (Williams 1975a). At the Monsanto Ponds, Maury County, 1 or 2 Common Moorhens were reported in early May during several years in the 1970s and 1980s; breeding was confirmed there during the Atlas period in June 1986, when a pair of adults was observed giving a distraction display near a raccoon (*Migrant* 57:107, Atlas results). Reports of adult moorhens observed between late May and early July exist from a few counties across the state (Robinson 1990, *Migrant*); some of these birds may have been nesting.

The Common Moorhen was reported from 6 Atlas blocks and confirmed in 4 of those blocks. The confirmed records were observations of a nest and fledglings at Reelfoot Lake, the pair performing the distraction display in Maury County mentioned above, and a brood in Meigs County on 7 August 1988. The possible records were of a moorhen observed at Cross Creeks NWR on 31 May 1986 and of a bird present at least a week in June 1986 at a pond in a Blount County pasture. This bird disappeared after cattle destroyed the grass and cattails bordering the pond.

Although the Common Moorhen breeds throughout much of eastern and southwestern North America and is legally hunted in many states including Tennessee, little is known of its population trends (Strohmeyer 1977). Because of the widespread loss of its preferred wetland habitats, its population has probably decreased in recent decades. The population at Reelfoot Lake appears reduced from that described by Ganier (1933b), a change probably due in part to the replacement of giant cutgrass with less suitable woody vegetation since water levels were stabilized in the 1940s (Henson 1990). The Knox County marsh where nesting occurred in 1970 has been reduced in area by filling for highway and commercial development. The Common Moorhen should be listed as In Need of Management in Tennessee.

Breeding Biology: Throughout its range, the Common Moorhen nests in marshes dominated by several different types of emergent vegetation (Strohmeyer 1977). In Tennessee, it has nested at Reelfoot in giant cutgrass (Ganier 1933b) and more recently in knotweed and yellow lotus (Nicholson pers. obs.), and in Knox County in blue flag and grasses (Alsop 1970, Williams 1975a). The Maury County site was dominated by cattails, and the Meigs County brood was observed in a shallow embayment of Chickamauga Reservoir choked with submerged and emergent vegetation (Atlas results).

Both adults participate in building the nest, which is usually placed low in emergent aquatic vegetation near

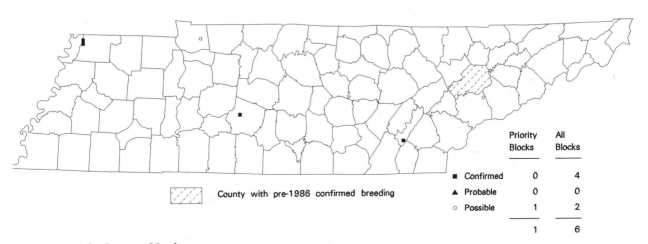

Distribution of the Common Moorhen.

open water (Strohmeyer 1977, Helm, Pashley, and Zwank 1987). The nest is a platform built of dead vegetation collected near the nest site. The Knox County nest was constructed of an aquatic grass (Alsop 1970) and recent Reelfoot nests were constructed of knotweed roots, stems and leaves (Nicholson pers. obs.). Nest construction probably peaks in late May, although it begins as early as late April. In addition to the nest where eggs are laid, the Common Moorhen also builds trial nests and nestlike brood platforms (Helm, Pashley, and Zwank 1987).

Two clutches of 8 eggs have been reported in Tennessee. Clutch sizes elsewhere range from 4 to 17 eggs and average about 9 eggs (Helm, Pashley, and Zwank 1987). Both adults incubate for about 3 weeks, beginning before the clutch is completed. The young are down covered at hatching and fed and brooded by their parents, often on the nest, a brood platform, or muskrat lodge for a few weeks after hatching. The young reach full size in 6 to 7 weeks (Fredrickson 1971); the age at which they become independent is poorly known. Populations elsewhere are sometimes double-brooded; too little information is available to determine the frequency of second broods in Tennessee.—*Charles P. Nicholson.*

American Coot
Fulica americana

The American Coot is a common migrant, a common winter resident, and a rare summer resident (Robinson 1990). Migrant and wintering birds are present from September through early May on lakes, ponds, and reservoirs. Summering birds occur in these habitats as well as in marshes. Although superficially resembling a duck, the American Coot is closely related to rails and gallinules and is the most aquatic member of that family. It is by far the most abundant member of that family occurring in Tennessee.

The first reference to the breeding of the American Coot in Tennessee was by Rhoads (1895a), who described it as breeding at Reelfoot Lake. The first documented nest record was also at Reelfoot Lake, where Ganier described it as a common summer resident on parts of the lake and found at least 4 nests with eggs on 17 June 1923 (Ganier 1933b). On 30 May 1932 Ganier (1933b) observed about 75 coots and several unfinished nests. The most recent Reelfoot nest record was in 1984 (Pitts 1985).

The few confirmed breeding records in other Tennessee locations have been from a variety of habitats. An adult with 5 young was observed in a Davidson County urban park on 11 May 1947 (*Migrant* 18:25). Records from large lakes include an adult with 8 half-grown young on Nickajack Reservoir, Marion County, on the unusually early date of 27 April 1971 (*Migrant* 42:46), an adult with 3 young on 26 June 1971 in Benton County (*Migrant* 42:68), a nest with eggs on Chickamauga Reservoir, Hamilton County, in 1982 (K. Dubke pers. comm.), and an adult with young in June or July 1984 on Old Hickory Reservoir, Sumner County (*Migrant* 55:91). Williams (1975b) found a nest with 4 eggs, disrupted by a muskrat, at Alcoa Marsh, Blount County, on 9 May 1973. Coots have been frequently reported in small numbers during the summer, without evidence of breeding, from lakes across the state for many years (*Migrant*). Some of these birds were probably injured during the previous waterfowl hunting season and unable to migrate; others have appeared uninjured. Some have also joined flocks of domestic ducks and readily accepted handouts from people (Nicholson pers. obs.). One member of the pair nesting in Hamilton County in 1982 was unable to fly, and the unusually tame pair nested on a docked boat (K. Dubke pers. comm.).

During Atlas fieldwork, American Coots were reported as possibly breeding in 14 blocks. Half of these records were in marshes and swamps of northwest Tennessee. The rest were from reservoirs in Middle and East Tennessee except for the Campbell County record, which was from a pond with several tame ducks at Indian Mountain SP. Only 1 bird was reported on a single miniroute.

The recent records of summering and nesting coots suggest that the nesting population at Reelfoot Lake has declined since the 1930s (Pitts 1985). The reasons for this are not clear, although it may be due to the aquatic vegetation changes since the 1930s (Henson 1990). Elsewhere in the state, summering coots have sporadically occurred without showing any population trends. The

American Coot. Elizabeth S. Chastain

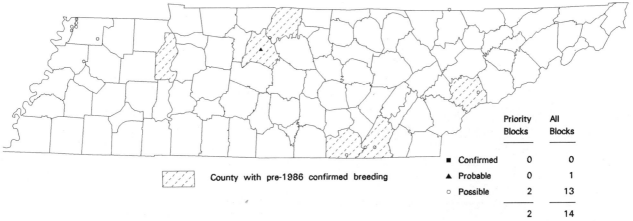

Distribution of the American Coot.

wintering coot population has shown an increase on Christmas Bird Counts since 1961 in West Tennessee and since 1971 in Middle and East Tennessee (Tanner 1985). At least some of this increase is due to the increase of Eurasian watermilfoil, an important food item, in the large reservoirs. The continental breeding population fluctuates with water conditions on the main northern prairie breeding areas, and like ducks nesting in that area, declined between 1955 and 1971 (Fredrickson et al. 1977). From 1966 to 1993, coot populations, as measured by rangewide BBS routes, showed no significant trend (Peterjohn, Sauer, and Link 1994).

The future of "wild" breeding populations of the American Coot in Tennessee is uncertain. The breeding population at Reelfoot Lake, once the largest in the state, has greatly decreased. Although reservoirs have provided abundant wintering habitat, they appear to be of little value to nesting coots, even when marsh vegetation lines the shore. Hunting probably has little or no effect on the local breeding population, whose preservation will require the management of large areas of wetlands with a mix of emergent vegetation and open water. This would also benefit the Least Bittern, Common Moorhen, and Purple Gallinule.

Breeding Biology: At Reelfoot Lake, American Coots nested, along with Common Moorhens and Least Bitterns, in areas of tall grass in water 0.6 to 1.8 m deep (Ganier 1933b, egg coll. data). The coot nests were partly floating platforms anchored in grass and a few cm above the water. The nest at Alcoa Marsh was made of cattails and built in a small stand of cattails in water about 30 cm deep (Williams 1975b). These nests are similar to coot nests elsewhere (Fredrickson et al. 1977). Both sexes build the nest as well as similar structures used for brooding the young.

Although the earliest record of a nest with eggs is on 9 May, the eggs of the brood of young observed on 27 April were laid by mid-March. Clutches of "wild" coots contained from 1 to 6 eggs. Only the 6-egg clutch showed signs of incubation, and therefore may have been complete. The normal clutch size is 9 or 10 eggs (Fredrickson et al. 1977). Incubation, performed by both adults, begins before the clutch is complete and lasts 21–25 days. The young hatch over a period of days and leave the nest soon after hatching. They are fed by their parents for a few days after hatching, and remain in their parent's territory at least 7 weeks. They begin to fly when about 10 weeks old (Fredrickson et al. 1977).
—*Charles P. Nicholson.*

Killdeer
Charadrius vociferus

The Killdeer is well known for its piercing call, from which it gets its name. This conspicuous plover is common in many bare-ground and short-grass areas in both wetland and upland settings. The Killdeer is the most widely distributed shorebird in North America, breeding from Alaska and Newfoundland to central Mexico and the Greater Antilles (AOU 1983). It is a common permanent resident throughout most of Tennessee and has one of the longest breeding seasons of the state's nesting birds.

The Killdeer is one of several open country birds that surely benefited from the clearing of the extensive eastern forests for agriculture by the early European settlers. Rhoads (1895a) found it at all of his stops in East and Middle Tennessee. Ganier (1933a) called it a common resident all across the state. Atlas and BBS results show that it remains a common species, and from 1966 to 1994, BBS results show a significant ($p < 0.05$) increase of 1.7%/year. Periodic, short-term declines have occurred following severe winters. Throughout its range,

Killdeer. Elizabeth S. Chastain

the Killdeer population showed a slight decreasing trend of 0.4%/year (p = 0.01) from 1966 to 1993 (Peterjohn, Sauer, and Link 1994).

The Killdeer was among the more widespread birds reported during the Atlas period; it was found in 82% of the completed priority blocks. The Atlas results are probably an accurate depiction of its distribution. It was confirmed breeding in all of the physiographic regions, but was sparsely distributed in some. The most notable gap was in the southern two-thirds of the Unaka Mountains, which are heavily forested. The relationship between the Killdeer's distribution and the percent of cropland by county is fairly strong. Killdeer were also found in some urban areas, such as industrial sites, shopping centers, and transportation corridors.

The average abundance of the Killdeer on Atlas miniroutes was 1.9 stops/route for the 360 routes (52%) recording it. By physiographic regions, its greatest abundance was in the Mississippi Alluvial Plain ($\bar{x}$ = 3.6 stops/route on 13 of 18 routes recording it) and the Loess Plain ($\bar{x}$ = 2.1 stops/route on 72/96 routes). The lowest abundance figures came from the Unaka Mountains ($\bar{x}$ = 1.2 stops/route on 5/47 routes) and Cumberland Mountains ($\bar{x}$ = 1.2 stops/route on 5/21 routes). The average relative abundance on Tennessee BBS routes during 1986–91, 3.9 stops/route (5.48 birds/route), was lower than the miniroute abundance; it was found on all but 2 of the BBS routes. No density information is available from plot censuses in Tennessee. Mace (1978) reported average densities of 30–33 pairs/100 ha in preferred Minnesota habitats.

Breeding was confirmed in almost one-third of the Atlas blocks recording the Killdeer. Of the confirmed records, 44% were of fledged young, 26% were distraction displays, and 22% were nests with eggs, which, with some patience, are often easy to find.

Breeding Biology: Killdeer breed in a wide variety of open areas with short or sparse vegetation, usually close to a conspicuous object such as a log or clump of taller vegetation. Heavily grazed pastures, crop stubble, recently plowed cropland, lake and pond margins, gravel road and railroad right-of-ways, parking lots, golf courses, ball fields, airports, and gravel rooftops are among the nest sites chosen. Factors in common for these areas include open space, in which the adults and young can forage, and gravel, wood chips, and other material for the nest site. Roof nesting was first reported in Tennessee in 1935 at Murfreesboro (Todd 1935). Subsequent published records of roof nesting are few: Nashville in 1953 (*Migrant* 24:56), Chattanooga in 1972 (*Migrant* 43:78), and Hawkins County in 1976 (Phillips and Alsop 1978). A Meigs County Atlas record was of an adult incubating on a rural school roof. Other records probably exist and should be published to document the extent and success of this behavior.

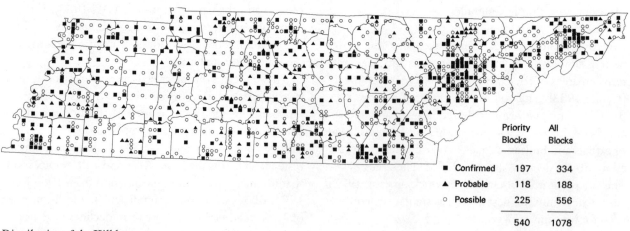

	Priority Blocks	All Blocks
■ Confirmed	197	334
▲ Probable	118	188
○ Possible	225	556
	540	1078

Distribution of the Killdeer.

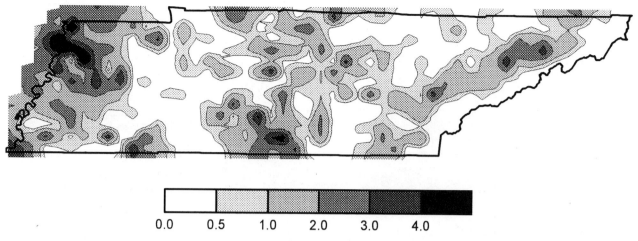

Abundance of the Killdeer.

The Killdeer nesting season in Tennessee extends from late winter through mid-summer. Two-thirds of the egg sets in the Todd collection (n = 179), mostly from Rutherford County, were found between 10 March and 20 April. The nest is a shallow scrape in the ground made by the male (Mundahl 1982) and lined with pebbles, shells, or wood chips. The eggs are buff colored, with numerous dark spots and scrawls providing protective markings. The normal clutch size is 4 eggs, but occasionally 3 or 5 are laid (Harrison 1975). Incubation, lasting 24–28 days, is shared by both parents during the day, but at night is carried out mainly by the male (Mundahl 1982). Killdeer may be double-brooded and Ganier (1934a) reported a case of apparent triple-brooding in Tennessee. The precocial young leave the nest within hours of hatching and catch their own insect food. They are brooded and guarded by 1 or both parents and fly in about 40 days (Harrison 1978). This species is well known for its broken-wing distraction display, used to lure predators away from the nest or fledged young. The males take a greater role in parental care, including distractions (Mundahl 1982). Males also show a greater tendency toward returning to the same territory in consecutive years (Lennington and Mace 1975). Although usually monogamous, at least 1 instance of sequential polyandry has been reported (Brunton 1988).—*Richard L. Knight.*

Black-necked Stilt
Himantopus mexicanus

The Black-necked Stilt, well-named for its disproportionately long, reddish legs, is a recently established, very local summer resident of the southwest corner of Tennessee and a very rare fall migrant elsewhere in the state (Robinson 1990). Stilts are usually present from early April through late September. The establishment of the Tennessee stilt population was described by Coffey (1985). The first state record was in northeast Shelby County in March 1981. In 1982, a pair nested at the sewage lagoons in Ensley Bottoms, in southwest Shelby County. Only a single stilt was observed in 1983, and two pairs nested in 1984. No evidence of nesting was found in 1985; stilts have nested successfully at Ensley each year since (*Migrant*, Atlas results).

The origin of the Tennessee stilt population is unknown. The North American stilt population nests in scattered locations throughout the West, east to Kansas, along the western Gulf coast, and along the southern Atlantic coast (AOU 1983). The establishment of the Tennessee population was coincident with an increase in the number of stilt records in Arkansas, where nesting first occurred in 1981 (James and Neal 1986, Parker 1991). The first Missouri nest records, in rice fields in the southeast corner of the state, were in 1990

Black-necked Stilt. Elizabeth S. Chastain

Confirmed Breeding Species

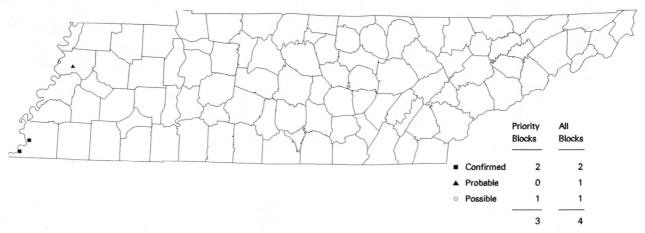

Distribution of the Black-necked Stilt.

(Jacobs 1991). The Black-necked Stilt has also recently extended its breeding range north along the Atlantic Coast to Pennsylvania, reclaiming some of the range from which they were extirpated early this century (Santer 1992).

Black-necked Stilts are typically loosely colonial, nesting in grassy marshes, shallow, frequently alkali ponds, and mud flats (Hamilton 1975). Most of the stilt nests at Ensley have been in 8–45 ha lagoons used for oxidizing sewage sludge prior to the sludge being spread on nearby fields (Coffey 1985, Waldron and Bean 1991). Nests have usually been on scattered raised hummocks supporting variable amounts of vegetation. The stilts forage in the sewage lagoons as well as in adjacent marshy areas and in puddles in nearby agricultural fields. The availability of marshy areas varies greatly from year to year, and stilts have also used them for nesting (*Migrant* 62:22).

The stilt population at Ensley has grown to 9–15 pairs in recent years (M. G. Waldron pers. comm.). In 1990, 2 stilts were also observed in northeast Dyer County on 6 June, and a single stilt was present in southwest Dyer County on 10 June (*Migrant* 62:22). None of these birds apparently nested; each is shown as a possible breeding record on the Atlas map. In 1991, stilts nested at the North Treatment Plant in Shelby County, about 17 km northeast of the Ensley site (Waldron pers. comm.). Fourteen adults, 3 nests and 4 young stilts were present there on 2 July.

Much of the Ensley site used by Black-necked Stilts, located within the Pidgeon Industrial Park, has recently become the Environmental and Resource Technology (EARTH) Complex, a long-term solid and organic waste reclamation and disposal area (Waldron and Bean 1991). Sixteen ha of the sewage lagoons will be managed to provide stilt nesting habitat, and the long-term continued use of adjacent areas for sludge disposal should provide foraging habitat. Much of the rest of the Ensley site is planned for industrial development. Other areas used by stilts in Tennessee presently receive little protection. Because of the vulnerability of its habitat, the Black-necked Stilt should be listed as In Need of Management in Tennessee.

Breeding Biology: Black-necked Stilts are monogamous and defend a territory containing the nest; neighboring nests are often spaced 20 m or more apart (Hamilton 1975). Fidelity to particular nest sites is fairly low, possibly because of the temporary nature of feeding sites (Sordahl 1984), and in Tennessee because of changes in the availability of nest sites in the lagoons. The nest itself is a simple scrape lined with variable amounts of twigs, weed stems or pebbles, placed in an open area. When threatened by rising water, the pair may rapidly raise the nest by adding material to it (Hamilton 1975). This behavior has not been documented in Tennessee.

Egg laying probably peaks in the second half of May. Because of the inaccessibility of the nests, clutch size information is very limited. The normal clutch size elsewhere is 4 eggs, less commonly 3 or 5 (Hamilton 1975). The eggs are buffy with numerous brown to black blotches and spots (Bent 1927). Both adults incubate for about 25 days (Hamilton 1975). The precocial young are active within hours of hatching, and the brood soon permanently abandons the nest site. The young are brooded and protected by the adults until fledging in 4–5 weeks (Sordahl 1984). The adults perform elaborate distraction displays when a nest or brood is threatened, and adults from nearby territories join together to mob a predator (Hamilton 1975). Of 11 nesting attempts at Ensley in 1989, 9 were successful, as were at least 4 of 6 nests in 1991 (Waldron pers. comm.).—*Charles P. Nicholson and Dianne P. Bean.*

Spotted Sandpiper
Actitis macularia

Tennessee is near the southern boundary of the Spotted Sandpiper's extensive breeding range (AOU 1983). Although this shorebird is a rare nesting species in the state, it occurs as a fairly common transient throughout Tennessee and is often seen on the banks of streams, rivers, lakes, and ponds during its spring and fall migration. Its habit of nervously bobbing its tail leads to its colloquial name of "teeter tail" and makes it readily recognizable as it feeds along the water's edge. Spotted Sandpipers typically arrive in mid-April and, except for a few nesting and non-nesting summering birds, depart by early June. Fall transients arrive in July and often linger into late October or early November. Wintering birds are rare (Robinson 1990).

The Spotted Sandpiper has been confirmed nesting in 6 locations across Tennessee. The earliest report is of a nest with 4 heavily incubated eggs on 10 June in northeast Tennessee (Lyle and Tyler 1934). The year of this nest is not given by Lyle and Tyler (contra Robinson 1990). The Lyle egg collection contains no record of a Spotted Sandpiper egg set collected in Tennessee, although it does contain a catalog entry of a set of 4 eggs, incubation advanced, collected on 10 June 1910 in Washington County, Virginia. This may be the egg set mentioned by Lyle and Tyler (1934). The next Tennessee breeding record, and the first with good location information, was the observation of an adult with 2 young on the banks of lower Sycamore Creek in Cheatham County on 26 June 1954 (Weise 1955). A nest with 2 young was reported at Austin Springs, Washington County, on 29 June 1971 (*Migrant* 42:71). The most frequently reported nest site is the Buena Vista Marsh/Metro Center area of Nashville, Davidson County. Nesting was first confirmed there in 1977 (*Migrant* 48:103) and also observed in 1979, 1980, 1981, 1983, 1984, and 1988 (*Migrant,* Vogt pers. obs.).

During the Atlas survey, nesting was confirmed at the Davidson County site in 1988 and at two West Tennessee locations. The first West Tennessee site was in southwest Shelby County in 1988 (*Migrant* 59:123, Atlas results) and the second in southern Lake County, where 3 adults and 2 young were observed on 6 July 1991 (*Migrant* 63:21). Spotted Sandpipers were reported with possible or probable breeding evidence from an additional 15 Atlas blocks across the state. The probable record was of a pair observed in late June 1987. The significance of the "possible" records is difficult to determine due to the species's late spring departure and early fall arrival. In fact, based on the dates of observations of young Spotted Sandpipers in Tennessee, egg laying and incubation begins while transient birds are still present, and some birds observed by Atlas workers in late May and considered migrants may have been nesting.

The Atlas results represent the largest number of breeding season records of the Spotted Sandpiper reported during a short time period. Whether this represents a population increase, however, is uncertain. Most of the recent breeding season records, both confirmed and unconfirmed, have been associated with human-made water bodies, and the increasing presence of these lakes and ponds may be contributing to the recent, relatively large number of records. The first breeding season records were by Rhoads (1895a), who described the species as numerous at low elevations across the state and rarer among the mountains. Based on Rhoads's itinerary, all of his Spotted Sandpiper observations from Chattanooga westward were probably of migrants. Lyle and Tyler (1934) described the Spotted Sandpiper as a common summer resident in northeastern Tennessee, a status not supported by other checklists from the area such as those by Ganier and Tyler (1934) and Herndon (1950b). With the exception of the Nashville area (Parmer 1985), other regional checklists have not listed the Spotted Sandpiper as a breeding species. Rangewide, the Spotted Sandpiper did not show a significant population trend on 1966–93 BBS routes (Peterjohn, Sauer, and Link 1994).

Breeding Biology: Spotted Sandpipers nest and raise their young in semi-open areas near water (Oring, Lank, and Maxson 1983). At least 4 of the 6 confirmed Tennessee nesting locations were close to water. These locations

Spotted Sandpiper. Elizabeth S. Chastain

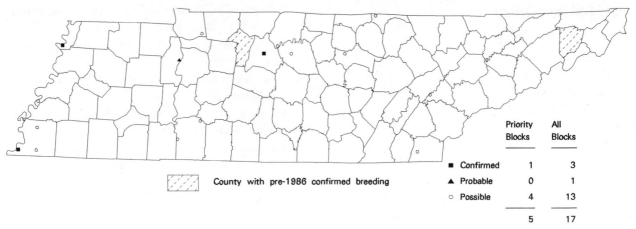

Distribution of the Spotted Sandpiper.

were near a bottomland stream, near a mill dam, near sewage settling ponds, and, at the Davidson County site used for several years, near a human-made lake. At least 4 of 5 nests were located within 30 m of water. The nest is minimal, consisting of a shallow depression in the ground lined with weed stems. It is generally situated in a weedy field, well concealed by the surrounding vegetation and the protective coloration of the eggs. One Tennessee nest was built in debris consisting of herbaceous material, driftwood, and a variety of litter; one of the eggs rested on a discarded liquor bottle (Vogt pers. obs.). All 4 Tennessee clutches contained 4 eggs (Lyle and Tyler 1934, Vogt pers. obs.), the normal clutch size for the species (Oring, Lank, and Maxson 1983); the eggs are buff and marked with dark, umber blotches of varying size and shape.

Among the most striking aspects of the Spotted Sandpiper's breeding biology is the polyandrous behavior of the female (Oring, Lank, and Maxson 1983). The female arrives at the breeding area first and courts males by strutting in an erect posture while spreading its tail and craning its neck. Both members of the short-term pair build the nest within a territory the male defends from other males. After egg laying the female abandons the male to mate with another male and lay another clutch. Although polyandry has not been confirmed in Tennessee, only 3 adults were observed in the vicinity of 2 simultaneously incubated nests at the Davidson County site (Vogt pers. obs.). This suggests these 2 clutches may have been laid by a single female.

The males incubate the eggs for about 3 weeks and care for the young for about 3 weeks. The young leave the nest within hours of hatching and quickly begin display tail-bobbing. When threatened, the male repeatedly gives a single loud "weet" note, which prompts the young to squat motionless on the ground where their dorsal coloration makes them virtually invisible (Vogt pers. obs.). The male may also feign injury in a manner reminiscent of the Killdeer's behavior. When the threat of danger has passed, the male gives a softer, high-pitched note that results in the young gathering and resuming normal feeding.—*David F. Vogt.*

American Woodcock
Scolopax minor

The American Woodcock is an uncommon statewide permanent resident. Unlike other members of the sandpiper family in Tennessee, it occurs in young, moist deciduous woodlands, where its mottled brown and black plumage blends with fallen leaves. It is most active at dusk and dawn, when it flies between woodlands used as daytime roosts and nearby fields used as nocturnal roosts and courtship areas. Its conspicuous courtship displays consist of loud, nasal "peent" calls given from the ground, and a high-pitched twittering song given during a high, spiral display flight. The woodcock is also a fairly popular game bird.

American Woodcocks are most numerous during spring migration, occurring in February and March, and fall migration, which peaks in November (Pitts 1978c, Roberts 1978). During the winter it is uncommon to rare, depending on the severity of the winter. Breeding woodcocks are uncommon in the eastern two-thirds of the state and rare in West Tennessee.

The first report of the American Woodcock was by Fox (1886), who observed it in March 1885 in Roane County. Rhoads (1895a) observed specimens at several unspecified locations across Tennessee and stated that it nested in Haywood County. Although Ganier (1917) described it as a very rare summer resident in Middle and East Tennessee, the first documented nest record is apparently a set of eggs collected by Monk in Davidson County on 15 March 1922 (WFVZ).

American Woodcock. Elizabeth S. Chastain

Since 1922, confirmed evidence of breeding has been reported from across the state. Whether the Shelby County nest (Roberts 1978) was in the Mississippi Alluvial Plain or Loess Plain physiographic region is unknown; nests or broods of young have been reported from all other physiographic regions and several summer records exist from elevations over 1524 m in the Unaka Mountains (e.g., Stupka 1963). Little information on the trends of either breeding, migrant, or wintering populations is available. The number of published records, particularly in the *Migrant*, peaked during the 1970s. Some of them, such as a winter 1973 report of over 50 birds in Hardin County (*Migrant* 44:22), suggested large local populations. This increased number of reports was probably due more to an increase in interest in the woodcock rather than a population increase. Surveys of singing males throughout the main breeding range, north of Tennessee, show a long-term population decrease (Artmann 1977, Tautin et al. 1983). This decrease is most evident in the eastern portion of the woodcock's range, and due mostly to loss of breeding habitat. Because of this decrease, the February hunting season was recently eliminated. This season, in effect since 1977, coincided with the spring migration and some local nesting; the present season is from October into December. The annual harvest is estimated to be about 12,000 woodcocks (TWRA 1992).

During the Atlas period, American Woodcocks were reported in about 5% of the completed priority blocks. This is probably a significant underestimate of their breeding distribution. Woodcocks are most conspicuous during their crepuscular courtship period, which peaks during February and March, a time of little Atlas fieldwork. Male woodcocks perform courtship displays while migrating (Sheldon 1967), and an unknown proportion of courting males in Tennessee are migrating birds (Pitts 1978b, Roberts 1978). Because of the overlap between migrating and locally breeding birds, observations before mid-April without stronger evidence of breeding than the presence of courting males were not included in the Atlas results. This conservative approach probably eliminated several observations of locally breeding woodcocks. Unless inadvertently flushed during the day, woodcocks were difficult for Atlas workers to observe during late spring and summer.

Breeding Biology: Courtship displays occur sporadically from mid-October into December, on a regular basis from late December into March, and then sporadically into late May (Pitts 1978c, Roberts 1978). The "singing grounds" used for courtship include heavily grazed pastures, old fields with small, grassy openings, cultivated fields, and young pine plantations (Pitts 1978b, Roberts 1978). They range in size from less than 0.1 to several hectares and are usually close to wooded diurnal habitat. Courtship typically occurs on evenings with

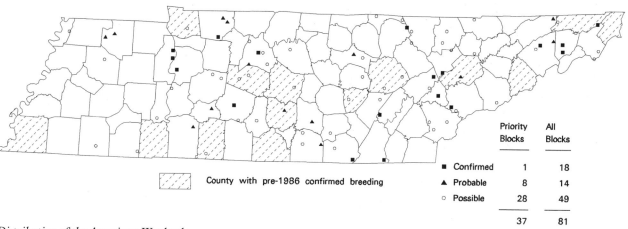
Distribution of the American Woodcock.

	Priority Blocks	All Blocks
■ Confirmed	1	18
▲ Probable	8	14
○ Possible	28	49
	37	81

Confirmed Breeding Species

temperatures above freezing, beginning about 15 minutes after sunset and lasting about half an hour (Roberts 1978). Females mate with the polygamous males on the singing grounds, and males take no further role in nesting or brood rearing (Sheldon 1967).

By mid-February, both males and females are physiologically capable of breeding (Roberts 1978), and records of eggs and young show a peak of egg laying during the first half of March. The eggs are laid on dead leaves in a slight depression with a few twigs or pine needles often added around the rim (Sheldon 1967). The nest is usually near the base of a small tree or shrub in second-growth hardwoods, close to a singing ground (Roberts 1978, Sheldon 1967). All clutches reported in Tennessee contained 4 eggs except for 1 fresh and possibly incomplete clutch of 3 eggs (Pitts 1978c, Roberts 1978, egg coll. data); the normal clutch size elsewhere is 4 eggs (Sheldon 1967). Incubation lasts about 21 days, and the precocial young leave the nest soon after hatching. They begin to fly in about 2 weeks and are almost fully grown in 4 weeks. Broods have been reported in young mixed hardwoods and conifers with both open and dense understories, a bottomland pine plantation with patchy honeysuckle thickets, bottomland hardwoods with open privet understory, brushy fields with scattered trees, and a recently cleared field with scattered slash (Roberts 1978, Nicholson pers. obs.). Renesting has been reported following loss of the initial clutch (Sheldon 1967) and suggested in Tennessee by Bamberg (1933). Only 1 brood is raised.—*Charles P. Nicholson.*

Least Tern
Sterna antillarum

The Least Tern is a locally common summer resident along the Mississippi River and a rare migrant elsewhere in the state. Spring migrants arrive in early May and are concentrated in West Tennessee (Robinson 1990). Fall migrants, present from July through early September, occur on lakes and large rivers throughout the state. The Least Tern feeds primarily on small fish caught in shallow water (Hardy 1957).

The first report of Least Terns in Tennessee was by Bartsch (1922), who observed them (as well as Black Terns, *Chlidonias niger*) carrying food at the junction of the then unimpounded Tennessee and Duck Rivers on 12 August 1907. He presumed they were nesting there. Least Terns, however, often leave the nest site soon after the young fledge, and the juveniles are fed by adults for several weeks (Hardy 1957). The terns Bartsch observed were probably wandering family groups, and it is unlikely they once nested along the lower Tennessee River.

Least Tern. Elizabeth S. Chastain

All definite nest records have been on sandbars in the Mississippi River, or mainland sandy areas up to 1.6 km from the river (*Migrant* 54:86, 55:88). The first nest record was in 1921, when Ganier (1930) observed a colony on Middle Bar in Lake County. Coffey observed a colony on Owen's Bar, Shelby County, in 1928 and 1929 (Ganier 1930). Although nesting probably occurred almost every year, colonies at these and other locations were only sporadically reported into the 1980s.

Following concern over low numbers and alteration of its riverine habitat, the inland nesting population of the Least Tern was listed as an endangered species in 1985 (USFWS 1990). Comprehensive counts of the population nesting along the lower Mississippi River, from Cape Girardeau, Missouri, to Vicksburg, Mississippi, including those nesting in Tennessee, were begun by the U.S. Army Corps of Engineers in 1985 (USFWS 1990). These surveys found a relatively stable population averaging 2363 adult birds from 1986 to 1988. The population adjacent to Tennessee consisted of 1212 birds at 24 nest sites in 1987 and 1135 birds at 23 sites in 1988 (J. P. Rumancik Jr., pers. comm.).

The amount of suitable nesting habitat is probably the main factor limiting inland Least Tern populations (USFWS 1990, Smith and Renken 1991). The area of sandbar and sand island habitat adjoining Tennessee declined 43% from 1937 to 1973 because of changes in the river flow regime (P. B. Hamel pers. comm.). The Least Tern appears to require sand islands and sandbars continuously exposed above water for at least 100 days between 15 May and 31 August (Smith and Renken 1991). These sites are the first exposed by receding water in early summer and are available in both high- and low-water years. The decline in nesting habitat is most ap-

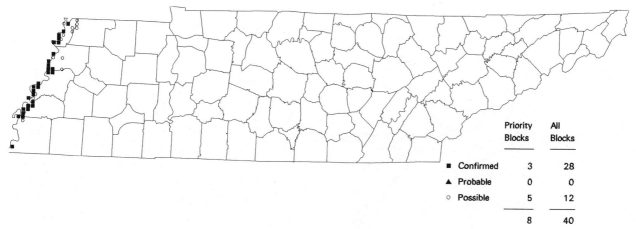

Distribution of the Least Tern.

parent in years of high water, when nests are more concentrated and subjected to higher human disturbance (Hamel pers. comm.). Other important nest site factors are vegetation cover of less than 10%, the presence of large amounts of driftwood, and little human disturbance. The greatest hazard to nesting terns is from natural flooding after nesting has begun, as occurred during the unusually wet summer of 1989 (Smith and Renken 1993). The gradual connection to the mainland of sandbars formed behind dikes is also detrimental because of the resultant increased predation by mammals and human disturbance (USFWS 1990, Smith and Renken 1993). The Least Tern Recovery Plan (USFWS 1990) and Smith and Renken (1991) both suggest methods of creating and enhancing nest habitat.

Since the listing of the inland population of the Least Tern as an endangered species in 1985, the tern has been the focus of much attention, including the development of a recovery plan (USFWS 1990), population censuses, and several research studies (e.g., Smith and Renken 1991). Compared to other inland populations, the lower Mississippi River population, including Tennessee, is relatively secure and nesting success is adequate to maintain the population (Smith and Renken 1993). Provided suitable nesting habitat is available, the greatest remaining threat to the Tennessee population is recreational use of islands and sandbars. Reducing this threat will take a combination of public education and law enforcement.

Breeding Biology: Least Terns are in the vicinity of potential nest sites by late May and remain in flocks awaiting the exposure of suitable nest habitat by dropping water levels (Hardy 1957). The nest consists of a shallow, unlined depression scraped in loose sand or gravel, and pairs begin several nest scrapes during courtship (Hardy 1957). Nests are often near driftwood, which may shelter nests from blowing sand and provides shade for tern chicks (Smith and Renken 1991). The number of nests in a colony varies greatly; adjacent to Missouri, it averaged 100 during a high-water year and 19 during a year of moderate river levels (USFWS 1990). Within a colony, nests are usually dispersed over an area of several hectares, and tern pairs defend the immediate area around their nests (Hardy 1957).

Clutch size is from 1 to 4 eggs, with 2 and 3 eggs most common (Ganier 1930, Hardy 1957). Clutch size from 1986 to 1989 at sites adjacent to Missouri, including many sites adjacent to Tennessee, averaged 2.4 eggs (USFWS 1990). The eggs are pale buff to olive buff with brown and gray spots and streaks (Hardy 1957) and blend well with their surroundings. Both adults incubate the eggs, usually for 20 to 25 days. The young wander from the nest site a few days after hatching and begin flying in about 3 weeks (Hardy 1957). The parents continue to feed them, often far from the nest site, until fall migration begins.—*Charles P. Nicholson.*

Rock Dove
Columba livia

The Rock Dove, more commonly known simply as the pigeon, is a common permanent resident of urban and agricultural areas across Tennessee. It is native to Eurasia and north Africa and has been introduced throughout the world (Goodwin 1983). Introduced populations are usually closely associated with humans, although their degree of dependence on humans is variable and often debatable. As a result of their origin from several different domestic strains, the plumage of Rock Doves in Tennessee varies greatly from almost pure white to gray to almost solid brown. The most common form is predominantly gray with 2 black wing bars, a black band at the end of the tail, a purplish green throat and neck, and a whitish rump, characteristics of the ancestral wild Rock Dove.

Rock Dove. Elizabeth S. Chastain

The ancestral wild Rock Dove nests on sheltered ledges or holes in cliffs or caves, usually in semidarkness (Goodwin 1983). This trait is evident in introduced populations, which nest in barns, on ledges of buildings and under bridges, and rarely on natural cliffs. In rural areas, the Rock Dove feeds in areas with short or no vegetation, such as farmyards, pastures, and plowed fields. Although flocks may fly up to a few kilometers to such sites, rural populations in Tennessee appear dependent on the presence of feeding areas within a kilometer or so of nest sites (Nicholson pers. obs.).

Unlike the other non-native bird species now established in Tennessee, the establishment and spread of feral Rock Dove populations in Tennessee and elsewhere is poorly known. French and British settlers brought domestic Rock Doves to North America in the early seventeenth century for use both as pets and food (Schorger 1952). The first year that settlers brought it to Tennessee is unknown; it was probably kept in most early-nineteenth-century settlements. Its numbers probably increased rapidly with the human population and growth of towns in the mid-nineteenth century, and again near the beginning of the twentieth century with the widespread construction of bridges.

The Rock Dove was first mentioned in the Tennessee ornithological literature by Ganier (1935c), who, after observing feral birds nesting in a Williamson County quarry, decided it was sufficiently established to be considered a part of the Tennessee avifauna. This observation, however, did not soon result in a change in the attitude of birdwatchers and ornithologists toward it. It was not reported on Christmas Bird Counts until 1973, or on Tennessee spring counts until 1974. Even then, some individual counts did not accurately report its numbers until well into the 1980s. The only reliable information on its population trend is from the Breeding Bird Survey, which shows no significant overall trend from 1966 to 1994, although it increased by 7.7%/year ($p < 0.01$) from 1966 to 1979. From 1966 to 1993, the Rock Dove population increased at an annual rate of 0.7% ($p < 0.04$) throughout North America (Peterjohn, Sauer, and Link 1994).

The Rock Dove is usually gregarious, occurring in flocks of up to a few dozen birds. Pairs often nest close to each other and only defend a small area around their nest. In addition to barns, urban buildings, and bridges, the Rock Dove in Tennessee occasionally nests in crannies on the cut rock faces of quarries (Ganier 1935c) and large railroad cuts (Nicholson pers. obs.), as well as natural cliffs. During atlas fieldwork, Rock Doves were found nesting on ledges adjacent to Fall Creek Falls in Bledsoe County (Nicholson pers. obs.). It is unlikely that the Rock Dove nested at Fall Creek Falls until resort facilities were constructed nearby in the 1960s.

Atlas workers found the Rock Dove in about 60% of the completed priority blocks. Because of its conspicuous behavior and easy identification, it is unlikely it was missed

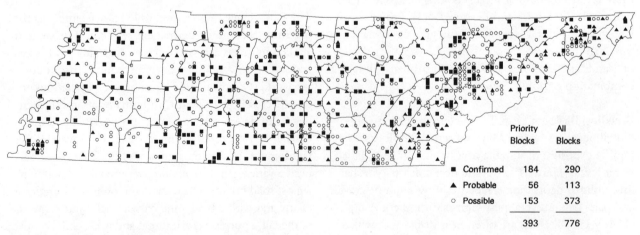

Distribution of the Rock Dove.

	Priority Blocks	All Blocks
■ Confirmed	184	290
▲ Probable	56	113
○ Possible	153	373
	393	776

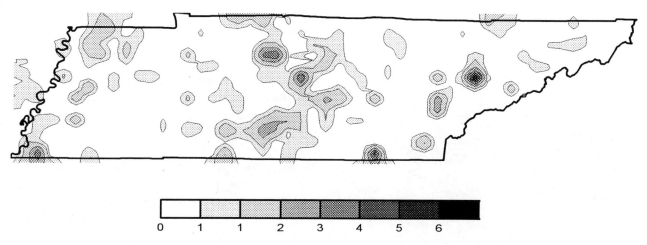

Abundance of the Rock Dove.

in many blocks. The Rock Dove was absent from parts of the Unaka Mountains, Cumberland Plateau and Mountains, and Highland Rim with low human populations and little farmland. It was also not found in several agricultural blocks where farmers probably eradicated it from barns and alternate nest sites were not available. The Rock Dove was recorded on about a quarter of the Atlas miniroutes at an average relative abundance of 1.5 stops/route. The highest abundance was generally in urban areas, although this was not evident in several cities across the state where miniroutes missed the built-up areas.

Breeding Biology: The ancestral wild Rock Dove is multiple-brooded and has been found nesting throughout the year, especially when food is abundant (Goodwin 1983). In Tennessee, the Rock Dove nests from at least January through August. Most Tennessee nest records are from April through early July, and the product of Atlas fieldwork. Nesting in the fall probably occurs, although populations elsewhere stop breeding while molting in early fall (Goodwin 1983). Confirming the breeding of the Rock Dove was usually easy for Atlas workers, and it was one of few species for which observing fledglings was harder than finding the nest. Over half of the confirmed records were of birds occupying nests. Because the Rock Dove often roosts at nest sites, a few of these birds may not have been actually nesting when observed.

The Rock Dove nest is a coarsely constructed platform of twigs and straws, with an unlined shallow bowl. The male brings most of the material to the nest site and the female arranges it (Goodwin 1983). The nest is often reused, and the droppings of the nestlings cement it into a rigid structure. The normal clutch size is 2 eggs, which are plain white. Both adults share incubation, which lasts 17–19 days. The young are fed crop milk at hatching and an increasing proportion of seeds as they mature. They leave the nest in about 25 days and remain in the nest area until the next brood hatches, when they are driven off by their parents (Goodwin 1983). The number of broods that pairs raise annually in Tennessee is unknown; second and third broods are probably common.—*Charles P. Nicholson.*

Mourning Dove
Zenaida macroura

The Mourning Dove is a common, widespread bird, well-known to most Tennesseans, especially in agricultural and suburban settings. Although present year-round and generally considered to be a permanent resident, northern populations migrate to and through the state, and some doves breeding here may leave during the winter. Much has been written about their biology, particularly their breeding chronology and population estimates for game management. Mourning Doves are a popular game bird in Tennessee—from 1981 to 1990, an average of 141,000 hunters harvested an average of almost 3 million doves per year, of which an estimated 93% were raised in the state (M. Gudlin pers. comm.).

Historical accounts (e.g., Rhoads 1895a) describe a range and status of the Mourning Dove similar to those of today. Presumably they were less common in prehistoric times and increased as the forests were cleared. Mechanical grain harvesters, which allow more acreage of crops but spill loose grain, have doubtless benefited dove numbers. Tennessee BBS routes show no significant population trends from 1966 to 1994. "Call counts," conducted yearly by TWRA personnel to aid in the establishment of hunting regulations, also show the population to be rather stable. The rangewide Mourning Dove population also does not show a significant overall trend from 1966 to 1993 (Peterjohn, Sauer, and Link 1994).

Confirmed Breeding Species

Mourning Dove. Elizabeth S. Chastain

The Mourning Dove was the fifth most frequently encountered species during the Atlas period. It was absent only from the higher elevations of the Unaka Mountains and some solidly forested blocks there and in the Cumberlands. Rhoads (1895a) described it as rare above 900 m, and that is generally true today; in the Great Smokies, it is infrequently found above 600 m (Stupka 1963). Heavy forest cover appears to be more of a limiting factor. In the northern half of the Unaka Mountains, Mourning Doves were found in several open sites—residential and farm clearings, roadsides, and some regenerating clearcuts—up to 1370 m (Knight pers. obs.). The most notable gap on the Atlas map is in the southern Unakas, primarily Monroe and Blount Counties. The Atlas map should reliably reflect the current distribution of Mourning Doves in Tennessee since the species is easily detectable.

Mourning Doves were well represented on the Atlas miniroutes, occurring on 96% of the routes at an average abundance of 6.4 stops/route. The areas of lowest abundance correspond well with areas of heavy forest cover and/or high elevation, while higher numbers are in agricultural or suburban areas. During the 1986–91 Atlas periods, Mourning Doves were recorded on all of the BBS routes, at an average relative abundance of 36.4 birds and 20.5 stops/50 stop route, nearly identical to the Atlas miniroute results.

Breeding Biology: The Mourning Dove has one of the longest breeding periods of all North American birds, with Tennessee nest records for 10 months. Courtship cooing and displays often begin during warm periods in January. Nest building has been reported as early as 21 January (*Migrant* 24:14), and there is an exceptional record of eggs laid in late December (*Migrant* 46:71). The peak of breeding runs from April to August and up to 6 nestings per year may occur; Monk (1949) felt doves regularly made 5 nesting attempts. Several time-saving adaptations, among them rapid construction of a small nest and frequent nest reuse, constant incubation and fast nestling growth, facilitate this multiple brooding (Monk 1949, Westmoreland, Best, and Blockstein 1986). Late nesting persists through September and occasionally into October. Monk (1949) found that 17% of all successful nests fledged in September, while Burch (1982) and Stogsdill (1983) reported that this proportion varied annually from 0 to 35%. This has caused controversy for federal and state wildlife agencies over the opening date for hunting, which traditionally has been 1 September in Tennessee. Geissler et al. (1987:23) claim that "it is unlikely that this early hunting has a significant effect on recruitment of fledglings into the dove population."

The nest of the Mourning Dove is a simple, flimsy-looking, small platform of sticks, often sparse enough for the eggs to be seen from beneath. Although some

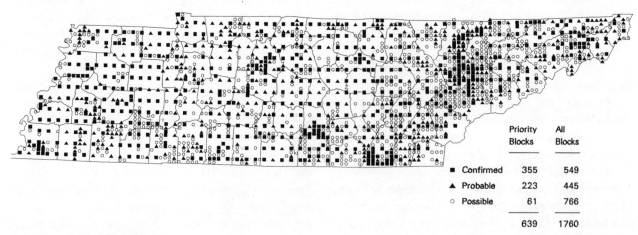

Distribution of the Mourning Dove.

	Priority Blocks	All Blocks
■ Confirmed	355	549
▲ Probable	223	445
○ Possible	61	766
	639	1760

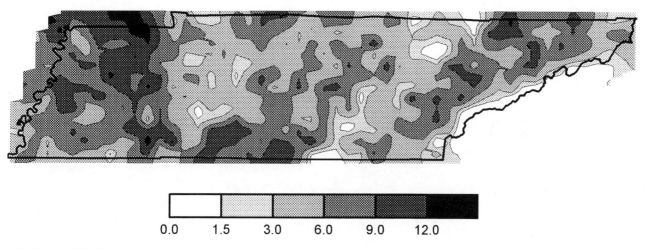

Abundance of the Mourning Dove.

are lost to storms, the nest is surprisingly strong (Monk 1949). The nest is built by the female, with the male carrying sticks to the site. Although usually built in the fork of a horizontal limb, the nest is sometimes placed on top of a deserted nest of another species, frequently a robin or thrasher. Occasionally it may be placed on a rock ledge or an artificial structure. Nests are often reused; Monk (1949) reported that 27% of nesting near Nashville involved reused nests, including one record of a nest used 4 consecutive times in a year. Studies elsewhere (in Westmoreland, Best, and Blockstein 1986) found nest reuse in up to 35–49% of nesting attempts. Evergreen trees and shrubs, especially red cedar, and deciduous trees with tangles of Japanese honeysuckle or grape vines are the most common nest supports. Nests are also commonly placed on horizontal branches of deciduous trees with dense branching structure. Most Mourning Dove nests found in Tennessee have been at heights of 2–5 m, but range from 0.6–11 m (Monk 1949, Burch 1982, NRC, egg coll. data).

The eggs of the Mourning Dove are white and unmarked. A clutch almost invariably consists of 2 eggs, but at least one clutch of 3 eggs is known from the state (Lyle egg coll.). Incubation is shared by both parents, the male incubating for much of the day and the female at night and part of the day. Incubation lasts 13–14 days. Both adults feed the young, at first solely from crop milk, then a mixture of crop milk and seeds, and by 6–8 days solely seeds (Westmoreland, Best, and Blockstein 1986). After fledging in 13–15 days, the young are tended by the male for up to a week.

Nesting success of Mourning Doves in Tennessee is generally good. Monk (1949) reported that 52% of nesting attempts succeeded, with 1.7 young/successful nest, in a primarily urban setting. Burch (1982) found 50% of rural nests, and 78% of urban nests were successful, with an overall production of 1.8 young/successful nest. In a rural area, Stogsdill (1983) recorded a 33% success rate, with 1.6 young/successful nest.—*Richard L. Knight.*

Black-billed Cuckoo
Coccyzus erythropthalmus

The Black-billed Cuckoo is an uncommon migrant and a rare summer resident of woodland and woodland edges in Tennessee. Spring migration occurs between late April and early June, and fall migration occurs between mid-August and early October (Robinson 1990). The timing of spring migration varies greatly between years.

Crook (1935b) summarized the status of the Black-billed Cuckoo in Tennessee through the mid-1930s. Rhoads (1895a) first reported it in the state and described it as one-fifth as common as the Yellow-billed Cuckoo. Rhoads did not list the locations at which he observed Black-billeds. By the mid-1930s, nests had been found in Davidson, Putnam, Grundy, and Washington Counties, a male with an incubation patch collected in Cheatham County, and possible breeding birds reported in a few other counties (Ganier 1926, Crook 1935b). Between the mid-1930s and the Atlas period, nests were reported from Madison and Campbell Counties (Roever 1951, Alsop 1971b), and a copulating pair was observed in Maury County (*Migrant* 55:91).

Stupka (1963) described the Black-billed Cuckoo as a fairly common summer resident in the Great Smoky Mountains, about as common as the Yellow-billed, and more frequently observed above 1067 m than the Yellow-billed. Annotated lists from Greene (White 1956) and Carter (Herndon 1950b) Counties both describe the Black-billed as a rare summer resident.

The few descriptions of Black-billed Cuckoo nesting habitat in Tennessee suggest the species shows little

Black-billed Cuckoo. Elizabeth S. Chastain

habitat preference. Nests have been found in a thicket, Cumberland Plateau mixed oak-pine woods, and woods at the top of a river bluff (Crook 1935b, *Migrant*, NRC). Atlas observations were from young deciduous trees with a dense understory, mixed oak-pine woods, rural deciduous woodlots, a fencerow with mature deciduous trees, and a pine plantation with trees about 4 m tall. Descriptions of its nesting habitat elsewhere in its range (e.g., Bent 1940, Eastman 1991) suggest a preference for areas with dense, shrubby vegetation.

Atlasers found Black-billed Cuckoos in only 14 Atlas blocks and confirmed evidence of breeding, a bird carrying food, was reported from 2 blocks in Cumberland County. The only miniroute observation was of 1 bird on a route in Scott County. It is likely that atlasers missed Black-billed Cuckoos in some Atlas blocks, and a few may have been misidentified as Yellow-billeds by atlasers not familiar with their song. These misses, however, would probably not have changed its status as a rare breeding species.

Because of the few historic and recent records of the cuckoo, it is difficult to determine its recent population trend. Comparison of the very few Atlas records from the Great Smoky Mountains with Stupka's (1963) description of its status there suggests a decline in its population. Black-billed Cuckoos have been recorded on too few Tennessee BBS routes to determine a significant population trend, and some of the birds recorded on routes run in late May and early June may have been migrants. Throughout its range, Black-billed Cuckoos showed no significant trend on BBS routes from 1966 to 1993, but significantly decreased ($p < 0.01$) by 3.0%/year between 1978 and 1988 (Peterjohn, Sauer, and Link 1994, Sauer and Droege 1992).

Breeding Biology: About a dozen Black-billed Cuckoo nest records are available from Tennessee. Nests have been reported from dogwood, small cedar, and pine trees, a buttonbush shrub, and a vine-covered dead sapling. Nest height ranges from 1.1 to 2.4 m, with an average of 1.9 m (s.d. = 0.4 m, n = 8), close to the average height reported elsewhere (Harrison 1975). The Black-billed Cuckoo's nest is a platform of loosely woven twigs, lined with bits of leaves, pine needles, and catkins, placed either near the tree trunk or on a branch some distance from the trunk (Spencer 1943). The birds continue to add nest material during incubation. Egg laying begins as early as late April and extends into mid-June. Elsewhere, nesting regularly extends into late summer. Clutches in Tennessee include 7 of 2 eggs, 3 of 3 eggs, 1 of 4 eggs, and 1 of 5 eggs. Because Black-billed Cuckoos begin incubation after the first egg is laid, some of these clutches, many of which were described as fresh, may

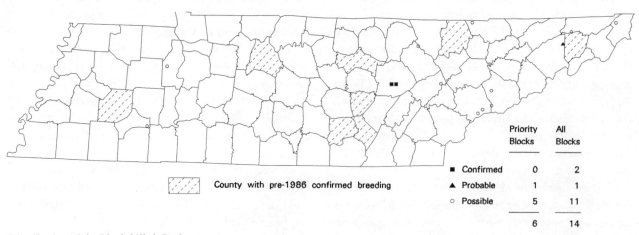

Distribution of the Black-billed Cuckoo.

have been incomplete. Studies elsewhere report the same range of clutch size and an average of just under 3 eggs (Spencer 1943, Sealy 1978a). The eggs are a dull greenish blue. Black-billed Cuckoos rarely lay eggs in the nests of Yellow-billed Cuckoos, as well as other species (Spencer 1943); this has not been reported in Tennessee.

Both adults incubate for 10–13 days (Spencer 1943, Sealy 1985). The newly hatched young have black skin and sparse, gray, hairlike feathers. They are fed by both adults, develop rapidly, and leave the nest when 7–9 days old (Spencer 1943, Sealy 1985). At this time, the young are fully feathered except for the flight feathers, which have not reached their full length. The fledglings are flightless for at least several days after leaving the nest and, when disturbed, assume a vertical pose with outstretched neck and upward pointed bill (Sealy 1985). Second broods have not been reported.

Studies north of Tennessee where Black-billed Cuckoos nest more regularly suggest that the density, timing, and clutch size of nesting cuckoos are influenced by the food supply available when the birds are settling on nesting territories (Nolan and Thompson 1975, Sealy 1978a). Large hairy caterpillars and cicadas are important items in their diet (Bent 1940). Following their spring migration, cuckoos are relatively nomadic, and their appraisal of local food conditions during this period influences settlement patterns (Hamilton and Hamilton 1965, Nolan and Thompson 1975). During periods of unusual insect abundance, such as tent caterpillar or periodic cicada outbreaks, they may nest in higher numbers and produce a larger clutch of eggs.

Insufficient information is available to determine whether Black-billed Cuckoo nesting in Tennessee is influenced by insect outbreaks. Dated observations of likely or confirmed nesting show few yearly peaks suggestive of a response to insect outbreaks, except for 1935, when several observations were reported (Crook 1935b). Nine of the Tennessee nests had eggs laid in mid-May or earlier, at which time tent caterpillars are frequently abundant (Nicholson pers. obs.). Sealy (1978a) suggests clutches of 4 and 5 eggs may be typical of nests during prey outbreaks; 2 such clutches have been reported in Tennessee.—*Charles P. Nicholson.*

Yellow-billed Cuckoo
Coccyzus americanus

The Yellow-billed Cuckoo is a fairly common to common summer resident of woodlands and woodland edges throughout Tennessee. It arrives in late April or May, and departs by mid-October. Its numbers, par-

Yellow-billed Cuckoo. Elizabeth S. Chastain

ticularly during the spring, are influenced by insect outbreaks, such as tent caterpillars or periodic cicadas.

Atlas workers found the Yellow-billed Cuckoo throughout the state and in almost 96% of the completed priority blocks. Although the cuckoo is often not easily seen, it is easily identified by its loud song and calls, which are given throughout the day and throughout the summer. It was absent from a few urban blocks and from some high elevation Unaka Mountain blocks. It was probably present, but missed by atlasers, in the completed blocks elsewhere in the state.

In the Unaka Mountains, the Yellow-billed Cuckoo becomes uncommon above about 1067 m and is absent from the spruce-fir forest (Stupka 1963, Kendeigh and Fawver 1981). It is also absent from the predominantly hardwood forests at the highest elevations of the Unicoi Mountains (Ganier and Clebsch 1946, Nicholson pers. obs.). At lower elevations elsewhere in the state, it occurs in virtually all forest types, except perhaps in pure stands of pole-sized or larger pines. It shows a preference for areas with dense saplings (Anderson and Shugart 1974) and is often found in open woodlands, forest edges, and old fields dominated by shrubs and saplings. The Yellow-billed Cuckoo has been recorded on many Tennessee census plots. Representative densities in different habitats include 4–12 pairs/100 ha in Ridge and Valley deciduous forest (Howell 1972, 1973), 4 to 12/100 ha in Cumberland Mountain deciduous forest–strip mine plots (Garton 1973, Yahner 1973, Nicholson 1980c), 8/100 ha in a field with hedgerows (Lewis 1975a), 5–10/100 ha in a Highland Rim forest and brushy field plot (Simmers 1979, 1982b), and 16/100 ha in Central Basin cedar and deciduous forests (Fowler and Fowler 1983a,

1984a). The highest density reported is 35/100 ha in a Ridge and Valley deciduous woodland (Smith 1975).

Atlasers reported the Yellow-billed Cuckoo on almost 80% of the miniroutes at an average relative abundance of 3.1 stops/route. The highest average abundance and the block with the highest abundance, 13 stops/route, were in the Mississippi Alluvial Plain. Other physiographic regions with high abundances were the Coastal Plain Uplands (4.5 stops/route), Loess Slope (3.4), and the Central Basin (3.4). These results support the findings of Ford (1990), who found the cuckoo to be among the most frequently occurring species in West Tennessee forested wetlands. The cuckoo also shows a tolerance of small forest areas (Robbins, Dawson, and Dowell 1989). Average abundances in the other physiographic regions were between 1.9 and 2.7 stops/route.

The Yellow-billed Cuckoo was probably an uncommon to fairly common bird in prehistoric Tennessee, most numerous around forest openings and in shrubby woodlands resulting from fire or other disturbances. Its numbers probably increased with European settlement, then declined with the near complete deforestation of parts of the state in the early twentieth century. Rhoads (1895a) first reported the cuckoo in Tennessee; his report suggests it was present throughout the state except on Roan Mountain. Ganier (1933a) described it as common across the state. From 1966 to 1994, the Yellow-billed Cuckoo population declined significantly ($p < 0.01$) by 1.8%/year on Tennessee BBS routes. The rate of decline has accelerated since 1980. The trend throughout its breeding range is similar, with an overall decline of 1.4%/year ($p < 0.01$), accelerating since 1978 (Sauer and Droege 1992, Peterjohn, Sauer, and Link 1994).

Breeding Biology: The breeding season of the Yellow-billed Cuckoo extends from late April through mid-September, one of the longest of any neotropical migrant species nesting in Tennessee. Other unusual aspects of its biology are a nomadic phase following spring migration, a very short nesting cycle, the rare parasitism of other cuckoos' and unrelated species' nests, and its response to outbreaks of insect prey (Hamilton and Hamilton 1965, Nolan and Thompson 1975). During periods of unusual prey abundance, it may nest in higher densities and lay larger clutches. Food supplies are apparently evaluated and settlement patterns determined during the post-migratory nomadic phase.

As a likely result of this nomadic behavior, cuckoo numbers in Tennessee vary from year to year during the late spring and early summer (spring count data, BBS results). Egg laying occurs from the end of April through late August, with a weakly defined peak in late May and early June, when observers are probably most active. Nolan and Thompson (1975) suggest that, in the absence of insect outbreaks, egg laying in southern Indiana peaked during late July and early August. Insufficient information is available to correlate cuckoo nesting with insect populations in Tennessee. Confirming breeding of the cuckoo was difficult for atlasers, and almost half of the confirmed records were of adults carrying food. The short nesting cycle and inconspicuous behavior of the fledglings contribute to the difficulty, and many cuckoos had probably not begun nesting when most Atlas fieldwork was conducted.

Both the male and female Yellow-billed Cuckoos build the nest, which is placed in a shrub or on a horizontal tree limb, often vine-covered. Tennessee nests vary from 0.9 to 6.7 m above ground, with an average height of 3.1 m (s.d. = 1.5, n = 52). The plant species most often used for nesting in Tennessee is hackberry; cedars, elms, willows, and pines are also frequently used. The nest is a platform of twigs, with a shallow cup lined with leaves, roots, pine needles, and frequently *Usnea* lichens (egg coll. data, Nicholson pers. obs.). The con-

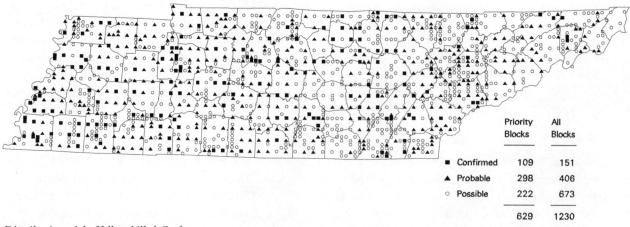

	Priority Blocks	All Blocks
■ Confirmed	109	151
▲ Probable	298	406
○ Possible	222	673
	629	1230

Distribution of the Yellow-billed Cuckoo.

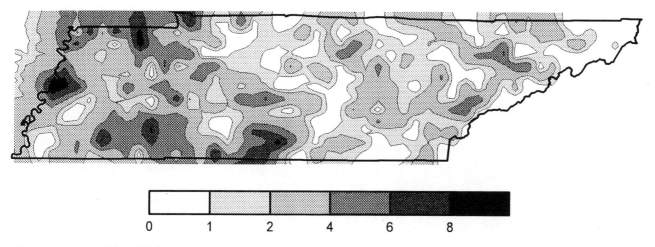

Abundance of the Yellow-billed Cuckoo.

struction of the nest varies from frail platforms with the eggs visible from below to substantial, well-built nests 30 cm or more in diameter (Bent 1940, Potter 1980).

Clutch size varies from 2 to 5 eggs, and in Tennessee averages 2.66 eggs (s.d. = 0.82, n = 51), similar to that reported elsewhere (Bent 1940, Nolan and Thompson 1975). The eggs are pale greenish blue. Incubation, carried out by both adults, begins after the first egg is laid and lasts 9–11 days (Hamilton and Hamilton 1965, Potter 1980). The young, naked and black-skinned when born, are fed by both parents and develop rapidly. At 6–7 days old, the long feather sheaths burst open, and by 9–10 days old the young leave the nest, although they can not yet fly (Bent 1940).

Because of its long nesting season, several authors have suggested that the Yellow-billed Cuckoo is double-brooded (e.g. Mengel 1965, Potter 1980); this has yet to be confirmed. Laying in nests of unrelated species has not been reported in Tennessee. A clutch described by Laskey (unpubl. notes) with eggs of different sizes and colors was likely the product of more than 1 female.
—Charles P. Nicholson.

Barn Owl
Tyto alba

The Barn Owl is one of the most widely distributed land birds, inhabiting most tropical and temperate regions where suitable habitat comprising grasslands, farmlands, and urban areas, is available (Bunn, Warburton, and Wilson 1982, AOU 1983). In Tennessee, it is a rare to locally uncommon permanent resident statewide, the least common of the 4 widespread breeding owls. Banding data show that young Barn Owls disperse long distances, sometimes hundreds of kilometers, and that more northerly populations may be at least partly mi-

gratory (Stewart 1952, Marti 1988). Therefore, while most of the state's breeding Barn Owls are apparently resident, there is some movement of other owls into and out of the state between autumn and spring.

The Barn Owl inhabits open or mostly open areas with low ground cover, mainly grasslands and marshes, where it hunts for small mammals, its major prey (Bunn, Warburton, and Wilson 1982, Marti 1988). Typical habitats in Tennessee are hayfields, lightly grazed pastures, wet meadows, and recently abandoned farmland, as well as urban habitats such as vacant lots, cemeteries, and parks. About two-thirds of prey items in regurgitated pellets from across the state consisted of voles (*Microtus* spp.); most of the rest were short-tailed shrews (*Blarina* spp.) and hispid cotton rats (*Sigmodon hispidus*) (Parmalee and Klippel 1991). Voles and cotton rats typically occur in grassy and shrubby habitats. Barn Owls also require relatively large cavities for roosting and nesting, although they are adaptable in the choice of cavities (Marti 1988).

Barn Owl. Elizabeth S. Chastain

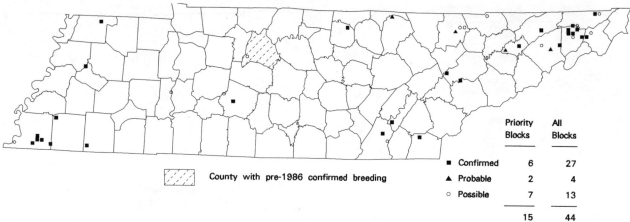

Distribution of the Barn Owl.

Hollow trees and caves in river bluffs are typical natural sites in Tennessee. Human-made sites include barns, silos, church steeples, abandoned buildings, under bridges, potholes in quarries, and nest boxes.

Little is known about the distribution of the Barn Owl in Tennessee prior to the twentieth century. Because of the limited extent of the open habitat, the Barn Owl was probably either very rare or absent from prehistoric Tennessee, and increased as European settlers cleared forests for agricultural uses.

Atlas workers reported the Barn Owl from several sites across the state; its mapped distribution, however, is probably not indicative of its true range. Barn Owls are highly nocturnal, shy, and lack a typical territorial call, and therefore are one of the most difficult birds to locate. Finding them usually required searching suitable roost and nest sites. Asking farmers about the presence of "monkey-faced owls" was also a useful source of records (Knight pers. obs.). The concentration of records in the northern Ridge and Valley was expected, as this area has a high proportion of grassland and, historically, relatively high Barn Owl numbers. The few records in the Central Basin, where the owl was regularly reported in the past *(Migrant),* are probably due to a lack of Atlas coverage. Heavily forested areas, such as the Unaka Mountains and Cumberland Plateau in the east, and farmlands dominated by row crops, such as much of West Tennessee, offer little or no suitable habitat.

The lack of recent Barn Owl records in areas where the owl previously occurred may also be evidence of a population decline, as has occurred in other parts of North America (Marti 1988). Because of concern over its status, the Barn Owl was listed as In Need of Management in Tennessee in 1976 (TWRA 1976), and has been on the National Audubon Society's Blue List since 1971 (Tate 1986). As suggested by Colvin (1985) in Ohio, the decline is probably due to the loss of grassland foraging habitat through more intensive farming practices and a general decline in farm acreage. The loss of nest sites may be a local factor in the decline, which can be remedied by providing nest boxes (Marti, Wagner, and Denne 1979). Recovery efforts for the Barn Owl in Tennessee have been minimal, mainly providing a few nest boxes and release of captive-reared young in the Memphis area. The success of this release program is unknown; large-scale restocking programs in the Midwest have shown little success (Marti 1988). Because of the Barn Owl's high reproductive capacity (see below) and dispersal ability, it is unlikely that restocking would be necessary, provided suitable habitat is present. Future management efforts should concentrate on a more thorough inventory and maintenance of foraging habitat, followed by nest boxes if nest sites prove to be limiting. For the time being, retention of the In Need of Management status is warranted.

Breeding Biology: Barn Owls have a flexible reproductive regime including monogamy and polygyny, a large range of clutch sizes, 2 broods per year if prey is abundant, and nesting at any time of year (Marti 1988). Eggs or nestlings have been reported in Tennessee every month except August. The peak of breeding activity is from April to June, with another small peak in September to December. No nest material is added to the nest site, although pellets and prey remains accumulate to form a nest substrate. Clutch size in Tennessee ranges from 2 to 7, with an average of 4.89 eggs (s.d. = 1.34, n = 27). The pure white eggs are laid at intervals of 2–3 days, and incubation begins with the laying of the first egg, resulting in a great size difference in the young (Marti 1988). This difference acts to adjust the number of surviving nestlings to the abundance of prey; when prey is abundant, most young would survive, while when prey is scarce, only the largest young would likely survive. The female incubates the eggs for about 30 days and is fed by the male during this period. Both adults later feed the nestlings, which fly in about 60 days (Marti 1988). The fledglings often return to the nest site to roost for several weeks after fledging. Two broods per year have been reported in the same nest site in Tennessee, with the interval from fledging to the next completed clutch as little as 10 weeks (*Migrant* 6:72, Jamison and Simpson 1940). Female Barn Owls may also begin a second clutch while the male is still feeding young from the first clutch (Marti 1988).—*Richard L. Knight.*

Eastern Screech-Owl
Otus asio

The Eastern Screech-Owl is a fairly common permanent resident, found at low elevations throughout the state. It is the most numerous owl and probably the most numerous of all raptors. It is easily identified by its small size, ear tufts, and distinctive trilled and descending whinny-like songs. Two color phases of the screech-owl occur in Tennessee, with the red phase outnumbering the gray phase by 1.7 to 1 among live birds and 3.4 to 1 among road-killed birds (Fowler 1985). An intermediate brown phase is rare.

The Eastern Screech-Owl is adaptable in its habitat use and occurs in a variety of habitats, including forests, rural agricultural areas, and suburbs. Use of deciduous and mixed forests is higher than coniferous forests (Smith and Gilbert 1984), and riparian woodlands are heavily used. The screech-owl roosts and nests in cavities that are either natural, excavated by a woodpecker, or human-made. It readily uses boxes erected for Wood Ducks, as well as boxes specifically built for owls. Fowler and Dimmick (1983) found screech-owl use of boxes for roosting and nesting in East Tennessee was higher in suburban and agriculturally dominated rural areas than in forested areas. These results suggest that suitable cavities are limiting in some habitats. Tennessee nests in natural cavities and old woodpecker holes have been from 2.4 to 16.8 m above ground, with no apparent preference for tree species (Duley 1979, egg coll. data). Belthoff and Ritchison (1990) found cavity depth, which averaged 31 cm for occupied nests, was a more important determinant of screech-owl cavity use than tree species, cavity height, and cavity orientation.

The Eastern Screech-Owl was probably common throughout prehistoric Tennessee, except for the higher elevations of the Unaka Mountains, where it infrequently occurs as high as 1219 m (Stupka 1963). It was first reported in Tennessee by Fox (1886) from Roane County. Langdon (1887) observed it in Blount County. Rhoads (1895a:480) did not observe the owl in Tennessee, but based on reports of local residents, described it as "common enough for the every day needs of the more superstitious natives." Ganier (1933a) described it as a common permanent resident throughout the state. Its population probably declined somewhat with widespread logging, and subsequent loss of cavities, in the late nineteenth and early twentieth centuries. Competition for nest sites from European Starlings, which were common statewide by the late 1930s, may affect owl numbers in agricultural and suburban landscapes (see Fowler and Dimmick 1983).

Little quantitative information is available on recent population trends of the screech-owl in Tennessee, and it is poorly sampled by BBS routes. Tanner (1985) found a significant increase in screech-owl numbers on Christmas Bird Counts since the early 1970s. Most of this increase, however, was probably due to increased efforts in detecting this species, rather than a real increase in numbers.

Atlasers found the screech-owl throughout the state in 40% of the priority blocks. Because it is nocturnal, observing it required special efforts by atlasers, and these efforts were not uniform across the state. The screech-owl sings little during April and May, but singing increases in June following fledging of the young (Ritchison et al. 1988). During June, it was usually not hard to record at dusk or shortly before dawn. It probably occurs in almost all of the low-elevation priority blocks, but may be absent or occur in very low numbers in some Middle and West Tennessee blocks with little woodland.

Confirming the breeding of the screech-owl was not easy for atlasers, and confirmed records made up about 15% of the total records. Over three-fourths of the confirmed records were of fledglings, and most of the probable records were of pairs or territorial birds.

Eastern Screech-Owl. Elizabeth S. Chastain

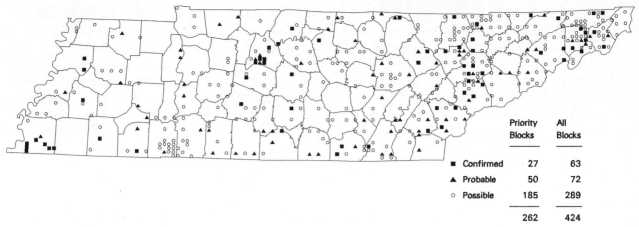

	Priority Blocks	All Blocks
■ Confirmed	27	63
▲ Probable	50	72
○ Possible	185	289
	262	424

Distribution of the Eastern Screech-Owl.

Breeding Biology: The Eastern Screech-owl begins courtship in late winter, and by the beginning of March the pair often roosts together. They add no nest materials to the nest cavity. Egg laying peaks during late March and early April (Duley 1979, egg coll. data). Clutch size in Tennessee ranges from 2 to 6 eggs. Duley (1979) found an average clutch size of 4.12 eggs (s.d. = 0.78, n = 25) during 1978 in East Tennessee; 44 other clutches from East and Middle Tennessee average 3.66 eggs (s.d. = 0.80). Clutches of up to 8 eggs have been reported elsewhere in the owl's range, and clutch size increases from south to north (Murray 1976). The eggs are elliptical to nearly spherical and pure white (Bent 1938).

The incubation period averages about 26 days (Bent 1938, Duley 1979). The female does most or all of the incubation, and both parents feed the young. The young leave the nest in about 31 days (Van Camp and Henny 1975) and are dependent on the parents a few weeks before beginning dispersal in July (Ritchison et al. 1988). Screech-owls are single-brooded, and they renest after the loss of a clutch early in the incubation period (Van Camp and Henny 1975, egg coll. data).—*Charles P. Nicholson.*

Great Horned Owl
Bubo virginianus

The Great Horned Owl, largest member of its family nesting in Tennessee, is an uncommon to fairly common permanent resident. It occurs throughout the state and is probably most common in areas of mixed fields and woodlands. The Great Horned Owl is easily identified by its large size, ear tufts, and white throat, and by very low-pitched 3- to 5-note hoots.

Although often described as occurring in heavily forested regions (Bent 1938, Ganier 1947), the Great Horned Owl occupies a variety of habitats. In Tennessee, it occurs in upland and bottomland forests, agricultural areas, and urban woodlands. Within the Unaka Mountains, it rarely occurs at high elevations (Stupka 1963). It may be most common in areas of mature (over 80-year-old) forests mixed with farmlands, as noted in southwest Virginia (McGarigal and Fraser 1984). In comparison with young forests, mature forests provide more large trees for nesting and a more open flight area beneath the forest canopy. Much of the owl's hunting is done in fields and along forest edges, where voles and rabbits, which make up much of its diet, are numerous (Petersen 1979).

The Great Horned Owl is probably more common now than in prehistoric Tennessee, when the open habitats in which it frequently forages were much less common. Its numbers probably increased with European settlement. The first published report from Tennessee was that of LeMoyne (1886), who, in contrast to later reports, described it as the most common owl in the Smokies. Langdon (1887) observed it at 1219 m in the

Great Horned Owl. Elizabeth S. Chastain

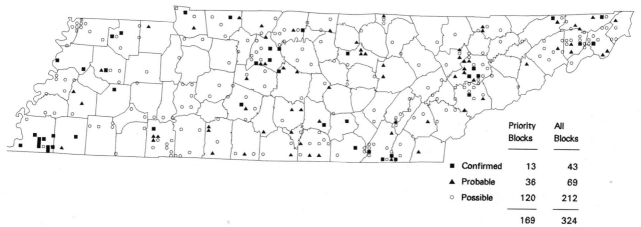

Distribution of the Great Horned Owl.

Smokies, and Rhoads (1895a) reported it from the Reelfoot Lake area, Shelby County, and on the Cumberland Plateau in Hamilton County. Ganier (1933a) described it as fairly common in West Tennessee and rare in Middle and East Tennessee. The more recent regional lists from the Nashville area (Parmer 1985) and Knox County (Howell and Monroe 1958) describe it as uncommon. Because the Great Horned Owl is poorly sampled by BBS routes, and efforts to record it on Christmas and spring bird counts have not been constant, little quantitative information is available on its population trends. The availability of suitable habitat has probably changed little in recent decades, and the Great Horned Owl is probably less persecuted by humans than a few decades ago.

Atlasers found the Great Horned Owl in about a quarter of the priority blocks. As with other nocturnal species, efforts to find it were not uniform across the state, and it was probably missed in many Atlas blocks. The proportion of priority blocks in which it was found was highest in the Ridge and Valley, probably due both to the abundance of woodland–field habitat and more effort to record it than elsewhere in the state. The owl was recorded in few heavily forested blocks in the Unaka Mountains, Cumberland Plateau, and Cumberland Mountains.

Confirming breeding of Great Horned Owl was frequently difficult; confirmed Atlas records about equally divided between nests with young and fledglings. Horned Owls nest early, and, by the time most Atlas work was done, most young had fledged. Most of the probable breeding records were of pairs, usually identifiable by the different pitches of their calls.

Breeding Biology: The Great Horned Owl nests in a variety of sites, including hollows in the broken-off tops of large trees, ledges on cliffs, and old nests of crows, squirrels, and hawks, particularly the Red-tailed Hawk.

Ganier (1947) felt that hollow trees and ledges were the owl's preferred nest sites. Nests in hollow trees have been from 4.6 to 18.3 m above ground, and nests in crow and hawk nests have been from 6.1 to 29.9 m above ground. Nests have been reported in many tree species, with little apparent preference other than the availability of a suitable nest site. Sites in the interior of woodlots appear preferable to more exposed nest sites, as noted elsewhere by Petersen (1979). The owl pair does no nest construction other than adding a few feathers.

The nesting season of the Great Horned Owl is the earliest of any species nesting in Tennessee. Courtship occurs in late fall and early winter, when the owl is most vocal, and eggs are usually laid in January. Incubating birds are occasionally covered by snow. Tennessee clutches include 22 sets of 2 eggs and 2 of 1 egg. Clutches of 3 eggs are uncommonly laid, as evidenced by 3 records of broods of 3 young. Throughout the owl's range, clutches of up to 5 eggs have been reported and clutches average 2.44 eggs (Murray 1976). The eggs are unmarked, dull white, and elliptical to almost spherically shaped.

The female does most or all of the incubation, which begins when the first egg is laid and lasts for 26–35 days (Johnsgard 1988). Both adults feed the young. Although they may leave the nest when about 5 weeks old, the young do not begin to fly well until about 9 weeks old (Johnsgard 1988). By the time the young fledge, the nest, especially if an old crow or squirrel nest, may have been disintegrated by the active nestlings. The young begin catching their own food 2–3 weeks after flying and remain with the parents for up to 3 more months (Petersen 1979). Most Great Horned Owls do not nest until almost 2 years old (Petersen 1979).—*Charles P. Nicholson.*

Barred Owl
Strix varia

The Barred Owl is an uncommon to fairly common permanent resident of wooded areas throughout Tennessee. It is easy to identify by its large size, barred brown plumage, brown eyes, and lack of ear tufts. It is more active during the day than the other owls nesting in Tennessee, and occasionally it gives its loud, distinctive "who cooks for you, who cooks for you-all" call during the day, especially in spring and summer.

The Barred Owl inhabits forests, and some authors (e.g., Robinson 1990) describe it as most common in bottomland and swamp forests. It is, however, also common in extensive, mature upland forests. It occurs at all elevations, and in the Smokies it is more common at higher than at low elevations (Stupka 1963). It nests in many forest types, including bottomland hardwoods and cypress, oak-hickory, oak-pine, mixed mesophytic, cove hardwoods, and spruce-fir. The Barred Owl has been recorded on Tennessee breeding bird censuses in virgin mixed mesophytic forest (Robertson 1979), second-growth mixed mesophytic forest (Smith 1977), and plots containing second-growth mixed mesophytic forest and strip mines (e.g., Yahner 1972) or old fields (Simmers 1980). Ford (1990) observed it on 34 of 59 West Tennessee forested wetland sites.

Although no quantitative studies of the Barred Owl's habitat use in Tennessee have been published, studies elsewhere suggest it requires large areas of mature forest. In southwest Virginia it occurred more often in mature (over 80-year-old) upland oak forests than in younger forests, and showed no preference for areas near farmland (McGarigal and Fraser 1984). Its home range averaged about 230 ha in Minnesota, where it made little use of old fields within its home range (Nicholls and Warner 1972).

The Barred Owl was probably common throughout prehistoric Tennessee except in the cedar glades of Middle Tennessee and the grasslands of Middle and northwest Tennessee. Among the owls, it was probably second in numbers only to the Eastern Screech-Owl. LeMoyne (1886) described it as common in the Smokies, and Rhoads (1895a) observed it in the Reelfoot Lake area and Shelby and Davidson Counties. Its numbers probably declined with agricultural clearing, and Ganier (1933a) described it as common in West Tennessee and fairly common in Middle and East Tennessee.

Tennessee BBS route results suggest an increasing population trend from 1966 to 1994; the proportion of routes showing increases is greater ($p < 0.05$) than the proportion with decreases. This trend is probably due to the overall increasing age and area of forest in the state, most pronounced in the low elevation of the Unaka Mountains, the Cumberland Plateau, Western Highland Rim, and Coastal Plain Uplands, all areas with high rates of farm abandonment in the 1920s and 1930s. Conversely, there have probably also been local declines as large areas of forests have been converted to pine plantations or bottomland soybean fields. The rangewide population did not show a significant trend from 1966 to 1991 (Peterjohn and Sauer 1993).

Atlasers found the Barred Owl in 40% of the completed priority blocks. Because it is relatively vocal during the day, the Atlas results are probably a better portrayal of its range than is the case with the other common owls. It was, however, probably still missed in many forested blocks. The Atlas results suggest it is absent or rare in many largely deforested blocks in Middle and East Tennessee. Many of its West Tennessee occurrences were in bottomland forests, where it shows some tolerance of fragmented forests. It was the only owl frequently reported by Atlas workers from the high elevations of the Unaka Mountains. The Barred Owl was recorded on 5% of the miniroutes, too few to show regional differences in abundance.

Atlasers confirmed the breeding of the Barred Owl in about 10% of the blocks, and almost all of the confirmed records were of fledglings. Fledgings have a loud, distinctive hissing call given throughout the day during the first few weeks after leaving the nest. The proportion of Atlas workers familiar with this call, however, was low.

Barred Owl. Elizabeth S. Chastain

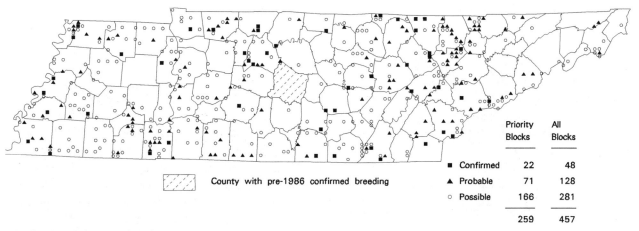

Distribution of the Barred Owl.

Breeding Biology: The breeding biology of the Barred Owl in Tennessee is poorly known and details of few nests are available. The Barred Owl nests in trees, either in open-topped cavities in large trees, the preferred site, or abandoned hawk, crow, and squirrel nests (Apfelbaum and Seelbach 1983, Bent 1938, Devereux and Mosher 1984). Seven Tennessee nests have been in tree cavities, 1 in an abandoned squirrel nest, and 1 in a nest box constructed from a section of a large, hollow log. The nests in natural cavities ranged in height from 4.3 to 12.2 m above ground. When nesting in a tree cavity, the owls normally do no nest construction. When nesting in an abandoned platform nest, they may scrape out a bowl and add green pine needles and a few sticks (Bent 1938). The same nest sites are frequently used for several consecutive years. In upland forest, nests tend to be near forest openings in sites with well-developed understories and large overstory trees (Devereux and Mosher 1984).

Egg laying occurs in early March. At least 4 clutches of 2 eggs and 2 of 3 eggs have been found in Tennessee. Clutches of 4 eggs also occur, as evidenced by a record of a nest with 4 young (Todd 1944). Clutch size elsewhere is 2 to 4 eggs, with an average of 2.4 throughout the species's range (Bent 1938, Murray 1976). The eggs are unspotted, dull white, and oval to elliptical in shape. Incubation lasts about 4 weeks and is carried out mostly or entirely by the female (Bent 1938). The young often leave the nest when 4 or 5 weeks old and fly when about 6 weeks old; their parents care for them until they are about 4 months old. When they leave the nest, the young are covered with gray to brown down, which is retained through late summer. The Barred Owl is single-brooded, and lays a second clutch after loss of the first (Bent 1938).
—*Charles P. Nicholson.*

Northern Saw-whet Owl
Aegolius acadicus

The Northern Saw-whet Owl is the smallest owl, indeed the smallest raptor, in eastern North America. This tame inhabitant of northern and western forests reaches its southeastern breeding limit in the mountains of Tennessee and North Carolina, where it is a rare and local summer resident. It is also a rarely encountered winter visitor across Tennessee, with just over a dozen records scattered from Shelby County to Washington County (Robinson 1990). The chocolate-brown juvenile Saw-whet is sufficiently different in appearance from the adult that it was once thought to be a separate species, the White-fronted Owl (Simpson 1970).

The first certain record of the Northern Saw-whet Owl in the southern Appalachian Mountains during the breeding season was on 21 June 1941, when one was heard calling on the North Carolina side of Clingman's Dome in the Great Smoky Mountains National Park (Stupka 1946). An earlier record of a very small owl seen briefly at dusk on 30 June 1933 on top of Mt. Le Conte in the Tennessee portion of the Smokies (Ganier 1946b), was almost certainly a Saw-whet; other possible records there extend back to around 1925 (Simpson 1968). Most subsequent records from the Smokies have been from the main crest between Clingman's Dome and Icewater Springs, along the Tennessee–North Carolina border (Stupka 1946, Savage 1965, Alsop 1991). Other records in the park have come from Mt. Le Conte (Tanner 1957). Prior to the Atlas, the only other Tennessee site with breeding season records was Roan Mountain, with calling owls noted in 1968 (Simpson 1968) and later beginning in 1981 (LeGrand 1982, *Migrant* 56:82). This site is also astride the state line. The first breeding evidence of the Northern Saw-whet Owl in the southern Appalachians was a juvenile seen 10 July 1965 in North

Northern Saw-whet Owl. Elizabeth S. Chastain

Carolina's Balsam Mountains (Peake 1965), southeast of the Smokies. Four subsequent records of fledged juveniles between 16 August and 2 September in western North Carolina have been reported (Simpson and Range 1974, McKinney and Owen 1989). The first nest for the two-state region was found in 1989 near Asheville, North Carolina (LeGrand 1990a). Only 2 instances of breeding have been documented in Tennessee, fledglings in 1988 and a nest in 1992, both discussed further below.

The Northern Saw-whet Owl was found in only 4 blocks during the Atlas, 3 of which were in traditional mountain areas. In the Smokies, calling owls were heard on the trail to the Mt. Collins shelter and near the Siler's Bald shelter, perhaps the westernmost record in the park. Another calling owl was reported from the Tennessee side of Roan Mountain. One of the most startling of all Atlas records was the discovery of a pair of adult Saw-whets with at least 2 recently fledged young on 8–11 May 1988 in the Ridge and Valley portion of Claiborne County (McKinney and Owen 1989). Whether this record represents "part of an unknown Cumberland Mountain foothills population or a one-time, accidental occurrence" (McKinney and Owen 1989:6) remains to be determined; a Saw-whet was heard calling at the same site in the late winter of 1989 (G. W. McKinney pers. comm.). Coverage for the Saw-whet during the Atlas was inadequate, as it was for most other nocturnal birds. Additional records from the Smokies would have been expected, as suggested by an owl calling in 1988 near Mt. Guyot a few hundred m into North Carolina (Nicholson pers. comm.). At least 2 calling owls, and later a nest, were found on Unaka Mountain in Unicoi County in 1992, a year after the Atlas. These were the first records for that site, which was not checked for nocturnal birds during the Atlas.

Very little data is available concerning the abundance and population trends of this species in the southern Appalachians, as most reports are anecdotal accounts of calling owls. Simpson (1972) calculated abundances of 1 "calling station" per 2.3 km (9 in 21 km) along a transect in the Great Balsam–Pisgah Ridge area of North Carolina and 1 per 2.2 km (5 in 11 km) along the road from Newfound Gap to Clingman's Dome in the Smokies. However, Alsop (1991) reported hearing as many as 11 calling Saw-whets in a single night at the latter location, which would translate to a density of 1 per 1.0 km. By comparison, a Wisconsin study reported an abundance of 1 per 0.4 km (Swengel and Swengel 1987). Since these figures are based on calling owls, some of the factors influencing their calling and the fieldworker's ability to hear them should be considered. Most sources (e.g., Simpson 1970, Alsop 1991) recommend clear nights, preferably with some moonlight, and little or no wind. April and May are the prime calling period, with Tennessee records from 22 February to 7 July. The owls may respond to an imitation of their distinctive "too-too-too . . ." call at any season and sometimes even in daylight.

Breeding habitat for Northern Saw-whet Owls is generally described as coniferous or mixed deciduous-coniferous forest. The southern Appalachian population is usually associated with the high-elevation spruce-fir belt, but Simpson (1972) stresses the importance of the transition area between this forest type and the adjacent northern hardwood forest. Most records have come from above 1600 m elevation, though some range down to 1400 m. The Unaka Mountain nest site was in a grove of mature spruce within the birch-maple forest at approximately 1350 m (Mayfield and Alsop 1992). The anomalous Claiborne County record was in a deciduous woodland containing patches of pines at 375 m elevation (McKinney and Owen 1989).

Breeding Biology: The cavity-nesting Northern Saw-whet Owl usually occupies old woodpecker holes, especially those of flickers, but natural cavities and nest boxes are accepted as well. No nest material is added, although prey remains and pellets may accumulate. Nest boxes should be lined with wood shavings. Nest height may vary considerably. Cannings (1987) found Saw-whets using nest boxes 2.6–6.1 m above ground in Canada and the Tennessee nest was in a nest box 4.6 m high (Mayfield and Alsop 1992). The white, nearly spherical eggs are laid at intervals of 1–3 days, with most clutches containing 5–6 eggs (Harrison 1975). Incubation is carried out mainly by the female, beginning with the laying of the first or second egg, resulting in asynchro-

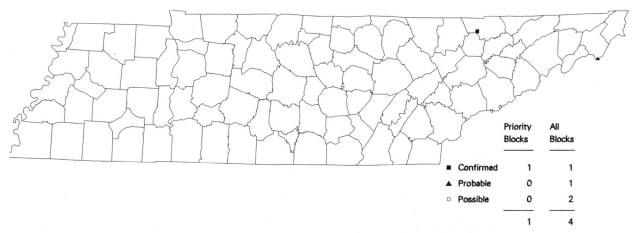

Distribution of the Northern Saw-whet Owl.

nous hatching. The Unaka Mountain nest contained 5 eggs in July, 4 of which hatched late that month, with 3 young fledging in late August (Mayfield and Alsop 1992). Even when considering the elevation difference, these dates are very late in comparison to the Claiborne County fledglings and seem to be very late for a species generally considered to be an early nester. Saw-whets are single-brooded and usually monogamous, although 2 cases of polygyny have been reported in the Pacific Northwest (Marks, Doremus, and Cannings 1989).—*Richard L. Knight.*

Common Nighthawk
Chordeiles minor

The Common Nighthawk is an uncommon to fairly common summer resident of Tennessee, usually present from late April through early October. During late August and early September, large flocks, sometimes numbering thousands of birds, migrate southward through the state (Robinson 1990). Breeding nighthawks occur in areas of exposed gravel or soil and few trees, and are most common in urban areas, Middle Tennessee cedar glades, and heavily agricultural areas.

Atlasers found Common Nighthawks in slightly less than 20% of the completed priority blocks. Nighthawks are most easily observed and identified by their erratic, stiff-winged flight, nasal "peent" calls, and the booming display of the male. They feed on flying insects, are most active at dawn and dusk, and only occasionally fly during the day. During the day, they usually roost on the ground, the roof of a building, or a tree limb. Their plumage usually blends well with the substrate, and they are difficult to observe. For these reasons, atlasers probably missed nighthawks in some blocks.

Nighthawks were recorded on only 10 of the Atlas miniroutes, too few to show regional differences in abundance. Using the proportion of completed priority blocks as a measure of abundance, they are most common in the Central Basin physiographic region, where atlasers found them in 38% of the priority blocks. Next in nighthawk abundance are the Loess Plain (34%) and the Ridge and Valley (23%). In all other physiographic regions, nighthawks were recorded in 15% or less of the blocks.

Before the settlement of Tennessee by Europeans, Common Nighthawks were probably most common in Central Basin cedar glades. Smaller numbers probably nested on the smaller glade areas on the Coastal Plain Uplands, Western Highland Rim, and Ridge and Valley. Nighthawks may also have nested in areas of mineral soil exposed by intensive forest fires, as they do elsewhere in their range (Nicholson pers. obs.). Suitable nesting habitat increased as European settlers cleared large areas for agriculture.

The first ornithologist to report nighthawks in Tennessee was Fox (1886) who observed them in Roane County in April 1884. The following year, Langdon (1887) observed 5 flying over Tuckaleechee Cove, Blount County, in mid-August; this is early for fall migrants, and they may have been locally breeding birds.

Common Nighthawk. Elizabeth S. Chastain

A few years later Rhoads (1895a:484) found nighthawks "in lowlands all across the State but not in the mountains." This suggests he found them on the Cumberland Plateau. At about the same time, Torrey (1896a) described them as common in the Chattanooga area. At the turn of the century, Gettys collected several sets of eggs from nests in McMinn County croplands (Gettys egg coll.). The flat, graveled roofs of large buildings also provided nesting habitat, and the first record of roof nesting was in Washington County in 1908 (Lyle egg coll.). Elsewhere in the nighthawk's range, roof nesting was first reported in Philadelphia in 1869 and was common in the Northeast by 1880 (Gross 1940).

Nighthawk numbers, as measured by Tennessee BBS routes, have shown no significant overall trend since 1966, although they declined by 8.3%/year (p < 0.10) from 1980 to 1994. The number of birds recorded on the routes, however, is small and the variance high. BBS routes throughout the species's range similarly show no significant change between 1966 and 1993 (Peterjohn, Sauer, and Link 1994). Because of concerns over a perceived decrease in numbers, however, the Common Nighthawk has been included on the *American Birds* Blue List since 1975 (Tate 1986).

Breeding Biology: As was true of the other members of its family, confirming the breeding of nighthawks was difficult for Atlas workers. Nighthawks lay their eggs directly on the substrate without building a nest, and the eggs, as well as the incubating adult, are cryptically colored. Checking potential rooftop nest sites usually required gaining access to roofs, which was often difficult. Several Tennessee egg collectors, however, were quite successful in finding their nests. H. O. Todd found at least 73 nests in Rutherford County (DeVore 1975), including 21 nests in 1936 (Todd 1944). Most of these nests were on rocky cedar glades. The tendency of nighthawks to return to the same nest site in consecutive years (Dexter 1961) probably helped Todd and others to build large collections.

Common Nighthawks are monogamous, and the male defends a territory that usually includes both nesting and feeding areas. (Armstrong 1965). The male's booming display is apparently used for both courtship and territorial defense (Weller 1958, Sutherland 1963). In this display, the male swoops downward and at the bottom of the dive, often only a few meters above the ground, flexes his wings. The loud booming sound is produced by air vibrating the wing feathers. Territory size varies; Armstrong (1965) found urban territories averaged 10.4 ha, and Sutherland (1963) found nests an average of 80 m apart. No information on territory size or breeding density is available from Tennessee.

Egg laying begins in early May and peaks during the second half of May. Nest sites include gravel roofs, exposed rock in cedar glades and pastures, and in plowed fields. The eggs are laid directly on gravel or soil, with no added nest material. When nighthawks nest on roofs, the eggs are usually laid near a parapet or other structure (Dexter 1961). The normal clutch size is 2 eggs (Gross 1940). A few clutches containing 1 egg have been reported in Tennessee; some of these may have been complete clutches. The eggs are elliptically shaped and pale gray, heavily spotted with fine brown and gray spots (Harrison 1975). The female incubates the eggs and broods of the nestlings while the male frequently roosts nearby (Weller 1958). Incubation lasts about 19 days. The young are covered with down when they first hatch and, within a day of hatching, follow the female to brood in shaded areas. Both adults feed the nestlings, most frequently at dusk and dawn (Sutherland 1963). The young begin flying when about 23 days old and remain dependent on the adults for at least another week (Weller 1958, Grazma 1967). The adults use an elabo-

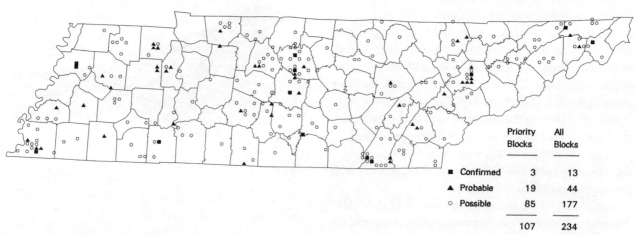

Distribution of the Common Nighthawk.

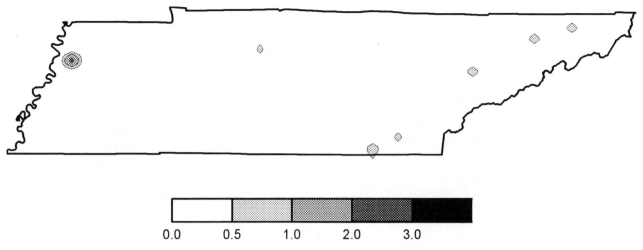

Abundance of the Common Nighthawk.

rate display to distract intruders from the eggs or nestlings. Two broods are occasionally raised in other parts of the nighthawk's range (Weller 1958), and the wide range of egg dates suggest this may occur in Tennessee.
—*Charles P. Nicholson.*

Chuck-will's-widow
Caprimulgus carolinensis

The Chuck-will's-widow is a fairly common, primarily nocturnal summer resident of Tennessee. It is most easily identified by its 4-syllable song, sung at dusk and dawn and throughout moonlit nights. Because of its loud voice, it is well known to many rural Tennesseans, although often mistaken for the Whip-poor-will, which has a 3-syllable song. During the day, Chuck-will's-widows roost in woodland on or near the ground, where their mottled plumage blends well with the forest floor. They arrive in Tennessee in mid-April, and most depart by late August (Robinson 1990); their fall migration is poorly known.

Atlasers found Chuck-will's-widows in about 40% of the priority blocks. Chuck-will's-widows are most vocal on moonlit nights, when they sing throughout much of the night (Cooper 1981). At other times, they sing briefly at dawn or dusk. Recording Chuck-will's-widows thus required special efforts by atlasers, and this effort was not uniformly applied across the state. Thus the species was not reported from many blocks in which it probably occurs. The concentration of records in southwest Tennessee is due to the special efforts of Ben and Lula Coffey, who surveyed them in this area for years (e.g., Coffey and Coffey 1980).

Historical changes in the distribution and abundance of Chuck-will's-widows in Tennessee are difficult to assess because of the paucity of early reports and standardized surveys, as well as the difficulty of precisely defining its habitat. Mengel (1965) described its habitat in Kentucky as farm woodlots, oak-hickory and pine groves, and old-field margins. Compared to the Whip-poor-will, the Chuck prefers more open habitat, where they often call from woodland edge and forage over fields (Bent 1940, Cooper 1981). In prehistoric Tennessee, Chuck-will's-widows probably occupied Middle Tennessee cedar forests, periodically burned pine and oak-pine forests in the southern Ridge and Valley and southern Coastal Plain Uplands, and upland forests of the Loess Plain. It was first reported in Tennessee by Wilson, who described it as rarely found north of Nashville, where he encountered it along the Cumberland River (Wilson 1811a). Rhoads (1895a) reported it from only Shelby County. Its simultaneous presence in southwestern Kentucky (Mengel 1965) suggests it occurred throughout upland western Tennessee. It was nesting in McMinn County in the southern Ridge and Valley about 1900 (Ijams and Hofferbert 1934).

Chuck-will's-widow. Elizabeth S. Chastain

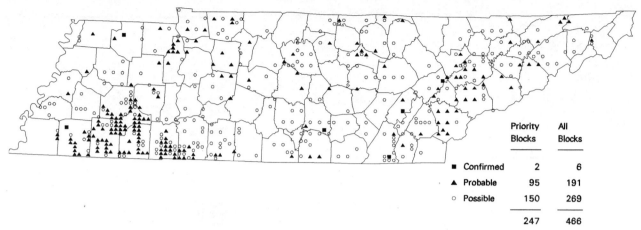

Distribution of the Chuck-will's-widow.

The Chuck-will's-widow's range probably expanded in Tennessee during the twentieth century, as suggested by Mengel (1965) for Kentucky. Ganier's (1933a) description of its distribution as common in West Tennessee, fairly common in the lowlands (Central Basin) of Middle Tennessee, and rare in the lowlands of East Tennessee, also suggests a more limited range than at present. It was a rare summer resident in upper East Tennessee in the early 1930s (Lyle and Tyler 1934) and present on the Highland Rim in Fentress County in 1937 (Ganier 1937a). It was also regularly reported from Knoxville in the 1930s *(Migrant)*. More recently, Tennessee BBS routes, which do not sample the species well, show no significant overall trend from 1966 to 1994, although it decreased by 7.1%/year (p < 0.01) from 1980 to 1994. Rangewide, Chuck-will's-widow numbers on BBS routes showed a declining trend of 1.2%/year from 1966 to 1993 (Peterjohn, Sauer, and Link 1994).

Because Atlas miniroutes were begun after most Chuck-will's-widows had ceased calling, they are a poor measure of regional differences in the species's abundance. Taking into account the differences in block coverage, the Atlas results suggest Chuck-will's-widows are presently most abundant in the central and southern Ridge and Valley, Eastern Highland Rim, southern Coastal Plain Uplands, and Loess Plain. They are virtually absent from the Unaka and Cumberland Mountains, rare to uncommon in the northern Ridge and Valley and Cumberland Plateau, uncommon in the Central Basin, Western Highland Rim, and northern Coastal Plain Uplands, and absent from the Mississippi Alluvial Plain.

Breeding biology: Chuck-will's-widows nest on the ground in wooded areas with an open understory. Nests in Tennessee have been reported in pine woods, mixed pine-hardwoods, and cedar-dominated woodlands. No nest is built, and the eggs are laid on dead leaves or pine needles. Individual females often renest in the same spot in consecutive years (Ganier 1964a). Because of the lack of a constructed nest and cryptic coloration of the birds, nests are difficult to find, and only about 20 detailed records are available from Tennessee. Confirmed records made up only 1.3% of the Atlas observations.

Egg laying in Tennessee begins in early May and appears to peak around the middle of May. The normal clutch size is 2 eggs, which are creamy white with blotches of reddish brown and gray (Harrison 1975). The female incubates for about 20 days (Hoyt 1953, Wilson 1959). At hatching, the young are covered with tan down and within a day are capable of running. Before they begin to fly at 14–16 days old, they may move several meters from where the eggs were laid (Wilson 1959, Nunley 1960). The feathers of the juvenile plumage emerge when the nestling is about a week old, and the molt to the basic plumage begins about the time the young begins to fly (Rowher 1971). The length of the fledging period is unknown, and likely around 2 weeks (Rowher 1971). The few available Tennessee nest records suggest that replacement clutches are laid and only 1 brood is raised; Rowher (1971) reached a similar conclusion.
—*Charles P. Nicholson and William B. Fowler.*

Whip-poor-will
Caprimulgus vociferus

The Whip-poor-will is a fairly common summer resident of Tennessee, most numerous in heavily forested areas. It arrives around the first of April and departs by early October. The Whip-poor-will is a nocturnal species and spends the day roosting on a tree limb or on the forest floor. It is much more easily heard than seen, and easily identified by its loud, distinctive, repetitive, 3-note song.

Whip-poor-wills occupy upland forests, usually near an opening, and frequently with an open understory. Mengel (1965) described the species in Kentucky as common in mixed mesophytic forests, rare in oak-hickory, and absent from bottomland forests. Cooper (1981) found the Whip-poor-will was more numerous in forested habitat than in open and suburban areas in Georgia. In Tennessee, it occurs in most upland deciduous and mixed forest types, and uncommonly in pine forests. Most Tennessee plot census studies recording Whip-poor-wills have included both forested and open areas. Densities on these plots are 8.5 pairs/100 ha in a Cumberland Mountain deciduous forest and strip mine plot (Nicholson 1979a), 3.3 to 5/100 ha in an Eastern Highland Rim mixed mesophytic woods and fields plot (Simmers 1979, 1980), 4.9/100 ha in a 6–7-year-old deciduous clearcut (Lewis 1981a, 1982), and 7.9/100 ha in a low-elevation Unaka Mountains mixed deciduous forest (Lewis 1983a). In the Unaka Mountains, Whip-poor-wills rarely occur above about 823 m (Stupka 1963).

The Whip-poor-will was first reported with certainty in Tennessee in the 1890s. Rhoads (1895a) observed it in the Reelfoot Lake area, presumably in the bluff forests, as well as in Shelby County, on the Cumberland Plateau in Fentress County and at Chattanooga. Torrey (1896a) found it on the Cumberland Plateau near Chattanooga, and Gettys collected its eggs in McMinn County (egg coll. data). Ganier (1933a) described the Whip-poor-will as rare in West Tennessee, fairly common in Middle Tennessee, and common in East Tennessee.

Atlasers found Whip-poor-wills in about one-third of the completed priority blocks. As with the Chuck-will's-widow, Whip-poor-wills are most easily observed by hearing their song at dusk and dawn, or throughout the night on moonlit nights. Recording them took special efforts by atlasers, and these efforts were not uniformly applied across the state. The concentration of records in southwest Tennessee is due to the special efforts of Ben and Lula Coffey, who surveyed them in this area for years (e.g., Coffey and Coffey 1980). Whip-poor-wills were recorded on only 11 Atlas miniroutes, too few to show regional differences in abundance. Atlas block results suggest that they are fairly common to common in the farmed valleys at the base of the Unaka Mountains, in the Cumberlands, Western Highland Rim, Coastal Plain Uplands, and along the northern Highland Rim escarpment.

Based on their apparent preference for forest openings, Whip-poor-wills were probably uncommon to fairly common in prehistoric Tennessee. Their numbers probably increased as settlers opened the forests, until deforestation reached the point where their numbers dropped. Reforestation of parts of the state since the 1930s has also affected their numbers. In primarily forested areas, the reforestation of former clearings may result in decreased numbers, as reported in the Smokies by Stupka (1963). Reforestation of marginal farmlands in primarily cleared areas may have resulted in local increases in numbers (e.g., Cooper 1981), and could account for the species's present abundance in the Coastal Plain Uplands. As noted above, Ganier (1933a) described it as rare in West Tennessee. During the Atlas period, Whip-poor-wills were not reported from a few West Tennessee counties where they were found in 1979 (Coffey and Coffey 1980), probably due to the lack of survey effort. Numbers on a 17 km roadside route in Sullivan County censused between 1959 and 1979 ranged from 1.9/km to 4.6/km, with no apparent trend (Hale 1979). Tennessee BBS route results show a significant decrease ($p < 0.05$) of 5.8%/year from 1966 to 1994. Rangewide BBS results show a slight overall declining trend during the same period (Peterjohn, Sauer, and Link 1994).

Confirming the breeding of Whip-poor-wills is difficult and usually results from the chance flushing of an adult from the nest. The percentage of confirmed Atlas records, 2%, is very low; the confirmed records included 1 distraction display, 4 fledglings, and 4 nests. Fewer than 2 dozen earlier Tennessee nest records are available.

Breeding Biology: Whip-poor-wills nest on the forest floor in open woodlands (Bent 1940). No nest is built, and the eggs are laid directly on dead leaves. The same nest site is often used in successive years, presumably by the same pair (Bent 1940, G. McKinney pers. comm.). Egg laying is synchronized with the lunar cycle so that

Whip-poor-will. Elizabeth S. Chastain

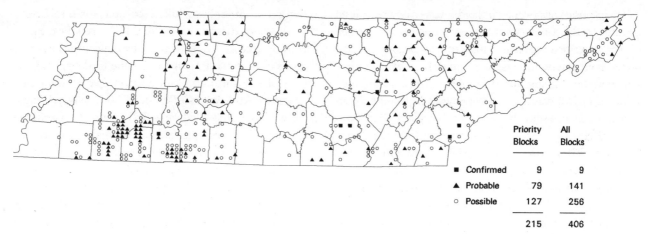

Distribution of the Whip-poor-will.

the eggs hatch during a young waxing moon, resulting in bright moonlight for efficient foraging while feeding nestlings (Mills 1986). Egg laying in Tennessee begins in late April and appears to peak during the first half of May. The normal clutch size is 2 eggs, which may be laid more than a day apart (Bent 1940, Goodpasture and Douglass 1964). The eggs are white with irregular gray and brown spots and blotches (Harrison 1975).

Incubation, usually undertaken by the female, lasts 19–20 days (Goodpasture and Douglass 1964, Ehrlich, Dobkin, and Wheye 1988). Newly hatched young are covered with cinnamon-colored down and may leave the nest site within a day or two. The female broods the nestlings during the day and both adults feed them. The young begin flying when about 19 days old and feed themselves when a month old (Mills 1986). Whip-poor-wills infrequently raise a second brood, which is begun while the young of the first brood are still dependent (Mills 1986). Although second broods have not been reported in Tennessee, the timing of a nest that hatched 12 July (NRC) is suggestive of a second brood.—*Charles P. Nicholson.*

Chimney Swift
Chaetura pelagica

The Chimney Swift is a common summer resident of Tennessee, present throughout the state but most numerous in urban and suburban areas. As its name implies, it nests and roosts primarily in chimneys. During the day it spends much of its time in flight, and it is a common site over settled areas. In flight it resembles a black cigar with bow-shaped wings. This distinctive silhouette and its loud, high-pitched, chattering call make it easily recognized and well known to many Tennesseans. The Chimney Swift arrives in Tennessee in late March or early April and departs by mid-October.

Before chimneys were available, the Chimney Swift nested and roosted in caves, sinkholes, and hollow trees. As chimneys became available, swifts moved into them, and their use was noted as early as 1671 in New England (Palmer 1949). Both Wilson and Audubon described them as nesting almost exclusively in chimneys in the early-nineteenth century (Tyler 1940). In addition to chimneys, swifts also less commonly nest on the inner walls of uninhabited buildings, in abandoned cisterns, and in wells. Each of these nest sites has been used in Tennessee (Stupka 1963, NRC).

Nesting or roosting in hollow trees has been reported several times in Tennessee. Tree roosts were found in the late nineteenth century in a large, hollow sycamore (Weakley 1941) and in a large tulip-poplar, "nearly six feet [1.8 m] through and at least fifty feet [18 m] up to the first limb," which the swifts entered through an 0.6 m hole where a large limb had broken off (McLaughlin 1926). Ganier (1962) observed a pair flying in and out

Chimney Swift. Elizabeth S. Chastain

of a hollow tree on 30 May 1925, high in the Smokies, and presumed it was their nest site. Wetmore (1939) reported, without details, that swifts were nesting in hollow trees on Mt. Guyot and 2 nearby peaks in the Smokies, and Ganier and Clebsch (1946) observed a swift trying to gather nest material and presumably nesting in a dead chestnut high in the Unicoi Mountains. A pair laid 4 eggs and raised 1 young in a nest in a hollow, open-topped, 4 m tall, 70 cm diameter dead oak in a mountainous part of Campbell County in 1979 (Nicholson pers. obs.). In late May 1981, a nest was reported in a hollow tree stub in wooded bottomlands of McNairy County (Nicholson 1984b), and, in July 1987, 3 birds were observed flying in and out of a 5.5 m tall, 55 cm diameter hollow snag near Clingman's Dome, the highest point in the state (Atlas results, *Migrant* 58:147). Tree nesting still rarely occurs in areas with numerous chimneys, as observed by Ferguson and Ferguson (1991) in Shelby County.

Atlasers found the Chimney Swift in almost all of the completed priority blocks, including many that are heavily forested with little or no human population. The swift was also recorded on 78% of the miniroutes, at an average relative abundance of 2.9 stops/block. Abundance was highest in urban areas, and the Ridge and Valley, with the highest human population density of any physiographic region, had the highest average swift abundance, 4.0/route. The miniroute results may underestimate swift numbers in some heavily forested blocks where flying swifts were difficult to see through the tree canopy.

The Chimney Swift population in Tennessee, as measured by BBS routes, has shown a declining annual trend ($p < 0.05$) of 1.5% from 1966 to 1994. Most of this decrease was from 1980 from 1994, when the trend was -2.3%/year ($p < 0.05$). Part of the decrease is due to the relocation of 2 routes since 1979, each of which then had many fewer swifts. The other reasons for the decrease are not clear. Its North American population declined by 0.9%/year ($p < 0.01$) from 1966 to 1993, although in the east it was stable during the latter half of this period (Peterjohn, Sauer, and Link 1994, Sauer and Droege 1992).

Atlasers often found it difficult to confirm Chimney Swift breeding, as inspection of nest sites was usually difficult. Most swift records were of birds in flight and assigned to the possible category. About 60% of the confirmed records were of swifts repeatedly entering nest sites and were coded ON.

Breeding Biology: After arriving in the spring, swifts roost in large numbers, and non-breeding birds continue to roost in groups throughout the summer. The same nest site is often used by the same pair of swifts for several consecutive years, and because the nest usually deteriorates during the winter, a new nest is normally built each year (Dexter 1969). Both adults construct the nest and gather nest material in flight by breaking twigs off trees with their feet. The nest is a half cup, made of twigs glued together and to the vertical surface with saliva. It is rarely closer than about 2 m to the opening of the chimney or tree (Dexter 1969). The adults often continue to add twigs to the nest after egg laying has begun.

Egg laying begins in mid-May. Clutch size in Tennessee is 3–6 eggs and averages 4.49 (s.d. = 0.74, n = 35). Tyler (1940) and Dexter (1981) reported similar clutch sizes elsewhere in the swift's range, with clutches of 3–5 eggs most common. The eggs are white, unmarked, and more cylindrical than those of most other birds. Both adults incubate the eggs for an average of 20 days (Dexter 1969). Both adults also feed the nestlings, which often leave the crowded nest in about 3 weeks and cling to the chimney wall for several days before flying from the nest site at 4 weeks old.

In addition to the nesting pair, swift nests are frequently attended by 1–3 additional unmated, usually immature

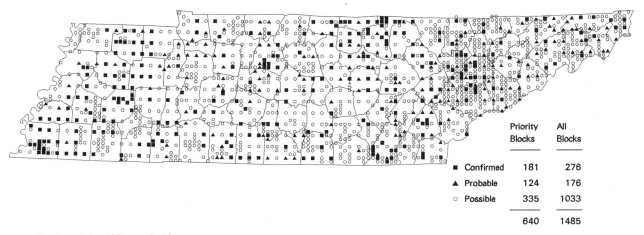

Distribution of the Chimney Swift.

	Priority Blocks	All Blocks
■ Confirmed	181	276
▲ Probable	124	176
○ Possible	335	1033
	640	1485

Confirmed Breeding Species

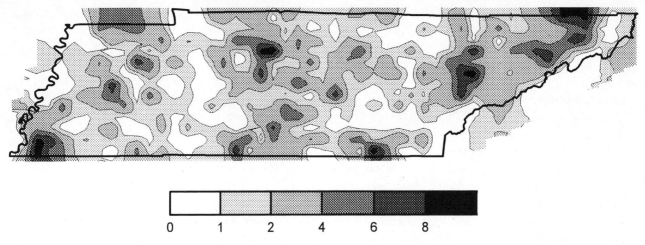

Abundance of the Chimney Swift.

swifts, termed "visitors" by Dexter (1981). These birds may assist with incubation and with feeding nestlings. The productivity of females nesting for the first time may be slightly increased when visitors are present; no increase was observed in the productivity of birds that had nested in previous years. Some of the visitors eventually nested in Dexter's Ohio study area.—*Charles P. Nicholson.*

Ruby-throated Hummingbird
Archilochus colubris

The Ruby-throated Hummingbird is the smallest of Tennessee's breeding birds. It occurs at all altitudes from the Mississippi River to the peaks of the Unaka Mountains and is becoming increasingly popular with Tennesseans as they discover the ease with which it can be attracted to flower gardens and especially hummingbird feeders. The hummingbird, a neotropical migrant, usually arrives by mid-April and departs by early October (Robinson 1990).

The Ruby-throated Hummingbird nests in a variety of forest and forest edge habitats. Its nest is usually built on a downward sloping tree limb near a woodland opening or edge, frequently over a stream or ravine (Johnsgard 1983b). It is dependent on flower nectar for much of its diet, and the availability of preferred flowers probably influences its local distribution, as it does the hummingbird's migration (Bertin 1982). Information on historical changes in hummingbird numbers is limited; its numbers, however, probably increased as the early clearing for agriculture by European settlers created more edge habitat. Rhoads (1895a) described the hummingbird as occurring throughout Tennessee except for the highest elevations of Roan Mountain. Ganier (1933a) described it as a fairly common summer resident in the 3 major divisions of Tennessee.

The Ruby-throated Hummingbird was found in 80% of the completed Atlas priority blocks and was relatively uniformly distributed throughout the state. Few common habitat features are obvious in the completed blocks where it was not found, and it was probably simply overlooked in those blocks. Away from flower gardens and feeders, the hummingbird is often difficult to observe because of its small size and quiet vocalizations. Its chance of detection also showed some increase with the hours a block was worked. The hummingbird was recorded on 22% of the miniroutes at an average density of 1.2 stops/route. It was least frequently reported in the Central Basin (on 8 of 85 routes); other regional differences are difficult to discern.

Ruby-throated Hummingbirds probably reach their highest density in riparian areas and bottomland forests. This was poorly reflected in the atlas miniroute results, which did not sample these habitats well. Ford (1990) found the Ruby-throated Hummingbird on 42 of 59 West Tennessee forested wetland sites at an average density of 47 birds/100 ha on transect counts. It

Ruby-throated Hummingbird. Elizabeth S. Chastain

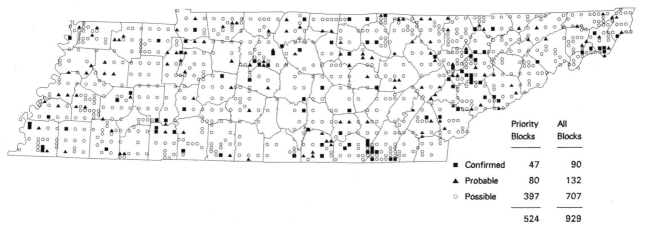

Distribution of the Ruby-throated Hummingbird.

was one of the more abundant and widely distributed species; habitat features correlated with high densities included the presence of tall snags and a tall herbaceous layer, both of which were probably related to the abundance of flowering plants. On mature upland forested plots elsewhere in the state, hummingbirds have been present in much lower densities or occasionally absent (e.g., Smith 1977, Robertson 1979, Turner and Fowler 1981a, Hardy 1991). An exception was a disturbed wooded ravine on the Eastern Highland Rim, where Simmers (1982a) recorded a density, by the spot-mapping technique, of 38 birds/100 ha.

Tennessee BBS routes show a slight, nonsignificant declining trend in hummingbird numbers from 1966 to 1994. No significant trend was evident in spring count results (Tanner 1986). The increasing area of forested land in recent decades and the increasing fragmentation of these forests have both probably been beneficial to hummingbirds, although they may have been more than offset by the detrimental aspects of increasing urbanization, changing agricultural practices, and clearing of bottomland forests. Rangewide, the Ruby-throated Hummingbird has shown an increasing trend of 1.6%/year ($p = 0.05$) (Peterjohn, Sauer, and Link 1994).

Atlas observers confirmed the breeding of the hummingbird in 10% of the Atlas blocks; backyard observers contributed several of these records, as the proportion confirmed in the priority blocks was a slightly lower 9%. Records of fledglings, many at feeders during midsummer, made up about one-third of the confirmed records.

Breeding Biology: Male hummingbirds establish a territory soon after their spring arrival. The probably promiscuous male displays in a pendulum-like flight that concludes in diving flights with the female and copulation. The male takes no further role in nesting. The nest, about 4 cm in outside diameter and built of plant down, lichens, and spider webs, often resembles a knot on a limb. Its small size and camouflaging make it difficult to find, although Gettys collected at least 67 hummingbird egg sets in McMinn County between 1897 and 1907 (Ijams and Hofferbert 1934, egg coll. data).

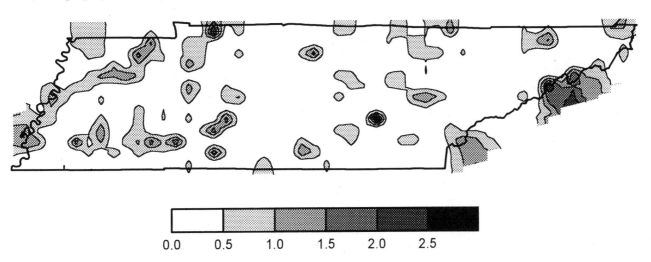

Abundance of the Ruby-throated Hummingbird.

Confirmed Breeding Species

Nests built in previous years may be refurbished and reused; Owen (1979b) described a nest, unusually located on a bracket attached to a carport, that was used 3 consecutive years. The nest is normally placed in a tree or large shrub; average height of 76 nests is 5 m (s.d. = 2.0 m, range 1.5–10.7 m). Nests have been reported in many species of deciduous trees, as well as rhododendron, pines, and hemlock. At least 15 of the nests found by Gettys were in pines.

Egg laying begins in late April and peaks during the second half of May. The normal clutch size is 2 eggs; a brood of 3 nestlings in Sevier County was very unusual (Stupka 1963). The eggs are pure white. Incubation lasts 11–14 days, and the young fly in 14–28 days (Johnsgard 1983b). The range of dates of nests with eggs and young in Tennessee suggests hummingbirds are sometimes double-brooded, as Johnsgard (1983b) suggested occurs elsewhere. During late summer and early fall, hummingbirds gather, rarely in numbers as high as 60 (Abernathy 1955), around patches of flowers, especially jewelweed and cardinal flower.—*Barbara H. Stedman.*

Belted Kingfisher
Ceryle alcyon

A flash of blue and white and a harsh, prolonged rattle drifting over water signal the Belted Kingfisher. This aggressive species is a permanent resident of Tennessee, although some withdraw to the south in harsh winters (Tyler and Lyle 1933). It occurs across the state wherever there are relatively clear waters with small fish, its principal food, and nearby vertical earth banks for nesting.

No evidence exists for significant change in the kingfisher's historical range in Tennessee. Cherokee Indians joked about this fish-gigging bird (Mooney 1972), and early naturalists knew it well. Fox (1886) found it to be fairly common on larger Roane County streams, and Langdon (1887) observed it at low elevations in the Great Smoky Mountains. Rhoads (1895a) reported it common across the state except at high elevations and on the top of the Cumberland Plateau. Ganier (1933a) described it as fairly common statewide in summer. Within the Unaka Mountains, it has nested up to 915 m and infrequently occurs above that elevation (Ganier 1926, Ganier 1962, Stupka 1963).

Atlasers found the Belted Kingfisher in all counties and in 68% of the completed priority blocks. It was found on 20% of the miniroutes at a relatively low, uniform density of 1.2 stops/route. Because of its conspicuous behavior and ease of identification, the Atlas results are probably an accurate portrayal of its distribution. Its lowest frequency of occurrence was in West Tennessee, the Central Basin, the Eastern Highland Rim, and the Cumberlands. It is probably scarce in much of West Tennessee because of limited feeding habitat due to the degradation of streams by channelization and siltation; in the Mississippi Alluvial Plain and Loess Plain, nest sites may be limited because of the flat topography. In the inner Central Basin, records were concentrated around the Cumberland River system impoundments. It was probably missed in a few Cumberland Plateau blocks where sizable perennial streams are in relatively inaccessible gorges and little other surface water exists. It is more common within the relatively well-watered Sequatchie Valley than in the adjacent southern Cumberland Plateau. It is also common in the Ridge and Valley, where surface water, in streams and reservoirs, is plentiful. Its status in other parts of the state without impoundments generally remains similar to earlier descriptions (i.e., West Tennessee: Calhoun 1941, Coffey 1944, 1975; Cumberland Plateau: Dubke and Dubke 1977, Ganier 1937a, Ganier and Clebsch 1940).

The numerous impoundments in Middle Tennessee have probably resulted in an increase in kingfisher numbers, especially in the inner Central Basin and north-central Highland Rim where, aside from the large rivers, surface water is scarce. Whether the impoundments in East Tennessee have resulted in increased kingfisher numbers is debatable. Studies elsewhere by Cornwell (1963) and Davis (1982) indicate that kingfisher densities are greater along clear streams rich in shoals and small fish than around lakes. Numerous nest sites are available in the vertical dirt banks surrounding much of the mainstream Tennessee River reservoirs, which show relatively little seasonal water-level fluctuation. This is reflected in the numerous confirmed records in Hamil-

Belted Kingfisher. Elizabeth S. Chastain

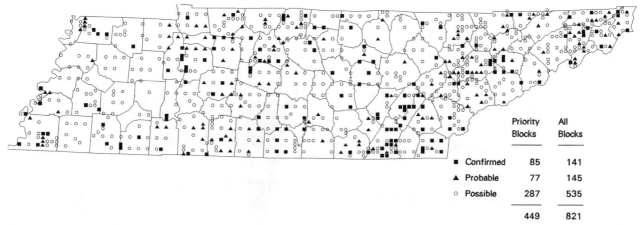

Distribution of the Belted Kingfisher.

ton, Meigs, and Rhea Counties. Nest sites are less common adjacent to the tributary reservoirs, which show great water-level fluctuation.

Fowler and Fowler (1985) found a breeding density of 1.1 kingfishers/km along the unimpounded Duck River in Maury County. No comparable density information is available from elsewhere in the state.

The breeding population has shown a significant ($p < 0.10$) decline of 2.0%/year on Tennessee BBS routes from 1966 to 1994. It has been recorded on all but 1 BBS route at an average density of 0.9/50 stop route, lower than the Atlas miniroute density. Likely reasons for this decline include loss of foraging habitat from stream pollution and channelization and loss of nest sites from revegetation of earthen banks and road cuts and riprapping of lakeshores. This loss of habitat may be more severe near the primary and secondary roads used by BBS routes than across Tennessee as a whole. The kingfisher's rangewide population showed a decreasing trend of 1.7%/year ($p < 0.01$) from 1966 to 1993 (Peterjohn, Sauer, and Link 1994).

Breeding Biology: The Belted Kingfisher begins breeding activities in March. The male selects a dirt bank suitable for nest excavation and later defines a feeding territory (Davis 1982). The nest site may be several hundred meters from the vigorously defended feeding territory. High circling flights with prolonged rattling calls are common during early spring and probably have both territorial and courtship functions (Bent 1940, Davis 1982). Nest sites used in Tennessee include railroad and road cuts, sandbanks, clay pits, and stream and lake banks. The soil must be soft enough for burrow excavation, and soils with a high sand and low clay content are preferred. Both adults dig the nest burrow with their bills and push the dirt out with their small feet (Bent 1940). The burrow entrance is usually about 10 cm in size and 0.5–1 m below the top of the bank; occupied burrows are easily recognized by the paired grooves scratched out by the bird's feet as it enters and leaves. The burrow is typically 1.2–1.8 m long, roughly horizontal, and ends in a chamber about 0.3 m diameter.

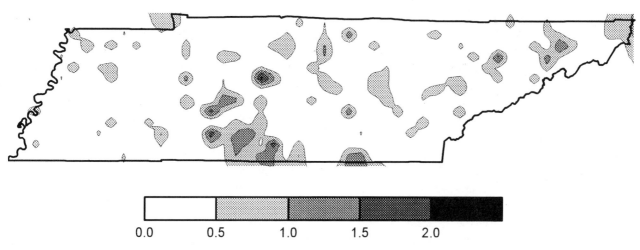

Abundance of the Belted Kingfisher.

Confirmed Breeding Species

Egg laying probably peaks in early April and extends well into May. The mean egg date is 1 May. Only a single brood is raised, and the late May and June egg dates were probably renests. Normal clutch size is 6–7 eggs (Bent 1940); Tennessee clutches average 6.41 eggs (s.d. = 1.01, n = 22, range 4–8). An incubated clutch of 2 eggs reported by Lyle and Tyler (1934) had probably lost eggs by predation or another cause. At least 4 broods of 7 nestlings and 1 of 6 have also been reported. Both adults incubate the eggs for 23–24 days; the male often brings food to the incubating female (Bent 1940, Cornwell 1963). Both adults feed the nestlings, which leave the nest in about 4 weeks. The adults teach the young to fish, and the young begin catching live fish in about 10 days. The young soon disperse and individual kingfishers hold separate territories during the fall and winter (Bent 1940).—*Ann T. Tarbell.*

Red-headed Woodpecker
Melanerpes erythrocephalus

The Red-headed Woodpecker is one of the most brilliantly colored and easily recognized birds nesting in Tennessee. Although present in the state throughout the year, a pronounced migration occurs during April and September and October. The size of the wintering population, probably composed mostly of birds nesting north of Tennessee, varies with the size of the acorn crop. The Red-headed Woodpecker is presently common to locally abundant in West Tennessee and declines in abundance to the east.

Nesting Red-headed Woodpeckers inhabit open areas with sparse ground cover and isolated trees or old utility poles, as well as woodland edges, mature bottomland hardwood forests, and parklike woodlands with large, tall trees and an open understory. A large portion of their foraging is by flycatching and stooping to take prey on the ground, an adaptation that enables them to nest in areas with very few trees (Jackson 1976). Prior to the European settlement of Tennessee, Red-headeds were probably common in the bottomlands of West Tennessee and in the barrens along the Kentucky border. Populations probably persisted for short periods in areas with standing dead trees resulting from disturbances such as beaver ponds and fires. Following European settlement, their numbers probably rapidly increased, as the "deadenings" created by girdling trees provided suitable habitat. They probably remained common around these farmsteads for many years, although their numbers probably declined somewhat as the dead trees fell. Langdon (1887) found them common around the clearings in the foothills of the Smokies.

Red-headed Woodpecker. Elizabeth S. Chastain

Rhoads (1895a) described Red-headed Woodpeckers as very abundant in West and Middle Tennessee and rare at higher elevations of East Tennessee. He also quoted Merriam, who described them as abundant in McMinn County. Ganier (1933a) described them as abundant in West and Middle Tennessee and fairly common in East Tennessee. Local declines in their numbers in Middle and East Tennessee have been reported in the *Migrant* since the late 1930s. In recent decades, nesting Red-headeds have virtually disappeared from areas of East Tennessee where they were once fairly common, such as Knox County (Howell and Monroe 1957). Concern over decreasing numbers led to them being listed by the TWRA as In Need of Management in 1976 (TWRA 1976). BBS route results, however, showed an increasing trend of 6.5% ($p < 0.05$) from 1966 to 1994, suggesting that the decrease occurred before 1966 or that it is localized. From 1966 to 1993, their numbers significantly ($p < 0.01$) decreased by 1.8%/year throughout North America (Peterjohn, Sauer, and Link 1994).

Two probable reasons for the decline in Red-headed Woodpecker numbers are loss of nesting habitat and competition for nest holes from European Starlings. Red-headeds most often nest in dead trees, from which the bark has fallen, located in open areas (Reller 1972, Jackson 1976, Ingold 1989). Dead limbs in live trees and wooden

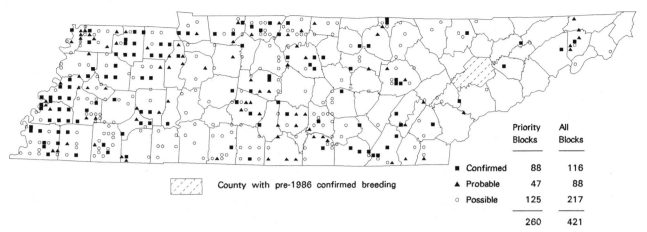

Distribution of the Red-headed Woodpecker.

utility poles are also used. Many of the early Tennessee nest records were in isolated trees in pastures. The recent trend toward improved pastures and removal of wooded fencerows between farm fields has probably been detrimental to their nest habitat. The tendency to trim dead limbs and remove dead trees in urban and suburban areas has also destroyed potential nest sites as well as active nests (Nicholson pers. obs.).

Competition with European Starlings for nest holes has been frequently mentioned in the Tennessee literature since the 1930s (e.g., Clebsch 1939), only a few years after starlings began nesting in Tennessee. Several reports exist of starlings occupying holes from previous seasons, as well as appropriating freshly excavated holes from Red-headeds, forcing the woodpeckers to dig additional cavities. Many of them eventually fledged young. Ingold (1989), in a study of woodpecker-starling competition in central Mississippi, found the rate of loss of freshly excavated Red-head cavities was very low because most starlings were already nesting by the time Red-headeds completed cavities.

Atlasers found Red-headed Woodpeckers in 40% of the completed priority blocks and recorded them on 89 of the miniroutes at an average density of 1.6 stops/route. By physiographic regions, the proportion of blocks in which they were found and their miniroute density decreased from west to east. Outside of the West Tennessee bottomlands, areas where they remain relatively common include the barrens area of the northern Highland Rim, heavily agricultural areas of southern Middle Tennessee, and the Crossville area on the Cumberland Plateau, where they have been common around golf courses since at least the 1950s. Slightly over one-quarter of the atlas records for Red-headed Woodpeckers were in the confirmed category, and slightly over half the confirmed records were of nests. This high proportion of nests is due to the Red-headed's tendency to nest in open areas.

Breeding Biology: The Red-headed Woodpecker nests later than any of the other widespread woodpeckers breeding in Tennessee. Migrant birds settle on nesting territories in late April or early May. An existing cavity

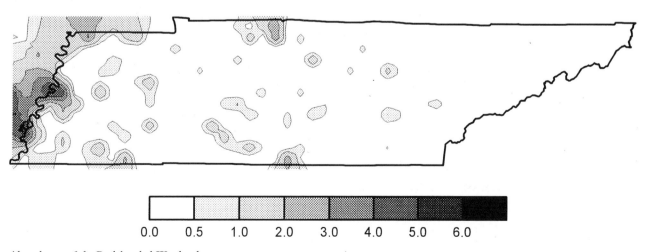

Abundance of the Red-headed Woodpecker.

Confirmed Breeding Species

may be renovated or a new cavity dug, often in the vicinity of the previous year's nest (Ingold 1991). The time required to dig the cavity varies and may be as short as 2 days (*Migrant* 22:43). Tennessee nests have been reported in both deciduous trees and yellow pines and range in height from 1.8 to 19.8 m above ground, with an average of 8.5 m (s.d. = 4.5 m, n = 19). Nests have been reported elsewhere as high as 24 m (Bent 1939). Clutches of 3 to 6 eggs have been reported in Tennessee, with 4 and 5 most common, and an average of 4.48 (s.d. = 0.68, n = 21), slightly less than the rangewide average of 4.8 (Koenig 1986). The eggs frequently rest on wood chips flaked off the cavity wall by the incubating adults (Jackson 1976).

Both the male and female incubate the eggs as well as brood and feed the nestlings (Jackson 1976). Incubation begins before the last egg is laid and lasts 12 to 13 days (Bent 1939, Reller 1972, Jackson 1976). The young leave the nest when about 27 to 30 days old. Red-headeds usually relay after loss of an early clutch, sometimes reusing the same cavity. Second broods are common elsewhere in their range (Reller 1972, Ingold 1989) and, based on the wide range of egg, nestling and fledgling dates, are probably common in Tennessee.—*Charles P. Nicholson.*

Red-bellied Woodpecker
Melanerpes carolinus

Red-bellied Woodpecker. Elizabeth S. Chastain

The Red-bellied Woodpecker is a fairly common to common statewide permanent resident. It is easily identified by its ladder back pattern and loud, distinctive voice. Because of the large red patch on its head, it is often misidentified as the Red-headed Woodpecker by Tennesseans, especially in those parts of the state where the true Red-headed Woodpecker is uncommon.

Atlasers reported Red-bellied Woodpeckers in 96% of the completed priority blocks, a frequency exceeded among the woodpeckers only by the Downy Woodpecker. The total number of Red-bellied records was the highest of all the woodpeckers. The few completed priority blocks where Red-bellieds were not found were in downtown Nashville, heavily wooded parts of the Cumberlands, and in the Unaka Mountains. Red-bellieds were reported on 86% of the miniroutes at an average density of 4.1 stops/route. They increased in abundance from eastern to western Tennessee and reached their highest average abundance, 5.6 stops/route, on the Loess Slope physiographic region of West Tennessee.

Red-bellied Woodpeckers occupy woodlands and occur in essentially all low-elevation forest types. Breeding densities, as measured on census plots, include 4.1 to 8.2 pairs/100 ha in Ridge and Valley hardwood forests (Howell 1972, 1973, Smith 1975), 17/100 ha in virgin Cumberland Plateau mixed mesophytic forest (Robertson 1979), 6.3/100 ha in Eastern Highland Rim mixed mesophytic forest (Simmers 1982a), and 16.4 to 32.8/100 ha in Central Basin upland hardwoods (Fowler and Fowler 1984a, 1984b). The low miniroute density in the inner portion of the Central Basin suggests that cedar-dominated forests are poor Red-bellied Woodpecker habitat. In the Unaka Mountains, Red-bellieds are uncommon and rarely occur above about 610 m (Stupka 1963, Nicholson pers. obs.). The same elevational limit holds true in the Cumberland Mountains. Quantitative studies of their habitat selection in deciduous forests of East Tennessee (Anderson and Shugart 1974) and Arkansas (James 1971) suggest they have rather generalized habitat requirements. Robbins, Dawson, and Dowell (1989) found their relative abundance increased with forest area, although they had a high tolerance of small woodlots. The ubiquitous distribution and abundance of Red-bellieds in sparsely forested West Tennessee supports their acceptance of small woodlots. They appear to show some aversion to areas of extensive forest with few openings, as noted by Mengel (1965) and evidenced by their low numbers in parts of the Cumberlands.

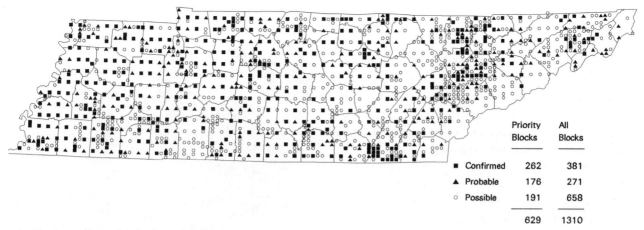

Distribution of the Red-bellied Woodpecker.

The distribution of Red-bellieds in Tennessee has probably not historically changed, although they were likely more numerous in the presently deforested areas of the state. At the end of the nineteenth century, Rhoads (1895a:483) found them abundant in the Reelfoot Lake and Memphis areas, and noted "several" at Bellevue, Chattanooga, and Harriman. He did not find them on the northern Cumberland Plateau, at Knoxville, or at Johnson City. Ganier (1933a) described the Red-bellied as common in West and Middle Tennessee and fairly common in East Tennessee. Lyle and Tyler (1934) found it rare in northeastern Tennessee; the Atlas results suggest it may now be more numerous there. Red-bellied numbers, as measured by BBS routes, did not show a significant trend from 1966 to 1994, although the proportion of routes on which their numbers have increased is greater than routes with decreases. Rangewide, their numbers increased by 0.6%/year ($p < 0.01$) from 1966 to 1993 (Peterjohn, Sauer, and Link 1994).

Atlasers confirmed the breeding of Red-bellied Woodpeckers in almost 30% of the blocks in which they were recorded, in 41% of the priority blocks, and in almost every county of the state. Slightly over three-quarters of the confirmed records were of fledglings or adults carrying food. Most of the remaining confirmed records were code ON, many of which were probably nests with eggs. Detailed descriptions of very few Red-bellied nests in Tennessee are available.

Breeding Biology: Red-bellied Woodpeckers are non-migratory and hold territories throughout the year (Kilham 1961). Courtship may begin as early as mid-winter and involves drumming, calling, raised-crest displays, and mutual tapping at potential nest sites. Excavation of the first nest probably occurs in late March and early April, as noted by Ingold (1989) in central Mississippi. The male selects the nest site and does most of the excavation (Kilham 1961). The nest cavity is most frequently excavated on the underside of a sloping dead limb, with bark still attached, of a live tree (Reller 1972, Ingold 1989). Red-bellieds also nest in dead trees and, uncommonly, in bird boxes after enlarging the entrance.

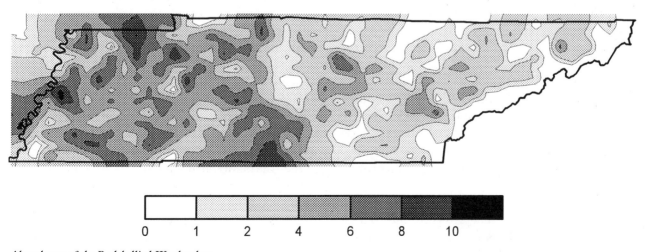

Abundance of the Red-bellied Woodpecker.

Confirmed Breeding Species

Nest trees are more often located in woodland than in an open area. Nests in Tennessee range in height from 1.5 to 12.2 m above ground, averaging 8.2 m (s.d. = 3.6, n = 13). Nests elsewhere have been reported as high as 24 m (Bent 1939). Trees used for nesting in Tennessee include yellow pine, oaks, maples, elm, hickory, and willow.

Egg laying in Tennessee probably peaks during the second half of April. Nine clutches of 4 eggs and 1 each of 3 and 5 eggs have been reported in Tennessee. Clutches of up to 8 eggs have been reported elsewhere (Bent 1939). Koenig (1986) found a rangewide average of 4.3 eggs, and Ingold (1989) found average sizes of 4.1 eggs for first clutches and 3.6 eggs for second clutches in Mississippi. The eggs are pure white. Both adults incubate the eggs for 12 to 14 days, with the average closer to 12 days (Bent 1939, Jackson 1976). Incubation may begin before the last egg is laid, resulting in the eggs hatching over a 2- or 3-day period. Both adults feed the nestlings, which leave the nest in about 26 days (Kilham 1961).

After loss of a clutch, Red-bellied Woodpeckers lay a replacement, sometimes in the same hole as the first clutch (Bent 1939, Ganier egg coll. data). Although Kilham (1961) found no evidence of second broods in Maryland, Ingold (1989) found them common in Mississippi. The numerous Tennessee records of nests with young and fledglings in July suggest that Red-belliedies raise second broods in Tennessee. Loss of nest cavities to European Starlings is common early in the nesting season, as both Red-belliedieds and Starlings begin nesting at the same time (Ingold 1989). Starling competition is much less during later Red-bellied nesting attempts.
—*Charles P. Nicholson.*

Yellow-bellied Sapsucker
Sphyrapicus varius

The Yellow-bellied Sapsucker is one of the rarest of Tennessee's breeding birds, presently restricted to a small area of high-elevation hardwoods in the Unicoi Mountains. Tennessee is at the southern limit of the sapsucker's eastern breeding range, and nesting sapsuckers have never been numerous. Ganier (1954) described them as rare, and Stupka (1963) described them as uncommon, localized, and rarer in Smokies than in the Unicois. All breeding records have been in hardwoods above 1150 m, and the breeding population has apparently declined in recent decades. Elsewhere in the state, sapsuckers are fairly common wintering birds, arriving in mid- to late September and departing by early May.

Sapsuckers were first reported as summer residents by Rhoads (1895a), who found a pair in June 1895 along

Yellow-bellied Sapsucker. Elizabeth S. Chastain

the Doe River at Roan Mountain, Carter County, at 1219 m and presumed they were nesting. Between 1920 and 1946, nesting sapsuckers were reported from the Great Smoky Mountains in Blount and Sevier Counties (Ganier and Clebsch 1938, Stupka 1963) and from south of the Smokies in the Unicoi Mountains in Monroe County (Ganier and Clebsch 1946). During this same period, the 1937 Smithsonian Expedition observed the species in the Holston Mountains above Shady Valley, Johnson or Sullivan County (Wetmore 1939).

Since the mid-1940s, there have been very few published records of probable nesting sapsuckers. These records are of an adult at Elizabethton, Carter County, elevation not given, in June and August 1955 (*Migrant* 26:51), 1 in the GSMNP on 21 July 1972 (*Migrant* 43:80), 1 at Stratton Meadows, Monroe County, on 6 July 1974 (*Migrant* 45:104), and 1 on Iron Mountain, Carter or Johnson County, on 24 June 1978 (*Migrant* 49:95). The reasons for this apparent recent decline are not clear. Lack of ornithological work may be part of the explanation in the Unicoi Mountains, but not in the Smokies and upper East Tennessee mountains. Sapsuckers have not been recorded on BBS routes in Tennessee, and there are no other censuses showing their population trends.

During Atlas fieldwork, sapsuckers were found in only 1 block, in the same area of the Unicois where Ganier and Clebsch (1946) reported them in the 1940s. At least 5 adult birds were observed in 3 locations in the same block in 1989. A sapsucker was recorded at 1 miniroute stop in this same block. These observations were in mature northern red oak, beech, and yellow birch at 1292 m, in large hardwoods near the edge of a recent clearcut at 1097 m, and in a sapling to pole-sized mixed stand of black birch, black cherry, ash, tulip-poplar, and hemlock at 1158 m. At the last site, I observed a pair repeatedly feeding at patches of holes chiseled in black birches. In mid-July 1991, I saw a pair with fledglings and at least 1 independent juvenile bird in this same area. These birds were also feeding on black birches; in Michigan, Tate (1973) noted a preference for feeding on birches during the summer.

With the exception of the observation in the mature oak-beech-birch stand, the recent reports were in areas with openings in the forest canopy. Stupka (1963) noted the species's preference for nesting in mature deciduous groves, with openings in the canopy resulting from logging, fire, blowdowns, and dying chestnuts. During the 1930s and 1940s, such canopy openings were common in the Smokies. The current rarity of sapsuckers there may be related to the reduced availability of high-elevation hardwoods with open canopies. This habitat may be maintained by timber harvesting in the southern portion of Cherokee National Forest. The apparent absence of sapsuckers from the northern portion of the national forest is difficult to explain.

The breeding sapsuckers in the southern Appalachians were described as a unique subspecies, *S. varius appalachiensis*, by Ganier (1954). Ganier distinguished the Appalachian sapsucker by its smaller size and darker coloration. His type specimen was a male collected 21 June 1946 at 1341 m in the Unicoi Mountains of Monroe County. This subspecies was recognized in the 1957 AOU Checklist (AOU 1957), but its identity has recently been questioned.

Although additional surveys—using broadcast recordings or imitations of its drumming—would be helpful, the Atlas results and the few other recent records suggest that the sapsucker no longer breeds in much of its former range. Current management practices in the Smokies are not conducive to maintaining the disturbed, open-canopy forests sapsuckers prefer, and its future probably depends on management practices in the Cherokee National Forest. Although the breeding population of the Yellow-bellied Sapsucker is listed as In Need of Management in Tennessee (TWRA 1976), a status of Threatened would be more appropriate.

Breeding Biology: Very little is known about the breeding biology of sapsuckers in Tennessee, as well as elsewhere in the southern Appalachians. There are no available records of eggs and only about 4 records of nestlings. Nests have been reported 12.2 m up in an "old yellow birch" (Stupka 1963:86) and 15.2 m up in a dead chestnut (Ganier and Clebsch 1944). Based on the dates of nests with young, egg laying occurs in mid-May.

In comparison to the southern Appalachians, the breeding biology of sapsuckers is well known in New England and eastern Canada (Kilham 1962, 1977, Lawrence 1967). Both the male and female show strong territorial fidelity year after year. The male arrives in the territory first in the spring and takes the lead in its establishment as well as most other nesting activities. A new cavity is excavated each year; digging it takes about 3 weeks. Most nest cavities are on a smooth, uncamouflaged section of tree trunk, averaging about 9 m above ground. The same tree is frequently used for more than 1 year. Live aspens (*Populus deltoides, P. tremuloides*) infected with fungal heart rot are most frequently used, although the nest may be in a dead stub or dead section of a live tree. Clutch size ranges from 3 to 7 eggs with 5 to 6 most common. Incubation lasts 12–13 days and the nestlings

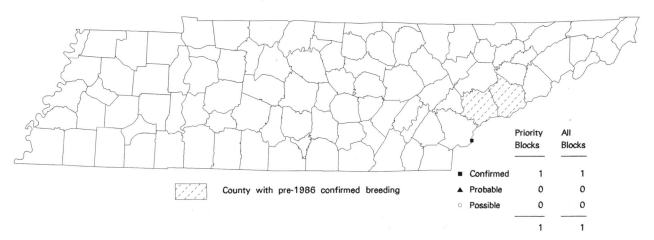

Distribution of the Yellow-bellied Sapsucker.

leave the nest after 25–29 days. The nestlings are fed insects, frequently mixed with sap, which the adult carries in its bill. The young begin feeding on sap soon after fledging and are independent in about 2 weeks. The family group may remain intact for several more weeks.—*Charles P. Nicholson.*

Downy Woodpecker
Picoides pubescens

The Downy Woodpecker is a common permanent resident found throughout Tennessee. Because it is relatively tame and frequents bird feeders, it is well known to many Tennesseans. The Downy is also probably the most numerous woodpecker in the state. Atlasers found it in more priority blocks than any other woodpecker; the higher miniroute density of the similarly widespread Red-bellied Woodpecker is probably due to the greater conspicuousness of the larger, louder Red-bellied.

Downy Woodpeckers occupy most forest types present in Tennessee, although they are probably less common in pine forests than in other low-elevation forest types. In the Unaka Mountains, it is uncommon at the highest elevations and rarely occurs in spruce-fir forests (Stupka 1963, Kendeigh and Fawver 1981, Nicholson pers. obs.).

Downy Woodpecker. Elizabeth S. Chastain

In Ridge and Valley forests, Downies prefer areas with numerous saplings in the understory (Anderson and Shugart 1974). In West Tennessee forested wetlands, they are the most common nesting woodpecker and are most abundant in forests with scattered large canopy trees and few shrubs (Ford 1990). Downies show little preference for large over small forest tracts (Robbins, Dawson, and Dowell 1989), and can nest in shrubby areas with few trees. Compared to other eastern woodpeckers, Downies tend to nest in less mature forests with smaller trees, and show little preference for dense forest over open areas (Conner and Adkisson 1977).

Densities of breeding Downy Woodpeckers on virgin forest census plots in Tennessee include 9 pairs/100 ha in Cumberland Plateau mixed mesophytic forest (Robertson 1979) and 11/100 ha in Great Smoky Mountain cove forest (Aldrich and Goodrum 1946). Densities in second-growth forests include 4 to 20/100 ha in Ridge and Valley hardwood forests (Howell 1973, Smith 1975), 10 and 12.5/100 ha in Cumberland Mountain mixed mesophytic (Smith 1977) and oak-maple forests (Turner and Fowler 1981a), 12.5/100 ha in Highland Rim mixed mesophytic forest (Simmers 1982a), and 0 and 12.5/100 ha in Central Basin cedar forests (Fowler and Fowler 1983b, 1983c).

Atlasers recorded Downy Woodpeckers on slightly fewer than three-fourths of the miniroutes at an average density of 2.2 stops/route. Average densities were lowest in the Unaka Mountains (1.6/route) and Cumberland Plateau (1.8/route) physiographic regions, attributable to the species's lower density at high elevations, and, on much of the plateau and southern Blue Ridge, lower density in pine forests. Much local variation in density occurred within physiographic regions, and densities were poorly correlated with regional differences in forest cover. Some of the local differences in density may have been due to the dates on which the routes were censused. Downies are less conspicuous in late May and early June when many are feeding nestlings than later in the summer.

As measured by Tennessee BBS routes, the Downy Woodpecker population did not change significantly from 1966 to 1994. The same was true of their range-wide population from 1966 to 1993 (Peterjohn, Sauer, and Link 1994). Tanner (1985) also found no significant change in numbers on Tennessee Christmas Bird Counts. During the Atlas period, Downies were recorded on all of the Tennessee BBS routes at an average density of 4.7 birds and 4.4 stops/route.

Of the Atlas records of Downy Woodpeckers, 28% were in the confirmed category, and slightly over two-thirds of the confirmed records were observations of

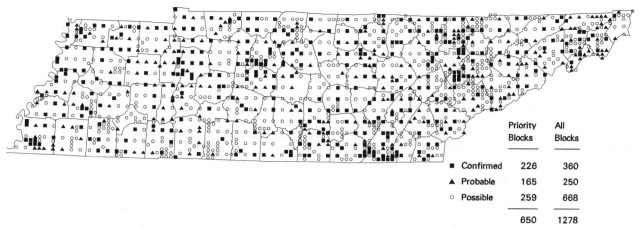

Distribution of the Downy Woodpecker.

fledglings. Most of the remaining confirmed records were observations of nests, and observations of nests with young outnumbered other nest records. Confirmed records were well distributed across the state and in virtually every county.

Breeding Biology: Male Downy Woodpeckers maintain their territories throughout the year, although territorial behavior is relaxed from the late summer through early winter (Lawrence 1967). Reproductive behavior begins in late winter, when the pair begins reciprocal drumming with one bird drumming after the other. This is followed by courtship displays, where the bird erects its crown feathers, spreads its wings and points with its bill (Lawrence 1967). The female usually chooses the nest site, and the male does most of the excavation of the nest cavity (Lawrence 1967).

The nests are excavated in the trunk of a dead tree or in the dead branch of a live tree. The cavity is excavated in wood that has been softened by fungi and is often near the broken-off end of the limb that may have been used in previous years (Lawrence 1967, Conner et al. 1975). Tennessee nests average 5.7 m above ground (s.d. = 2.9 m, n = 30), with a range of 1.2–15.2 m.

Egg laying in Tennessee peaks in mid- to late April and clutches range from 3 to 6 eggs, averaging 4.6 eggs (s.d. = 0.67, n = 22). Koenig (1986) found a slightly larger rangewide average clutch size of 4.8 eggs. The male incubates the pure white eggs at night, and both adults share incubation duties during the day; incubation lasts about 12 days (Lawrence 1967). Both adults also feed and brood the young until they fledge in 20–22 days. The nestlings give soft calls until they are almost ready to fledge, when their loud calls make locating the nest much easier. Adults carry food to the young in their bills; records of this accounted for 9% of the confirmed Atlas records. The fledglings fly well when they first leave the nest. They begin feeding themselves within 3 weeks of fledging, and the family remains together for 4–6 weeks after the young fledge (Lawrence 1967). No evidence of second broods exists in Tennessee.—*Charles P. Nicholson.*

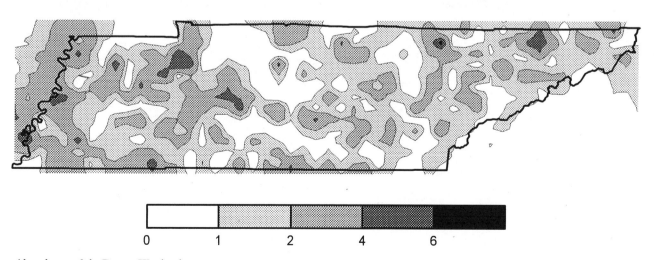

Abundance of the Downy Woodpecker.

Hairy Woodpecker
Picoides villosus

Over the state as a whole, the Hairy Woodpecker is an uncommon permanent resident. It occupies larger and more mature woodlands than the more familiar, smaller, and similar appearing Downy Woodpecker. In some heavily forested parts of the state, such as the Cumberlands and Unaka Mountains, the Hairy Woodpecker is a fairly common to common species.

Atlas workers reported Hairy Woodpeckers from about two-thirds of the completed priority blocks and in every county. Because nesting Hairies are usually quiet and inconspicuous until they have large nestlings, they were probably missed in some blocks where suitable habitat was limited and access to it difficult. Robbins, Dawson, and Dowell (1989) found the Hairy Woodpecker's probability of occurrence in a forest increases with the forest area up to about 200 ha, where it levels off. The Atlas results show a gross relation with the percent of forest cover. The woodpecker's presence in several West Tennessee counties with little forest area shows its ability to inhabit farm woodlots and the remaining bottomland forests.

Hairy Woodpeckers were recorded on only one-fifth of the miniroutes at an average density of 1.3 birds/route. Little pattern is evident in their density, other than high numbers in the southern Coastal Plain Uplands, parts of the Cumberlands, and the Smokies.

Hairy Woodpeckers occur in most forest types present in Tennessee. In the Unaka Mountains, they occur at all elevations, and Stupka (1963) found them outnumbering Downy Woodpeckers by a ratio of 4 to 1 above 1067 m elevation in the Smokies. The reverse situation was true at lower elevations. In Ridge and Valley deciduous forests, Hairy Woodpeckers show a preference for areas with large numbers of tall trees and a well-developed forest canopy (Anderson and Shugart 1974). They do, however, nest in forests with a wide range of basal area, canopy height, and tree density (Conner and Adkisson 1977). Breeding densities reported in Tennessee include 4.2 pairs/100 ha in a Ridge and Valley hardwood forest (Howell 1971, 1972. 1973, 1974), 5/100 ha in a Cumberland Mountain mixed mesophytic forest (Smith 1977), 9/100 ha in a virgin Cumberland Plateau mixed mesophytic forest (Robertson 1979), and 4/100 ha in a low-elevation Unaka Mountain deciduous forest (Lewis 1983a).

Prior to the widespread nineteenth-century clearing of forests, Hairy Woodpeckers were probably fairly common and, except for the barrens area along the Kentucky

Hairy Woodpecker. Elizabeth S. Chastain

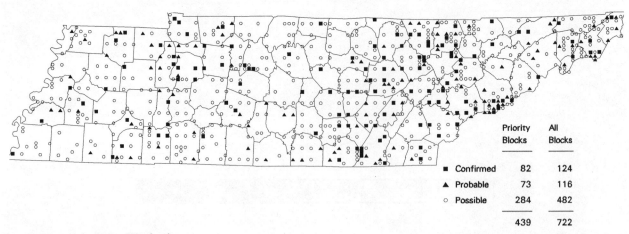

	Priority Blocks	All Blocks
■ Confirmed	82	124
▲ Probable	73	116
○ Possible	284	482
	439	722

Distribution of the Hairy Woodpecker.

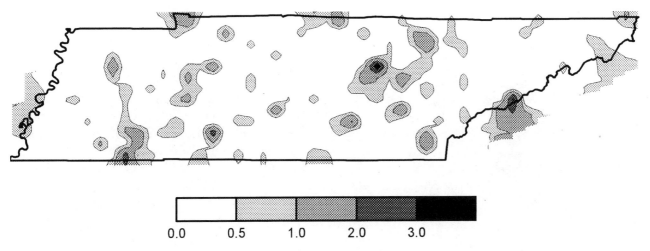

Abundance of the Hairy Woodpecker.

border and parts of the Central Basin, continuously distributed across the state. Rhoads (1895a) described it as a common bird all across the state, and Ganier (1933a) described it as fairly common in West, Middle, and East Tennessee. Suitable Hairy Woodpecker habitat has probably increased in recent decades with the increase in forest area and the maturing of the forests. This habitat trend, however, has not been reflected in Hairy Woodpecker numbers, which, as measured by BBS routes, did not show a significant trend from 1966 to 1994. During the years of Atlas fieldwork, Hairy Woodpeckers were recorded on 32 of 42 BBS routes, at average densities of 1.4 stops and 1.4 birds/route. Tanner (1985) also failed to find a significant change in numbers on Tennessee Christmas Bird Counts. Throughout North America, Hairy Woodpeckers showed an increase of 1.1%/year (p = 0.01) between 1966 and 1993 (Peterjohn, Sauer, and Link 1994).

Breeding Biology: Few details of Hairy Woodpecker nests in Tennessee are available, and the breeding biology of this species is the least known of the widespread Tennessee woodpeckers. Atlasers confirmed the Hairy Woodpecker in only about 17% of the blocks, the lowest rate of any of the small woodpeckers. Almost 80% of the confirmed records were of relatively conspicuous fledglings or adults carrying food and the remaining confirmed records were observations of nests.

Hairy Woodpecker pairs are territorial throughout the year and remain mated for several consecutive years (Lawrence 1967, Kilham 1960). Courtship begins in the fall and lasts throughout the winter; it includes drumming and display flights. Nest site selection and the start of nest cavity excavation probably occur in March in Tennessee. The pair usually excavates a new nest cavity each year, and the male leads in selecting the nest site and excavating the nest (Lawrence 1967). Nests reported in Tennessee were between 4.6 and 12.2 m above ground, averaging 6.1 m (s.d. = 3.6, n = 6). Nest heights elsewhere range from 1.5 to 19.8 m and averaged 8.8 m in southwest Virginia (Bent 1939, Conner et al. 1975). Trees used for nesting in Tennessee include elm, ash, boxelder, sweetgum, and black and white oaks, and cavities have been excavated in dead trees and dead sections of live trees.

Most Hairy Woodpecker clutches in Tennessee are probably laid in mid- to late April. The few egg records include 1 clutch of 2 eggs (incubated 3 days), 4 of 3, and 2 of 4 eggs. Clutches of 3 to 6 eggs have been reported elsewhere with 3 and 4 egg clutches most common (Bent 1939). Incubation lasts 11 to 12 days and is carried out by both adult woodpeckers, with the male incubating more than the female (Lawrence 1967). Both adults feed and brood the young which leave the nest in 28 to 30 days. During the latter part of the nestling period the young call almost constantly. They fly well when they first leave the nest and become independent about 2 weeks after fledging (Lawrence 1967). Hairy Woodpeckers are single brooded.—*Charles P. Nicholson.*

Red-cockaded Woodpecker
Picoides borealis

The Red-cockaded Woodpecker was, until very recently, a rare permanent resident of Tennessee, found in mature, open pine forests. It historically nested in the southern Unaka Mountains, at several sites on the Cumberland Plateau, and in the southern end of the Coastal Plain Uplands physiographic regions. Its population has drastically declined in the last 25 years and by 1991 consisted of a single bird in the southeast corner of the state. By the end of 1994, none were known to occur in the state.

Red-cockaded Woodpecker. Elizabeth S. Chastain

The Red-cockaded Woodpecker is most abundant in the longleaf and loblolly pine forests of the southeastern Coastal Plain (USFWS 1985). In Tennessee, near the northern limit of the woodpecker's range, it occupied shortleaf, Virginia, and pitch pines. The Red-cockaded is unique among woodpeckers in excavating a nest cavity in old, mature living pines, usually infected with the fungus *Phellinus pini*. The breeding pair and often 1 or more unmated helpers, collectively known as a clan, occupy 1 to several cavity trees, usually within a 230 m radius, throughout the year. This area, known as the cavity tree cluster, is made up of mature pine trees, at least 80–90 years old, with an open, parklike understory. Most foraging is on pines. The average annual home range of Coastal Plain clans is 87 ha (Hooper et al. 1982); nothing is known of the home range of the species in Tennessee.

Audubon (1839) made the first reference to the Red-cockaded Woodpecker in the state, suggesting that it existed as far inland as Tennessee. Whether he actually observed the bird in Tennessee is unknown. The first documented specimens were collected by Fox in Roane County in 1884 (Fox 1886); a few years later Rhoads (1895a) found it in Fentress, Scott, and Morgan Counties. Gettys reported a nest in the limb of an oak in McMinn County in 1901 (Ijams and Hofferbert 1934); because of the uniqueness of this nest site, however, this record is questionable. There have been no other reports of nesting Red-cockadeds in the Ridge and Valley province of Tennessee, although it occurred in Georgia just south of Chattanooga until about 1964 (Nicholson 1977).

Between the 1930s and 1960s, additional clusters were reported in the western Great Smoky Mountains National Park, the southern Coastal Plain Uplands, and the Cumberland Plateau at Pickett SP, Savage Gulf State Natural Area, and Catoosa WMA, and a pair observed in Prentice Cooper State Forest, Marion County (Nicholson 1977). The largest known population was at Catoosa, where 6 occupied clusters probably occurred during the 1960s and early 1970s (Nicholson 1977, C. Watson pers. comm.). No more than 1 or 2 clusters were known from the other areas. The only region of the state where the species has been reported without evidence of nesting is in Stewart County, where a bird was observed on 30 October 1937 (Wetmore 1939). Little pine has historically occurred there.

Nicholson (1977) estimated a state population of between 6 and 25 birds, with the low estimate based on the then active sites at Catoosa, and the higher estimate assuming woodpeckers were still present at Pickett, Campbell County, and the Great Smokies. Based on the results of intensive efforts during the late 1970s and 1980s to locate additional clusters (Alsop 1979b, Dimmick, Dimmick, and Watson 1980, C. Watson pers. comm., Nicholson unpubl. data), at least 11 clusters were probably active at Catoosa, Savage Gulf, Campbell County, the Great Smokies, and Cherokee National Forest in Polk County during the 1970s. By the mid-1980s, however, the Tennessee population had crashed. The last report of the Red-cockaded Woodpecker at Catoosa was in 1985, at Pickett in 1971, at Savage Gulf in 1980, in the Smokies in 1985, and in Campbell County in 1983 or 1984 (*Migrant* 56:80, Nicholson 1977, unpubl. data). The only location where the species occurred during Atlas fieldwork was the Polk County site, first discovered in 1986. By fall 1989, a single male was present at this site, and this bird disappeared in late 1994.

Because of its declining population, the Red-cockaded Woodpecker was listed as an endangered species by the U.S. Fish and Wildlife Service in 1968 and by the state of Tennessee in 1975. The primary cause of the decline, in Tennessee as elsewhere in the bird's range, has been loss of old-growth pine habitat through logging and, in the absence of periodic fires, succession to hardwoods (USFWS 1985). Succession to hardwoods has caused most of the recent decline in Tennessee, almost all of which was on public lands. Recovery efforts in Tennessee included searching for clusters (Alsop 1979b, Dimmick, Dimmick, and Watson 1980, L. Mitchell pers. comm.), describing cluster sites (C. Watson unpubl. data) and limited clearing of hardwoods from cluster sites. These efforts were most intensive at the Polk County site, where they also included protection of existing cavity trees, excavating artificial cavities, creation of a pine stand for future occupancy, and 2 unsuccessful attempts to introduce a female (L. Mitchell, U.S. Forest Service, pers. comm.).

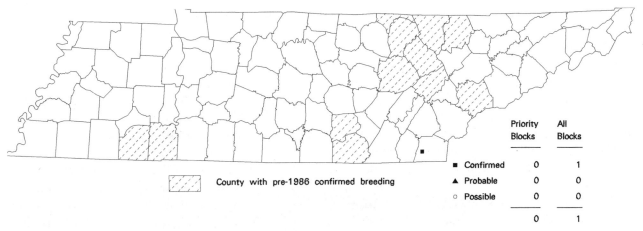

County with pre-1986 confirmed breeding

	Priority Blocks	All Blocks
■ Confirmed	0	1
▲ Probable	0	0
○ Possible	0	0
	0	1

Distribution of the Red-cockaded Woodpecker.

A long-term commitment to intensive habitat management across a large area will be necessary if the Red-cockaded Woodpecker is to return to Tennessee. Savage Gulf may be the most suitable area because of the large acreage of pine and flat terrain, which facilitates controlled burning. Catoosa may also be a suitable site for such efforts, although the current contiguous acreage of mature pine there may be less than at Savage Gulf. Management efforts necessary to reestablish and maintain a Red-cockaded Woodpecker population in Tennessee include providing suitable nest sites by releasing suppressed pines, inoculating them with *P. pini*, and excavating artificial cavities (Locke, Conner, and Kroll 1983, Copeyon 1990). Hardwoods should be controlled within colony sites and foraging habitat by cutting or poisoning and periodic burning, and young pine stands, preferably shortleaf rather than Virginia pines, should be established nearby for future habitat. Because of the lack of colonies close to potential recovery areas, translocation of birds from healthy populations will probably be necessary.

Breeding Biology: Egg laying in Tennessee, as determined primarily by dated observations of nestlings, occurs between the last week of April and the third week of May. Although no Tennessee records of complete clutches are available, observations of nestlings suggest a clutch size of from 2 to 4 eggs, as reported elsewhere (USFWS 1985). Both adults incubate the eggs for about 10 days, and the young fledge in 24 to 29 days (Ligon 1970, USFWS 1985). After fledging, the juveniles remain in the natal territory into late fall, and young males may remain longer as helpers. There are no unequivocal records of helpers in Tennessee (Nicholson 1977).
—*Charles P. Nicholson.*

Northern Flicker
Colaptes auratus

The Northern Flicker, more commonly known by its former name, Yellow-shafted Flicker, or colloquially as the yellowhammer, is a fairly common permanent Tennessee resident. Because of its large size, loud and frequent calls, distinctive plumage with bright white rump, and frequent habit of foraging on the ground, it is probably the most conspicuous and easily recognized woodpecker nesting in the state. Migrant birds from north of Tennessee are present during the winter, when the flicker is probably more numerous than during the summer. Most flickers nesting in Tennessee are permanent residents (Moore 1995).

Northern Flicker. Elizabeth S. Chastain

Northern Flickers occupy open woodlands and forest edges and nest in almost all forest types found in Tennessee. In the Unaka Mountains, they have nested at elevations as high as 1700 m (Stupka 1963), but are generally rare above 1525 m (McNair 1987c, Nicholson pers. obs.). At their upper elevational limit, they are most common at the edges of balds (Ganier and Clebsch 1946, Stupka 1963). In Ridge and Valley forests, Anderson and Shugart (1974) found them most common in areas with many large trees and a well-developed canopy and subcanopy. Their nests, however, tend to be in trees standing in open areas or at the edge of woodlands (Conner and Adkisson 1977). The nearby presence of open ground for foraging is probably an important component of their nesting habitat.

Northern Flickers were probably uncommon in presettlement Tennessee and most numerous on the Cumberland Plateau and Eastern Highland Rim where fire-maintained, open forests with grassy understories were common. Their numbers probably increased rapidly with European settlement, as clearings with numerous standing dead trees were created. In the late nineteenth century, Rhoads (1895a:483) described the flicker as "nowhere . . . as abundant as in the Middle States, but it was seen in all localities." Ganier (1933a) described it as common across the state.

Atlasers found Northern Flickers in 89% of the completed priority blocks. It was absent from some heavily forested blocks in the Cumberlands and Unaka Mountains, probably due to the lack of forest openings. Its absence from blocks elsewhere in the state is difficult to explain, and atlasers may have missed it in some of these blocks. Flickers were reported on about half of the miniroutes, at an average relative abundance of 1.6 birds/route. The highest abundance, reported on several blocks across the state, was 5. The average abundance and proportion of blocks on which flickers were reported varied little among physiographic regions.

Northern Flicker numbers have declined in recent decades. On BBS routes, flickers declined between 1966 and 1994 at the rate of 2.8%/year (p < 0.01). This Tennessee trend follows that of the Yellow-shafted race on rangewide BBS routes, which show a decline of 2.9%/year (p < 0.01) from 1966 to 1993 (Peterjohn, Sauer, and Link 1994). Competition with the European Starling for nest holes is probably a reason for these declines (Moore 1995).

Changing land use practices are probably also a factor in Tennessee. In the Smokies and parts of the Cumberland Plateau where starlings are virtually absent, flicker numbers have declined since the early 1940s (Ganier 1937a, Ganier and Clebsch 1940, Wilcove 1988, Atlas results). At that time, these areas were recently logged or recently abandoned farmland; numerous standing dead chestnut trees were also present. The reforestation of these areas since then has probably made them less suitable for flickers. As with the Red-headed Woodpecker, the recent trend toward improved pastures and removal of wooded fencerows between farm fields has probably eliminated nest sites. Widespread fire control has probably also resulted in a reduction of flicker nest habitat.

Breeding Biology: Northern Flickers begin courtship as early as midwinter when they are nonmigratory or, in the case of migratory populations, quickly after returning to their breeding territories (Kilham 1959, Lawrence 1967). Courtship includes drumming and calling and displays where the male and female face each other and bob, spread their wings and tail to show the yellow feather shafts, and wave their bills in a circular motion. Flickers frequently return to the previous year's nest cavity and renovate it or dig a new one nearby (Lawrence 1967). The male leads in selecting the nest site and excavating the nest cavity, as well as in most other nesting activities.

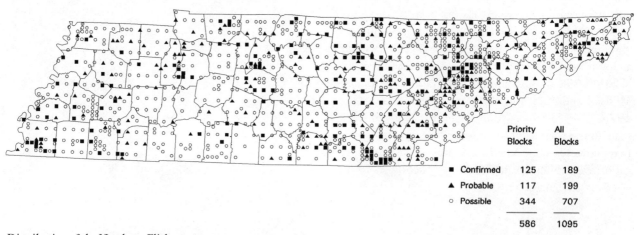

	Priority Blocks	All Blocks
■ Confirmed	125	189
▲ Probable	117	199
○ Possible	344	707
	586	1095

Distribution of the Northern Flicker.

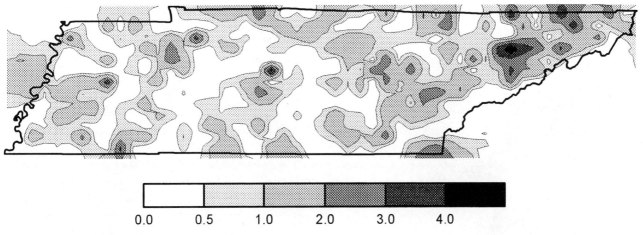

Abundance of the Northern Flicker.

Flicker nest cavities are most often excavated either in a dead tree or the dead limb of a live tree, in wood that has been softened by fungal decay (Conner et al. 1975). They also dig cavities in wood utility poles and fence posts, into knotholes in live trees, and infrequently used bird boxes, sawdust piles, and dirt banks. The height of the cavity entrance above ground varies greatly; in Tennessee, nests range from 1.2 to 13.7 m high and average 5.6 m (s.d. = 3.4 m, n = 45). Nests on the ground, uncommon elsewhere, have not been reported in Tennessee. Flickers nest in a variety of hardwood tree species, as well as in pines, and show little apparent preference. They also commonly nest in dead snags standing in young clearcuts (Conner et al. 1975, Nicholson unpubl. data).

The time required to excavate the nest depends on whether an existing cavity is renovated or a new cavity dug; Lawrence (1967) found an average of 12 days was required. Egg laying extends over a long period in Tennessee, with a poorly defined peak in early May. Competition for cavities from starlings is common (Nicholson unpubl. data), and loss of cavities to starlings delays egg laying. Clutches in Tennessee vary from 4 to 8 eggs with an average of 5.6 (s.d. = 1.0, n = 41). One clutch of 10 eggs, laid by 2 females, has been reported (Ganier egg coll. data). Koenig (1986) found a rangewide clutch size of 6.5 eggs and noted that clutch size increased with latitude; at the latitude of Tennessee, it averaged 6.2 eggs. Incubation lasts about 12 days, and the male incubates more than the female (Lawrence 1967). Both adults feed the young by regurgitating food; the young fledge in about 26 days. After fledging, the brood may soon leave the nest area, and the young remain with the parents for a few weeks (Lawrence 1967). The range of egg and nestling dates suggests the flicker is occasionally double-brooded in Tennessee.—*Charles P. Nicholson.*

Pileated Woodpecker
Dryocopus pileatus

The Pileated Woodpecker is an uncommon to locally common permanent resident of Tennessee. Because of its large size, distinctive plumage, and frequent lack of wariness, it is familiar to rural Tennesseans. It inhabits woodlands with enough large trees to provide nesting and foraging sites.

Atlasers recorded Pileated Woodpeckers throughout the state in 84% of the completed priority blocks and on slightly fewer than half of the miniroutes. It was absent from much of the inner Central Basin, the north-central portion of the Highland Rim, and much of the Loess Plain. All of these areas have a relatively low proportion of forest cover. Its near total absence from Rutherford County, which has a higher proportion of forest than other inner basin counties, is probably due to the unsuitability of cedar forests, a dominant type in the county, as Pileated Woodpecker habitat, as well as the small size class of the forests (Vissage and Duncan 1990). The species occurs throughout the elevational range of the hardwood forests in the Unaka Mountains, becoming less common at the higher elevations and rare in spruce-fir (Stupka 1963).

Pileated Woodpecker pairs maintain a strong pair bond and are territorial throughout the year (Hoyt 1957, Kilham 1979). Territory sizes reported in eastern North American forest types range from 43 ha in Louisiana mixed bottomland forest (Tanner 1942) to an average of 87 ha in Missouri oak-hickory forest (Renken and Wiggers 1989). Territory size is inversely related to the proportion of canopy cover made up by overstory trees, the volume of logs and stumps, and the proportion of the forest consisting of trees larger than 30 cm diameter (Renken and Wiggers 1989). The Pileated Woodpecker is a forest interior species,

Pileated Woodpecker. Elizabeth S. Chastain

and their probability of occurrence increases with forest area (Whitcomb et al. 1981, Robbins, Dawson, and Dowell 1989). As they will readily fly across open areas and are somewhat tolerant of forest fragmentation, their occurrence is more dependent on the regional forest area than on the size of individual forest tracts. No Tennessee studies of the territory size or specific habitat requirements of Pileated Woodpeckers are available.

Atlasers recorded Pileated Woodpeckers on 337 of the miniroutes at an average density of 2.1 stops/route. The density was highest in the Unaka Mountains (3.3 stops/route), followed by the Cumberland Plateau (2.9), the Cumberland Mountains (2.2) and the Mississippi Alluvial Plain (2.2), and lowest on the Loess Plain (1.2). The individual routes with the highest densities were almost all in blocks dominated by mature deciduous forests.

Pileated Woodpeckers were probably common in Tennessee prior to the nineteenth-century agricultural clearing and timber harvesting. Their numbers decreased as the area of mature and old-growth forests decreased. At the end of the century, most of the virgin forest outside of the Unaka Mountains had been cut, and many second-growth woodlands had not matured enough to provide suitable habitat. Rhoads (1895a) observed Pileated Woodpeckers at 6 of the 9 areas he visited and gave no other indication of its status. By the early twentieth century, many ornithologists (e.g., Pearson 1917), alarmed over decreasing Pileated Woodpecker numbers, feared the species was unable to adapt to second-growth forests and would become extinct. Pileated Woodpeckers, however, have survived and adapted to second-growth forests.

Pileated Woodpecker numbers in Tennessee, as measured by BBS routes, do not show a significant trend from 1966 to 1994. From 1966 to 1993, they showed a significant ($p < 0.01$) increase of 2.0%/year throughout North America (Peterjohn, Sauer, and Link 1994). Their numbers on Tennessee Christmas Bird Counts have shown a significant ($p < 0.05$) increase (Tanner 1985). These increases are likely due to both the increasing forest area and the increasing size of the trees in the forests.

Breeding Biology: Pileated Woodpeckers were difficult for atlasers to confirm, and nest records accounted for only 37% of the confirmed records. Relatively few detailed Tennessee nest records are available; the Tennessee studies of Sharp (1932), Humphrey (1946), and McGowan (1967) provide information on nest trees and nesting behavior.

Pileated Woodpeckers excavate their nest hole in large dead trees or dead sections of live trees. The nest hole is typically located in the main trunk of a tall, dead stub

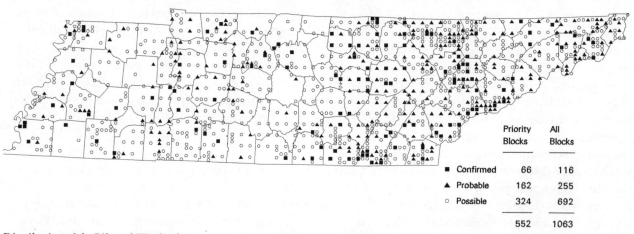

	Priority Blocks	All Blocks
■ Confirmed	66	116
▲ Probable	162	255
○ Possible	324	692
	552	1063

Distribution of the Pileated Woodpecker.

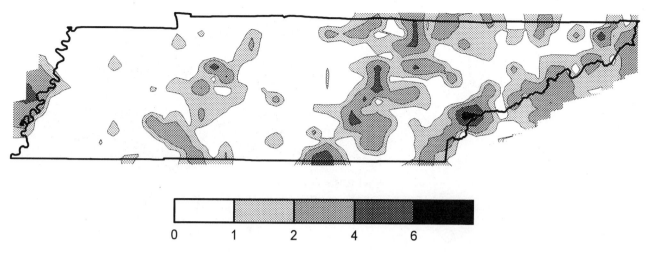

Abundance of the Pileated Woodpecker.

weakened by fungus and with the top broken off (Conner et al. 1975). The average diameter at breast height of nest trees in Virginia and Missouri is about 54 cm (Conner et al. 1975, Renken and Wiggers 1989). In a study near Memphis, McGowan (1967) reported a 63 cm average diameter of trees containing roost and nest holes and an average cavity height of 12.4 m. Nests elsewhere in the state range in height from 5.5 to 25.9 m, averaging 10.8 m (s.d. = 4.7 m, n = 16). Conner et al. (1975) reported average nest heights of 13.6 m in forested areas and 7.5 m in clearcuts and nonforested areas. Hoyt (1957) and Conner et al. (1975) noted a tendency for nests to be near water, which may be due to the greater likelihood of suitable large trees being on moist sites than on less productive upland sites. Nests in Tennessee have been reported from several species of deciduous trees, including beech, sycamore, sweetgum, sugar maple, ash, tulip-poplar, and oaks. Use of coniferous trees, including hemlock and pines, has been reported elsewhere (Hoyt 1957, Conner et al. 1975, Kilham 1979). A nest tree may be used several consecutive years, although a new cavity is usually dug each year below previous cavities (Sharp 1932, Hoyt 1957).

Both male and female Pileated Woodpeckers participate in all nesting activities; the male often spends more time excavating the cavity, which may be up to 66 cm deep, and incubating the glossy white eggs than the female (Hoyt 1957). Egg laying in Tennessee occurs from early April to early May with no well-defined peak. Clutch size ranges from 3 to 5 eggs with 4 the most common size (Hoyt 1957). Tennessee clutches, not including 2 possibly incomplete fresh clutches of 2 eggs, average 3.8 eggs (s.d. = 0.7, n = 23, range 3–5). Incubation lasts 18 days, and the young leave the nest in about 28 days (Hoyt 1957). After leaving the nest, the young are fed by the adults for several days, and the family group may stay together for up to 3 months.—*Charles P. Nicholson.*

Olive-sided Flycatcher
Contopus cooperi

The Olive-sided Flycatcher was formerly an uncommon local summer resident of high-elevation coniferous or mixed forests of the Unaka Mountains. It occurred in areas with many standing dead trees, where it was usually seen perched on branches in treetops or on tall snags, uttering its far-carrying and easily recognized "quick-three-beers" song. Breeding pairs usually arrived by mid-May and departed by mid-September. Otherwise, this species is a rare migrant throughout Tennessee (Robinson 1990).

Breeding of the Olive-sided Flycatcher has been confirmed only in the Great Smoky Mountains National Park (Williams 1976); Roan Mountain is the only other Tennessee location where the species has probably bred. The flycatcher was first reported in the state by Rhoads (1895a), who found 2 birds in hemlock ravines at 1220 m in Rock Creek valley on the northwest side of Roan

Olive-sided Flycatcher. Elizabeth S. Chastain

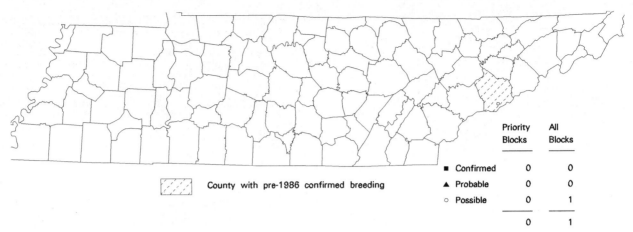

Distribution of the Olive-sided Flycatcher.

Mountain, Carter County, during mid-June for the only probable breeding record from this locality. The most recent possible breeding records from Roan Mountain were during 1977, when 2 birds were observed on 17 June and 1 on 19 June in an area of sapling and scattered mature spruce and fir trees on the North Carolina side of the mountain (Phillips 1979b).

Other likely breeding records have been from the Great Smokies. Burleigh (1935b) reported the first breeding season records there—a singing male on 10 July 1932 near Newfound Gap, Sevier County (1525 m) and a probable pair at the same site in 1933. Subsequently, Stupka (1963) stated this flycatcher was an "uncommon summer resident" in the Great Smokies above 1068 m from 1934 to 1961. However, it is unclear if this species was regularly present at more than just a few breeding sites in the spruce-fir forest, usually from Newfound Gap to Clingman's Dome, along the Alum Cave Bluff Trail, and along the Sawteeth. Outside these areas, Wetmore (1939) discovered single birds during mid-June 1937 on Cosby and Inadu Knobs at 1525–1740 m. The highest count in the Great Smokies was 8 birds at 1068–1830 m from 13–20 June 1938 (Ganier and Clebsch 1938).

The Olive-sided Flycatcher was reported but once in suitable breeding habitat during the Atlas period, in June 1989 along the Alum Cave Bluff Trail near where Williams (1976) found a nest in 1974 and observed a pair in July 1975. Other post-1975 reports exist from this area, and from near Newfound Gap, but with only possible breeding evidence (see McNair 1987c). Low-elevation reports by Atlas workers during late May and early June in Johnson, Cocke, and Monroe Counties (*Migrant* 59:130, 60:109, Atlas results) were likely migrants and are not mapped as possible breeders.

The Olive-sided Flycatcher has continued its long-term decline in the southern Appalachian Mountains and elsewhere throughout much of its eastern breeding range since the beginning of this century (see McNair 1987c; Sauer and Droege 1992, Peterjohn and Sauer 1993). The small, disjunct populations in the southern Appalachian Mountains of Tennessee and North Carolina at the southeastern extremity of its breeding range have now mostly become extirpated (Atlas data; McNair 1987c; *contra* LeGrand 1990b, Robinson 1990). The reasons for its virtual disappearance are unknown. Recent heavy mortality of mature fir and lighter spruce mortality have resulted in increased forest edge habitat at least temporarily suitable for occupation by breeding Olive-sided Flycatchers (see also LeGrand 1990b). Consequently, the Olive-sided Flycatcher should now be state-listed as a Threatened instead of the In Need of Management status it was given in 1994 (TWRA 1994a).

Breeding Biology: The only confirmed breeding record in the Great Smokies was the discovery of an active nest at 1350 m along the Alum Cave Bluff Trail from 30 June–5 July 1974 (Williams 1976). The nest was located on a horizontal branch near the top of a red spruce 30 m above ground and contained a nestling and 2 infertile eggs on 5 July. The breeding biology of the Olive-sided Flycatcher is also poorly known elsewhere in its range (Ehrlich, Dobkin, and Wheye 1988). The nest is typically high in a conifer, on a horizontal limb far from the trunk. Clutches of 3 eggs are probably most common; the female incubates them for about 14 days, and the young fledge in about 3 weeks.—*Douglas B. McNair.*

Eastern Wood-Pewee
Contopus virens

The Eastern Wood-Pewee is a common statewide summer resident. It sings its distinctive, plaintive whistled song throughout the day and is much more easily heard than seen. Although the number of Atlas blocks in which

Eastern Wood-Pewee. Elizabeth S. Chastain

the Eastern Kingbird and Eastern Phoebe were reported was higher than for the pewee, the pewee was found in more of the priority blocks. This, and the fact that the pewee's density on miniroutes was the highest of any flycatcher, suggests that the Eastern Wood-Pewee is the most abundant flycatcher nesting in Tennessee. Pewees arrive in Tennessee in late April and depart by early October.

Eastern Wood-Pewees occupy woodlands and woodland edges and feed high in tree canopies (Hespenheide 1971). Within woodlands, they typically occur in areas with an incomplete canopy. In Ridge and Valley deciduous forests, Anderson and Shugart (1974) found pewees selected habitat based on several variables, among them a low number of small trees and a high number of mid-sized and large trees. Pewees also occupy pine woods. Their occurrence is not related to the area of forest (Robbins, Dawson, and Dowell 1989), and in Tennessee they frequently occur in wooded suburbs, wooded fencerows, and isolated groves of trees. In the Unaka Mountains, they commonly occur in hardwoods up to about 1525 m and occasionally in the spruce-fir forest, although there is no evidence of nesting in this habitat (Stupka 1963, Nicholson pers. obs.). In West Tennessee wetland forests, Ford (1990) found pewees to be one of the most frequently occurring species, present on 54 of 59 study sites.

Eastern Wood-Pewees have probably always been common in Tennessee. They have adapted to many of our modifications of the landscape, and their numbers have probably increased in areas dominated by small fields and mature, parklike woodlands. Conversely, their numbers have probably decreased where most of the forest has been cleared, such as parts of the Loess Plain, Central Basin, and Ridge and Valley. Historic Tennessee sources such as Rhoads (1895a) and Ganier (1933a) give no indication of changes in pewee numbers. Since 1966, their population, as measured by Tennessee BBS routes, has not changed. They have been reported on all of the Tennessee routes, at an average relative abundance of 7.8 stops and 8.4 birds/50-stop route during the Atlas period. Throughout their North American breeding range, their population declined significantly ($p < 0.01$) by 1.7%/year between 1966 and 1993 (Peterjohn, Sauer, and Link 1994).

Atlasers found pewees on almost 80% of the miniroutes, at an average relative abundance of 3.1 stops/route. The highest abundance was 11, and numbers often varied greatly among nearby blocks. Physiographic regions with the highest average abundances were the Coastal Plain Uplands and Western and Eastern Highland Rims. The Cumberland Mountains, Ridge and Valley, and Unaka Mountains had the lowest average abundances. Pewee numbers were not closely related to the proportion of forested land.

Eastern Wood-Pewees have been recorded at measurable densities on relatively few of the breeding bird censuses in Tennessee. In virgin cove hardwoods in the Smokies, Aldrich and Goodrum (1946) found a density

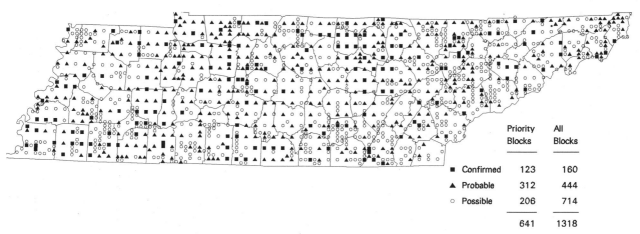

Distribution of the Eastern Wood-Pewee.

		Priority Blocks	All Blocks
■	Confirmed	123	160
▲	Probable	312	444
○	Possible	206	714
		641	1318

Confirmed Breeding Species

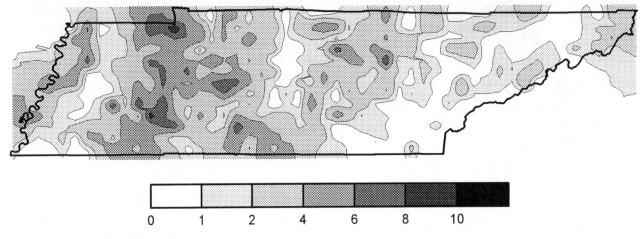

Abundance of the Eastern Wood-Pewee.

of 11 pairs/100 ha. Density in a 2–4 year old clearcut with scattered large trees was 5/100 ha (Lewis and Smith 1975, Lewis 1977, Lewis 1978). Garton (1973) found a density of 8.8/100 ha in a Cumberland Mountain deciduous forest and strip mine plot, and Simmers (1978) recorded a density of 23/100 ha in a pasture with wooded fencerows. Stedman and Stedman (1992) found a density of 14.7/100 ha in a Highland Rim woodlot.

Confirming the breeding of pewees was difficult for atlasers, and only 12% of Atlas records were confirmed. Although pewee nests are usually in an exposed position far out on a tree branch, the nest is small and high in the tree, and the adult is inconspicuous in its vicinity. Nests accounted for slightly less than half of the confirmed records. Most of the nest records were of the code ON, as atlasers were unable to determine the contents of the inaccessible nests.

Breeding Biology: Eastern Wood-Pewees are late nesters, rarely beginning nest construction before the middle of May. They are usually monogamous, although polygyny has been reported (Eckhardt 1976). The female builds the nest (Harrison 1975), which is a compact, well-built, shallow cup. The outer cup is constructed of bits of plant fibers, moss, short grass, and weed stems, bound together with spider webs. The nest is lined with fine grasses and hair and camouflaged on the outside with lichens. From below, the nest often resembles a knot. The nest is usually placed on a horizontal limb far from the trunk, either straddling the limb or on a fork. Nests are occasionally built on dead limbs. Pewee nest heights in Tennessee range from 1.8 to 13.7 m above ground, with an average height of 6.0 m (s.d. = 2.4 m, n = 42), considerably lower than the average height of 10.7 m given by Harrison (1975). Most Tennessee nests have been built in oaks and pines.

Egg laying in Tennessee peaks during the last week of May and first third of June. Clutches of 2–4 eggs have been reported in Tennessee, with a mode of 3 and average of 2.85 (s.d. = 0.41, n = 48). The eggs are creamy white, marked with brown and lilac blotches in a wreath around the larger end. The female incubates the eggs for 12 to 13 days. Both adults feed the nestlings, which fledge in 15–18 days.

The detailed pewee nest records available from Tennessee, most dating before 1940, include no records of parasitism by Brown-headed Cowbirds. The only available evidence of cowbird parasitism is 2 observations during the Atlas period of pewees feeding fledgling cowbirds (Nicholson pers. obs.). Cowbird parasitism of pewees is uncommon elsewhere (Friedmann 1963). —*Charles P. Nicholson.*

Acadian Flycatcher
Empidonax virescens

This small flycatcher is usually first identified by its explosive "spit-chee" song or by the high-pitched twittering call given as it flies from one perch to another. It is a fairly common statewide summer resident of low elevations. Acadians arrive in late April and depart by late September (Robinson 1990).

Acadian Flycatchers have relatively specialized habitat requirements consisting of moist, predominantly deciduous woodland. In most of the state, they are found along wooded streams and in moist ravines. In West Tennessee forested wetlands, Acadians are one of the most abundant species, and they prefer forests with a tall, dense canopy and few shrubs (Ford 1990). Fowler and Fowler (1985) also found it to be one of the most abundant species along a portion of the Duck River in Maury County. In the eastern mountains, Acadians also

Acadian Flycatcher. Elizabeth S. Chastain

occupy streamside hemlocks, up to a maximum elevation of about 1070 m (Stupka 1963). Acadian Flycatchers are more frequently found in large tracts of forest than in small woodlots (Robbins, Dawson, and Dowell 1989).

Rhoads (1895a) was the first to describe the Acadian Flycatcher's occurrence in Tennessee, and he found it statewide below 1070 m and abundant in West and Middle Tennessee. Its overall range has not changed since then, although, as shown by the Atlas results, it is no longer abundant in parts of West and Middle Tennessee. This is no doubt due to the extensive forest clearing in these areas.

Atlasers found Acadian Flycatchers in about two-thirds of the completed priority blocks. They were most frequently reported in the heavily wooded Blue Ridge, Cumberlands, and Western Highland Rim. Atlasers probably missed Acadians in some blocks in the Ridge and Valley and southern Cumberland Plateau where access to suitable habitat was difficult.

Acadian Flycatchers were recorded on about 40% of the miniroutes, at an average relative abundance of 2.6 stops/route. The regions of highest abundance were in the forested sections of the Mississippi Alluvial Plain, where the highest count of 13 was made, the Western Highland Rim, and southern Blue Ridge. The highest densities reported on census plots are 88 pairs/100 ha in a mixed mesophytic, deciduous ravine in Jackson County on the Eastern Highland Rim (Simmers 1982a), and 60 to 71 pairs/100 ha in hemlock-deciduous forest in the Great Smokies (Wilcove 1988).

From 1966 to 1994, the Acadian Flycatcher population showed no significant overall trend on Tennessee BBS routes, where its average relative abundance was 2.7 birds/route on 40 of the 42 routes. Throughout North America, Acadian Flycatcher populations likewise showed no significant trend from 1966 to 1993 (Peterjohn, Sauer, and Link 1994). They reach their highest numbers, according to BBS results, in the Cumberland Plateau physiographic province (including the Cumberland Mountains) (Robbins, Bystrak, and Geissler 1986).

Breeding Biology: In contrast to most other small passerines, most of the confirmed records of Acadian Flycatchers were actual observations of nests. A few Atlas workers became proficient at finding the nests, and these individuals reported a disproportionate number of the confirmed records. Most of these were of the breeding code ON, reflecting the difficulty of examining the nest contents. Acadian Flycatchers nest on horizontal tree limbs, usually far from the trunk. The nest is usually near or at the lower edge of the tree canopy with an open space below (Mumford 1964). Nests are often near the bottom of ravines and sometimes over water. Nest heights in Tennessee averaged 3.9 m (s.d. = 2.0, n = 38), and ranged from 1 from 12 m. Nests have been reported from a variety of deciduous trees, with beech,

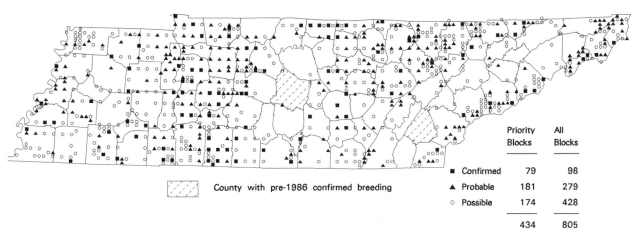

Distribution of the Acadian Flycatcher.

Confirmed Breeding Species

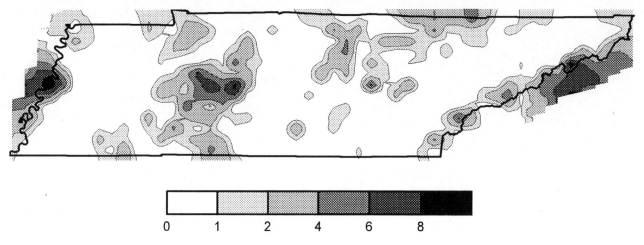

Abundance of the Acadian Flycatcher.

dogwood, and sweetgum most often used. Nests have also been reported from hemlock and shortleaf pine trees.

The cup-shaped nest, built by the female, is suspended from a fork of small branches (Mumford 1964). It is built of dried grasses, small twigs, grapevine, and fine bark strips, held together with spider webs. Stupka (1963) reported a nest in the Great Smoky Mountains constructed of *Usnea* lichens. The nest often has little lining, and pieces of nest material hang several cm below the nest. Completed nests appear frail and unfinished, and it is often possible to see the eggs through the bottom of the nest. These features make the nest easy to identify. Despite their frail appearance, Acadian nests may last up to 3 years; nests built during the previous season are occasionally reused (Mumford 1964).

Clutch sizes of 2 and 3 have been reported in Tennessee, with an average of 2.73 (s.d. = 0.44, n = 33). Clutches of 4 have been reported elsewhere, although they are uncommon (Mumford 1964, Harrison 1975). The eggs are creamy white and marked with a few brown spots near the large end. The female incubates the eggs for 13 to 15 days; both parents feed the nestlings, which leave the nest in 13 to 14 days (Mumford 1964).

Studies elsewhere have found that Acadian Flycatchers sometimes renest after fledging young from the first nest (Mumford 1964, Walkinshaw 1961). This renesting has not been proven by studies of banded birds in Tennessee. Egg dates, which range from 20 May to 8 July, with a strong peak in late May and a lesser peak in late June, suggest renesting occurs. Mumford (1964) observed adults alternately feeding nestlings and building a second nest, and feeding fledglings while incubating a second clutch.

Few Tennessee records of cowbird parasitism of Acadian Flycatchers are available. A set of 2 flycatcher eggs and 1 cowbird egg was collected in Washington County in 1954, and atlasers reported 1 instance of flycatchers feeding a fledging cowbird. Elsewhere in their range, Acadian Flycatchers are a fairly common cowbird host, and local populations may be heavily parasitized (Friedmann, Kiff, and Rothstein 1977, Walkinshaw 1961). Although the flycatcher may occasionally build a new nest floor over the cowbird eggs, most parasitized nests fail to fledge flycatcher young (Walkinshaw 1961).—*Charles P. Nicholson.*

Alder Flycatcher
Empidonax alnorum

The Alder Flycatcher is a drab, neotropical migrant that is confusingly similar to other members of its genus. It is best identified by its song, a burry "fee-beeo," which it seldom sings in migration. As a result, very few spring records exist from across Tennessee, although many must surely pass through. It was lumped with the Willow Flycatcher as a single species until 1973, further clouding the status of both (AOU 1973). The Alder Flycatcher is the last species to arrive on its Tennessee breeding grounds, appearing in mid-May and departing in August.

The Alder Flycatcher has only recently extended its breeding range southward into the southern Appalachian Mountains (Scott 1982, McNair 1987c, Hall 1989). The first breeding season report in Tennessee came from Roan Mountain in 1977, but was not verified until the following year when successful breeding was confirmed there through the observation of recently fledged young (LeGrand 1979, Lura, Schell, and Wallace 1979). Alder Flycatchers have been present and breeding on Mount Rogers in southwestern Virginia since 1974 (Scott 1982). The first summer report in North Carolina came from near Blowing Rock in 1972 (Teulings 1972); since then, isolated populations have been reported from 5 western North Carolina counties, with breeding confirmed as far south as Haywood County (McNair 1987c).

Alder Flycatcher. Elizabeth S. Chastain

In Tennessee, the Alder Flycatcher was reported from 3 blocks during the Atlas period, with the only confirmed breeding at Roan Mountain. The other records were from new locations in Johnson County. A singing male was found in a streamside thicket in early June 1989 at Mountain City (D. B. McNair pers. comm.). A territorial male was present in a shrubby bog in Shady Valley throughout June 1990–91 (Knight and J. Shumate pers. obs.); this habitat was destroyed in 1992. A few other birds elsewhere could have been missed, as some isolated patches of suitable habitat may have been overlooked, such as on Unaka and Big Bald Mountains in Unicoi County, Camp Creek Bald in Greene County, and some Smoky Mountain balds (Knight pers. obs.). Alder Flycatchers have a brief period of singing and are also rather elusive, staying low in thickets. None were recorded on Atlas miniroutes.

The habitat of Alder Flycatchers in the southern Appalachians is dense, scrubby, deciduous thickets in either upland or slightly boggy settings at high elevations, mostly above 1460 m. These thickets consist of a variety of deciduous shrubs and small trees, especially green alder, hawthorn, mountain-ash, and fire cherry, with a few evergreens (mainly red spruce) sometimes present. Blackberry is often an important component (McNair 1987c, Hull 1990). At Roan Mountain, these thickets are primarily on the fringes of grassy balds at elevations of 1675–1860 m. The 2 Johnson County sites were at unexpectedly lower elevation (701 and 853 m); both were deciduous thickets, with blackberry prominent, and adjacent to water. Density figures for the southern Appalachian breeding population generally have not been recorded, but McNair (1987c) found 17 singing males within a 3 km radius near Shining Rock Wilderness Area, North Carolina.

The Roan Mountain population, which straddles the Tennessee–North Carolina border, has remained relatively stable for several years at 6–8 territorial males (Knight pers. obs.). Its further growth is probably limited by the availability of suitable habitat. At lower elevations, competition with the Willow Flycatcher may limit the Alder's distribution. When sympatric, the Willow Flycatcher is often dominant over the Alder and may usurp the Alder's territory (Prescott 1987). Willow Flycatchers were present on adjacent territories at both Johnson County Alder Flycatcher sites; these populations should be closely monitored in the future.

Breeding Biology: No Alder Flycatcher nests have yet been found in Tennessee, although recently fledged young have been reported several times from the Tennessee side of Roan Mountain (Lura, Schell, and Wallace 1979, Knight and F. J. Alsop pers. obs.). Hull (1990) found a nest containing 4 eggs on 23 June 1988 on the North Carolina side of Roan Mountain. Three of these eggs hatched by 3 July, and young fledged on 13 July.

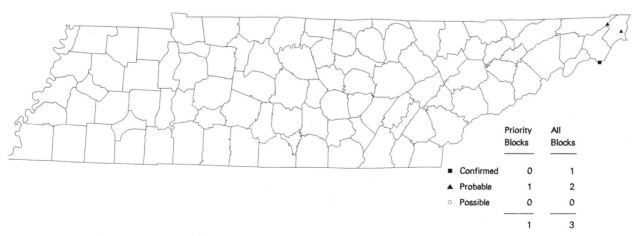

	Priority Blocks	All Blocks
■ Confirmed	0	1
▲ Probable	1	2
○ Possible	0	0
	1	3

Distribution of the Alder Flycatcher.

This nest was attached to 3 blackberry stems 84 cm above the ground and woven entirely of grasses and sedges, some of which hung 38 cm below the nest (Hull 1990). Nests elsewhere average about 60 cm above ground, are typically untidy, and hold a clutch of 3–4 eggs (Harrison 1975). The female builds the nest and incubates the eggs for 12–14 days; both adults feed the nestlings, which fledge in about 2 weeks (Bent 1942, Harrison 1975). Cowbird parasitism of the Alder Flycatcher has not been reported in the southern Appalachians.—*Richard L. Knight*.

Willow Flycatcher
Empidonax traillii

The Willow Flycatcher, a small, uncommon summer resident, is much more easily identified by its "fitz-bew" song than its drab plumage. It inhabits brushy areas, and is present in Tennessee from early May until September; few spring records exist outside of suitable breeding habitat. Until 1973, it and the Alder Flycatcher were considered different forms of a single species, the Traill's Flycatcher. The two species are most easily distinguishable by song.

The Willow Flycatcher is a relatively recent addition to the state's nesting avifauna. Nesting was first confirmed in 1958 in Carter County (Herndon 1958), although a few earlier late May and early June records exist. For over a decade, all breeding records were in Carter and nearby Johnson County. Then, within the space of a few years, its known breeding range greatly increased. Nesting was observed in Meigs County in 1969 (*Migrant* 40:70), in Blount and Knox Counties in 1970 (Alsop 1971a, Williams 1975b), and in Davidson County in 1971 (Goodpasture and Alsop 1972). In 1973, 6 birds were present in suitable breeding habitat in Benton County (Pitts 1982c). Territorial birds were first present in eastern Obion County in 1976 and at Reelfoot Lake in 1982 (Pitts 1982c). During the late 1970s and early 1980s, Willows were reported from many new locations within the overall range outlined by the earlier records. During the Atlas period, the species extended its range to the extreme southwest corner of the state. The Willow Flycatcher has been recorded on too few Tennessee BBS routes, however, to detect a significant increase.

The spread of the Willow Flycatcher into Tennessee is part of a broad expansion of this species from the midwestern prairies into the eastern United States (Stein 1963). Although the Willow Flycatcher has nested in the Grand Prairie area of east-central Arkansas, about 160 km from Memphis, since the time of Audubon (James and Neal 1986), there is no evidence that this population spread to Tennessee. The spread of the Willow Flycatcher into Tennessee was apparently from the north, first into the northeast corner of the state, then, about 1970, across a broad front.

Most Tennessee records of Willow Flycatchers, for which detailed habitat descriptions are available, have been in moist, shrubby areas or shrub swamps, with willow trees the dominant woody plant (e.g., Herndon 1958, Goodpasture and Alsop 1972, Williams 1975b). Territories may be in extensive shrub swamps or in narrow bands of willows along streams or lakeshores, usually adjoining grassy areas. Willow Flycatchers also less commonly occur in drier thickets in pastures or hayfields, where taller vegetation consists of species such as osage orange, plum, and blackberry. Such habitat has been used in Williamson (Stedman 1987), Washington (R. L. Knight pers. comm.), and Campbell and Knox (Nicholson pers. obs.) Counties. This drier habitat is frequently used by Willows elsewhere in their range (Stein 1963). The Willow Flycatcher occurs frequently at elevations to 975 m, and rarely to 1100 m (Knight pers. comm.).

Atlasers found Willow Flycatchers throughout the state, in a total of 79 blocks, and about 5% of the completed priority blocks. It is likely they were missed in some blocks where suitable habitat was limited and not near roads. None were reported in some areas of the state where seemingly suitable habitat occurs, such as the southeastern Highland Rim. They were found on 11 miniroutes, at an average relative abundance of 1.1 stops/route; because of the low numbers, the miniroute results do not portray regional differences in abundance well. Areas of the state where Willow Flycatchers are locally common include Cross Creeks National Wildlife

Willow Flycatcher. Elizabeth S. Chastain

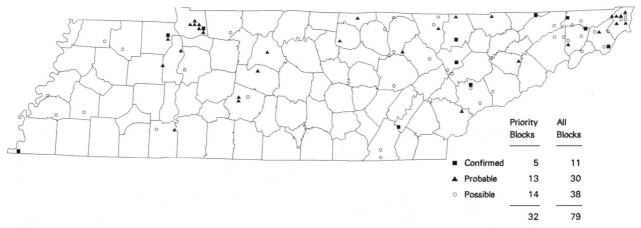

Distribution of the Willow Flycatcher.

Refuge and Johnson County, where they occurred in several blocks, and McNair (pers. comm.) found 19–20 singing males in an Atlas block. Four or more singing males also occurred in blocks in Washington, Fentress, and Campbell Counties (Knight pers. comm., Nicholson pers. obs.). Williams (1975b) observed 4 pairs in a 7 ha Blount County marsh. Walkinshaw (1966a) found an average territory size of 0.7 ha in southern Michigan.

Breeding Biology: Willow Flycatchers are relatively late nesters in Tennessee, with nest building beginning in late May and egg laying peaking in mid-June. Although most often monogamous, they are occasionally polygynous, with 2 females nesting within the territory of a single male (Sedgwick and Knopf 1989). The nest is usually built in the upright crotch of a small tree or bush or on a horizontal branch at the junction of an upright branch, to which it is fastened (Walkinshaw 1966a). Tennessee nests have been of both types and average 1.8 m (s.d. = 0.7 m, n = 23) above the ground, ranging from 1.1 to 4.6 m. Nests have been most often reported in willow, buttonbush, and elm saplings; other species used include St. John's wort, blackberry, and alder. The nest is a compact cup built by the female of down from milkweeds, cattails, and other plants, as well as small leaves, hair, and feathers bound together with spider webs and lined with fine grasses (Walkinshaw 1966a, Goodpasture and Alsop 1972, Nicholson unpubl. data).

Clutches in Tennessee range from 2 to 4 eggs, with 3 and 4 most common and an average size of 3.46 (s.d. = 0.66, n = 13). Walkinshaw (1966a) found an average clutch size of 3.7 and a rare clutch of 5 eggs. The eggs are creamy white with fine black or brown spots at the larger end. The female incubates the eggs for 13–15 days; both adults feed the nestlings, which fledge in 12–15 days and are then fed by the parents for about 10 more days (Walkinshaw 1966a). Cowbird parasitism, although fairly common to the north and west (Friedmann 1963), has yet to be reported in Tennessee.—*Charles P. Nicholson.*

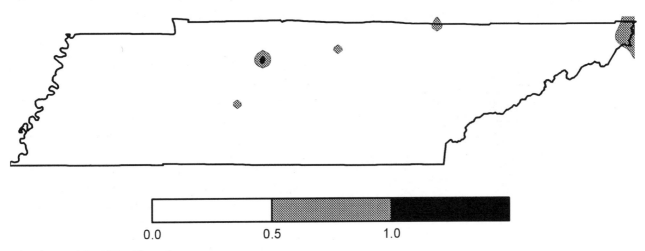

Abundance of the Willow Flycatcher.

Least Flycatcher
Empidonax minimus

The Least Flycatcher is another of the small, drab, confusingly similar *Empidonax* flycatchers, best identified by its song, an emphatic "che-bek." It is the most persistent singer of the 4 *Empidonax* breeding in Tennessee. Because it also sings en route, it is the most often encountered spring migrant of this genus, occurring all across the state. Most fall records come from banding stations. The Least Flycatcher is a rare to locally fairly common summer resident of the state's eastern highlands, arriving in late April or early May and departing in August or September. Tennessee is near the southeastern limit of its breeding range (AOU 1983).

The earliest mention of nesting Least Flycatchers in Tennessee is by Rhoads (1895a:486), who stated, "It breeds at Roan Mountain Station and thence up the Doe River Valley to near the limit of deciduous trees." Ganier (1917, 1933a) called it a very rare to rare summer resident of the mountains. Ganier and Tyler (1934:22) found it "fairly common, chiefly in the woodlands along the creek" in Shady Valley, Johnson County. Stupka (1963) considered it an uncommon summer resident in the eastern half of the Great Smoky Mountains National Park, probably nesting for several years during the early 1950s at Gatlinburg. Scattered, inconsistent summer reports have come from the northern Cumberlands since 1970 (Campbell and Howell 1970), but breeding has yet to be confirmed there (C. P. Nicholson pers. comm.).

The distribution of the Least Flycatcher during the Atlas period was similar to its historic range in the state. Reports came from just 18 blocks, mostly in the Unaka Mountains and Cumberland Mountains/Plateau, with breeding confirmed only in the Unakas. The species is fairly common in the immediate Roan Mountain vicinity and uncommon in Johnson and northern Carter Counties. These areas remain as its stronghold. Recent reports from the Smokies have been very scarce, with a record (albeit of 6 singing males and a fledgling) from only 1 Atlas block. Least Flycatchers were present in the Unicoi Mountains at Whigg Meadow, Monroe County, in 1981–82 (McConnell and McConnell 1983), but could not be found there in 1989 or 1991 (Nicholson pers. comm.). Five Atlas reports from the Cumberland Mountains/Plateau were of solitary singing males, all observed in early June. Some of these were perhaps late migrants, but enough sightings have come from that region in the last 20 years to indicate some level of summer residency. More fieldwork is needed there for clarification. A territorial male in the northern Sequatchie Valley, a bird on the Eastern Highland Rim, and a pair on the eastern edge of the Ridge and Valley during the Atlas present anomalies.

Least Flycatchers in Tennessee have not been adequately surveyed by standard methods to determine their population trend; they were recorded on just a single miniroute stop during the Atlas. Anecdotal reports from areas with historic populations seem to suggest stable numbers at Roan Mountain, a slight decrease in Johnson County, and a significant decline in the Smokies. Rangewide BBS results show a decrease of 0.9%/year (p = 0.03) from 1966 to 1993 (Peterjohn, Sauer, and Link 1994).

Habitat of the Least Flycatcher is usually described as open, deciduous woods or woodland edge, but groves of deciduous or coniferous trees in open surroundings are also suitable. Breckenridge (1956) found that an open subcanopy and understory was an important feature of Least Flycatcher habitat. Males defend small territories (Davis 1959), and they frequently occur in loose colonies at high densities. Least Flycatchers forage very energetically in the subcanopy, taking insects in flight and from leaves (Holmes, Black, and Sherry 1979). Where its range overlaps that of the Acadian Flycatcher, the Acadian prefers a denser understory (Hespenheide 1971). In Tennessee, Least Flycatcher habitat includes open beech-maple forest, mature beech forest, forest edge of beech-maple-cherry, an open grove of Virginia pine, a grove of shortleaf pine, an open stand of locust and sumac, a mature cedar grove in a pasture, and an apple orchard. It usually occurs at elevations of 750–1600 m.

Breeding Biology: Least Flycatcher breeding biology information from Tennessee is scant, with few nests and fewer nest contents having been described. The nest of

Least Flycatcher. Elizabeth S. Chastain

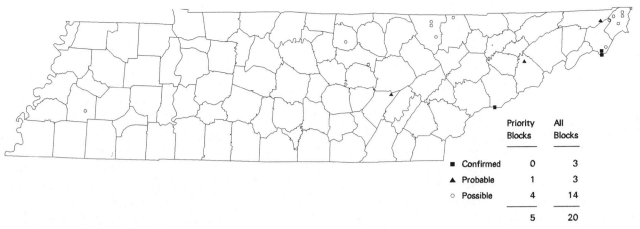

Distribution of the Least Flycatcher.

the Least Flycatcher is a compact cup made of bark, weed stems, and grasses, lined with plant down and hair, and built by the female in 6–8 days (Walkinshaw 1966b, Harrison 1975). Nest construction has been reported as early as 7 May (Herndon 1950a), and nests have been found in pine and maple trees. Nests are usually placed in the fork of a limb 3–6 m above the ground, and the highest Tennessee nest was 11 m above ground (Knight pers. obs.). Clutch size is 3–6 creamy white eggs and averages about 4; they are incubated by the female for 13–14 days (Walkinshaw 1966b). Although there are no dated Tennessee egg records, apparently incubating birds have been observed from 1–24 June (Knight pers. obs., F. J. Alsop III pers. comm.). Three nests with young each contained 3 nestlings. Both parents tend the young, which fledge in 12–16 days (Walkinshaw 1966b). The very late observation of an adult "feeding several young" on 2 September 1937 in the Smokies (Bellrose 1938:2) suggests that second broods are sometimes raised, as reported elsewhere (Briskie and Sealy 1987). The Least Flycatcher is usually monogamous, although polygyny has been reported (Briskie and Sealy 1987). Cowbird parasitism is uncommon overall (Friedmann and Kiff 1985) and unreported in Tennessee.—*Richard L. Knight.*

Eastern Phoebe
Sayornis phoebe

The Eastern Phoebe is the familiar, plain, tail-flicking flycatcher that sings its name "fee-bee" and is appreciated for its captivating behavior in Tennessee, where it frequently places its mud and moss nest in barns and outbuildings and occasionally under eaves of occupied houses. It breeds throughout the state and is present throughout the year, though its numbers sharply diminish in winter (Robinson 1990).

The Eastern Phoebe is an uncommon to common breeder in Tennessee, in general increasing from west to east (Atlas results). It was found in 96% of the completed priority blocks and on 64% of the miniroutes at an average density of 2.0 stops/route. It is most numerous in rolling to mountainous, semi-open hardwood or mixed forest, containing numerous streams, rock walls, and human-made nest sites from the Eastern Highland Rim eastward to the lower elevations of the Unaka Mountains, where it has nested up to 900 m (Stupka 1963). In North Carolina, phoebes have nested up to 1490 m in the Great Smokies (Stupka 1963) and 1684 m elsewhere in the Blue Ridge (McNair 1987c). The areas of greatest concentration in Tennessee are the northern half of the Cumberlands and Eastern Highland Rim. The phoebe is also numerous in portions of the Western Highland Rim, in the hillier south, and along the bluffs of the Tennessee River. It is least numerous in

Eastern Phoebe. Elizabeth S. Chastain

Confirmed Breeding Species

West Tennessee, the Central Basin, and portions of the Ridge and Valley (Atlas miniroute data), especially in areas of flat terrain or limited surface drainage. These areas also generally have the scantiest forest cover in Tennessee; phoebes will occupy agricultural areas with scattered trees and bushes, though it is less-favored habitat (see Weeks 1979, also Hill and Gates 1988). The scarcity of suitable nest sites probably limit its occurrence in the Mississippi Alluvial Plain, the only region where it was not widespread.

The Eastern Phoebe was probably uncommon in prehistoric Tennessee, nesting on rocky bluffs and in cave mouths in the eastern two-thirds of the state. Rhoads (1895a) recorded it only from the Cumberland Plateau eastward and south to Chattanooga, although Pindar (1889) listed it as a common summer resident in Fulton County, Kentucky, just north of Reelfoot Lake. It nested in a cave and old cabin in extreme northeastern Mississippi in 1904 (Allison 1907, Coffey 1943b), and Howell (1910) found it in Lawrence County in early September 1908, before fall migration begins. Ganier (1933a) stated that breeding phoebes were rare in West Tennessee and fairly common in Middle Tennessee. Its spread throughout the state is poorly documented, and the first actual nest record southwest of Nashville was in Wayne County in 1938 (Tanner 1988). Wetmore (1939) recorded likely breeding phoebes in Fayette, Lake, Obion, and Wayne Counties in 1937, and Calhoun (1941) found phoebes during late summer 1939 in Hardeman and McNairy Counties. All evidence considered, it is likely that at least some scattered pairs of phoebes were established in the Coastal Plain Uplands region and the southern portions of Middle Tennessee prior to 1933 as Ganier stated. Coffey discovered the first nest in the Memphis area in 1942 (Tanner 1988; see also Jeter 1957), which suggests that phoebes were becoming established throughout West Tennessee. By the mid-1960s, phoebes were uncommon to common throughout Tennessee (BBS results, spring counts in *Migrant*).

The phoebe's breeding-range expansion in West Tennessee was undoubtedly a response to the availability of human-made nesting structures, primarily bridges and culverts. This is also true of the southeast coastal plain outside Tennessee, where the phoebe has continued to modestly expand its breeding range (Jeter 1957, Jackson and Weeks 1976, Jackson et al. 1976, McNair 1984a, 1990). The expansion and consolidation of the phoebe's breeding range in the coastal plain of Tennessee agrees with this general population increase, and is also consistent with statewide Tennessee BBS results from 1980 to 1994, which show a highly significant ($p < 0.01$) increase of 4.2%/year. From 1966 to 1994, its Tennessee population showed no significant overall trend; rangewide, it increased by 0.7%/year ($p = 0.03$) from 1966 to 1993 (Peterjohn, Sauer, and Link 1994).

The Eastern Phoebe had the highest frequency (62.7%) of confirmed breeding of any common species in Tennessee; in priority Atlas blocks only, the frequency of confirmed breeding rose to 84.6%. Most confirmed records were active or used nests. Nest-site tenacity is strong (Bent 1942, Hull 1983) and used nests, which frequently last a year and may be reused within a season or again the next year (Weeks 1978, 1979), accounted for 20–25% of all confirmed breeding records.

Breeding Biology: Nests are either placed on a horizontal support (statant) or attached to a vertical surface (adherent) and built by the female of mud, grasses, weeds, and moss lined with fine grasses or hair. In natural habitats, phoebes nest on rocky bluffs or at cave mouths where overhanging ledges provide protection from the weather. Even today, nests occur in about three-quarters of all cave mouths examined (C. Nicholson pers. comm., NRC). Rarely, phoebe nests in natural habitats may be

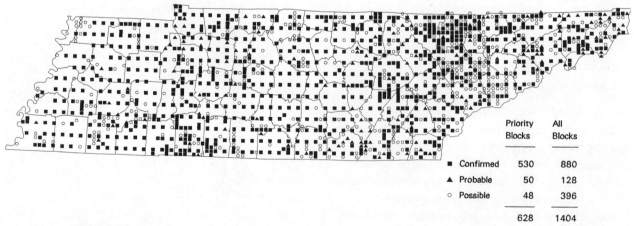

Distribution of the Eastern Phoebe.

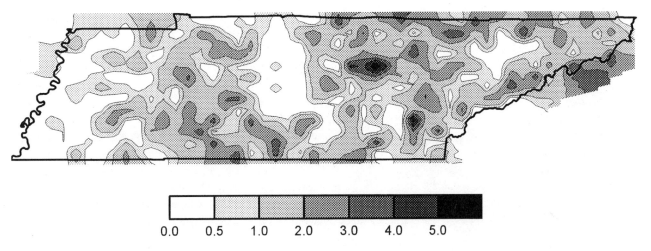

Abundance of the Eastern Phoebe.

placed under streambanks and not depend upon the presence of rock (Cuthbert 1962); one nest was found attached to roots at the top of an eroded dirt slope about 2 m above a road in Anderson County (C. Nicholson pers. comm.). These natural nest sites and habitats continue to be important today in many areas of Middle and East Tennessee, though the adoption of human-made nesting structures allowed phoebes to increase considerably into anthropogenic habitats after European settlement (e.g., Coffey 1963). These human-made nesting structures in disturbed habitats also include rock quarries, entrances to mineshafts, strip mine banks, road cuts, and old wells, the equivalents to natural nest sites during the pre-European settlement period.

Exceptionally, egg laying may begin in early March (Stupka 1963), but normally extends from late March to late June (Atlas results). The phoebe is typically double-brooded (Weeks 1978). The modal clutch size is 5, and the average size of unparasitized clutches in Tennessee is 4.61 (s.d. = 0.69, range 3–6, n = 120), which is similar to published data elsewhere (Weeks 1978, Faanes 1980, Conrad and Robertson 1993). The female incubates the usually unspotted, white eggs for about 16 days, and the nestling period is 16–18 days (Bent 1942, Coffey 1963, Klaas 1975, Weeks 1978, 1979, Hill and Gates 1988, Conrad and Robertson 1993). Predation is the main cause of nest failure (Bent 1942, Klaas 1975, Weeks 1979, Hill and Gates 1988 and others). The average clutch size is lowest in new adherent nests (Weeks 1978). The benefits of larger clutch size may outweigh the energetic investment of new nest construction and the threat of predation, which was greater in statant than in adherent nests (see Weeks 1979). Losses due to cowbird parasitism range from light to heavy (Klaas 1975, Weeks 1979, Friedmann and Kiff 1985, Hill and Gates 1988); the primary impact is on first broods. The parasitism rate in Tennessee is very low—1 of 78 nests studied by Coffey (1963) was parasitized, as were 3 of 156 nests described by egg collection and NRC records.—*Douglas B. McNair.*

Great Crested Flycatcher
Myiarchus crinitus

A loud "wheep" and a flash of yellow and rufous from an exposed perch distinguish the Great Crested Flycatcher. This large flycatcher is a fairly common summer resident of low elevations across Tennessee, arriving in mid-April and departing by mid-September (Robinson 1990). The Great Crested Flycatcher occurs in a wide range of rural and suburban forest and forest edge habitats containing cavities suitable for its use as nest sites.

Great Crested Flycatchers were probably common in most of prehistoric Tennessee, occurring in riparian and floodplain forests and in small openings in upland forests. Exceptions were probably the prairie areas of north Middle and northwest Tennessee and the cedar glades of Middle Tennessee, where nest cavities may have been limiting. Their numbers probably decreased slightly with nineteenth-century agricultural clearing. The first breeding season records were by Rhoads (1895a:485), who recorded them as "everywhere" except the highest elevations of Roan Mountain, and abundant in West and Middle Tennessee. Ganier (1933a) described the flycatcher as fairly common across the state. Since then, Great Crested Flycatchers have probably declined somewhat from competition with the European Starling for nests; the magnitude of this decrease, however, is poorly known. Tennessee BBS results do not show a significant trend from 1966 to 1994. Their rangewide population likewise showed no significant change from 1966 to 1993 (Peterjohn, Sauer, and Link 1994).

Great Crested Flycatcher. Elizabeth S. Chastain

Great Crested Flycatchers are presently distributed widely in Tennessee, occurring in 93% of completed Atlas blocks and on 55% of the Atlas miniroutes. For those miniroutes on which Great Crested Flycatchers occurred, they averaged 1.9 stops/route. Their relative abundance was highest in the Coastal Plain Uplands, Mississippi Alluvial Plain, and Eastern Highland Rim, although regional differences were not dramatic. Their miniroute abundance was considerably higher than their abundance on Tennessee BBS routes, where they were recorded at an average of 3.2 stops and 3.5 birds/50 stop route from 1986 to 1991. They were found on all of the 42 routes. Crested Flycatchers are uncommon in the eastern mountains, where they rarely occur above 1220 m (Stupka 1963, Eller and Wallace 1984). Many of the priority blocks in which they were not found were in heavily forested parts of East Tennessee.

Other Tennessee studies have found Great Crested Flycatchers to be widespread at low densities. They were among the most widely distributed birds in the Central Basin, they occurred on 25 of 31 study sites and averaged 12.5 individuals per sq km (Ford and Hamel 1988). In forested wetlands in West Tennessee, these flycatchers occurred on 53 of 59 randomly selected study sites and averaged 3.9 individuals per 1 km transect (Ford 1990). Breeding Bird Census results include 17 pairs/100 ha in a Ridge and Valley mixed deciduous woodlot (Smith 1975) and 33/100 ha in a Central Basin upland hardwood forest (Fowler and Fowler 1984a). They were found on few other Breeding Bird Census plots.

Great Crested Flycatchers are generally associated with floodplain forests, small woodlots, suburban forests, and forest edges in Tennessee, which is typical for most of their breeding range. They usually forage high in trees, and their summer diet includes fruit as well as insects and other invertebrates (Johnston 1971). In extensive forests, they frequently occur around standing dead trees in canopy gaps. Robbins, Dawson, and Dowell (1989) found their probability of occurrence increased with forest area up to about 72 ha; in larger tracts, this probability decreased somewhat. In forested wetlands of West Tennessee, highest densities of Great Crested Flycatchers occurred in forests with a low density of subcanopy trees (Ford 1990).

Breeding Biology: The proportion of Atlas records in the confirmed category, 12%, was relatively low compared to other cavity nesters and most other flycatchers. Nests were often difficult to locate and accounted for about one-third of the confirmed records. Great Crested Flycatchers nest in either natural cavities or cavities excavated by other species. Human-made sites such as Purple Martin houses, bluebird boxes, and rural mailboxes are also used; these sites are usually close to wood-

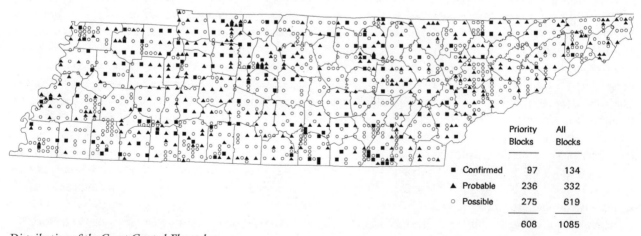

Distribution of the Great Crested Flycatcher.

		Priority Blocks	All Blocks
■	Confirmed	97	134
▲	Probable	236	332
○	Possible	275	619
		608	1085

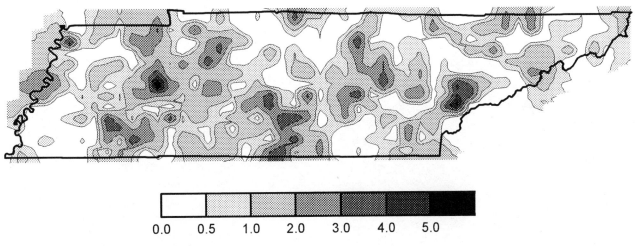

Abundance of the Great Crested Flycatcher.

lands. The height of nests varies with available cavities; nests in natural cavities and woodpecker holes averaged 4.7 m high (range 2.0–15.2 m, s.d. = 3.6, n = 14). Nests in bird boxes have been as low as 1 m. No preference for cavities in particular tree species is evident.

Both the male and female vigorously defend the nest area; the female builds the nest in 4–5 days (Taylor and Kershner 1991). The nest itself is made of leaves, pine needles, grasses, and bark strips, and lined with hair and feathers. It usually contains a snakeskin (Taylor and Kershner 1991, egg coll. data).

Egg laying peaks during the second half of May. Tennessee clutches average 4.28 eggs (range 3–5, s.d. = 0.64, n = 21); clutches of 4–8 eggs have been reported elsewhere, with 5 eggs most common (Harrison 1975, Taylor and Kershner 1991). Eggs are buff colored, blotched, and lined with dark brown and lavender (Harrison 1975). The female incubates the eggs for 13–15 days; both adults feed the nestlings, which fledge in 13–15 days (Taylor and Kershner 1991). The family group remains together for up to 3 weeks after fledging. Only a single brood is raised, although renesting occurs if the first nest is destroyed (Taylor and Kershner 1991).

Parasitism by the Brown-headed Cowbird is rare (Friedmann and Kiff 1985) and has not been reported in Tennessee. Ganier (1955) reported an instance of nest competition with European Starlings, in which flycatchers threw a starling egg out of the nest; neither species remained to nest. Most starlings are well into their nest cycle by the time Great Crested Flycatchers begin nesting.—*Robert P. Ford and Charles P. Nicholson.*

Eastern Kingbird
Tyrannus tyrannus

The Eastern Kingbird is a common summer resident of low-elevation open areas throughout Tennessee. Kingbirds usually arrive in early April and depart during August and early September. Because of their size, bright contrasting black-and-white plumage, use of exposed perches, and aggressive chasing of larger birds, kingbirds are conspicuous and easily identified.

Eastern Kingbirds occupy a wide variety of habitats, with the common feature being the absence of a forest canopy. Typical habitats in Tennessee include fencerows, grasslands with scattered trees, orchards, young clearcuts, lakeshores, and recently burned areas. Within the eastern mountains, it occurs up to about 840 m (Atlas results), somewhat higher than the limit given by Stevenson and Stupka (1948). Atlasers found kingbirds in 92% of the completed priority blocks; most of the blocks where they were not found were heavily forested and in the Unaka Mountains, Cumberland Mountains, and Cumberland Plateau.

Before European settlement, kingbirds were probably an uncommon bird in Tennessee, nesting along the large rivers, in areas of recent, severe forest fires, around fields maintained by Indians, and in cedar glades. The largest population was probably in the barrens area of north-central and northwest Tennessee. The first state record was by Michaux, who found it, then known as *Lanius Tyrannus,* in the Nashville area on 21 June 1795 (Williams 1928). Kingbird populations probably increased rapidly with the clearing of the landscape. Rhoads (1895a) described them as abundant throughout the state except for the Roan Mountain area. Their population probably peaked around 1940, when the proportion of the state in farmland was highest.

Eastern Kingbird. Elizabeth S. Chastain

Since 1966, Eastern Kingbird numbers on Tennessee BBS routes have shown a decreasing trend of 1.4%/year ($p < 0.10$). Throughout their extensive North American breeding range, kingbird numbers did not significantly change between 1966 and 1993 (Peterjohn, Sauer, and Link 1994).

Atlas workers recorded Eastern Kingbirds on 450 miniroutes at an average relative abundance of 2.4 stops/route. By physiographic regions, abundance was lowest in the Mississippi Alluvial Plain, an area dominated by extensive croplands and bottomland forests, with little woodlot or fencerow habitat. The next lowest numbers were in the Unaka and Cumberland Mountains and the highest abundance was in the Central Basin.

Breeding Biology: Atlasers found it easy to confirm kingbird breeding. Kingbird nests are often conspicuous, and the noisy fledglings are attended by the parents for several weeks. Slightly less than half of the confirmed records were of nests, and most of the other confirmed records were of fledglings and adults carrying food. Eastern Kingbirds build a bulky nest in a tree or shrub adjacent to an open area or over water. The nest is often exposed, and other supports, such as tree stumps in water and utility poles, are occasionally used. The female, escorted by the male, builds the nest (Morehouse and Brewer 1968), which is usually placed on the fork of a limb far from the trunk. The outer shell of the relatively large nest is built of weed stems, grasses, twigs, and often bits of leaves, rags, string, paper, or cotton. The cup, about 8 cm across by 4 cm deep, is lined with rootlets and fine grasses.

Tennessee nests average 6.1 m above ground or water (s.d. = 3.9 m, n = 60), with a range of 1–20 m, the highest nest being on a crosspiece of a high-voltage electrical transmission tower (Nicholson pers. obs.). Nests in upland areas are usually higher than those over water, as noted elsewhere by Blancher and Robertson (1985). Kingbird nests have been found in many species of deciduous trees, with little apparent preference. Nests are also frequently built in pines and infrequently in cedars.

Egg laying peaks during the last third of May. Studies elsewhere (e.g., Murphy 1983) have observed that egg laying may be delayed for several days after the nest is completed during cool or wet weather. The frequency of this delay in Tennessee is unknown. Clutch size in Tennessee ranges from 2 to 5, with a mode of 3 and an average size of 3.43 (s.d. = 0.77, n = 41). Other studies north of Tennessee have found the same range of clutch sizes and a similar average clutch size of 3.39 (Murphy 1983, Blancher and Robertson 1985). Eastern Kingbird eggs are creamy white with a variable amount of brown, black, and lavender spotting (Harrison 1975).

The female incubates for 14–16 days (Morehouse and Brewer 1968, Murphy 1983). Both adults feed the young, which grow slowly and leave the nest in 14–17 days (Morehouse and Brewer 1968). The young fly

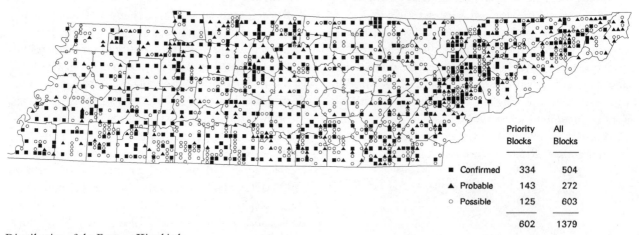

	Priority Blocks	All Blocks
■ Confirmed	334	504
▲ Probable	143	272
○ Possible	125	603
	602	1379

Distribution of the Eastern Kingbird.

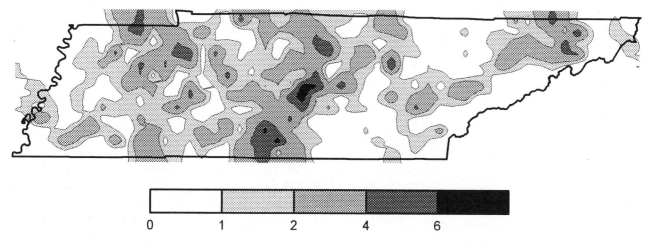

Abundance of the Eastern Kingbird.

poorly at fledging and do not begin flycatching for over a week. They continue to be fed by their parents for almost 5 weeks after fledging (Morehouse and Brewer 1968). Kingbirds are single-brooded. Following the loss of a nest, they usually renest if the loss occurred early in the nesting cycle or early in the season (Murphy 1983).

Eastern Kingbirds are fairly common hosts of Brown-headed Cowbirds, and Murphy (1986) found 9% of frequently inspected nests in Kansas were parasitized. Kingbirds, however, usually eject cowbird eggs; consequently, parasitism is rarely reported. No records of cowbird parasitism of kingbirds in Tennessee are available.
—*Charles P. Nicholson.*

Scissor-tailed Flycatcher
Tyrannus forficatus

The Scissor-tailed Flycatcher is a rare vagrant and a rare summer resident of Tennessee. This strikingly plumaged bird, easily recognized by its long tail feathers, was first reported in the state on 26 April 1964 in Shelby County (Robinson 1990). Since then it has been recorded each month from April to November; most of the records are in the fall, and reports have recently increased in frequency (Robinson 1990, *Migrant*). It has nested in the state on at least 3 occasions, 2 of them during the Atlas period.

The first nest of the Scissor-tailed Flycatcher was found in July 1978 in Rutherford County (*Migrant* 49:92). Only 1 adult was observed at the nest containing 3 small young, none of which fledged. At least 1 adult returned to this site each year through 1985. Two juveniles were observed in late summer, 1983; again, only 1 adult was present (*Migrant* 55:26). A single adult was observed nest-building during the summer of 1985 (*Migrant* 56:111). Unfortunately, many details of these observations remain unpublished.

In the autumns of 1983 through 1986, a Scissor-tailed Flycatcher was observed in the same area of southern Meigs County (Harris 1984, *Migrant* 57:33, 58:28). Whether this bird (or birds) nested in the area is unknown; the fact that it appeared in the same area 4 consecutive autumns raises the possibility.

During Atlas fieldwork, Scissor-tailed Flycatcher nest attempts were reported in Franklin County in 1989 and Hardin County in 1990. At the Franklin County site, a large (more than 40 ha) pasture on the Eastern Highland Rim, a bird was first observed on 13 May and a pair present by 2 June. In mid-June, the female was observed placing nest material in a fork of an isolated tree in the pasture. The outcome of this nest attempt is unknown (D. Davidson pers. comm.).

In late May 1990, at least 1 adult Scissor-tail was found in an open field with scattered small trees at Pickwick Dam, Hardin County. On 21 June the female of the pair

Scissor-tailed Flycatcher. Elizabeth S. Chastain

Confirmed Breeding Species

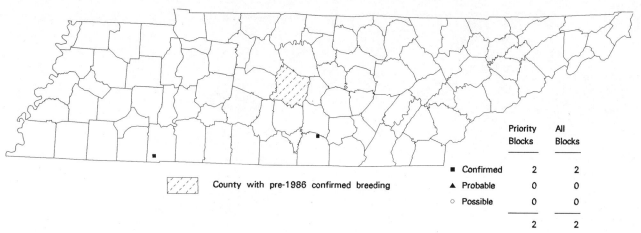

Distribution of the Scissor-tailed Flycatcher.

was incubating 2 eggs in a nest in a loblolly pine. One egg was in the nest on 25 June, and in early July the nest was empty and no adults present (D. Simbeck pers. comm.).

Six other Scissor-tailed Flycatchers, all apparently vagrants and not mapped, were observed during the Atlas period within the species's breeding period. In 1987, single birds were observed on 1 June at LBL, Stewart County (Robinson 1990), and on 30 June in northern Lake County (*Migrant* 58:137). The 1988 observations were on 30 May in south-central McNairy County (*Migrant* 59:95) and on 20 June in southwest Shelby County. On 4 June 1989 a bird was observed in Cades Cove, GSMNP, Blount County (Alsop 1990), and on 26 July 1991 a bird was present in Franklin County (Atlas results). The Franklin County observation was about 13 km southwest of the location of the 1989 nesting attempt.

These Scissor-tailed Flycatcher observations suggest an expansion of the species's breeding range to east of the Mississippi River. The historic breeding range of the Scissor-tailed Flycatcher was in the south-central United States east and north to southwest Missouri and central Arkansas (Fitch 1950). Within this range, it originally occupied the prairie-deciduous forest ecotone of dry grasslands, scattered trees, and scrub and has since occupied pastures and other agricultural lands with scattered trees. It has recently extended its range throughout much of Arkansas (James and Neal 1986) and into central Missouri (Robbins and Easterla 1992). Isolated breeding outside this range has been reported from Indiana in 1974 (Mumford and Keller 1984), northeast Mississippi in 1975 and Iowa in 1979 (AOU 1983), northwest Alabama in 1990 (Kittle and Patterson 1990), and Tennessee. The flycatcher's population, as measured by BBS routes, showed no significant overall trend from 1967 to 1993 (Peterjohn, Sauer, and Link 1994). From 1967 to 1978, however, it greatly declined and since then it has partially recovered (Sauer 1990).

Observers should be alert for future occurrences of the Scissor-Tailed Flycatcher in Tennessee and publish full details of all nesting observations.

Breeding Biology: The Scissor-tailed Flycatcher is monogamous and single-brooded (Fitch 1950, Murphy 1988). The female builds the nest, usually low in an isolated tree near the end of a horizontal branch. Little preference is evident in the tree species used, and human-made structures are occasionally employed. The Hardin County nest was about 2.4 m up in a 4 m loblolly pine (Simbeck pers. comm.). The nest is bulky, with an outer shell of plant stems, twigs, string, and other materials, and a lining of grass blades, bark, feathers, or wool (Fitch 1950). The normal clutch size is 3–5 eggs, with 5 most frequent (Murphy 1988). The female incubates the eggs for 2 weeks, and both adults feed the nestlings, which fledge in about 16 days. Of the 6 Tennessee nest attempts, only 1, in 1983, was known to be successful.—*Charles P. Nicholson.*

Horned Lark
Eremophila alpestris

The Horned Lark is a permanent resident of Tennessee, increasing in abundance from east to west. It occupies sparsely vegetated croplands and extensive areas of short grass interspersed with bare areas. During the winter, the Horned Lark often occurs in large flocks, many of which are probably the northern race *E. a. alpestris,* although specimen data is limited. Nesting birds, which are probably present throughout the year, are the race *E. a. praticola,* commonly known as the Prairie Horned Lark (Wetmore 1939).

Prior to the beginning of European settlers' cultivation of the Midwest, the breeding range of the Prairie Horned Lark was apparently restricted to the midwestern tall-grass prairie (Hurley and Franks 1976). The south-

Horned Lark. Chris Myers

eastern edge of the lark's range was in central Illinois and Missouri. Following cultivation of the prairies and the opening of the forests to the east, the lark rapidly spread eastward, reaching New York by 1880 (Hurley and Franks 1976). Within its original range, it also shifted in its habitat preference from grasslands to cultivated fields (Graber and Graber 1963). Its spread into the southeast was apparently somewhat slower than its spread to the northeast. The first evidence of Horned Larks nesting in Tennessee was in 1925 in northwestern Davidson County (Ganier 1931d). It was probably present somewhat earlier, as the first Kentucky breeding season record was in Logan County, adjacent to Robertson County, Tennessee, in 1904 (Mengel 1965). The first West Tennessee breeding record was in 1932 (*Migrant* 3:36), and the first for East Tennessee was about 1934 (Lyle and Tyler 1934).

Since the early 1930s, Horned Larks have been reported during the breeding season from numerous counties across the state. Although their numbers on most BBS routes have been low and highly variable, BBS results show a declining trend of 7.9%/year ($p < 0.10$) from 1966 to 1994. A pattern of local declines is probably most pronounced in urban areas, as noted around Nashville (Parmer 1985, *Migrant* 57:108), although local declines have also occurred in rural areas in Middle and East Tennessee. These declines are probably due to a loss of habitat from changing land use practices. Throughout North America, their numbers decreased by 0.8%/year from 1966 to 1993; during the first half of this period, they decreased in the Northeast and increased in the Southeast (Robbins, Bystrak, and Geissler 1986, Peterjohn, Sauer, and Link 1994). They also increased throughout the Mississippi Alluvial Plain because of the increase in cropland there since the 1960s.

Atlas workers found Horned Larks in 18% of the priority blocks and on 7% of the miniroutes at an average relative abundance of 2.2 birds/route. Their abundance was highest in the Mississippi Alluvial Plain, where larks was found on all but 1 miniroute and averaged 4.1 birds/route. Within the Loess Plain region, it was more numerous toward the north, where the proportion of cropland is higher. Other concentrations occurred in the extensive alluvial agricultural areas of McNairy and Hardin Counties in the Coastal Plain Uplands and in plowed portions of the Highland Rim Barrens of Coffee County. The Horned Lark's absence from some areas of seemingly suitable habitat, such as the extensive, flat agricultural lands in southern Lincoln and Giles Counties, is puzzling. In East Tennessee, up to 40 birds were present in eastern Jefferson County in the early 1980s (*Migrant* 54:90); smaller numbers were present during Atlas fieldwork. Although the male's flight song is conspicuous and easily identified, the lark is often inconspicuous when on the ground. They could have been missed in some blocks in which they occurred in low numbers and which were not worked early in the season.

The Horned Lark nests in extensive areas of short grass or bare soil with scattered low vegetation. Nests and fledglings have been found in pastures, grasslands surrounding airport runways, rowcrop stubble, and freshly plowed fields. The highest elevation breeding record is at 1800 m on Roan Mountain (Tyler 1936). It was not found at high elevations during Atlas fieldwork, probably because, with the recent absence of grazing, the vegetation there is too tall and dense. The only high-elevation summer record from the Smokies is a bird on 8 May 1960 at the Clingmans Dome parking lot, 1923 m elevation, and a short distance into North Carolina (Stupka 1963). Some observations on the Cumberland Plateau were in extensive recent clearcuts and large strip mine areas with sparse grasses (Nicholson pers. obs.); an adult with fledglings was present in a Cumberland County clearcut in 1978 (*Migrant* 49:94).

Breeding Biology: The Horned Lark is among the earliest nesting passerines, laying eggs as early as the first half of March. The nest is built by the female in a shallow depression excavated in the soil and constructed of grass stems with a lining of rootlets, fine grasses, and feathers or wool (Pickwell 1931, Goodpasture 1950). There is usually no overhead cover. Small pebbles are usually placed on one side of the nest rim, sometimes forming a windbreak. Clutches of 2 to 4 eggs have been found in Tennessee; the average size is 3.57 eggs (s.d. = 0.65, n = 14). The female incubates the eggs for 11 days, and the young leave the nest in 9 to 12 days (Pickwell 1931,

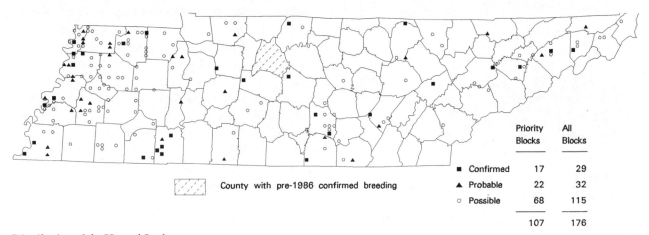

Distribution of the Horned Lark.

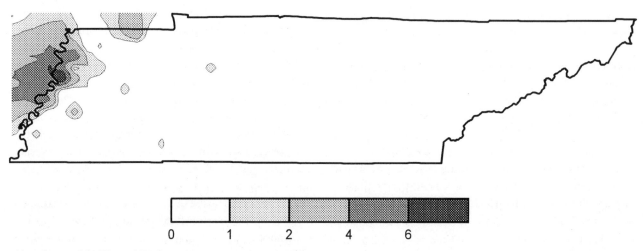

Abundance of the Horned Lark.

Goodpasture 1950). The Horned Lark is usually double-brooded. Goodpasture (1950) observed 4 nesting attempts by a single pair, only 1 of which was successful. Causes of nest losses include trampling by cattle (Tyler 1936) and flooding (*Migrant* 29:60); some are also probably destroyed by spring and summer plowing. Maturing crops probably make many fields unsuitable for second broods. Parasitism by Brown-headed Cowbirds, infrequent elsewhere (Friedmann, Kiff, and Rothstein 1977, Friedmann and Kiff 1985), has not been reported in Tennessee.—*Charles P. Nicholson.*

Purple Martin
Progne subis

The Purple Martin is a fairly common to common summer resident of towns, suburbs, and farmsteads. It nests exclusively in "martin houses" and gourds hung from poles, and its local distribution is dependent on the presence of these artificial nest sites. The earliest spring migrants are usually present by the first of March; their arrival is earlier during unusually warm, late-winter temperatures. The majority of the population, however, arrives in late March and April and departs the state by early September.

The Purple Martin nests in cavities, normally in colonies. It originally depended on abandoned woodpecker cavities. Although nests in woodpecker cavities were reported in several eastern states during the nineteenth and early twentieth centuries (Allen and Nice 1952), they have not been definitely reported in Tennessee. Native Americans began the practice of providing nest structures for martins. Wilson (1812) described martins as numerous in Kentucky and Tennessee and noted that Choctaw and Chickasaw Indians hung hollowed calabash gourds from saplings near their cabins to serve as martin houses. European settlers also provided nest structures, both gourds and multiple compartment wooden houses. Audubon (1831) observed that these martin boxes were common at country taverns and towns.

The Purple Martin population in Tennessee increased with European settlement. Torrey (1896a) described it

Purple Martin. Elizabeth S. Chastain

as common at Chattanooga in the spring of 1894. Rhoads (1895a) found it numerous at low elevations and uncommon on the Cumberland Plateau. Ganier (1933a) described it as a common summer resident in each of the 3 major divisions of the state. From 1966 to 1994, the martin population, as measured by Tennessee BBS routes, showed an increasing trend ($p < 0.05$) of 3.2%/year. Throughout its extensive North American range, the martin population did not show a significant trend from 1966 to 1993 (Peterjohn, Sauer, and Link 1994).

In addition to gourds and martin houses, Purple Martins have been found nesting in a few other human-made structures. The old Fayetteville courthouse, prior to burning in the 1930s, had hundreds of martins nesting in its cornices (*Migrant* 9:65). Several pairs nested in cylindrical blinders on traffic lights in Ashland City, Cheatham County, in 1962 (*Migrant* 35:117).

During Atlas fieldwork, the Purple Martin was recorded in three-quarters of the completed priority blocks. Martin houses and gourds were usually easily observed from roads, and martin breeding was confirmed in almost three-quarters of the priority blocks, mostly with the code ON. Birds in flight, often heard calling just before dawn, were observed in a few heavily forested blocks with little or no human population. These birds were probably nesting in nearby blocks.

The Purple Martin was recorded on 42% of the Atlas miniroutes at an average relative abundance of 1.8 stops/route. Its abundance was, as expected, very low in the Unaka Mountains, where it was found on only 2 routes. The low average abundance in the heavily settled and agricultural Central Basin, where it was found on about one-third of the routes at 1.2 stops/route, was unexpected. The proportion of routes on which it was recorded and its abundance in the other heavily settled or agricultural physiographic regions were at least equal to the statewide averages.

Breeding Biology: The first Purple Martins to arrive in the spring are mostly adult males; they soon select nest sites, which they defend from other males (Allen and Nice 1952). Female martins usually arrive a few days to a few weeks after the males; they also soon select a nest compartment and with it a mate. Nest construction is often delayed for up to a few weeks after the arrival of the adults and occurs from mid-April through early June. Both adults build the nest, which is crudely constructed of grasses, weed stems, and leaves. A wall of mud is sometimes built inside cavities with large entrances, presumably for protection from rain (Allen and Nice 1952) and has been reported in a few Tennessee nests (egg coll. data, *Migrant* 14:41–42). Fresh green leaves are added during incubation.

For such a common species, relatively few detailed Purple Martin nest records are available from Tennessee. Clutches of eggs have been reported from mid-May to

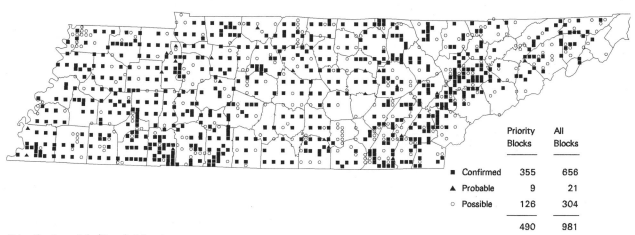

Distribution of the Purple Martin.

		Priority Blocks	All Blocks
■	Confirmed	355	656
▲	Probable	9	21
○	Possible	126	304
		490	981

Confirmed Breeding Species

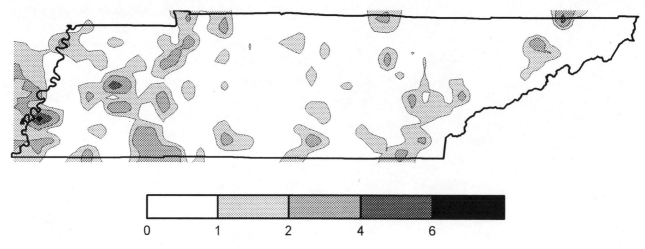

Abundance of the Purple Martin.

mid-June, although records of nests with young indicate that egg laying occurs as early as late April. Clutches average 4.91 eggs (s.d. = 0.79, range 3–6, n = 12). Clutches of 4 and 5 are most common elsewhere (Allen and Nice 1952). The female incubates for 15–16 days. Both adults feed the nestlings, which fledge in about 28 days. Purple Martins raise a single brood; renesting occurs after loss of a clutch (Allen and Nice 1952). After the young have fledged, adult and juvenile martins gather in large roosts, often located in deciduous trees in towns and numbering thousands of birds (Nicholson pers. obs.).

The Purple Martin is presently the only bird nesting in the state that is totally dependent on human-made nest sites. Its future, therefore, depends on the supply of martin houses and gourds. Martin houses and gourds are sold in many cities, and guidelines for their construction and management are readily available. The European Starling and House Sparrow are common competitors for martin nest sites (Jackson and Tate 1974); proper martin colony management includes control of these competing species.—*Charles P. Nicholson.*

Tree Swallow
Tachycineta bicolor

The Tree Swallow is an uncommon summer resident that nests in cavities near water across Tennessee. It is usually present in the state from mid-March into October. Migrants are fairly common during the spring and fall, when they are occasionally locally abundant, especially near the Mississippi River (Robinson 1990). The Tree Swallow became established as a nesting species in 1968, and it continues to expand its nesting distribution.

The early history of Tree Swallow nesting in Tennessee was described by Nicholson and Pitts (1982). The first record was in 1918 at Reelfoot Lake. Until 1968, there were no additional nest records, although possibly breeding Tree Swallows were reported during the early summer from Reelfoot Lake and East Tennessee. In 1968, it nested in Anderson and Maury Counties. Between 1968 and 1982, nests were reported almost every year, and the number of counties where it had nested rose to 13. In the early 1980s, the largest local populations of Tree Swallows were probably at Reelfoot Lake and along Kentucky Reservoir near the mouth of the Duck River.

The Tree Swallow continued to spread after 1982, and during Atlas fieldwork several new county nest records were established. By the end of the Atlas period, Tree Swallows had nested in at least one-third of the counties in the state. Although they are usually conspicuous, some local populations were probably missed because of the swallow's frequent early departure from nest sites.

The reasons for this relatively rapid range expansion into and throughout Tennessee are not clear. The Tree Swallow breeds throughout northern North America; until recently, breeding south of the Ohio River was uncommon and sporadic (AOU 1983). During the period of its recent expansion in Tennessee, similar expansions occurred in other southeastern states including Arkansas (James and Neal 1986), Kentucky (Monroe, Stamm, and Palmer-Ball 1988), and North Carolina (Duyck 1981). Breeding habitat has increased in Tennessee with the construction of ponds and large reservoirs; most of this habitat increase, however, occurred during the 1930s and 1940s, well before the swallow's recent expansion. Although the availability of nest sites has recently increased in some areas as waterfront homeowners have erected bird boxes, cavities in dead trees near water have probably always been available. Between 1965 and 1979 Tree Swallow numbers significantly in-

Tree Swallow. Elizabeth S. Chastain

Breeding biology: Confirming Tree Swallow breeding was usually easy, and nest records made up 82% of confirmed Atlas records. Tree Swallows are usually monogamous and defend a small territory around the nest site (Kuerzi 1941). They are unique among North American songbirds in that the female requires two years to attain its adult plumage, which resembles that of the male (Tyler 1942). First-year females, with dull brown back plumage, are capable of breeding and have been reported at several Tennessee nests (Nicholson and Pitts 1982, Atlas results).

The Tree Swallow begins nesting activities in April, when migrating swallows are still present. Most nests have been in cavities in dead trees, usually in woodpecker holes up to 12.2 m high, and in bird boxes. Tree Swallows made little use of boxes erected for Prothonotary Warblers along Kentucky Reservoir, choosing instead natural cavities that were generally higher and in a more open situation than the warbler boxes (Petit et al. 1987, L. Petit pers. comm.). The nest, constructed mostly by the female, is loosely built of dead grasses shaped into a cup lined with feathers (Kuerzi 1941). The presence of feathers and the pure white coloration of Tree Swallow eggs make nest identification easy.

Egg laying probably peaks during mid-May, although several clutches were not begun until well into June. These late clutches were likely either renests following destruction of the first nest or from pairs that were late in beginning nesting. Unusually high water levels may delay nesting along Kentucky Reservoir (Petit pers. comm.). Tennessee clutches range from 4 to 6 eggs and average 4.92 eggs (s.d. = 0.67, n = 12), similar to those reported elsewhere. The female incubates for 13–16 days; both adults feed the nestlings, which fledge in about 20 days (Kuerzi 1941, Nicholson and Pitts 1982). Family groups may leave the nest area a day after the young fledge, and large flocks occur in early July (Nicholson and Pitts 1982).—*Charles P. Nicholson.*

creased ($p < 0.05$) in eastern and central North America (Robbins, Bystrak, and Geissler 1986). Their rangewide population showed a slight increasing trend from 1966 to 1993 (Peterjohn, Sauer, and Link 1994).

All of the Tree Swallow nest records for which details are available have been over or within about 400 m of a body of water. The water bodies range in size from 0.1 ha farm ponds to Kentucky Reservoir; the ponds have been surrounded by grassland. Some sites with small populations have been only used for 1 year or 2, without obvious changes in habitat or cavity availability (Nicholson and Pitts 1982, Atlas results). Other sites with large populations have been used continuously for over a decade. The short-term use of some sites may be the result of short adult life spans and low tendency for Tree Swallows to nest near their place of birth. Based on banding studies north of Tennessee, the annual adult mortality of Tree Swallows is about 61%, and fewer than 10% of nestlings return to their natal areas (Butler 1988).

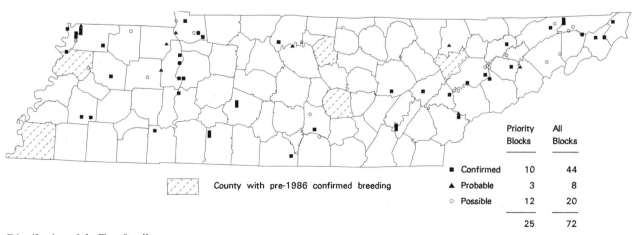

Distribution of the Tree Swallow.

Northern Rough-winged Swallow
Stelgidopteryx serripennis

The Northern Rough-winged Swallow, named after the serrations on its outer wing feathers, is an uncommon summer resident across the state of Tennessee. This drably marked species arrives in late March and early April. Fall migratory flocks are present as early as mid-July, and most have departed the state by late September. The Northern Rough-winged Swallow nests in natural and human-made cavities and crevices in dirt banks and rock walls; the availability of nest sites is probably the major factor determining its distribution.

Prior to the beginning of widespread forest clearing in the early nineteenth century, the Northern Rough-winged Swallow was probably an uncommon nesting species, found along large streams and rivers that provided nest sites and adequate open area for foraging. Clearing of forests along smaller streams made the streambanks accessible to nesting swallows, and rock quarries, railroad cuts, and highway cuts also created nest sites. The response by the swallow population is poorly documented, both because of the few nineteenth-century records available and because some early naturalists (e.g., Gettys in Ijams and Hofferbert 1934) confused the Northern Rough-winged Swallow with the similarly colored Bank Swallow, which has a much more restricted breeding range in Tennessee. The Rough-winged was first reported in the state by Fox (1886), who described it as the most common swallow in Roane County in April. Rhoads (1895a) described it as abundant everywhere from Samburg to Johnson City. At that time it was, with the exception of the Purple Martin, the most widely distributed member of its family.

Northern Rough-winged Swallow. Elizabeth S. Chastain

Ganier (1933a) described it as common in West Tennessee and fairly common in Middle and East Tennessee.

Atlasers reported Rough-winged Swallows in about two-thirds of the completed priority blocks. They were fairly common in Middle and East Tennessee and reported from three-quarters or more of the blocks in the Central Basin, Cumberland Plateau and Mountains, and Ridge and Valley regions. The lack of suitable nest sites probably limited their occurrence in much of West Tennessee, especially in the Loess Plain region. The numerous records at Reelfoot Lake, most in the possible category, are probably of birds feeding over the wetlands; Rough-wingeds are not known to nest in the immediate lake area (Pitts 1985). In the Unaka Mountains, Rough-winged Swallows are probably limited at high elevations by the lack of nest sites. The highest elevation nest record in Tennessee is at 1615 m in a road cut in the Smokies (Savage 1964). Elsewhere in the southern Blue Ridge, nesting has been reported as high as 1708 m (McNair 1987c).

Rough-winged Swallows were recorded on only 22% of the miniroutes, at an average relative abundance of 1.3 stops/route and a maximum of 3 stops/route. The mapped miniroute results show little trend except for an area of high abundance in the Cumberland Mountains and the adjacent portion of the Ridge and Valley.

Rough-winged Swallow numbers, as measured by Tennessee BBS routes, do not show a significant trend from 1966 to 1994. During the Atlas period, they were recorded at an average of 2.0 stops/route on 37 of the 42 50-stop BBS routes. Tanner (1986) did not find a significant trend in their numbers on Tennessee spring counts, which are usually held following arrival of nesting birds but while migrants may still be present. Throughout North America, Rough-winged numbers did not show a significant trend from 1966 to 1993; they increased significantly ($p < 0.05$) from 1966 to 1979 (Robbins, Bystrak, and Geissler 1986, Peterjohn, Sauer, and Link 1994). From 1966 to 1979, however, their numbers decreased throughout the Cumberland Plateau and Blue Ridge physiographic regions. The decrease in the Cumberlands seems contrary to the habitat trend from the 1950s through the late 1970s, when the increase in coal strip mining, which created numerous road cuts and clifflike highwalls, resulted in much suitable habitat (Nicholson pers. obs.). This habitat trend was at least partially reversed by reclamation laws in the late 1970s. Rural road improvement programs, involving road widening and the sloping and planting of road banks, have also resulted in a loss of nesting habitat in recent decades, probably greatest in West Tennessee (cf. Coffey and Coffey 1980).

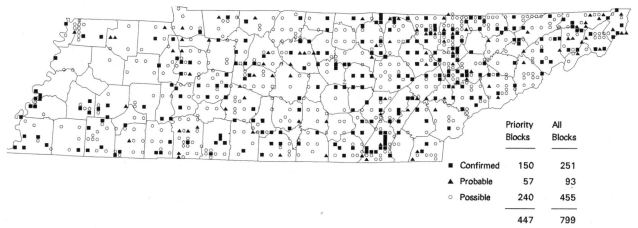

	Priority Blocks	All Blocks
■ Confirmed	150	251
▲ Probable	57	93
○ Possible	240	455
	447	799

Distribution of the Northern Rough-winged Swallow.

Breeding Biology: Natural nest sites in Tennessee include abandoned Belted Kingfisher burrows, crevices in rock bluffs and cave mouths, and tunnels along streams and lakes formed by rotted tree roots. Where their range overlaps with that of Bank Swallows, Rough-wingeds commonly nest in Bank Swallow burrows (Lunk 1962); Rough-wingeds have been found in the vicinity of Bank Swallow colonies in Tennessee. Human-made nest sites commonly used in Tennessee include drain holes in bridges and retaining walls and crevices in road cuts and quarries. Among the more unusual sites are a traffic light (Farrell 1964) and holes in semi-trailers (Stedman and Simbeck 1988). Although many reports (e.g., Dingle 1942) state that Northern Rough-winged Swallows often dig their own burrows, Lunk (1962) found no definitive evidence of this.

Northern Rough-winged Swallows appear at nest sites quickly after their arrival in the spring (Lunk 1962), and nest building occurs as early as 3 April in Tennessee. The pair defends a small territory around the entrance of the nest burrow, and the male perches near the nest entrance during much of the nest cycle. The female builds the nest in about 5 to 7 days (Lunk 1962). The nest is rather loosely constructed of weed stems, grass, and twigs and is lined with grass. Nests are placed from 0.4 to 2 m from the tunnel entrance, with the average less than a meter, and as low as 0.3 m above water. Loss of nests due to flooding is not unusual (Nicholson pers. obs.).

Egg laying is often delayed for several days after the nest is completed (Lunk 1962) and in Tennessee peaks during the first half of May. Clutch size ranges from 4 to 8, with 5 to 7 eggs most common; Lunk (1962) found an average of 6.3 eggs in southern Michigan. Thirty-one apparently complete Tennessee clutches averaged 6.10 eggs (s.d. = 1.14) and range from 4 to an unusually large 9 (Crook collection, WFVZ). The female incubates the eggs for an average of 16 days, and both adults feed the nestlings, which leave the nest in about 19 days (Lunk 1962). The family group often leaves the nest area soon after the young fledge. Northern Rough-winged Swallows are normally single-brooded and usually renest after loss of a nest, occasionally reusing the same hole (Lunk 1962).—*Charles P. Nicholson.*

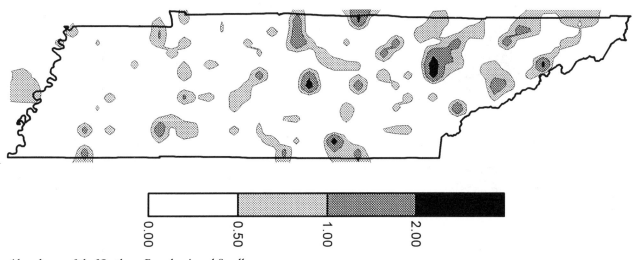

Abundance of the Northern Rough-winged Swallow.

Bank Swallow
Riparia riparia

The Bank Swallow is an uncommon migrant and a very localized, uncommon summer resident. As its scientific name implies, it occurs along rivers, traditionally nesting in colonies in burrows excavated into riverbanks. It has nested along the Mississippi River since at least the early nineteenth century and in recent decades at human-made sites in East Tennessee. It is usually present in Tennessee from late April through mid-September (Robinson 1990).

Evidence of the Bank Swallow in Tennessee was first reported by Audubon, who in late 1820 observed "thousands" of nest holes in Second Chickasaw Bluff, on the Mississippi River near present-day Randolph, Tipton County (Audubon 1929). Rhoads (1895a), who visited the Reelfoot Lake area, stated it bred abundantly along the Mississippi River bluffs. It is unclear whether Rhoads actually observed the Bank Swallow or was paraphrasing Audubon. Ganier and Weakley (1936), either unaware of the earlier reports or relying only on their personal discoveries, stated that the Bank Swallow had not been found nesting in Tennessee. Over the next 50 years, new county nest records were reported in Dyer (*Migrant* 25:52), Lake (Pitts 1972), Lauderdale (*Migrant* 47:5), and Shelby Counties (Waldron 1981). All West Tennessee colonies have been either overlooking the Mississippi River or less than a km upstream from the mouth of the Wolf (Waldron 1981) and Hatchie Rivers (*Migrant* 55:88). Reports from individual sites during this period were infrequent.

Bank Swallow. Elizabeth S. Chastain

Nesting outside of the Mississippi Alluvial Plain was first reported at Austin Springs, Washington County, in 1969 (*Migrant* 43:68, 71). The nest burrows were in a vertical face of a pile of rocks, sand and silt dredged from the adjacent Watauga River and awaiting crushing (R. L. Knight pers. comm.). Tanner (1974) reported colonies active in 1973 in zinc mine tailings at Mascot, Knox County, and near New Market, Jefferson County. The Mascot colony had existed for several years. A colony in a fly ash pile at Kingston Steam Plant, Roane County, was first reported in 1984 (*Migrant* 55:94); swallows were also present at this site the previous year, but no nests observed. A single pair was reported at a hole in a streambank in Blount County on 22 June 1984 (*Migrant* 55:94). Solitary nesting by the Bank Swallow is very unusual (Peterson 1955). The only Middle Tennessee nest records are along Carson Creek near Culpepper, Cannon County, where 32 nest holes, but no attendant Bank Swallows, were observed on 22 May 1984 (R. P. Ford pers. comm.), and in Lawrence County along Shoal Creek a short distance from the Alabama line in the early 1980s (*Migrant* 57:108, D. J. Simbeck pers. comm.). A Bank Swallow was observed near this site in June 1986.

Atlas workers found at least 12 Bank Swallow colonies. The only colony in a previously unreported area was in the fly ash pile at John Sevier Steam Plant, Hawkins County, in 1988. Fly ash piles at some of the other steam plants were not thoroughly checked for swallow colonies. Additional colonies may have been present at them as well as along unworked sections of the Mississippi River. The possible breeding records, coded as "Observed," were apparently wandering birds; few of them were near potential breeding habitat.

Breeding Biology: Bank Swallows excavate their own nest burrows in vertical banks of relatively coarse-textured, sandy soils. Nest chambers tend to be deeper in soils with higher sand content (Peterson 1955). Soils with high-clay content, such as occur along most Middle and East Tennessee rivers, are presumably not friable enough for Bank Swallows to nest in them. The zinc mine tailings at the Knox and Jefferson County colonies were predominantly finely ground dolomite limestone with the consistency of fine sand (Tanner 1974). Steam plant fly ash also has a sandy texture (Nicholson pers. obs.). Colonies are usually in banks at least 1.8 m tall and burrows at least 0.4 m below the top of the bank. Without continued slumping of the bank face from wave action or human activities, which maintains the bank's vertical profile, colony sites are rarely suitable for nesting more than a couple seasons (Freer 1979). Bank Swallows have adapted to the ephemeral nature

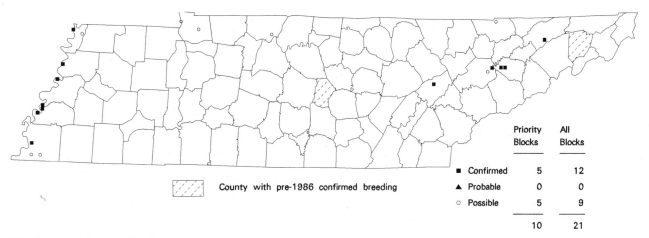

Distribution of the Bank Swallow.

of their nest sites by having a low degree of site faithfulness compared to other swallows (Freer 1979).

The largest Bank Swallow colony reported in Tennessee was Audubon's report of "thousands" of burrows. Colonies reported since then have had from 4 to 600 nest burrows and average 108 burrows (s.d. = 124.0, n = 38). Too few thorough counts have been made in consecutive years at individual colony sites to show population trends. Channel maintenance activities along the Mississippi River, such as placing riprap on banks, have made some previous colony sites unsuitable for nesting (Coffey 1976). Similarly, changes in gravel-crushing operations at Austin Springs and in fly ash disposal at Kingston Steam Plant made those sites periodically unsuitable. Following restoration of suitable habitat, Bank Swallows are capable of quickly reoccupying colony sites.

Most reports of nesting Bank Swallows in Tennessee have only mentioned the number of burrows and swallows present; nesting details are very limited. Partially completed burrows have been reported as early as 29 April (*Migrant* 59:101). A nest in Lauderdale County was constructed of straws and feathers and contained 5 eggs on 18 June 1990 (Atlas results). Clutches elsewhere are usually 4 to 6 eggs, with 5 most common (Peterson 1955). Most young are probably fledged by early July, as several colony sites have had few swallows present by the second week of July. Bank Swallows typically raise a single brood and readily renest after loss of a nest (Peterson 1955). The colony at Kingston Steam Plant apparently renested in new burrows in 1985 following the early June collapse of their nest site (*Migrant* 56:114).

The future of the Bank Swallow population along the Mississippi River, although affected by channel maintenance and erosion control activities, is probably secure. The population in East Tennessee, however, is dependent on human activities. The colonies in zinc tailings piles will probably flourish as long as tailings disposal practices continue as they have in the last few years. Bank Swallows have not nested at Austin Springs since 1980, when gravel-crushing operations ceased (Knight pers. comm.). The future of colonies in steam plant ash piles is questionable due to changing ash disposal methods.—*Charles P. Nicholson.*

Cliff Swallow
Petrochelidon pyrrhonota

The Cliff Swallow is a fairly common migrant and a locally common summer resident in Tennessee. Most birds arrive in early April and depart by early September. It normally nests in colonies, most of which are under bridges and dams along the Tennessee River, the lower Cumberland River, and their large tributaries. The mud nest, in the shape of a gourd or inverted igloo, is one of the most distinctive of any Tennessee bird.

Emlen (1954) listed essential features of Cliff Swallow nesting habitat as an open foraging area, a vertical substrate with an overhang for nest attachment, and a supply of mud suitable for nest construction. Prior to European settlement, Cliff Swallows were restricted to nesting on cliffs or at the mouths of caves (Gross 1942). Following settlement, they began nesting on buildings and, later, on bridges and dams. The newly cleared land also provided foraging habitat. In Tennessee, bridges and dams with concrete surfaces suitable for nest attachment were rare until early this century. Suitable habitat greatly increased with the construction of large dams, beginning in the 1930s.

The first published accounts of Cliff Swallows nesting in Tennessee were in 1936, when nests were reported on 3 locks on the lower Cumberland River, Stewart and Montgomery Counties, and on Swallow Bluff on the Tennessee River, Decatur County (Ganier and Weakley 1936, Weakley 1936). According to lock

Cliff Swallow. Elizabeth S. Chastain

tenders, the birds began nesting at Lock D near Dover shortly after the lock was built in 1916. The fact that a bluff, island, and landing were named after the birds suggests that the Swallow Bluff site was occupied much earlier. Since 1936, the range and numbers of Cliff Swallows has greatly increased in Tennessee. Alsop (1981) describes this increase and the species's Tennessee distribution through the mid-1970s. The increase continues, and several new county nest records were discovered by Atlas workers. The only counties where swallows previously nested, but were not reported by atlasers, were Shelby and Hardeman in West Tennessee. Single nests were reported from each of these counties during the 1970s (Alsop 1981).

Cliff Swallows have been reported on too few BBS routes in Tennessee to show a significant trend; more routes, however, have shown increases than decreases. Rangewide, Cliff Swallow populations showed an increasing trend of 1.1%/year (p = 0.03) from 1966 to 1993 (Peterjohn, Sauer, and Link 1994).

Atlas workers found Cliff Swallows in all physiographic regions except the Cumberland Mountains, where nesting habitat is limited. Suitable nest sites are also limiting in West Tennessee, as many bridges there are unsuitable wood and steel. Locating and confirming breeding of Cliff Swallows was easy for atlasers because of the swallows' colonial breeding, distinctive nest, and preference for human-made structures in open areas. It is unlikely that the species was missed in priority blocks; more work in nonpriority blocks may have resulted in additional records along rivers such as the Duck. Too few mini-routes recorded Cliff Swallows to show regional differences in abundance.

The Cliff Swallow subspecies breeding from the southern Appalachians to central Texas was recently described as *P. albifrons [pyrrhonota] ganieri*, named after Tennessee ornithologist Albert F. Ganier (Phillips 1986). The type locality of this race is Swallow Bluff, Tennessee.

Breeding Biology: The Cliff Swallow's practice of nesting on or in barns, common north of Tennessee (e.g., Gross 1942, Samuel 1971), has only been reported from the northeast portion of Tennessee (Herndon 1947). Besides Swallow Bluff, nesting on natural substrates has occurred at Marvin's Bluff in 1945 (Weakley 1945) and a bluff near Peter's Landing (Atlas results), both in Perry County, and Gray Cliff and the mouth of Nickajack Cave, both in Marion County, in 1960 (West 1961). Swallow nests were present at Swallow Bluff in 1983 (Nicholson 1984a). I saw no evidence of Cliff Swallow nests on Gray Cliff in 1987, but they still nest in the mouth of Nickajack Cave, as well as on nearby bridges. The largest colony reported in the state was about 2500 nests in Decatur County in 1958 (Nicholson 1980b), and the average colony size is probably between 100 and 200 nests (Alsop 1981). Nest heights range from about 2 to over 20 m above ground or water. Nests are usu-

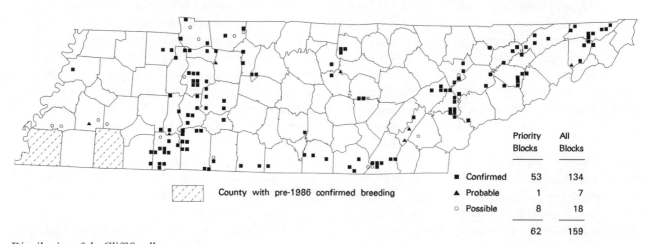

Distribution of the Cliff Swallow.

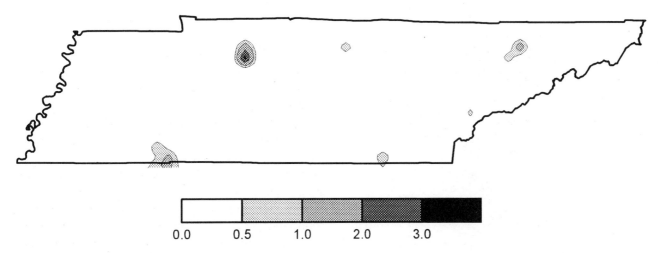

Abundance of the Cliff Swallow.

ally densely packed in the colony, and adjacent nests may share walls. Because of their shape and frequent inaccessibility, the contents of Cliff Swallow nests are difficult to examine, and relatively little is known about their breeding biology in Tennessee. Most Atlas records were confirmed by watching the adults at nests.

Cliff Swallow nests from previous years are often repaired and reused, although infrequently by their previous users (Mayhew 1958). Pair formation and copulation take place at the nest site (Emlen 1954), and females often copulate with neighboring males. Both members of the pair construct the nest, which requires up to 2 weeks (Samuel 1971). Flocks of colony members usually gather mud together, and the nest is constructed from up to 1200 mud pellets (Emlen 1954). The nest chamber is about 15 cm in diameter with an entrance tunnel up to 15 cm long (Harrison 1975). The chamber is lined with dried grasses and, less commonly, feathers. Egg laying often occurs before completion of the nest.

Dates of nests with young and the few Tennessee egg records suggest that most egg laying occurs in the first half of May. Tennessee clutches include 1 with 3 eggs, 5 with 4 eggs, and 1 with 5 eggs. Studies in Virginia and West Virginia found clutch sizes of 2–6 eggs, with an average of 3.3 (Grant and Quay 1977, Samuel 1971). The eggs are white to pale pink with speckles or blotches of brown, often concentrated around the larger end. Both sexes incubate the eggs for about 14–16 days, and both feed the nestlings, which leave the nest in about 24 days (Mayhew 1958, Samuel 1971). The fledged young remain near the nest for 2–3 days, then join flocks with their parents. Cliff Swallows occasionally raise a second brood (Samuel 1971). All of the birds depart the colony en masse, sometimes by the end of June (Nicholson pers. obs.), and sometimes before second broods are fledged (Samuel 1971).

Brown-headed Cowbird parasitism of Cliff Swallows is very rare (Friedmann 1963). Other parasites, however, affect their reproductive success and site tenacity. Parasitic insect populations can become high enough to cause nest or colony site abandonment (Brown and Brown 1986). House Sparrows lay eggs in swallow nests, and the swallows often raise the sparrow young, to the detriment of their own young (Stoner 1939). House Sparrows also build their nests in Cliff Swallow nests, and the sparrow populations may build up to the point that the swallows abandon the site (Gross 1942), as observed by Patterson (1966) in Hardin County.—*Charles P. Nicholson.*

Barn Swallow
Hirundo rustica

The Barn Swallow, a common, widespread summer resident, is well known by Tennesseans for its sociable nest habits and distinctive forked tail. It is the most numerous member of its family in the state and usually found in cleared areas, where it forages low over fields and water. It nests in barns, under overhangs of other rural buildings, and under bridges. The Barn Swallow is present in Tennessee from late March through early October (Robinson 1990).

The prehistoric breeding distribution of the Barn Swallow is poorly known (Mengel 1965); it originally nested in the mouths of caves and crevices in cliffs (Snapp 1976). Although these potential nest sites occur across the eastern two-thirds of Tennessee, they are rarely, if ever, used today. Wilson (1812) specifically mentioned not observing the Barn Swallow while traveling through the state in April and May 1810. It was first reported in Tennessee by Fox (1886), who noted its spring arrival in April 1884 and 1885. Merriam (in Rhoads 1895a) observed it in McMinn County in late

Barn Swallow. Elizabeth S. Chastain

July 1887, when nesting is often still underway. The first nest was reported by McEven (1894) in a barn at Bell Buckle, Bedford County, in 1893. Rhoads (1895a) described the Barn Swallow as not abundant and reported it from near Memphis, Nashville, Chattanooga, Knoxville, Greeneville, and Johnson City. Based on the dates of Rhoads's visits to these areas, the swallow was probably nesting from at least Nashville eastward. Gettys (egg coll. data) collected Barn Swallow egg sets in McMinn County in 1901 and 1902, and the first actual Nashville nest record was in 1902 (Ganier 1922). Three decades later, it was considered a rare summer resident in Middle Tennessee and a fairly common summer resident in East Tennessee (Ganier 1933a). The first West Tennessee nest record was in 1935 at Memphis (*Migrant* 7:69). The range and number of Barn Swallows rapidly increased, and by 1966 they were present throughout the state (BBS results). At least part of the reason for this increase is probably due to the replacement of many wooden rural bridges with concrete bridges, as noted by Jackson and Burchfield (1975) in Mississippi.

Since 1966, the Barn Swallow population, as measured by BBS routes, has shown a significant ($p < 0.01$) decreasing trend of 2.5%/year. The rate of decrease was greatest (2.8%/year) between 1980 and 1994. This decrease is probably due to the decrease in farmland and the subsequent loss of barn nesting habitat. Many wooden barns have also been replaced with metal farm-equipment storage buildings, which offer fewer swallow nest sites. The recent accelerated replacement of many steel and wooden rural bridges with concrete culverts has probably been partly responsible for slowing the rate of decrease. Throughout its large North American breeding range, the Barn Swallow population showed no significant overall trend from 1966 to 1993, although it decreased during the last part of this period (Peterjohn, Sauer, and Link 1994, Sauer and Droege 1992).

The Barn Swallow was found in 95% of the completed priority blocks; the total number of blocks in which it was reported was ninth highest of all species. Because of the ease in finding the Barn Swallow, it is unlikely it was missed in completed priority blocks. It was absent from blocks in the heavily wooded parts of the Unaka Mountains, Cumberland Mountains, Cumberland Plateau, and Mississippi Alluvial Plain, where neither bridges nor buildings suitable for nesting occurred. Barn Swallows occasionally forage at high elevations in the eastern mountains, at some distance from known nesting areas (Robinson 1990, Nicholson pers. obs.); this may have been true of some of the high elevation records in the "Possible" category. It commonly nests up to about 1070 m in northeast Tennessee and occasionally as high as 1490 m on the North Carolina side of the Smokies (Stupka 1963).

Atlasers recorded the Barn Swallow on 83% of the miniroutes at an average relative abundance of 2.8

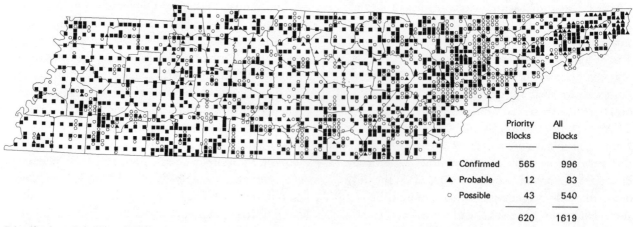

		Priority Blocks	All Blocks
■	Confirmed	565	996
▲	Probable	12	83
○	Possible	43	540
		620	1619

Distribution of the Barn Swallow.

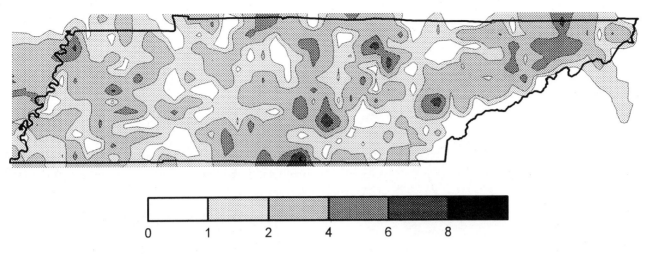

Abundance of the Barn Swallow.

stops/route. Its average abundance and proportion of routes recording it were lowest in the Cumberland Mountains (x̄ = 1.8, freq. = 57%) and Unaka Mountains (2.1, 57%). Numbers were higher than average in the Ridge and Valley, Eastern Highland Rim, Central Basin and Mississippi Alluvial Plain.

Breeding Biology: Confirming breeding of the Barn Swallow was easier than for most other species, due both its use of buildings and bridges as nest sites and the conspicuous behavior of fledglings. Two-thirds of the confirmed records were of nests, which are easily identified and often conspicuous. The Barn Swallow nests solitarily or in small colonies of rarely more than about 20 active nests (Nicholson pers. obs.). Within a colony, breeding synchrony is low, pairs defend an area around the nest, and nests are dispersed (Snapp 1976). The size of the colony increases with the size of the structure and the number of entryways.

The nest is typically placed on a ledge, a vertical wall, or at the corner of 2 vertical walls, a short distance below a horizontal surface. On smooth concrete or painted surfaces, the nest is often placed on mud-dauber nests (Jackson and Burchfield 1975, Nicholson pers. obs.). Nests under bridges are usually on inner spans, sheltered from blowing rain. The height of the nest varies with the height of the nest support, and Barn Swallow nests are rarely lower than about 1.7 m. Both adults build the nest in slightly less than a week (Samuel 1971). The cup-shaped nest is constructed of pellets of mud and pieces of straw and lined with feathers. Many nests are reused for second clutches, and nests from previous years are often refurbished (Samuel 1971).

Egg laying begins in late April and for first clutches peaks about 10–15 May. Clutch size is 3–6 eggs, with 5 the most frequent number and an average of 4.87 (s.d. = 0.67, n = 68). Clutches of 7 were reported elsewhere by Samuel (1971), who also found 5 eggs to be the most common size of first clutches and 4 eggs most common in second clutches. Tennessee clutches begun before 20 May averaged 4.91 eggs (s.d. = 0.70, n = 35), not significantly different from later clutches (x̄ = 4.82, s.d. = 0.64, n = 33).

The female carries out most of the incubation, which lasts about 17 days (Samuel 1971). Both adults feed the nestlings, which fledge in about 21 days. After fledging, the juveniles perch in the open near the nest site, where they are fed for several days. No records of its parasitism by Brown-headed Cowbirds are available from Tennessee, and such parasitism is infrequent elsewhere (Friedmann and Kiff 1985). House Sparrows and Eastern Phoebes occasionally build their nests on Barn Swallow nests, although this does not appear to effect the swallow's productivity (Nicholson pers. obs.).—*Charles P. Nicholson.*

Blue Jay
Cyanocitta cristata

Blue Jays are colorful, loud, and common birds, familiar to most Tennesseans. They are partially migratory, with many birds born in Tennessee migrating southward in the fall and additional birds from north of the state present in the winter (Laskey 1958). The fall migration lasts from mid-September into early November; spring migrants pass through from late March into early May. Migrating birds are often conspicuous as they fly during the day in loose flocks just above the treetops.

Blue Jays nest throughout the state and breeding has been confirmed up to an elevation of at least 1638 m in the Great Smoky Mountains (Stupka 1963). Atlasers found Blue Jays in all priority blocks except for a few at high elevations in the Unaka Mountains and found them to be common in both suburban and rural areas. They

Blue Jay. Elizabeth S. Chastain

were recorded on 652 miniroutes at an average relative abundance of 4.4 stops/route. Their abundance in the Mississippi Alluvial Plain, Cumberland Mountains, and Unaka Mountains was somewhat lower than average; in the Coastal Plain and Ridge and Valley Provinces it was higher than average. With the exception of the Mississippi Alluvial Plain, their abundance shows a gross inverse relationship with the proportion of forest cover.

Blue Jays occupy a variety of forest types and occur in both the forest interior and forest edge (Kroodsma 1984, Whitcomb et al. 1981). In Ridge and Valley deciduous forest, Anderson and Shugart (1974) found jays selected areas with a dense understory and large canopy trees. The highest densities on census plots have been in Central Basin mixed deciduous forests, 33 to 49 pairs/100 ha (Fowler and Fowler 1984a, 1984b), and in Ridge and Valley mixed hardwood forests, 33/100 ha (Howell 1974) and 22/100 ha (Smith 1975). Blue Jays were one of the most common species in these plots, all of which were in regions dominated by woodlots and farmland. Numbers in mixed mesophytic forests on the Eastern Highland Rim and Cumberland Mountains have been much lower, generally less than 5/100 ha (e.g., Smith 1977, Simmers 1982a, 1991), although Robertson (1979) reported 17/100 ha in a virgin mixed-mesophytic forest. Wilcove (1988) found densities of 9 to 42 pairs/100 ha in Great Smoky Mountain oak forests.

Blue Jays were probably fairly common in the widespread oak forests of prehistoric Tennessee. Their numbers may have initially declined somewhat with the widespread forest clearing associated with European settlement. Rhoads (1895a) found the Blue Jay present throughout the state and most abundant in the Mississippi River bottoms. Torrey (1896a) described the Blue Jay as scarce in the Chattanooga area during the spring of 1894. Their numbers apparently increased as they adapted to humans. Bock and Lepthien (1976a) noted a 30% increase in numbers on rangewide Christmas Bird Counts during the 1960s, at least partly attributable to increased winter bird feeding. The increase in area and maturation of Tennessee forests in recent decades has probably benefited wintering Blue Jays, which feed heavily on acorns. Numbers of wintering jays on CBCs have increased in the Unaka Mountains and at Nashville (Tanner 1985). Some breeding populations have also increased. Wilcove (1988) noted an increase on Great Smoky Mountains census plots between 1947 and 1983. Statewide, however, their numbers on BBS routes significantly decreased ($p < 0.01$) by 2.7%/year from 1966 to 1994. Blue Jay numbers on eastern North American BBS routes significantly decreased ($p < 0.05$) from 1966 to 1987 (Robbins et al. 1989b), as did the continental population from 1966 to 1993 (Peterjohn, Sauer, and Link 1994). The reasons for these declines are not clear.

Slightly over one-third of all Atlas records and over half of the records from priority blocks were of confirmed

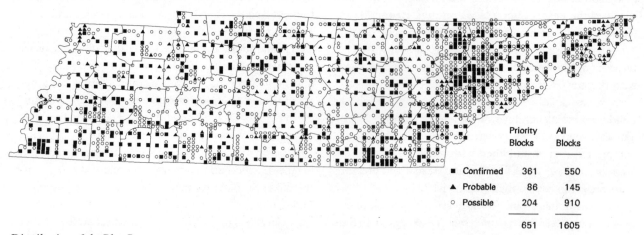

	Priority Blocks	All Blocks
■ Confirmed	361	550
▲ Probable	86	145
○ Possible	204	910
	651	1605

Distribution of the Blue Jay.

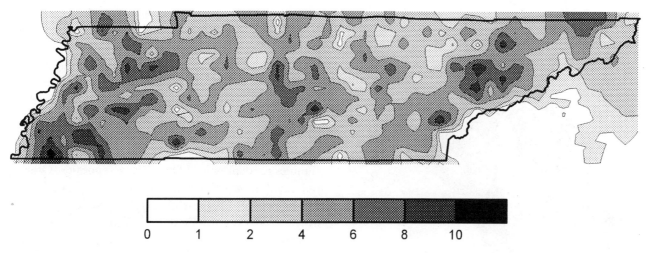

Abundance of the Blue Jay.

nesting. Less than one-third of the confirmed records, however, were of nests. The nest is often well hidden, and the adults, especially early in the nest cycle, are secretive in its vicinity.

Breeding Biology: Blue Jays begin breeding activity in late winter, forming courtship flocks of a female and a few male birds (Hardy 1961). These flocks have a distinctive display flight: each bird, with its tail spread, alternates slow, extended wing beats and gliding. The female eventually mates with one of the males. Late in the spring, these flocks are made up mostly of yearling birds, many of which do not nest. Blue Jays are monogamous, and the pair defends a small area around the nest from individual jays whose behavior suggests they are potential rivals (Hardy 1961). After mating, the pair begins nest building, often building a "false nest" when the male gives sticks to the female, who places them at a potential nest site. This false nest is usually not completed (Hardy 1961). In her study at Nashville, Laskey (1958) noted a yearling bird complete and lay in a false nest her parents had built a few weeks earlier.

Both the male and female Blue Jay build the true nest in about 3–5 days, although the female does most of the work (Hardy 1961). The outer portion of the bulky nest is made of sticks, weed stems, and often paper or similar items. This is usually lined with mud; the cup is then finished with a lining of rootlets. The nest is built in an upright crotch or on the horizontal branch of a tree or, less commonly, a large shrub. Tennessee nests average 5.5 m (s.d. = 2.6, n = 119) above ground and ranged from 1.7 to 13.7 m high. Nests have been reported in a variety of tree species, with oaks and maples probably used most often; nests are also commonly built in cedars and pines.

Egg laying peaks in late April. The normal clutch size is 3 to 5 eggs; Tennessee clutches average 4.3 eggs (s.d. = 0.63, n = 87). Clutches of 6 eggs have been reported elsewhere (Goodwin 1976). The eggs are olive green to buff or bluish, marked with dark brown and gray spots. Both green and blue eggs occasionally occur in the same clutch (Nicholson pers. obs.). The female performs most of the incubation, which lasts 16 to 18 days (Goodwin 1976). During this period, as well as during nest construction, the female receives most of her food from the male (Hardy 1961). The young leave the nest in 17 to 20 days (Laskey 1958, Hardy 1961). Fledglings begin foraging for themselves in about 3 weeks, and may be fed by the parents for up to 2 months after fledging. Blue Jays frequently raise 2 broods (Laskey 1958). Cowbird parasitism of Blue Jays has not been reported in Tennessee. It is very rare elsewhere, and Blue Jays usually remove cowbird eggs experimentally placed in their nests (Friedmann, Kiff, and Rothstein 1977).—*Charles P. Nicholson.*

American Crow
Corvus brachyrhynchos

The American Crow is a common and conspicuous permanent resident of Tennessee that is found throughout the state and was the fourth most frequently reported species during the Atlas survey. Crows occur on dispersed nesting territories in the breeding season. In the fall and winter, however, crows usually occur in flocks and form communal roosts sometimes numbering several thousand birds (e.g., Robinson 1990). Migrant crows from north of Tennessee are also present during the winter.

American Crows typically occur in open country, with a mix of farmland and woodland. They are omnivorous and feed primarily on the ground; corn and other grains are often a large portion of their diet (Goodwin 1976). Crows have a long history of being persecuted by humans and as a result are usually wary

American Crow. Elizabeth S. Chastain

of them, particularly in rural areas. This persecution has often been organized; for example, during the 1930s major newspapers and the state game and fish agency sponsored crow killing contests (Anon. 1934). Year-round hunting to control crow depredations is still legal, although the "sport-hunting" season has recently been restricted to the fall and winter. No estimate of the number of crows killed annually is available, and no significant long-term effects on crow numbers are evident as a result of this hunting. In urban and suburban areas, where crows are much less persecuted, crows have recently become habituated to human presence and are now more common (Nicholson pers. obs., Knight, Grout, and Temple 1987).

At the time of the first European settlements in Tennessee, the crow was probably uncommon to rare in the state and probably associated with Indian settlements and open-glade and prairie habitats. Alexander Wilson (1811a) noted that crows were outnumbered by Common Ravens along the Natchez Trace southwest of Nashville in 1810. Large settlements did not then exist along the Tennessee portion of the Natchez Trace. Agricultural clearing soon accelerated and crow numbers increased. At the end of the century, Rhoads (1895a:487) described the crow as "not abundant" but occurring statewide and up to the spruce-fir zone in the Roan Mountain area. A few decades later, Ganier (1933a) described the crow as a common summer resident in West Tennessee and a fairly common summer resident in Middle and East Tennessee. Other published information from that period (e.g., Calhoun 1941), however, suggests that crows were less common in West Tennessee than in the rest of the state, as is presently the case. During the late 1930s and early 1940s, crows were rare or absent from some relatively remote, heavily forested areas such as Pickett Forest, Fall Creek Falls, and mid- to upper elevations of the Great Smokies (Ganier 1937a, Ganier and Clebsch 1940, Wilcove 1988). Since then, their numbers have greatly increased in all these areas (Wilcove 1988, Atlas results). These local increases are probably due to increased human development and less persecution within and adjacent to these areas. From 1966 to 1994, crow numbers, as measured by Tennessee BBS routes, decreased by 0.6%/year ($p < 0.10$); the decline was concentrated in the first half of this period. The North American crow population showed an increasing trend of 0.7%/year ($p < 0.01$) on BBS routes from 1966 to 1993 (Peterjohn, Sauer, and Link 1994).

The American Crow was the only species found in all completed Atlas priority blocks. It was recorded on all but 10 of the completed miniroutes at an average relative abundance of 7.3 stops/species. Their abundance generally increased from west to east, except that it was lower in the Unaka Mountains than in the adjacent Ridge and Valley. In West Tennessee, they were more common in the more heavily wooded Coastal Plain Uplands than in the less forested Mississippi Alluvial

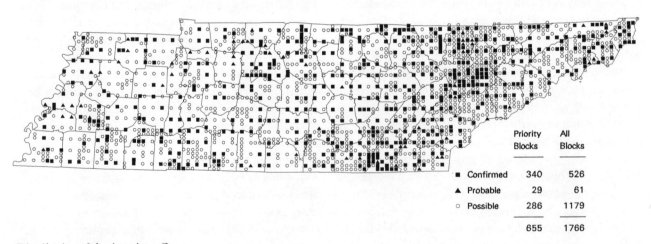

	Priority Blocks	All Blocks
■ Confirmed	340	526
▲ Probable	29	61
○ Possible	286	1179
	655	1766

Distribution of the American Crow.

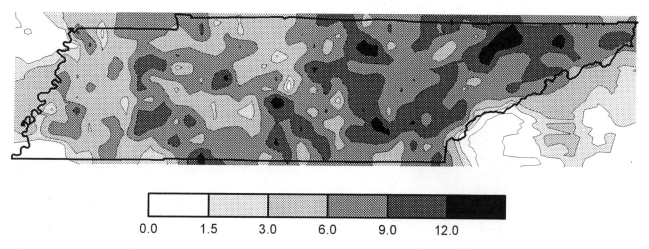

Abundance of the American Crow.

Plain and Loess Plain; limited wooded nesting habitat may account in part for their lower density in the latter 2 regions. In the Unaka Mountains, crows occur regularly up to the highest elevations. The highest elevation at which breeding has been confirmed, however, is at about 900 m (Wilcove 1988).

Breeding Biology: Atlasers found it fairly easy to confirm the breeding of American Crows, although only 9% of the confirmed records were of nests. Most Atlas fieldwork occurred after crows fledged, and 80% of the confirmed records were of fledglings. Although there are numerous nest records of crows available from Middle and East Tennessee, many details of its breeding biology are poorly known, as is true in much of the rest of its range (Kilham 1989). Recent studies, for example, have found that crow populations in Florida and Massachusetts usually breed cooperatively in groups of up to 10 birds composed of a breeding pair and their offspring from previous years (Chamberlain-Auger, Auger, and Strauss 1990, Kilham 1989).

Nest construction begins in March. The nest is a well-made cup about 0.6 m in diameter, with an outer shell of sticks and twigs. Inside this is frequently a layer of mud, then a lining of shredded bark, hair, moss, grasses, and other material. Nest construction takes up to 2 weeks, and crows may begin and abandon nests before completing one (Chamberlain-Auger, Auger, and Strauss 1990, Kilham 1989). The nest is built in a tree, either in a central crotch or on a lateral branch. Nest trees are located in woods or, less commonly, stand alone in an open area. Evergreens, when available, seem to be preferred, perhaps because they offer better concealment early in the nesting cycle. All but 1 of 27 nests found by Todd in Rutherford County, in the Central Basin, were in eastern red cedars (egg coll. data). Almost 40% of all Tennessee nests were in cedars and about 20% in pines; the remainder were in a variety of deciduous trees. Nest height ranges from 3 to 21.3 m and averages 9.9 m (s.d. = 4.1, n = 77).

Egg laying peaks during late March and early April and appears to be a few days later in East Tennessee than in Middle Tennessee. Clutch size in Tennessee ranges from 3 to 6, with an average of 4.40 (s.d. = 0.84, n = 87). The eggs are bluish to pale green with brown and gray spots and blotches (Goodwin 1976). The female incubates the eggs for about 18 to 19 days, and during this period she gets much of her food from the male (Kilham 1989). The female also broods the nestlings, which both adults feed. The young fledge in about 35 days and are dependent on the parents for at least 2 weeks (Kilham 1989). The family group may remain intact into the winter (Knopf and Knopf 1983). American Crows will often renest after the loss of a clutch, and the late April and May egg dates are probably second clutches. They are single-brooded (Kilham 1989).—*Charles P. Nicholson.*

Fish Crow
Corvus ossifragus

The Fish Crow is a locally uncommon permanent resident of West Tennessee, found near the Mississippi River and along its direct tributaries. It is endemic to the eastern United States, where its primary range is southern tidewater areas (Bent 1946, Johnston 1961). Its ancestral habitats in tidewater areas are riverine and broken forests, and use of these habitats has probably preadapted them to use open forests in river valleys and along lakes in the interior, where they have dramatically increased their range and abundance for over the last 25 years (McNair 1987b, Wells and McGowan 1991, Peterjohn and Sauer 1993).

Fish Crow. Elizabeth S. Chastain

The first record of the Fish Crow in Tennessee did not occur until 1931, near the Mississippi River in Shelby County (Coffey 1942b). The crow gradually expanded its range northward along the Mississippi. It was first reported in southwest Kentucky in 1959 and regularly reported there by the mid-1960s (Mengel 1965, Able 1967). The first southeast Missouri report was also in the mid-1960s (Robbins and Easterla 1992). Shelby County has since remained the crow's center of abundance in Tennessee (Tanner 1985, Robinson 1990). Fish Crows are still generally scarce during winter throughout the remainder of the Mississippi floodplain (though see Robinson 1990 for counts of recent large winter flocks in Lauderdale County), where they increase during spring migration from mid-March through April. Atlas block data clearly show that breeding-season Fish Crows are now widely distributed in all counties adjoining the Mississippi River, despite this region's extensive reduction in forest cover for agricultural uses in recent decades.

Prior to the Atlas period, reports of Fish Crows were restricted to Shelby County and to the Mississippi Alluvial Plain, except for an extralimital record of a flock of 4 vagrants on 10 April 1979 in Hamilton County at the southern end of the Ridge and Valley Province (Dubke 1982). Atlas block data have documented a modest expansion of breeding-season Fish Crows along several major Mississippi River tributaries into the Loess Plain region, though substantiated breeding evidence is not documented for this area. This modest recent range expansion is consistent with the pattern of Fish Crow range expansion elsewhere (McNair 1987a, Wells and McGowan 1991).

Almost all breeding-season records in Tennessee pertain to possible breeding only (31 of 35 blocks, 89%). Many of these individuals may not have bred in the block where observed. Breeding Fish Crows forage peacefully with other Fish Crows in a large home range (Bent 1946, McNair pers. obs.), frequently forage away from the nest site, usually reach nest sites by flying over woods, and vigilantly defend an area confined to the immediate nest site (McNair 1984c), which is apparently why observers found scant solid breeding evidence in Tennessee. The scarcity of suitable forested habitats in many blocks may also reduce breeding opportunities for Fish Crows.

Fish Crows have not expanded their range in Tennessee into nonriparian habitats away from the immediate vicinity of water except for residential areas of Shelby County (Atlas data). Fish Crows are tamer than American Crows and are much more tolerant of disturbance near their nests. These characteristics would facilitate colonization of residential areas by Fish Crows. The reasons for the lack of a more widespread expansion into nonriparian habitats are obscure. Expansion into open pine groves or woods away from water, for example, has been widespread in many Southeast states such as Georgia and the Carolinas (McNair 1987b), though this habitat is limited in western Tennessee within the present range of the Fish Crow.

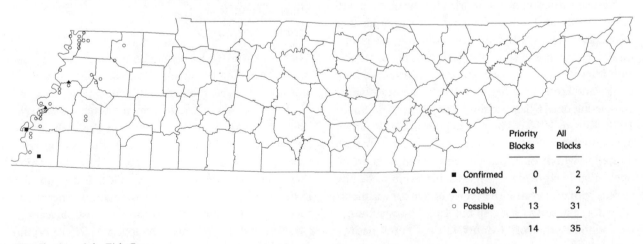

Distribution of the Fish Crow.

	Priority Blocks	All Blocks
■ Confirmed	0	2
▲ Probable	1	2
○ Possible	13	31
	14	35

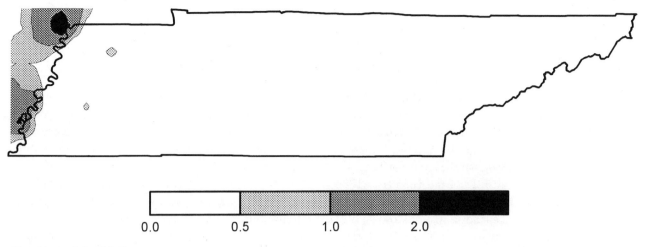

Abundance of the Fish Crow.

Fish Crows were recorded on 8 Atlas miniroutes, at a mean abundance of 1.4 stops/route, significantly less than the 4.6 stops/route for the American Crow on the same 8 miniroutes. Miniroute and Atlas block data within the entire Mississippi Alluvial Plain also reveal that Fish Crows were much less widely distributed than American Crows. However, despite their greater distribution and abundance in the Mississippi Alluvial Plain compared to Fish Crows, American Crows are less numerous and widely distributed in this region compared to any other in Tennessee. American Crows may be relatively less abundant here because the habitat is less favorable, or because the more competitive Fish Crow is increasing at their expense in the area of sympatry. The clear geographical pattern of Fish Crow distribution and abundance in West Tennessee in comparison with the American Crow suggests that the true status of the Fish Crow is not obscured by any mistakes in identification (cf. McNair 1987b, Wells and McGowan 1991).

Fish Crows are highly adaptable. Given recent population trends, we expect a further consolidation and probable expansion of breeding populations in West Tennessee.

Breeding Biology: The first Fish Crow nests were not discovered in Tennessee until 1980, by B. B. Coffey Jr. in Shelby County (Robinson 1990). The only 2 confirmed breeding records during the Atlas period were also discovered in Shelby County, one in a residential area, the other in bottomland forest. One active nest with young and another nest with contents unseen but with adults present throughout the breeding season were located in large cottonwood and sycamore trees, 11.5 and 21.5 m above ground. The majority of nest sites used in the interior of the United States have been in the crowns of pines (Bent 1946, McNair 1987b). However, pines are generally absent from the range of the Fish Crow in Tennessee, except for some widely scattered, planted groves. No Fish Crow clutches have yet been described in Tennessee; nest building has been observed during the first week of April, and nestlings have been observed during late June.—*Douglas B. McNair.*

Common Raven
Corvus corax

Because of its great size, acrobatic flight, large repertoire of calls, and frequent preference for wilderness areas, the Common Raven is much sought after by naturalists. Although once found throughout Tennessee, most recent observations have been from the Unaka Mountains, where it is an uncommon permanent resident. Ravens are occasionally reported from the northeastern end of the Ridge and Valley, especially in winter.

Common Raven. Elizabeth S. Chastain

The Common Raven was first reported in Tennessee by Alexander Wilson (1811a), who observed it along the Natchez Trace southwest of Nashville, where it then outnumbered American Crows. Its range contracted with the spread of European settlers, who persecuted it because of its taste for poultry and livestock (e.g., Ganier 1973). By the late nineteenth century, its range was restricted to the Cumberlands and Unaka Mountains. Fox (1882) observed flocks of 8 or 10 ravens on Lookout Mountain near Chattanooga in the spring of 1882 and found it in Roane County in 1884 and 1885 (Fox 1886). The only documented twentieth-century observation of the raven in the Cumberlands is that of Howell (1924), who observed a pair flying over the Tennessee River at the Alabama state line in April 1914. Both Howell (1910) and Ganier (1923a, 1973) gave secondhand reports of its occurrence elsewhere in the Cumberlands, and Ganier found old nests there he believed were built by ravens. The Common Raven probably disappeared from the Cumberlands by 1920.

The earliest record from the Unaka Mountains is from Roan Mountain in 1895 (Rhoads 1895a). By the mid-1940s, it had been reported from all counties bordering North Carolina. Alsop (1979b) summarizes these reports, which suggest that ravens occurred on most mountain ranges reaching elevations of at least 915 m from the Smokies northeast to Virginia. Most of the reports are from accessible parts of the Great Smoky Mountains National Park and from Roan Mountain. Southwest of the Smokies, a raven was reported from Big Frog Mountain, Polk County, in July 1937 by Wetmore (1939). James T. Tanner and I observed one there in May 1983. Ganier and Clebsch (1946) observed ravens in the Unicoi Mountains of Monroe County near Stratton's Bald in the mid-1940s, and ravens were reported several times near there through the early 1980s (Alsop 1979b, McConnell and McConnell 1983).

Although Common Ravens have recently occurred throughout the Unaka Mountains, many of the few known nests have been in the Smokies. The best-known site is on Peregrine Ridge near Alum Cave Bluffs. Young birds were reported there in 1944 and a nest first observed in 1960 (Stupka 1963). The site has been regularly used since then (Williams 1980b, Nicholson pers. obs.). Stupka (1963) also reports a nest near Roaring Fork in the 1930s. Other Smokies sites where nesting has probably occurred are just east of Charlies Bunion and on Greenbriar Pinnacle (Stupka 1963, Atlas results). Outside of the Smokies, 2 nests have been reported from Carter County. An adult was observed on 29 April 1982 at a large nest built of sticks on a cliff in the Doe River Gorge at about 640 m elevation; no eggs or young were seen (*Migrant* 53:70, P. Range pers. comm.). A raven was observed in this area in 1991, and a territorial pair was present in 1992 (R. L. Knight pers. comm.). A nest was reported, without details, near Ripshin Lake, elevation 1,067 m, on 27 April 1985 (*Migrant* 56:82).

Atlasers found the Common Raven in 36 Atlas blocks, and 14 of them completed priority blocks. The only confirmed record was from the Alum Cave site in the Smokies; pairs or territorial birds were reported from Greenbriar Pinnacle, Unaka Mountain, Iron Mountain, and Roan Mountain. Most of the other observations were of birds in flight, and many of them were probably non-breeding juveniles. Common Ravens begin nesting when 3 or 4 years old (Heinrich 1991). Once mated, the pair bond is permanent, and the territory is occupied year-round and defended against other ravens. Juvenile birds wander extensively, sometimes in flocks.

The Common Raven was listed as Endangered in Tennessee in 1975 (TWRA 1975). The Atlas results suggest that its population is relatively stable in the Smokies and northeast Tennessee. Little is known of the nesting success of this population, which may be limited by food supplies, as Hooper (1977) suggests for birds nesting at high elevations in Virginia. The increasing recreational use of the informal trail along the crest of Peregrine Ridge in the Smokies is probably detrimental to the ravens nesting there, and this trail should be closed from February through May. Southwest of the Smokies, the single report from the Unicoi Mountains suggests a population decrease, although the breeding status of the birds formerly occurring there is unknown. As the population is in no present danger of extinction, its status was lowered to Threatened in 1994 (TWRA 1994a); an even lower status of In Need of Management is probably justified.

The raven population in western Virginia has recently increased, and, after an absence of several decades, a raven nest was discovered in Kentucky, about 48 km north of Tennessee, in 1984 (Fowler et al. 1985). Ravens have also been reported from the Kentucky–Virginia portions of Cumberland Gap National Park. If this population expands, ravens should repopulate nearby parts of Tennessee. Numerous cliffs, some probably suitable for nesting, occur on Pine Mountain, as well as on Clinch Mountain and adjacent ridges.

Breeding Biology: The Common Raven begins nesting in late January or February. The few Tennessee nests have all been on narrow ledges of cliffs, protected by overhanging rock, typical of most nest sites elsewhere in the southern Appalachians (e.g., Hooper 1977). Suitable cliff nest sites are absent from some areas where ravens have been regularly reported; birds in these areas

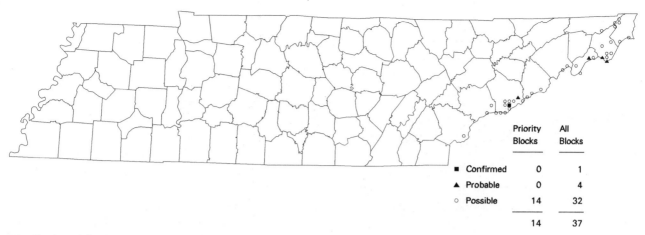

Distribution of the Common Raven.

may nest in trees, as occurs infrequently in western Virginia and more commonly further north (Hooper 1977). The nest is built of sticks with a well-formed cup; Williams (1980b) described a nest in the Smokies built of laurel branches. Eggs are probably laid in late February or early March. Williams (1980b) reported a clutch of 7 eggs. Clutches of 3–7 eggs have been reported elsewhere, with 4–6 eggs most common (Goodwin 1976, Stiehl 1985). From 1 to 3 young have been found in Tennessee nests. The female incubates the eggs for 18–21 days; both adults feed the nestlings which fledge in about 7 weeks (Goodwin 1976, Stiehl 1985).—*Charles P. Nicholson.*

Black-capped Chickadee
Poecile atricapillus

The Black-capped Chickadee is a locally common permanent resident with relic, disjunct populations at high elevations of the southern Blue Ridge mountains. In Tennessee, the only convincingly documented population is in the Great Smoky Mountains. Black-cappeds have occasionally been reported from the Unicoi Mountains immediately south of the Smokies and from a few locations to the northeast (Robinson 1990, Atlas results, Mitchell and Stedman 1993). The identification of Black-cappeds at these locations, however, is questionable, and it is likely these birds were Carolina Chickadees, whose vocalizations can closely resemble those of Black-cappeds (Robbins, Braun, and Tomey 1986).

The Black-capped Chickadee was first reported from the southern Blue Ridge by Brewster (1886) in 1885 (see Simpson 1977), and then from the Smokies in Blount County, Tennessee by LeMoyne (1886), who reported observing 2 nests, but gave no further details. Ganier (1926) reported it from the high elevations of the Smokies, but little was known about the Tennessee population until the study of Tanner (1952).

Tanner (1952) found Black-capped Chickadees generally distributed throughout the Smokies at elevations above 1220 m during the nesting season. The lowest confirmed nesting is at 1036 m (Tanner 1952) and territorial birds occur at elevations as low as 750–900 m on the north-facing slopes of cove hardwood forests (Kendeigh and Fawver 1981, Wilcove 1988). Black-cappeds nest upslope of Carolina Chickadees, separated by an apparent elevational gap of about 180 m. In southern Blue Ridge mountain ranges where Black-cappeds are absent, Carolina Chickadees nest in low density through about the lower half of the elevational range occupied by Black-cappeds in the Smokies (Tanner 1952). During the winter, Black-cappeds occur at lower elevations and occasionally mix with Carolina Chickadees (Tanner 1952).

Of the 19 Atlas reports of Black-capped Chickadees, 17 were from the Smokies; Black-cappeds were also found on 4 miniroutes, all in the Smokies, at an average relative

Black-capped Chickadee. Elizabeth S. Chastain

abundance of 5.3 stops/route. The 2 Atlas reports outside of the Smokies were in the Big Fodderstack area of the Unicoi Mountains. Mitchell and Stedman (1993) also reported a pair of Black-cappeds, and no Carolinas, in 1992 on a 10 ha census plot in northern hardwood forest at 1400 m near Whigg Meadow, a few km south of Big Fodderstack. No chickadees were observed in this area in 1993 (L. Mitchell pers. comm.). Decades earlier, Ganier and Clebsch (1946, see also Tanner 1952) collected Carolina Chickadees in the vicinity of Stratton Gap, (1310 m), between Whigg Meadow and Big Fodderstack; McConnell and McConnell (1983) also reported only Carolinas in this area. During Atlas fieldwork, Carolinas were again found in the Stratton Gap area. We believe additional study, including measurement of captured birds and analysis of recorded vocalizations (e.g., Robbins, Braun, and Tomey 1986) is required to verify the identification of Black-cappeds in the Unicois.

The Black-capped Chickadee occurs in all forest types within its elevational range. Tanner (1952) stated that the presence of yellow birch trees for nest sites may be an important habitat component and described the Black-capped's comparative abundance during the nesting season as 64% in spruce-fir, 29% in northern hardwoods and hemlock, and 7% in lower-elevation forest types. Densities on censuses in undamaged virgin spruce-fir forest, 17.5–25 pairs/100 ha (Kendeigh and Fawver 1981, Alsop and Laughlin 1991), have been relatively high compared to most other habitats, with the possible exception of high-elevation beech gap forests, where Wilcove (1988) reported 43 pairs/100 ha at one site. Rabenold (1978) reported an anomalous high density of 125 pairs/100 ha in mature spruce-fir, probably an artifact of the elongated shape of his census plot. Average densities in other forest types are 17/100 ha in mixed hemlock-deciduous forest, 16/100 ha in northern red oak, 9/100 ha in pine heath, 7/100 ha in cove hardwoods, and 4/100 ha in chestnut oak (Kendeigh and Fawver 1981). Wilcove (1988) reported 43 pairs/100 ha in high-elevation beech gap forest.

Limited quantitative information is available on Black-capped Chickadee population trends. Wilcove (1988) found no clear trend on hardwood and hemlock plots censused in 1947–48 and 1982–83. Alsop and Laughlin (1991) found a decrease from 17.5 pairs/100 ha in 1967 to 2.5/100 ha in 1985 on a spruce-fir plot. Between these 2 censuses, however, most of the overstory trees had been killed by the balsam woolly adelgid. If this decrease occurred throughout the spruce-fir, which seems likely in the higher-elevation fir-dominated stands, a substantial recent reduction in the southern Black-capped Chickadee population has occurred.

Breeding Biology: Few details of Black-capped Chickadee nesting are available from Tennessee (or North Carolina). Tanner (1952) observed excavation of nest cavities in late April and early May; the 4 nests he found were in dead yellow birches, from 1.5 to 18.3 m above ground. Ganier (egg coll. data) found a nest with 6 eggs, incubated 7 days, in a dead stump on 30 May 1925. All of the confirmed Atlas records were families with dependent fledglings, observed between 27 June and 9 July.

The chickadee's breeding biology is well known from studies elsewhere in its range and here summarized from Smith (1991). Pair formation occurs within the winter flock, which breaks up in late winter, and the pairs then defend breeding territories within the flock's range. The female builds the nest in a cavity, which is usually excavated by both adults in a rotten stub; woodpecker holes, natural cavities, and bird boxes are less frequently used. The nest is constructed of bark strips, pine needles, moss, and hair. Most clutches contain 6–8 eggs, which the female incubates 12–13 days. Both adults feed the nestlings,

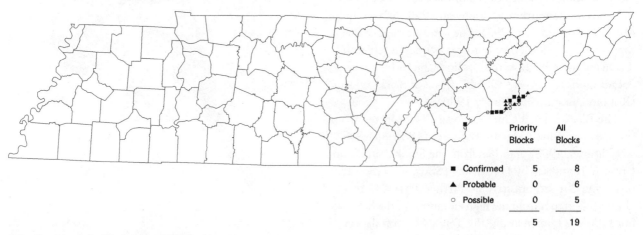

Distribution of the Black-capped Chickadee.

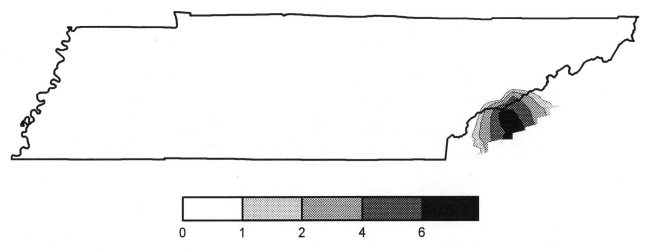

Abundance of the Black-capped Chickadee.

which leave the nest, almost fully feathered, in about 16 days. The family group stays together 2–4 weeks; following its breakup, the young rapidly disperse from the nest area, sometimes for distances of several kilometers. Second broods are occasionally raised. Tove (1980) observed a recently fledged young chickadee he described as a Black-capped on 9 August in the North Carolina mountains, suggesting second broods are at least occasionally raised in the southern Blue Ridge. Whether this chickadee was a "pure" Black-capped, however, is questionable.—*Charles P. Nicholson and Douglas B. McNair.*

Carolina Chickadee
Poecile carolinensis

Carolina Chickadees occur throughout Tennessee except at the highest elevations of the eastern mountains. If trees are present, even in suburban and urban areas, chickadees are likely to be present. Chickadees are small, but they can be conspicuous, especially in yards with bird feeders offering seeds or suet. Some chickadees become tolerant of people and may even perch on an outstretched hand that is holding sunflower seeds. The species most likely to be confused with the Carolina Chickadee is the Black-capped Chickadee, which occurs only in the high elevations of the eastern part of the state. Carolina Chickadees are permanent residents in Tennessee.

Carolina Chickadees were probably abundant in Tennessee when European settlers arrived. The subsequent clearing of forests reduced the amount of chickadee habitat. However, few areas became totally unsuitable for chickadees. They were found in every priority Atlas block except a few at the highest elevations of the Unaka Mountains. Carolina Chickadees were recorded on 88% of the miniroutes censused during the Atlas project at an average relative abundance of 3.3 stops/route. The number of chickadees per route generally increased from west to east across the state. In the Mississippi Alluvial Plain and Loess Plain, chickadees were detected at 2.4 and 2.1 stops/route, respectively, while in the easternmost Unaka Mountains they were recorded at 4.3 stops/route.

The number of chickadees is probably directly correlated with the amount of forest present. Carolina Chickadees occupy both deciduous and coniferous forests, with the highest densities apparently occurring in areas having mixed stands of deciduous trees and conifers. The only wooded areas in the state without Carolina Chickadees are the high elevations above approximately 1200 m in the Appalachians where Black-capped Chickadees are often present (Tanner 1952). From 1966 to 1994, Carolina Chickadees did not show a significant overall population trend on Tennessee BBS routes, although they increased by 2.2%/year ($p < 0.05$) from 1980 to

Carolina Chickadee. Elizabeth S. Chastain

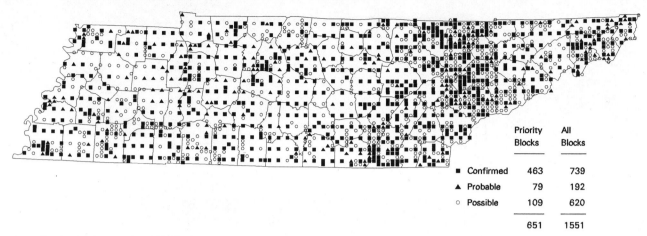

	Priority Blocks	All Blocks
■ Confirmed	463	739
▲ Probable	79	192
○ Possible	109	620
	651	1551

Distribution of the Carolina Chickadee.

1994. Rangewide, they decreased by 0.7%/year (p = 0.02) from 1966 to 1993 (Peterjohn, Sauer, and Link 1994).

Carolina Chickadees are monogamous, and mates may remain together on the same territory for more than one nesting season (Pitts pers. obs.). Few chickadees live more than 4 or 5 years, but one of my banded adults lived for 10 years. During the winter, chickadees live in flocks that may also contain Tufted Titmice, Downy Woodpeckers, and other species. In late winter the flocks break up as each pair begins territorial defense. Males are especially conspicuous as they patrol their territories and sing repeatedly. Following territory establishment and prior to fledging of the young, chickadees are less conspicuous and may be underestimated on miniroutes censused early in the season.

Breeding Biology: Carolina Chickadees nest in cavities that they either excavate or find. They select rotten snags for excavation. In Michigan, nest snags were 11–13 cm in diameter, and the entrance to the nest was 1.6–2.0 m above the ground; cavities excavated by chickadees averaged 6.5 cm in diameter (Brewer 1961). In Tennessee, 55 nests in natural cavities ranged from 0.5 to 3.1 m and averaged 1.4 m (s.d. = 0.6 m) above ground. Some of these nests were in hollow fence posts, and most of the others were in dead snags of various species. The percentage of chickadees that excavate their nest cavity is not known. Chickadees will use nest boxes, and, if successful, they may reuse the same box a second or third year. When given a choice of nest boxes with floor sizes of 71.5 sq cm or 143 sq cm, chickadees always selected the smaller box. When given a choice of nest boxes with floor size of 36 sq cm or 71.5 sq cm, chickadees again selected the smaller box (Pitts unpublished data). Based on Drury (1958) and a small number of observations, Pitts (1978a) suggested that nest boxes containing sawdust may be more attractive to chickadees. Most or all of the sawdust will be removed by the chickadees before they begin nest construction. The bottom part of the nest is constructed of green moss. The nest cup is lined with either mammal hair or thin strips of plant fibers. During the egg-laying period, eggs are usually covered with a flap of fur while the female is away from the nest.

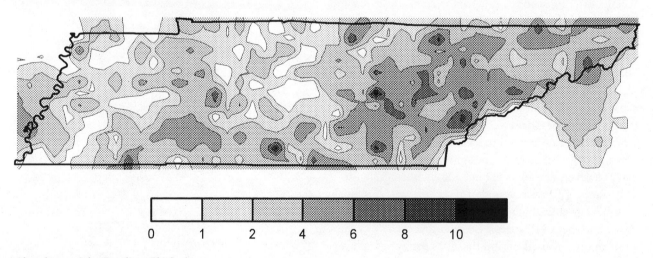

Abundance of the Carolina Chickadee.

Egg laying in northwest Tennessee typically begins in mid-March and peaks in early April but may be delayed by a week or more following harsh winters (Pitts unpublished data). In nest boxes, clutches of 5–6 are most common with an average of 5.7 (s.d. = 1.31, n = 148) (Pitts unpublished data). The largest clutches (7–8 eggs) are laid early in the year by experienced females, while the smallest clutches (2–4) are usually laid late in the nesting season. Nests in natural cavities also averaged 5.7 eggs (s.d. = 1.01, n = 70). Each pair normally has only one nest each year. However, I have one record of a female that produced two successful broods (with a different male at the second nest) and one record of a pair that successfully re-nested after the loss of their first nest. While some of the late nests are second attempts, I suspect that many of these nests are produced by pairs that were not able to find and defend suitable nesting sites earlier in the year.

Only the female incubates the eggs. The male frequently delivers food to the incubating female. Incubation usually requires 14–15 days. Both adults feed the young, which remain in the nest for 17–18 days. Caterpillars and other small animals are the most common foods of nestlings. If trapped inside a nest cavity, adults frequently perform a "snake display" (Sibley 1955) consisting of a slow swaying of the body followed by a lunge toward the intruder; this sudden movement is accompanied by an explosive hissing sound. This behavior resembles the actions of a snake and is thought to startle or repel potential predators, such as rodents, that might attempt to enter the nest cavity. After fledging, the young remain with their parents for 2 weeks or more. During this time, typically late May and early June, the broods are especially conspicuous. The mortality rate for fledglings is high, and surviving young attempt to nest their first year after hatching.

Nest success varies greatly. In years with frequent rains, eggs may fail to hatch and young may die in soggy nests. Predation by snakes is relatively uncommon because many of the early nests are completed before the peak of snake activity. Other cavity-nesting species such as Tufted Titmice and Eastern Bluebirds sometimes usurp active chickadee nests and use the cavity for their own nests. Two of 148 nests in boxes received a single Brown-headed Cowbird egg (Pitts unpublished data); neither nest was successful. In both cases the nest box entrance had been enlarged by woodpeckers. None of the 70 nests in natural cavities had cowbird eggs. Approximately 60% of the nests are parasitized by bird blowflies (*Protocalliphora deceptor*), whose larvae feed on the nestling chickadees (Pitts unpublished data). Blowflies rarely kill the nestlings, but parasitized nestlings are weakened and are more vulnerable to food shortages and adverse weather.
—*T. David Pitts.*

Tufted Titmouse
Baeolophus bicolor

One of the earliest signs of the approach of spring in Tennessee is the singing of Tufted Titmice. The "peto peto" mate-calling songs are commonly heard in mid-January, long before the actual beginning of nest construction. During the non-breeding season, groups of 2–4 titmice commonly move about with flocks of Carolina Chickadees and Downy Woodpeckers and readily use bird feeders located in or adjacent to woodlands.

Tufted Titmice are common permanent residents throughout Tennessee, except for the highest elevations of the Unaka Mountains. They were probably common throughout the extensive forests of prehistoric Tennessee. Rhoads's (1895a) comment that they are "abundant everywhere" is essentially still true, except in extensively deforested areas and above about 1525 m in the Unakas (Stupka 1963, Atlas results). Because of their loud, easily recognized, and frequently repeated songs, their presence is usually easy to document. They were found in all but 3 of the completed Atlas priority blocks, and the total number of blocks in which they were reported was the tenth-highest of any bird. They were also found on 94% of the miniroutes at an average relative abundance of 4.4 stops/route. With the exception of the Unaka Mountains, where they become less common at higher elevations, their abundance generally correlated well with regional proportions of forested area, although regional differences were not great. Their numbers were also low in several of the major metropolitan areas. Based on winter trapping and banding in Weakley County, I expected Carolina Chickadees to be more abundant than titmice on the miniroutes, but the average chickadee count was lower. The more conspicuous songs and notes of the titmouse may result in a larger percentage of them being recorded.

Tennessee BBS routes show no significant overall trend in titmouse numbers from 1966 to 1994. During the latter decade of this period, however, their population increased significantly ($p < 0.05$) at the rate of 1.9%/year. Their numbers on Tennessee Christmas Bird Counts also show no long-term trend (Tanner 1985). Range-wide, their population increased by 0.6%/year ($p = 0.01$) from 1966 to 1993 (Peterjohn, Sauer, and Link 1994).

The highest Tufted Titmouse densities on census plots, from 29 to 52 pairs/100 ha, have generally been in Ridge and Valley deciduous forests (Howell 1971, 1972, 1973, 1974, Smith 1975). Densities in mixed mesophytic forests include 22/100 ha in a Cumberland Mountains plot (Smith 1977), 17/100 ha in a virgin Cumberland Plateau plot (Robertson 1979), and 9/100 ha in a disturbed

Tufted Titmouse. Elizabeth S. Chastain

Highland Rim plot (Simmers 1982a). Densities in Central Basin upland hardwoods were 16–33/100 ha (Fowler and Fowler 1984a, 1984b) and in Central Basin cedar-dominated forests, 4–49/100 ha (Fowler and Fowler 1983a, 1983b, 1983c). Mitchell and Stedman (1993) found a density of 10/100 ha in mature northern hardwoods at 1500 m elevation in the Unicoi Mountains. Ford (1990) found titmice ubiquitous in West Tennessee forested wetlands, where their density averaged 27 pairs/100 ha on transect counts. Their density correlated with the number of snags and trees in the subcanopy. Somewhat similarly, Anderson and Shugart (1974) found that titmice, although relatively abundant throughout their Ridge and Valley deciduous forest sample plots, tended to select stands with an open understory and a well-developed subcanopy.

Breeding Biology: Titmice, like their relative the Carolina Chickadee, nest in cavities, either naturally occurring or nest boxes. While chickadees occasionally excavate their nest cavity, titmice apparently do not (Harrison 1975). Of 77 Tennessee nest records, only 20 were in natural cavities. This does not mean that titmice readily nest in boxes. Over a 20-year period when I observed over 1100 nests of Eastern Bluebirds and 150 nests of Carolina Chickadees in nest boxes, I found only 24 titmice nests. Natural nest cavities used vary from 1 to 7 m above ground and average 3.7 m (s.d. = 2.0, n = 17).

Titmice are monogamous, and a pair may use the same nest cavity for more than one year (Pitts pers. obs.). They construct their nests of a variety of materials, including dry leaves, green and dead moss, fragments of snake skin, and numerous kinds of animal fur. Some titmice do not hesitate to remove hair from live animals, including humans (Goertz 1962, Eshbaugh and Eshbaugh 1979). Eggs were laid in 48 of 57 nests (84%) in April; the earliest laying date was 2 April. The average clutch size of 65 Tennessee nests was 5.62 (s.d. = 1.18) with a range of 3–8. Clutches of 5–7 eggs were most common. As in many other species, the largest clutches are normally laid early in the year and by experienced females (Pitts pers. obs.). Clutches are also larger (t-test, p = 0.0001) in nest boxes ($\bar{x}$ = 5.83, s.d. = 1.13, n = 57) than in natural cavities ($\bar{x}$ = 5.05, s.d. = 1.16, n = 20); the reasons for this are not clear. The few nests initiated in May and early June are mostly attempts to replace earlier nests that were destroyed; only rarely does a pair have 2 successful nests in a year. The female incubates the eggs, and during this time she may be fed by the male. Incubation usually lasts 13–14 days (Laskey 1957).

Both adult titmice feed the nestlings, and occasionally a third adult will help care for the young (Tarbell 1983). The young typically remain in the nest for 17–18 days. Young titmice remain with their parents for several weeks after fledging (Van Tyne 1948) and sometimes throughout the winter (Laskey 1957). When the female is incu-

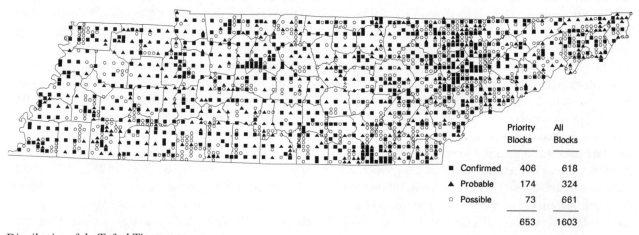
Distribution of the Tufted Titmouse.

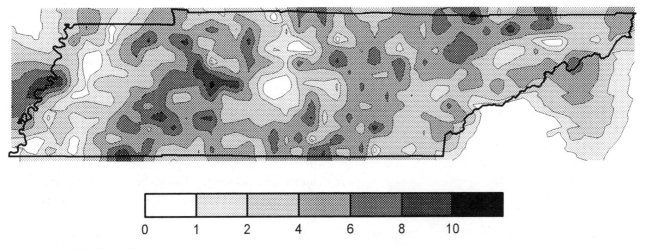

Abundance of the Tufted Titmouse.

bating eggs or brooding young, she is reluctant to flush and, as noted by several observers, may have to be lifted off the nest in order to determine the nest contents. She will also repeatedly give the "snake display" (see Carolina Chickadee account). Females that have been handled early in the nest cycle are more likely to abandon the nest (Pitts pers. obs.). Nest success in Tennessee is high: 75% of 293 eggs hatched and 61% of the eggs produced fledglings. In nests where the fate of the eggs was known, only 10 of 48 nests (21%) did not produce any fledglings. Causes of nest failure include predation by rat snakes and raccoons and occasional usurpation by bluebirds. Bird blowflies laid eggs in at least 21% of titmouse nests in bluebird boxes (Pitts pers. obs.). Cowbird parasitism of titmice has been documented in other states (Friedmann 1963) but not in Tennessee.—*T. David Pitts.*

Red-breasted Nuthatch
Sitta canadensis

The Red-breasted Nuthatch is a fairly common permanent resident of the Unaka Mountains and an irruptive winter resident elsewhere in Tennessee, especially in the east (Tanner 1985). Within the Unaka Mountains, it breeds in mature, high-elevation mixed and coniferous forests, where it is one of the most characteristic bird species. Outside of the mountains, the Red-breasted Nuthatch is usually present from September into early May (Robinson 1990), in numbers varying greatly from year to year depending on conifer seed crops (Larson and Bock 1986).

The first Tennessee record of the Red-breasted Nuthatch was by Fox (1882), who observed it in early spring 1882 in Hamilton County. The first breeding season report was by Rhoads (1895a) from Roan Mountain. The largest population is in the Great Smokies, where the first nest in the state was found in 1925 (Ganier 1926, Stupka 1963). Eight of the 13 Atlas block records and all of the 4 miniroutes recording the nuthatch, at an average relative abundance of 2 stops/route, were in the Smokies. A small population in the Unicoi Mountains southwest of the Smokies was reported by Ganier and Clebsch (1946), who found 2 birds in virgin hemlocks at 1340 m, a short distance into North Carolina. A pair was found in the same area on 17 June 1981 by McConnell and McConnell (1983); whether these birds were in Tennessee or North Carolina is uncertain. The Atlas observation in the Unicois was in large hemlocks at the Falls Branch Scenic Area, about 6 km west of the state line (Nicholson pers. obs.).

Red-breasted Nuthatch. Elizabeth S. Chastain

Breeding Red-breasted Nuthatches occur in spruce-fir forest, where they are generally most numerous, as well as hemlock, mixed hemlock-northern hardwood forest, and high-elevation pine forests (Stupka 1963, Kendeigh and Fawver 1981, McNair 1987c). They normally occur at elevations above about 1070 m, although individuals have been reported as low as 885 m during June (Stupka 1963). Following completion of breeding, nuthatches may move downslope as low as 550 m by early August (Nicholson pers. obs.). The only proven breeding record for the Red-breasted Nuthatch outside the Unaka Mountains was in pine woods in the Ridge and Valley at Knoxville at around 300 m, where recently fledged young were observed in early June 1977 (Owen 1979a).

Most Tennessee Red-breasted Nuthatch census data are from various habitats in the Great Smoky Mountains where peak breeding densities in virgin spruce-fir forest reached 50–78 pairs/100 ha (Fawver 1950, Kendeigh and Fawver 1981); Rabenold (1978) reported an unusually high density of 100 pairs/100 ha, an estimate that probably needs to be confirmed. Following a severe balsam woolly adelgid infestation, a replicated census in virgin spruce-fir forest (dominated by Fraser fir) on Mt. Guyot determined that the nuthatch density increased to 25 pairs/100 ha in 1985, compared to a slightly lower density of 20 pairs/100 ha in 1969 (Alsop and Laughlin 1991). The remaining scattered standing mature red spruce, an abundance of standing dead firs for nest-sites, and an apparent bountiful supply of insects were sufficient to support a substantial number of breeding Red-breasted Nuthatches. The nuthatch population there will probably decrease to below pre-infestation levels in the near future when the dead trees fall, as occurred on Mt. Mitchell, North Carolina (Adams 1959, Hammond and Adams 1986). Red-breasted Nuthatch densities in other habitats include 20 pairs/100 ha in second-growth red spruce (Mayfield 1993), 18/100 ha in Table Mountain-pitch pine heath (Kendeigh and Fawver 1981), 21–27/100 ha in oak-dominated forests (Wilcove 1988), and 13–42/100 ha in hemlock-northern hardwood forests (Wilcove 1988).

Both census and anecdotal data indicate that numbers of breeding nuthatches may fluctuate greatly, even in the absence of apparent habitat change (see McNair 1987c). The size of breeding populations, which primarily feed upon insects (Anderson 1976), may be partially determined by food supply during the previous fall and winter when nuthatches are heavily reliant upon conifer seed crops (cf. Larson and Bock 1986). The absence of conifer seeds during some winters would account for the simultaneous absence of nuthatches in the Great Smokies (Stupka 1963) and the nearby Black Mountains of North Carolina (Burleigh 1941). However, breeding nuthatches are regularly present in large areas of mature spruce-fir forest in the Great Smokies and Black Mountains, even following their absence the previous winter (Burleigh 1941, Stupka 1963, Nicholson pers. obs.), which underscores their lack of dependence on conifer seeds during the breeding season.

In contrast to the Great Smokies, the spruce-fir forest on Roan Mountain is much less extensive and was heavily logged by the late 1930s. Red-breasted Nuthatches have occasionally been absent there during the breeding season (Phillips 1979b), although the majority of breeding-season surveys there have found them in rather low numbers (e.g., Rhoads 1895a, Bruner and Field 1912, Ganier 1936).

Breeding Biology: Active nests of Red-breasted Nuthatches in Tennessee have been described in spruce-fir forest on Roan Mountain (Ganier 1936) and in the Great Smokies (Ganier 1926). Of the 6 confirmed Atlas observations, 5 were of adults with fledglings; the

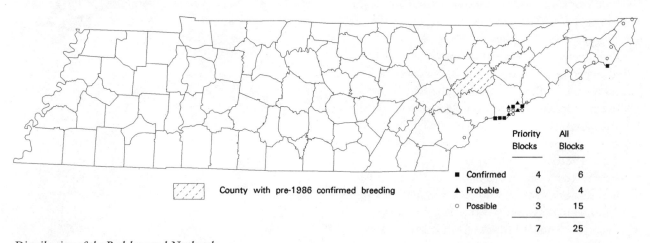

Distribution of the Red-breasted Nuthatch.

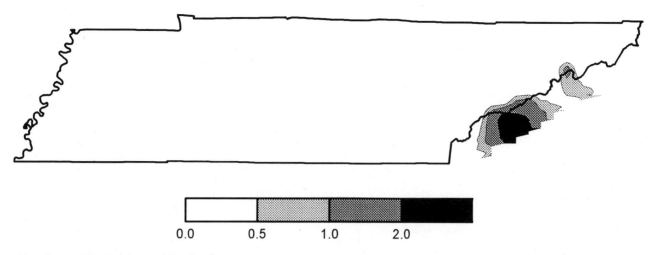

Abundance of the Red-breasted Nuthatch.

other confirmed record, from Roan Mountain, was of an adult carrying food. The Red-breasted Nuthatch usually excavates its nest cavity in a snag, at heights of 2.1 to 6.7 m ($\bar{x}$ = 4.3, s.d. = 2.2, n = 4) in Tennessee. The nest is built primarily of shredded bark within the cavity, and the cavity entrance is usually smeared with sap (Ganier 1936, Tyler 1948b). Based on observations of fledglings, egg laying in the Unaka Mountains begins during the last third of May. The pair that nested at Knoxville probably laid eggs by early May. The 2 clutches reported in Tennessee contained 5 and 6 eggs. The incubation and nestling periods are given by Tyler (1948b) as 12 days and 14–21 days, respectively, although both are poorly studied.—*Douglas B. McNair and Charles P. Nicholson.*

White-breasted Nuthatch
Sitta carolinensis

The White-breasted Nuthatch is a chunky, mostly gray-and-white bird with the unusual habit of frequently hitching headfirst down tree trunks. It frequents bird feeders, where it is easily recognized by its acrobatics. It is a fairly common, permanent resident occurring across Tennessee in mature, primarily deciduous woodlands, where it nests in cavities and feeds heavily on acorns and other nuts during the winter.

Historically, the White-breasted Nuthatch was probably a common resident of the primeval forests that once covered most of the state. Its numbers would have decreased with the widespread forest clearing begun by eighteenth- and nineteenth-century European settlers. It was known by the Cherokee Indians (Mooney 1972:281), and in the late nineteenth century Rhoads (1895a) described it as sparingly distributed across the state except for the high-elevation coniferous forests of Roan Mountain, where it was absent. Its population probably reached its lowest point during the first half of the twentieth century, when the area of forest was lowest. Ganier (1933a) described it as fairly common in West, Middle, and East Tennessee.

The White-breasted Nuthatch population has shown a recent increase, probably in response to recent increases in both the forested area and the age of the forests. BBS routes recorded a significant (p < 0.01) increase of 8.3%/year from 1966 to 1994. This is one of the greatest increases of any Tennessee bird. Christmas Bird Counts also show increased numbers (Tanner 1985). Its range-wide population has also significantly (p < 0.01) increased, at an average rate of 2.1%/year, from 1966 to 1993 (Peterjohn, Sauer, and Link 1994).

Atlasers found the White-breasted Nuthatch in 77% of the completed priority blocks. No major change from

White-breasted Nuthatch. Elizabeth S. Chastain

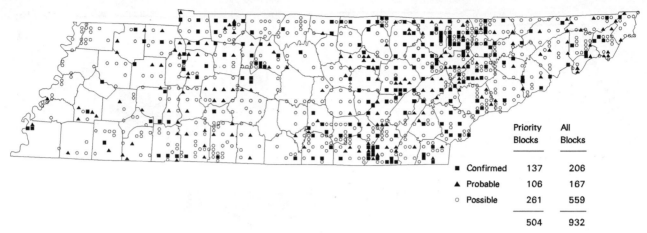

		Priority Blocks	All Blocks
■	Confirmed	137	206
▲	Probable	106	167
○	Possible	261	559
		504	932

Distribution of the White-breasted Nuthatch.

its historical range is evident, although it is now absent or rare in much of the Loess Plain and Central Basin. It is also uncommon in parts of the Coastal Plain Uplands and Ridge and Valley. Because its calls and movements on tree trunks are easily detected, Atlas results probably give an accurate picture of its distribution. It was probably missed in some blocks where the limited forested area was not easily accessible. The nuthatch was recorded on 44% of the miniroutes at an average relative abundance of 1.9 stops/route. Its highest local abundance was on the southern Cumberland Plateau, an area of extensive oak-hickory forests. The physiographic regions with the highest average abundance and highest proportion of routes reporting the nuthatch were the Cumberland Plateau ($\bar{x} = 2.7$, on 62 of 71 routes) and the Cumberland Mountains ($\bar{x} = 2.2$, 12/16). Numbers in the heavily forested Unaka Mountains were somewhat lower than average, due in part to its reduced numbers at high elevations. It occurs throughout the elevational range of the oak forests (Stupka 1963).

The probability of the White-breasted Nuthatch occurring in a given woodland increases with the woodland area until it reaches a maximum at about 300 ha (Robbins, Dawson, and Dowell 1989). Its probability of occurrence drops rapidly in woodlands less than 3 ha in area. This tolerance of moderately small woodlands helps explain its occurrence, although in very low numbers, in several counties with a low proportion of forest. The White-breasted Nuthatch also appears to avoid cedar forests, as suggested by the few Atlas records from the inner Central Basin and its absence in the cedar forests censused by Fowler and Fowler (1983a, 1983b, 1983c). Densities in other forested census plots in Tennessee, expressed as pairs/100 ha, include 11 in virgin, mid-elevation Unaka Mountain cove hardwoods (Aldrich and Goodrum 1946), 12 in each of 2 Cumberland Mountain plots dominated by maple, gum, and hickory (Smith 1977) and by oak and maple (Turner and Fowler 1981a), and 7 in Central Basin oak-maple (Fowler and Fowler 1984a). Butts (in Tyler 1948a) gave typical breeding territory sizes of 10 to 12 ha elsewhere, similar to the Tennessee densities.

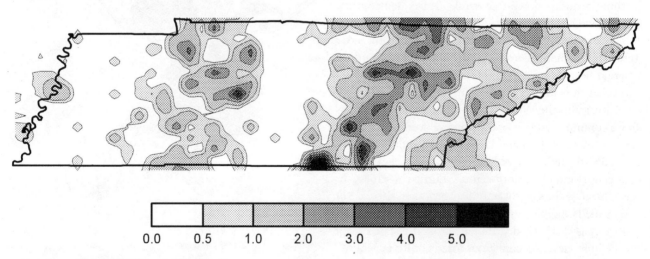

Abundance of the White-breasted Nuthatch.

Breeding Biology: The White-breasted Nuthatch breeds in monogamous pairs and begins breeding activities in late winter with loud calling, singing, bowing displays, and later mate-feeding by the male (Kilham 1972). The female selects a natural cavity, old woodpecker hole, or occasionally a nest box and packs it with twigs, fur, hair, feathers, and bark shreds, beginning as early as 6 March. One Tennessee nest was constructed of litter from a gray squirrel *(Sciurus carolinensis)* nest (Tarbell pers. obs.). The few Tennessee records of nests in sites other than boxes include knotholes in oak and hackberry trees 10.7 m high and an abandoned Red-cockaded Woodpecker cavity 7 m up in a Virginia pine (Duyck, McNair, and Nicholson 1991, Nicholson pers. comm.). Nests as high as 18 m have been reported elsewhere (Tyler 1948a, Ritchison 1981).

Detailed records of completed nuthatch clutches are limited and include 1 set of 6 eggs and 2 sets each of 7 and 9 eggs. Clutch size elsewhere ranges from 5 to 10 eggs, with 8 the most common number (Tyler 1948a). The eggs are white with rufous spotting. First laying appears to peak in late March to early April. Parasitism by Brown-headed Cowbirds, rare elsewhere (Friedmann, Kiff, and Rothstein 1977), has not been reported in Tennessee. The female incubates the eggs for 12 to 14 days; during this period she is often fed at the nest by the male (Tyler 1948a, Ritchison 1981). Both parents feed the nestlings and fledglings, which are conspicuous and are fed by the parents for at least a couple of weeks after fledging. The nuthatch is single-brooded but renests after nest failure, as noted in Tennessee by Patterson following predation of a nest in a box by a corn snake *(Elaphe guttata)* (NRC). The family often remains together into the fall (Tyler 1948a).—*Ann T. Tarbell.*

Brown-headed Nuthatch
Sitta pusilla

Brown-headed Nuthatches are permanent residents of southeastern pine forests from eastern Texas to Florida, north to Delaware and west to Arkansas (Bent 1948, Norris 1958). Until recently, their range did not include Tennessee. Prime habitat is mature longleaf pine forest which does not occur in the state (Bent 1948, Norris 1958). Brown-headed Nuthatches are secondarily associated with mature, less than 35-year-old loblolly pine forest (Myers and Johnson 1978). In Tennessee, the original, native range of loblolly pine extends northward in the southern part of the Ridge and Valley province to central Rhea, Meigs, and McMinn Counties, though shortleaf and Virginia pines are often dominant in early successional stands in this transitional oak-pine region. It is within this native range of loblolly pine that the Brown-headed Nuthatch has recently established itself as a locally uncommon resident, derived from established populations in north Georgia (Burleigh 1958, Tanner 1988). In Tennessee, they prefer open loblolly and shortleaf pine forest, though they also occasionally occur in Virginia pines (Haney 1981, Atlas results).

Brown-headed Nuthatches use open pine stands with minimal encroachment of a hardwood midstory (Norris 1958, O'Halloran and Conner 1987). In Tennessee, Brown-headed Nuthatches occur in human-modified habitats, including residential areas landscaped with sufficient pines, managed pine forests at recreation areas where the pines have been selectively thinned, and pine stands that are grazed and periodically burned (Haney 1981, Atlas results). These habitats are probably inferior to natural, unmodified habitats, though the open understory would be a feature of fire-maintained natural stands (see McNair 1984d).

The first Brown-headed Nuthatch in Tennessee was recorded in 1968 in southeastern Hamilton County (Basham 1969), which has continued to be the nucleus of their Tennessee range. Breeding populations became established by the late 1970s; the first nest record was discovered at McDonald in 1977 (Haney 1981). An active nest was found in adjacent Bradley County in 1985 (Robinson 1990; C. P. Nicholson, in litt.), where breeding was also confirmed during the Atlas. In Polk County, breeding was confirmed (dependent fledglings) for the first time near Parksville on 18 June 1990 (Atlas results). Brown-headed Nuthatches have also been recorded during early winter in southern Meigs County (Tanner 1988). The only extralimital record of this species is 5 birds on 29 July 1974 at Fall Creek Falls SP at 550 m in Van Buren County, on the Cumberland Plateau

Brown-headed Nuthatch. Elizabeth S. Chastain

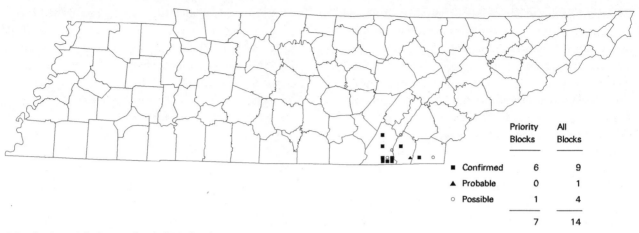

	Priority Blocks	All Blocks
■ Confirmed	6	9
▲ Probable	0	1
○ Possible	1	4
	7	14

Distribution of the Brown-headed Nuthatch.

(Haney 1981), where they have not been reported again. During Atlas fieldwork, the species was found in 14 blocks and at 1 stop of 1 miniroute; the growing breeding population is now entrenched at elevations of 220 to 300 m (Haney 1981, Tanner 1988, Atlas results).

The timing of the breeding range extension to Tennessee (Haney 1981) coincided with a reinvasion of Brown-headed Nuthatches to other areas, including the French Broad River Valley of Buncombe County in the Blue Ridge Mountains of North Carolina, which were reoccupied after a lapse of about 75 years (Simpson 1969, Whitehurst 1986, Duyck and McNair 1990). The range-wide population, however, did not show a significant trend on 1966–93 BBS routes (Peterjohn, Sauer, and Link 1994).

The range of the Brown-headed Nuthatch is more restricted in Tennessee than the historical range of another pine-obligate species, the Red-cockaded Woodpecker, which has primarily occurred in shortleaf pine forest. It is puzzling that Brown-headeds have not occupied at least some of these sites, because they can inhabit smaller, younger stands than Red-cockaded Woodpeckers. Despite numerous searches, Brown-headed Nuthatches have never been found in Hardeman and McNairy Counties, where highly favorable shortleaf pine habitat exists, even though they are present as close as 16 km south of Tennessee in Mississippi where loblolly pine is present (Haney 1981). Native loblolly pine may once have extended into southwest Tennessee (Defebaugh 1906), though this fact is disputed; planted loblolly pine is now widespread there. In contrast, Brown-headed Nuthatches have dispersed to Buncombe County, North Carolina, where shortleaf pines are present and native loblolly pines are also absent. Evidently Brown-headed Nuthatches are dependent upon the proximity of loblolly pine forest and a corridor of continuous yellow pine habitat for effective dispersal to other sites, even where loblolly pine may be locally absent.

We do not expect the Brown-headed Nuthatch to expand its range much beyond the native limit of loblolly

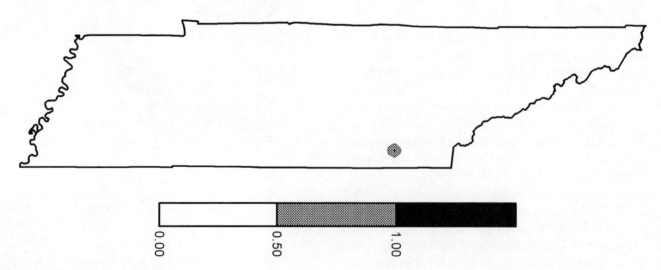

Abundance of the Brown-headed Nuthatch.

pine forest in Tennessee. Within their present range, local extinctions (Haney 1981, Tanner 1988) and occupation of new sites will probably occur, depending upon the availability of suitable nesting habitat. While a future breeding range extension to McMinn, Meigs, or Rhea Counties would not be surprising, the presence of a pair with fledglings near Kingston, Roane County, in 1995 was unexpected.

Breeding Biology: Breeding of Brown-headed Nuthatches is relatively easy to confirm. Cavity excavation is a conspicuous activity, for Brown-headeds usually excavate their own cavities, and nest-sites can be located in open areas some distance from foraging areas (McNair 1984d). However, they prefer nesting within intact forest in partially rotted pine stumps or posts (see McNair 1984d). The mean height of the cavity entrance at rather diverse nest sites in Tennessee was 3 m (s.d. = 2.6 m, n = 7) (Haney 1981); this mean included 1 nest box but did not include a pair nest building 9 m up in a loblolly pine in an old woodpecker hole, an unusual nest site (see McNair 1984d). Nuthatches commence breeding quite early; nest construction has been observed as early as 21 February and clutches from mid-April to mid-May. They are usually single-brooded, though renesting may occur occasionally, and second broods have been reported (McNair 1984d). In Tennessee, clutches have been of 4, 4, and 6 eggs (Haney 1981). The incubation period lasts 14 days, and the young fledge in about 18.5 days (McNair 1984d).—*Douglas B. McNair.*

Brown Creeper
Certhia americana

Brown Creepers are small, brown birds of mountain, swamp, and bog forests throughout northeastern and western North America. They are silent and methodical while foraging and, as a result, often inconspicuous. Suddenly, however, a creeper may flutter like a falling leaf to the base of a nearby tree trunk, or sing its beautifully wild song which slips away "in an indescribable plaintive cadence like the soft sigh of the wind among the pine boughs" (Brewster in Pearson 1917). Although a common winter resident across Tennessee, nesting creepers are common at high elevations in the eastern mountains and rare and occasional elsewhere in the state.

Traditionally, the creeper's breeding stronghold in Tennessee has been in the Blue Ridge, where it rarely occurs below about 1000 m elevation during the summer (McNair 1987c) and is most numerous in the spruce-fir forest. Confirmed nesting in Tennessee was first described by LeMoyne (1886), who discovered a nest in the Smokies

Brown Creeper. Elizabeth S. Chastain

in 1885. The first breeding season report outside of the Unaka Mountains was in 1937, when Pickering (1937) observed a bird at Reelfoot Lake on 10 May, unusually late for a migrant. Creepers had regularly bred in southeast Missouri at the turn of the century (Widmann 1895, Robbins and Easterla 1992). Breeding outside of the Unakas was confirmed in 1976 at Radnor Lake, Davidson County (Bierly 1978). Robertson (1979) found a territorial male, also observed carrying food, during June 1977 in a virgin mixed-mesophytic forest in Savage Gulf, Grundy County, on the Cumberland Plateau. Criswell (1979) observed a pair nest building in 1977 and a nest with young in 1978 in Dyer County.

While conducting an extensive survey of breeding birds at 59 West Tennessee forested wetland sites, Ford (1987, 1990) found Brown Creepers at 2 sites in 1985. In 1986, creepers were observed at both 1985 sites as well as an additional 3 sites. Brown Creepers were also found at 2 sites in Tipton County in 1989 (Atlas results). These observations suggest creepers occur more frequently in West Tennessee than had been previously assumed. They may also be evidence of an increase in the breeding Brown Creeper population in the Mississippi Valley, further supported by recent observations in southern Illinois (*Amer. Birds* 38:1025), western Kentucky (Monroe, Stamm, and Palmer-Ball 1988), and in southeast Missouri, where, after an absence of over 70 years, creepers were again reported in the mid-1980s (Robbins and Easterla 1992). Some of these observations, however, are the result of the recent increase in ornithological study in Mississippi Valley forests.

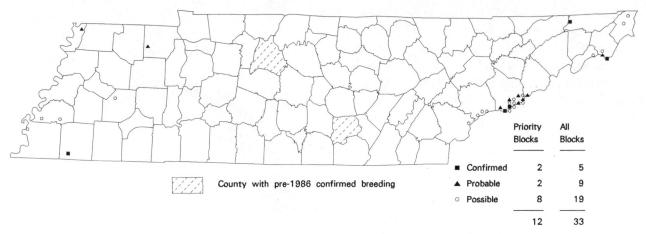

Distribution of the Brown Creeper.

Aside from the West Tennessee records, the only Atlas report of Brown Creepers outside of the Unaka Mountains was in the Ridge and Valley in Sullivan County, where a pair nested in 1987 in mixed pine-hardwoods on Bays Mountain, elevation 572 m. Outside of the Smokies, the few nests reported contained young in mid-May, before most Atlas work was conducted. Because of their inconspicuous song and calls and early nesting, it is possible that creepers were missed in a blocks outside of the Unaka Mountains. Even in the Unakas, where they are relatively common, they were not found on any Atlas miniroutes.

The highest Brown Creeper breeding densities, measured on census plots, have been in virgin spruce-fir forests. Kendeigh and Fawver (1981) found a density of 95 pairs/100 ha in Fraser fir-dominated forest on Clingman's Dome; other densities in undisturbed spruce-fir were 60/100 ha (Rabenold 1978) and 25/100 ha (Alsop and Laughlin 1991). Wilcove (1988) reported densities of 11–15 pairs/100 ha in hemlock-deciduous forest at 1300 m elevation. Most West Tennessee observations have been in open-canopied forests, flooded by beavers or other causes, with numerous dead standing trees (Criswell 1979, Ford 1987). The Davidson County nest was in a grove of boxelder, black willow, sycamore, and hackberry trees (Bierly 1978); both it and the 1987 Sullivan County nest were within a few meters of lakes.

Limited information is available on Brown Creeper population trends. The recent death of the mature fir does not seem to have greatly impacted the Unaka Mountain population. Alsop and Laughlin (1991) found an increase from 25 pairs/100 ha in 1967 to 40 pairs/100 ha in 1985, while Rabenold (1978, unpubl. ms.) found a decrease from about 60 pairs/100 ha in 1974 to about 50 pairs/100 ha in 1986. Throughout North America, Brown Creepers did not show a significant population trend from 1966 to 1991 (Peterjohn and Sauer 1993).

Breeding Biology: Brown Creepers nest under a slab of peeling bark on the trunk of a dead tree (Tyler 1948c, Davis 1978). The nests are often difficult to find, and confirming breeding was difficult for Atlas workers. Tennessee nests have been in red spruce, Fraser fir, Virginia pine, and unidentified hardwood trees, at heights of 2.0–12.2 m ($\bar{x}$ = 4.4 m, s.d. = 3.7, n = 7). The nest, constructed by both the female and male, is built in the gap between the bark slab and the tree trunk, its size conforming to the size of the gap (Davis 1978). It is constructed of fine twigs, strips of dead wood, and spider webs, lined with fine shredded barks and frequently feathers, and sometimes completely hidden behind the bark slab (Ganier and Clebsch 1938, Davis 1978). Nest building in the Unaka Mountains has been observed from late April through mid-June. Outside of the Unakas, it probably begins in late March.

Egg laying probably begins in early April outside of the Unakas and in late May in the mountains. Clutches typically contain 4–8 eggs, most frequently 5–6 (Tyler 1948c, Davis 1978). Tennessee observations include 3 completed clutches of 5 eggs and 1 of 4 eggs, as well as a brood of 6 nestlings. The eggs, white and sparsely marked with reddish brown, often in a wreath, are incubated by the female for 14–17 days. Both adults feed the nestlings, which leave the nest in 13–16 days. After fledging, the brood roosts in a tight circle with their heads inward (Davis 1978). Cowbird parasitism has not been reported in Tennessee and is very rare elsewhere (Friedmann and Kiff 1985).—*Robert P. Ford and Charles P. Nicholson.*

Carolina Wren
Thryothorus ludovicianus

The Carolina Wren is a common to abundant permanent resident of low-elevation woodlands and woodland edges across the state. Although it usually stays in or close to thick brush, the Carolina Wren gives its loud distinctive songs and calls throughout much of the year and is familiar to most Tennesseans. It was the sixth most frequently reported species during Atlas fieldwork.

The Carolina Wren was probably a common bird throughout the low elevation of prehistoric Tennessee. The first reference to it in the state was by Wilson (1811), who described it as abundant along his route from Pittsburg to New Orleans. Fox (1886) found it common in Roane County, and Langdon (1887) described it as common up to 915 m in the Smokies foothills of Blount County. Rhoads (1895a) described it as very abundant across the state except for above 1070 m on Roan Mountain.

Atlasers found the Carolina Wren in all but 6 of the completed priority blocks; the blocks where it was not found were all at high elevations of the Unaka Mountains. Within the Unaka Mountains, territorial birds have been reported as high as 1280 m (Stupka 1963); it decreases in abundance, however, above about 780 m in the Smokies and in the Cumberland Mountains (Wilcove 1988, Nicholson unpubl. data, Atlas results).

The Carolina Wren was recorded on 94% of the Atlas miniroutes at an average relative abundance of 4.8 stops/route. The highest average abundance, 7.2/route, occurred in the Ridge and Valley physiographic region, and the lowest, 3.4/route, on the Eastern Highland Rim. All of the other regions were between 4.0 and 4.7 stops/route.

Carolina Wrens occupy habitats with at least some tree cover. Within heavily forested areas, they are more common in disturbed areas such as near fallen trees and along streams (Anderson and Shugart 1974). Ford (1990) found them on 57 of 58 plots distributed across a wide range of West Tennessee forested wetlands; their average density on transect counts was 9 pairs/100 ha. Densities on census plots include between 58 and 132 pairs/100 ha in a Ridge and Valley deciduous forest (Howell 1971, 1974), 9/100 ha in virgin Cumberland Plateau mixed

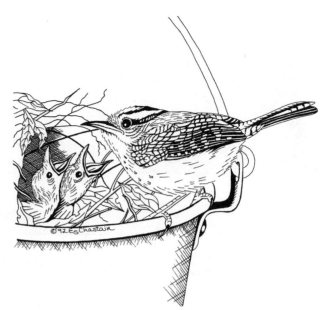

Carolina Wren. Elizabeth S. Chastain

mesophytic forest (Robertson 1979), 25/100 ha in disturbed Eastern Highland Rim mixed mesophytic forest (Simmers 1982a), 45/100 ha in Eastern Highland Rim mixed pine and mature deciduous forest (Stedman and Stedman 1992), 0 and 33/100 ha in Central Basin deciduous forest (Fowler and Fowler 1984a, 1984b), 8/100 ha in a field with hedgerows (Lewis 1975), and 25/100 ha in a year-old low elevation Unaka Mountain deciduous clearcut (Lewis and Smith 1975). Central Basin cedar forest plots had either a partial territory or no wrens (Fowler and Fowler 1983a, 1983b, 1983c).

The Carolina Wren population showed no significant overall trend on Tennessee BBS routes from 1966 to 1994 and averaged 11.2 birds/route. Its population, however, varies greatly in relation to the severity of the winter. It steadily increased to 14.7/route in 1976, then, following 2 severe winters, crashed to 5.6/route in 1978. By 1984 it recovered to slightly higher than the long-term average,

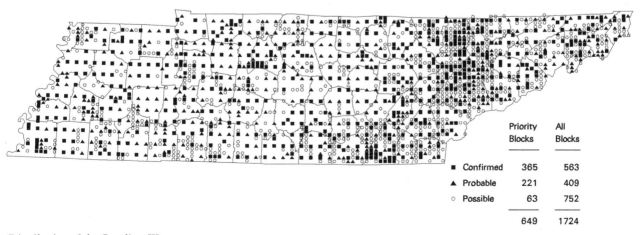

	Priority Blocks	All Blocks
■ Confirmed	365	563
▲ Probable	221	409
○ Possible	63	752
	649	1724

Distribution of the Carolina Wren.

Confirmed Breeding Species

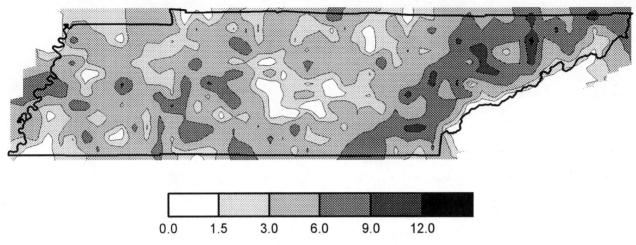

Abundance of the Carolina Wren.

then decreased to 7.5/route in 1985. Since then it increased to 19.1/route in 1991, almost twice the long-term average. The earliest such decrease noted in the Tennessee literature followed the winter of 1939–40 (Coffey 1942a). These population fluctuations occur throughout much of the species's range, often resulting in a temporary retraction of the northern range limits (Bent 1948, Robbins, Bystrak, and Geissler 1986). Stupka (1963) felt that a similar retraction of the upper limits of its elevational range occurred in the Smokies. The wren's rangewide population showed an increasing trend of 0.9%/year ($p < 0.01$) from 1966 to 1993 (Peterjohn, Sauer, and Link 1994).

Breeding Biology: The Carolina Wren has a long nesting season, normally from late March into August. Second broods are common, and a third brood is occasionally raised in Tennessee (Laskey 1939). Confirming the breeding of the Carolina Wren was fairly easy for atlasers, as evidenced by the high proportions of confirmed breeding. About half of the confirmed records were observations of fledglings, which are conspicuous and more easily observed than nests, which accounted for about one-third of the confirmed records.

The Carolina Wren nest, built by both adults, is a large, domed mass with a side entrance. It is constructed of leaves, grasses, weed stems, twigs, moss, and bark strips and is lined with feathers, rootlets, and fine grasses. Construction of the first nest of the season may begin 2 weeks or more before egg laying. The nest is usually placed in an enclosed area such as in exposed tree roots, thick shrubbery and vines, brushpiles, crevices in embankments, tree cavities, and bird boxes. They also nest on ledges and nooks of outbuildings and abandoned houses and in hanging flowerpots.

Egg laying normally begins in late March. An exception was a nest in Hamilton County with 2 young and 1 egg found abandoned in late December 1967 during unusually warm weather (DeVore 1968b). The egg was estimated to have been laid about 13 December. No clear peak of egg laying is evident. Clutch size ranges from 3 to 7 eggs, with 5 eggs most frequent and a mean of 4.73 (s.d. = 0.78, n = 103). An exceptional clutch of 12 eggs, probably the product of 2 females, was found in Grundy County on 19 May 1952 (Caldwell 1952). This nest was abandoned before the eggs hatched. There is little difference in the size of clutches initiated during March and April ($\bar{x}$ = 4.83, s.d. = 0.71, n = 40), presumably first clutches, and those of clutches initiated after the first of May ($\bar{x}$ = 4.70, s.d. = 0.80, n = 61). The eggs are white to pale pinkish with scattered brown spots, usually forming a wreath around the large end (Bent 1948). The female incubates the eggs for about 14 days, and both adults feed the nestlings, which fledge in 12 to 15 days (NRC data). A brood of fledglings at Nashville remained with its parents for 17 days after fledging (Monk in Coffey 1942a). Laskey (*Migrant* 4:37) observed a pair fledge 5 young on 17 July and complete a clutch of 4 eggs on 28 July.

Parasitism by Brown-headed Cowbirds is rare (Friedmann and Kiff 1985). The few Tennessee reports include 3 wren and 1 cowbird egg, all collected before hatching, in a relatively conspicuous nest placed in a tree crotch (Crook 1934) and 6 wren and 2 cowbird eggs in a nest in a hanging basket, from which only the cowbirds fledged (NRC data).—*Charles P. Nicholson.*

Bewick's Wren
Thryomanes bewickii

The Bewick's Wren is a rare permanent resident, less common during the winter than in other seasons. It was formerly a fairly common bird of brushy thickets in areas of otherwise sparse vegetation, rural farms, and suburban areas throughout most of the state. Its numbers have greatly declined since the 1940s, and it presently nests only in Middle Tennessee and the adjacent Coastal Plain Uplands of West Tennessee.

Although the Bewick's Wren underwent a northward range expansion in the Midwest in the late nineteenth century (Bent 1948), there is no evidence of a historic expansion of its range in Tennessee. In prehistoric Tennessee, it was probably found in brushy areas around burned areas and clearings maintained by Native Americans. Its numbers probably increased somewhat with European settlement. It readily nested around buildings; because of this habit, it was colloquially known as the "House Wren" in the South. Brewster (1886) described it as one of the most abundant birds in western North Carolina towns, nesting in nearly every outbuilding at Asheville.

The Bewick's Wren was first reported in Tennessee by Fox (1886), who found one in Roane County in April 1885. Rhoads (1895a) described it as local but present in every county he visited, and present up to 1220 m on Roan Mountain. Within the Unaka Mountains, its breeding was confirmed at 1006 m (Wetmore 1939), and singing birds have occurred up to 1490 m on the North Carolina side of the Smokies (Stupka 1963). Ganier (1933a) described it as fairly common in West Tennessee, common in Middle Tennessee, and fairly common in East Tennessee.

A decline in the wren's numbers apparently began a few years later (Coffey 1942a). By 1948, it was rare in northeastern Tennessee (Tyler 1948), and by 1965 it had almost disappeared from formerly occupied areas in Nashville (Laskey 1966). The decline was described in more detail by Alsop (1979b). Between 1966 and 1994, the Bewick's Wren significantly ($p < 0.01$) decreased by 22.0%/year on Tennessee BBS routes. This decline, the largest of any Tennessee bird, has also been widespread in eastern North America (Robbins, Bystrak, and Geissler 1986). The Appalachian population, often considered a distinct subspecies, *T. bewickii altus*, was, until recently, a candidate for Federal listing as Endangered or Threatened (USFWS 1994, 1996). The statewide Bewick's Wren population is listed as Threatened in Tennessee (TWRA 1975).

Several reasons have been suggested for the decline in Tennessee, including competition from the House Wren (Tyler 1948), severe winter weather (Alsop 1979b), and loss of habitat (Laskey 1966). Although competition with the House Wren does occur, as witnessed by B. P. Tyler (1948), it is inadequate to explain the widespread, general decline, as the decline was underway in

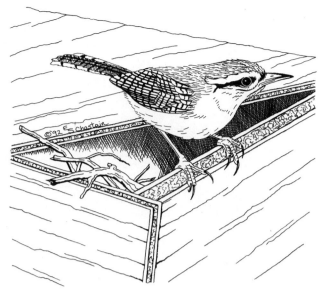

Bewick's Wren. Elizabeth S. Chastain

much of the state prior to the expansion of the House Wren's breeding range into Tennessee (e.g., Laskey 1966). The House Wren is also not yet established in some parts of the state where Bewick's Wrens no longer occur. While declines occurred after the severe winters of 1939–40 and 1950–51 (Alsop 1979b), it is unlikely they were a major factor in the long-term decline. Nesting habitat, in the form of brushpiles and unkempt outbuildings, has decreased in urban and suburban areas (e.g., Laskey 1966), but is still common in rural areas. The major causes of the wren's decline remain unknown.

During Atlas fieldwork, the Bewick's Wren was found in 36 blocks, about half of them completed priority blocks. The Central Basin and some Western Highland Rim records were in the suburban and rural farmyard habitats where the wren has typically been reported in the past. Seven singing males were found in this habitat in a Houston County block. Several other Western Highland Rim and Coastal Plain Uplands records were in young clearcuts, where the wrens occupied windrowed slash piles, most 1 to 3 years old, resulting from preparation of the site for pine planting (Robinson 1989, Atlas results). Bewick's Wrens were found in this habitat in Stewart, Montgomery, and Wayne Counties. The only East Tennessee report was of a bird singing in similar clearcut habitat on 7 June 1988; it was not observed the following day. Unlike elsewhere in the southern Appalachians (Adkisson 1991, Simpson 1978b), the Appalachian population does not seem to have persisted longer in Tennessee at high elevations than at low elevations.

Since the Bewick's Wren population in Tennessee was listed as Threatened in 1975, it has continued to decline. It appears to be extirpated from the Appalachians and persists in very low numbers in the Central

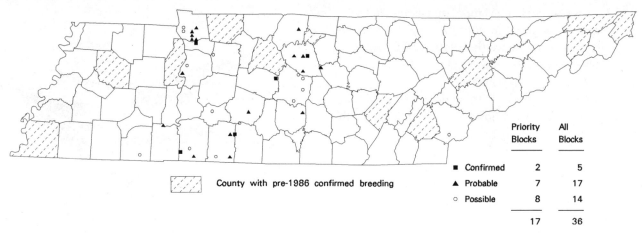

Distribution of the Bewick's Wren.

Basin and Western Highland Rim. The recent discovery of its use of slash piles in clearcuts (Robinson 1989) is the only hopeful sign for the future of this species. Research on its use of clearcuts, including detailed descriptions of occupied sites, comparisons of occupied and unoccupied sites, reproductive success, and dispersal are urgently needed. Until these studies are carried out and their results integrated into large-scale forest management plans, the Bewick's Wren should be considered Endangered in Tennessee.

Breeding Biology: Some nesting activities begin as early as late February; most birds, however, begin nesting in late March (Laskey 1946). Both birds build the nest, which is usually placed in a cavity or recess such as a bird box, mailbox, natural tree cavity, junked car, brushpile, or sheltered ledge in a building. A nest reported during Atlas fieldwork was in thick vines and saplings near the base of a low rock bluff. The nest is usually globular shaped with a side opening, constructed of sticks, grasses, and weed stems, and lined with moss, hair, and feathers.

Clutch size ranges from 4 to 9 and averages 6.09 eggs (s.d. = 1.16, n = 96). A single, unusually large (Bent 1948) clutch of 12 eggs (Todd egg coll.), perhaps laid by 2 females, is not included in this average. The female incubates the eggs for 12 days, and both adults feed the nestlings, which fledge in 13 to 14 days (Laskey 1946). At least 2 broods are normally raised. Parasitism by the Brown-headed Cowbird has been reported once; a 1966 nest in Hardin County contained 5 wren and 1 cowbird egg (NRC, Lemon 1969). Only the cowbird fledged. The Bewick's Wren is rarely parasitized by cowbirds elsewhere (Friedmann, Kiff, and Rothstein 1977).—*Charles P. Nicholson.*

House Wren
Troglodytes aedon

The House Wren is a fairly common migrant, an uncommon summer resident, and a rare winter resident of Tennessee. Spring migrants usually arrive in mid-April, and most birds have departed the state by mid-October. Migrant birds are usually present in areas of dense shrubbery; nesting birds occur in urban and suburban areas, as well as around rural homes. The House Wren is easy to identify by its loud, bubbly song, and it sings into mid-summer.

The status of the House Wren in Tennessee has changed dramatically during the past century. Rhoads (1895a) did not observe it in Tennessee. Four decades later it was a rare migrant in Middle and East Tennessee (Ganier 1933a) and was extending its breeding range southward in Kentucky and North Carolina (Odum and Johnston 1951, Mengel 1965). Herndon (1956) summarized Tennessee records of the House Wren through 1955. The first Tennessee nest record was in 1913 in Johnson City. There were apparently no other summer records for almost 20 years, when a pair was present at Johnson City in 1932 and singing birds observed on top of Roan Mountain in July 1934 (Lyle and Tyler 1934). An unmated male built a nest at Knoxville in 1944, and by the late 1940s the species was nesting in Carter, Greene, Johnson, and Washington Counties and was common in parts of Sullivan County. Successful nesting occurred at Knoxville in 1950. This East Tennessee population apparently spread into the state from southwest Virginia. Migrant House Wrens were also reported with increasing frequency elsewhere in the state during this period.

An apparently unmated House Wren was present in Nashville during the summer of 1956, and the first successful Middle Tennessee nesting was at Nashville in 1957 (Laskey 1966). These birds probably invaded

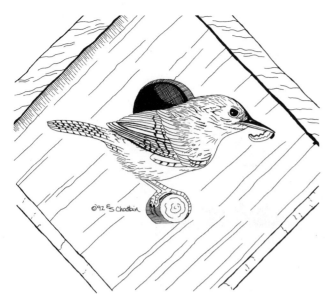
House Wren. Elizabeth S. Chastain

Tennessee from Kentucky, as the species was nesting in south-central Kentucky by the late 1940s (Mengel 1965). The Middle Tennessee population apparently increased slowly until the 1970s, and by the mid-1980s it was nesting in several towns in the northern half of Middle Tennessee (*Migrant* 56:111). The first West Tennessee nest records were from Dyer County in 1976, and nest building was reported at Memphis in 1979 (*Migrant* 50:86). The wren's spread in the Ridge and Valley southwest of Knoxville was slow to be documented, probably due in part to the few observers in that area. Nest building was first reported in the Chattanooga area in 1978 (*Migrant* 49:94).

Atlasers found the House Wren to be common in northeastern Tennessee and fairly common in the Knoxville area and north-central Tennessee. It was reported in 11% of the completed priority blocks and on 5% of the mini-routes, at an average relative abundance of 1.7 birds/route. Few were reported outside of the general range known prior to 1986, and its spread appears to have slowed or halted. Observers reported decreased numbers in some East Tennessee locations in the early 1980s (*Migrant* 54:91). Odum and Johnston (1951) hypothesized that high temperatures may limit its egg production in Georgia, and Kendeigh (1963) noted its southern range limit may be limited by May temperatures high enough to discourage egg laying. This may be limiting its establishment in south-central and southwestern Tennessee.

The House Wren is found on too few Tennessee BBS routes for its population trend to be reliably analyzed. During the 1986–91 Atlas period, it was reported from 9 routes at an average abundance of 1.9 stops and 2.3 birds/route. Local variation in its population appears high, as only 2 of these 9 routes recorded it during at least 3 of the 6 years. Throughout the North American portion of its range, the House Wren increased ($p < 0.01$) by 1.5%/year from 1966 to 1993 (Peterjohn, Sauer, and Link 1994).

At present, nesting House Wrens most frequently occur in urban and suburban areas with shade trees, shrub thickets, and expanses of mowed lawn. Nesting regularly occurs around houses up to 1372 m on Roan Mountain, and in 1987 a pair nested there at Carver's Gap, elevation 1680 m (*Migrant* 58:147). Singing birds were reported at Whigg Meadow, in the Unicoi Mountains at 1513 m elevation, in 1980 and 1981 (McConnell and McConnell 1983). The species was not found there during Atlas fieldwork. In western North Carolina, nesting has been confirmed as high as 1708 m (McNair 1987c).

Breeding Biology: The male House Wren arrives before the female in the spring, usually returning to the territory occupied the previous year (Gross 1948). He selects potential nest sites and begins filling one or more of them with sticks. The House Wren typically

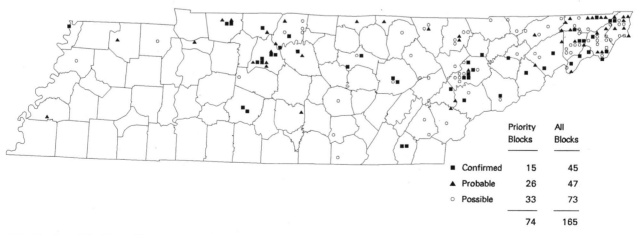

Distribution of the House Wren.

	Priority Blocks	All Blocks
■ Confirmed	15	45
▲ Probable	26	47
○ Possible	33	73
	74	165

Confirmed Breeding Species

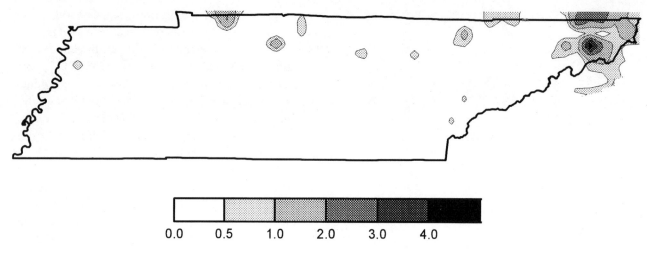

Abundance of the House Wren.

nests in a cavity, such as a natural cavity in a tree, old woodpecker hole, or bird house. Other fully or partially enclosed sites are occasionally used. Descriptions of few nests in Tennessee are available, and most of them have been in bird houses; other sites used include old woodpecker holes, the space between logs forming the walls of a cabin, and between the studs of a garage wall. The highest Tennessee nest was 4.6 m above ground. After the female arrives and selects a nest site, both adults complete construction of the nest (Gross 1948). The cavity is almost filled with twigs, among which a nest cup, lined with feathers, grasses, and other materials, is built. Laskey (1966) counted 922 twigs in a wren nest in a bluebird box. The male House Wren is occasionally polygamous, with more than 1 female nesting on a territory (Gross 1948), possibly a function of the male's habit of building several nests.

Egg laying in Tennessee begins in early May and extends into July. Apparently completed clutches include 4 of 4 eggs, 1 of 5, 2 of 6, 1 of 7, and 1 of 10. At least 1 brood of 8 young has also been reported. Most clutches reported elsewhere are of 6–8 eggs (Gross 1948). The female incubates the eggs for an average of 13 days, and both adults feed the young, which fledge in 15–16 days. The fledglings remain with the parents about 12 days (Gross 1948). The House Wren is typically double-brooded, and second broods have been reported in Tennessee, often in the same box as the first brood (Laskey 1966, Nicholson pers. obs.). Nest parasitism by Brown-headed Cowbirds is very rare (Friedmann and Kiff 1985) and has not been reported in Tennessee.

The House Wren has long been known to occasionally destroy the eggs or young of other open and cavity-nesting species, including other House Wrens (Belles-Isles and Picman 1986). This behavior is most pronounced in unmated males and females that have not yet begun clutches, and it may function to reduce competition for food and nest sites. Other than attacks on Bewick's Wren nests (see Bewick's Wren account), this behavior has not been reported in Tennessee.—*Charles P. Nicholson.*

Winter Wren
Troglodytes troglodytes

The winter wren is a small, dark, mouselike bird that is much more often heard than seen. It nests throughout much of boreal North America and Eurasia and is best known for its song, a long, series of warbles and trills, one of the most complex, repeatable songs of any songbird (Kroodsma 1980). The Winter Wren is a locally common permanent resident of the high elevations of the Unaka Mountains and an uncommon winter resident elsewhere in Tennessee. Although a few wrens remain at high elevations throughout the winter, most migrate to lower elevations (Stupka 1963). Wintering birds are present throughout the state from October into early April (Robinson 1990).

Winter Wrens nesting in the southern Appalachians were described as a distinct subspecies, the Southern Winter Wren, *T. troglodytes pullus,* by Burleigh (1935a). The Winter Wren occurs in cool mesic forest and reaches its highest density in spruce-fir. The presence of rotten stumps, fallen trees, and upturned roots, used for foraging and nesting, is an important habitat feature (Bent 1948). The wren is common from the highest elevations down to about 1220 m and uncommon as low as 730–800 m (Stevenson and Stupka 1948, Tanner 1955, Wilcove 1988). At the lower elevations, it is usually found in hemlock along streams. Densities on census plots in deciduous and mixed forests include 12–44 pairs/100 ha in cove forest, 24–50/100 ha in hemlock-deciduous forest (Kendeigh and Fawver 1981, Wilcove 1988), and

Winter Wren. Elizabeth S. Chastain

The Winter Wren was found on 5 Atlas miniroutes at an average relative abundance of 5.4 stops/route. Three of these routes were in the Great Smoky Mountains; it was found at all stops on a route in the Mt. Guyot area entirely within spruce-fir forest.

Information on population trends of the Winter Wren is limited, as it is not recorded on Tennessee BBS routes. Wilcove's (1988) census results suggest an increase in the Smokies from the late 1940s to early 1980s. Alsop and Laughlin (1991) found a decrease from 83 to 70 pairs/100 ha in a virgin spruce-fir stand following death of most of the mature fir. Wren numbers, both in the Smokies (Stupka 1963) and elsewhere (Robbins, Bystrak, and Geissler 1986), often decrease for a year or more following unusually cold winters. The wren's North American population did not show a significant long-term trend from 1966 to 1993 (Peterjohn, Sauer, and Link 1994).

5/100 ha in mature northern hardwoods (Mitchell and Stedman 1993). Within spruce-fir forests, reported densities have been 7.5/100 ha in early successional stands, 73/100 ha in mid-successional stands, and 83–85/100 ha in climax stands (Kendeigh and Fawver 1981, Alsop and Laughlin 1991).

During Atlas fieldwork, the Winter Wren was found in 33 blocks. Because the male sings loudly and frequently, the Atlas results are probably an accurate reflection of its breeding distribution. The Winter Wren was not found on Big Frog Mountain in Polk County, where Nicholson and Tanner (unpubl. data) found a singing bird in late May 1983. Whether this bird was part of a local breeding population is unknown; Winter Wrens were not found there in 1937 (Wetmore 1939). Otherwise, there is no evidence of a change in the wren's breeding distribution since it was first reported during the summer by Rhoads (1895a).

Breeding Biology: The breeding biology of the Winter Wren in the southern Appalachians is poorly known because of its secretive habits and well-concealed nest. All 3 of the confirmed Atlas records were of fledgling wrens, and only 1 Tennessee nest has been described (Ganier 1962).

Winter Wrens usually return to high elevations by late April and males sing vigorously into July (Stupka 1963). Males in Eurasia are highly polygynous (Armstrong 1955); whether this behavior occurs in eastern North America is unknown (Kroodsma 1980). The nest is built into a hollow or cavity in upturned tree roots and rotten stumps (Bent 1948). The nest described by Ganier (1962) on Mt. Le Conte was tucked among moss-covered tree roots on the slope of a ravine. The male constructs several rudimentary nests with a base of fine twigs, leaves, and moss (Bent 1948, Armstrong 1955). The female completes the nest with a dome of moss and other plant material and a lining of fur or feathers. Ganier's (1962) nest was lined with Ruffed Grouse feathers.

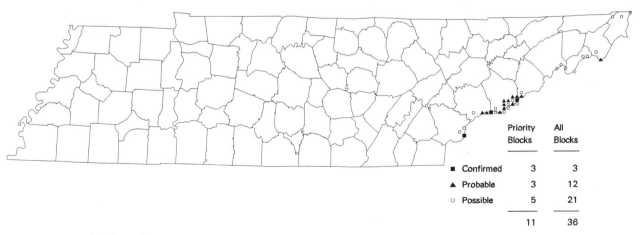

Distribution of the Winter Wren.

	Priority Blocks	All Blocks
■ Confirmed	3	3
▲ Probable	3	12
○ Possible	5	21
	11	36

Confirmed Breeding Species

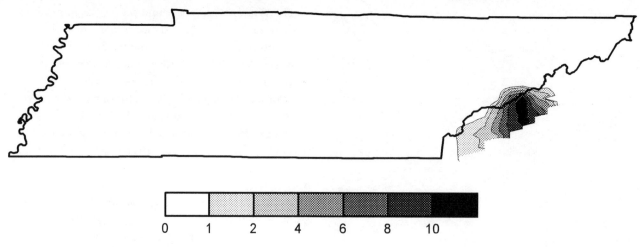

Abundance of the Winter Wren.

Clutches usually contain 5–6 eggs, pure white or marked with fine brown spots (Bent 1948). Ganier (1962) observed 5 nestlings, which fledged on 21 June; egg laying had probably begun about 15 May. Lyle (egg coll. data) observed a brood of at least 6 fledglings. The female incubates the eggs for 14–17 days; both adults feed the nestlings, which fledge in about 19 days (Bent 1948, Armstrong 1955).—*Barbara H. Stedman.*

Golden-crowned Kinglet
Regulus satrapa

The Golden-crowned Kinglet is a tiny but hardy breeding bird of northern and montane coniferous forests. Its southeasternmost breeding limit is reached in the mountains of Tennessee and the Carolinas, where it is a locally common summer resident and an uncommon winter resident of high-elevation coniferous forests. This species is also a common winter resident statewide, arriving in October and departing by April. The Golden-crowned Kinglet is very tame and easily identified during the winter by its high-pitched call notes. During the breeding season, however, it is often difficult to observe and its song, although distinctive, is unfamiliar to many Tennessee birdwatchers.

Probably the earliest recorded proof of breeding by Golden-crowned Kinglets in the state is that by Sennett (1887), who collected a fledged young from a brood near the summit of Roan Mountain on the Tennessee–North Carolina border on 23 July 1886. Although Sennett described it as uncommon there, Brewster (1886:178) called it "one of the most numerous and characteristic birds" of the upper slopes of the nearby Black Mountains of North Carolina. Rhoads (1895a) apparently did not find the kinglet on Roan Mountain in June 1895, although it was present there in 1911 (Bruner and Field 1912). It was not found there in June 1936 (Ganier 1936), perhaps temporarily eliminated by the logging of the spruce-fir forest a few years earlier. It was present there in the 1950s and is common there today (Eller and Wallace 1984, Atlas results). Golden-crowned Kinglets are common in the higher portions of the Great Smoky Mountains (Wetmore 1939, Stupka 1963, Alsop 1991), although the earliest forays there did not record them (Ganier 1926). This species is also common on the summit of Unaka Mountain, Unicoi County (Knight pers. obs., Mayfield 1993). This kinglet is common in high-elevation coniferous forests in western North Carolina (Simpson 1976b, McNair 1987c).

Most Atlas records were from traditional areas in the high elevations of the Smokies, Roan and Unaka Mountains. The lowest observation was at Rock Creek Park in Unicoi County, elevation 720 m, where it was first reported

Golden-crowned Kinglet. Elizabeth S. Chastain

in June 1980 (*Migrant* 51:96). New locations were in Shady Valley, Johnson County, at 850 m and 2 blocks in the Unicoi Mountains, Monroe County, at 1100–60 m. While the high-pitched vocalizations and treetop nature of this species may have resulted in a few isolated populations being missed, most atlasers working within its range were familiar with it, and the Atlas map is a good representation of the species's distribution in Tennessee.

The observations of the Golden-crowned Kinglet in Shady Valley and the Unicoi Mountains may be evidence of a recent range expansion, as each of these areas has received intensive study in the past (e.g., Ganier and Tyler 1934, Ganier and Clebsch 1946, McConnell and McConnell 1983). The Great Smoky Mountains census results of Wilcove (1988) also suggest that both an increase in range and numbers has occurred there. The Golden-crowned Kinglet has, in recent decades, expanded its breeding range into maturing low-elevation spruce plantations in the Northeast (Mulvihill 1992) and was recently reported during the breeding season for the first time in South Carolina (McNair 1987c).

Golden-crowned Kinglets breeding in the southern Appalachians primarily inhabit the high elevation (greater than 1350 m) spruce-fir forests; they also occur in hemlock or white pine forests, usually at moderately high elevation (Simpson 1976b). The Roan Mountain population and the majority of the Smokies population breed in the spruce-fir. The forest atop Unaka Mountain is all spruce. On Greenbrier Pinnacle, in the Smokies, kinglets were found in hemlocks and in a stand of Table Mountain and pitch pines at elevations of 1310–70 m (Knight pers. obs.). The kinglets in Shady Valley, Rock Creek Park, and the Unicoi Mountains were in hemlocks.

The Golden-crowned Kinglet is usually described as common in the spruce-fir belt (e.g., Stupka 1963); this description is supported by several censuses, which have found the kinglet to be the second or third most abundant species, usually only exceeded by the Dark-eyed Junco. A 1977 Roan Mountain census recorded 149 males/100 ha (Hale 1980), while 101 males/100 ha were found on nearby Unaka Mountain in 1992 (Mayfield 1993). Densities in climax GSMNP spruce-fir plots have been 95–125 pairs/100 ha (Alsop 1969, Rabenold 1978, Kendeigh and Fawver 1981) and in late successional spruce-fir, 13/100 ha (Kendeigh and Fawver 1981). Following the death of most of the mature firs due to the balsam woolly adelgid, the kinglet density on Mt. Guyot dropped from 124 to 41/100 ha (Alsop and Laughlin 1991).

The Golden-crowned Kinglet was recorded on 6 of 15 Atlas miniroute stops on Mt. Guyot, the only miniroute entirely within spruce-fir, and at 2 stops on 1 other Smokies miniroute. Its density in other habitats has been lower than in spruce-fir: 13 to 56 males/100 ha in hemlock-deciduous forest plots and 14/100 ha in beech gap forest, which presumably contained a few conifers (Wilcove 1988).

Breeding Biology: Nests of the Golden-crowned Kinglet are very difficult to find and few have been located in the southern Appalachians, although numerous reports of fledged young exist. Most of the confirmed Atlas records were also of recently fledged young. Five nests from Roan Mountain (*Migrant* 51:96, 58:147; Knight 1987) are apparently all that have been discovered in the spruce-fir belt of Tennessee and North Carolina.

The nest of the Golden-crowned Kinglet is an oblong, globular mass of woven moss, lichens, bark strips, and spider webs with a deep cup lined with fine plant fibers, hair, and feathers (Galati and Galati 1985). The nest is mainly built by the female in a conifer, most often a spruce, and may be placed near the trunk or suspended beneath a branch away from the trunk. In either case, it is in the tree's crown and sheltered by overhanging foliage. Nest

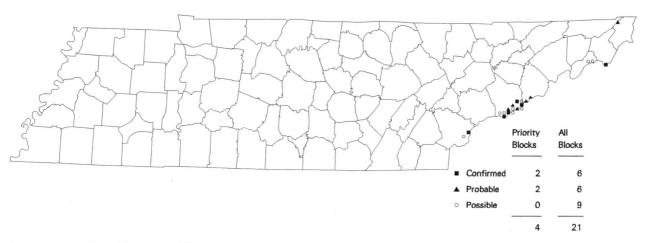

Distribution of the Golden-crowned Kinglet.

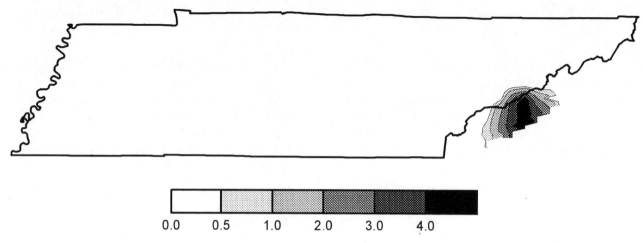

Abundance of the Golden-crowned Kinglet.

height ranges from 1.8 to 19.6 m, but usually is 9–15 m (Harrison 1975, Galati and Galati 1985). Of 5 Roan Mountain nests, 1 was about 4 m high, the others 6.5–9.6 m high (Knight 1987, F. J. Alsop III pers. comm.). No Tennessee clutch information is available; clutches elsewhere usually contain 8 or 9 eggs, a large number for such a small bird (Galati and Galati 1985). The white to cream-colored eggs, with various amounts of brown spots, are incubated by the female alone for about 15 days. Both parents tend the young, which fledge in 16–19 days; the female, however, may start building a second nest before the first brood fledges, in which case the male does most of the feeding (Galati and Galati 1985). Insufficient breeding records are available to determine whether second broods are raised in the southern Appalachians. Cowbird parasitism is uncommon (Friedmann and Kiff 1985), and unknown in Tennessee.—*Richard L. Knight.*

Blue-gray Gnatcatcher
Polioptila caerulea

The Blue-gray Gnatcatcher occurs in a variety of habitats throughout much of temperate and tropical North America. In Tennessee, it is a fairly common summer resident of woodlands and woodland edges. It is one of the first migrant songbirds to return in the spring, usually arriving by late March; most depart by late September (Robinson 1990).

Gnatcatchers nest throughout the state at elevations up to around 760 m, although their upper elevational limit is poorly documented (Stupka 1963). They are among the smallest of Tennessee's nesting birds and spend much of their time high in trees. Their frequently given and easily identified "spee" calls and high activity level, however, increase their conspicuousness, and they were probably missed in few priority Atlas blocks.

Rhoads (1895a:500) described the gnatcatcher as breeding "all across the state," and there is no evidence of a change in its range since 1895. It has probably decreased in numbers in parts of the Ridge and Valley, Central Basin, and West Tennessee because of agricultural and urban development. From 1966 to 1994, its population, as recorded on Tennessee BBS routes, did not show a significant overall trend. Rangewide, it increased by 1.0%/year ($p = 0.07$) from 1966 to 1993 (Peterjohn, Sauer, and Link 1994).

Atlasers recorded gnatcatchers on 466 (two-thirds) of the miniroutes at a mean relative abundance of 3.1 stops/route. During the same period they were recorded on all of the state's BBS routes at an average of 7.8 birds and 6.2 stops/route. The highest miniroute abundance was in the heavily forested Coastal Plain Uplands (3.7/route) and Western Highland Rim (4.7/route) physiographic regions. The lower abundance in the heavily

Blue-gray Gnatcatcher. Elizabeth S. Chastain

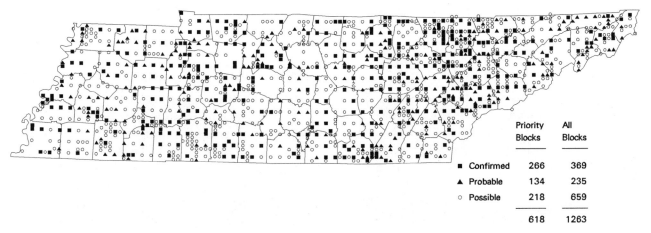

Distribution of the Blue-gray Gnatcatcher.

forested Unakas (2.5/route) is in part due to the higher average elevation of that region; elevation may also partly explain their lower numbers in the Cumberlands.

Although gnatcatchers occur in both woodland and woodland edge habitats, they are sensitive to forest area, occurring more frequently in larger forests (Whitcomb et al. 1981, Robbins, Dawson, and Dowell 1989). Quantitative habitat studies in upland deciduous forests show they are more often found in forests with open understories and large trees forming a high canopy (Anderson and Shugart 1974, Robbins, Dawson, and Dowell 1989). In West Tennessee forested wetlands, where Ford (1990) found them to be the most abundant species, they are most numerous in forests with a high basal area and few shrubs.

Breeding densities, as recorded on upland forest census plots in Middle and East Tennessee, include 10–16 pairs/100 ha in Ridge and Valley deciduous forests (Howell 1973, 1974, Smith 1975), 4–20 pairs/100 ha in Cumberland Mountain deciduous forests (Yahner 1972, Garton 1973), and 19–23 pairs/100 ha in Eastern Highland Rim mixed mesophytic forests (Simmers 1979, 1980, 1982a, 1982b, 1983, 1984). Densities in Central Basin forests include 0–16 pairs/100 ha in both cedar-dominated and upland hardwood forests (Fowler and Fowler 1983a, 1983b, 1983c, Fowler and Fowler 1984a, 1984b). Ford (1990) found an average density of 150 pairs/100 ha on transect counts in West Tennessee forested wetlands. Their highest rangewide breeding densities are also in southeastern United States bottomland forests (Ellison 1992).

Breeding Biology: Gnatcatchers have a long nesting season, raise more than 1 brood, frequently call when near the nest, and are fearless when visiting the nest. These factors contributed to the high proportion of confirmed Atlas records, 29% of all blocks and 42% of completed priority blocks, unusual for a small, canopy-dwelling songbird. Observations of nests composed 43% of the confirmed records.

Gnatcatchers arrive in Tennessee well before deciduous trees leaf out and quickly begin defending a territory. At this time, they sing their lengthy, high-pitched, relatively

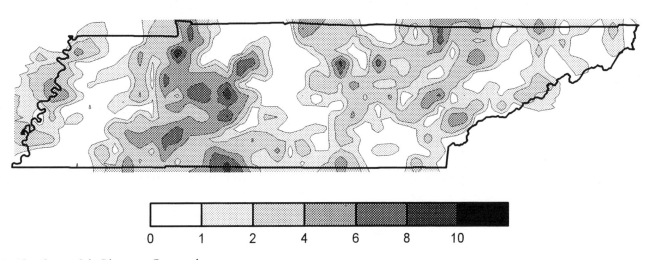

Abundance of the Blue-gray Gnatcatcher.

Confirmed Breeding Species

quiet song, consisting of variable warbles, whistles, and the more familiar "spee" calls. This song is primarily associated with courtship activities (Root 1969) and infrequently sung after early May (Nicholson pers. obs.). Unmated males frequently inspect potential nest sites and lead the female to these sites during courtship (Root 1969).

Both adults construct the nest, which is a small cup about 6 cm wide and 6 cm tall and constructed mostly of fine dead grasses and plant fibers held together with spider silk and lined with plant down, feathers, or hair. Bits of lichen are attached to the outside of the nest; these usually help camouflage the nest. Construction of the first nest of the season takes about 8 to 14 days (Ellison 1992). Construction of later nests is much faster because gnatcatchers have the unusual habit of tearing apart a previously used nest and reusing parts of it in a new nest (Ellison 1992, Nicholson pers. obs.).

The nest is usually saddled on a horizontal tree limb far from the trunk, either directly on the limb or at the junction with an upright or horizontal branch. Less frequently the nest is placed in the upright fork of a vertical limb (Ellison 1992). Gnatcatcher nests have been reported in many species of large and small trees in Tennessee. Pines (including shortleaf and Virginia), elms (including winged elm), oaks, and sweetgum are among the most frequently used trees. Todd (1944) also reported frequent use of tall, slender cedars in the Murfreesboro area. Nest height in Tennessee ranges from 1.8 to 13.7 m above ground and averages 6.4 m (s.d. = 3.1, n = 52).

Egg laying begins in early April. Clutches of 3–5 eggs have been reported in Tennessee, with an average of 4.47 (s.d. = 0.70, n = 43), very close to the range-wide average of 4.5 eggs/clutch (Ellison 1992). The eggs are pale blue to bluish white and spotted with brown. Both adults and feed the nestlings; incubation lasts an average 13 days, and the young leave the nest in about 13 days (Ellison 1992). Renesting may begin the day after the young leave the nest, with the female caring for the brood while the male constructs the new nest (Root 1969).

Blue-gray Gnatcatchers are a fairly common cowbird host (Friedmann 1963), although there are few Tennessee records. At least 2 nests containing cowbird eggs have been reported (*Migrant,* egg coll. data), and I have twice observed gnatcatchers feeding cowbird fledglings, in Anderson and Macon Counties.—*Charles P. Nicholson.*

Eastern Bluebird
Sialia sialis

Eastern Bluebirds are common permanent residents in Tennessee. While bluebirds from states farther north migrate through Tennessee in fall and early spring, evidence of northern bluebirds overwintering in the state is limited to 2 winter recoveries of birds banded north of Tennessee (Pinkowski 1971). Likewise, few bluebirds nesting in Tennessee leave the state; Laskey (1956) reported only a single out-of-state recovery from thousands of nesting bluebirds she banded in Tennessee.

The habitat preference of bluebirds has played a key role in the association of bluebirds with humans, both prehistorically and recently. The primary summer foods of bluebirds are insects, spiders, and other small animals; these prey items can be found and captured by bluebirds most efficiently in short, sparse vegetation, such as heavily grazed pastures. Other open areas such as lawns and golf courses are also commonly used. Prior to the arrival of European settlers in Tennessee the forests were dotted with openings due to fires, wind, and insect damage, beaver ponds, Indian villages, and prairies.

Early naturalists, such as Catesby (1771), commented on the abundance of bluebirds, both in rural and urban areas, throughout the eastern United States. As forested areas were cleared, bluebird numbers increased. Much of the early clearing was by girdling trees and planting crops and pasture under the standing dead trees. This created an especially favorable habitat for bluebirds with the combination of open areas, short vegetation, and many nest sites produced by woodpeckers in the dead trees. Langdon (1887) noted bluebirds around these "deadenings" in coves of the Unaka Mountains. Brewster (1886) considered bluebirds common in the mountains of western North Carolina up to elevations of about 1220 m; he also noted that in many places bluebirds used boxes erected for them. Rhoads (1895a) observed bluebirds throughout the state up to 1220 m on Roan Mountain and described them as frequently seen in parts of West and Middle Tennessee.

Bluebirds are often described in the popular press as a rare species that has greatly declined in recent decades. While this decline is not well documented, some writers (e.g., Zeleny 1976) have estimated declines as high as 90%. The main reasons for the decline are changing land use practices and nest cavity competition. Late-nineteenth- and early-twentieth-century farms, with a combination of woodlots, pastures, and small, cropped areas separated by fencerows, provided high-quality bluebird habitat. The transition from mules to tractors resulted in a reduction of pasture and hayfields and an increase in row-crops,

Eastern Bluebird. Elizabeth S. Chastain

94% of the completed priority blocks. Their total number of records was the eighth highest of any species, and they were found throughout the state except for some urban areas and very heavily wooded or high-elevation areas in the Cumberlands and the Unaka Mountains. Despite their wide distribution, their density is not particularly high; they were recorded on 79% of the Atlas miniroutes at an average relative abundance of 2.9 stops/route. By physiographic regions, their average abundance was lowest in the Mississippi Alluvial Plain, Cumberland Mountains, Unaka Mountains, and Loess Plain, and highest in the Western Highland Rim and Coastal Plain Uplands.

Winter weather is the factor that most dramatically influences the number of bluebirds in Tennessee. During the late 1970s, their breeding population plummeted as a result of the record-breaking cold winters of 1976–77 and 1977–78 (Pitts 1981). The average number per BBS route declined from 7.3 in 1976 to 1.9 in 1979. By the early 1980s the population had recovered in most areas, and from 1980 to 1994 it grew steadily at a rate of 7.3%/year ($p < 0.01$). Sauer and Droege (1990) found bluebird populations in 1987 to be at approximately the same level as in the mid-1960s; the 1991 Tennessee population was somewhat higher.

Breeding Biology: Bluebirds nest in a cavity in or adjacent to an open area. Naturally occurring cavities such as abandoned woodpecker holes are commonly used, as are large numbers of nest boxes erected for bluebirds. The female lines the nest cavity with dead vegetation such as grass, pine needles, or rootlets, often forming a loosely built cup. She lays 4–6 pale blue or occasionally pure white eggs at the rate of one per day. Of 800 nests in boxes in northwest Tennessee, 4% had 6 eggs, 55% had 5 eggs, 35% had 4 eggs, and 5% had 3 eggs (Pitts unpubl. data). Clutches as large as 7 and as small as 2

which provided little, if any, bluebird habitat. Fencerows were eliminated as fields became larger, and in many remaining fencerows, wooden posts, which had often provide nest cavities, have been replaced with more durable metal fenceposts unsuitable as nest sites. Competition for nest cavities from the introduced European Starling and House Sparrow is also detrimental. The many thousands of bluebird nest boxes erected across the state have helped offset this decline and often resulted in high local bluebird densities. Bluebird nest boxes should have an entrance hole 3.8 cm in diameter to exclude starlings; these boxes, however, are still accessible to the similarly sized House Sparrow. Placement of boxes away from barns and other House Sparrow concentrations can reduce box use by House Sparrows.

Due to their preference for open areas, their bright colors, and their easily recognized songs, bluebirds were readily detected by Atlas workers and were found in

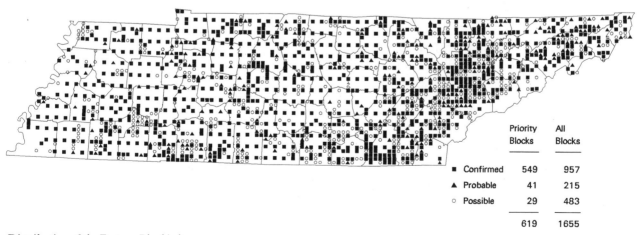

	Priority Blocks	All Blocks
■ Confirmed	549	957
▲ Probable	41	215
○ Possible	29	483
	619	1655

Distribution of the Eastern Bluebird.

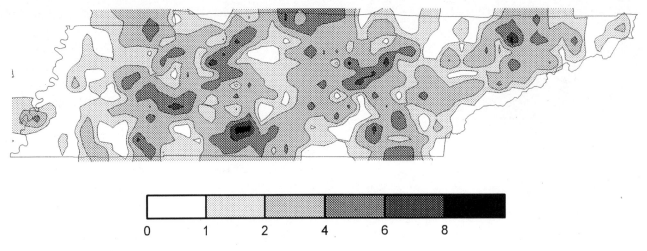

Abundance of the Eastern Bluebird.

occur at rare intervals. The female incubates the eggs for 12–14 days. The young remain in the nest cavity for 15–18 days and are fed by both parents. More than 85% of the eggs usually hatch if the nest is not destroyed by a predator. In a sample of 855 nests, 572 (67%) successfully produced at least one fledgling (Pitts unpubl. data). However, most of these boxes were placed on metal pipes, accessible to rat snakes but less accessible to raccoons than nests would be in most natural cavities. Cowbird parasitism is rare, only 1 of 855 nests; this parasitized nest was in a box with an enlarged entrance.

The nesting season of bluebirds in Tennessee is long. Most pairs have 2 nests per year, and many pairs have 3. If late winter weather is mild, eggs are commonly laid in March. Eggs are deposited in the last nests in July and August. Fledglings are dependent on their parents for food for about a month following departure from the nest. The young birds then form flocks that roam the countryside. Some of these flocks may have 30 or more birds; flocks of 100 have been seen in Tennessee (P. Dietrich pers. comm.) but such large flocks may consist partly or entirely of migrating birds from northern states. Bluebirds begin nesting the year after they are hatched if they are successful in locating and defending a suitable nest site. Most adults survive for only one nesting season.—*T. David Pitts.*

Veery
Catharus fuscescens

The Veery is a common summer resident at the high elevations of the East Tennessee, where its evening chorus is one of the most beautiful and memorable song performances of any Tennessee bird. Veeries arrive on their breeding grounds between the end of April and mid-May and depart by mid-September. Elsewhere in the state, they are an uncommon migrant, present from late April to late May, and late August to early October.

The Veery was first recorded in Tennessee during the breeding season by Rhoads (1895a), who found it abundant on Roan Mountain between 915 and 1525 m. Rhoads also observed Veeries at Chattanooga in late May and suspected them of breeding; these birds were most likely migrants. The presence of Veeries in the Cumberland Mountains during the breeding season was not discovered until the mid-1970s; prior to this time, little fieldwork had been done on the peaks of the Cumberland Mountains.

With their distinctive song and call notes, Veeries were probably not often missed by atlasers, unless the block was only visited at midday, when they rarely sing. As shown on the Atlas map, Veeries occur at high elevations throughout the Unaka Mountains and on a few of the highest peaks of the Cumberland Mountains. Within the Blue Ridge, Veeries typically occur at elevations above 915 m and occupy both deciduous and coniferous forest types. The population is probably more continuous along the state line north of the Smokies than shown on the map. To the south of the Smokies, Veeries occur in the Whigg Meadow-Stratton Gap-Whiteoak Flats area of Monroe County and on Big Frog Mountain, Polk County. The Big Frog population, isolated by lower elevations from other Tennessee populations, probably extends southward along the Blue Ridge into Georgia.

In the Cumberlands, Veeries occur on Cross Mountain and Frozen Head Mountain at elevations above 854 m (Nicholson 1987). Each of these mountains reaches a maximum height of over 1000 m and has a forested area above 915 m of several hundred hectares. A few inaccessible peaks in the Cumberlands were not surveyed, and Veeries may occur on some of them. Coal surface mines have probably reduced the amount of

Veery. Elizabeth S. Chastain

Veeries occupy a variety of mesic forest types, including cove and northern hardwoods, spruce-fir, mixed hemlock, and beech gaps (Stupka 1963, Wilcove 1988). Exposed, xeric forest types, such as chestnut oak, are avoided. Wilcove (1988) found the highest density of Veeries in the Great Smoky Mountains, 61 pairs/100 ha, in a hemlock-deciduous forest with a birch understory at about 1300 m elevation. Mitchell and Stedman (1993) found a very high density of 200 pairs/100 ha in a mature maple-beech-birch forest at 1500 m in the Unicoi Mountains. On Frozen Head Mountain, Veeries occupy forest dominated by buckeye, tulip-poplar, black cherry, and northern red oak, with an understory dominated by *Hydrangea arborescens* (Nicholson 1987). Robertson (1979) found a density of 17 pairs/100 ha in Cumberland Plateau virgin mixed mesophytic forest.

In the Unaka Mountains, Wood Thrushes occur sympatrically with Veeries near the lower limit of the Veery's altitudinal range (Stupka 1963, Noon 1981). Along a north to northwest slope in the Smokies, Noon (1981) found the 2 species overlapped along a small proportion of their elevational ranges. Veeries reached their highest abundance near and above the Wood Thrush's upper elevational limit. Veeries occupied territories with a lower canopy height, fewer tree species, and higher tree and shrub densities than Wood Thrushes; these differences are typical of the higher altitude forests inhabited by Veeries (Noon 1981). In the Cumberlands, Wood Thrushes occur throughout the range of the Veery (Nicholson 1987, pers. obs.).

suitable habitat on several peaks there. Robertson (1979) found 2 territorial Veeries at 470 m elevation in virgin mixed mesophytic forest on the Cumberland Plateau in Savage Gulf, Grundy County, in June 1977. Whether these birds successfully nested is unknown. They were not reported there by atlasers; if they were part of a very localized population, it is likely they were simply missed.

Veeries were recorded on 10 miniroutes, at an average of 4.6 stops/route. The highest relative abundance was 10, recorded on 2 routes. Only 1 bird was recorded on 1 route in the Cumberlands. On a nonrandom miniroute at Frozen Head, the 2-year average abundance was 5 stops/route (Nicholson unpubl. data). Veeries have been found on too few Tennessee BBS routes to establish a population trend. On plot censuses conducted in the Great Smoky Mountains, Wilcove (1988) did not detect a significant population change between 1947 and 1983. Rangewide, Veeries declined significantly ($p < 0.01$) at a rate of 1.4%/year from 1966 to 1993; the rate accelerated during the latter part of this period (Sauer and Droege 1992, Peterjohn, Sauer, Link 1994).

Breeding Biology: Veeries begin nest building in late May and lay eggs during early June (Ganier 1962, egg coll. data). Nest building is performed by the female and requires 6 to 10 days (Harrison 1975). Nests are rarely more than 1 m above the ground and often on the ground. Rhoads (1895a) reported a pair building

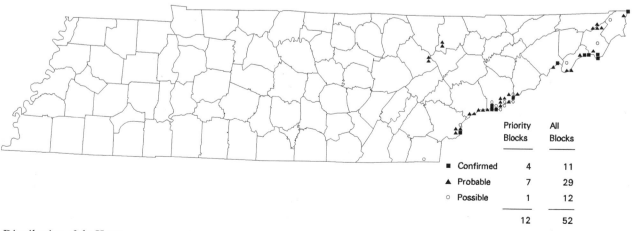

Distribution of the Veery.

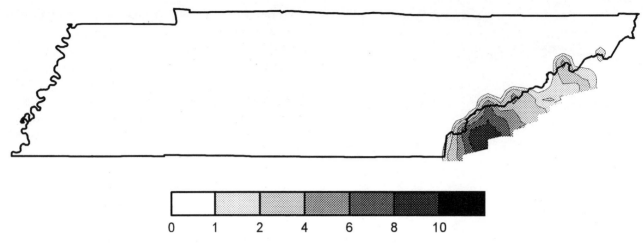

Abundance of the Veery.

12 m up in a maple on Roan Mountain and noted that it was an unusual nest location. The nest is usually well hidden and may be placed on a low, horizontal conifer limb, in thick shrub stems or stump sprouts, or on a low stump or brushpile hidden by thick vegetation (Ganier 1962, egg coll. data). Only 2 of the confirmed Atlas records were of nests. The foundation of the relatively large nest is made of dead leaves, with a cup formed of twigs, weed stems, and grape vines and lined with bark strips, rootlets, and grasses. The 3 to 4 eggs are pale blue, normally without spots. Incubation, carried out by the female, requires 10–12 days. Both adults feed the young, which leave the nest in 10–12 days (Harrison 1978). Whether Veeries in the southern Appalachians are double-brooded, as suggested by Ehrlich, Dobkin, and Wheye (1988), is unknown; egg collection data and dates of confirmed records reported by atlasers suggest they are single-brooded. No reports of Brown-headed Cowbird parasitism of Veeries in Tennessee are available. Cowbirds are rare or absent in most of the Veery's breeding range in and south of the Smokies. To the north of the Smokies, where forests are more fragmented, and in the Cumberlands, Veeries are probably vulnerable to cowbird parasitism. Populations north of Tennessee are often parasitized (Friedman 1963, Friedman, Kiff, and Rothstein 1977); Veeries readily accept cowbird eggs.—*Charles P. Nicholson.*

Wood Thrush
Hylocichla mustelina

The Wood Thrush is a common summer resident of woodlands throughout the state. Most birds arrive during the first half of April and depart by mid-October (Robinson 1990). Because of their loud and distinctive flutelike song and habit of frequently nesting in wooded residential areas, Wood Thrushes are well known to many Tennesseans.

The Wood Thrush was found in about 92% of the completed Atlas priority blocks, and it is unlikely they were overlooked in more than a very few blocks. The few completed blocks where Wood Thrushes were not found were either at the highest elevations in the Unaka Mountains or blocks with very little woodland. Stupka (1963) found Wood Thrushes to be common in the Great Smoky Mountains up to the lower limits of the spruce-fir zone, which varies between 1070 and 1525 m. Nest records exist as high as 1372 m in the Smokies (NRC), and the highest is at Cloudland on Roan Mountain, 1850 m (Lyle egg collection). Wetmore (1939) reported a recently fledged juvenile at 1532 m in the Smokies. At the upper limit of its range, Wood Thrushes are often sympatric with Veeries, in both the Unaka and the Cumberland Mountains (Stupka 1963, Noon 1981, Nicholson 1987).

The Wood Thrush breeds throughout most of the eastern North American deciduous forest. Within this large area, it occurs in many types of deciduous as well as mixed deciduous-coniferous forest. The different forest types all have a well-shaded understory, small trees with low, exposed branches, and a fairly open forest floor with decaying leaf litter (James et al. 1984). Robbins, Dawson, and Dowell (1989) found that the probability of detecting Wood Thrushes increased with the size of the forest and reached a maximum at about 500 ha; they are also frequently found in forests of 1 ha or less. The highest densities on Tennessee census plots have been in upland mesic forests: 129 pairs/100 ha in a virgin cove hardwood forest in the Smokies, 60/100 ha in a virgin Cumberland Plateau mixed mesophytic forest, 63/100 ha in an Eastern Highland Rim mixed mesophytic forest, and 55/100 ha in a Cumberland Mountain mixed mesophytic forest (Aldrich and Goodrum 1946, Robertson 1979, Simmers 1982a, Smith 1977).

Wood Thrush. Elizabeth S. Chastain

Densities in Central Basin and Ridge and Valley hardwood forests have been considerably lower (Fowler and Fowler 1984a, 1984b, Howell 1965, 1971, 1972, 1973, 1974, Smith 1975). Ford (1990) found Wood Thrushes on 45 of 59 West Tennessee forested wetland sites at an average density, measured by transect counts, of 9.6/100 ha.

Wood Thrushes were recorded on two-thirds of the miniroutes, at an average relative abundance of 2.8 stops/route and a maximum of 10 stops/route. Their abundance on miniroutes was closely related to the area of forest cover. The Cumberland Mountains had the highest abundance of any physiographic region with an average of 4.6 stops/route. The Cumberland Plateau (including the Cumberland Mountains) is among the physiographic areas with the highest numbers of Wood Thrushes, as measured by BBS routes throughout eastern North America (Robbins, Bystrak, and Geissler 1986).

Although deforestation and urban development have apparently eliminated Wood Thrushes from a few local areas where they probably once occurred, no major change in their distribution is evident. Rhoads (1895a) found Wood Thrushes at low elevations across the state, except on the Cumberland Plateau. He probably overlooked them on the Plateau, where later observers have found them common (e.g., Ganier 1937a). They have probably always been uncommon in the inner portion of the Central Basin due to the scarcity of moist deciduous forests. Wood Thrushes declined on Tennessee BBS routes at a significant ($p < 0.01$) rate of 2.3%/year from 1966 to 1994. Rangewide, their population declined at an overall rate of 1.9%/year ($p < 0.01$) from 1966 from 1993; the rate of decline accelerated during the latter part of this period, as was also the case in Tennessee (Sauer and Droege 1992, Peterjohn, Sauer, and Link 1994).

Breeding Biology: Wood Thrush nests are often easy to find, and nests accounted for 61% of the confirmed Atlas records. About half the nest records were of used nests, which weather slowly and are easy to identify. Wood Thrushes begin nesting soon after their spring arrival. The nest is built by the female in 5 to 6 days (Brackbill 1958) of dead leaves and grasses, held together with mud and lined with rootlets and grasses. It is usually placed in the fork of a horizontal or upward-sloping tree branch or adjacent to the stem of a sapling. The nest is fairly exposed, often on the lowest branch of a tree (Brackbill 1958, James et al. 1984). Nests are usually built in deciduous trees, most often in dogwood, beech, hackberry, maples, and elms in Tennessee. More Tennessee nests have been reported in hemlocks than other conifers. Tennessee nest heights range from 1.4 to 7.6 m, with an average of 3.0 m (s.d. = 1.3, n = 38), within the range given by studies elsewhere (Brackbill 1958, Longcore and Jones 1969).

Clutch size is usually 3 or 4 eggs and clutches of 2 and 5 are rare (Brackbill 1958, Longcore and Jones 1969).

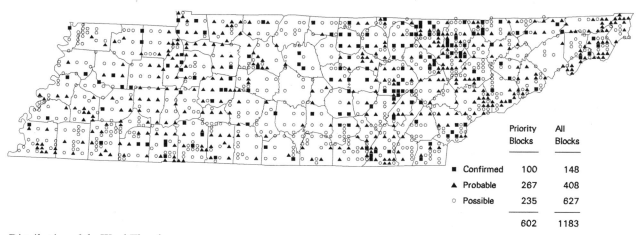

	Priority Blocks	All Blocks
■ Confirmed	100	148
▲ Probable	267	408
○ Possible	235	627
	602	1183

Distribution of the Wood Thrush.

Confirmed Breeding Species

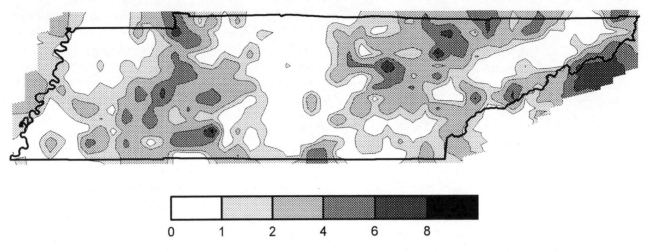

Abundance of the Wood Thrush.

Fifty-two Tennessee clutches averaged 3.81 eggs (s.d. = 0.53, range 2–5). The eggs are pale bluish to bluish green, unmarked, and slightly smaller and paler than an American Robin's eggs (Harrison 1975). The female alone incubates the eggs for 12–13 days (Brackbill 1958, Nolan 1974). Both adults feed the young, which leave the nest in 12–14 days and become independent when 3–4 weeks old. Second broods are common, and the success of late season nests is often higher than that of nests begun in May (Longcore and Jones 1969). Tennessee egg dates show a peak during the second 10 days of May and a lesser peak in mid-June.

Only 2 Tennessee records of Brown-headed Cowbirds parasitism of Wood Thrushes predate 1965—Todd's (1936) report of a nest with 1 cowbird and 3 thrush eggs, and a 1945 egg collection record of 2 cowbird and 4 thrush eggs. Eight of 20 nests reported since then contained cowbird eggs or nestlings (NRC, Nicholson unpubl. data). Several observations of Wood Thrushes feeding fledgling cowbirds were reported by atlasers. Elsewhere in its range, the Wood Thrush is frequently parasitized by cowbirds, and half or more of the nests have been parasitized in some local areas (Friedmann 1963, Friedmann, Kiff, and Rothstein 1977). Wood Thrushes rarely reject cowbird eggs.—*Charles P. Nicholson.*

American Robin
Turdus migratorius

The American Robin, a common permanent resident, is well known by almost all Tennesseans. It occurs at all elevations throughout the state and is most numerous in suburban communities, where it forages for earthworms and other invertebrates on mowed lawns and nests in trees and large shrubs. It was one of the species most frequently reported by Atlas workers.

Prior to the widespread alteration of the environment by European settlers, robins were probably uncommon and restricted to the eastern half of Tennessee, occupying grassy woodland openings, forested areas with an open, grassy understory, and the moss or grass-carpeted understory of the high-elevation spruce-fir and beech forests. During the late eighteenth century, an increase in their numbers probably began with the forest clearing by European settlers. Michaux collected the species near Nashville in June 1795 (Williams 1928), where it was probably nesting. A century later Rhoads (1895a) described the robin as rare in West and Middle Tennessee and not abundant in East Tennessee. At that time Tennessee was near the southern edge of the species's breeding range; robins were also absent west of the Tennessee River in Kentucky (Mengel 1965).

In the early twentieth century, the breeding range of the American Robin spread southward in Georgia (Odum and Burleigh 1946) and, although less studied, westward in Tennessee. The rapid increase in the urban human population during this period, with the resultant landscaped lawns, provided new robin habitat. The robin probably first colonized towns and later the intervening rural areas in the state, as occurred in Georgia (Odum and Burleigh 1946). By 1942, the breeding robins had been common at Nashville for at least 25 years (Monk 1942), and the 39 nests reported at Memphis in 1948 (Coffey 1948) suggest it was common there.

The increase in the robin breeding population is still continuing. On Tennessee BBS routes, the robin population significantly ($p < 0.01$) increased at the rate of 2.9%/year between 1966 and 1994. Its population in eastern North America also increased significantly ($p < 0.01$) between 1966 and 1987 (Robbins et al. 1989), as it did across North America through 1993 ($p < 0.01$) (Peterjohn, Sauer, and Link 1994).

American Robin. Elizabeth S. Chastain

Atlasers found the robin in 95% of the completed priority blocks. It was not found in a few blocks with little or no suitable habitat in the Mississippi Alluvial Plain, the Cumberland Plateau and Mountains, and in the southern Unaka Mountains. It was recorded on 86% of the miniroutes, at an average relative abundance of 4.2 stops/route. By physiographic regions, the Ridge and Valley had the highest average abundance (5.7 stops/route and on all but 2 routes); numbers were lowest in West Tennessee. Most urban areas had high robin numbers; this is especially evident in West Tennessee. Its abundance was also high around the central Cumberland Plateau resort communities.

Limited information on robin territory size and density on census plots is available for Tennessee. Pitts (1984) found an average territory size of 0.42 ha (range 0.12–0.84 ha, n = 62) on a northwest Tennessee college campus, considerably larger than that reported by Howell (1942) in New York. Densities in Smoky Mountain spruce-fir forests ranged from 8 to 20 pairs/100 ha (Kendeigh and Fawver 1981, Alsop and Laughlin 1991). The population in spruce-fir, however, has recently declined as brushy growth increased following death of most of the mature firs (Alsop and Laughlin 1991). Average density in Smoky Mountain beech gap forest was 13/100 ha and did not change between 1947 and 1982 (Wilcove 1988).

Breeding Biology: The breeding season of the American Robin is long, as it is normally double-brooded and occasionally triple-brooded (Howell 1942). Atlasers found confirming the breeding of the robin to be easy; 56% of records from all blocks were confirmed, and 82% of the records from priority blocks were confirmed. Robin nests are usually easy to locate and used nests easily identified; 37% of the confirmed records were of nests. Adults carrying food and fledglings are also conspicuous, and these categories comprised 58% of the confirmed records.

The American Robin begins breeding activities early in the spring, with the actual date of nest initiation dependent on local temperatures (James and Shugart 1974). The earliest date of nest construction in Tennessee is 13 February 1991 (Atlas results); most pairs, however, do not begin nest construction until late March or early April. The robin is normally monogamous, and the male establishes a territory that both adults defend and later includes the nests. They may leave the territory to feed, and the males occupy communal roosts early in the nesting season (Howell 1942, Pitts 1984).

Both adults construct the nest, which is placed in a crotch of a large shrub, sapling, or tree, or on the fork of a horizontal tree branch. Nests are occasionally built on ledges or other parts of buildings and on shelf-like nest boxes. Nests on the ground have been reported elsewhere in the species's range (Howell 1942). The nest is constructed of dead grasses and often twigs, weed stems, bits of paper, and string held together with mud and

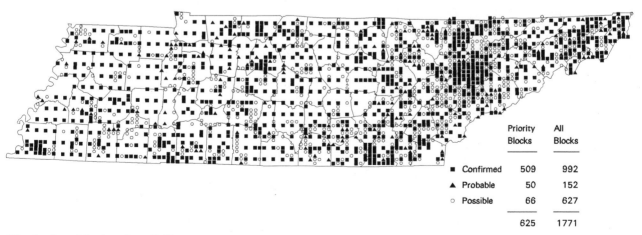

Distribution of the American Robin.

	Priority Blocks	All Blocks
■ Confirmed	509	992
▲ Probable	50	152
○ Possible	66	627
	625	1771

Confirmed Breeding Species

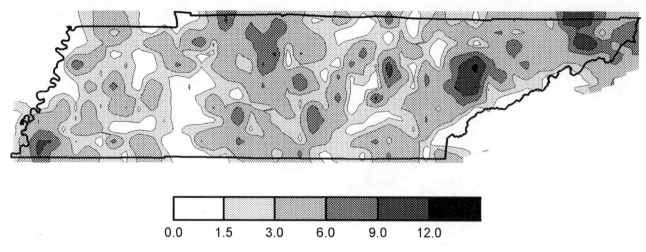

Abundance of the American Robin.

lined with grass. Tennessee nests range in height from 0.9 to 12.2 m and average 3.4 m (s.d. = 1.8 m, n = 102). Coffey (1948) reported an average height of 5.8 m for 39 Memphis nests. Robins nest in a wide variety of tree and a few shrub species. Cedars and pines are frequently used early in the nesting season, when most deciduous trees are not yet leafed out, as also reported by Howell (1942). Other frequently used species are hackberry, maples, and, at Memphis, elms (Coffey 1948). Ganier (1926) found nests in stunted beech and on low limbs of spruce and firs at high elevations in the Smokies.

Egg laying, as determined from a sample of 92 dated observations, peaks 11–15 April; no peak in second clutches is evident. Clutch size in Tennessee ranges from 3 to 5 eggs, with 4 most frequent and an average of 3.87 (s.d. = 0.44, n = 82). The female incubates the eggs for an average of 12.5 days, and the young leave the nest in about 13 days (Young 1955). Both adults feed the nestlings. The fledglings become independent about 2 weeks after leaving the nest; second broods are frequently begun before the young of the first brood reach independence (Howell 1942, Young 1955). Robins normally eject cowbird eggs laid in their nests, and no records of cowbird parasitism of robins in Tennessee are available.—*Charles P. Nicholson.*

Gray Catbird
Dumetella carolinensis

The Gray Catbird, named for its loud, catlike calls, is an uncommon to fairly common summer resident across the state of Tennessee. It usually arrives in mid- to late April and departs by late October. Many winter records exist, mostly on Christmas Bird Counts; whether these birds successfully overwinter is poorly known (Robinson 1990).

Catbirds occur in a variety of forest edge habitats dominated by shrubs and low trees. They also show a preference for moist habitats (Robbins, Dawson, and Dowell 1989). In Tennessee, typical low-elevation catbird habitats include streamside thickets and towns and suburbs with thick shrubbery. In the Unaka Mountains, catbirds occur up to the highest elevations, where they nest in shrubby openings in the spruce-fir forests and on balds (Stupka 1963, Ganier 1936, Ganier and Clebsch 1946, Atlas results). Within the Smokies, they are most numerous at high elevations in the western part of the park, where the forests are predominantly deciduous and balds are more common (Stupka 1963, Atlas results). Catbirds do not seem to have responded to the die-off of the fir forest (Alsop and Laughlin 1991).

The few Tennessee breeding bird censuses in relatively uniform habitats that have recorded catbirds are all from the Unaka Mountains. Kendeigh and Fawver (1981) recorded average densities of 12.5 pairs/100 ha in young pine-oak forests, 5/100 ha in red oak and pine heath forests, 6.3/100 ha in beech gap forests, and 12.5/100 ha in a mature spruce-fir forest. Lewis (1980, 1981a, 1982, 1983b) found 5 pairs/100 ha in a 6- to 8-year-old, low-elevation deciduous clearcut. Densities in higher elevation clearcuts of similar age in northern hardwoods are probably higher (Nicholson pers. obs.).

Because of their preference for dense shrubbery, catbirds are often difficult to see. Their song and loud calls, however, are distinctive, and it is unlikely that atlasers missed catbirds in many priority blocks. The Atlas results show catbirds to be most common in the Unaka Mountains, the Cumberlands, and the Highland Rim. They have probably declined in abundance in the rest of the state. Rhoads (1895a) described the catbird as occurring statewide in all localities he visited, and Torrey (1896a) described it as very common at Chattanooga. Ganier (1933a) described it as common in West and East Tennessee and abundant in Middle Tennessee. This

Gray Catbird. Elizabeth S. Chastain

is clearly no longer the case, especially in West Tennessee. Statewide, catbirds were recorded in less than two-thirds of the completed priority blocks and were found on 189 miniroutes at an average relative abundance of 1.5 stops/route. During the Atlas period, catbirds were recorded on 36 of the 42 BBS routes in Tennessee, at an average relative abundance of 3 birds and 2.7 stops/route.

Tennessee BBS results also show that Gray Catbird numbers have decreased at the average rate of 5.0%/year ($p < 0.01$) from 1966 to 1994. This is one of the greatest declines of any Tennessee bird, and it was most pronounced (-10.2%/year) from 1980 to 1994. Throughout their range, catbirds showed a decreasing trend of 0.4%/year ($p = 0.03$) from 1966 to 1993 (Peterjohn, Sauer, and Link 1994). In eastern North America, catbirds increased at a significant ($p < 0.05$) annual rate of 0.6%/year between 1966 and 1978, and declined at an annual rate of 1.4%/year ($p < 0.05$) from 1978 to 1987 (Robbins et al. 1989). The reasons for the decline are not clear. Catbirds are considered to be generalists in their winter habitat requirements, and thus may not be impacted by changes on wintering grounds. Suitable catbird habitat in Tennessee has probably decreased due to the general maturing of the forests and decreased acreage of sapling and pole-sized stands, as well as the trend toward cleaner farming with elimination of much fencerow habitat and brushy pastures. Catbirds do not seem to have benefited from the increasing area of pine plantations, as these young woodlands are probably too dry.

Breeding Biology: Male catbirds begin defending a territory shortly after their arrival in the spring. Territorial defense is apparently limited to the vicinity of the nest, and catbirds may leave the territory to feed with other catbirds in undefended areas (Nickell 1965). Following pairing, the male assists in the selection of a nest site by placing a few sticks in a suitable site and displaying to the female in the vicinity of this false nest. The female, who does most of the nest construction, either accepts this nest site or begins building at another nearby site (Nickell 1965).

The nest is usually well concealed in thick shrubbery, vines, or a small tree. Catbird nests in Tennessee range in height from 0.3 to 5.2 m and average 2.2 m (s.d. = 1.0, n = 73) above ground. Nests have been reported in many plant species, including hawthorns, osage orange, eastern red cedar, blackberry, forsythia, privet, and saplings of willow, oak, and maple, with little apparent preference for any individual species. The bulky nest is placed in a vertical fork of the supporting plants, or less commonly on a fork of a horizontal limb. It is composed of a foundation of twigs, grasses, and weed stems, an inner basket of bark strips, leaves, or paper, and a lining of rootlets (Layne 1931, Nickell 1965).

Egg laying in Tennessee peaks in mid-May and extends into early July. Gray Catbirds in Tennessee, as

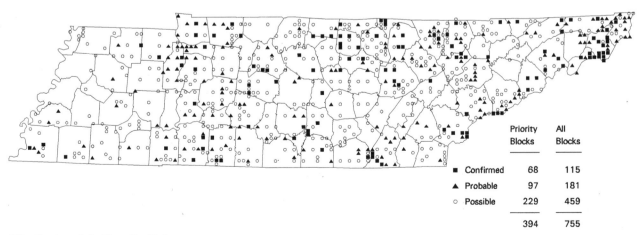

Distribution of the Gray Catbird.

		Priority Blocks	All Blocks
■	Confirmed	68	115
▲	Probable	97	181
○	Possible	229	459
		394	755

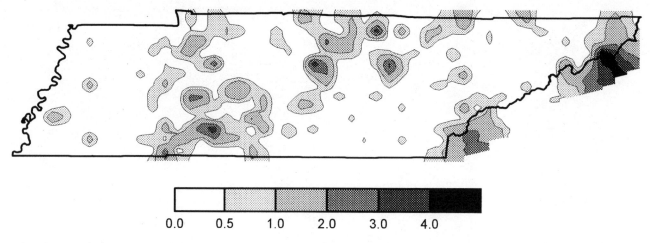

Abundance of the Gray Catbird.

elsewhere, renest after loss of a nest and often raise 2 broods (Swindell 1962, Layne 1931, Nickell 1965). The normal clutch size is 3–5 eggs with an average size of 3.8 (s.d. = 0.58, n = 72) in Tennessee, close to the average elsewhere (Nickell 1965, Scott, Lemon, and Darley 1988). First clutches are larger than later clutches (Nickell 1965, Scott, Lemon, and Darley 1988); this is also true in Tennessee, although the sample size of June and July clutches is small. Catbird eggs are a distinctive, unmarked, deep greenish blue.

The female incubates the eggs for about 14 days; during this period she is often fed by the male (Nickell 1965). She also broods the nestlings, which both adults feed. During the incubation and early nestling period, the male remains close to the nest, aggressively guarding it (Slack 1976). The young leave the nest in about 11 days, and the adults care for them for at least 2 weeks (Layne 1931, Nickell 1965).

At least 1 Tennessee record exists of a Brown-headed Cowbird egg in a catbird nest. This nest contained 1 cowbird and 3 catbird eggs when found; the following day, there were 4 catbird and no cowbird eggs (NRC). Gray Catbirds usually eject cowbird eggs and consequently rarely raise cowbird young (Friedmann, Kiff, and Rothstein 1977, Friedmann and Kiff 1985). No Tennessee records of catbirds raising cowbird young are available.—*Charles P. Nicholson.*

Northern Mockingbird
Mimus polyglottos

The Northern Mockingbird is a common permanent resident found throughout most of the state. It is one of the birds best known by Tennesseans because of its designation as the state bird, its conspicuousness around human dwellings, and its ability to mimic the songs of numerous other birds. It occurs in areas of short grass or herbs and scattered trees and shrubs and is common in towns, suburbs, farmyards, and fencerows.

The Northern Mockingbird was probably uncommon in prehistoric Tennessee, nesting in the natural grasslands along the north-central and northwestern border, around Indian villages, and in the open areas of cedar glades. Its numbers increased with European settlement, although it apparently was still uncommon in the late nineteenth century. The first published record of it in Tennessee is apparently that of Fox (1886), who observed a single bird in Roane County in April 1885. Torrey (1896a) found it rare at Chattanooga in the spring of 1894. Rhoads (1895a) also described it as rare and did not record it on the Cumberland Plateau. Rhoads's descriptions imply that the mockingbird was less numerous and less widespread than both the Gray Catbird and the Brown Thrasher, a reverse of the present situation.

Northern Mockingbird. Elizabeth S. Chastain

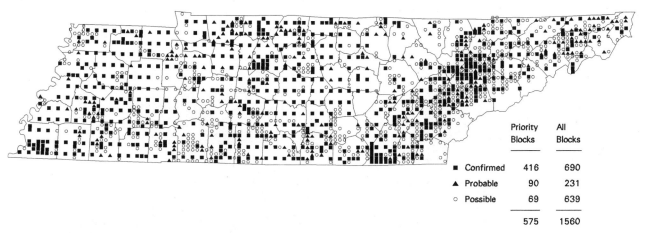

Distribution of the Northern Mockingbird.

	Priority Blocks	All Blocks
■ Confirmed	416	690
▲ Probable	90	231
○ Possible	69	639
	575	1560

Northern Mockingbirds have increased since the 1890s and probably peaked during the 1930s when the forested area reached a minimum. From 1966 to 1994, its population, as measured by Tennessee BBS routes, decreased by 1.1%/year (p < 0.01). This decrease is probably due to the increase in forest area as well as cleaner farming techniques, which have eliminated much of the fencerow habitat used by the mockingbird, and was concentrated in the first half of the survey period. Its rangewide population decreased by 1.0%/year (p < 0.01) from 1966 to 1993 (Peterjohn, Sauer, and Link 1994).

The Northern Mockingbird, however, remains a common bird and was one of the most frequently reported species during Atlas fieldwork. It was found in 87% of the completed priority blocks and occurs statewide except for heavily forested blocks in the Unaka Mountains and the Cumberlands. It was also absent from a few heavily forested blocks elsewhere in the state. It may be limited in its dispersal abilities across extensive areas of forest, as it was often absent from isolated farmsteads in forested areas (Nicholson pers. obs.).

Atlasers recorded the Northern Mockingbird on 78% of the miniroutes at an average relative abundance of 5.47 stops/route. It was most common on the Loess Plain, where it was recorded on all but 1 miniroute and averaged 7.8 stops/route. The Central Basin had the next highest abundance, 6.8 stops/route. It is also one of the most numerous native birds in areas of dense human population.

Breeding Biology: Confirming breeding of the Northern Mockingbird was easy for atlasers, and its proportion of confirmed records is among the highest of open-nesting species. Almost half of the confirmed records were of fledglings, which are usually easy to observe; 30% were of nests. The mockingbird's breeding biology in the Nashville area was described by Laskey (1962), the source of most of the breeding biology information presented here. Its breeding season extends from late March into August, and it annually rears up to 4 broods.

The Northern Mockingbird nests in small evergreens, deciduous trees and shrubs with dense foliage or branching

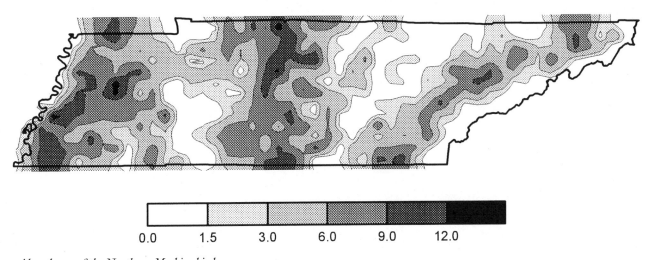

Abundance of the Northern Mockingbird.

Confirmed Breeding Species

structure, and vine thickets. Laskey found over half of all nests in evergreens and one-third of them in red cedar. The male places nest material in possible nest sites during courtship and both adults build the nest. The nest is constructed of a heavy outer layer of twigs, an inner layer of leaves, plant stems, paper, or other materials, and a lining of rootlets. Laskey reported nests from ground level to 16 m above ground; 76% of 189 nests she examined were from 1 to 2 m above ground. A separate sample of 78 nests, mostly in Middle and East Tennessee (NRC, egg coll. data) averaged 1.8 m above ground (s.d. = 0.8).

Egg laying begins in March or early April and is earlier in years with above normal late winter temperatures. Of 235 clutches observed by Laskey, 2% in March, 40% were laid in April, 23% in May, 24% in June, and 11% in July. A few clutches have been laid in early August. Clutch size is normally 3 to 5 eggs; a single record of 7 eggs exists (Laskey 1963). The most common clutch size is 4 eggs. Laskey found an average clutch size of 3.91 eggs (s.d. = 0.61, n = 182); other clutches (NRC, egg coll. data) averaged 3.89 eggs (s.d. = 0.64, n = 96). The eggs are pale blue and green with heavy brown blotches and spots. The female incubates the eggs for 12–12.5 days. Both adults feed the young, which usually fledge in 12 days. Pairs make several nesting attempts in a season and usually fledge 2 broods. Banded nestlings have dispersed as far as 320 km from Nashville. The Northern Mockingbird normally removes Brown-headed Cowbird eggs from its nest (Friedmann and Kiff 1985); its parasitism by cowbirds has not been reported in Tennessee.—*Charles P. Nicholson.*

Brown Thrasher
Toxostoma rufum

The Brown Thrasher is a statewide resident of Tennessee, fairly common in summer and uncommon in winter. A pronounced migration occurs in the fall and spring. During the winter, it is most common in southwestern Tennessee and decreases in numbers toward the north and east (Tanner 1985).

Brown Thrashers occupy forest edge habitats, usually with thick shrubbery and scattered trees. Typical habitats include wooded fencerows, brushy pastures, open cedar forests, dense thickets of multiflora rose and other shrubs, young clearcuts, and wooded residential areas. On the Cumberland Plateau, the Highland Rim, and the Coastal Plain Uplands, it often nests in thickets of farkleberry and deciduous holly in the understory of dry deciduous forests (Nicholson pers. obs.). Compared to the other members of the family Mimidae in Tennessee, with which thrashers frequently coexist, thrashers are

Brown Thrasher. Elizabeth S. Chastain

less dependent on areas of short grass than Northern Mockingbirds and less restricted to moist areas than Gray Catbirds.

Rhoads (1895a) found the Brown Thrasher present across the state at elevations below 914 m. Their breeding distribution has apparently changed little since then. At the higher elevations of the Unaka Mountains, they uncommonly occur on grassy balds at elevations up to at least 1585 m (Stupka 1963). Prior to European settlement, thrashers were probably fairly common across the state, occurring around Indian towns and in cedar glades and in frequently burned woodlands. BBS results show a significant ($p < 0.01$) rangewide decreasing trend of 1.3%/year in Brown Thrasher numbers from 1966 to 1993 (Peterjohn, Sauer, and Link 1994). In Tennessee, it decreased by 1.0%/year ($p < 0.10$) from 1966 to 1994; most of the decline occurred in the second half of this period.

Brown Thrashers become conspicuous in March, when they reoccupy breeding territories, and the males sing vigorously from exposed, high perches. By May, however, their rate of singing decreases, and by July they have virtually ceased singing (Nicholson pers. obs., Howell and Monroe 1957). Despite the low occurrence of singing at the time when most Atlas fieldwork was done, thrashers remained conspicuous enough to be recorded on most of the completed priority blocks, and the Atlas map is an accurate representation of their distribution in Tennessee. Most of the completed blocks where they were not found were either in very heavily wooded parts of the Unaka Mountains and Cumberland Plateau or in West Tennessee bottomlands.

Atlasers recorded Brown Thrashers on about 70% of the miniroutes, at an average relative abundance of 2.1

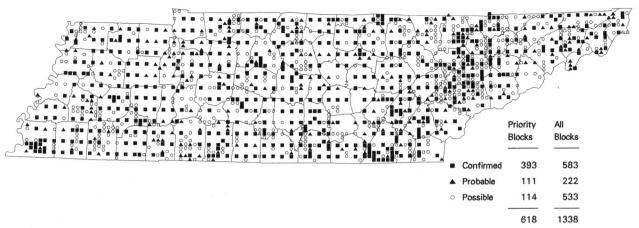

	Priority Blocks	All Blocks
■ Confirmed	393	583
▲ Probable	111	222
○ Possible	114	533
	618	1338

Distribution of the Brown Thrasher.

stops/route. Its numbers were uniformly low across the state, with no great differences between physiographic areas. Mengel (1965) noted a similar situation in Kentucky. Only 8 Tennessee routes reported it from more than 5 stops/route and the maximum count was 10, on a route in the Central Basin. During the Atlas period, Brown Thrashers were recorded on all of the Tennessee BBS routes and averaged 5.3 birds and 4.6 stops/route. Densities in the few breeding bird censuses in relatively uniform plots include 2.5 pairs/100 ha in a grazed, Ridge and Valley hardwood forest (Smith 1975), 16.5 pairs/100 ha in Central Basin cedar glades (Fowler and Fowler 1983a, 1983b), and 8.2 and 33 pairs/100 ha in Central Basin abandoned farmland (Fowler and Fowler 1984c, 1984d).

Breeding Biology: Atlasers found that confirming breeding of Brown Thrashers was relatively easy; almost two-thirds of the records from priority blocks were of confirmed breeding. Brown Thrashers often attempt to raise at least 2 broods (Erwin 1935, Murphy and Fleischer 1986) and have a long nesting season. Most of the confirmed records were of fledglings or of adults carrying food, both of which are often conspicuous. Its breeding biology in Tennessee is fairly well known due to the study of Erwin (1935) at Nashville and numerous unpublished nest records.

Brown Thrashers are apparently monogamous and begin defending territories during the second half of March. Nest construction, carried out by both the male and female, is usually underway by early April (Erwin 1935, Nicholson pers. obs.). The nest is loosely constructed with a foundation of twigs, an inner cup of dead leaves, bark, and weed and grass stems, and lined primarily with rootlets. Erwin (1935) found early nests were usually built in 4 to 5 days, while later nests, which were more lightly constructed, were usually built in 3 days.

The nest is usually built in thick shrubbery, vines, or a small tree. It is also occasionally built on the ground in thick vegetation, in a brushpile, or on a low horizontal branch of a large tree. Nests are built in many plant species, with eastern red cedar and blackberry the most frequently reported. Brown Thrashers also nest in several

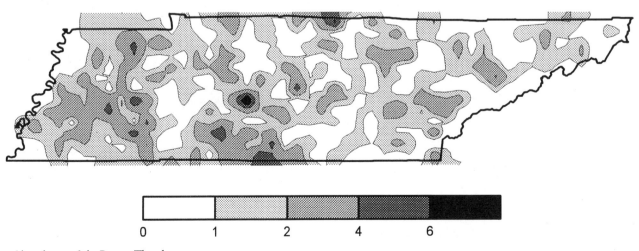

Abundance of the Brown Thrasher.

species of naturalized shrubs and vines, especially multiflora rose, Japanese honeysuckle, and privet. Tree nests have been in both deciduous trees and pines. Aboveground nests range in height from 0.3 to 4.9 m. Erwin (1935) reported an average nest height of 1.5 m (n = 59), and 138 other nests also averaged 1.5 m (s.d. = 0.82 m) high. Erwin (1935) found no ground nests; only 1 of 109 Tennessee nests found by Todd (in Bent 1948) was on the ground, and 3 of a separate Tennessee sample of 132 nests were on the ground.

Egg laying peaks in mid- to late April, when most first clutches are laid. Clutches of 4 eggs are most common, 3 and 5 eggs also common, and 2 eggs rare. Erwin (1935) reported an average clutch size of 3.9 eggs (s.d. = 0.66, n = 50), and 147 other Tennessee clutches average 3.8 eggs (s.d. = 0.70). The latter sample include 4 clutches of 2 eggs being incubated. Murphy and Fleischer (1986) reported an average clutch size of 4.1 eggs in Kansas and found no 2-egg clutches. Burleigh (1958) reported that 2-egg clutches were not infrequent in late nests in Georgia. Erwin (1935) noted a decrease in the size of late season clutches. Eggs are very pale blue to white and covered with very small reddish-brown spots.

Both adults incubate the eggs for 11–14 days, with average periods given as 12.6 days (Erwin 1935) and 13.6 days (Murphy and Fleischer 1986). Both adults also feed the young, which grow rapidly and leave the nest in 11 days (Erwin 1935, Murphy and Fleischer 1986). The female may begin building another nest within 3 days of the young fledgling, while the male cares for the young for 2 weeks or more (Erwin 1935). About 38% of the nests in Erwin's urban Nashville study area fledged young. The Brown Thrasher is at least an infrequent cowbird host and usually ejects cowbird eggs (Friedmann 1963, Friedmann, Kiff, and Rothstein 1977). The only Tennessee incident of cowbird parasitism was reported by Laskey (1943); the incubated cowbird egg disappeared before hatching.—*Charles P. Nicholson.*

Cedar Waxwing
Bombycilla cedrorum

The Cedar Waxwing is present in Tennessee throughout the year, usually in nomadic flocks of a few dozen birds that feed primarily on berries. Its numbers are greatest during the fall and spring, when it occurs in flocks of up to a few hundred birds. Waxwing numbers decrease during May and are lowest during the summer. Nesting waxwings are most common from the Cumberland Plateau eastward. Prior to Atlas fieldwork, nesting had been reported once from each of 3 Middle Tennessee counties and from 15 East Tennessee counties. During Atlas fieldwork, confirmed breeding was reported from an additional 23 counties across the state, probably the largest increase in the known breeding distribution of any Tennessee bird.

The first breeding season report of the Cedar Waxwing was by LeMoyne (1887), who collected a bird in immature plumage in mid-August 1886 at 915 m in Blount County. Rhoads (1895a) described its Tennessee distribution as universal, although based on our present knowledge of its breeding chronology, most of his observations west of Knoxville could have been of migrant birds. Gettys collected a clutch in McMinn County in 1898 (egg coll. data). Ganier (1933a) described it as a very rare summer resident in Middle Tennessee and a common summer resident in East Tennessee. Within East Tennessee, it has, at least until recently, been more numerous in the Unaka Mountains, where it nests at all elevations (Stupka 1963, Atlas results), than in the Ridge and Valley or Cumberlands (e.g., Lyle and Tyler 1934, Bierly 1980).

The Cedar Waxwing was found in 35% of the completed Atlas priority blocks, and its breeding status was either probable or confirmed in 42% of those blocks. Many of the possible records, especially those in early summer, could have been of birds that did not nest locally; with the exception of the southwest corner of the state, however, the distribution of the confirmed records generally encompassed the possible records. The first confirmed breeding in Stewart, Unicoi, and Hawkins Counties was in 1986 and in Bledsoe, Scott, and Jackson Counties in 1987. Waxwings greatly expanded their range in 1988, when breeding was confirmed in a total of 18 counties, including 8 for the first time. The first West Tennessee breeding record was of fledglings in a bottomland forest in Crockett County on the unusually

Cedar Waxwing. Elizabeth S. Chastain

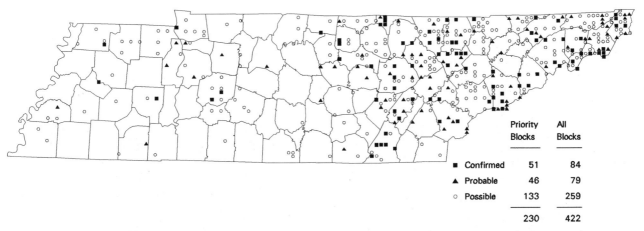

Abundance of the Cedar Waxwing.

early date of 13 June 1988. The number of confirmed records dropped greatly in 1989, increased again in 1990, and dropped somewhat in 1991.

Although new county nest records for the Cedar Waxwing were probably an inevitable result of the Atlas field-work, the great increase in its breeding range was unexpected. Tennessee BBS results also show a recent increase, indicative of a real change in the waxwing population. From 1966 to 1994, waxwing numbers increased by 14.6%/year ($p < 0.01$). This increase began about 1985 and accelerated in 1988, when waxwings were reported at 20 stops of 7 BBS routes, an increase over the 1966–87 average of 7.1 stops on 3.8 routes. Their numbers dropped slightly in 1989 and returned to record high levels in 1990. Their rate of increase is one of the highest of any native bird. From 1966 to 1993, their rangewide population increased ($p < 0.01$) at the rate of 2.0%/year (Peterjohn, Sauer, and Link 1994).

The waxwing was found on 12% of the Tennessee miniroutes, at an average relative abundance of 1.6 stops/route. Its highest numbers were in the Unaka Mountains, northern Ridge and Valley, and northern Cumberlands, all historic breeding areas. Had all of the miniroutes been censused during the years of high waxwing numbers, the waxwing's abundance would probably have been higher in other parts of the state.

Breeding Biology: Most Tennessee waxwing nests have been in woodland edges, frequently near water, and in parklike areas with scattered shrubs and large trees. Unusual nest locations were a cypress swamp in Obion County (Atlas results) and a heavily forested mountainside in Campbell County, about 15 m from a narrow gravel road (Nicholson pers. obs.). In other parts of their range, waxwings are often loosely colonial (Rothstein 1971); more than 1 simultaneously active nests less than 50 m apart have occurred in at least 4 Tennessee counties (Atlas results, Nicholson pers. obs.). At a Smith County site, where 2 active nests were 25 m apart in June 1988, at least 7 waxwing nests were evident in the same area following leaf-drop that fall. Most Tennessee observations, however, have apparently been of isolated pairs.

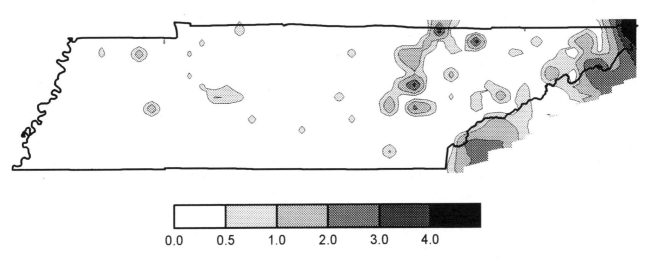

Distribution of the Cedar Waxwing.

Confirmed Breeding Species

Waxwings only defend a small area, which includes the nest and a conspicuous elevated perch used by the male (Putnam 1949); nesting birds often forage in small flocks.

Confirmed Atlas records are about equally divided between observations of nest building, fledglings, and records of nests, usually coded ON because of the difficulty of inspecting nest contents. As elsewhere in its range (Leck and Cantor 1979), the Cedar Waxwing nests relatively late in Tennessee. The earliest nest was observed by Stupka (1963) on 15 May 1946. Nest construction peaks during the first third of June and extends into August. Although second broods, begun before the first fledge, are common elsewhere (Putnam 1949), they have not been definitely reported in Tennessee. The nest is placed on the fork of a horizontal tree limb or less frequently in a tall shrub. Nests have been in many species of deciduous and coniferous trees, with about 40% in the 4 common species of pine. The average height of 54 nests was 7.9 m (s.d. = 3.4, range 2.3–16.8 m), higher than the normal range of 1.5–6.1 m reported elsewhere (Putnam 1949). Both adults construct the nest in about 6 days of loosely woven twigs, grasses, weed stems, and bark strips lined with moss, rootlets, or fine grasses (Putnam 1949, Nicholson pers. obs.).

Clutch size in Tennessee averages 4.13 eggs (s.d. = 0.64, range 3–5, n = 8); Leck and Cantor (1979) give a rangewide average of 4.2 eggs. The eggs are pale bluish gray with sparse brown and black spots. The female incubates the eggs for 12 days, while the male perches conspicuously nearby and feeds the female at or near the nest (Putnam 1949). Both adults feed the nestlings, for the first few days with insects and then largely with fruit. The young leave the nest in 16 days and are independent about 10 days later (Putnam 1949). Parasitism of waxwings by the Brown-headed Cowbird has not been reported in Tennessee. It is uncommon elsewhere, in part because waxwings usually reject cowbird eggs (Friedmann, Kiff, and Rothstein 1977).—*Charles P. Nicholson.*

Loggerhead Shrike
Lanius ludovicianus

The Loggerhead Shrike occurs across Tennessee in low-elevation grasslands, croplands, and old fields with scattered shrubs, small trees, or fencerows. It is unusual among songbirds both in feeding on small mammals, birds, and reptiles as well as insects, and in its habit of impaling prey on thorns and barbed wire. The shrike is present throughout the year and most common during the winter. Its breeding population increases in abundance from the east, where it is rare to uncommon, to West Tennessee, where it is fairly common to common.

Loggerhead Shrike. Elizabeth S. Chastain

The Loggerhead Shrike was probably an uncommon and locally distributed species in prehistoric Tennessee, occurring in the prairies of north-central and northwest Tennessee and large Indian-maintained savannas and old fields elsewhere in the state. It was first reported by Fox (1886), who found it in Roane County in March 1885, when migrant birds could have been present. Rhoads (1895a) observed it only in the Reelfoot Lake area and in Shelby County, where he found it breeding. These reports, other early annotated lists, and its absence from the McMinn County egg collection of Gettys all suggest that nesting shrikes were rare in Middle and East Tennessee in the late nineteenth and early twentieth centuries. Ganier (1933a) described its summer status as common in West Tennessee and rare in Middle and East Tennessee.

Literature records suggest shrike numbers began increasing in Middle and East Tennessee during the 1940s, an increase that continued into the 1960s. White (1956) described it as uncommon in Greene County, and Howell and Monroe (1958) described it as fairly common in Knox County. The first nest record for Carter County was in 1964 (*Migrant* 35:66), in Sullivan County in 1965 (*Migrant* 36:67), and in Johnson County in 1966 (*Migrant* 37:24). The reasons for the increase during this period are not clear, as suitable habitat in the form of farmlands and hedgerows peaked about 1940.

Tennessee BBS results show a significant (p < 0.01) decrease in shrike numbers of 7.0%/year from 1966 to 1994, one of the largest of any species nesting in the state. Annual BBS averages reached their lowest levels in the early 1980s and then slightly increased. Rangewide, the shrike decreased at the average rate of 3.5%/year (p < 0.01) from 1966 to 1993 (Peterjohn, Sauer, and Link 1994). Causes of this decrease are the loss of farmland to reforestation and urban development and changing farming practices toward larger fields with fewer hedgerows and improved pastures. Other more subtle causes, such as pesticides (Anderson and Duzan 1978, Busbee 1977), may also be involved, as the rate of the shrike's decline was faster than that of its gross habitat. Much seemingly suitable habitat is unoccupied, particularly in East Tennessee.

The Loggerhead Shrike was found in 57% of the completed Atlas priority blocks, most frequently in the Ridge and Valley, Central Basin, north-central and south-central Highland Rim, and the Loess Plain. Some local concentrations in Middle Tennessee were the result of special searches for the shrike (*Migrant* 57:109). It was absent from the Cumberland Mountains as well as most of the Unaka Mountains and Cumberland Plateau, all regions with limited suitable habitat. The presence of suitable habitat in the Sequatchie Valley portion of the plateau is reflected by the shrike records in Marion, Sequatchie, and Bledsoe Counties. The Loggerhead Shrike was recorded on 19% of the miniroutes at an average relative abundance of 1.3 birds/route, indicative of the low density of this bird that often frequents roadsides. It was found on more than one-quarter of the routes in the Mississippi Alluvial Plain, Loess Plain, and Central Basin.

Because of its population decline, the subspecies of Loggerhead Shrike found in the northeastern U.S. and the eastern portion of Tennessee, *L. l. migrans,* was considered a Category 2 candidate for federal listing as Endangered or Threatened until that designation was abolished (USFWS 1994, 1996). Its Tennessee population apparently increased from the 1940s to the 1960s and then declined in the 1970s. Atlas results show it to be widespread, and recent BBS results suggest a slight increase. A state listing such as endangered or threatened is not presently justified. There is, however, a need for detailed study of the shrike's habitat use and nest success in different habitats.

Breeding Biology: Confirming the breeding of the Loggerhead Shrike was fairly easy for Atlas workers because of the conspicuousness of fledglings, which accounted for almost two-thirds of the confirmed records. Shrike nests, although usually moderately well hidden, are with practice fairly easy to find and identify; they accounted for slightly over one-fifth of the confirmed records. The number of detailed nest records available, however, is surprisingly low.

Compared to other passerines, the Loggerhead Shrike nests fairly early; males begin occupying territories and singing in late winter. The nest site fidelity of males is considerably higher than that of females (Kridelbaugh 1983). The relatively large nest, well made and bulky, is built by both adults of sticks, twigs, weed stems, and often human-made materials such as string. The cup is lined with grasses, feathers, and frequently cotton. The nest is placed in a tree or shrub, and the most frequently reported species is red cedar. Other commonly used species are hackberry and multiflora rose. When in a deciduous tree, the nest is usually in a tangle of vines, such as Japanese honeysuckle, or on a lower horizontal limb with densely branching twigs. Other studies have noted the frequent use of red cedar and found nest success is often higher for nests in red cedar than in other species such as multiflora rose (Kridelbaugh 1983, Gawlik and Bildstein 1990). Nest height in Tennessee averages 3.0 m (s.d. = 1.5, n = 45, range 1.2–9.1 m).

Egg laying begins in late March and peaks in early to mid-April. Shrikes frequently raise a second brood (Gawlik and Bildstein 1990), although the peak of initiation of

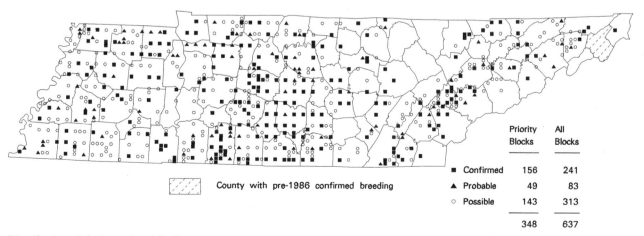

Distribution of the Loggerhead Shrike.

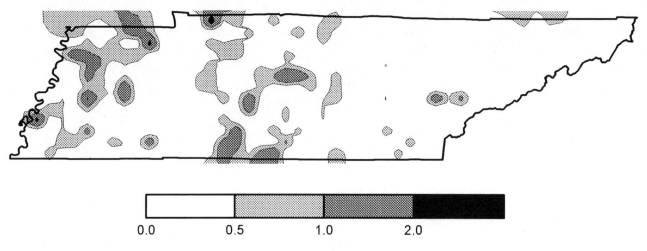

Abundance of the Loggerhead Shrike.

second clutches is not evident in the limited Tennessee data. Clutches of 5 and 6 eggs are most common, with an average of 5.32 eggs (s.d. = 0.94, n = 34, range 3–7). Average clutch sizes in studies elsewhere are from 4.4 to 6.4 eggs and increase with latitude (Kridelbaugh 1983, Gawlik and Bildstein 1990). The eggs are dull white to buffy with brown and gray spots, often concentrated at the larger end (Harrison 1975). The female incubates the eggs for about 17 days, beginning before the clutch is complete; during this period she is fed by the male (Kridelbaugh 1983). Both adults feed the nestlings, which leave the nest in about 19 days. The fledglings remain near the nest for 2 or 3 days and are dependent on the adults for 3 or 4 weeks (Kridelbaugh 1983).—*Charles P. Nicholson*.

European Starling
Sturnus vulgaris

The European Starling is an abundant permanent resident of the developed portions of Tennessee. It occurs in large flocks most of the year, feeding in urban areas, mowed lawns, and agricultural areas. During the winter, migrants from north of Tennessee join blackbirds and the resident starlings in large nocturnal roosts that may number in the hundreds of thousands or millions of birds.

As its name implies, the European Starling is not native to North America. Following unsuccessful introduction attempts, it became established in the New York City area as a result of birds released in 1890 (Wing 1943). From New York it spread rapidly to the south and southwest. The normal pattern of its spread was for small flocks of winter stragglers to first appear in an area, followed by nesting about 5 years later. After initial nesting, their numbers usually increased slowly the first few years, then increased rapidly until stabilizing in 25 to 30 years (Wing 1943).

The route of the starling's spread into Tennessee was through the Great Valley of Virginia and perhaps from central Kentucky into Middle Tennessee. The first Tennessee records were in December 1921 at Nashville and in Bluff City, Sullivan County (Tyler 1922, Ganier 1924). Nesting was reported at Bristol and Knoxville in 1925, at Athens in 1926, and at Nashville in 1928 (Tanner 1988). A pair was present at Woodbury, Cannon County, in June 1927 and nested there in 1928 (Ganier 1928). The first known West Tennessee nestings were at McKenzie in 1933 and at Memphis in 1935 (Tanner 1988).

Following the initial nestings, the European Starling population increased rapidly. In several areas, they became one of the most abundant, or at least most conspicuous, birds (e.g., Knox County, Howell and Monroe 1957). Although there is limited supporting quantitative data, the starling has been detrimental to several native bird species with which it competes for nest cavities.

European Starling. Elizabeth S. Chastain

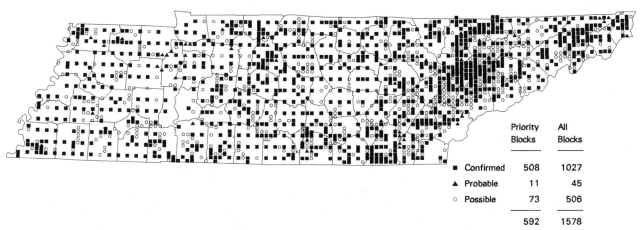

Distribution of the European Starling.

These species include the Northern Flicker, Red-headed and Red-bellied Woodpecker, Great Crested Flycatcher, Purple Martin, and Eastern Bluebird. The starling also competes with several species for wild fruits in the fall and winter. Roosts of the starling and other blackbirds occasionally cause public health problems, resulting in local roost control efforts. These measures have little apparent effect on the statewide starling population.

Atlasers found the European Starling easy to observe and recorded it in 90% of the completed priority blocks. It was absent from many blocks in the Unaka Mountains, as well as a few in the Cumberlands and the Western Highland Rim that were heavily forested and had a very low human population density. Its absence from priority blocks in West Tennessee is difficult to explain, as virtually all of those blocks have some farmland and human inhabitants. Elevation is probably not a limiting factor for the starling, as it was found in settled blocks at elevations above 975 m.

The starling population, as measured by Tennessee BBS routes, has shown a nonsignificant overall trend from 1966 to 1994, although it decreased from 1966 to 1979 and increased from 1980 to 1994. Its rangewide population decreased by 1%/year ($p < 0.01$) from 1966 to 1993 (Peterjohn, Sauer, and Link 1994). The starling is an abundant species and the second most abundant species reported on Tennessee BBS routes. It was reported on 73% of the Atlas miniroutes, at an average relative abundance of 3.9 stops/block. By physiographic region, its highest average abundance, 5.4 stops/route, was in the Ridge and Valley, where it occurred on all but 6 routes. Next highest were the Central Basin (4.3) and the Eastern Highland Rim (4.1). All of the other areas has averages of 3.4 or less. Its numbers were also frequently, but not always, high in urban areas.

Breeding Biology: Atlasers found it easy to confirm breeding of the European Starling and the proportion of confirmed records was one of the highest of any species. The starling starts nesting activities in late winter, when it begins visiting and roosting in potential nest sites (Kessel 1957). The winter flocks break up during

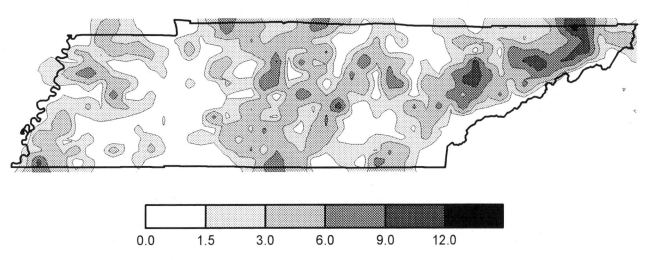

Abundance of the European Starling.

March, and most birds are paired and occupying nest sites by early April. They are usually monogamous, although polygyny has been reported several times (Kessel 1957). The male usually first occupies the nest site, which it aggressively defends from other birds. The male sings and displays by fanning its wings and erecting its iridescent throat feathers while it is perched near the nest hole. Territorial defense is limited to the immediate vicinity of the nest site.

Starlings are very adaptable in their choice of nest sites and nest in woodpecker and natural cavities in trees, holes in buildings, crevices in rock faces, and bird boxes. The male often begins nest construction before he is mated; following mating, both birds build the nest (Kessel 1957). They fill much of the nest cavity with grasses and weed stems and make a cup in the top of the mass. The cup may be lined with feathers, grasses, or other materials. Construction of the first nest may take up to a month and later nests may be built in a day or 2 (Kessel 1957).

Egg laying in Tennessee peaks during the second and third weeks of April. Clutch size is 3–7 eggs, with clutches of 4 and 5 most common throughout its range (Crossner 1977). Tennessee clutches, excluding 1 example of 9 eggs, which may have been the product of 2 females, average 4.94 eggs (s.d. = 0.79, n = 53). The eggs are pale bluish green to almost white and usually unmarked. Both adults incubate the eggs for an average of 12 days and both feed the nestlings, which leave the nest in about 21 days (Kessel 1957). At the time of fledging, they are almost fully feathered and fly well. The young are independent of the adults 3 to 4 days after fledging and form flocks with other juveniles. Renesting after loss of a clutch or brood is common and studies north of Tennessee show second broods are common (Kessel 1957). The chronology of nest records and confirmed Atlas observations suggest that second broods are uncommon in Tennessee.
—*Charles P. Nicholson.*

White-eyed Vireo
Vireo griseus

A common migrant and summer resident, the White-eyed Vireo arrives by early April and remains through mid-October; some late migrants linger well into November (Robinson 1990). It inhabits old fields and second-growth areas dominated by a variety of shrubs, young trees, and brier thickets.

The White-eyed Vireo, a persistent singer, is often heard long before it is seen; its song can be heard as soon as it arrives in the spring, throughout the summer, and even into September. In a study spanning 20 states, Bermuda, and the Bahamas, Borror (1987) documented

White-eyed Vireo. Chris Myers

614 different song types for this species; songs typically last slightly over a second and are given at the rate of about 13 times per minute. However, this vireo occasionally sings an extended version of the song, lasting several seconds and consisting of a wild medley of buzzes, whistles, and other harsh sounds; this song is reminiscent of the Gray Catbird and is referred to as a "chattering" song by Bradley (1980).

White-eyed Vireos usually sing songs of one type for a while before changing to another song type; most birds have a repertoire of 10–14 song types (Borror 1987). The White-eyed Vireo may mimic other species (Bent 1950, Adkisson and Conner 1978), but it is not as accomplished in this behavior as the Northern Mockingbird.

White-eyed Vireos have a statewide distribution and were found in 88% of the completed priority blocks. They become less common at higher elevations in the eastern part of the state, especially in the Unaka Mountains where birds are typically recorded below an elevation of 1000 m (Eller and Wallace 1984, Robinson 1990). Because it is highly vocal and thus easily recognized, the Atlas data most likely present a reliable reflection of the true distribution of the species within the state.

Atlasers found White-eyed Vireos at an average of 2.8 stops on the 418 miniroutes (60%) recording the species. Birds were most frequently recorded in the Coastal Plain Uplands and Western Highland Rim physiographic regions. They occur in conspicuously low numbers in local areas of the state managed intensively for grasslands, pastures, and grazing; such areas often lack the shrubby vegetational structure required by this species. Recent density estimates reported on Breeding Bird Census plots include 46 pairs/100 ha in a brushy pasture on the Eastern Highland Rim in Putnam County

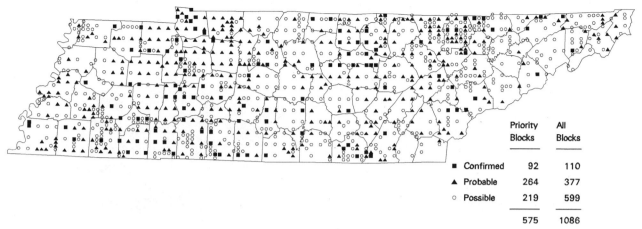

Distribution of the White-eyed Vireo.

	Priority Blocks	All Blocks
■ Confirmed	92	110
▲ Probable	264	377
○ Possible	219	599
	575	1086

(Simmers 1978) and 35/100 ha in a 5-year-old deciduous clearcut in Washington County (Lewis 1979).

From 1986 to 1991, an average of 7.0 birds per route was recorded on the Tennessee BBS routes; the species was present on 39 of Tennessee's 42 routes. Tennessee BBS results for 1966–94 period show a declining trend of 1.2%/year ($p < 0.10$). The declining trend was 3.3%/year ($p < 0.01$) from 1980 to 1994. Probable reasons for this decline include the elimination of brushy habitat and hedgerows in and around farmlands and a general shift in Tennessee forests toward sawtimber-size stands since 1980 (May 1991). Rangewide its population did not show a significant overall trend from 1966 to 1993, although it declined by 3.0%/year from 1978 to 1988 ($p < 0.01$) (Peterjohn, Sauer, and Link 1994, Sauer and Droege 1992).

Breeding Biology: The White-eyed Vireo utilizes a monogamous breeding strategy. Courtship involves the male displaying before the female, at which time he may fluff his feathers and spread his tail while giving a snarling call (Terres 1980, Ehrlich, Dobkin, and Wheye 1988). Both sexes build the nest, which is characteristically placed on the twig of a small tree, shrub, or bush, usually within 2.5 m of the ground. The nest is suspended from the twig and is often pointed or tapered at the bottom. External nest materials include mosses, lichens, pieces of wasp nests, leaves, small sticks, and bark shreds; internally, the nests are lined with bark strips, fine grasses, plant fibers, plant down, and spider's silk (Bent 1950, Harrison 1978). At the nest site, White-eyed Vireos are typically fearless of humans (Bent 1950, Ehrlich, Dobkin, and Wheye 1988).

In Tennessee, White-eyed Vireos place their nests in a wide variety of shrubs and small trees, including rhododendron, buckbush, and beautyberry shrubs, as well as dogwood, hickory, elm, boxelder, walnut, tulip-poplar, willow, white oak, hawthorn, crabapple, and cedar trees. The height of 36 nests averaged 0.9 m (s.d. = 0.4 m) and ranged from 0.3 to 2.1 m. Descriptions of nest materials found in Tennessee match those given by Bent (1950) and Harrison (1978) above.

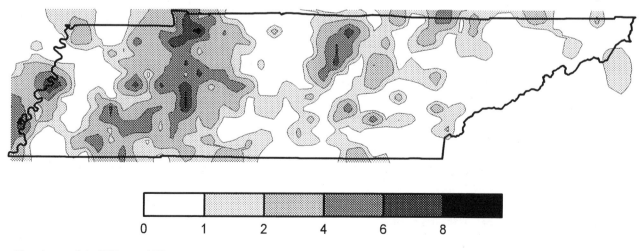

Abundance of the White-eyed Vireo.

Confirmed Breeding Species

Clutch size reported for 48 nests in Tennessee was usually 3–4, although one nest had 5 eggs ($\bar{x}$ = 3.75, s.d. = 0.60, range 2–5). Elsewhere, clutch sizes of 3–5 are widely reported (Bent 1950, Harrison 1978). The eggs are white with tiny black speckling or spotting. The species is commonly parasitized by the Brown-headed Cowbird. At least 5 nests in Tennessee were parasitized by cowbirds, including one Warren County nest containing a recently hatched cowbird young in 1980. Incubation is by both sexes and lasts about 12–15 days (Harrison 1978). The length of the nestling period is poorly documented. Both sexes feed the young (Ehrlich, Dobkin, and Wheye 1988).

Although it is apparently double-brooded in the southern part of its range, Tennessee confirmed breeding records for this vireo suggest that only one brood is usually raised in the state. Nests with eggs are rarely found in July and likely represent abandoned nest attempts or reinitiation of nesting after an earlier failure. The median date of 60 observations of nests with eggs is 23 May, suggesting that most eggs are usually laid from about early May to early June; recently fledged young may be observed into early August.—*John C. Robinson.*

Solitary Vireo
Vireo solitarius

The Solitary Vireo is a fairly common summer resident of the mountainous areas of East Tennessee. Within its breeding habitat, it is one of the first migrants to return in the spring, arriving in mid- to late March. In the rest of the state, it is an uncommon migrant, with a prolonged spring migration lasting into mid-May. Fall migration occurs from late September to early November. There are several winter records, mostly from West Tennessee (Robinson 1990).

Solitary Vireos breeding in the southern Appalachians were described as a distinct subspecies, the Mountain Solitary Vireo, *V. s. alticola*, by Brewster (1886). The first Tennessee records of this subspecies were by Fox from Hamilton and Roane Counties (Fox 1886, 1887, Wetmore 1939), and by Langdon (1887) from Chilhowee and Defeat Mountains, Blount County. Langdon's specimens, collected in August 1886, were in molting summer plumage, suggesting the birds had bred nearby.

Solitary Vireos have a loud, distinctive voice, more musical than the other forest-dwelling vireos within its range. The Mountain Solitary's song has also been described as louder and more musical than that of the northeastern race *V. s. solitarius* (Brewster 1886, Bent 1950). No quantitative studies have been published describing a distinct southern dialect.

Solitary Vireo. Chris Myers

As shown by the Atlas results, the Solitary Vireo is distributed throughout the Unaka Mountains, more locally distributed in the Cumberlands, and very local in the northern Ridge and Valley. The Atlas results are probably an accurate portrayal of its range, although Solitaries were probably missed on some high ridges, such as Clinch Mountain, in the northern Ridge and Valley. Stupka (1963) reported Solitary Vireos regularly breed from the highest elevations of the Great Smokies down to about 610 m and listed several summer records at somewhat lower elevations. During Atlas fieldwork, however, it was found as low as 320 m in the southern Unaka Mountains. In the Cumberlands, it regularly occurs from the highest elevations down to about 520 m, and the lowest nest record is at 610 m (Nicholson 1987). It occasionally occurs at lower elevations in the Ridge and Valley—the probable record shown in central Anderson County was a territorial male present at least 2 years at 275 m elevation near Norris (Nicholson pers. obs.).

There has been no apparent change in the Solitary Vireo's Tennessee range in the last century. Rhoads (1895a) did not report it from the Cumberland Plateau; it is presently uncommon in the plateau areas he visited, and he probably overlooked it. The records of Fox (1886, 1887) suggest it was nesting on the plateau at that time. Solitaries have expanded their range elsewhere in the Southeast (Odum 1948).

On the Cumberland Plateau and low elevations of the Unaka Mountains, Solitary Vireos usually occur in yellow pines or mixed hemlock-hardwoods on northern slopes and in ravines. At higher elevations, it occurs in all forest types, including spruce-fir. Where the elevational ranges of the Solitary and Red-eyed Vireos overlap, the two species are often sympatric; above the limit of the Red-eye's range, the Solitary is the only nesting

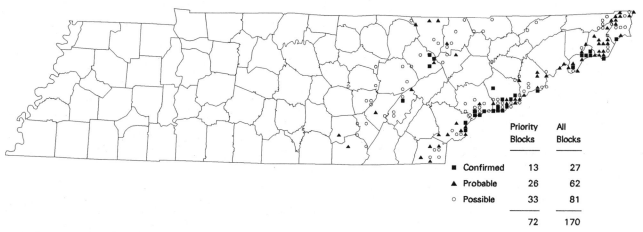

Distribution of the Solitary Vireo.

vireo. Kendeigh and Fawver (1981) found its highest densities (in 1947–48) in climax spruce-fir forest (average of 60 pairs/100 ha), cove forest (62.5/100 ha), and hemlock-deciduous forest (90/100 ha). Robertson (1979) found a density of 17/100 ha in a Cumberland Plateau virgin mixed mesophytic forest. Solitaries have been reported in densities of fewer than 12 pairs/100 ha in the Cumberland Mountains (e.g., Smith 1977, Turner and Fowler 1981b).

Solitary Vireos were recorded on 41 Atlas miniroutes, at an average relative abundance of 2.9 stops/route. Its highest counts were in the Great Smoky Mountains, where it was found at 15 stops on one route and 14 stops on another. It was found on 11 routes in the Cumberlands, at a maximum of 3 stops/route.

Solitary Vireos have been found on too few Tennessee BBS routes to detect a population trend. Wilcove (1988), in a recensus of plots in the Smokies first censused 35 years earlier, found no clear change in numbers and much variation in consecutive years. Range-wide, Solitary Vireos increased 3.2%/year ($p < 0.01$) on BBS routes from 1966 to 1993 (Peterjohn, Sauer, and Link 1994). The rate of increase has slowed since 1978, possibly due to changes in Central American and Caribbean wintering habitats (Robbins et al. 1989). In contrast with more northern and western populations, the Mountain Solitary Vireo winters in the southeastern U.S. (Bent 1950, AOU 1957) and would be subject to different pressures during the winter.

Breeding Biology: Many details of the breeding biology of the Mountain Solitary Vireo are poorly known. Males of the northeastern race begin building nests prior to the females' arrival, and they display at partially built nests or suitable nest sites (James 1978). Both birds build the nest, with the female eventually doing most of the building; the nest is completed in about 8 days. Nests are suspended from a forked branch in the lower portion of a tree canopy or in an understory shrub or sapling. The nest is a well-made, thick-walled cup built of pieces of grass, bark fibers, and moss and lined with fine weed stems, rootlets, or hair and decorated

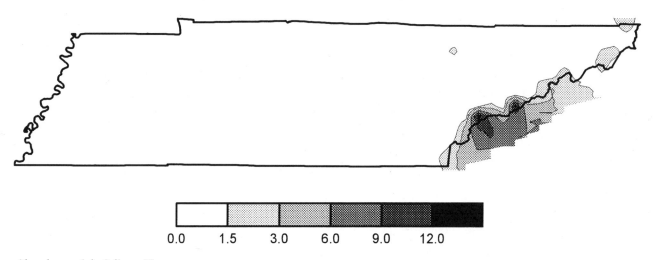

Abundance of the Solitary Vireo.

Confirmed Breeding Species

on the outside with bits of lichen, paper from wasp nests, and spider egg cases (Bent 1950, egg collection data). Tennessee nests range from 0.3 to 9 m above ground and average 3.8 m high (s.d. = 2.0 m, n = 16). The most commonly reported tree used for nesting is the sugar maple; other species include yellow birch, hydrangea, oaks, and hemlock. Bent (1950) noted an apparent preference for nesting in yellow birches in the southern Appalachians.

Clutches of 3 and 4 eggs have been reported from the southern Appalachians (Bent 1950); 11 of 16 apparently complete Tennessee clutches contain 4 eggs and the remainder 3 eggs. A possibly incomplete clutch of 2 eggs was reported in mid-July (NRC). The eggs are creamy to pinkish white, lightly spotted with brown. Both adults incubate the eggs, and the male may sing while on the nest (James 1984). The adults are very approachable when at the nest, occasionally allowing an observer to touch them. The incubation and nestling periods in the southern Appalachians are unknown.

Bent (1950) presents evidence of Mountain Solitary Vireos raising 2 broods. The available nest records suggest this occurs in Tennessee, especially at lower elevations. The earliest nest record is of an almost completed nest on 6 April 1974 at 670 m in the Smokies (M. D. Williams and Nicholson, pers. obs.), and Stupka (1963) reported an incubating bird on 23 July. Most egg dates have been from late May and June. The only record of cowbird parasitism available from Tennessee is of a nest with 1 cowbird and 3 vireo eggs (NRC). Solitary Vireos are an uncommon cowbird host (Friedmann, Kiff, and Rothstein 1977).—*Charles P. Nicholson.*

Yellow-throated Vireo
Vireo flavifrons

The hoarse-voiced Yellow-throated Vireo is a fairly common to uncommon summer resident, found in open deciduous and mixed woodlands at low elevations across Tennessee. It usually arrives in early April and departs by mid-September (Robinson 1990).

The distribution of the Yellow-throated Vireo in Tennessee is very similar to that of the Red-eyed Vireo, except that the Yellow-throated is absent from elevations above about 915 m in the Unaka Mountains (Stupka 1963). Yellow-throated Vireos are less persistent singers than Red-eyeds; this and their habit of foraging high in tree canopies (James 1976, Williamson 1971) make them more difficult to observe. Atlasers found Yellow-throated Vireos in about two-thirds of the completed priority blocks, including most priority blocks in the Coastal Plain Uplands, Western Highland Rim, Cumberland Plateau, and Cumberland Mountain regions, all areas with a high proportion of forest.

Yellow-throated Vireos occur in hardwood or mixed pine-hardwood forests with, in comparison to the other low-elevation vireos, larger, taller trees and more open understory (James 1971, Robbins, Dawson, and Dowell 1989). Studies of their minimum forest area requirements have reached different conclusions (Whitcomb et al. 1981, Robbins, Dawson, and Dowell 1989). Yellow-throated Vireos utilize both forest interior and edge habitats and often nest along a woodland edge (Whitcomb et al. 1981, Nicholson pers. obs.). They typically occupy large territories at low densities, especially in comparison with the Red-eyed Vireo (Williamson 1971). They have been recorded on several Breeding Bird Census plots in hardwoods or mixed pine-hardwoods; the highest density, 22 pairs/ 100 ha, was in a Cumberland Mountains maple-gum-hickory forest at middle elevations (Smith 1977). Most other Tennessee censuses have reported densities of 12/100 ha or fewer. Mengel (1965) reported relatively uniform densities of 12–18/100 ha in several forest types on the Cumberland Plateau of Kentucky.

Yellow-throated Vireos were found on 37% of the Atlas miniroutes at an average relative abundance of 1.7 stops/route. Its distribution on miniroutes generally correlates well with Atlas block results except in the northern Loess Plain, where it was found on few miniroutes. The greatest abundance, averaging 3.2 stops/ route, was in the Cumberland Mountains, one of the most heavily forested parts of the state. Rangewide BBS results show the vireo reaches its highest average abundance in the Cumberlands and Blue Ridge (Robbins, Bystrak, and Geissler 1986).

Yellow-throated Vireos were probably fairly common throughout most of prehistoric Tennessee, and

Yellow-throated Vireo. Chris Myers

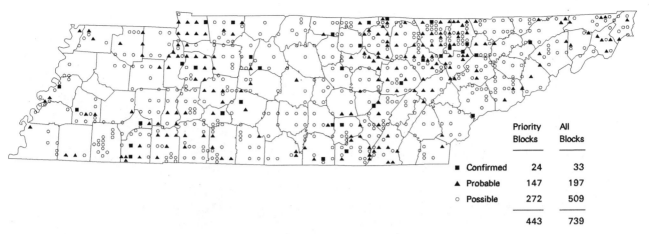

Distribution of the Yellow-throated Vireo.

nineteenth-century deforestation would have resulted in local declines. Rhoads (1895a) recorded them at all low-elevation areas he visited. Ganier (1933a) described the species as fairly common across the state. Yellow-throateds are now uncommon in the Ridge and Valley, rare in inner portion of the Central Basin, and uncommon and locally distributed in West Tennessee. Tennessee BBS results show no significant population trend from 1966 to 1994; rangewide, it increased by 1.0%/year from 1966 to 1993 (Peterjohn, Sauer, and Link 1994). During the Atlas period, Yellow-throated Vireos were recorded on Tennessee BBS routes at an average of 3.8 birds and 3.7 stops/50-stop route.

Breeding Biology: Confirming the breeding of Yellow-throated Vireos was difficult for atlasers, and few nests have been reported in the state. Male Yellow-throateds establish territories shortly after their spring arrival. They sing frequently and may begin building one or more nests (James 1978). After the female's arrival, the male displays to her at the unfinished nests. Following mating, the female does most of the nest building, completing the nest in about a week. The cup-shaped nest is suspended from the fork of a tree limb, sometimes as low as 1 m above the ground, but more typically high in a tree in the inner portion of the canopy (Bent 1950, James 1976). Ten Tennessee nests ranged from 3 to 18.3 m high and averaged 9.5 m above ground (s.d. = 5.1 m) within the range reported by Bent (1950), which is slightly higher than the average reported by Williamson (1971) and somewhat lower than the average height reported by James (1976). Most Tennessee nests have been in hardwoods, including white oak, tulip-poplar, and sycamore; 2 McMinn County nests were in pines. The nest is built of bark strips, plant fibers, and grasses held together with spider webs and lined with fine grasses or hair. The outside is decorated with lichens, moss, and masses of spider webbing, and Bent (1950) described it as the handsomest nest of any vireo.

The few Tennessee nest records show egg-laying peaks in early May. Clutches of 3 and 4 eggs have been

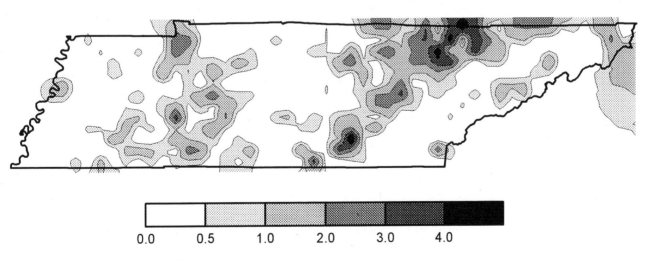

Abundance of the Yellow-throated Vireo.

Confirmed Breeding Species

reported in Tennessee; elsewhere, clutch size ranges from 3 to 5, with 4 the most common (Bent 1950). The eggs are white to cream-colored, with prominent brown or black spots concentrated around the large end. Both the male and female incubate the eggs for about 2 weeks, and the male may sing infrequently while incubating (Bent 1950, Smith, Pawlukiewicz, and Smith 1978). Both adults feed the young, and following fledging the brood is divided among the adults, who care for them for an undetermined time period.

The Yellow-throated Vireo is frequently parasitized by Brown-headed Cowbirds and may occasionally build a new nest floor over a cowbird egg (Friedmann 1963). Tennessee records of cowbird parasitism are few. I have 3 times observed adult vireos feeding fledgling cowbirds, in Cumberland, Van Buren, and Smith Counties. On 28 June 1986 a pair of vireos was seen fighting off a female cowbird attempting to approach the vireo's nest (*Migrant* 57:109).—*Charles P. Nicholson.*

Warbling Vireo
Vireo gilvus

The Warbling Vireo is an uncommon summer resident of Tennessee, found mostly in the western half and the northeastern corner of the state. It usually arrives by mid-April and departs by mid-September; migrants are infrequently reported outside of nesting areas. Although it is drably plumaged and spends most of its time in tree canopies, the Warbling Vireo is easily identified by its melodic warbled song, which it sings almost incessantly until late summer.

The Warbling Vireo occupies open deciduous woodlands and parklike open areas with scattered trees, frequently close to a stream (James 1971, 1976). In Tennessee, it has been found in rural and suburban yards with large shade trees, urban parks, cemeteries, riparian woodlands, and bottomland forests. Breeding density information is very limited. Ford (1990) found a single Warbling Vireo on 3 of 59 1-km-long transects in West Tennessee forested wetlands. In prehistoric Tennessee, it probably occurred in riparian woodlands and the savannah-like areas in the north-central and northwest parts of the state. Its numbers and range probably increased as European settlers opened the forests, creating suitable habitat in upland as well as riparian areas.

The Warbling Vireo was first reported in Tennessee by Rhoads (1895a), who observed it at all localities he visited across the state except for the Roan Mountain area. Gettys considered it a rare breeder in McMinn County around 1900 (Ganier 1935a). Ganier (1933a) described it as a fairly common summer resident in each

Warbling Vireo. Chris Myers

of the 3 major divisions of the state. Herndon (1950b) described it as a rare summer resident in Carter County, and Howell and Monroe (1957) described it as an uncommon summer resident in Knox County.

In recent decades, the Warbling Vireo has disappeared from parts of its range in the eastern half of the state. None have been reported during breeding season from the Cumberland Plateau, Chattanooga area, or McMinn County since Rhoads's and Gettys's observations. The last such observation in the Knoxville area was in 1972 (*Migrant* 43:54). The reasons for this decrease are unclear, as no change in habitat is obvious. The Warbling Vireo is recorded on too few Tennessee BBS routes to analyze its population trends, and virtually all routes recording it are in the western half of the state. From 1966 to 1988, the Warbling Vireo significantly ($p < 0.05$) increased by 0.9%/year on eastern North American BBS routes, although the increasing trend reversed during the latter half of this period (Sauer and Droege 1992).

The Warbling Vireo was found in 10% of the completed Atlas priority blocks. It occurred in the highest proportion of blocks in the Mississippi Alluvial Plain (10 of 14), followed by the Western Highland Rim (24 of 113). Each of these regions has extensive riparian forests along the major rivers and their tributaries. Most records elsewhere in the state were also in riparian areas. The Warbling Vireo was also reported on 27 miniroutes, at an average relative abundance of 1.6 stops/route. The highest regional average, 2.7 stops/route, was in the Mississippi Alluvial Plain. Although it was probably missed in a few blocks where access to riparian forests was difficult, the Atlas results are probably a good portrayal of its general range in the state.

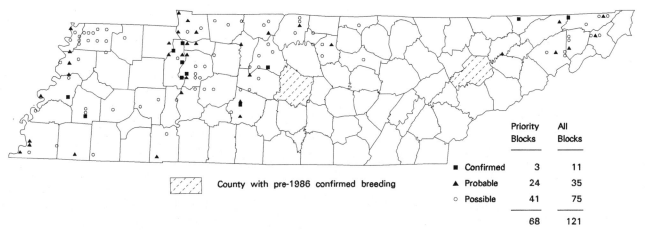

Distribution of the Warbling Vireo.

Breeding Biology: Confirming the breeding of the Warbling Vireo was difficult, due to its low numbers and its habit of frequently nesting high in tree canopies. The male's habit of singing while near and on the nest (Howes-Jones 1985) should make finding nests somewhat easier. Only 4 of the confirmed records, however, were of nests. All of the other confirmed records were of fledglings.

The Warbling Vireo often begins nest construction in Tennessee by the end of April, and most clutches are laid during the first half of May (Monk 1934). The nest is suspended from the crotch of the horizontal limb of a deciduous tree, usually near the outer edge of the canopy (Tyler 1950b). It is usually protected by the overhead canopy, but visible from below (Monk 1934). Both adults build the cup-shaped nest from bark strips, plant fibers, grasses, and small leaves, bound together with spider webs. The lining is made of fine grasses and hair. Thirty-five nests observed by Monk (1934) in the Nashville area were between 2.4 and 12.2 m above ground, with most below 7.6 m. These nests were in 9 deciduous tree species. Eight other nests averaged 6.6 m above ground (range 3.4–10.7 m, s.d. = 2.9). Nests elsewhere have been as high as 27.4 m (Tyler 1950b). More Tennessee nests have been reported in sycamores than in any other tree species.

Clutches of 3 and 4 eggs have been reported in Tennessee, with an average size of 3.6 (s.d. = 0.52, n = 9). Both adults incubate the eggs for about 12 days, as well as feed the nestlings, which leave the nest in about 2 weeks (Monk 1934, Tyler 1950b). The fledglings are cared for by the parents for at least 2 weeks (Monk 1934). The Warbling Vireo is single-brooded in Tennessee; renesting occurs following destruction of a nest (Monk 1934).

Parasitism of the Warbling Vireo by the Brown-headed Cowbird has been reported at least twice in the Nashville area (Monk 1936). One of 35 nests contained 1 cowbird and 4 vireo eggs, and adult vireos were also observed tending a fledgling cowbird. Elsewhere in eastern North America, the Warbling Vireo is a frequent cowbird host (Friedmann 1963).—*Charles P. Nicholson.*

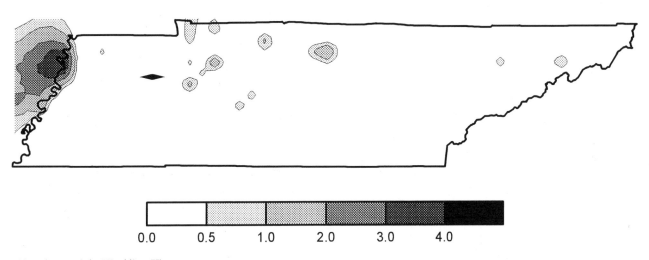

Abundance of the Warbling Vireo.

Confirmed Breeding Species

Red-eyed Vireo
Vireo olivaceous

The Red-eyed Vireo is a common summer resident of hardwood and mixed forests throughout Tennessee. It usually arrives in mid-April. Fall migration begins in mid-August (Nicholson unpubl. data), and most birds have departed by early October. From its spring arrival until mid-summer, male Red-eyes sing most of the day, more persistently than other forest species.

The Red-eyed Vireo is one of the most common and widely distributed forest-dwelling species, found in 92% of the completed priority blocks. Because of the vireo's conspicuous song, it is unlikely that it was overlooked by atlasers. Robbins, Dawson, and Dowell (1989) found Red-eyed Vireos somewhat sensitive to the size of the forest, with their probability of occurrence dropping below 50% in forests of less than 2.5 ha. Several blocks

Red-eyed Vireo. Elizabeth S. Chastain

in the Western Coastal Plain and Central Basin apparently lacked sufficient contiguous deciduous forest cover to provide habitat for Red-eyed Vireos. Prior to widespread forest clearing, Red-eyed Vireos probably occurred throughout the state except for the highest elevations of the Unaka Mountains, where they are scarce above 1525 m (Stupka 1963).

Red-eyed Vireos breed in a variety of forest habitats, including mature and second-growth forests, and prefer stands with numerous saplings in the understory (Lawrence 1953). Pure coniferous forests are avoided. Although several authors (e.g., Tyler 1950a) have noted their occurrence in suburban, parklike areas with large shade trees, this habitat appears to be infrequently occupied in Tennessee (Nicholson pers. obs.). The highest densities of Red-eyed Vireos reported on Breeding Bird Census plots are 190 pairs/100 ha in a mixed-mesophytic ravine forest on the Eastern Highland Rim (Simmers 1982a) and 108/100 ha in a mixed deciduous forest in the Cumberland Mountains (Yahner 1973).

Atlasers recorded Red-eyed Vireos at an average of 4.8 stops on the 439 miniroutes recording the species. They were most abundant in the Cumberlands, where they were recorded at 14 or 15 miniroute stops in several blocks, and in the Unaka Mountains. A third and lesser center of abundance is in the Western Highland Rim and adjacent southeastern portion of the Coastal Plain Uplands. These areas are all relatively heavily forested. The lower abundance along parts of the Tennessee–North Carolina border is due to their scarcity at high elevations.

The Red-eyed Vireo population, as measured by Tennessee BBS routes, showed a slight declining trend from 1966 to 1994. During the last half of this period, it increased by 2.1%/year ($p < 0.01$). This recent increase is probably due to the increases in forest area and forest

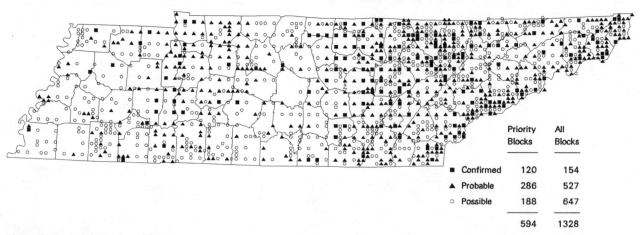

	Priority Blocks	All Blocks
■ Confirmed	120	154
▲ Probable	286	527
○ Possible	188	647
	594	1328

Distribution of the Red-eyed Vireo.

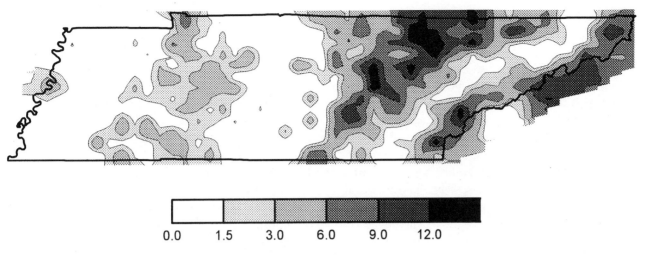

Abundance of the Red-eyed Vireo.

age. During the Atlas period, they were recorded on 39 of 42 BBS routes at averages of 14 birds and 9.3 stops/50 stop route. Rangewide, Red-eyed Vireo numbers significantly ($p < 0.01$) increased by 1.1%/year from 1966 to 1993 (Peterjohn, Sauer, and Link 1994).

Breeding Biology: The male Red-eyed Vireo arrives on the breeding territory several days before the female. Pairing occurs soon after the arrival of the female and courtship includes chases, attacks on the female by the male, and occasional courtship-feeding of the female (Lawrence 1953, Barlow and Rice 1977). The female, often accompanied by the frequently singing male, selects the nest site and builds the nest in 4–5 days (Tyler 1950a, Lawrence 1953). The nest is a rather elaborate cup suspended from the fork of a horizontal branch (Tyler 1950a). It is built of fine grasses, rootlets, and strips of grapevine bark, held together and to the branch with spider webs. The outside is often decorated with bits of bark, paper from wasps' nests, and lichens, and the nest is lined with dried grasses, fine weed stems, grapevine bark, and hair. Nests have been reported in many shrub and deciduous tree species; most frequently reported were oak, maple, hickory, and dogwood. Only one nest has been reported in a coniferous tree—a pine (Gettys egg collection). The height of 32 nests in Tennessee averaged 2.7 m (s.d. = 1.5) and ranged from 0.8 to 7.6 m. Elsewhere, nest heights of up to 18.3 m, but usually less than 5 m, have been reported (Tyler 1950a, Lawrence 1953). Nests built for second nestings tend to be higher than those of first nestings (Lawrence 1953).

Although the nest may be fairly conspicuous, few Tennessee atlasers were adept at finding them, and nest records accounted for only about 2% of Atlas records. The proportion of confirmed records, 12% (19% in more thoroughly worked priority blocks), was relatively low. Over 80% of these records were of fledglings or adults carrying food.

Clutch size in Tennessee is 3 or 4 eggs ($\bar{x} = 3.25$, s.d. = 0.44, n = 28). Clutch sizes of 3–5, with 4 the most common size, have been reported elsewhere (Tyler 1950a). The eggs are pure white with a few fine markings of reddish or dark brown or black, concentrated near the large end. The female incubates the eggs for 12–14 days and may be fed by the male during this period (Lawrence 1953). Both parents feed the nestlings, although the female brings food to the nest more often than the male. The young open their eyes when 6–7 days old and leave the nest 10–11 days after hatching. The young, poor flyers when they first leave the nest, are attended by the parents for up to a month (Lawrence 1953). Although the plumage of the fledglings is inconspicuous, their begging call, similar to some chickadee calls, is given repeatedly and loudly. Tennessee egg dates, most from late May, suggest one brood is normally raised. Lawrence (1953) found that Red-eyed Vireos are single-brooded and renest once or twice following an unsuccessful attempt.

The Red-eyed Vireo has been described as one of the most frequent hosts of the Brown-headed Cowbird (Friedmann 1963, Mayfield 1965), and there are several records of this parasitism in Tennessee. None of 14 sets collected by Gettys in McMinn County between 1900 and 1908 were parasitized. Five of 20 later egg sets, most collected in the 1930s, contained up to 3 cowbird eggs. Half of the more recent Nest Record Cards described parasitized nests, and atlasers reported several instances of Red-eyeds feeding fledgling cowbirds. Red-eyed Vireos have little defense against cowbird parasitism except for occasionally abandoning a nest containing only cowbird eggs (Mayfield 1965, Friedmann, Kiff, and Rothstein 1977).—*Charles P. Nicholson.*

Blue-winged Warbler
Vermivora pinus

The Blue-winged Warbler is a rare to fairly common summer resident of Tennessee, nesting primarily in the middle part of the state. It usually arrives during mid-April and departs by early October. The Blue-winged Warbler is easily identifiable by its bright yellow head and throat, black eyeline, and unmusical "bee-buzz" territorial song. It occurs in shrubby, second-growth areas with scattered taller trees, where it is more easily heard than seen. Blue-winged Warblers occasionally hybridize with the closely related Golden-winged Warbler; these hybrids are described in the miscellaneous species section at the end of this chapter.

According to Gill (1980), in the mid-1800s the Blue-winged Warbler probably occurred in most of Middle Tennessee and in East Tennessee west of the Unaka Mountains. The first published report from Tennessee was by Rhoads (1895a), who found it uncommon from Shelby County eastward to Chattanooga and Knoxville. His Shelby County observations were made in the second week of May and could have been of migrants; it has not recently bred there (Waldron 1987). Torrey (1896a) observed a single bird at Chattanooga and a pair on nearby Missionary Ridge in early May 1894. Ganier (1933a) described it as a fairly common summer resident in West and Middle Tennessee and a very rare summer resident in the east. Since the early twentieth century, the Blue-winged Warbler has been only sporadically reported from East Tennessee. Ganier and Clebsch (1940) found a nest at about 530 m on the Cumberland Plateau in Van Buren County in 1940, and Stupka (1963) lists reports of Blue-wingeds below 500 m in the Smokies during 3 different years from 1934 to 1958. A Blue-winged was observed for several years during the late 1970s on a BBS route in eastern Scott County. The reference to an adult feeding young in Blount County in 1976 (*Migrant* 47:102, Robinson 1990) is erroneous; the birds were in Georgia a short distance south of Chattanooga (K. Dubke pers. comm.).

BBS routes suggest a decline in the statewide Blue-winged Warbler population from 1966 to 1991, although the trend is not significant because of the small number (around 6) of routes recording it each year. From 1982 to 1991, the proportion of Tennessee routes showing decreases was significantly ($p < 0.01$) greater than those with increases. Rangewide, its population did not show a significant overall trend from 1966 to 1993 (Peterjohn, Sauer, and Link 1994).

Atlas workers found the Blue-winged Warbler in about 13% of the priority blocks. Because it stops regular singing fairly early in the summer and becomes inconspicuous in late June (Goodpasture 1954, Nicholson pers. obs.), the Blue-winged may have been missed in some blocks. Some Atlas workers were also probably unfamiliar with the longer, more variable "nesting song" given more frequently than the territorial song once nesting has begun. The Atlas results, however, are probably a good representation of its overall range in Tennessee. Blue-winged Warblers were recorded on 38 miniroutes at an average relative abundance of 1.7 stops/route; the highest numbers were on the Western Highland Rim.

Most of the Atlas records were from the Eastern and Western Highland Rims and the Coastal Plain Uplands, where there has been little change in its historic range. The East Tennessee records suggest a slight increase in numbers there, although most sites were only occupied for a year or 2. The Polk County location, where 4 singing males occurred in 1986, is adjacent to the only North Carolina counties where Blue-wingeds have occurred (Potter, Parnell, and Tuelings 1980). No Blue-wingeds were found during more intensive Atlas work in Polk County in 1989–91.

The Blue-winged Warbler occurs in early to mid-successional habitats resulting from reversion of abandoned farmland, fires, logging, or other disturbances. Within suitable habitat, shrubs and herbaceous vegetation, and to a lesser extent trees, usually occur in discrete patches (Confer and Knapp 1981). Suitable habitat occurs over a broad range of succession, from lightly grazed or recently abandoned brushy pastures through areas with an almost unbroken forest canopy. Its highest relative abundance on the Western Highland Rim is probably due to the region's high proportion of forest interspersed with numerous small farms and recently logged areas. Within the northeastern United States, the Blue-winged Warbler usually occurs in loose colonies in tracts of 10 ha or more of suitable habitat (Confer and

Blue-winged Warbler. Elizabeth S. Chastain

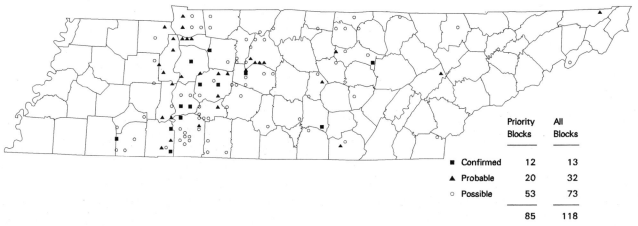

Distribution of the Blue-winged Warbler.

Knapp 1981); this colonial nature is less apparent in Tennessee. Habitat around Tennessee nests has been described as a rolling hillside covered with patches of brush, blackberries, and broom sedge (Crook 1933), an 0.13 ha blackberry and broom sedge field surrounded by woods (Monroe 1937), and among blackberries, wild plum, and coral berry shrubs at the edge of a meadow and woodlands (Goodpasture 1949a). The only Tennessee plot censuses to record the Blue-winged Warbler are those of Simmers (1980, 1983), who found a maximum of 1 pair on a 27 ha area of mixed woods and old fields on the Eastern Highland Rim.

Confirming the breeding of the Blue-winged Warbler was difficult for most Atlas workers, and about half of the confirmed records were of fledglings. Finding its nest is difficult and very few detailed nest records are available from the state.

Breeding Biology: The Blue-winged Warbler begins nest building as early as late April, following a very short 1- to 2-day courtship period (Morse 1989). The female constructs the bulky nest, sometimes assisted by the male (Harrison 1975). The nest is a narrow, deep cup, built on or close to the ground in the base of a grass clump of among weed stems. The outer portion of the nest is made of dead leaves with some leaf tips arching over the nest (Bent 1953). A cup is woven within the leaves from coarse grasses and bark strips and lined with grass, fine pieces of bark, and hair.

Tennessee nest records show a slight peak in egg laying during the first third of May. One clutch of 3 eggs, 2 of 4 eggs, 4 of 5 eggs, and 1 of 6 eggs have been reported. Elsewhere clutches of up to 7 eggs occur, with 5 the most common size (Bent 1953). The eggs are white with sparse, fine brown spots. The female incubates the eggs for 10 to 11 days; both adults feed the young, which fledge in 8 to 10 days (Bent 1953). The only available record of its parasitism by Brown-headed Cowbirds in Tennessee is of a pair of Blue-wingeds feeding a cowbird fledgling in Perry County (Nicholson pers. obs.). The Blue-winged Warbler is a fairly frequent cowbird host elsewhere (Friedmann 1963).—*Charles P. Nicholson.*

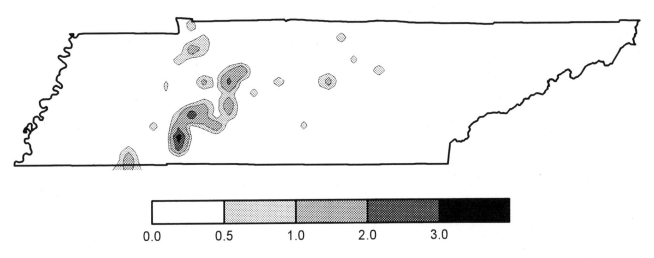

Abundance of the Blue-winged Warbler.

Golden-winged Warbler
Vermivora chrysoptera

The Golden-winged Warbler is an uncommon summer resident of the Cumberland Plateau, Cumberland Mountains, and Unaka Mountains. It usually arrives in the state during the second half of April and departs by early October. Outside of its breeding range, spring migrants usually depart by mid-May. The numbers and breeding range of the Golden-winged Warbler have decreased in recent decades, and it is presently one of the least common warblers nesting in the state. Golden-wingeds occasionally hybridize with the closely related Blue-winged Warbler; these hybrids are described in the miscellaneous species section at the end of this chapter.

The Golden-winged Warbler was considered a rare bird by early ornithologists north of Tennessee; it increased in numbers in the late nineteenth century, probably in response to the availability of abandoned farmlands (Confer 1992). A similar pattern probably occurred in Tennessee, where it was first recorded by Lemoyne (1886), who found it nesting in the Smokies. Torrey (1896a) described it as common in broken woods north of Chattanooga in May 1894. The following year Rhoads (1895a) described it as uncommon on the Cumberland Plateau in Hamilton and Fentress Counties, in Roane County, and at Knoxville. Around the turn of the century Gettys reported it as tolerably common and breeding in McMinn County (Ganier 1935), but apparently did not collect its eggs.

By the early 1980s, the Golden-winged Warbler had been reported throughout much of the Cumberlands, in the northern Unaka Mountains, and in the Smokies. Stupka (1963) described it as fairly common at low and middle elevations of the Smokies, and rarely occurring up to 1600 m. The only report from the Unaka Mountains south of the Smokies was in the Unicoi Mountains of Monroe County in June 1992 and 1993 (J. G. Bartlett pers. comm.). A territorial bird was present on the Western Highland Rim in Lewis County in the late 1970s (*Migrant* 50:88), and singing birds have occasionally appeared on the Eastern Highland Rim (e.g., Simmers 1982b, Atlas results). Late July records in Middle Tennessee (Robinson 1990) have probably been of early migrants.

Quantitative information on Golden-winged Warbler population trends in Tennessee is limited. It has been recorded on 7 of the 42 BBS routes, and these results suggest a declining trend; since 1986, it has been found on only 2 routes. Rangewide, its 1966–93 population trend suggests a decline, although it is not statistically significant (Peterjohn, Sauer, and Link 1994). This apparent decline was greatest from 1966 to 1978 (Sauer and Droege 1992). The decrease in Tennessee is probably attributable to changes in breeding habitat. North of Tennessee, its decline has been attributed to habitat loss and competition with the Blue-winged Warbler (Gill 1980, Morse 1989).

The Golden-winged Warbler occurs in a narrow range of habitats, typically second-growth areas with clumps of shrubs, scattered trees, grassy ground cover, and a forested edge (Confer 1992). In the Tennessee Cumberlands, typical habitats are abandoned farms, young pine plantations, and strip mines reclaimed with fescue and locust trees. In the Smokies it occurred in old fields, forest clearings growing up in brush and young trees, and on dry brushy hillsides (Stupka 1963). In 1947 and 1948, Kendeigh and Fawver (1981) recorded an average density of 30 pairs/100 ha on low-elevation seral pine-oak census plots in the Smokies. These plots had trees 1.8–6 m tall scattered among greenbriars, sumac, and blackberries; a luxuriant herb growth covered most of ground. The Golden-winged Warbler has been reported on several forest/strip mine census plots in the Cumberland Mountains (Yahner 1972, 1973, Garton 1973, Nicholson 1979a, 1980c).

Golden-winged Warbler habitat probably peaked in the first half of this century, when much farmland was abandoned in the Unaka Mountains and in the Cumberlands. The scarcity of recently abandoned farmlands in the Unakas probably accounts for the birds' scarcity there. Although early successional habitat is being created there by hardwood clearcutting, these areas rarely provide suitable habitat, as noted by Confer and Knapp (1981) in the northeastern United States. Beginning in the 1950s, suitable habitat was created in the Cumberlands

Golden-winged Warbler. Elizabeth S. Chastain

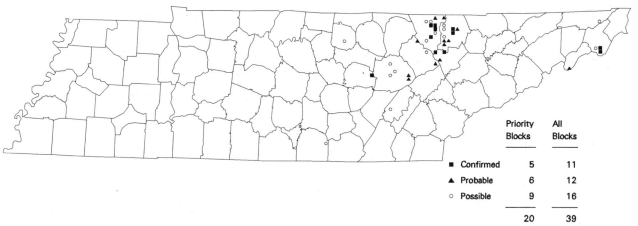

Distribution of the Golden-winged Warbler.

by pine plantations and strip mining; this has at least partially offset the loss of abandoned farmland habitat. Plantations, however, are probably suitable for up to 5–6 years, and strip mines a decade or more (Nicholson pers. obs.). Although the amount of habitat created by mining has greatly decreased in the last decade, mines could probably provide habitat for a longer period if reclaimed with grasses and widely spaced hardwoods, rather than thick plantings of pines and shrubs. While competition with the Blue-winged Warbler, which is less restricted to early successional habitat than the Golden-winged, has been attributed to declines in the Golden-winged Warbler north of Tennessee, there is little evidence of this in Tennessee. The Blue-winged Warbler has historically been rare in East Tennessee, and there is little evidence of a recent increase in its numbers there.

Atlas workers found the Golden-winged Warbler in 20 priority blocks, most of them in the Cumberland Mountains. Because its singing is sporadic once nesting has begun, it was probably missed in a few blocks. Confirmed records, most of fledglings, made up a fairly high proportion (28%) of the Atlas records. The warbler was also recorded on 6 miniroutes, at an average relative abundance of 1.7 stops/route. The highest miniroute count, 5 stops, was in a block with large areas of young pine plantations, as well as strip mines and old farms. It was not found in the Smokies, where most areas that were once suitable habitat have grown into mature forest. It was also not found on parts of the southern Cumberland Plateau where it formerly occurred, and there has been no evidence of breeding in the Ridge and Valley since early this century.

Breeding Biology: The Golden-winged Warbler has a very short, 1- to 2-day courtship period (Morse 1989). The nest, built by the female, is placed on or very close to the ground, usually at the base of a tall herb or shrub stem and sometimes in a clump of grass. It is coarsely made of grasses and bark strips, placed on a base of dead leaves, and lined with hair and grasses (Confer 1992). It is usually well hidden, somewhat unkempt in appearance, and often concealed from above.

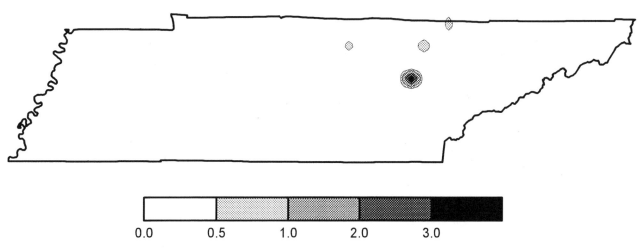

Abundance of the Golden-winged Warbler.

The few Tennessee egg records indicate a peak of egg laying in the second half of May. Three records of clutches of 5 eggs and 2 of 4 eggs exist; clutches elsewhere range from 3 to 6 and average 5 eggs (Confer 1992, Harrison 1975). The eggs are white to creamy white with variable brown spots, concentrated around the larger end. The female incubates the eggs for 10 to 11 days, and the young leave the nest in about 10 days (Confer 1992). Parasitism by the Brown-headed Cowbird has not been reported in Tennessee, but is occasional to fairly common elsewhere (Friedmann 1963, Friedmann, Kiff, and Rothstein 1977).—*Charles P. Nicholson.*

Northern Parula
Parula americana

The rising, buzzy trill, snapping at the end, of the Northern Parula is a typical sound of bottomland and ravine forests across the state. The Northern Parula is an uncommon to fairly common summer resident in Tennessee, arriving in early April and departing by late September (Robinson 1990).

Northern Parulas forage high in trees at the tips of branches (Morse 1967); consequently, they are often difficult to see. Their primary song, however, is loud and distinctive, and it is unlikely Northern Parulas were missed in many priority Atlas blocks. Atlasers found them across the state, from the Mississippi River to the Blue Ridge, where they are absent above about 1585 m (Stupka 1963). In West Tennessee, they occur in bottomland forests and riparian woodlands. On the Highland Rim, they occur in riparian woodlands. In the Cumberlands and Blue Ridge, they most frequently occur in ravines, often in hemlock trees.

Northern Parula. David Vogt

Northern Parulas were recorded on 147 miniroutes at an average relative abundance of 1.8 stops/route. The Mississippi Alluvial Plain contained the block with the highest count, 9 stops/route, and had the highest average of any physiographic province. This abundance agrees with the 1975 Lauderdale County foray results, where the Parula was the third most common woodland warbler (Coffey 1976), and with Ford (1990), who found them to be the second most abundant warbler in West Tennessee forested wetlands. The next highest miniroute averages were in the Western Highland Rim and Cumberland Mountains.

Rhoads (1895a) described Parulas as abundant statewide except at high elevations in the Unaka Mountains. They are presently abundant in few areas. The decline has probably been greatest in the Central Basin and Ridge and Valley, due in part to forest clearing and the destruction of riparian woodlands by reservoir construction. Northern Parulas no longer occur in Knox County in the numbers given by Howell and Monroe (1957). They have not been recently found in the Norris, Anderson County, area where they formerly nested (*Migrant* 7:73); this area remains relatively heavily forested. Although present in most West Tennessee priority blocks, Parulas are common there in only a few heavily wooded areas. Their occurrence in many West Tennessee Atlas blocks with a low percentage of forest area suggests they have some tolerance for small forests. Robbins, Dawson, and Dowell (1989) found that the probability of detection increased with forest area, but had too few samples to give a minimum area requirement.

Tennessee BBS results show a nonsignificant increasing trend in Northern Parula numbers from 1966 to 1994; during the last half of this period, the increase was significant ($p < 0.05$). Rangewide BBS routes do not show a significant overall trend from 1966 to 1993; between 1978 and 1988, however, they significantly ($p < 0.01$) declined (Peterjohn, Sauer, and Link 1994, Sauer and Droege 1992).

The distribution of Northern Parulas has often been tied to that of pendant "old man's beard" (*Usnea*) lichens and the superficially similar Spanish moss (*Tillandsia*), in which Parulas commonly nest. Spanish moss does not occur in Tennessee, and pendant *Usnea* lichens are uncommon outside of the eastern mountains. Ganier and Tyler (1934) described the Northern Parula as the most abundant bird in a coniferous bog, with abundant *Usnea*, in Johnson County. Northern Parulas, however, are absent from the high-elevation spruce-fir forests, where pendant *Usnea* is common.

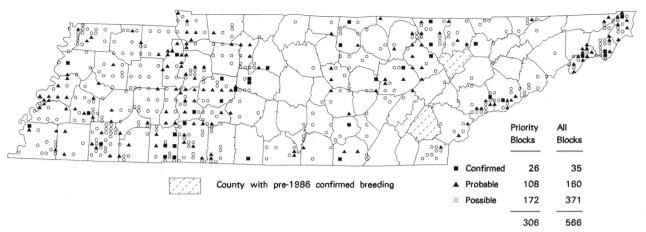

	Priority Blocks	All Blocks
■ Confirmed	26	35
▲ Probable	108	160
○ Possible	172	371
	306	566

Distribution of the Northern Parula.

Breeding Biology: Northern Parulas build their nests in hanging strands of lichens, in clusters of evergreen needles or deciduous leaves, or in flood debris deposited on tree limbs. Nests in flood debris have not been reported in Tennessee. Nests built in pendant *Usnea* have been found in Johnson and McMinn Counties (Ganier and Tyler 1934, Ijams and Hofferbert 1934) and consist of a cup woven into hanging strands of lichen, with an opening at the side or top. Jones (1931) described several nests in the northeast corner of the state built at the ends of hemlock branches. These nests were constructed mostly of *Usnea,* along with a few pieces of fine grasses, woven among hemlock twigs and needles. A few hairs and spider cocoons were added to the lining. Such nests are rarely visible from the ground. Other Tennessee nest records include one built of *Usnea* in a yellow pine in Pickett County, and one "in thick *Usnea*" in an red cedar in Cheatham County (egg coll. data); it is not clear if these nests were in pendant *Usnea* or built among the evergreen needles. No nest records are available from West Tennessee.

Nest building in Tennessee begins as early as the end of April. Most of the few egg records, however, are from June. Most Tennessee nests have been in hemlock trees; other trees used include pine, cedar, and oak. They range from 2.7 to 29 m in height and average 15.5 m (s.d. = 9.4 m, n = 9) above ground. Of 8 egg sets reported in Tennessee, 5 contained 3 eggs and the rest 4 eggs. Elsewhere clutches of 4 or 5 eggs are most common, and up to 7 eggs have been reported (Bent 1953). The female incubates the eggs for 12–14 days (Harrison 1978). Both adults feed the nestlings; the age at which the young leave the nest is not known. Lyle and Tyler (1934) stated that 2 broods are raised in northeast Tennessee. This has apparently not been verified by others.

There are no reports of cowbird parasitism of Northern Parulas in Tennessee, which may be due in part to the few available breeding records of Parulas. Cowbird parasitism of Parulas, however, is rare elsewhere in their range (Friedmann 1963).—*Charles P. Nicholson.*

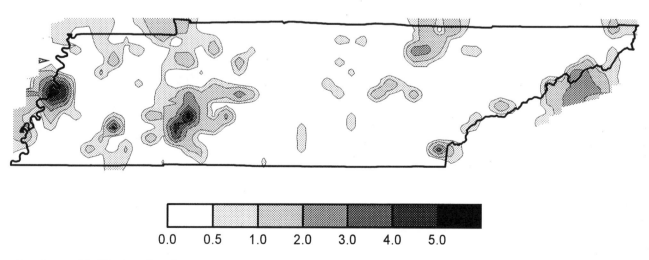

Abundance of the Northern Parula.

Confirmed Breeding Species

Yellow Warbler
Dendroica petechia

The aptly named Yellow Warbler is a fairly common summer resident of East and Middle Tennessee and a rare summer resident of West Tennessee (Robinson 1990). It arrives in early April and departs in August and early September. During the spring and early summer, it is conspicuous in its edge habitat, and its song is loud and usually easily identified.

Yellow Warblers are most common in moist, shrubby thickets adjoining streams and lakes. Use of suburban habitats, urban parks, orchards, and dry upland thickets, common farther north (Morse 1989), is at present uncommon in Tennessee. Historic descriptions of nest sites (egg. coll. data) suggest these habitats were more commonly used earlier this century. The reasons for this change are unclear. Plot census results for Tennessee include 13 pairs/100 ha in young, low-elevation, Smoky Mountain pine-oak forest (Kendeigh and Fawver 1981) and 22/100 ha in riverside field and hedgerow habitat (Lewis 1975).

In prehistoric Tennessee, the Yellow Warbler was probably uncommon and found in riverbank thickets and bottomland fields abandoned by Native Americans, especially in the eastern two-thirds of the state. As elsewhere, its numbers probably greatly increased with the increase in early second-growth habitats resulting from European settlement (Morse 1989). It was first reported in Tennessee by Fox (1886), who found it rather common in April in Roane County and observed nest building there. Rhoads (1895a) found it uncommon at Nashville, present on the Cumberland Plateau, and numerous in East Tennessee up to about 1070 m. He did not record it in West Tennessee, where the first probable breeding records were near the Tennessee River in the 1950s (Nicholson 1982b). Although not reported from northwest Tennessee until the 1970s, it was present in western Kentucky in the late nineteenth and early twentieth centuries (Nicholson 1982b). Its presence in Kentucky may have been the basis for Ganier's (1933a) premature description of it as a rare summer resident in West Tennessee. Its distribution in West Tennessee has changed little in the last decade.

The Yellow Warbler was found in about 30% of the completed Atlas priority blocks, and it was probably missed in some blocks where suitable habitat was limited and difficult to access. Atlas results show the warbler to be uncommon in most of Middle and East Tennessee, except for the Western Highland Rim, where it is fairly common, and in the Cumberland Mountains and northeastern corner of the state, where it is common. Its abundance in the latter 3 areas is probably due to the numerous permanent streams bordered by small farms with shrubby field borders and abandoned fields in each of these predominantly forested areas. In Johnson County, it is common in Shady Valley at elevations of 910 m (Atlas results), and it occurs up to about 1525 m on Roan Mountain (Eller and Wallace 1984). Stupka (1963) reported the Yellow Warbler to be common at low elevations in the Smokies, nesting up to about 955 m. Probable nesting birds have occurred at 1490 m on the North Carolina side of the Smokies (Stupka 1963), but the Yellow Warbler has not been reported from balds with seemingly suitable habitat at similar elevations along the state line.

The Yellow Warbler was reported on only about 10% of the Atlas miniroutes at an average relative abundance of 1.4 stops/route. The proportion of routes on which it was recorded was highest in the Cumberland Mountains.

Tennessee BBS routes show a declining trend in Yellow Warbler numbers of 3.0%/year ($p < 0.05$) from 1966 to 1994. Most of this decline has occurred since 1979 and is due in part to urban development along routes as well as the maturing of forests on farms abandoned a few decades ago. In the Smokies, Yellow Warblers now rarely occur away from maintained low-elevation openings. In Shady Valley, Johnson County, it may have increased since Ganier and Tyler (1934) reported 6 or 8 birds in June 1934. Throughout both its extensive North American range and eastern North America, the Yellow Warbler population has significantly ($p < 0.05$) increased since 1966 (Sauer and Droege 1992, Peterjohn, Sauer, and Link 1994).

Tennessee Atlas workers, in contrast to those in states farther north where the Yellow Warbler is much more

Yellow Warbler. David Vogt

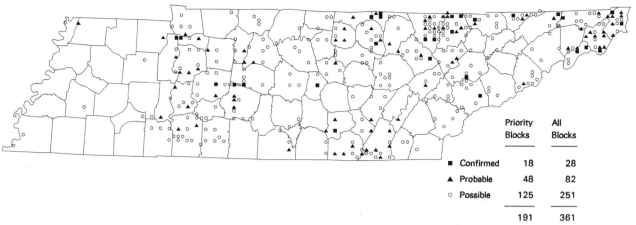

Distribution of the Yellow Warbler.

common (e.g., Andrle and Carroll 1988), had difficulty confirming Yellow Warbler breeding, and confirmed records made up only 8% of all records. About two-thirds of the confirmed records were of fledglings or adults carrying food. Most Tennessee nest records of the Yellow Warbler are from Nashville and northeastern Tennessee and date from the 1920s and 1930s.

Breeding Biology: Yellow Warblers are typically monogamous, although a male may rarely be mated to 2 females (Hobson and Sealy 1989). They often nest in dense colonies, with territories as small as one-twentieth ha (Hobson and Sealy 1989). The nest, built by the female in about 4 days, is placed in the fork of a shrub or sapling, usually near an opening (Bent 1953). It is constructed of woven plant fibers and grasses and lined with plant down, hair, and feathers. Tennessee nests were built 1.2–4.9 m above ground, with an average height of 3.2 m (s.d. = 1.2 m, n = 18). More nests were built in elm saplings than any other plant species; other plants used were fruit trees, ornamental shrubs, ash, maple, hawthorn, and osage orange.

Egg laying begins in late April and peaks during the first half of May. Tennessee clutches all contain 4 or 5 eggs, with an average of 4.6 eggs (s.d. = 0.5, n = 21). Clutches of 3 to 6 eggs have been reported elsewhere, with 4 or 5 eggs most common (Bent 1953). The eggs are pale grayish to greenish white, with variable brown and gray markings, usually forming a wreath around the large end. The female, frequently fed by the male, incubates the eggs for 11–12 days and both adults feed the nestlings, which leave the nest in 8–10 days (Schrantz 1943). Yellow Warblers are normally single-brooded (Clark and Robertson 1981), and the available nest records suggest this is true in Tennessee.

The first records of cowbird parasitism on Yellow Warblers in Tennessee were by Monk (1936), who found 2 parasitized nests at Nashville in the 1920s, each with 1 cowbird and 2 warbler eggs. Since then, additional parasitized nests with cowbird eggs and young, and adult warblers feeding cowbird fledglings, have been reported several times. Local cowbird parasitism rates elsewhere are as high as 41%; when parasitized early in

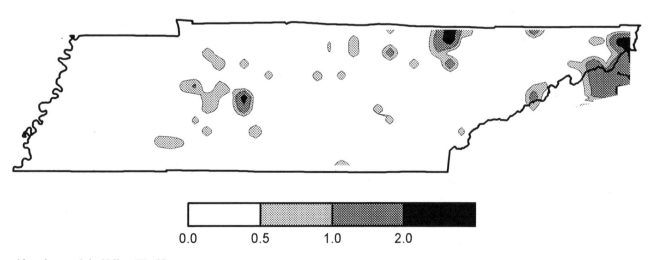

Abundance of the Yellow Warbler.

the nest cycle, with no more than 1 warbler egg present, the warbler usually builds a new nest floor that buries the eggs or, less commonly, deserts the nest (Clark and Robertson 1981).—*Charles P. Nicholson.*

Chestnut-sided Warbler
Dendroica pensylvanica

The Chestnut-sided Warbler is a common summer resident in the Unaka and Cumberland Mountains and a rare summer resident on the Cumberland Plateau. Elsewhere in Tennessee, it is a fairly common migrant, present from late April to mid-May and late August to mid-October (Robinson 1990). Chestnut-sideds are occasionally reported from low elevations outside their known breeding range, especially during early June (Robinson 1990, *Migrant*); these birds are probably late migrants or wandering birds. The Atlas records from Davidson and western Pickett Counties are probably in this category.

The breeding population of Chestnut-sided Warblers in the southern Appalachians was first reported from western North Carolina by Brewster (1886:174). He described them as "generally distributed between 2000 and 4000 feet [610–1220 m], but nowhere really numerous. Indeed, I rarely saw more than one or two in a single day." A decade later, Rhoads (1895a) found them breeding on Roan Mountain, but gave no description of their abundance. Earlier in the century, the Chestnut-sided Warbler had been a rare bird in the eastern United States; following the clearing of the original forest and a cycle of farm abandonment, its numbers rapidly increased in the late 1800s to the point where it became one of the commonest breeding birds in the Northeast (Morse 1989).

Chestnut-sided Warbler. David Vogt

No information is available on the status of breeding Chestnut-sided Warblers in the southern Appalachians prior to their population explosion farther north. Suitable habitat was undoubtedly present on heath balds and in the dense understory of stunted, mountaintop oak forests, as well as temporary openings resulting from fire and other disturbances. The regrowth from the large-scale logging of the southern Appalachians in the early twentieth century created much habitat suitable for Chestnut-sided Warblers. Additional habitat was created following the chestnut blight in the 1930s, as American chestnut was one of the dominant trees in much of the southern Appalachians (Braun 1950, Whittaker 1956). Brooks (1947) described the Chestnut-sided Warbler as the most common and characteristic warbler of the chestnut sprout association in the central Alleghenies. Ganier and Clebsch (1946) found the Chestnut-sided to be the fourth most common species in the Unicoi Mountains.

Stupka (1963) described the Chestnut-sided Warbler as common above 915 m in the Great Smoky Mountains. In the Smokies, it occurs in several forest types, including heath balds, chestnut oak, red oak, pine heath, beech gap, and spruce-fir, where it occurs in early successional stands or openings with deciduous saplings (often mountain-ash) in mature stands (Kendeigh and Fawver 1981). Kendeigh and Fawver (1981) found it at its highest density, 353 pairs/100 ha, in early successional spruce-fir. In the Unicoi Mountains, southwest of the Smokies, it also occurs in sapling-sized clearcut areas and in the dense understory of maple-beech-birch stands (e.g., Mitchell and Stedman 1993).

In the Cumberland Mountains, Chestnut-sided Warblers are most common above about 825 m (Nicholson 1987 pers. obs.). The lowest elevation at which breeding has been confirmed is 640 m (Nicholson pers. obs.). Several records exist of singing males, some territorial, from as low as 550 m on the Cumberland Plateau. In the Cumberlands, Chestnut-sideds occur in cutover woodland, young pine plantations, shrubby strip mines, and in the shrubby understory of mountaintop forests (Nicholson pers. obs.).

Chestnut-sided Warblers were recorded on 16 miniroutes at an average relative abundance of 4.4 stops/route. The highest counts, 13 and 11 stops/route, were in a Smokies spruce-fir block with abundant sapling growth following fir die-off and in a Unicoi Mountains block with recent timber harvest areas and a thick shrub layer in much of the remaining forest. Relatively little information is available on recent population trends of Chestnut-sided Warblers in Tennessee. They are found on too few BBS routes to show a trend. Wilcove

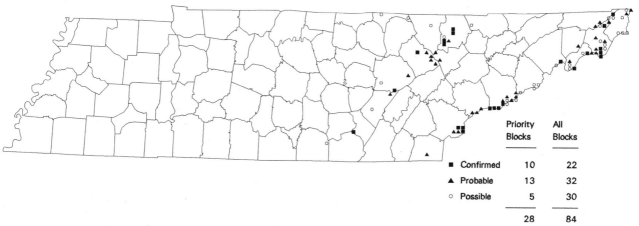

Distribution of the Chestnut-sided Warbler.

(1988) did not find a significant change in numbers on plots censused in the Smokies in 1947 and 1982. Chestnut-sideds were absent in 1982 from some plots where they occurred in 1947; this may have been due to maturation of the forest. Alsop and Laughlin (1991) observed an increase from zero to 15 pairs/100 ha between 1967 and 1985 on a spruce-fir plot in the Smokies; the increase was due to the increased sapling growth following death of the mature firs. On rangewide BBS routes, Chestnut-sided Warblers decreased by 0.8%/year (p = 0.09) from 1966 to 1993 (Peterjohn, Sauer, and Link 1994).

Breeding Biology: Nesting activities begin in Tennessee in mid-May in the Cumberlands and somewhat later at the high elevations of the Unaka Mountains. The female builds the nest, which is thin walled and loosely constructed (Bent 1953, Ganier 1956). The outer shell is made of grasses, weed stems, bark strips, plant down, and spider webs. The nest is lined with hair, rootlets, or fine grasses. Twenty Tennessee nests were built between 0.3 and 1.5 m above ground, averaging 0.8 m (s.d. = 0.4). Most Tennessee nests have been built in blackberries; other species used include crabapple, blueberry, and maple, beech, and birch sprouts.

Three or 4 eggs are usually laid, and the average clutch size in Tennessee is 3.6 (s.d. = 0.49, n = 29) eggs. Bent (1953) gives 4 as the most common clutch size and notes that 5 eggs are occasionally laid; this has not been reported in Tennessee. The eggs are creamy white with variable markings of browns and gray. The female incubates the eggs for 12–13 days, and the young leave the nest in 10–12 days (Bent 1953). Both adults care for the young. The dates of nests with eggs and young suggest that only 1 brood is raised in Tennessee. Parasitism by Brown-headed Cowbirds on Chestnut-sided Warblers has not been reported in Tennessee, although cowbirds regularly occur within the Chestnut-sided's range in the Cumberlands and in northeast Tennessee. Chestnut-sided Warblers are frequently parasitized by cowbirds elsewhere in their range (Friedmann 1963).
—*Charles P. Nicholson.*

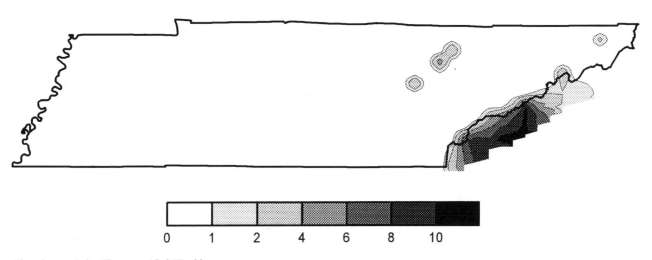

Abundance of the Chestnut-sided Warbler.

Black-throated Blue Warbler
Dendroica caerulescens

This strikingly plumaged blue and black warbler is a common nesting species in high-elevation forests of the eastern mountains. In contrast to the male, the female is drab green, and the species has one of the highest levels of sexual dimorphism of any warbler. Outside of the eastern mountains, it is a rare to uncommon migrant, occurring from mid-April into mid-May and from early September to mid-October.

The southern Appalachian population of this northern-breeding species was long ago described as a separate subspecies, *D. caerulescens cairnsi* (Coues 1897). Male "Cairn's Warblers" were distinguished by having back feathers spotted with or primarily black, instead of blue, as in the nominate race. Blue-backed males, however, are not uncommon in the southern Appalachians, and a reappraisal of *cairnsi*, partly based on Tennessee specimens, is underway (Gary Graves pers. comm.).

Male Black-throated Blue Warblers, with their conspicuous plumage and distinctive song, sung well into the summer, were easily observed by atlasers; their mapped distribution is probably an accurate reflection of their Tennessee range. They occur throughout the higher elevations of the Unaka Mountains and in an isolated population in the Cumberland Mountains on Frozen Head Mountain (Nicholson 1987). The Cumberland Mountain population was probably more widespread in the past, as Howell (1910) collected birds he identified as *cairnsi* about 25 km northeast of Frozen Head at 1036 m on Cross Mountain, Campbell County, in mid-August 1908. The collection dates, before fall migrants normally appear in the state, suggest the birds bred locally. Although other sympatric northern species, such as Veeries, occur on Cross Mountain, there are no recent records of breeding Black-throated Blues there. It may have been eliminated by the forest clearing associated with coal mining.

In 1922, Ganier (1923a:29) collected an adult and nest in "thick second growth in a thin plateau woods" in Grundy County. The Cumberland Plateau in this region ranges from about 550 to 580 m elevation, below the species's current altitudinal distribution. No other Black-throated Blue Warblers were observed by Ganier in the area; another anomalous Plateau record is a bird at Fall Creek Falls SP on 3 June 1975 (*Migrant* 46:88–89). LeMoyne's nesting record from 1311 m elevation in Roane County (LeMoyne 1886) is erroneous, as the maximum elevation in Roane County, along the western edge of the Ridge and Valley, is about 610 m.

On Frozen Head Mountain, Black-throated Blues occur down to about 790 m elevation (Nicholson 1987).

Black-throated Blue Warbler. David Vogt

In the Smokies, they are common above 850 m in most forest types (Stupka 1963) and common down to about 725 m in cove hardwoods (Wilcove 1988). They reach their highest densities in cove hardwoods, where Kendeigh and Fawver (1981) noted that Black-throated Blues, at populations of up to 290 pairs/100 ha, made up 41% of the total bird population. This high density was attributed to their exploitation, without competition from other species, of the *Rhododendron maximum* understory, a habitat absent from mixed mesophytic forests outside the southern Appalachians. Wilcove (1988) thought the Black-throated Blue Warbler was the second most abundant bird in the Smokies, exceeded only by the Dark-eyed Junco. Their abundance is probably underestimated by the Atlas miniroutes, which recorded them on 12 routes at an average relative abundance of 4.0 stops/route. These routes did not sample the isolated Frozen Head population; 7 birds were recorded at 5 stops of a 15-stop, nonrandom (not shown on map) miniroute there (Nicholson pers. obs.). Black-throated Blue Warblers are absent from mountains with maximum elevations of around 910 m, such as Oswald Dome in Polk County, presumably because such mountains lack sufficient area of high-elevation forest types.

Limited information is available on population trends of Black-throated Blue Warblers, as they are not sampled by Tennessee BBS routes. Wilcove (1988) found no clear trend on plots in the Smokies censused in both the late 1940s and early 1980s. Rangewide BBS routes do not show a significant trend from 1966 to 1993 (Peterjohn, Sauer, and Link 1994).

Although only a little over one-third of all Atlas records of this species were in the confirmed category, observing fledglings or adults carrying food was not difficult. The nest, however, is well hidden, and only a single nest record was reported.

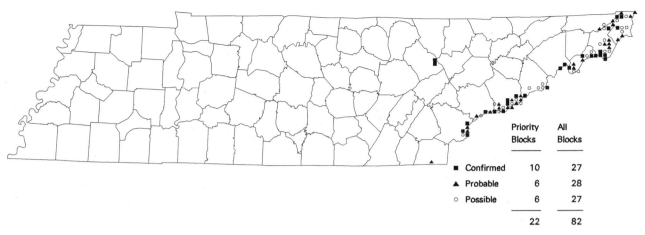

Distribution of the Black-throated Blue Warbler.

Breeding Biology: Black-throated Blue Warblers begin nesting shortly after their arrival on the breeding grounds and nesting may extend into August. The nest is placed in thick understory vegetation, frequently rhododendron, mountain laurel, or dog-hobble (Bent 1953, Ganier 1956, Burleigh 1927, NRC). Other species used in the southern Appalachians include blue cohosh, viburnum, and saplings of deciduous and coniferous trees. Nest building is by the female and takes about 4–6 days (Cairns in Bent 1953, Holmes, Black, and Sherry 1979); the male may rarely help (Harding 1931). Nest height ranges from about 0.2 to 1.5 m, with most nests between 0.5 and 1.2 m above ground (Ganier 1956, NRC). The nest is constructed of bark strips and bits of rotten wood held together with spider webs and lined with black rootlets, hairlike moss, fine grasses, or hair (Bent 1953, Ganier 1956).

Egg laying begins around the first of May at lower elevations and peaks in late May and early June. The eggs are white to creamy white, marked with brown speckles or blotches with gray undertones and concentrated at the large end, often forming a wreath or brown cap (Bent 1953). Three or 4 eggs are laid, and of 8 Tennessee clutches half were of each size. Larger samples elsewhere had average clutch sizes closer to 4 (Harding 1931, Holmes, Black, and Sherry 1979, Holmes et al. 1992). Incubation, carried out by the female, lasts 12–13 days (Holmes, Black, and Sherry 1979). The female broods the young, which fledge in 10–12 days (Harding 1931, Holmes, Black, and Sherry 1979). In the New Hampshire population studied by Holmes et al. (1992), about 10% of the males were polygynous and almost half of the females attempted to raise a second brood.

No records of Brown-headed Cowbird parasitism of Black-throated Blue Warblers are available from Tennessee. Most of the population in the Smokies and Unakas south of the Smokies is probably not exposed to cowbirds, which occur in the rest of the warbler's Tennessee range. Cowbird parasitism is infrequent elsewhere in the warbler's range (Friedmann, Kiff, and Rothstein 1977).—*Charles P. Nicholson.*

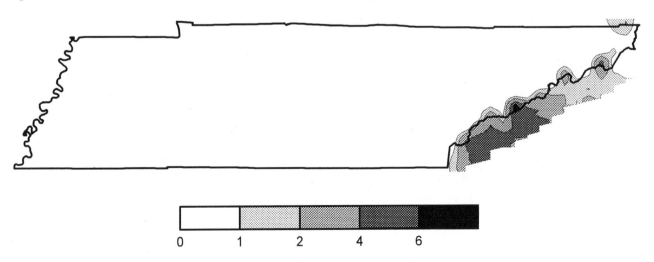

Abundance of the Black-throated Blue Warbler.

Confirmed Breeding Species

Black-throated Green Warbler
Dendroica virens

The strikingly plumaged Black-throated Green Warbler is usually observed high in the outer canopy of a large tree, foraging and singing its distinctive song. It is fairly common to common throughout the Blue Ridge and locally common in the Cumberlands. Elsewhere in the state, it is a fairly common migrant. Its migration periods are among the longest of any warbler, lasting from about the end of March to mid-May in the spring and late August to late October in the fall.

Black-throated Green Warblers were first reported breeding in the state by Rhoads (1895a), and the species's breeding range has changed little since then. In his first checklist, Ganier (1917) listed it as a very rare summer resident in Middle Tennessee, an erroneous status later corrected (Ganier 1933a). He also underestimated its abundance in the eastern mountains, where he listed it as rare. Wetmore (1939:224) corrected this, noting they were commoner "than has been supposed."

The Atlas results show the species distributed almost continuously throughout the Unaka Mountains and scattered throughout the Cumberlands. The only records from outside these physiographic regions were an out-of-range territorial male singing on 3 and 10 June 1991 in oak-hickory woods on the Western Highland Rim (P. B. Hamel, Atlas results), and 2 records from high ridges in the Ridge and Valley Province at 518 and 701 m in mixed pine-hardwoods. Atlasers may have missed the species at other locations in the Ridge and Valley because of difficult access to the high ridges (e.g., Clinch, Powell, and Bays Mountains) where suitable habitat occurs. Rhoads (1985b) listed Knoxville among the locations where he observed the species breeding. The species is not presently known to breed in the Knoxville area, although its presence on the high ridges northeast of the city is possible. In 1979, a territorial male was present in white pines at an elevation of 424 m near Norris, Anderson County (Nicholson pers. obs.).

Stupka (1963) described the Black-throated Green Warbler as one of few species occurring at all elevations within the Smokies. On a statewide basis, it nests from 2012 m, the highest elevation in the Smokies, to 240 m in gorges dissecting the Cumberlands. This range may be exceeded by only one other warbler, the Common Yellowthroat. Within its North American breeding range, Black-throated Green Warblers occupy a wide variety of forest types (Collins 1983, Morse 1993). The same is true in Tennessee. It occurs in spruce-fir, hemlock, and mixed hemlock-deciduous forests, white and Virginia pines, cove hardwoods, mixed mesophytic, oak-hickory, and tulip-poplar-oak forests. At lower elevations, it is usually found in ravines or gorges. In the Cumberland Mountains, it is often found in hardwoods on steep slopes. It occupies relatively mature forests in all of these habitats. In the Smokies, Black-throated Green Warblers reach their highest density, up to 290 pairs/100 ha, in hemlock-deciduous forests (Kendeigh and Fawver 1981, Wilcove 1988), where it is one of the most common species. Robertson (1979) found a density of 170/100 ha in virgin Cumberland Plateau mixed mesophytic forest.

Black-throated Green Warblers were recorded on 48 miniroutes at an average relative abundance of 2.8 stops/route. They reached their highest abundance, up to 9 stops, in the southwest portion of the Smokies and the southern portion of Cherokee National Forest. Black-throated Green Warblers are found on too few Tennessee BBS routes to reliably analyze their population trend, although the limited available results suggest a population decline. Wilcove (1988), however, found it was the only neotropical migrant with a widespread population increase on census plots in the Smokies between 1947 and 1982. Its population may vary greatly from year to year, as Wilcove (1988) noted a 50% decrease between 1982 and 1983 on the census plots. Alsop and Laughlin (1991) found a decrease from 30 to 2.5 pairs/100 ha in spruce-fir forest following death of the mature firs. Throughout its range, it showed no significant trend on BBS routes from 1966 to 1993; between 1978 and 1988, however, its numbers declined significantly ($p < 0.01$) by 3.2%/year (Peterjohn, Sauer, and Link 1994, Sauer and Droege 1992).

Atlasers confirmed breeding of the Black-throated Green Warbler in 34% of the priority blocks, relatively high for a bird that spends most of its time in tree canopies. All but 3 of these records, however, were of fledglings or adults carrying food. Whether the birds outside of the Blue Ridge and Cumberlands successfully breed remains unknown.

Black-throated Green Warbler. David Vogt

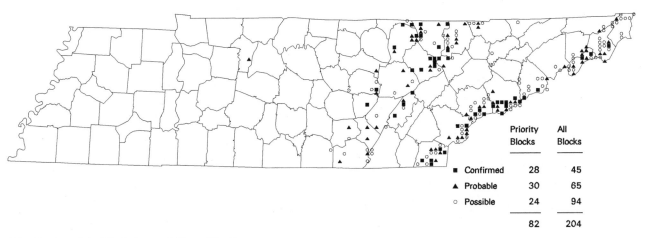

Distribution of the Black-throated Green Warbler.

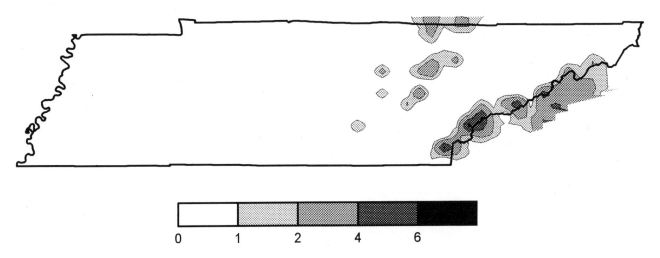

Abundance of the Black-throated Green Warbler.

Breeding Biology: The breeding biology of the Black-throated Green Warbler in Tennessee is poorly known. Nest building has been observed in mid-April, and the earliest records of fledglings are in late May. The few reported nests have been in red spruce, hemlock, buckeye, black birch, yellow birch, sugar maple, box elder, and basswood trees at heights of 4.9–10.6 m ($\bar{x}$ = 8.5 m, s.d. = 2.1 m, n = 8). The nest is placed in a crotch against the main trunk or on a branch a short distance from the trunk. Nicholson (in press) gives further information on Tennessee nests. Elsewhere in the warbler's range, it nests at heights of from near ground level to 22 m, with most nests 1–3 m high (Morse 1993). Both adults build the nest from twigs, grass stems, and bark strips and line it with hair, fine stems, or rootlets (Morse 1993). Most nests are well built and tightly woven, although some have been so lightly constructed that light shone through the walls (Morse 1993).

The only available Tennessee egg record is a clutch of 4 (Nicholson in press). Normal clutch size elsewhere is 3–5 eggs, with 4 most common (Morse 1993). The female incubates the eggs for 12 days, and the young leave the nest in about 9–10 days. Both parents feed the young and remove fecal sacs from the nest.

The Black-throated Green Warbler is a regular but usually infrequent cowbird host (Friedmann and Kiff 1985). One of 2 Tennessee records of nests with known contents was parasitized by cowbirds; this nest, in Scott County, contained a single large cowbird, and no young warblers, on 1 June 1995 (Nicholson in press). —*Charles P. Nicholson.*

Blackburnian Warbler
Dendroica fusca

The brilliantly colored Blackburnian Warbler is a summer resident of the high elevations of East Tennessee and, according to Atlas results, is the rarest warbler regularly nesting in the state. In the Unaka Mountains, it usually occurs above about 915 m elevation, most often in coniferous and mixed forests (Stupka 1963). In the Cumberland Mountains, it occurs in oak-dominated forests

Blackburnian Warbler. Elizabeth S. Chastain

above about 640 m (Nicholson 1987). Outside of its breeding range, it is an uncommon migrant, present from late April through late May and from late August to mid-October (Robinson 1990).

The first breeding season report of the Blackburnian Warbler in Tennessee, aside from the questionable records of LeMoyne (1886), was by Rhoads (1895a), who found it on Roan Mountain. By the late 1930s, most of the spruce-fir forest there had been heavily logged and the Blackburnian has not been reported during more recent studies (e.g., Ganier 1936, Phillips 1979b, Atlas results). It has been regularly reported from the Smokies since the 1930s (Stupka 1963). South of the Smokies, the only known Tennessee breeding location is in the Unicoi Mountains of Monroe County, where it was first reported by Ganier and Clebsch (1944, 1946) and present during Atlas fieldwork.

The only extant Tennessee population outside of the Unaka Mountains is in the Cumberland Mountains, where it was first reported by Howell and Campbell (1972), who observed a territorial male in June 1971 at 650 m in Campbell County. This area had received little previous ornithological study; Blackburnians were probably present before then, as they have been found in eastern Kentucky since early this century (Mengel 1965). The Cumberland Mountains population is small, and most reports are of singing males, often not present in the same area the following year. The most stable population, and only locality where breeding has been confirmed, is at Frozen Head State Natural Area, Morgan County (Nicholson 1987, pers. obs.). Farther south, an "active nest" was reported from the Cumberland Plateau at Fall Creek Falls SP, Van Buren County,

on 16 June 1970 (*Migrant* 41:69). Unfortunately, no other details of this observation are available; the species was also reported there in 1971 and 1972 (*Migrant* 42:70, 43:77).

The Blackburnian Warbler may have been overlooked in a few Atlas blocks, as it has a high-pitched, relatively quiet song, rarely sung after late June, when much of the fieldwork in the Unaka Mountains was done. It also occupies mature forests, where it uses the tops of large trees for foraging, singing, and nesting (Morse 1994), further adding to observation difficulties. However, with the exception of the Fall Creek Falls population, unreported since 1972, the Atlas results suggest little recent change in the warbler's distribution and are probably an accurate portrayal of its range. Blackburnians were recorded at a single stop of 3 miniroutes, all in the Unakas and probably an underestimate of its abundance. On a nonrandom 15-stop miniroute (not mapped) censused during 2 years at Frozen Head, Blackburnians were found at an average of 1.5 stops (Nicholson pers. obs.).

The only quantitative information on population trends of the Blackburnian Warbler is from replicated censuses in the Smokies, most of which suggest a population decline. Wilcove (1988) found lower numbers in 1982–83 on 2 of 3 relatively unchanged hemlock-deciduous forest plots that had average and maximum densities of 103 and 150 pairs/100 ha in 1947–48. Their numbers in the spruce-fir forest have also declined, largely attributable to the death of overstory firs from the balsam woolly adelgid infestation. Alsop and Laughlin (1991) reported a decline from 8 pairs/100 ha in the virgin Mt. Guyot spruce-fir forest in 1967 to none in a post-infestation 1985 census. A similar decline occurred on Mt. Mitchell, North Carolina (McNair 1987c). Rangewide BBS results, however, show no significant overall population trend from 1966 to 1993 (Peterjohn, Sauer, and Link 1994).

These census results, as well as other information in McNair (1987c), indicate that Blackburnian Warbler populations have fluctuated greatly within relatively unchanged habitats and between years. This is probably especially true in spruce-fir, as both Kendeigh and Fawver (1981) and Rabenold (1978) failed to record Blackburnians in undisturbed Smokies spruce-fir forests. In contrast, they have been regularly reported from and reach their highest density in mature hemlock and hemlock-deciduous forests, probably their optimal habitat in the southern Appalachians. In the Unicoi Mountains, they were found in virgin hemlocks in the 1940s and in both mature hemlocks and mixed hemlock and northern hardwoods during Atlas fieldwork.

In the Cumberland Mountains, Blackburnian Warblers occur in deciduous forests. The bird reported by Howell

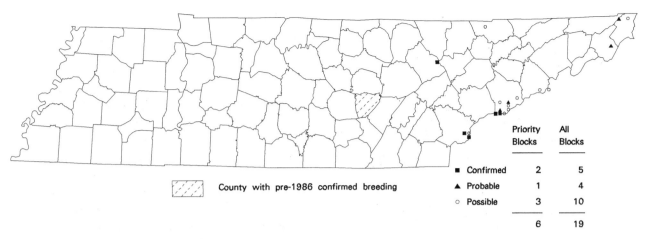

Distribution of the Blackburnian Warbler.

and Campbell (1972) was in forest dominated by chestnut oak, black locust, and tulip-poplar. At Frozen Head, Blackburnians occur between 795 and 945 m on southeast slopes dominated by northern red oak, tulip-poplar, and black cherry (Nicholson 1987, pers. obs.). The Atlas observation from Claiborne County was in large oaks. Mengel (1965) found Blackburnians in eastern Kentucky above 850 m in maple-beech-basswood forests, moderately disturbed areas containing black locust, and in drier forests of oak-chestnut. Blackburnians also occur at low densities in similar forest types near the lower limit of their elevational range in the southern Blue Ridge (Kendeigh and Fawver 1981, McNair 1987c).

Breeding Biology: Other than the Fall Creek Falls nest, the only Tennessee nest record of the Blackburnian Warbler is Stupka's (1963) report of an adult bringing food to a nest on 27 June 1952 in the Smokies. No other details of this record are available. Based on reports of fledglings in late June and early July (Stupka 1963, Nicholson 1987, Atlas data), nest construction and egg laying probably occur in mid- to late May. Elsewhere in its range, Blackburnians usually nest in conifers, especially hemlocks if present, on a limb some distance from the trunk (Morse 1994). Nest height varies from 1.5 m to over 25 m and averages over 10 m. Two nests in north Georgia were near the ends of limbs of large white oaks, 12 and 17 m above ground (Burleigh 1958). The cup-shaped nest is densely built by the female of twigs, bark, and rootlets and lined with grasses, hair, and moss. Normal clutch size is 3–5, with 4 eggs most common. The female incubates the eggs for 12–13 days, and both parents feed the young (Morse 1994). The length of the nestling period is not known. There are no reports of the Blackburnian being parasitized by cowbirds in Tennessee and few elsewhere (Friedmann 1963).—*Charles P. Nicholson.*

Yellow-throated Warbler
Dendroica dominica

A fairly common migrant and summer resident, the Yellow-throated Warbler arrives by late March and remains through mid-September; some late migrants linger into November (Robinson 1990). It is usually found in mixed deciduous-coniferous woodlands and in riparian habitats or similar habitat near water, where it most frequently associates with pine, sycamore, or bald cypress trees. Because it appears before full leaf development and the arrival of most other migrant bird species, its distinctive song and bright yellow throat aid identification and make it easy to find in the spring.

The Yellow-throated Warbler has a statewide distribution and was found in 51% of the completed Atlas priority blocks. It was not found in many areas in the Ridge and Valley and Central Basin. It was also rare at high elevations in the east. The densest populations occur in the Cumberland Mountains, Cumberland Plateau, and Eastern and Western Highland Rims. Except for its abundance in the Cumberlands, these findings are consistent with the general description of this species's distribution given by Robinson (1990), who summarized the upper elevational breeding limit of this species in the eastern mountains to be about 910 m. Although Ganier (1933a) once listed this species as a rare summer resident in eastern Tennessee, it is doubtful that he had then had much field experience in East Tennessee outside of the Smoky Mountains.

In the western part of the state, the Yellow-throated Warbler is most closely associated with the Mississippi, Obion, and Hatchie River valleys; Ford (1990) found it in 36 of 59 West Tennessee forested wetland sites, at an average density of 10.3 pairs/100 ha on line transect counts. A few records in the northern and extreme southern Coastal Plain Uplands and the northern Loess Plain

Yellow-throated Warbler. David Vogt

were in pine groves; the predominance of nonforested areas probably accounts for its localized distribution elsewhere in this part of the state. The species is also scattered and locally distributed in extreme northeast Tennessee, where less habitat exists. Its relative absence from most of the Central Basin may be the result of several interrelated factors, including lack of suitable habitat, inadequate survey coverage, and perhaps other ecological site factors.

Atlasers found Yellow-throated Warblers at an average of 2.1 stops on the 173 miniroutes recording the species. Birds were most frequently recorded on the Cumberland Plateau, the Western Highland Rim, and along the Tennessee River in West Tennessee; at these locations the species was recorded on 2 to 4 stops/route in several blocks. The center of abundance is the Cumberland Mountains where it was found on 12 of 16 routes at an average of 3.4 stops/route.

The Yellow-throated Warbler was first reported during the breeding season by Merriam (in Rhoads 1895a) in 1887 in McMinn County. Rhoads (1895a) described it as breeding in Davidson, Fentress, Knox, Greene, and Washington Counties in 1895. Since then pine plantings have probably increased its range and abundance, although these gains may have been partially offset by the destruction of forested wetlands and riparian areas. BBS routes show a significant ($p < 0.10$) annual increase in Yellow-throated Warbler numbers in Tennessee of 2.0% from 1966 to 1994. The species has been found on 30 of Tennessee's 42 BBS routes. Between 1966 and 1979, the Highland Rim and Cumberland Plateau were areas of major population increases, and Yellow-throated Warblers were more abundant during that period in Tennessee than in any other state or province (Robbins, Bystrak, and Geissler 1986). Rangewide, their population showed no significant overall trend from 1966 to 1993 (Peterjohn, Sauer, and Link 1994).

Breeding Biology: Male Yellow-throated Warblers arrive first in the spring and immediately begin to establish and defend territories (Ganier 1953). The species utilizes a monogamous breeding strategy, and both sexes are involved in building the nest (Ehrlich, Dobkin, and Wheye 1988). Nests are usually built high (6 to 15 m, but sometimes as high as 36 m above ground) in sycamore, pine, or cypress trees; nests below a height of 4.5 m are uncommon (Terres 1980, Ehrlich, Dobkin, and Wheye 1988). Ganier (1953) described a Tennessee nest in a willow tree and inferred that nests may also occur in oak trees on the Cumberland Plateau. In the Nashville area, nests were historically placed primarily in sycamore trees (Goodpasture 1949b).

Nests are typically placed on a horizontal branch and consist of a cup of fine grasses, weed stems, shreds of bark, plant fibers and down, hair, dead leaves, and caterpillar or spider webs; plant down or feathers may be used to line the nest (Ganier 1953, Harrison 1978).

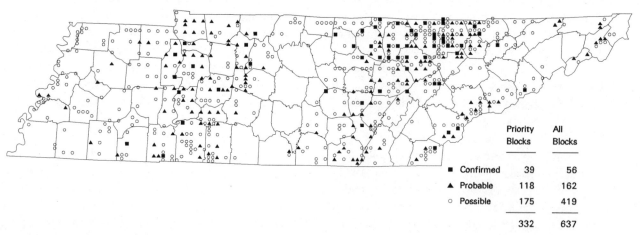

	Priority Blocks	All Blocks
■ Confirmed	39	56
▲ Probable	118	162
○ Possible	175	419
	332	637

Distribution of the Yellow-throated Warbler.

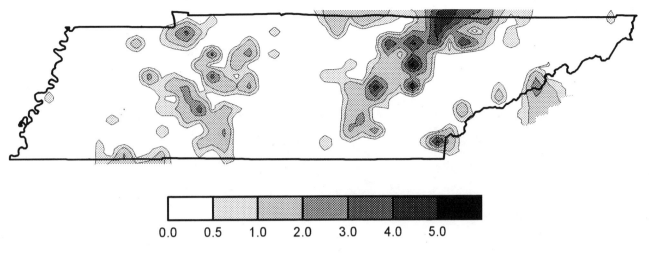

Abundance of the Yellow-throated Warbler.

The height of 12 nests in Tennessee averaged 10.2 m (s.d. = 3.9) and ranged from 4.5 to 18 m; each was built on the branch of a pine or sycamore tree. Nests built early in the spring may be placed lower in the tree than nests built later in the season, which may often be attached to a small, nearly vertical limb in the upper tree canopy (Ganier 1953).

In Tennessee, the clutch size averages 4.0 eggs (n = 17, s.d. = 0.4, range 3–5). A clutch size of 3 to 4 (usually 4, rarely 5) is reported widely for this species (Harrison 1978, Griscom and Sprunt 1979, Terres 1980). The eggs are pale greenish or grayish white and speckled, spotted, or blotched with a variety of darker colors (Harrison 1978). Information on incubation and the nestling period is scant; the incubation period probably lasts about 12 to 13 days (Harrison 1975). Second broods may be fairly common for Yellow-throated Warblers in Tennessee, as suggested by Ganier (1953) and in the Southeast by Bent (1953). Egg dates and observations of recently fledged young suggest that egg laying peaks in late April to early May. Ganier (1953) reported nest building on 11 July and recently fledged young are observed into late July.

Cowbird parasitism of the Yellow-throated Warbler is rarely documented; four known records are summarized by Friedmann and Kiff (1985). Records of cowbird parasitism in Tennessee are equally rare (a record in June 1991 of adult warblers feeding a cowbird fledgling exists for Robertson County, fide P. B. Hamel); although this may be an artifact of the paucity of nesting data for the Yellow-throated Warbler. This warbler is probably vulnerable to cowbird parasitism throughout its Tennessee breeding range.—*John C. Robinson.*

Pine Warbler
Dendroica pinus

The Pine Warbler is a fairly common summer resident and a rare to uncommon winter resident of Tennessee. As its name suggests, it inhabits pine and mixed pine-hardwood woodlands, and its occurrence, particularly during the summer, is closely tied to these forest types. A large proportion of its population migrates south from Tennessee in the fall. This fall migration is indistinct, and most birds probably leave by early November. Numbers begin increasing in February, and the Pine Warbler is usually present and singing regularly by the second half of March.

The Pine Warbler is a habitat specialist, occupying pure stands of Virginia, shortleaf, and loblolly pine, and mixtures of these species and deciduous trees. It also occurs in white pines, although at lower densities (Hardy 1991). The warbler selects stands with open overstories of trees with large canopies and sparse understory (Anderson and Shugart 1974). Such conditions are often typical of mature, natural pine stands but not necessarily of commercial plantations. In mixed stands of pines and deciduous trees, the Pine Warbler nests in stands with as little as 10% pine, and its density varies with the proportion of pine, in which it forages almost exclusively (Morse 1989). Its relative abundance is not significantly ($p > 0.05$) related to the forest area (Robbins, Dawson, and Dowell 1989).

In prehistoric Tennessee, the Pine Warbler was probably fairly common throughout the range of pines and most common in East Tennessee on the Cumberland Plateau, the southern Unaka Mountains, and Ridge and Valley. In West Tennessee, it was likely concentrated in the Coastal Plain Uplands. Fox (1886) found it common in Roane County in late March 1885, but did not

Pine Warbler. Elizabeth S. Chastain

observe it the following month. Rhoads (1895a) found it in Shelby, Davidson, and Fentress Counties, and described it as uncommon. Gettys collected at least 6 sets of Pine Warbler eggs in McMinn County at the turn of the century (Ijams and Hofferbert 1934). Given the inconspicuousness and frequent inaccessibility of its nests, these sets suggest it was fairly common there. Ganier (1933a) described it as a rare summer resident in West, Middle, and East Tennessee, likely an underestimate of its true numbers in the east. In Middle Tennessee, it was restricted to the few small areas of pines, such as at Craggie Hope, Cheatham County, where Ganier found it nesting in the 1920s (egg collection data). In the Unaka Mountains, it rarely occurs above 763 m, and the highest elevation at which it breeds elsewhere in the southern Appalachians is about 988 m (McNair 1987c).

Atlasers found the Pine Warbler in slightly over half of the priority blocks. It was present in most of the low-elevation priority blocks from the Cumberland Plateau eastward, except for the northeastern Ridge and Valley. Its occurrence on much of the Highland Rim and Coastal Plain Uplands, as well as scattered occurrences elsewhere in West and Middle Tennessee, is evidence of a fairly recent range expansion, attributable to the widespread planting of pines, particularly loblolly pine, since the 1930s for erosion control, landscaping, and lumber and pulpwood production. The recent maturation of this planted pine has also resulted in a significant ($p < 0.01$) annual increase of 9.8% on Tennessee BBS routes from 1966 to 1994, the largest increase of any species of warbler. An even greater increase has occurred throughout much of the Southeast (James, McCulloch, and Wolfe 1990). B. B. Coffey Jr. (1966) described the warbler's occurrence in Natchez Trace SP, Henderson County, following planting of pines in 1938. The first 2 were present in 1950, about 6 were present from 1956 to 1963, and they were fairly common in 1965. This pattern seems typical of the Pine Warbler's occupancy of planted pines elsewhere in the state (Hardy 1991) and the Southeast (Hamel 1992). Its rangewide population increased by 1.8%/year ($p < 0.01$) from 1966 to 1993 (Peterjohn, Sauer, and Link 1994).

The Pine Warbler was recorded on one-quarter of the Atlas miniroutes, at an average of 2.5 stops/route. The areas of high relative abundance generally correspond with the natural range of pine forests; blocks with the highest miniroute counts, 8 to 10 stops/route, were in the southern Unaka Mountains and northern Cumberland Plateau. The low numbers on the southern Cumberland Plateau are difficult to explain.

On the few Tennessee spot-map censuses that have reported the Pine Warbler, its densities were 45 males/100 ha in a low-elevation Smokies pine-oak forest (Kendeigh and Fawver 1981) and 5/100 ha in a Highland Rim hardwood-dominated mixed forest (Stedman and Stedman 1992). Hardy (1991), using transect counts, found average densities over 2 years of 118 pairs/100 ha and 8/100 ha in 35-year-old plantations of loblolly and white pine, respectively.

Almost one-fifth of the Pine Warbler Atlas records were of confirmed breeding, a fairly high proportion for a tree-nesting warbler. Over three-fourths of the confirmed records were of fledglings, which are conspicuous as they move through the pines, calling frequently. Most of the other confirmed records were of adults carrying food.

Breeding Biology: The Pine Warbler is the earliest nesting warbler in Tennessee, with nest construction begun as early as mid-March. The nest is placed on a pine limb, usually in a cluster of needles. It is compact and well built, probably by the female, of fine strips of bark, pine needles, weed stems, and grasses, and lined with hair, plant down, and feathers (Bent 1953). Tennessee nests range in height from 4.3 to 18.3 m and average 10.8 m above ground (s.d. = 4.3, n = 15). Nest heights elsewhere are from 2.4 to 24 m, and rarely to 41 m (Bent 1953). All Tennessee nests for which the tree species were identified were in shortleaf or Virginia pines. Most of these nest records are from the first half of this century, before mature loblolly pines were widespread.

Egg laying begins as early as late March, as indicated by a clutch with advanced incubation collected on 6 April (Gettys egg coll.) and probably peaks in early April. Tennessee clutches average 3.6 eggs (s.d. = 0.5, n = 10, range 3–4). Clutches throughout the warbler's range average 3.73 eggs, and clutches of 5 eggs occasionally occur

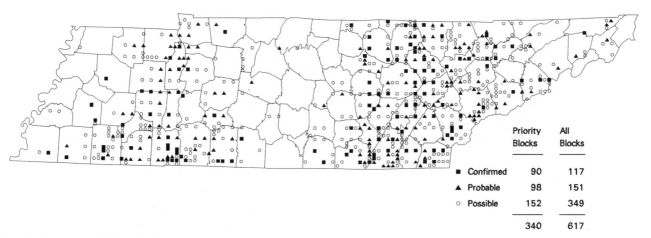

Distribution of the Pine Warbler.

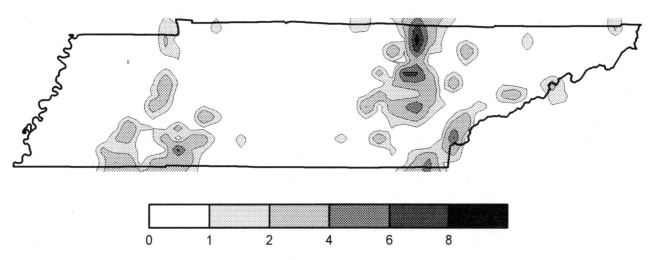

Abundance of the Pine Warbler.

(McNair 1987a). The eggs are white to grayish white with brown stops and blotches, usually forming a wreath around the large end. The length of the incubation and nestling periods are poorly known; incubation is probably solely by the female, and both birds feed the nestlings. In Tennessee, the Pine Warbler is probably double-brooded, as nest building and fresh eggs have been reported as late as early June and fledglings into August.

The Pine Warbler is rarely parasitized by the Brown-headed Cowbird (Friedmann, Kiff, and Rothstein 1977), and no records of such parasitism are available from Tennessee.—*Charles P. Nicholson.*

Prairie Warbler
Dendroica discolor

The diminutive Prairie Warbler is an uncommon to locally common summer resident in most of Tennessee. It arrives during the first half of April and departs by mid-September. Prairie Warblers are found in low-elevation, dry, second-growth habitats with a dense ground cover, numerous shrubs, and an open canopy.

The name "prairie" was given to this species by Alexander Wilson, who found it in 1810 in south-central Kentucky in savannah-like habitat then called prairie, and now often referred to as barrens (Nolan 1978). This habitat extended into the northern portions of Middle and West Tennessee (Braun 1950), where Prairie Warblers presumably occurred. Other habitats used prior to European settlement probably included cedar glades, most common in the Central Basin, as well as rocky, unforested cliff edges and forest openings in the Cumberlands (Mengel 1965), brushland resulting from activities of the Indians, and second growth resulting from forest fires, insect outbreaks, and other natural disturbances (Nolan 1978). The last two of these habitat types were only suitable to the warblers for a few years, while the other types were relatively stable over time. Prairie Warblers were probably uncommon and locally distributed during this presettlement period.

Prairie Warbler. David Vogt

Prairie Warbler habitat greatly increased following settlement and the resultant cutover forest, brush pastures, and periodically abandoned croplands. Rhoads (1895a) found them numerous between Nashville and Knoxville. Their habitat may have peaked following the wave of farm abandonment in the 1930s and 1940s; early in that period, Ganier (1933a) described them as fairly common across the state. They were the most abundant species in the Pickett Forest area and common at Fall Creek Falls (Ganier 1937a, Ganier and Clebsch 1940). Each of these areas on the Cumberland Plateau had been recently logged or abandoned by farmers.

Prairie Warblers occurred in 63% of the completed Atlas priority blocks throughout the state, except for the Mississippi Alluvial Plain and the higher elevations of the Unaka Mountains. They are easily identified and conspicuous in their habitat, and it is unlikely they were missed in priority blocks. Prairie Warblers were recorded on one-third of the miniroutes at an average of 2.3 stops/route.

Their relative abundance often varied greatly between nearby blocks because of local land use practices. They were most abundant on the Cumberland Plateau, the southern end of the Cumberland Mountains, the Western Highland Rim, and the Coastal Plain Uplands, all areas with recent, extensive logging and young pine plantations. Strip mines also provide habitat in the Cumberlands. Prairie Warblers are rare in upper East Tennessee, as was the case earlier this century (Herndon 1950b, White 1956).

Tennessee BBS results show a decreasing trend of 3.6%/year ($p < 0.01$) in Prairie Warbler populations from 1966 to 1994. Much of this decrease is due to the loss of breeding habitat through urbanization, cleaner farming, and maturation of the forests. Stupka (1963) described the Prairie Warbler as a fairly common summer resident at elevations up to 820 m in the Smokies area. Because of the present lack of low-elevation, early successional habitat, it is now very rare or absent from the Smokies. Throughout their breeding range, Prairie Warbler populations declined 2.9%/year ($p < 0.01$) from 1966 to 1993; as in Tennessee, the rate of decrease was greatest early in this period (Peterjohn, Sauer, and Link 1994, Sauer and Droege 1992).

Breeding Biology: The breeding biology of the Prairie Warbler is well known through the work of Nolan (1978) in Indiana, and a few dozen Tennessee nest records exist. Prairie Warblers are primarily monogamous, although a small percentage of the males mate with more than 1 female. The male arrives first in the spring, usually reclaiming the territory used the previous year. Courtship begins with the male attacking and chasing the female; later, the male performs a mothlike courtship flight. The female selects the nest site and constructs the nest in a small tree or shrub, either in the upright fork of a branch, against the trunk at the base of a twig, or on a horizon-

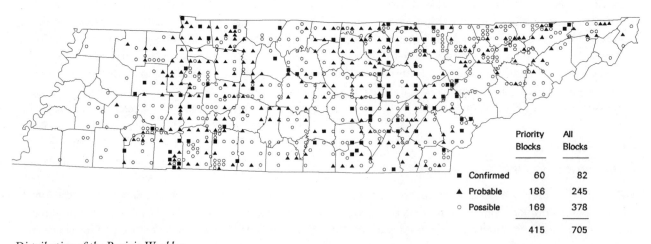

	Priority Blocks	All Blocks
■ Confirmed	60	82
▲ Probable	186	245
○ Possible	169	378
	415	705

Distribution of the Prairie Warbler.

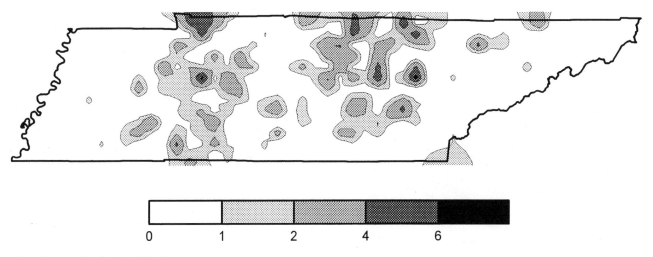

Abundance of the Prairie Warbler.

tal or diagonal branch. The nest is a well-made cup, constructed of down from milkweed and other plants, bits of shredded bark, dead leaves, grass, and spider webs, woven around the supporting branches. It is lined with fine grasses, hair, feathers, and rootlets. Twenty-five nests in Tennessee averaged 1.8 m above ground (range 0.8–4.0 m, s.d. = 0.7). Nolan (1978) found an average nest height of 2.3 m. Tennessee nests have been built in a variety of shrubs and trees, including elm, oak, red cedar, pine, red maple, and blackberry. Most Middle Tennessee nests have been in elms. Nolan (1978) noted disproportionate use of elms in his southern Indiana study area and attributed it to the elm's frequent forking and many crooked branches and twigs, which offer many nest locations.

Tennessee clutches contain 3–5 eggs and average 3.82 eggs (s.d. = 0.6, n = 28). Nolan (1978) gave an average of 3.89 for 188 clutches. Egg laying in Tennessee peaks in mid-May. Prairie Warbler eggs are pale cinnamon to grayish or creamy white and spotted with brown, usually in a wreath near the large end or, less frequently, forming a cap on the large end. The female incubates the eggs for an average of 12 days, and most young leave the nest 9 to 10 days after hatching. The male feeds the incubating female, and both adults feed the nestlings. Shortly after leaving the nest, the brood divides between the male and female, who care for the young for about a month. Nest mortality is high, and Prairies readily renest; second broods are rare.

Few incidents of Brown-headed Cowbird parasitism of Prairie Warblers have been reported in Tennessee. One out of 29 egg sets collected from 1900 to 1937 contained a cowbird egg, and at least 2 fledgling cowbirds attended by Prairie Warblers were reported by atlasers. Nolan (1978) found slightly over one-quarter of the nests he examined in Indiana contained cowbird eggs.

Almost half of the parasitized nests were deserted by the warblers. Because of this high rate of abandonment, the frequent asynchronous hatching of the cowbird eggs, and high predation rate, few cowbirds are successfully raised by Prairie Warblers.—*Charles P. Nicholson.*

Cerulean Warbler
Dendroica cerulea

Cerulean Warblers are sky blue birds of mature, generally moist hardwood forests. They live in the upper canopy of these forests and often strain the patience, vision, and neck muscles of even the most experienced student of birds. Cerulean Warblers are summer residents, occurring across the state in scattered populations, where it is at times locally common. They arrive in Tennessee in mid-April and depart by early September (Robinson 1990).

Cerulean Warblers were probably fairly common and occurred in much of prehistoric Tennessee, except for xeric pine and pine-oak forests across the state and the cedar-dominated forests of the Central Basin. They were probably rare in the Unaka Mountains and absent from the highest elevations there, as they have been historically (e.g., Ganier and Clebsch 1946, Stupka 1963). Their numbers undoubtedly declined as widespread forest clearing for agriculture occurred in the nineteenth century.

Rhoads (1895a) described the Cerulean Warbler as breeding in the Reelfoot Lake area and in Shelby and Davidson Counties; he admitted he was unable to identify it with certainty elsewhere, a problem he probably shared with his contemporaries and some more recent naturalists. Early-twentieth-century reports (e.g., Ganier 1933a) described the Cerulean as rare in the state. Later regional lists often reported higher numbers. Parmer (1985) described it as uncommon in the Nashville area,

Cerulean Warbler. Elizabeth S. Chastain

and in the early 1970s it was fairly common in Campbell County (Alsop 1971b), uncommon in Lawrence County (Alsop and Williams 1974), and uncommon in Lauderdale County (Coffey 1976).

Cerulean Warblers are locally distributed in Tennessee, occurring in 14% of the completed priority Atlas blocks. Noticeable gaps in its distribution are in the Ridge and Valley, Central Basin, and uplands of the Coastal Plain. Large tracts of mature, mesic forests are rare habitats in each of these physiographic regions. On the Cumberland Plateau, it was restricted to wooded gorges, although it was not found in Savage Gulf, where it was present in 1977 (Robertson 1979). Within the Ridge and Valley, it was not reported from McMinn County, where it bred in 1904 (Ijams and Hofferbert 1934), in Knox County, where it had formerly bred (Howell and Monroe 1957), and the Oak Ridge Reservation, Anderson–Roane Counties, where Anderson and Shugart (1974) had found several. While a few of these gaps may have resulted from failure to work suitable habitat in nonpriority blocks, the Atlas results are probably a fairly accurate portrayal of the Cerulean's distribution. Atlas results also show the Cerulean to be more widespread in the Unaka Mountains than suggested by the literature, although some of these records could have been misidentified Black-throated Blue Warblers.

Cerulean Warblers occurred on 6% of the Atlas miniroutes at an average relative abundance of 2.1 stops/route. The only area where they were common was in the Cumberland Mountains, where they occurred on average at 3.8 stops/route on 10 of 16 miniroutes. Their abundance in this physiographic region, dominated by mixed mesophytic forest, is shown by studies at Frozen Head State Natural Area where Nicholson (1987) recorded 38 singing males on 13 km of trails and more recently found an average of 21 birds at 11 of 15 stops of a nonrandom (not mapped) miniroute (Nicholson pers. comm.). Results from Breeding Bird Censuses in this area also show high densities, up to 110 pairs/100 ha (Smith 1977). Throughout its range, the highest density of the Cerulean Warbler is in the Cumberlands from West Virginia to Tennessee (Robbins, Fitzpatrick, and Hamel 1992).

Other notable populations occurred at lower densities on the Highland Rim and in West Tennessee, especially in the Mississippi Alluvial Plain. Most of the West Tennessee records were in extensive floodplain forests, either publicly owned or managed by private hunt clubs. Ford (1990) found Cerulean Warblers on 15 of 59 randomly selected forested wetlands in this region, at an average of 2.7 birds/1 km transect. Simmers (1982a) found 19 pairs/100 ha in an Eastern Highland Rim wooded ravine.

Cerulean Warblers typically inhabit mature and old-growth deciduous forest, particularly in floodplains or other mesic areas (e.g., Lynch 1981, Hands, Drobney, and Ryan 1989, Robbins, Fitzpatrick, and Hamel 1992). In Middle and West Tennessee, they foraged primarily in the upper half of the canopies of trees of larger diameter than average and were restricted to the larger tracts of forest (Robbins, Fitzpatrick, and Hamel 1992). Robbins, Dawson, and Dowell (1989) also found a preference for large tracts of mature deciduous forest in Maryland and estimated that maximum densities occurred in tracts of at least 3000 ha. In Missouri, optimal habitats contained a large number of large trees (more than 30 cm dbh), and a high (more than 18 m) closed canopy (Kahl et al. 1985). Detailed habitat studies have not yet been conducted in the Cumberlands.

Tennessee BBS results suggest a large, although statistically nonsignificant, decline in Cerulean Warbler numbers from 1966 to 1994. Between 1966 and 1979, it declined by 5.1%/year ($p < 0.01$), and has been comparatively stable since then. During the Atlas period, it was found on one-quarter of the routes at an average of 4.0 birds and 3.5 stops/route. Rangewide, it declined by 4.2%/year ($p < 0.01$) from 1966 to 1993; the rate of decline slowed during the latter part of this period (Peterjohn, Sauer, and Link 1994, Sauer and Droege 1992). Because of their rapid decline, the Cerulean Warbler was considered a Category 2 candidate species for federal listing as Endangered or Threatened until that designation was abolished (USFWS 1991, 1996). The reduced area and fragmentation of mature forest, shorter rotation cycles of commercial forests, changes in tree species composition,

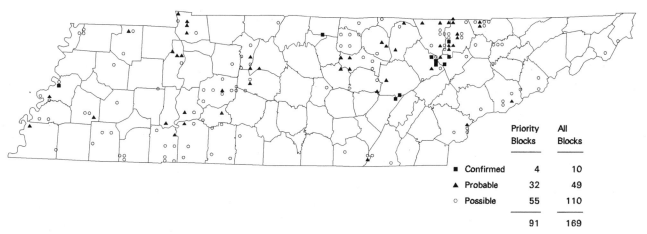

Distribution of the Cerulean Warbler.

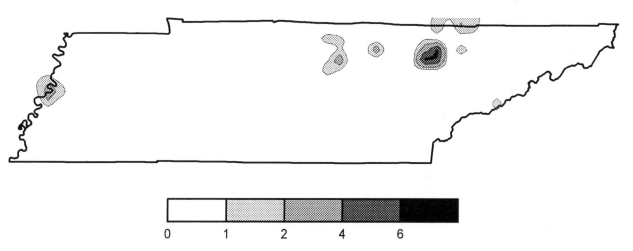

Abundance of the Cerulean Warbler.

and nest parasitism by Brown-headed Cowbirds have been suggested as major limiting factors on the breeding grounds (Robbins, Fitzpatrick, and Hamel 1992). A combination of these factors are probably limiting Cerulean Warbler distribution and density in Tennessee.

Breeding Biology: Relatively little is known about Cerulean Warbler life history, and all of the confirmed Atlas records were of fledglings. Females typically build the nest, which ranges in height from 4.5 to 27 m, with most in the 9–12 m range; because of the difficulty in detecting nests, the average is probably higher (Robbins, Fitzpatrick, and Hamel 1992). Tennessee nests have been at heights of 9.1, 15.2, 18.3, and 19.8 m in tulip-poplar, American elm, basswood, and white oak trees (Ganier 1951c, Goodpasture 1951, Campbell and Howell 1970). The nest is a relatively shallow cup placed on a branch far from the tree trunk, made of bark strips, moss, and lichens, covered with gray plant materials and spider webs (Bent 1953, Harrison 1975). Although no Tennessee clutch information is available, the few records of nestlings and fledglings suggest that egg laying peaks about the middle of May. Clutch size elsewhere is 3–5 eggs, with 4 probably most common (Harrison 1975). Length of the incubation and nestling periods are poorly known. Cowbird parasitism, not reported in Tennessee, is uncommon to fairly common elsewhere (Friedmann 1963, Robbins, Fitzpatrick, and Hamel 1992).—*Robert P. Ford and Paul B. Hamel.*

Black-and-White Warbler
Mniotilta varia

The Black-and-White Warbler is a fairly common summer resident of upland deciduous and mixed forests across Tennessee. The first birds are usually present by the end of March, making it one of the first migrant warblers to arrive in the spring. It usually departs by early October (Robinson 1990). The Black-and-white Warbler is easily identified by its clear, whistled song, black-and-white striped plumage, and nuthatch-like foraging on tree trunks and large limbs.

Black-and-White Warbler. David Vogt

The Black-and-White Warbler was probably abundant in the upland forests of prehistoric Tennessee. Although its numbers would have decreased with forest clearing as European settlers spread across the state, it remained numerous into the twentieth century. The first report of it in the state was by Fox (1886) who noted its spring arrival in Roane County and described it as common. Langdon (1887) described it as very common in the foothills of the Smokies in mid-August, after its nesting season. Rhoads (1895a) found it abundant across the state and up to 1067 m on Roan Mountain.

By the time of Ganier's 1933 checklist, the Black-and-white Warbler was absent from much of West Tennessee; he described it as rare there and fairly common in Middle and East Tennessee. This description was probably an underestimate of its abundance in much of the Cumberlands and the Unaka Mountains. Between 1966 and 1994, its population on Tennessee BBS routes decreased by 5.5%/year ($p < 0.01$). Its continental population showed no significant trend between 1966 and 1993 (Peterjohn, Sauer, and Link 1994).

Atlasers found the Black-and-White Warbler in 37% of the completed priority blocks and on 19% of the miniroutes, at an average relative abundance of 1.9 stops/route. Its overall distribution and abundance correlated well with the proportion of forested land. Robbins, Dawson, and Dowell (1989) considered the Black-and-White Warbler to be area-sensitive, rarely occurring in isolated forests smaller than about 200 ha. Miniroute results showed it to be most abundant in the Cumberland Mountains, where it occurred on all but 3 miniroutes at an average of 2.4 stops/route. It was also widespread, but somewhat less abundant, in the Unaka Mountains and Cumberland Plateau. Farther west, it was concentrated in the southern portions of the Coastal Plain Uplands and Western Highland Rim. The low number of reports from heavily wooded areas of the northern Western Highland Rim are puzzling. The scarcity of records from the Loess Plain and Mississippi Alluvial Plain are indicative of the limited extent of upland forests there.

Although the Black-and-White Warbler has been reported from all elevations of the Unaka Mountains, it rarely breeds above about 1525 m and does not regularly occur in the spruce-fir forest type (Kendeigh and Fawver 1981, Stupka 1963, Tanner 1955). Breeding densities in the Unaka Mountains include 15–39 pairs/100 ha in cove forests, 12–42/100 ha in oak forests, and 12/100 ha in a low-elevation, red maple–dominated forest (Wilcove 1988, Lewis 1983a). Robertson (1979) found 34 pairs/100 ha in a Cumberland Plateau virgin mixed-mesophytic forest and Simmers (1982a) reported 13/100 ha in a Highland Rim disturbed mixed-mesophytic forest. Densities on Cumberland Mountain census plots are 30/100 ha in mixed-mesophytic forest (Smith 1977) and 13/100 ha in oak-maple forests (Turner and Fowler 1981a, 1981b).

Breeding Biology: Nests of the Black-and-White Warbler are usually very well hidden, and only 3 were reported by atlasers. The majority of the confirmed Atlas records were

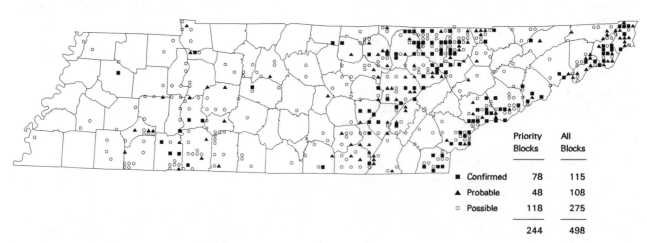

Distribution of the Black-and-White Warbler.

	Priority Blocks	All Blocks
■ Confirmed	78	115
▲ Probable	48	108
○ Possible	118	275
	244	498

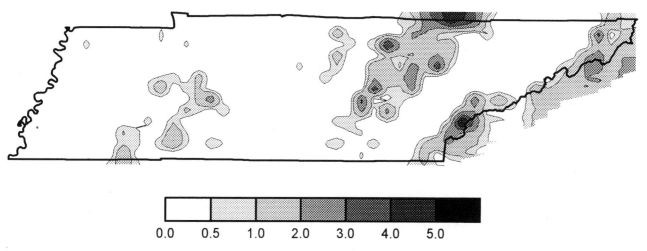

Abundance of the Black-and-White Warbler.

of fledglings, which are conspicuous. The female builds the nest on the ground or, less commonly, on a low stump (Harrison 1984). It is usually near the base of a tree, shrub, or rock, or under a fallen limb, concealed from above by dead leaves. The nest is built of dead leaves, grasses, and bark fibers and lined with fine grasses, rootlets, and hair. Nests are often on a hillside or slope of a ravine.

The available nest records show a peak in egg laying in early May. Goodpasture (1954) reported fledglings on 1 May, which, if not an error, would have come from an unusually early clutch. The normal clutch size is 4 or 5 eggs (Tyler 1953); Tennessee clutches average 4.25 eggs (s.d. = 0.45, n = 12). The eggs are white or creamy white and covered with small brown spots concentrated around the large end. The incubation and nestling periods are poorly known and given by Tyler (1953) as 10 days and 8–12 days, respectively. Adults perform a distraction display when flushed from the nest (Harrison 1984, Nicholson pers. obs.). Probably only 1 brood is raised.

Two reports of Brown-headed Cowbird parasitism are available: a nest with 3 warbler eggs and 1 cowbird egg (NRC) and adults feeding a fledgling cowbird (Goodpasture 1954). The Black-and-White Warbler has been described as an uncommon (Friedmann 1963) and common (Harrison 1984) cowbird host elsewhere.
—*Charles P. Nicholson.*

American Redstart
Setophaga ruticilla

American Redstarts are stunningly beautiful and active birds, often noticed for their flycatching or ritualized territory defense. They are uncommon to locally common summer residents across Tennessee, found along edges of woodlots and in mature, moist forests. They usually arrive in mid-April and depart by early October (Robinson 1990). They have also experienced one of the most precipitous population declines of any wood warbler nesting in the state.

American Redstarts were probably fairly common in mesic woodlands throughout prehistoric Tennessee. They were probably absent from cedar-dominated parts of the Central Basin, as well as the xeric oak-pine woods of the Coastal Plain Uplands and eastern one-third of the state. Redstarts were first reported in Tennessee by Fox (1886), who found them common in April in Roane County. Langdon (1887) observed several at low elevations of the Smokies in mid-August 1886; these birds had probably nested locally. Rhoads (1895a:498) described it as occurring "in favorable localities throughout the entire route" from Reelfoot Lake to Roan Mountain. Ganier (1933a) described it as rare in West Tennessee and fairly common in both Middle and East Tennessee.

Atlas workers found American Redstarts locally distributed across Tennessee, occurring in 13% of the completed priority blocks and on 4% of the Atlas miniroutes. These results suggest that redstarts have greatly declined in their distribution and abundance. They were not reported in some East Tennessee counties where nesting was historically confirmed, such as Sullivan and Washington (Lyle egg coll.). They were virtually absent from the Great Smoky Mountains, where Stupka (1963) described them as fairly common below 762 m. Redstarts may have been overlooked in a few Atlas blocks because the males sing much less after early June; this factor, however, is insufficient to account for the great difference in the current and historical distributions.

Additional evidence for a decline in redstart numbers is provided by Tennessee BBS results, which show a significant ($p < 0.01$) declining trend of 6.4%/year from 1966 to 1994. BBS results for the redstart from 1980 to 1994 suggest the rate of decline has slowed, and during

American Redstart. David Vogt

the Atlas period redstarts were found on 8 of 42 routes at an average of 5 birds and 4 stops per 50-stop route. The American Redstart has also shown a rangewide declining trend from 1966 to 1993 and a significant ($p > 0.10$) decline of 1.5%/year from 1978 to 1988 (Peterjohn, Sauer, and Link 1994, Sauer and Droege 1992).

Specific reasons for these population declines are unclear. Redstarts do not appear to be especially sensitive to forest fragmentation (Robbins, Dawson, and Dowell 1989), and in Tennessee, suitable habitats seem to be plentiful. Sherry and Holmes (1992) concluded that events on the breeding grounds, especially weather conditions, predation and parasitism, are contributing to the decline through low fledging success, possibly more than are events in wintering areas. More research is needed on this species in Tennessee, in coordination with research in wintering areas.

Redstarts are associated with damp woodlands and riparian zones throughout most of their breeding range, although they are not restricted to moist forested habitats. Mengel (1965) described the habitat in Kentucky as slashings and small openings along willow and sycamore communities near streams, and at the edges of mature floodplain forests and/or cottonwood and willow communities. Tennessee Atlas results show them to be most abundant in the primarily mixed mesophytic forests of the Cumberland Mountains and the floodplain forests of the Mississippi Alluvial Plain. In the Cumberland Mountains, redstarts usually occur in small gaps in the canopy of relatively mature forest (Mengel 1965, Nicholson pers. comm.). Breeding Bird Census plot densities in mixed mesophytic forests have reported up to 130 (Smith 1977) and 147 pairs/100 ha (Yahner 1973). The largest local population is probably in Frozen Head State Natural Area, Morgan County, where Nicholson (1987) reported 40 singing males along 13 km of trails; on a nonrandom, 15-stop miniroute (results not mapped) there, 21 birds at 14 stops and 30 birds at 15 stops were recorded over 2 years (Nicholson pers. comm.). Redstarts were among the most abundant of all birds in these studies. In West Tennessee forested wetlands, redstarts occurred on 37 of 59 randomly selected sites, and averaged 2.8 individuals per 1 km transect (Ford 1990). Redstarts foraged and sang most frequently 10 m or lower in the understory maples of more mature floodplain forests (Ford unpubl. data).

Breeding Biology: Nests of American Redstarts are typically in an upright fork of a deciduous tree or shrub, often partly supported by vines. Tennessee nests average 4.6 m (s.d. = 1.3, n = 7) above ground, with a range of 2.1–6.1; trees used for nesting include tulip-poplar, American beech, winged elm, red maple, and sycamore. Nests elsewhere have been as low as 0.8 m (Sturm 1945). The nest, built by the female in about 3 days, is a compact woven cup of plant down, lined with fine grasses, hair and sometimes feathers (Sturm 1945, Ficken 1963).

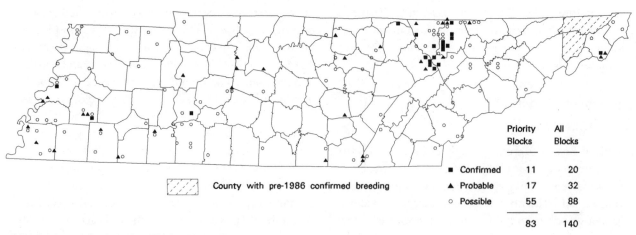

Distribution of the American Redstart.

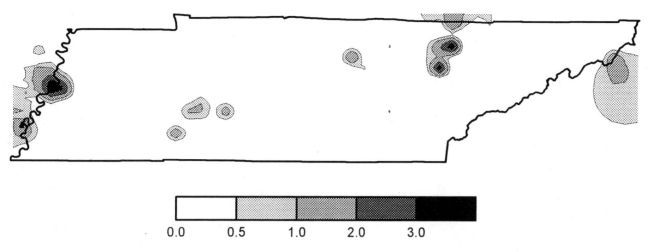

Abundance of the American Redstart.

Egg laying in Tennessee peaks during the second half of May. Clutches in Tennessee average 3.5 eggs, equally divided between 3- and 4-egg clutches (n = 6). Clutches of 2–5 eggs occur elsewhere, with 4-egg clutches most common (Bent 1953). The eggs are creamy white, with blotched or dotted brown and gray at the larger end (Harrison 1975). The female incubates the eggs for 11–12 days, and both adults birds feed the nestlings, which fledge in 8–9 days (Sherry and Holmes 1992). American Redstarts are single-brooded (Sherry and Holmes 1992), although they will renest if the first nest is destroyed.

The frequency of Brown-headed Cowbird parasitism varies greatly from common to rare (Friedmann 1963, Sherry and Holmes 1992). The only Tennessee record is of a redstart feeding a fledgling cowbird in Knox County (Howell and Monroe 1957).—*Robert P. Ford.*

Prothonotary Warbler
Protonotaria citrea

Once called the "Golden Swamp Warbler," the Prothonotary Warbler is a common bird of cypress swamps and river bottomlands and is one of the most colorful and intriguing of our summer residents. This neotropical migrant winters from Nicaragua to Venezuela (AOU 1983) and breeds throughout the eastern United States. Its first state breeding records were at Reelfoot Lake and Shelby County (Rhoads 1895a). Spring migrants arrive in Tennessee during the last week of March and early April (Robinson 1990); along Kentucky Reservoir, earliest arrival of breeding males is between 4 and 12 April (Petit unpubl. data). The warblers leave Tennessee breeding areas in late July to early August, but late migrants are occasionally recorded into October (Robinson 1990).

Prothonotary Warblers are unique among the eastern wood warblers because they nest in holes in trees. Most nests are placed in abandoned woodpecker holes or in cypress "knees," but the warblers have been known to nest in a variety of strange places (Bent 1953, Harrison 1984, Petit and Petit 1988) and will readily accept nest boxes (Ijams 1937, Meyer and Nevius 1943, Fleming and Petit 1986, Petit et al. 1987). Apart from suitable nest holes, the presence of standing water is the single most important habitat feature for Prothonotary Warblers, as they are rarely found more than 100 m from water (Kahl et al. 1985, Walkinshaw 1979). General habitat requirements include flat topography and vegetation features characteristic of mature bottomland forests and swamps (Kahl et al. 1985). Ford (1990) found their density in West Tennessee was highest on sites dominated by water tupelo with numerous snags and a low herbaceous layer.

Because of their specific habitat requirements and occasional difficult access to those habitats, Prothonotary Warblers occurred in only 28% of the completed priority blocks, primarily in the western half of the state where suitable habitat is relatively abundant. The largest concentrations were found along the floodplains of the Mississippi and Tennessee Rivers. Accuracy of Atlas results for this species is probably high, because it is unlikely to remain undetected if suitable habitat was searched. Males are extremely vocal and conspicuous throughout the day, and nest holes are relatively easy to locate. In fact, the percentage of records of confirmed breeding (26%) was relatively high compared with other warblers, probably because of the greater conspicuousness of the birds and their nest sites.

Prothonotary Warblers were recorded at an average of 1.7 stops on the 67 (10%) Atlas miniroutes on which the species was present. In a census on the Duck River in Middle Tennessee, Fowler and Fowler (1985) found

Prothonotary Warbler. Elizabeth S. Chastain

densities of 0.9 birds per 0.8 km segments of river length. In appropriate habitat along the western arm of the Tennessee River, it occurs in densities of 3 pairs/km of river (Petit 1986). Ford (1990) found it on 57 of 59 West Tennessee forested wetland sites at an average density of 54 pairs/100 ha.

Between 1966 and 1994, Prothonotaries did not show a significant population trend on Tennessee BBS routes. Throughout its breeding range, the Prothonotary Warbler exhibited an annual decreasing trend of 1.5% (p = 0.02) from 1966 to 1993, although it apparently increased from 1966 to 1978 (Peterjohn, Sauer, and Link 1994, Sauer and Droege 1992). Reasons for the decrease include the declines in Tennessee's and the nation's wetland areas.

Breeding Biology: Interestingly, most of the research on breeding ecology of Prothonotary Warblers has been conducted in Tennessee, primarily at Reelfoot Lake (Walkinshaw 1941) and on the Tennessee River in Benton County (Petit 1989, 1991a, 1991b, 1991c). Males arrive on breeding areas several days to a week before females and establish territories around 1 or more suitable nest sites, marking each nest cavity by placing pieces of live moss inside. Amounts of moss can vary from a few bits to several centimeters in depth. Females are responsible for building the nest. During nest building, the male follows the female closely as she gathers moss, small rootlets, and grape vine bark for nesting materials. Along the Tennessee River, fishing line is often incorporated into the nest lining as well.

The average height of 27 natural nest cavities used by the warblers in Tennessee was 205 cm (range 67–450 cm; Petit unpubl. data). Optimum diameter at breast height of the cavity tree is approximately 20 cm (Kahl et al. 1985; Petit unpubl. data). Time required for nest building is 3–5 days (Petit unpubl. data), although Walkinshaw (1941) noted an average of 8.8 days was required.

Prothonotary Warblers are almost exclusively monogamous, although if males are able to defend multiple nest sites (e.g., if nest boxes are provided) they may acquire more than 1 mate (Petit 1991b). Earliest egg-laying dates in Tennessee are usually between 18 and 23 April (Petit 1989, 1991a), but Walkinshaw (1941) noted a nest at Reelfoot Lake that contained the first egg on 6 April 1939. Approximately 45% of females in Tennessee make 2 nesting attempts per season, 30% attempted a second nest after a successful nest, and 20% successfully raise 2 broods per season (Petit 1989 and unpubl. data).

Median clutch size of Prothonotary Warblers in Tennessee is 5 eggs ($\bar{x}$ = 4.6, range: 3–6, n = 123; Petit 1989). Eggs are white with lavender-brown spots distributed over the entire surface, often densest around the larger end

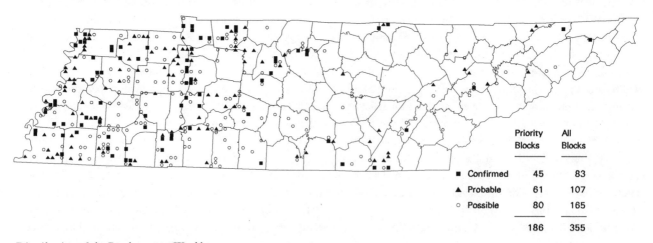

Distribution of the Prothonotary Warbler.

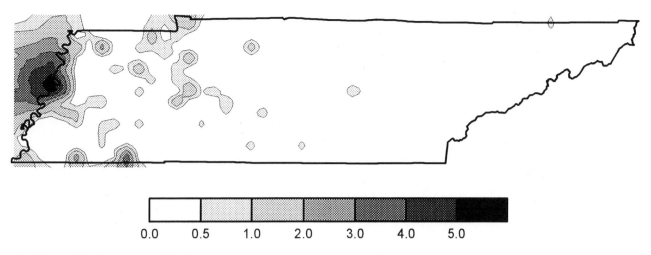

Abundance of the Prothonotary Warbler.

(Bent 1953). Females incubate the eggs for 12–13 days and males will occasionally feed the female on the nest. Nestling care is shared almost equally by both parents, but the male often brings food more frequently than the female. The nestling period is 10–11 days (Walkinshaw 1941, Petit 1989), and fledglings are dependent upon the parents for 3–4 weeks after leaving the nest. Parents often divide the brood after fledging so that each adult cares for only 2–3 of the young. However, the male cares for all young if the female renests.

Despite its hole-nesting habit, the Prothonotary Warbler can suffer a relatively high rate of brood parasitism by Brown-headed Cowbirds (Friedmann 1963). One of the earliest reports of cowbird parasitism in Tennessee was of a parasitized Prothonotary nest in Shelby County in 1921 (Ganier 1921). Frequency of parasitism of Prothonotary nests along the Tennessee River can be especially high (21% compared to an average of 13% in other geographic regions), although parasitism does not occur every year in this area (Petit 1991c). Prothonotary Warblers often respond to parasitism by abandoning the nest; this tendency, however, depends on whether another nest site is available for renesting (Petit 1991c).—*Lisa J. Petit.*

Worm-eating Warbler
Helmitheros vermivorus

The Worm-eating Warbler is an uncommon to fairly common summer resident of East Tennessee and the Western Highland Rim and a rare summer resident elsewhere in the state. It is usually present in the state from mid-April through early September (Robinson 1990). Within its woodland habitat, this drably plumaged warbler is usually inconspicuous except when singing its unmusical, flat, trilled songs or defending its young.

The Worm-eating Warbler was first reported in Tennessee by Fox (1886) during April of 1884 and 1885 in Roane County. Langdon (1887) found it August 1886 in the foothills of the Smokies, and Rhoads (1895a) observed migrants in West Tennessee and none elsewhere in the state. Ganier (1917), presumably based on records in surrounding states, listed it as a very rare summer resident in West Tennessee, a rare summer resident in Middle Tennessee, and a fairly common summer resident in East Tennessee. The first unequivocal breeding record is that of McNish (1923), who found its nest near Nashville. It was not found in the Smokies in the early 1920s (Ganier 1926), on the Cumberland Plateau at Pickett (Ganier 1937a), nor in Fall Creek Falls SP (Ganier and Clebsch 1940). Ganier's revised annotated list (Ganier 1933a) described the bird as a rare summer resident in both Middle and East Tennessee.

The scarcity of early nest and breeding season records of the Worm-eating Warbler, in contrast with its present distribution and abundance, suggest either that its numbers have greatly increased in the last few decades or that it was unfamiliar to and regularly overlooked by early ornithologists. No apparent change in the warbler's habitat has occurred that would result in a great increase in numbers, and such an increase has not been reported in adjoining states (e.g., Mengel 1965). The Worm-eating Warbler has shown a declining trend of 2.6%/year ($p < 0.10$) on Tennessee BBS routes from 1966 to 1994. Throughout its eastern North American range, the Worm-eating Warbler population did not change significantly from 1966 to 1993; from 1978 to 1988, however, it significantly ($p < 0.10$) declined (Peterjohn, Sauer, and Link 1994, Sauer and Droege 1992).

The Worm-eating Warbler occupies forested hillsides and ravines with numerous saplings and shrubs in the understory. In Tennessee, it occurs in the mixed mesophytic

Worm-eating Warbler. Elizabeth S. Chastain

and cove forest types, chestnut oak, oak-hickory, and oak-pine forests, especially with a mountain laurel understory, and hemlock-deciduous forest, especially with a rhododendron understory. It is sensitive to forest fragmentation; Robbins, Dawson, and Dowell (1989) found its probability of occurrence dropped below 50% when forest tract size fell below 150 ha.

Atlas results show the Worm-eating Warbler to be widespread in the Unaka Mountains, Cumberland Mountains, the Western Highland Rim, and parts of the Cumberland Plateau and Eastern Highland Rim. Most of the West Tennessee records are from the Coastal Plain Uplands and Mississippi River bluffs; wooded hillsides and ravines are rare elsewhere in West Tennessee. Most of the records in the Ridge and Valley were from wooded ridges in the northern part of the region. The Worm-eating Warbler was reported on 18% of the miniroutes at an average of 1.9 stops/route. The highest average relative abundance, 2.7 stops/route, was in the Cumberland Mountains, where it was also reported on the highest proportion of miniroutes (13 of 16) and in all priority blocks. Next in abundance were the Cumberland Plateau (2.3 stops/route) and the Unaka Mountains (2.1 stops/route). In the Unaka Mountains, it infrequently occurs above 915 m (Stupka 1963). It occurs throughout the elevational range of the Cumberland Mountains (Nicholson pers. obs.).

Highest densities of the Worm-eating Warbler on census plots have been 89/100 ha in an Eastern Highland Rim mixed mesophytic ravine (Simmers 1982a) and 45 pairs/100 ha in Cumberland Mountain mixed mesophytic forest (Smith 1977). Mengel (1965) similarly noted its highest densities in mixed-mesophytic forests in Kentucky, and this preference for mixed-mesophytic forests as well as steep slopes partially accounts for its abundance in the Cumberland Mountains. Other plot census results include 12–37/100 ha in high elevation, Cumberland Mountain oak-maple forests (Turner and Fowler 1981a, 1981b), 8/100 ha in a Unaka Mountain low-elevation mixed deciduous forest (Lewis 1983a), and average densities of 2.6 and 25/100 ha in Smoky Mountain cove and chestnut oak forests (Kendeigh and Fawver 1981). The Worm-eating Warbler may also occur in regenerating deciduous clearcuts; Lewis (1981a, 1981b, 1982, 1983b, 1984) found between 2.5 and 12/100 ha in low elevation Unaka Mountain clearcuts between 2 and 9 years old.

Breeding Biology: Confirming the breeding of the Worm-eating Warbler was often easy for Atlas workers, and the proportion of confirmed records in priority blocks is high for a small forest insectivore. Almost all of the confirmed records were of fledglings and adults—both of which are frequently conspicuous and vocal—carrying food. Few nests have been reported in Tennessee and only 4 during the Atlas period. The nest is well concealed on the ground, usually under a sprig of fallen

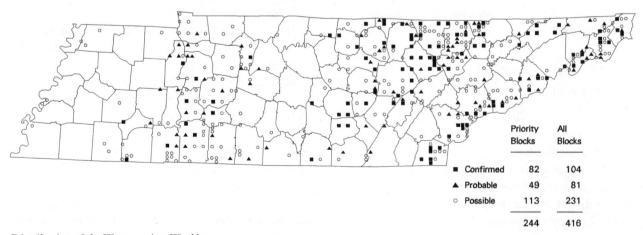

	Priority Blocks	All Blocks
■ Confirmed	82	104
▲ Probable	49	81
○ Possible	113	231
	244	416

Distribution of the Worm-eating Warbler.

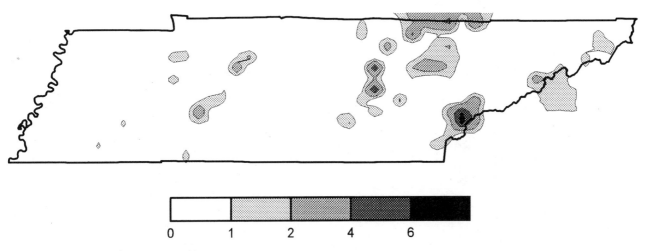

Abundance of the Worm-eating Warbler.

dead leaves at the base of a shrub or sapling on a hillside. The nest is built of skeletonized dead leaves and lined with fine moss stems (Bent 1953).

Egg laying begins as early as the last week of April and, based on numerous dated observations of fledglings, probably peaks in early May. Three nests with 5 eggs and 1 with 4 eggs have been reported in Tennessee. Clutches of 3–6 eggs have been reported elsewhere, with 4 and 5 eggs most common (Bent 1953). The eggs are white with a variable amount of brown spotting, heaviest around the large end. The female incubates the eggs for about 13 days, and the young leave the nest in about 10 days (Bent 1953). The number of broods raised is not known. The only report of Brown-headed Cowbird parasitism of the Worm-eating Warbler in Tennessee is an observation of adult warblers feeding a fledgling cowbird (*Migrant* 35:69). Cowbird parasitism is uncommon elsewhere in the warbler's range (Friedmann 1963).—*Charles P. Nicholson.*

Swainson's Warbler
Limnothlypis swainsonii

The Swainson's Warbler is an uncommon summer resident of wooded bottomlands and ravines in Tennessee. It spends most of its time on the ground and low in thick shrubs and saplings and is much more easily identified by its distinctive, loud, ringing song than by sight. It arrives in its breeding habitat in mid- to late April and probably departs in mid-September. Its fall migration is poorly known, and there are few spring records outside of suitable breeding habitat.

Swainson's Warblers presently nest in 3 distinct areas in Tennessee—in the West Tennessee bottomlands, in the Cumberlands in ravines, and in the Unaka Mountains in ravines. Within each of these areas, they occupy forested areas with a thick, shrubby understory. This habitat has traditionally been considered to include cane in West Tennessee and rhododendron in the mountains (Harrison 1984, Robinson 1990). In West Tennessee, they also frequently occur in bottomlands with dense shrubs, deciduous saplings, and no cane. In East Tennessee, they also occur in areas with dense deciduous and coniferous (frequently hemlock) saplings and occasionally cane (Howell and Campbell 1972, Nicholson pers. obs.). They have been found up to 915 m elevation and regularly occur in Johnson County at 825 m (Wetmore 1939, Atlas results). In Tennessee, as elsewhere (Brown and Dickson 1994), they occupy a variety of bottomland forest types with the apparent requirements of a partially open canopy and dense, woody understory. Maintenance of suitable habitat, at least in the Coastal Plain, requires some periodic disturbance of the forest from timber harvesting or natural causes.

Swainson's Warblers were first reported in Tennessee by Rhoads (1895a), who found them along the Wolf River in Shelby County. Rhoads predicted it would be found in most counties between the western valley of the Tennessee River and the Mississippi River, a prediction that has since proved true. Prior to the 1930s, the Swainson's Warbler was thought to be restricted to the coastal plain of the southeastern United States (Meanley 1971). In 1932, F. M. Jones of Bristol found it nesting in southwestern Virginia near the present South Holston Lake. A few years later, the Wetmore expedition found several birds, and collected specimens in breeding condition, in Johnson County, Tennessee (Wetmore 1939). Swainson's Warblers were not reported from the Cumberland Mountains until 1970, when summer fieldwork began there (Campbell and Howell 1970). Ganier (1937a) did not report it from Pickett SP, where it has regularly occurred since the late 1970s.

Confirmed Breeding Species

Swainson's Warbler. David Vogt

W. R. Gettys collected a nest and set of 3 eggs he identified as those of the Swainson's Warbler in 1902 from "near an old pond in some low wet woods" in McMinn County (Ijams and Hofferbert 1934:3). This record, if correctly identified, is the only nest record from the Ridge and Valley. Ganier (1934b) examined the nest and eggs, noted their resemblance to the nest and eggs of the Indigo Bunting, but concluded they were correctly identified. However, because of the uncertain identification and lack of other records from nearby in the Ridge and Valley, I feel it best to consider Gettys's record to be more likely a record of the Indigo Bunting. The only other possible breeding record from the Tennessee Ridge and Valley is an Atlas record of a nonsinging bird in late June near the Powell River, where the habitat more closely resembles that of the nearby mountain populations.

Although Swainson's Warblers are often difficult to see, they sing well into July (Meanley 1971, Nicholson pers. obs.). Because of the difficulty in exploring their habitat, however, atlasers probably missed Swainson's Warblers in some priority blocks. They were recorded on only 7 miniroutes, at an average relative abundance of 1.3 stops/route.

Swainson's Warblers have been recorded on too few BBS routes in Tennessee to show a significant recent population trend. Coffey (1941a) suggested Swainson's Warbler numbers had recently increased in the Memphis area. Much suitable habitat has been lost from drainage and clearing of West Tennessee bottomlands in recent decades. In the early 1940s, Swainson's Warblers were found nesting in wooded swamps and riverbank canebrakes along the Cumberland and Red Rivers in Montgomery and Cheatham Counties (Ganier 1940b, Clebsch 1942). There are no recent records from Montgomery County, probably due to habitat loss from agricultural clearing and reservoir construction. A territorial bird was present in Davidson County in 1975 (Parmer 1985). At several locations in the Cumberland Mountains and Blue Ridge, Swainson's Warblers have been found for one or two summers, and not in following years, despite no obvious change in the habitat (Nicholson pers. obs.). During the late 1970s, at least 6 Swainson's Warblers occurred along Cove Creek in Campbell County; few have occurred there in recent years (Nicholson pers. obs.). Because of concern over their numbers, Swainson's Warblers were listed as In Need of Management by the Tennessee Wildlife Resources Agency in 1976; this listing is still warranted. Throughout their range, Swainson's Warblers did not show a significant population trend from 1966 to 1993 (Peterjohn, Sauer, and Link 1994).

Breeding Biology: The breeding biology of the Swainson's Warbler in Tennessee is poorly known, and there were

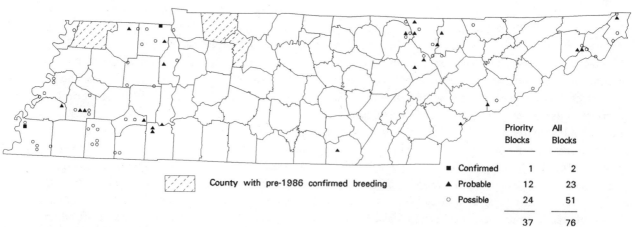
Distribution of the Swainson's Warbler.

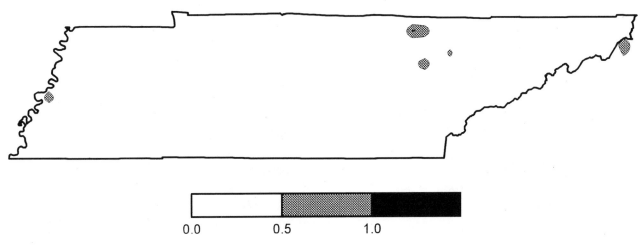

Abundance of the Swainson's Warbler.

only 2 confirmed Atlas records. The nest, although large and bulky, is often difficult to find; none have been reported from East Tennessee. The few Tennessee nests for which details are available were between 0.9 and 1.5 m above ground, averaging 1.2 m (s.d. = 0.3 m, n = 5). Five nests were in cane, 1 in blackberries, and 1 in thick vines. The nest, built by the female, consists of a large outer shell (up to 15 cm diameter) of dry leaves and sticks, and an inner cup of leaf skeletons and dead grasses (Ganier 1940b, Meanley 1971). It is often placed on dead leaves lodged in the understory and, upon completion, resembles fallen leaves or flood debris. The few egg records indicate laying occurred in the second half of May, and 2 complete clutches each contained 4 eggs. Unlike other warblers nesting in Tennessee, the eggs are usually pure white. The female incubates the eggs for 13–15 days, and both parents feed the young, which fledge in 10–12 days (Meanley 1971). Parasitism by Brown-headed Cowbirds has not been reported in Tennessee, although it is common in other parts of the warbler's range (Meanley 1971, Brown and Dickson 1994).—*Charles P. Nicholson.*

Ovenbird
Seiurus aurocapillus

The Ovenbird is a relatively large warbler found in upland forests. Its loud, distinctive "teacher, teacher, teacher" song is familiar to most birders, and, because of its cryptic plumage and ground-dwelling habitats, it is heard much more often than seen. Ovenbirds are present in Tennessee from mid-April to mid-October (Robinson 1990).

Ovenbirds are common in East Tennessee, uncommon in Middle Tennessee, and rare in West Tennessee. They reach their greatest abundance in the Cumberlands and in the low to middle elevations of the Unaka Mountains. Their distribution coincides well with that of extensive upland deciduous or mixed deciduous-coniferous forests. Ovenbirds are sensitive to forest fragmentation, and their probability of being detected in a forest drops as forest area decreases (Whitcomb et al. 1981, Robbins, Dawson, and Dowell 1989). Even when present in small (less than 140 ha) isolated forests, Ovenbirds may not successfully breed, as the proportion of paired males is much lower than in larger forest tracts (Gibbs and Faaborg 1990). In the Smokies, Ovenbirds occur commonly up to about 1280 m elevation, and reach 1650 m in gray beech forests (Stupka 1963, Kendeigh and Fawver 1981). A few summer records in spruce-fir forests exist (Stupka 1963), but nesting in this habitat has not been confirmed. In Ridge and Valley forests, Ovenbirds show a preference for areas with a closed canopy, large trees, little ground cover, and few shrubs (Smith and Shugart 1987).

The Ovenbird was found in 42% of the completed Atlas priority blocks. Because of its loud, conspicuous song, it was probably rarely missed by atlasers, and the Atlas map is an accurate portrayal of its range. It was reported on 26% of the miniroutes at an average relative abundance of 3.4 stops/route. The reasons for their low numbers on the heavily forested Western Highland Rim are unclear.

The Ovenbird is often one of the most abundant members of its family. Turner and Fowler (1981a, 1981b) recorded densities of 68 and 86 pairs/100 ha in mature, high-elevation oak-maple forests in the Cumberland Mountains; the Ovenbird was the most abundant species on these plots. At a density of 47 pairs/100 ha in a maple-dominated, middle-elevation hardwood forest in the northern Unaka Mountains, the Ovenbird was the second most abundant species (Lewis 1983a, 1984a). Densities in the Smokies include 54–105 pairs/100 ha on oak-dominated plots, 15–47/100 ha in cove hardwoods,

Ovenbird. David Vogt

Ovenbird numbers, as measured by BBS routes, declined by 2.3%/year (p < 0.01) from 1966 and 1994. Most of the decrease occurred after 1980, and occurred despite the increase in acreage of mature deciduous forest. This suggests other factors, such as forest fragmentation and changes in their Central and northern South American wintering habitat, are responsible. During the Atlas period Ovenbirds were found on one-third of the BBS routes at an average of 6.9 stops and 8.0 individuals/50-stop route. Throughout their breeding range, Ovenbirds increased by 0.5%/year (p = 0.03) from 1966 to 1993; most of this increase, however, was in the early part of this period (Sauer and Droege 1992, Peterjohn and Sauer 1993).

Breeding Biology: The male Ovenbird arrives in the spring about a week before the female and promptly begins defending a territory. Nest construction begins around the first of May, is done by the female, and takes about 5 days (Hann 1937). The nest, placed on the ground, resembles an old-fashioned domed Dutch oven and is the source of the species's name. Nests are often placed near an opening in the forest, such as along the edge of a trail. The entrance, a small slit on the side of the dome, usually faces downhill. The nest is built of dry grass and leaves and lined with hair, rootlets, fine plant stems, and moss.

and 51–270/100 ha on hemlock-deciduous forest plots (Wilcove 1988). Densities in mixed-mesophytic forests on the Cumberland Plateau, Cumberland Mountains, and Eastern Highland Rim, up to 24 pairs/100 ha (Robertson 1979, Smith 1977, Garton 1973, Simmers 1982a), are much lower than those in the ecologically similar cove and hemlock-deciduous forest plots in the Smokies. Smith and Shugart (1987) found territory size, which averaged 0.26 ha, varied inversely with the abundance of invertebrate prey.

In addition to those parts of Tennessee where they are presently common, Ovenbirds probably once occurred throughout the Ridge and Valley, southern Highland Rim, and upland areas of West Tennessee prior to widespread agricultural clearing. Rhoads (1895a) did not observe Ovenbirds in West Tennessee or in the Nashville area; farther east, he described them as "always present" on the Cumberland Plateau, Ridge and Valley, and lower elevations of the Unaka Mountains.

Egg laying peaks during the first half of May. Clutch size in Tennessee ranges from 3 to 5 and unparasitized clutches average 4.5 eggs (s.d. = 0.6, n = 29). Clutches of 6 and a most common size of 5 eggs have been reported elsewhere (Hann 1937). Ovenbirds are occasionally double-brooded, and the second clutch may be smaller. The eggs are white with gray to reddish brown spotting, concentrated in a wreath around the large end. Incubation, carried out entirely by the female, lasts 12–13 days (Hann 1937). When incubating and brooding

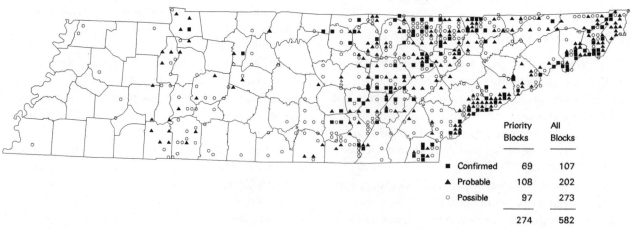

	Priority Blocks	All Blocks
■ Confirmed	69	107
▲ Probable	108	202
○ Possible	97	273
	274	582

Distribution of the Ovenbird.

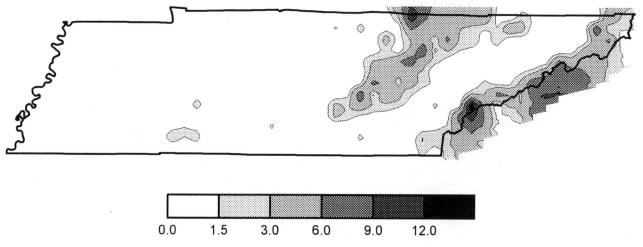

Abundance of the Ovenbird.

newly hatched young, the female sits tightly. Once flushed, she feigns injury, dragging her wings and tail while walking about. Both adults feed the nestlings. The young leave the nest relatively early, about 8 days after hatching. At this point their plumage is not fully developed, and they do not begin flying for several days. The parents care for the fledglings for 3–4 weeks after they leave the nest (Hann 1937). Observing fledglings was much easier than finding nests for atlasers, and fledglings accounted for two-thirds of the confirmed Atlas records.

None of 16 nests reported before 1930 in Tennessee was parasitized by Brown-headed Cowbirds. Of 19 nests reported since 1930, 7 contained cowbird eggs. These parasitized nests include 1 with 4 cowbird eggs and 2 with 3 cowbird eggs. No Tennessee records of Ovenbirds feeding fledgling cowbirds are available. Numerous studies elsewhere, summarized in Friedmann (1963), have shown that from one-third to half of all Ovenbird nests are parasitized by Brown-headed Cowbirds. Although Ovenbirds accept cowbird eggs, the survival rate is low, as the nestling Ovenbirds fledge faster than the nestling cowbirds (Hann 1937).—*Charles P. Nicholson.*

Louisiana Waterthrush
Seiurus motacilla

Louisiana Waterthrushes are uncommon to fairly common summer residents, found across the state along woodland streams. They are one of the earliest migrant warblers to arrive in the spring and one of the first to depart in the fall and are normally present in Tennessee from late March to mid-August.

The song of waterthrushes is loud and easily identified, making singing birds conspicuous; atlasers probably found them in most priority Atlas blocks with accessible woodland streams. Robbins, Dawson, and Dowell (1989) found forest area to be a significant predictor of the probability of occurrence of Louisiana Waterthrushes and that probability is reduced by half in forests less than 350 ha in area. The lack of adequate woodland stream habitat probably accounts for their absence from much of West Tennessee, the Central Basin and south-central Highland Rim, and the Ridge and Valley. The limited surface drainage by creeks and small streams contributes to their low numbers in the Central Basin, and a shortage of suitable, elevated nest sites may contribute to their low numbers along West Tennessee streams. The lack of records from parts of the Cumberland Plateau is probably due to the inaccessibility (to Atlas workers) of streams in many plateau blocks. In the Unaka Mountains, waterthrushes regularly occur up to about 1070 m (Stupka 1963).

Waterthrushes defend linear territories along streams and confine most of their activities to this narrow corridor. Territories are vigorously defended from other waterthrushes by singing and rapidly chasing intruders. Eaton (1958) found territories in New York state averaged 400 m in length, and Craig (1984) found territories in Connecticut averaged 0.67 ha. Simmers (1982a) found 2 waterthrush territories along a 518 m stretch of stream in an Eastern Highland Rim ravine; no other published information on territory sizes in Tennessee is available.

Atlasers recorded Louisiana Waterthrushes on 96 of the miniroutes, at an average of 1.4 stops/route. The highest relative abundance was on the Eastern and Western Highland Rims and in the Cumberland Mountains, probably due to the tendency of roads to follow streams in these regions as well as their high proportion of forest.

In the spring of 1810, Alexander Wilson found waterthrushes "among the mountain streams in the state of Tennessee . . . pretty numerous" (Wilson 1811a). Although Wilson, as well as other ornithologists of that

Louisiana Waterthrush. Elizabeth S. Chastain

period, did not distinguish between the Louisiana and Northern *(S. noveboracensis)* Waterthrushes, other parts of his description suggest he was describing Louisianas. Fox (1886) listed Louisiana Waterthrushes as rather common in Roane County in April 1885, and a decade later Rhoads (1895a) described them as a cosmopolitan summer resident, numerous across the state. At that time, their numbers, particularly in the Ridge and Valley, were probably somewhat reduced from presettlement levels due to clearing along streams. Twentieth-century population reductions have probably been most severe in West Tennessee. From 1966 to 1994, waterthrush numbers declined by 2.3%/year ($p < 0.10$) on Tennessee BBS routes. They were recorded on 18 of the 42 routes during the Atlas period, at an average of 1.9 stops and 2.1 birds/50-stop route. No significant change occurred in their rangewide population from 1966 from 1993 (Peterjohn, Sauer, and Link 1994).

Breeding Biology: Male waterthrushes occupy territories immediately after their spring arrival and sing throughout the day from tree perches several meters above the ground (Eaton 1958). The females arrive a few days after the male. Following mating, song frequency is much reduced. Early in the nesting season, most of their foraging occurs in and along the stream (Eaton 1958, Craig 1984). They wade in shallow water and walk along rocks, logs, and the stream bank, capturing submerged and floating prey as well as picking prey off dead leaves they pull from the water. Most of their aquatic prey consists of larvae of aquatic insects; small fish and salamanders, isopods, snails, and adult aquatic insects are also eaten (Craig 1984). Later in the season, they forage more on dry ground and tree limbs; this may be due in part to the declining populations of aquatic prey following the spring emergence of adult insects.

Louisiana Waterthrushes nest in a small hollows under overhanging roots or other vegetation on a steep bank within a few meters of a stream. Occasionally the nest is built in a rock crevice alongside the stream. Both adults build the nest, although the female does most of the work (Eaton 1958). After excavating a depression in the soil within the hollow, the birds drag dead, wet leaves into it and place them to form the nest cup and an often relatively conspicuous pathway extending downhill from the nest. The nest bowl is formed from grass or weed stems, pine needles, and twigs, lined with rootlets, hair, fine grasses, or moss. Nest construction takes 3–5 days (Eaton 1958).

Egg laying peaks during mid-April, and waterthrushes are single-brooded (Eaton 1958). The few late May and June egg records are probably replacement clutches. Normal clutch size is 4–6 eggs (Bent 1953); 41 Tennessee clutches averaged 4.7 eggs (s.d. = 0.54). Eaton (1958) found an average clutch size of 5.8 eggs in New York; his larger average may be due to a latitudinal gra-

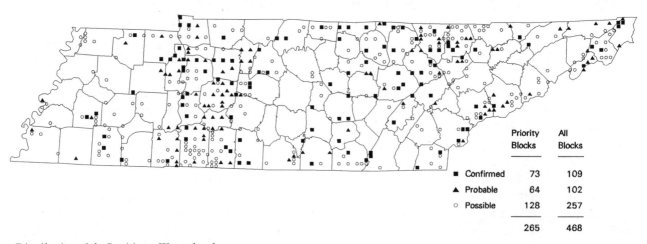

Distribution of the Louisiana Waterthrush.

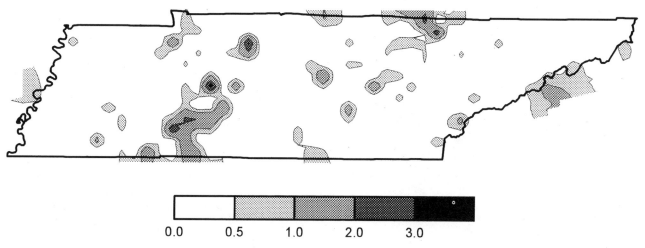

Abundance of the Louisiana Waterthrush.

dient in clutch size. The female incubates the eggs for 12–14 days (Bent 1953, Eaton 1958). Both adults feed the young, which leave the nest in 10 days. When flushed from the nest or in the vicinity of recently fledged young, the adults feign injury, dragging their extended wings or tail along the ground. After the nestlings are about 4 days old, the adults carry the fecal sacs from the nest and drop them in the stream, where they may accumulate in shallow water (Eaton 1958). These accumulated fecal sacs can be helpful in finding the nest (Nicholson pers. obs.). After fledging, the young are tended by the adults for 3–4 weeks (Eaton 1958). The family may wander some distance from the stream and is fairly conspicuous (Nicholson pers. obs.); records of fledglings accounted for over half of all confirmed Atlas records.

Louisiana Waterthrushes are rather frequently parasitized by Brown-headed Cowbirds in Tennessee, as elsewhere in their range (Friedmann 1963). The first report of parasitism was in 1928 (Vaughan egg coll.), and about 12% of the nests reported since 1950 have been parasitized. Cowbird young are often fledged by waterthrushes, and atlasers made several observations of fledged cowbirds being fed by waterthrushes.—*Charles P. Nicholson.*

Kentucky Warbler
Opornis formosus

The Kentucky Warbler is more often heard than seen. Its persistent song rings through woodlands with dense understories, but the shy warbler may offer only a glimpse of its olive and yellow colors and black sideburns. It is a summer resident, arriving in Tennessee in April and departing from late August through October.

Chiefly a southeastern species (Bent 1953), the Kentucky Warbler's range in Tennessee appears unchanged. Wilson (1811a) described it as particularly numerous south of Nashville in primeval forest, and Rhoads (1895a:491) listed it as "abounding all across the state from the Mississippi Bluff to the foothills of the Great Smoky Mountains." As forests were cleared, its numbers declined, and Ganier (1933a) described it as common in West and Middle Tennessee and fairly common in the east. Continued deforestation in recent decades in West Tennessee and urban areas has probably resulted in further declines.

Atlasers found the Kentucky Warbler throughout Tennessee; it was the most frequently reported warbler occurring in forested habitats. It was easily identified by its loud, distinctive song, and the Atlas map is likely an accurate portrayal of its distribution. It was found in 74% of the completed priority blocks and on 38% of the miniroutes at an average relative abundance of 1.9 stops/route. Centers of abundance were in the Mississippi Alluvial Plain, particularly Lauderdale County, and in the Cumberland Mountains. Its average abundance in these regions were 3.3 stops/route on 6 of 15 routes, and 3.5 stops/route on 14 of 16 routes, respectively. Its abundance was higher than average in the Western Highland Rim and Cumberland Plateau. Within the Unaka Mountains, it occurred on one-third of the routes at 1.5 stops/route. The lower abundance in this heavily forested region is due in part to its rare occurrence above 1070 m (Stupka 1963). It is also uncommon in the relatively xeric pine and mixed pine forests common in the southern Unaka Mountains and southern Cumberland Plateau.

Between 1966 and 1994, the Kentucky Warbler population showed a decline of 3.0%/year ($p < 0.05$) on Tennessee BBS routes. This decline was greatest before 1980. Rangewide, its population declined by 0.8%/year ($p = 0.07$) from 1966 to 1993; from 1978 to 1988, however, it decreased by 2.0%/year ($p < 0.05$) (Sauer and Droege 1992, Peterjohn, Sauer, and Link 1994).

Kentucky Warbler. Elizabeth S. Chastain

The Kentucky Warbler has been classified as a forest interior species, with its probability of occurrence increasing with the size of the woodland (Whitcomb et al. 1981, Robbins, Dawson, and Dowell 1989). It is, however, also adapted to smaller woodlands, as its probability of occurrence peaks at a forest area of about 300 ha. Gibbs and Faaborg (1990) also found little difference in its density and mating success between fragmented (less than 140 ha) and extensive (more than 500 ha) forests in Missouri. Its ability to inhabit fragmented forests accounts for its occurrence in much of West Tennessee, parts of the Highland Rim, and the outer portion of the Central Basin, all areas where forests are relatively fragmented.

The Kentucky Warbler shows a preference for moist woodlands with a relatively open overstory of large trees and dense, low understory foliage (Anderson and Shugart 1974, Robbins, Dawson, and Dowell 1989). It frequently occurs along streams and ravines and in laurel and rhododendron tangles. Densities on Tennessee census plots include 5 pairs/100 ha in virgin hardwoods at 900 m in the Smokies (Aldrich and Goodrum 1946); from 66/100 ha in 1965 to 4/100 ha in 1972 in Ridge and Valley hardwoods (Howell 1972); 75/100 ha in Cumberland Mountain maple-gum-hickory forest (Smith 1975); 17/100 ha in Cumberland Plateau virgin mixed-mesophytic forest (Robertson 1979); and 76/100 ha in a disturbed deciduous ravine forest on the western edge of the Eastern Highland Rim (Simmers 1982a). In the Central Basin, 2 Maury County cedar-deciduous plots had densities of 12 and 16/100 ha (Fowler and Fowler 1983a, 1983c); it was absent from nearby upland hardwood plots (Fowler and Fowler 1984a, 1984c). Lewis (1977–1984b), studying a low-elevation Unaka mountains clearcut planted with white pine, first found Kentucky Warblers 2 years after cutting at a density of 10 pairs/100 ha; they peaked at 30/100 ha 5 years after harvest and declined to 10/100 ha 9 years after harvest. Ford (1990) found the Kentucky Warbler widespread in West Tennessee forested wetlands; it occurred on 46 of 59 sites at an average density, measured by line transects, of 7 pairs/100 ha.

Breeding Biology: Although the adults readily show agitated behavior and give distraction displays near the nest, the nest is often difficult to find, and only 9% of the confirmed Atlas records were of nests. The total number of Tennessee nest records, however, is probably higher than for any other woodland warbler except the Prothonotary. Breeding activity begins in early May; the male actively defends the territory, and the female chooses the nest site and builds the nest (Bent 1953).

The nest is a well-made, bulky cup built of weed, vine, and bark shreds, lined with black rootlets, grass, and hairs, and placed on a base of dead leaves, often

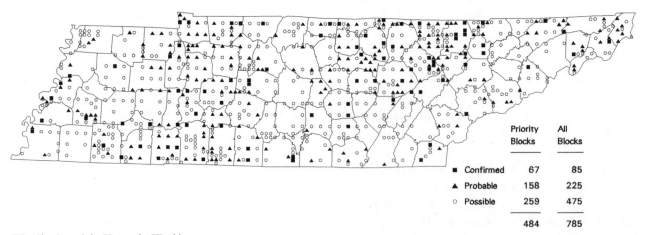

Distribution of the Kentucky Warbler.

	Priority Blocks	All Blocks
■ Confirmed	67	85
▲ Probable	158	225
○ Possible	259	475
	484	785

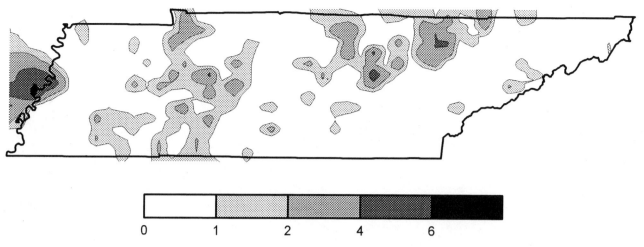

Abundance of the Kentucky Warbler.

several centimeters tall. It is often near the edge of the woods or a small opening, on the ground or up to 15 cm high, and well hidden in a low crotch, under a bush, or in thick herbaceous vegetation. Concealing vegetation includes American chestnut sprouts, mountain laurel, poison ivy, wild hydrangea, woodbine, and asters.

Tennessee clutches average 4.4 eggs (n = 48, s.d. = 0.82, range 2–6), very close to the Kentucky Warbler's range-wide average of 4.5 eggs (Morse 1989). The eggs, white with gray and brown markings concentrated around the large end, are incubated by the female for 12–13 days (Bent 1953). Laying peaks in the second and third weeks of May. Both parents attend the young, and 1 brood is raised (Bent 1953). Care of the fledglings usually ends in mid-July, and local dispersal occurs in August (Goodpasture 1977b).

The Kentucky Warbler is a common victim of the Brown-headed Cowbird. Of 70 nests in Tennessee, 8 (11%) contained cowbird eggs or young. The earliest record, from Knox County in 1924, contained 1 cowbird and 2 warbler nestlings and an addled warbler egg (Ganier unpubl. notes, Univ. Tennessee Library Special Coll., Knoxville). In Williamson County, 2 of 9 nests (22%) in 1949–53 held cowbird eggs (Goodpasture 1954), and a 1970 nest held 3 cowbird and 2 warbler eggs (*Migrant* 41:70). There are several records of Kentucky Warblers feeding cowbird fledglings (e.g., Ganier and Clebsch 1942, Nicholson pers. comm.).—*Ann T. Tarbell.*

Common Yellowthroat
Geothlypis trichas

The Common Yellowthroat is a common summer resident and one of the most abundant of the wood warblers nesting in Tennessee. It usually arrives in mid-April and departs by late October. Several winter records, most during December, exist (Robinson 1990). The Common Yellowthroat nests at all elevation throughout the state in brushy, early successional habitats.

The Common Yellowthroat was probably an uncommon bird in prehistoric Tennessee, nesting in the prairie areas of north-central and northwest Tennessee, along the edges of rivers and wetlands, in areas of recent severe fires, and around openings maintained by Native Americans. Its numbers probably increased rapidly as European settlers cleared the forest. In the late nineteenth century, Rhoads (1895a) found it abundant at low elevations throughout the state. Its numbers have probably fluctuated with the area of cleared land, peaking in the 1930s and 1940s. Since then, yellowthroat numbers have probably decreased with the area of brushy habitat due to urban development, cleaner farming practices, and the maturation of the forests.

The yellowthroat population in Tennessee decreased by 1.2%/year ($p < 0.05$) on BBS routes from 1966 to 1994; most of this decline was in the second half of this period. Throughout its extensive North American breeding range, its population decreased by 0.5%/year ($p < 0.01$) from 1966 to 1993 (Peterjohn, Sauer, and Link 1994).

No major change in the breeding distribution of the Common Yellowthroat is evident in the Tennessee literature. It has, however, apparently spread to the highest elevations of the Unaka Mountains since the 1920s (Stupka 1963). Brewster (1886) did not observe it above 640 m in western North Carolina, Rhoads (1895a) found it up to 915 m on Roan Mountain, and Ganier (1926) reported it only from low elevations in the Smokies. It was still apparently absent from high elevations of Roan Mountain in 1936 (Ganier 1936). Stupka (1963) lists several high-elevation summer records from the Smokies, beginning about 1935, and the yellowthroat presently nests high on Roan Mountain. This spread into the high

Common Yellowthroat. Elizabeth S. Chastain

elevations may have been due to the great increase in young, brushy habitat in the early 1900s due to logging of the high-elevation forests (Stupka 1963) and the somewhat later death of the American chestnut, as noted in West Virginia by Brooks (1940).

The Common Yellowthroat nests in brushy habitats, among them brushy pastures and hayfields, fencerows, grassy marshes, and heavily cutover woodlands. Densities on census plots in various habitats are 22 pairs/100 ha in an ungrazed field with hedgerows (Lewis 1975a), 49 and 33/100 ha in abandoned agricultural fields (Fowler and Fowler 1984c, 1984d), and 2/100 ha in a 5-year-old clearcut (Lewis 1980). Kendeigh and Fawver (1981) found densities in the Smokies of 30/100 ha in a young, open pine-oak forest with thick shrubby understory, and 68/100 ha in a recently burned, shrubby spruce-fir forest.

Breeding Biology: Male yellowthroats establish their territories quickly after spring arrival and defend them into late summer. They sing throughout much of the summer, occasionally performing flight songs by flying as high as 30 m and singing a jumble of notes, including bits of song, then silently descending. Males are occasionally polygynous, with more than one female nesting in the territory (Stewart 1953).

The nest is placed in low, thick vegetation and is difficult to locate. Observations of nests made up about 10% of the confirmed Atlas observations. Fledglings and adults carrying food are fairly easy to observe and these codes made up most of the confirmed records.

The female constructs the nest, which has an outer layer of leaves and coarse grasses and weeds, an inner layer of finer leaves and stems, and a lining of fine grasses and hair (Stewart 1953). The nest is usually built in a clump of grass, sedges, or weeds within a few centimeters of the ground. The supporting vegetation often forms a canopy over the nest. Nests are also occasionally built in low shrubs and blackberries up to 75 cm above ground (egg coll. data).

Egg laying peaks during the second half of May. Clutches of 3 to 5 eggs, with a mode of 4 and average of 4.2 (s.d. = 0.53, n = 33), have been reported in Tennessee. Clutch size elsewhere ranges from 3 to 6 eggs, with 4 and 5 most common (Stewart 1953). Yellowthroat eggs are creamy white with fine brown to black dots, usually forming a wreath around the large end. The female incubates the eggs for about 12 days, and both adults feed the young, which leave the nest in 8 or 9 days (Stewart 1953). The adults take care of the fledglings for at least 2 weeks. They are usually double-brooded (Stewart 1953), and available nest records suggest this is the case in Tennessee.

The Common Yellowthroat is commonly parasitized by the Brown-headed Cowbird. The proportion of nests

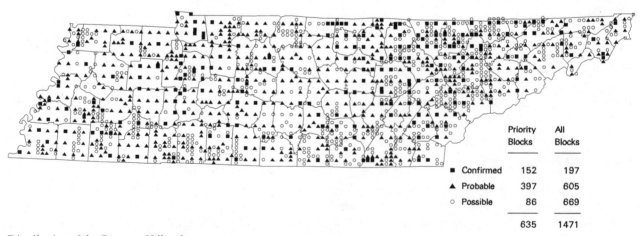

	Priority Blocks	All Blocks
■ Confirmed	152	197
▲ Probable	397	605
○ Possible	86	669
	635	1471

Distribution of the Common Yellowthroat.

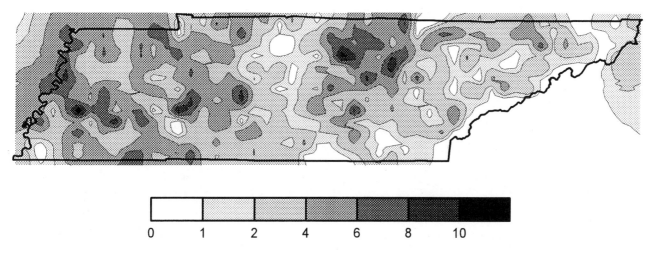

Abundance of the Common Yellowthroat.

found parasitized in different studies in eastern North America ranges from 7 to 46% (Friedmann 1963), and the yellowthroat usually accepts cowbird eggs. There are several Tennessee records of cowbird eggs in yellowthroat nests, as well as of adults feeding cowbird fledglings, dating to at least 1920 (Monk 1936).—*Charles P. Nicholson.*

Hooded Warbler
Wilsonia citrina

The Hooded Warbler is a summer resident, common in the wooded uplands of East Tennessee and uncommon elsewhere in the state. It arrives in mid-April and departs by mid-October. Although often difficult to see, it is easily identified by its loud, distinctive song, which it sings well into July.

The Hooded Warbler inhabits woodlands with a thick, shrubby understory. In the east, it is common in cove and mixed mesophytic forests and also occurs in relatively xeric, ridgetop pine-hardwood forests with a thick understory of laurel or blueberry. Farther west, it more often occurs in moist forested ravines. In West Tennessee, it occurs in swamp forests and in the moist Mississippi River bluff forests. The Hooded Warbler occurs throughout the elevational range of the Cumberland Mountains (Nicholson pers. obs.). In the Unaka Mountains, it rarely breeds above about 1220 m; the highest breeding record is at 1497 m (Stupka 1963). It shows a moderate sensitivity to forest fragmentation (Whitcomb et al. 1981, Robbins, Dawson, and Dowell 1989).

The Hooded Warbler is often one of the most abundant warblers and sometimes among the most numerous of all bird species. Densities on census plots in Cumberland Mountain hardwood forests range from 25 pairs/100 ha in an oak-maple forest (Turner and Fowler 1981b) to 155/100 ha in a maple-gum-hickory forest (Smith 1977). Simmers (1982a) recorded 44/100 ha in an Eastern Highland Rim mixed mesophytic forest. The Hooded Warbler also occurs in deciduous clearcuts; in a low-elevation Unaka Mountains clearcut, its density increased from 5/100 ha 2 years after harvest to 111/100 ha 9 years later (Lewis and Smith 1975, Lewis 1984a). During the latter years of this study, the Hooded Warbler was the second most abundant species. Average densities in Smokies census plots include 25 pairs/100 ha in xeric pine-oak forest and 80/100 ha in chestnut oak forest with thick sapling and shrub understory resulting from the death of American chestnuts (Kendeigh and Fawver 1981). Anderson and Shugart (1974) found it to be the most common warbler on their Ridge and Valley study area. Ford (1990) found it on 20% of the forested wetland sites sampled in West Tennessee.

Hooded Warbler. Elizabeth S. Chastain

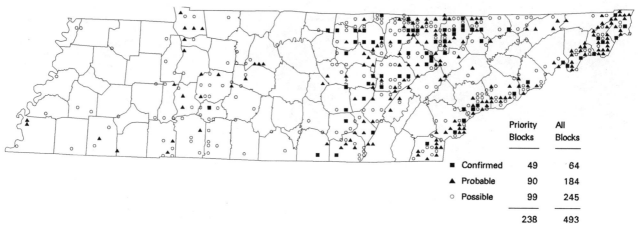

Distribution of the Hooded Warbler.

Atlasers recorded the Hooded Warbler in slightly over one-third of the completed priority blocks. Because of its loud voice and persistent singing, it is unlikely that it was missed in many priority blocks in the eastern two-thirds of the state. In West Tennessee, where it is more restricted to swamp forests, it was probably missed in some blocks where access to its habitat was difficult. The Hooded Warbler was found on one-fifth of the mini-routes, at an average relative abundance of 2.7 stops/route. It was most abundant in the Cumberland Mountains (4.6/route), the Unaka Mountains (3.5/route), and the Cumberland Plateau (2.8/route).

The Hooded Warbler was probably a common bird throughout much of prehistoric Tennessee. Alexander Wilson observed it during his trip through Middle Tennessee in 1811 and noted that it was abundant in cane thickets (Wilson 1811a). Its numbers probably declined with the forest clearing following European settlement. In the late nineteenth century, Rhoads (1895a:497) described it as the "most thoroughly representative and evenly distributed summer warbler of Tennessee." Within the range of the chestnut in Middle and East Tennessee, its numbers probably increased during the 1930s and 1940s in response to the thick sapling growth following death of the chestnuts (cf. Brooks 1940). Between 1966 and 1994, its population on Tennessee BBS routes decreased by 1.5%/year ($p < 0.01$). Howell (1972) noted its disappearance on a Knoxville census plot during the late 1960s. Rangewide, its population increased by 1.6%/year ($p = 0.05$) from 1966 to 1993 (Peterjohn, Sauer, and Link 1994).

Breeding Biology: Confirming the breeding of the Hooded Warbler was not easy for many Atlas workers. The nest is usually well hidden in shrubs or saplings, and records of nests made up less than one-fifth of the confirmed records. The Hooded Warbler begins nesting activities soon after its arrival in Tennessee. The female constructs the nest, which is placed in an upright fork or between parallel stems of a shrub, saplings, or a stout herbaceous plant. The compact nest is placed on a platform of dead leaves, and the cup is woven of plant

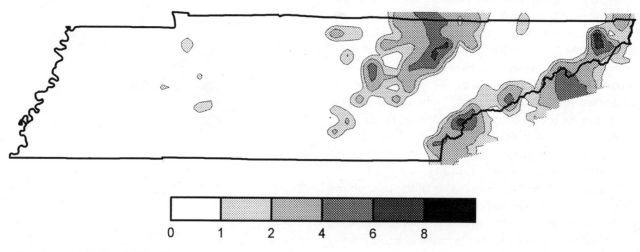

Abundance of the Hooded Warbler.

fibers and strips of bark from grape or other species, held together with spider webs (Bent 1953). The nest is lined with fine grasses, plant fibers, and hair. Tennessee nests range from 40 to 180 cm above ground, with an average of 85 cm (s.d. = 33 cm, n = 18). Nests elsewhere are most often 60–90 cm above ground (Bent 1953). In Tennessee, nests have been in sprouts and saplings of several deciduous tree species, as well as hydrangea, blueberry, cane, and cohosh plants.

Clutch size is usually 3 or 4, and rarely 5 eggs (Bent 1953). In Tennessee, 13 clutches were all of 3 or 4 eggs and averaged 3.7 eggs (s.d. = 0.48). The eggs, which the female incubates for about 12 days, are creamy with reddish brown spots, usually confined to the large end (Bent 1953). Both adults feed the young, which leave the nest in about 9 days. The Hooded Warbler readily renests after a nest is destroyed and may occasionally raise 2 broods (Bent 1953). The range of dates of nests with eggs and young suggests that the species is double-brooded in Tennessee.

The Hooded Warbler is a common host of the Brown-headed Cowbird (Friedmann, Kiff, and Rothstein 1977). In Tennessee, no nests reported before 1940 were parasitized; at least 3 of 10 more recent nests contained cowbird eggs or young. Young cowbirds are frequently fledged by Hoodeds, and Hooded Warblers feeding fledgling cowbirds have been reported at least twice in Tennessee.—*Charles P. Nicholson.*

Canada Warbler
Wilsonia canadensis

The Canada Warbler is a common summer resident of the highest mountains in East Tennessee, where it occupies dense shrub thickets in both deciduous and coniferous forests. Although much of its biology is poorly known, the Canada is very active and easily identified by its ebony necklace draped across a golden breast, as well as its yellow eye rings, gray back, and staccato song. Outside of the eastern mountains, it is an uncommon migrant, present from late April through late May and mid-August to late September (Robinson 1990).

The first evidence of the breeding of the Canada Warbler in the southern Appalachians was Brewster's (1886) description of it as abundant above 915 m in western North Carolina. LeMoyne (1896) first reported it from Tennessee and mentioned finding several nests, although details of his observations are limited. Rhoads (1895a) described it as breeding on Roan Mountain from 915 to 1220 m. Ganier (1926:37) described it as "perhaps the most common warbler" at high elevations of the Smokies. Although known as a summer resident

Canada Warbler. David Vogt

on Black Mountain, Kentucky, since early in the century (Mengel 1965), it was not found during the summer in the Cumberland Mountains of Tennessee until the 1980s (Nicholson 1987).

Atlas workers found the Canada Warbler to be widespread in the Unaka Mountains and very local in the Cumberland Mountains. In the Cumberlands, it was uncommon in the 2 blocks making up most of Frozen Head State Natural Area, and a single territorial bird was present in a priority block on Cross Mountain, Campbell County (Atlas results). This was the first summer report from the mountains of Campbell County, which have probably received more ornithological work than any other part of the Cumberland Mountains of Tennessee. The Canada Warbler was found on 7 Atlas miniroutes at an average relative abundance of 2.3 stops/route; all of these routes were in the Unaka Mountains. These results, however, may underestimate its abundance, as its rate of singing is low in late June and early July, when most high-elevation routes were censused. On a nonrandom miniroute (not included on density map) at high elevations on Frozen Head, it occurred at an average of 1.5 stops/route over 2 years (Nicholson unpubl. data).

In the Unaka Mountains, the Canada Warbler is common in moist forests with a thick understory of rhododendron or other shrubs above 1036 m (Stupka 1963). It occurs somewhat lower, to about 900 m, along streams in hemlock and cove forests. At lower elevations, it is frequently sympatric with the closely related Hooded Warbler (Kendeigh and Fawver 1981). Breeding densities on plots censused in the Smokies (Kendeigh and Fawver 1981, Wilcove 1988) were highest (average of 80 pairs/100 ha) in hemlock-deciduous forests, followed by mid- and late successional spruce-fir forests

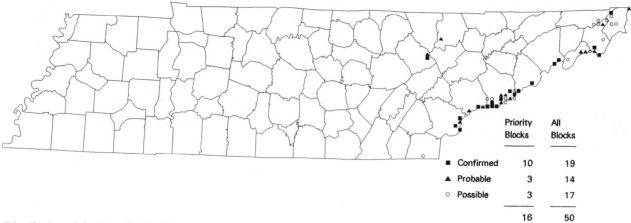

Distribution of the Canada Warbler.

(55 and 53/100 ha). Average densities in other forest types were 16/100 ha in cove forest, 13/100 ha in red oak forest, and 6/100 ha in beech gap forest. No Canada Warblers were found in climax spruce-fir plots censused in 1947–48 between Newfound Gap and Clingman's Dome (Kendeigh and Fawver 1981) or on Mt. Guyot in 1967 (Alsop and Laughlin 1991). Following death of most of the fir trees, however, the Mt. Guyot plot had 18 Canada Warbler pairs/100 ha in 1985 (Alsop and Laughlin 1991).

On Frozen Head, in the Cumberland Mountains, the Canada Warbler ranges from the highest elevations down to about 833 m (Nicholson 1987). It occurs in the *Hydrangea*-dominated understory of the mixed mesophytic forest as well as black locust-black cherry-northern red oak forest (Nicholson 1987, Nicholson pers. obs.). The bird found during Atlas work on Cross Mountain was at the edge of an abandoned strip mine at about 747 m. The mine was mostly revegetated with black locust, red maple, and blackberry; the adjacent forest was a mix of oak, hickory, maple, and basswood.

Robbins, Dawson, and Dowell (1989) found the probability of occurrence of the Canada Warbler increased with the area of the forest and suggested a minimum area requirement of 400 ha. This minimum area requirement could explain its absence from several isolated mountains in both the Cumberlands and Unakas, which reach a maximum elevation of about 915 m and therefore may lack sufficient high-elevation habitat.

Information on population trends of the Canada Warbler in Tennessee is limited, as it is not censused by BBS routes. Wilcove (1988) found no noticeable differences in its density on hemlock and hardwood-dominated plots in the Smokies censused in 1947–48 and 1982–83. Its population has increased within the spruce-fir forest, as suitable habitat was created following the death of mature firs. Throughout its breeding range, its population declined by 2.1%/year ($p = 0.07$) from 1966 to 1993 (Peterjohn, Sauer, and Link 1994).

Breeding Biology: The breeding biology of the Canada Warbler in the southern Appalachians is relatively poorly

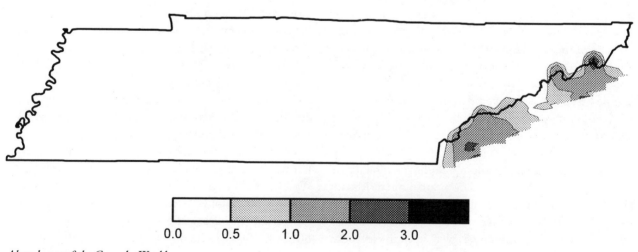

Abundance of the Canada Warbler.

known; the only Tennessee nest records available are those of LeMoyne (1886). The Atlas records of confirmed breeding were of an adult carrying nest material on 20 May, a distraction display, adults carrying food (8 records), and fledglings (9 records). The Canada Warbler builds a fairly bulky nest on a base of dead leaves, placed on or near the ground, typically in upturned tree roots, a mossy hummock or stump, or a hollow in a bank (Bent 1953). The nest cup, of bark, grass, and weed stems, is lined with fine rootlets and hair. It is usually concealed by overhanging grasses or ferns. Egg laying probably extends from mid-May through late June. Clutches of 3–5 eggs, most frequently 4, have been reported elsewhere (Bent 1953). The eggs are white or creamy, marked with chestnut, brown, and purple, usually wreathed near the large end. The incubation and nestling periods are poorly known.

The Canada Warbler is infrequently parasitized by the Brown-headed Cowbird (Friedmann and Kiff 1985), and no incidences of this parasitism have been reported in Tennessee. Cowbirds rarely occur within the Canada Warbler's range in the Smoky and Unicoi Mountains. Parasitism is more likely in the Cumberlands and northern Unaka Mountains, where cowbirds are more common.
—*Ronald D. Hoff and Charles P. Nicholson.*

Yellow-breasted Chat
Icteria virens

The Yellow-breasted Chat is a common summer resident in low-elevation brushy areas throughout Tennessee, present from mid-April to late September (Robinson 1990). Although chats spend much of their time in low, dense vegetation, their song is loud and distinctive, and they were the second most frequently reported wood warbler. The song, often given at night as well as throughout the day, is a series of squawks, whistles, and chattering and mewing notes, occasionally including mimicked notes of other birds. The Yellow-breasted Chat differs from other wood warblers in its large size as well as several anatomical and behavioral traits (Ficken and Ficken 1962). Because of this, its relationship to other wood warblers has long been in question. Recent classifications, based on DNA studies, show that it is more closely related to the wood warblers than any other group of birds and continue to classify it with the warblers in the tribe Parulinae (Sibley and Ahlquist 1990).

At the time of European settlement, Yellow-breasted Chats were probably restricted to the brushy areas associated with cedar glades and barrens, widespread natural forest disturbances such as fires, and brushy areas resulting from Native American activities. With the wide-

Yellow-breasted Chat. David Vogt

spread forest clearing accompanying European settlement, their numbers greatly increased. Rhoads (1895a) described the species as abundant in all of the areas he visited except Roan Mountain. In recent decades, chat numbers have declined with the decreased area of early successional forest and agricultural trends toward larger fields and improved pastures. Tennessee BBS routes show a declining trend of 4.2%/year ($p < 0.01$) from 1966 to 1994. Rangewide chat numbers decreased by 0.6%/year ($p = 0.07$) from 1966 to 1993 (Peterjohn, Sauer, and Link 1994). On a rangewide basis, the Cumberland Plateau and Highland Rim (including the Central Basin) are among the physiographic provinces with the highest numbers of chats (Robbins, Bystrak, and Geissler 1986).

Yellow-breasted Chats were recorded in 96% of the completed Atlas priority blocks. They were not found in a few priority blocks in urban areas, in 1 block in the Big South Fork area that lacked early successional habitat, and in high-elevation blocks in the Blue Ridge. Stupka (1963) gave summer records of chats at elevations over 1525 m in the Smokies without evidence of breeding. None were reported at such elevations by atlasers, and Stupka's high-elevation records may have been wandering birds. The upper elevational limit for nesting chats in Tennessee is probably around 975 m. None were present in seemingly suitable habitat above that elevation in the Unicoi Mountains in Monroe County (Nicholson pers. obs.). The absence of chats in several rural priority blocks in Middle and West Tennessee is difficult to explain.

Chats were recorded on 81% of the miniroutes, at an average relative abundance of 3.5 stops/route. Three routes recorded it at the highest abundance of 13 stops/route. The chat's abundance often varied greatly between adjacent blocks, averaging highest in the Coastal Plain

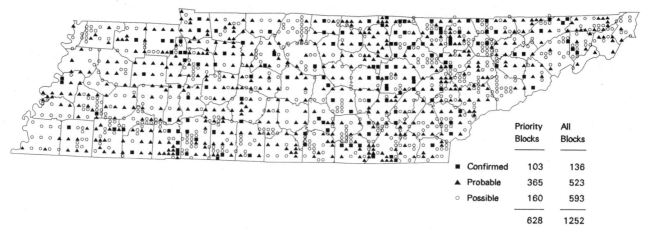

	Priority Blocks	All Blocks
■ Confirmed	103	136
▲ Probable	365	523
○ Possible	160	593
	628	1252

Distribution of the Yellow-breasted Chat.

Upland, Highland Rim, and Cumberland physiographic regions, and lowest in the Ridge and Valley and Unaka Mountains.

Because few plot censuses have been conducted in relatively homogeneous, early successional habitats, breeding density information is limited. Kendeigh and Fawver (1981) reported a density of 60 males/100 ha in young, low-elevation pine-oak forest in the Smokies in 1947–48. The chat population in a clearcut area in Washington County rose from 20/100 ha the first year following harvest to a peak of 79/100 ha the sixth year after harvest (Lewis and Smith 1975, Lewis 1981a).

Breeding Biology: Breeding Yellow-breasted Chats are territorial and usually monogamous (Thompson and Nolan 1973). The males' displays include a flight song, given in boundary disputes and other contexts. The male, while singing, flies upward, hovers with deep, exaggerated wingbeats and legs dangling, then descends to a perch. In a courtship display, the male fluffs his neck feathers and sways side to side (Ficken and Ficken 1962).

The relatively large, bulky nest is usually built low in a thicket of blackberry or other dense shrubs. Only 12% of the confirmed Atlas records were of nests; the majority of confirmed records were of adults carrying food. Chats, however, are well represented in Tennessee egg collections. The female builds the nest of leaves, grass, and weed stems and lines it with fine grasses and roots. Sixty-six Tennessee nests averaged 92 cm above ground (s.d. = 28), and ranged from 30 to 170 cm high. Nests have been most frequently reported in blackberries, and other species used include buckthorn, cane, rose, vine-covered sumac, and shrubby hackberry, elm, and sumac. Several Middle Tennessee nests have been in buckthorn, elm, and hackberry; this may be due as much to the nests being easier to find in these species as to regional preferences.

Egg laying in Tennessee peaks during the second half of May. Clutch size is 3–5 eggs, and averages 3.84 (s.d. = 0.56, n = 78) in Tennessee. The eggs are creamy white, with brown, gray, or purple spots spread over most of the egg. Incubation, carried out by the female, has been

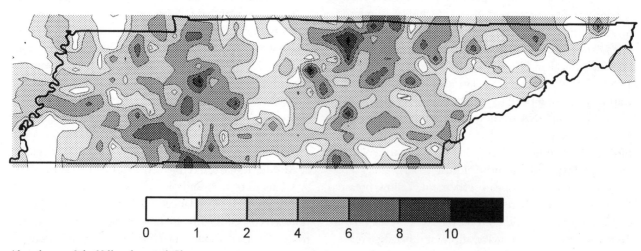

Abundance of the Yellow-breasted Chat.

given as 11 (Petrides 1938) to 14 days (Thompson and Nolan 1973). Both adults feed the young, which leave the nest in about 10 days (Thompson and Nolan 1973). Egg dates in Tennessee suggest that one brood is normally raised. Thompson and Nolan (1973) found a high rate of nest failure in their Indiana study area and noted frequent renesting after nest failure. Later nests tended to be more successful than early nests.

Of 70 sets of Tennessee chat eggs, mostly collected between 1900 and 1940, only 1 contained a Brown-headed Cowbird egg. Since then, there have been at least 3 records of cowbird young in chat nests and 1 record of chats feeding a fledgling cowbird. The frequency of parasitism seems to vary elsewhere in the chat's range and may be locally common (Friedmann 1963).—*Charles P. Nicholson.*

Summer Tanager
Piranga rubra

The "summer redbird" is an uncommon to fairly common summer resident, occurring statewide in low-elevation woodlands. It arrives in Tennessee about the third week of April and departs by early October (Robinson 1990). Males do not attain their fully red plumage until their second fall; first-year breeding birds may be a mixture of yellow-orange with red patches (Isler and Isler 1987). The greenish yellow females occasionally have varying amounts of red plumage.

Summer Tanagers occur in open woodland of a variety of types, including oak-hickory and other hardwoods, mixed oak-pine, and, less commonly, pine-dominated types. They also occur in wooded residential areas and bottomland forests. Breeding densities, as recorded on census plots, are relatively low and include 16 pairs/100 ha in Central Basin upland hardwoods (Fowler and Fowler 1984a, 1984b), and 8 pairs/100 ha in Ridge and Valley hardwood forests (Howell 1972, Smith 1975).

Early accounts described the Summer Tanager as occurring statewide and less numerous in the eastern part of the state than elsewhere (Rhoads 1895a, Ganier 1917). This description applies to its present distribution and abundance. The Summer Tanager was found in 84% of the completed Atlas priority blocks and in every county except Johnson in the extreme northeast corner of the state. It was also found on 55% of the miniroutes, at an average relative abundance of 2.4 stops/route. It was most abundant in the Western Highland Rim and Coastal Plain Uplands regions, where it occurred on over half the routes at averages of over 3 stops/route. It is generally absent above 610 m in the Unaka Mountains and from heavily wooded parts of

Summer Tanager. David Vogt

the Cumberlands (Atlas results, Stupka 1963). Summer Tanagers sing well into July and were probably rarely overlooked by atlasers. Although their song may occasionally be confused with the Scarlet Tanager's, they give their distinctive "pick-a-tuck" calls often enough that misidentification is probably infrequent.

In parts of the state where extensive forest clearing has occurred, such as the Loess Plain, Central Basin, and Ridge and Valley, the tanager's density is low and probably decreased from decades ago. Some local decreases in density have apparently also occurred in areas where the amount of forest has increased in recent decades. Ganier and Clebsch (1940) described it as fairly common at Fall Creek Falls SP. At that time, much of the park consisted of recently abandoned farmlands; few were found there during Atlas fieldwork, when most of the area was again forest. Robbins, Dawson, and Dowell (1989) found the tanager's probability of occurrence in Maryland and adjacent states increased with forest tract size, and dropped below 50% in forests of less than 40 ha. The tanager's wide distribution in Tennessee, especially in the Central Basin and Loess Plain—regions with little forested area—suggests a tolerance of small woodlots. Summer Tanagers frequently occur in forest edge habitats (Whitcomb et al. 1981, Nicholson pers. obs.).

Tennessee BBS routes do not show a significant trend in the Summer Tanager population from 1966 to 1994, although the proportion of routes with decreases was significantly ($p < 0.05$) greater than those with increases. Their population also did not show a significant rangewide overall trend from 1966 to 1993 (Peterjohn, Sauer, and Link 1994). Summer Tanagers have been recorded on 41 of the 42 BBS routes in Tennessee, at an average of 4.7 birds and 4.3 stops/50-stop route during the Atlas period.

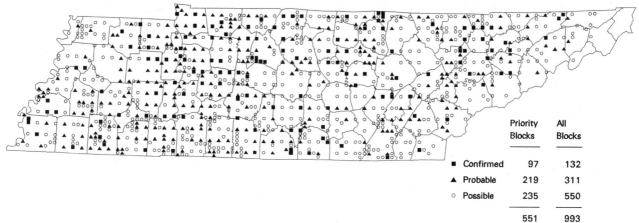

	Priority Blocks	All Blocks
■ Confirmed	97	132
▲ Probable	219	311
○ Possible	235	550
	551	993

Distribution of the Summer Tanager.

Summer Tanagers feed heavily on insects and are well known for eating wasps and honeybees (Isler and Isler 1987). Several accounts of this behavior in Tennessee, including observations of them tearing apart paper wasp nests, have been published. They frequently catch flying insects and then conspicuously perch on a fence, powerline, or low tree limb while feeding.

Breeding Biology: In contrast to the Scarlet Tanager, there are many Summer Tanager nest records from Tennessee. This is due, in part, to the Summer's habit of nesting at the edge of clearings, over roads, and in rural yards. Although the percentage of confirmed Atlas records is relatively low, about half of all confirmed records were records of nests. Nest construction begins in early May. The nest is built entirely by the female, who is often accompanied by the male (Isler and Isler 1987). The nest is a shallow cup, about 10 cm in outer diameter, placed on a horizontal branch, often at a fork far from the trunk. It is often flimsily constructed, and the eggs may be visible through the nest from below. The nest is built of dried weed stems, dried or fresh grass stems, and occasionally pine needles, lined with fine grass stems or rootlets. Tennessee nests averaged 4.0 m above the ground, with a range of 1.2–9.1 m (n = 50, s.d. = 1.6 m). Nests have been reported from at least 19 tree species in Tennessee, including pines and red cedar. The most frequently reported species are white and post oaks, elm, maple, and hickories.

Egg laying peaks during the second half of May, and second broods are probably sometimes raised, as several late summer nest records exist, including a nest with recently hatched young on 19 August (Anon. 1933). Mengel (1965) and Potter (1985) give evidence of second broods elsewhere. Clutch size ranges from 3 to 5, with an average of 3.62 (s.d. = 0.56, n = 34). Bent (1958) gave 4 eggs as the most common clutch size, and Mengel (1965) gave an average of 3.2 for 14 Kentucky nests. The eggs are pale blue to pale green, with brown spots or blotches over much of the egg and concentrated at the large end. Both the nests and eggs are similar enough to the Scarlet Tanager's that, in areas where both species occur, identification of unattended nests is difficult.

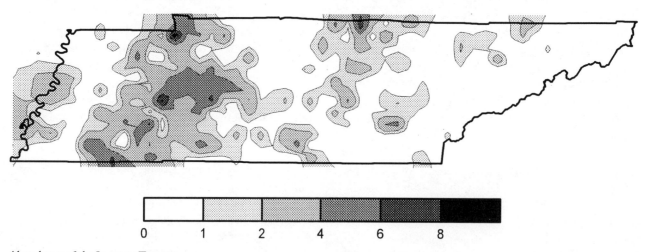

Abundance of the Summer Tanager.

The female incubates the eggs for 12–14 days, and the incubating female is frequently fed by the male, either on the nest or on a nearby limb (Bent 1958, Fitch and Fitch 1955, Potter 1985, Nicholson pers. obs.). Both adults feed the nestlings. The length of the nestling period is poorly known; Potter (1985) observed a brood leave the nest in 11 days. Fledglings remain in the parents' territory about 3 weeks (Fitch and Fitch 1955).

Thirteen of 49 (27%) Tennessee Summer Tanager nests contained up to 3 Brown-headed Cowbird eggs or young, indicating that cowbird parasitism is common. None of the 10 egg sets in the Gettys collection, collected in McMinn County from 1900 to 1905, were parasitized. The first case was reported in 1920 (Monk 1936), and 2 of 8 sets in the Todd collection, mostly from the late 1930s in Rutherford County, were parasitized. The rate has since continued to increase. Friedmann and Kiff (1985) described the Summer Tanager as a regular cowbird host, noting that the parasitism rate was underestimated in earlier summaries. There is no evidence of tangers rejecting cowbird eggs, and several Tennessee records of tangers feeding cowbird fledglings exist.
—*Charles P. Nicholson.*

Scarlet Tanager
Piranga olivacea

This brilliantly colored species is one of 2 members of the very large, primarily tropical, tanager family nesting in the Tennessee. Scarlet Tanagers are a summer resident, arriving by mid-April and usually departing by mid-October (Robinson 1990). They spend most of their time in treetops and are most easily seen early in the spring, before deciduous trees are fully leafed out.

The Scarlet Tanager occurs throughout most of the state in mature hardwood and mixed hardwood-pine forests. Stupka (1963) described it as a common summer resident between 450 and 1525 m in the Smokies, and it occurs at all elevations in the Cumberlands. It was found in 61% of the completed Atlas priority blocks and is present in almost all priority blocks in the Blue Ridge, Cumberland Mountains, and Cumberland Plateau, and in most blocks on the Western Highland Rim. It is uncommon and local in the Central Basin and West Tennessee, which may be partially due to its preference for extensive forests. Robbins, Dawson, and Dowell (1989) found its probability of occurrence increased with forest area, reaching a maximum in forests over 3000 ha and reduced below 50% in forests less than 12 ha. It was found on 37% of the Atlas miniroutes at an average relative abundance of 3.0 stops/route. Its miniroute abundance correlates well with the distribution of extensive deciduous forests and was highest in the Cumberland Mountains where it was found on all routes at an average of 4.7 stops/route.

Scarlet Tanager. David Vogt

In the Smokies, Kendeigh and Fawver (1981) and Wilcove (1988) found Scarlet Tanagers in all forest types censused at elevations below 1525 m, including cove hardwoods, northern red and chestnut oak, mixed hemlock-deciduous, pine-oak, and pine heath. The highest density, an average of 38 pairs/100 ha, occurred on chestnut oak census plots (Kendeigh and Fawver 1981). Densities of approximately 40 pairs/100 ha have been reported from Cumberland Mountain mixed mesophytic forests (Yahner 1972, Smith 1977).

The overall breeding range of the Scarlet Tanager in Tennessee has probably changed little in historic times, although local declines in abundance have undoubtedly resulted from deforestation. It was first reported in Tennessee by Rhoads (1895a), who described it as occurring statewide and more abundant in the east than elsewhere; his West Tennessee observations may have included migrating birds. Ganier (1917, 1933a) described it as very rare in the west, rare in Middle Tennessee, and fairly common in the east. The difference between Ganier's description and the tanager's present abundance on the Western Highland Rim is more likely due to an underestimate by Ganier than a change in tanager numbers. It has probably increased recently in the Ridge and Valley; Howell and Monroe (1957), for example, described it as rare in Knox County. The Scarlet Tanager population shows an increasing trend of 2.2%/year ($p < 0.10$) on Tennessee BBS routes from 1966 to 1994. Range-wide, its population did not show a significant overall trend from 1966 to 1993, although it decreased by 1%/year ($p < 0.05$) from 1978 to 1988 (Sauer and Droege 1992, Peterjohn, Sauer, and Link 1994).

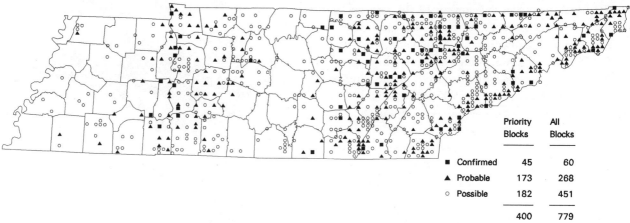

Distribution of the Scarlet Tanager.

	Priority Blocks	All Blocks
■ Confirmed	45	60
▲ Probable	173	268
○ Possible	182	451
	400	779

The range of the Scarlet Tanager in Tennessee overlaps that of the closely related Summer Tanager, and the song of these species are similar enough that atlasers probably misidentified a few of each. Stupka (1963) described the Scarlet Tanager as more often occurring above about 450–605 m, while the Summer Tanager occurs at lower elevations. Scarlet Tanagers, however, occur as low as 90 m along the Hatchie and Obion River drainages. Both species were found together in about half the priority Atlas blocks and both have been reported from several census plots (e.g., Smith 1975, Nicholson 1979a, Simmers 1982b). Shy (1984) found that they partition the habitat into non-overlapping territories: the Scarlet Tanager occupies forests with higher and denser canopy cover and more large trees. In the absence of the other species, the Scarlet Tanager occupied somewhat more open habitat and the Summer Tanager occupied somewhat denser habitat. They countersing when sympatric, and the aggressive response to the other's songs maintains the habitat segregation.

Breeding Biology: A very low proportion of Scarlet Tanager Atlas records was in the confirmed category; two-thirds of these were of fledglings or adults carrying food. Their nests are difficult to locate and in areas where both tanager species occur, unattended nests are difficult to identify. The few Tennessee nest records show nest construction begins about the first of May. Nests have been reported from tulip-poplar, hickory, black and northern red oaks, hickory, maple, and dogwood trees. Seven nests ranged in height from 2 to 12 m, averaging 7.1 m (s.d. = 3.7 m). The female tanager chooses the nest site and constructs the nest, a shallow cup on a horizontal limb, often far from the tree trunk (Prescott 1965). The nest is built with weed stems, twigs, grass, and rootlets and lined with fine grasses and plant fibers.

The only Tennessee egg records are during May and show an average clutch size of 3.8 (s.d. = 0.45, range 3–4, n = 5). A larger sample would likely include June egg dates, as dates of fledglings and adults carrying food extend into late July. Clutches of 4 eggs are most com-

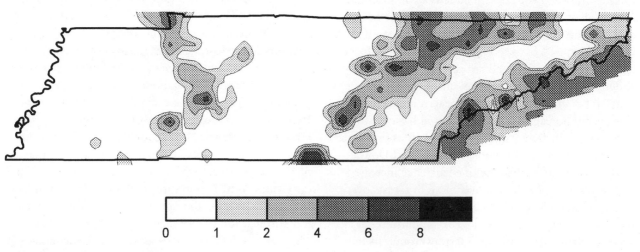

Abundance of the Scarlet Tanager.

mon elsewhere, with clutches of 5 occasional (Tyler 1958b, Prescott 1965). The eggs are pale blue to green, with brown spots or splotches concentrated at the large end. The female incubates the eggs for 13–14 days and is often fed by the male. Both parents feed the nestlings insects and fruit, and the young fledge in about 15 days (Tyler 1958b, Prescott 1965).

Of 8 Tennessee Scarlet Tanager nests with known contents, 1 was parasitized by Brown-headed Cowbirds. This nest held 1 cowbird egg and no tanager eggs. One instance of an adult tanager feeding a cowbird fledgling was reported during Atlas fieldwork. Friedmann (1963) described the Scarlet Tanager as the most heavily parasitized tanager, with some local populations at times heavily parasitized. Its true status as a cowbird host, however, is not well understood because of the low number of nest records available (Friedmann, Kiff, and Rothstein 1977).—*Charles P. Nicholson.*

Northern Cardinal
Cardinalis cardinalis

The Northern Cardinal, a permanent statewide resident, is familiar to almost everyone because of its handsome plumage, cheerful whistled song, and frequent visitation of bird feeders. Abundant in all seasons, the "redbird" is found in urban, suburban, and rural regions from bottomlands to mountainsides.

Because the Northern Cardinal is primarily a bird of forest edges, it was probably uncommon to fairly common in prehistoric Tennessee, most numerous in prairie edges, tree-fall gaps, forest burns, and around Native American towns. Cherokee Indian legends tell of cardinals in prehistoric times (Mooney 1972), and all early naturalists traveling through the South knew it well. Michaux's 1795 Nashville record (Williams 1928) is probably the earliest published Tennessee record. Rhoads (1895a:491) found it "very abundant everywhere" from Reelfoot Lake to 1220 m on Roan Mountain. Its numbers undoubtedly increased as early European settlers cleared the forests, and early settlement practices of small farms, numerous fencerows, and woodlots would have provided abundant cardinal habitat.

Northern Cardinals are not migratory; Tennessee banding studies have shown most spend their lives within a few kilometers of their hatching sites (Laskey 1944). An unusually long-lived male nested for 13.5 years in the same suburban Nashville yard (Ganier 1937c). Nevertheless, some young birds wander, and since 1886, when it was rarely seen north of the Ohio River Valley, it has expanded its range northward to become a permanent resident in Canada and the Northeast (Beddal

Northern Cardinal. David Vogt

1963, Bent 1968). Its range in Tennessee has remained static. It occurs statewide and is common up to about 1070 in the Smokies and regular to 1370 on Roan Mountain (Eller and Wallace 1984). There is no evidence of it moving upslope into forest openings created by the widespread death of mature Fraser firs (e.g., Alsop and Laughlin 1991).

It would be difficult to overlook this conspicuous resident, and Atlas records of the Northern Cardinal were the second most numerous of all species. It was found in all completed priority blocks except for 7 at high elevations in the Unaka Mountains. The Atlas map is undoubtedly an accurate portrayal of its distribution. Confirmation of breeding was high everywhere, due to its long breeding season and multiple broods. Of all confirmed records, over half (51%) were of fledged young and 25% were of nest building and active nests with eggs or young.

Atlas workers recorded Northern Cardinals on all but 10 of the miniroutes at slightly over half the census stops ($\bar{x}$ = 8.0 stops/route), the second highest relative abundance of any widespread species. The highest average abundance occurs in the heavily agricultural Loess Plain ($\bar{x}$ = 10.6 stops/route), followed by the Mississippi Alluvial Plain (8.9), Coastal Plain Uplands (8.8), Western Highland Rim (8.4), and Ridge and Valley (8.3). Lowest numbers were in the heavily forested Unaka Mountains (4.9) and Cumberland Mountains (5.2).

Tennessee BBS routes show a significant ($p < 0.01$) decline in the Northern Cardinal of 1.2%/year from 1966 to 1994. Most of this decline occurred before 1980. Although the trend of increasing forest area since the 1930s has likely increased cardinal numbers, this effect is probably short-lived and decreases as the forests mature. The maturing of forests, as well as changing

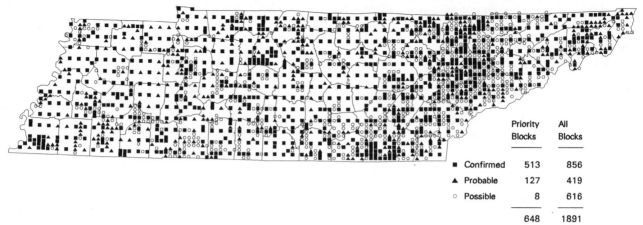

Distribution of the Northern Cardinal.

agricultural practices including removal of fencerows and cleaner pastures, probably account for the recent decline. Throughout North America, the cardinal population did not show a significant long-term trend from 1966 to 1993 (Peterjohn, Sauer, and Link 1994).

Northern Cardinal habitat requirements include open patches of ground for feeding, trees for singing perches, and dense, low growth for nesting. Occupied habitats include open woodland with a thick understory, forest edges, second growth, swamps, streamsides, hedgerows, and shrubbery around homes, farms, and parks (Bent 1968). Plot census results from Ridge and Valley woodlands found the cardinal to be the most abundant species. In a suburban woodland, Howell (1971, 1972, 1973, 1974) found cardinal densities averaging 142 pairs/100 ha. A partially cutover and grazed rural woodland had 110/100 ha (Smith 1975). Within Central Basin cedar forests broken by brushy glade openings, probably characteristic of much of the region's original forest cover, Northern Cardinals averaged 36/100 ha, first or second in rank (Fowler and Fowler 1983a, 1983b, 1983c). Densities in closed canopy upland forests are much lower. Within West Tennessee forested wetlands, Ford (1990) found the cardinal to be the second most abundant breeding species.

Breeding Biology: Detailed descriptions of the breeding of the Northern Cardinal in Tennessee have been published by Shaver and Roberts (1930), Ganier (1941), and Laskey (1944). In late February and March, cardinals form stable, monogamous pair bonds. Both the male and female defend the territory with song, displays, and mild combat. Courtship feeding by the male occurs regularly. The female picks the nest site and builds the nest with occasional help from the male, usually beginning in April, but as early as 17 March (*Migrant* 6:13). The first nest is a carefully built cup of layers of thin sticks and plant stems, leaves, paper, bark strips, and a lining of fine grasses and rootlets; later nests are often flimsy. The nest is well hidden on a crotch or branch in evergreens,

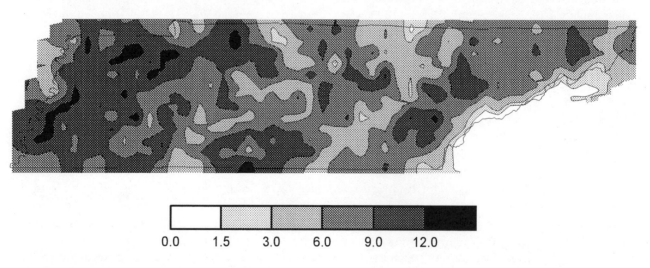

Abundance of the Northern Cardinal.

bushes, and saplings, often supported by vine tangles. Nest heights average 1.5 m (n = 104, s.d. = 0.6 m, range 0.3–6.1 m); Laskey (1944) reported 103 nests between 0.76 and 3.6 m high, with most 1.2–1.5 m. Favored plants are red cedar, hawthorns, blackberry, roses, and Japanese honeysuckle.

Egg laying peaks about 17 April to 4 May, but extends throughout the summer. The average clutch size is 3.1 eggs (n = 97, s.d. = 0.65, range 1–5), with 66% of the clutches containing 3 eggs, and 13% and 19% with 2 and 4 eggs respectively. Ganier (1941) and Laskey (1944) found 3-egg clutches common and fewer nests with 4 eggs. Cardinal eggs are pale grayish to greenish white with variable amounts of brown and gray spots and blotches. The female incubates the eggs for 12–13 days; both adults feed the nestlings, which leave the nest in 7–11 days, most often in 9–10 days (Laskey 1944). Four to 5 nestings are common (Shaver and Roberts 1930, Laskey 1944); elsewhere 8 renests with no successes have been reported (Kinser 1973). Time from nest failure to the first egg in a new nest is from 3 to 16 days (Shaver and Roberts 1930, Scott, Lemon, and Darley 1987).

The Northern Cardinal is frequently parasitized by the Brown-headed Cowbird. Rangewide, Friedmann, Kiff, and Rothstein (1977) noted about 10% of nests were parasitized. The earliest reports in Tennessee were 2 nests each with 2 cowbird and 3 host eggs in 1932 (Woodring 1932, Monk 1936). Of 97 nest records available through 1990, 10 held 1–3 cowbird eggs. This 10% rate may be an underestimate, as cowbird eggs are sometimes difficult to distinguish from cardinal eggs. Laskey (1944) found 6 of 17 (31%) suburban nests parasitized in 1942. There are numerous reports of cardinals feeding cowbird fledglings.—*Ann T. Tarbell.*

Rose-breasted Grosbeak
Pheucticus ludovicianus

The Rose-breasted Grosbeak is an uncommon to fairly common summer resident of the East Tennessee mountains. In the rest of the state, it is a fairly common migrant, present from mid-April through mid-May and mid-September through mid-October (Robinson 1990).

Probable breeding Rose-breasted Grosbeaks have been found in the Cumberland Mountains in deciduous forests at elevations as low as about 640 m (Howell and Campbell 1972), although the lowest confirmed breeding is at 730 m (Nicholson pers. obs.). In Frozen Head State Natural Area, they occur most frequently above 885 m (Nicholson 1987).

In the Unaka Mountains, Rose-breasted Grosbeaks occur above about 915 m and are most common be-

Rose-breasted Grosbeak. David Vogt

tween 975 and 1525 m (Stupka 1963). There are several records, without evidence of nesting, from lower elevations in the Smokies (Stupka 1963) and in several other counties in and adjacent to the Unaka and Cumberland Mountains *(Migrant)* during late June and July. These birds have presumably wandered from nearby high-elevation areas. The "possible" Atlas records from western Monroe County and southeastern Polk County may have been of wandering birds. The record from Hancock County, at 550 m on Newman Ridge in late June, was of a male rapidly moving through the area, with no evidence of nesting (Nicholson pers. obs.). Nearby ridges reach elevations of 730 m, and there are no known nesting populations in the area. There is no evidence of change in the grosbeak's historic Tennessee breeding range.

The area of forest above a certain minimum elevation is probably important in determining the occurrence of Rose-breasted Grosbeaks in the Cumberland Mountains and on isolated peaks in the Unaka Mountains. They were not found on peaks with maximum elevations of less than about 915 m. Additional research is necessary to refine their area requirements in Tennessee. Robbins, Dawson, and Dowell (1989) noted that grosbeak abundance was positively correlated with forest area in Maryland and adjacent states, but had too few samples to estimate minimal area requirements.

Rose-breasted Grosbeaks are most common in deciduous forests with a moderately dense to dense understory. In the Smokies, Kendeigh and Fawver (1981) found average densities in 1947–48 of 6.3 pairs/100 ha in mixed hemlock-deciduous forest, 3.3/100 ha in mature red oak forest, and 6.3/100 ha in beech gap forest. In 1982, Wilcove (1988) found a much higher density of 73 pairs/100 ha in a chestnut oak forest.

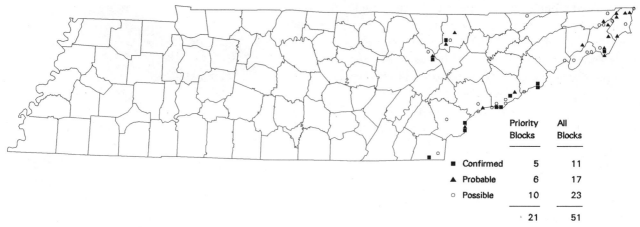

Distribution of the Rose-breasted Grosbeak.

Rose-breasted Grosbeaks have not been found on other census plots in the Unaka Mountains or on the few census plots at high elevations in the Cumberlands.

Rose-breasted Grosbeaks are found on too few BBS routes in Tennessee to show population trends. Wilcove (1988) found no clear trend in numbers on plots censused in the Smokies in 1947–48 and 1982–83. Throughout their breeding range, Rose-breasted Grosbeaks have shown no significant overall population trend from 1966 to 1993, although they showed an increase until 1978 and then decreased from 1978 to 1988 (Sauer and Droege 1992, Peterjohn, Sauer, and Link 1994). They were found on 10 Atlas miniroutes in Tennessee at an average relative abundance of 1.7 stops/route. Because grosbeaks sing infrequently after mid-June (Nicholson pers. obs.), they probably were underestimated on the miniroutes, as most of those in the Unaka Mountains were censused in late June or early July.

Breeding Biology: The breeding biology of Rose-breasted Grosbeaks in Tennessee is poorly known, and few detailed nest records are available. They are apparently relatively early nesters, beginning as early as the second week of May, shortly after their spring arrival (Ganier and Clebsch 1946, Nicholson 1987). The female—and sometimes the male (Bent 1968)—builds the nest from twigs, pieces of vine, and weed stems. Grasses, rootlets, and hair are used in the nest lining. The nest is often so lightly constructed that the eggs are visible from below. It is built in a variety of shrubs and small trees or occasionally high on the branch of a deciduous tree. Plant species holding nests are rhododendron, hydrangea, mountain laurel, black locust covered with grape vines, and large beech and tulip-poplar trees. Eleven nests ranged in height from 1.2 to 12.2 m, with an average of 4.9 m (s.d. = 3.9 m). Ganier (1962) reported an average height of 2.4 m for an unspecified number of nests in the Smokies. Nests have been reported from heath balds, shrubby abandoned strip mines, and mature forest.

Three presumably complete clutches, each containing 3 eggs, have been reported in Tennessee. The few broods of young reported have contained 2 or 3 nest-

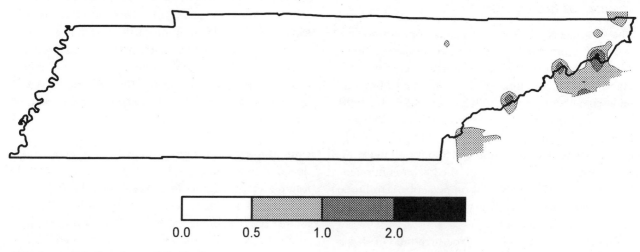

Abundance of the Rose-breasted Grosbeak.

lings. Elsewhere, clutch size ranges from 3 to 6, with 4 most common (Harrison 1975). The eggs are pale blue to green and marked with brown spots concentrated at the large end. Both the male and female incubate the eggs for 12–14 days, and the male often sings while incubating (Bent 1968). As in closely related species, the female also sings while nest building, incubating, brooding, and feeding the young (Ritchison 1986). Both adults feed the nestlings, which leave the nest in 9–12 days, and are dependent on the adults for about 3 more weeks (Harrison 1978). Second broods have been reported elsewhere (Harrison 1975), although there is no evidence of them in Tennessee.

Brown-headed Cowbird parasitism of Rose-breasted Grosbeaks has not been reported in Tennessee. This is no doubt due to the few available records of grosbeak nests and fledglings, as cowbirds occur within the grosbeak's range in the Cumberland Mountains and northeast Tennessee. Farther north, cowbird parasitism is fairly frequent (Friedmann 1963).—*Charles P. Nicholson.*

Blue Grosbeak
Guiraca caerulea

Blue Grosbeak. David Vogt

The Blue Grosbeak is a fairly common summer resident, occurring in brushy fields and hedgerows adjoining grasslands and croplands across Tennessee. It usually arrives by the end of April and departs by the end of September. The present distribution and abundance of this primarily southern species, first found nesting in the state in 1945, is evidence of one of the most dramatic recent increases by a native bird species.

The first Tennessee record of the Blue Grosbeak was in May 1929 at Memphis (Coffey 1955). During the next 16 years, there were a few other reports, mostly of spring migrants, from across the state. Ganier (1933a) described it as a rare summer resident in West Tennessee and a very rare summer resident in Middle Tennessee, although there were no published records supporting these designations. Then, in 1945, a small breeding population and a nest were discovered in southern McNairy County (Warriner 1945). Warriner also observed an apparent territorial male in Hardin County. By 1947, another population was established in southwest Hardeman County (Coffey 1955). An adult carrying food was observed in Knox County in 1948, and a nest was found there in 1951 (Howell 1951b). The first summer record in the Nashville area was in 1950 (Goodpasture 1968), and in 1953 there were summer records from near Chattanooga as well as a nest reported just south of the state line in Georgia (West 1953). By the mid-1960s, Blue Grosbeaks were established throughout the state. From 1966 to 1994, their numbers have increased on Tennessee BBS routes at an average rate of 3.6%/year ($p < 0.01$). This increase was much more pronounced from 1966 to 1979 (8.1%/year) than from 1980 on (2.8%/year). Their rangewide population also increased significantly ($p < 0.01$) by 1.5%/year from 1966 to 1993 (Peterjohn, Sauer, and Link 1994).

The route of this spread into the state is conjectural. Suitable habitat had been present for decades. By the time of the first Tennessee breeding records, the grosbeak was established in southern Illinois (Ridgway 1914a in Mengel 1965), West Virginia (Bent 1968), most of North Carolina (Pearson, Brimley, and Brimley 1942), northern Georgia (Burleigh 1958), and northern Mississippi (Warriner 1945), but apparently not in Kentucky (Mengel 1965). The early West Tennessee records were probably of birds spreading into Tennessee from the south. This may have also been true in Middle and East Tennessee, where the first breeding records were well to the north near Nashville and Knoxville. Fieldwork during the 1930s and 1940s (based on published material in the *Migrant*) was apparently limited south of Nashville (excluding the Murfreesboro area) and in the Ridge and Valley south of Knoxville.

Blue Grosbeaks occupy early successional habitats such as brushy pastures and abandoned fields with numerous shrubs and saplings. They also occur in brushy hedgerows adjoining hayfields and fields of small grains such as wheat, as well as extensive young, clearcut timberlands (Burleigh 1958, Nicholson pers. obs.). Unlike the similarly plumaged Indigo Bunting, with which it often occurs, the Blue Grosbeak does not occur in small woodland openings. Blue Grosbeaks occur on the

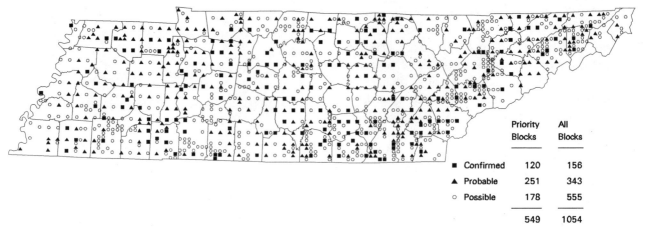

Distribution of the Blue Grosbeak.

Cumberland Plateau at elevations to about 610 m; in the northern Unaka Mountains, atlasers found them in the cleared valleys at elevations up to about 855 m.

The Blue Grosbeak was found in 84% of the completed Atlas priority blocks. Its song, a series of rich, clear to slightly burry warbles rising and falling in pitch, is occasionally confused with the song of the Orchard Oriole and Summer Tanager by novice birders and, presumably, Atlas workers. The song is also infrequently sung during midday; when not singing or feeding young, grosbeaks can be quite secretive. It is thus likely that atlasers missed Blue Grosbeaks in some West Tennessee, Highland Rim, and Central Basin priority blocks where suitable habitat occurred. Grosbeaks were not found in predominantly wooded blocks in the Cumberlands and Unaka Mountains, as well as in a few urban priority blocks, where there was little suitable habitat.

Atlasers found Blue Grosbeaks on 61% of the miniroutes at an average relative abundance of 2.5 stops/route. Their abundance and the proportion of the miniroutes on which they were recorded was greatest in the Ridge and Valley Province ($\bar{x}$ = 2.9, 94/130 routes). During the Atlas period, they were recorded on 40 of the 42 BBS routes at an average of 5.0 stops and 5.4 birds/route. The only plot census results from relatively uniform grosbeak habitat, in Central Basin, 6-year-old abandoned hayfields, showed densities of 8 pairs/100 ha (Fowler and Fowler 1984c, 1984d).

Breeding Biology: Many details of the breeding biology of Blue Grosbeaks are poorly known, and details are available for few nests in Tennessee. Nest building begins in late May. Blue Grosbeaks nest in shrubs, saplings, and vines, and the nest is usually well concealed. Tennessee nests average 1.0 m above ground (s.d. = 0.6, n = 16, range 0.37–2.7). Nests have been most frequently reported in blackberries; other species used are pokeberry, privet, elderberry, kudzu, and saplings of sweetgum, elm, persimmon, hackberry, willow, red oak, and cedar. Saplings used for nesting are often covered with honeysuckle or other vines. The nest is built of weed stems and heads, bark fibers, and bits of leaves

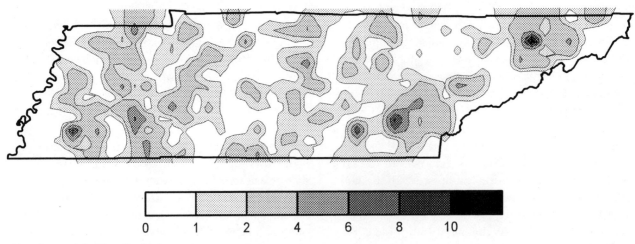

Abundance of the Blue Grosbeak.

and lined with fine weed stems, grasses, or hair. The outer shell usually contains pieces of snake skin, paper, or plastic (Bent 1968, Nicholson pers. obs.). The female apparently does most of the building.

Clutch size is most commonly 4 eggs, with clutches of 2, 3, and 5 also reported (Bent 1968). Tennessee clutches include 4 records of 3 eggs and 3 records of 4 eggs; nests containing 2 eggs have also been reported, but it is unclear whether these clutches were complete. The eggs are pale blue to white and usually unspotted. The incubation period and role of the male in incubating are poorly known; Bent (1968) gives the period as about 11 days. Both adults feed the nestlings, and the female makes more feeding visits than the male (Bent 1968, Alsop 1979a). The age at which the young leave the nest has been given as 9–10 days (Ehrlich, Dobkin, and Wheye 1988) and about 13 days (Bent 1968). Praying mantises are a locally important food for nestlings in Tennessee (Alsop 1979a). Blue Grosbeaks commonly raise 2 broods elsewhere (Bent 1968); the range of dates of nests with eggs and young suggests this is true in Tennessee.

The Blue Grosbeak is a fairly frequent host of the Brown-headed Cowbird and is heavily parasitized in some areas (Friedmann 1963, 1971). Two records of cowbird parasitism are available in Tennessee: a nest with 2 grosbeak and 2 cowbird eggs (Nicholson unpubl. data) and a nest with 4 grosbeak eggs and 1 cowbird egg that was buried in the nest lining (NRC by M. D. Williams). The behavior of the grosbeak burying the cowbird egg in the nest lining has apparently not been previously reported.—*Charles P. Nicholson.*

Indigo Bunting
Passerina cyanea

The Indigo Bunting is one of the most abundant and widely distributed birds nesting in Tennessee and probably the most abundant neotropical migrant. It occurs in brushy areas and woodland edges at all elevations, but, because it infrequently occurs in manicured urban and suburban habitats, it is unfamiliar to many Tennesseans. Indigos are a summer resident, generally arriving in late April and early May and departing in mid-October.

Indigo Buntings occupy a variety of habitat types with the common features of open areas with dense, low vegetation for feeding and nesting and high singing perches (Taber and Johnston 1968). These habitats include brushy fields, fencerows, forest edges, and both coniferous and deciduous forests. In mature forests, the species occupies small openings with brushy understories resulting from fallen trees, logging roads, and other natural or human-made disturbances.

Indigo Bunting. David Vogt

No significant change in the historical range of the Indigo Bunting has been reported. Rhoads (1895a:491) described it as "very abundant over my entire route, and one of the few lowland birds that breed on the summit of Roan Mountain." Its occurrence at the highest elevations may vary from year to year, and many of these reports from above 1525 m are of apparently unmated males (Stevenson and Stupka 1948, Stupka 1963). Indigo Buntings were probably widespread in lower numbers in prehistoric Tennessee, occurring in the glades of Middle Tennessee, the prairies of north-central and northwest Tennessee, and forest openings throughout the state created by fires, blowdowns, and other natural disturbances as well as the activities of Native Americans. Early authors (in Taber and Johnston 1968) noted its abundance in southeastern canebrakes, once widespread in stream bottoms throughout the state. Its numbers probably increased with the clearing of forests by settlers. An additional widespread short-term increase probably occurred in the first half of this century as buntings occupied the temporary forest openings created by the die-off of American chestnut (cf. Brooks 1940).

The total number of Atlas blocks in which Indigo Buntings were found was the highest of any species; similarly, they were found in all but 2 completed priority blocks. Because the males sing from conspicuous perches, frequently along roadsides and throughout the breeding season, the Atlas results are an accurate reflection of its distribution. Indigos were also recorded on all but 7 miniroutes at an average relative abundance of 8.9 stops/route, again the highest proportion of routes and greatest abundance of any species. Indigos occurred in low numbers in some high-elevation blocks in the Unaka Mountains, some heavily agricultural blocks in West Tennessee,

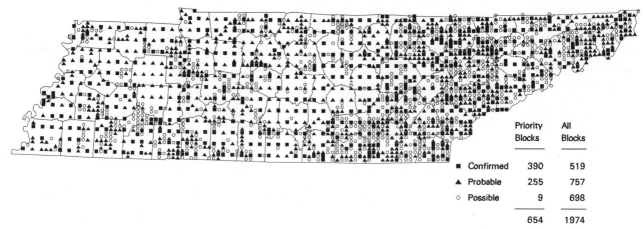

Distribution of the Indigo Bunting.

	Priority Blocks	All Blocks
■ Confirmed	390	519
▲ Probable	255	757
○ Possible	9	698
	654	1974

and some urban blocks. Their average abundance did not vary greatly between physiographic regions and ranged from 7.6 stops/route in the Central Basin to 10.6 stops/route in the Cumberland Mountains.

Despite its present abundance, the Indigo Bunting population, as measured by Tennessee BBS routes, declined at the average annual rate of 1.8%/year ($p < 0.01$) from 1966 to 1994. This decline is probably caused by the loss of early successional habitats due to increasing urbanization and changing agricultural practices. Most of the decline was before 1980, as from 1980 to 1994 there was no significant trend. Throughout its range the Indigo Bunting population decreased by 0.6%/year from 1966 to 1993 (Sauer and Droege 1992, Peterjohn, Sauer, and Link 1994). During the early part of this period, the Indigo population reached its greatest rangewide abundance in the Lexington Plain and Highland Rim physiographic provinces, including the Highland Rim and Central Basin of Tennessee (Robbins, Bystrak, and Geissler 1986).

Confirming Indigo Bunting breeding was not difficult for Atlas workers, and 57% of the priority block records were confirmed. Observations of adults carrying food or fecal sacs and attending fledglings made up 61% of the confirmed records. Although the nest is often well-concealed, a relatively high 30% of confirmed records were of nests.

Breeding Biology: Indigo Buntings often occur in small flocks for a short period after their spring arrival. Once they settle on territories, males vigorously defend their territories throughout the breeding season, often engaging in extended chases of other males. A small percentage are polygamous, with more than one female nesting on their territories (Carey and Nolan 1979, Payne 1982). Unlike most songbirds, many males in their first breeding season retain a variable amount of brown in their plumage. These birds appear mixed brown and blue, often with a white belly, and usually breed, although they are less successful than older males (Taber and Johnston 1968, Payne 1982).

Reproductive activities often begin soon after the female settles on a territory, although Nolan (1978) found nesting in southern Indiana was occasionally postponed

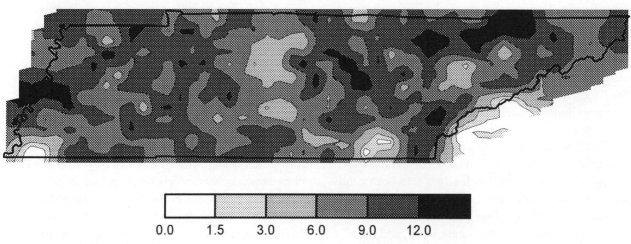

Abundance of the Indigo Bunting.

until the first of June. The female chooses the nest site and builds the nest (Payne 1990). It is a usually distinctive, well-made cup of dead leaves, dry grass, bark strips, weed stems, and twigs, lined with fine grasses and rootlets. Tennessee nests average 1.0 m (s.d. = 0.9, n = 129, range 0.3–3.7) above ground; only 2 nests were higher than 2 m. The nest is placed in the crotch of a sapling or shrub, or in a dense clump of blackberry or tall herbs such as goldenrod. Other commonly used plant species include saplings of winged elm, red and sugar maple, and flowering dogwood, occasionally partly supported by Japanese honeysuckle or grape vines. Hydrangea and buffalo nut are frequently used in the eastern mountains.

Indigo Buntings are usually double-brooded, and some females lay up to 4 clutches; the renesting interval is longer early in the breeding season (Payne 1990). No peaks in egg laying in Tennessee are noticeable. Tennessee clutches average 3.4 eggs (s.d. = 0.56, n = 86, range 2–5), with clutches of 2 and 5 eggs very rare, as reported elsewhere (Taber and Johnston 1968, Payne 1990). The female incubates the white to very pale bluish, usually unmarked eggs for 12–13 days, and the young fledge in about 10 days. Most of the care of the nestlings and fledglings is by the female (Payne 1990).

Indigo Buntings are commonly parasitized by Brown-headed Cowbirds (Payne 1990), and 21 of 123 Tennessee nests, dating from 1901, contained cowbird eggs. The earliest parasitized nest was found in 1918; the parasitism rate has increased in recent decades. The largest number of cowbird eggs reported in a nest was 3, in a nest which also held 1 bunting egg.—*Charles P. Nicholson.*

Painted Bunting
Passerina ciris

The Painted Bunting is a rare and very local summer resident of extreme southwestern Tennessee. It is usually present from late April to late July, though many birds may not arrive until mid- to late May (Coffey 1933, Robinson 1990). The multicolored, polygynous male is conspicuous as he sings throughout most of the breeding season, or as he engages in frequent combative interactions with other males (Parmelee 1959). The winter range of the Tennessee population is unknown, but probably lies within southern Texas, Tamaulipas, and northern Yucatan, Mexico (see Thompson 1991b).

Coffey (1933) reported the first record of the Painted Bunting in Tennessee in the Memphis area on 26 May 1929, soon after intensive ornithological surveys began in this region. The species was irregular until 1933, when breeding was conclusively documented at 2 sites. A maximum of about 6 pairs has apparently occupied

Painted Bunting. David Vogt

Shelby County in any year; the highest count for a single site is 10 birds at President's Island in late May 1981 (Robinson 1990).

Atlas results show that the present breeding range in Tennessee is similar to that of the 1930s, though breeding was not confirmed during this recent 6-year period. The current population may be a bit larger than suggested by Atlas results, since several areas where Painted Buntings could occur were not surveyed. These birds have been suspected of breeding occasionally in Covington in adjacent Tipton County since first detected there in 1975 (Robinson 1990). No birds were found in Tipton County during the Atlas period.

Painted Buntings occupy habitats with a predominant old-field component along with a smaller woodland edge component (see Norris and Elder 1982). Breeding territories in Tennessee are located in these partly open habitats, such as thick weedy and scrubby areas containing scattered thickets and some trees along railroad tracks, near abandoned warehouses and other buildings, alongside the Mississippi River levee, and, at least formerly, in residential areas several kilometers from the river. The area where the Painted Bunting has been most regularly observed in recent years, near the TVA Allen Steam Plant, has a ground cover of rank grasses and herbs, including Johnson grass, sweet clover, vetch and trumpet vine, patches of bare, sandy soil, and thickets of sumac, cottonwood, elm, and shrubby poison ivy (C. Nicholson pers. comm.). The easternmost Atlas record was at a large nursery and greenhouse complex, where territorial buntings occurred for several years until 1987.

The most critical habitat feature is probably extensive grassy areas where Painted Buntings forage on the ground (Norris and Elder 1982), which is why they may be highly tolerant of human-made habitat disturbance.

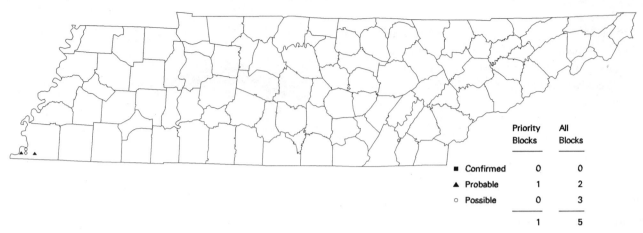

Distribution of the Painted Bunting.

Whether the bunting's prehistoric breeding range extended into Tennessee is, of course, unknown; early explorers (in Williams 1928) reported grassy areas along the Mississippi River, which were probably suitable Painted Bunting habitat.

The Painted Bunting breeds in 2 disjunct North American ranges, one along the southeast Atlantic coast and the other from southwest Alabama to southwest Missouri, Texas, and New Mexico (Thompson 1991b). Thompson (1991b) proposed that these populations may actually be two distinct species, which he called the eastern Painted Bunting and western Painted Bunting. The breeding birds in Tennessee are part of the western Painted Bunting population. This population differs from the eastern population in the timing and pattern of its molt and migration. No information on its molt in Tennessee is available, but the lack of August and September records roughly corresponds to their expected departure from breeding areas about 2 months earlier than birds from the eastern Painted Bunting population because of different molt-migration strategies (see Thompson 1991a, 1991b). This apparent early departure, however, is more likely due to an inherited migration schedule rather than local food scarcity, as is the case elsewhere in the western population. Additional information on the bunting's departure from Tennessee would be valuable.

Despite the availability of apparently suitable habitat, the peripheral population of the western Painted Bunting remains small, rather stable, and restricted to extreme southwest Tennessee. Individual birds have been found during the breeding season in Lauderdale and Obion County (Robinson 1990), but there has been no evidence of territorial birds or breeding north of Tipton County. The bunting is similarly absent from southeast Missouri, although its breeding range in central and western Missouri is apparently limited by factors other than habitat (Norris and Elder 1982). Consequently, the rare peripheral breeding population of the western Painted Bunting in Tennessee should be monitored because it is potentially vulnerable, but because evidence of a decrease is lacking, a special protective status is not warranted at this time.

Breeding Biology: The only Painted Bunting nest reported in Tennessee was located 1 m above ground in an elm bush alongside an old path in a brushy area in Memphis (Coffey 1933). The small, cuplike nest was built of grasses and weeds. Three large young fledged on 19 July. Clutch size elsewhere is usually 3–4 eggs, which the female incubates for 11–12 days. Both adults feed the nestlings, which fledge in 12–14 days (Ehrlich, Dobkin, and Wheye 1988).—*Douglas B. McNair.*

Dickcissel
Spiza americana

The Dickcissel is a summer resident of Tennessee, common in the west, and decreasing in abundance toward the east. It usually arrives from early to mid-May and occasionally in April. Male Dickcissels are conspicuous throughout the breeding season because singing, their major activity (Schartz and Zimmerman 1971), continues through mid-July near the conclusion of the breeding season. They also frequently sing their simple, loud, distinctive song from roadside habitats in agricultural landscapes. Consequently, despite the Dickcissel's sporadic occurrence throughout the periphery of their normal breeding range, the mid- to tall-grass prairie (Gross 1968, Fretwell 1986, Robbins, Bystrak, and Geissler 1986), their distribution and abundance in Tennessee is very faithfully represented by Atlas data. The species is rare in winter (Robinson 1990).

Atlas workers found the Dickcissel in one-quarter of the completed priority blocks and on 19% of the mini-

Dickcissel. David Vogt

routes at an average relative abundance of 3.0 stops/route. It is common and widespread in the heavily agricultural Gulf Coastal Plain province and most abundant in the northern Loess Plain region, not the Mississippi Alluvial Plain (*contra* Robinson 1990).

The distribution and abundance of the Dickcissel sharply declines eastward of West Tennessee. In Middle Tennessee, it is most common on the heavily agricultural, former prairie region of the Pennyroyal Plateau section of the West Highland Rim. Other concentrations on the Highland Rim occur in the vicinity of Cross Creeks NWR in Stewart County and on the Barrens portion of the Eastern Highland Rim, centered around Coffee County. It is uncommon in rural areas of the Central Basin, more numerous in the Outer Basin than in the thin-soiled Inner Basin. Their low numbers in the Nashville area during the Atlas period is consistent with Parmer's (1985) evaluation of a general decline of the Dickcissel since at least 1975 in this area, likely due to urban development.

In East Tennessee, virtually all Dickcissel records were from the Ridge and Valley and concentrated in the most extensive agricultural area, in the northeast, though it is more widely distributed than suggested by Robinson (1990). Dickcissels were, however, absent from urban Knox County, where they were formerly rare to uncommon (Howell and Monroe 1957). The concentration of records in the northeast today is consistent with the distribution of other grassland species in this region, e.g., Grasshopper Sparrow.

Throughout Tennessee, breeding Dickcissels are found in restricted habitats, primarily old fields, with specific microhabitat characteristics (see Zimmerman 1982). The densely vegetated, level to gently rolling old fields vary in their species composition and age (1–6 years), are at least 1 m high, and are frequently bordered by shrubby ditches, hedges, and thickets. This "briar patch" is a component of old-field habitat along roadsides and is frequently suitable for occupation by Dickcissels (Meanley 1963), even when contiguous habitat is not old fields but cropland. In West Tennessee, they are also common on the grass-vetch slopes of Mississippi River levees and in the stubble of bottomland fields left fallow until early summer. Successful reproduction in stubble habitat, however, is probably rare.

The switch from cotton to soybeans after World War II in West Tennessee has probably favored an increase of Dickcissels since they do occasionally nest in soybean fields, after the plants are large enough to support their nests (Best 1986). The amount of cropland has dramatically increased in West Tennessee since the 1960s, despite an overall decline of about 30% in the amount of cropland throughout Tennessee since 1940. This increase in the amount of cropland in West Tennessee would also favor an increase in Dickcissels, as has the recent creation of old-field habitat through the Conservation Reserve Program. Statewide BBS results show a slight decreasing trend from 1966 to 1994; variance in these results is high, especially in Middle and East Tennessee. From 1966 to 1979, the decrease was significant (8.9%/year, $p < 0.01$), while from 1980 to 1994 the Dickcissel population increased by 3.2%/year ($p < 0.10$). Though probably less widely distributed than when the area of farmland peaked in the early 1900s, Dickcissels are perhaps more numerous today in Tennessee than in the past. Rhoads (1895a), for example, stated that Dickcissels were abundant from the Mississippi River to the western escarpment of the Cumberland Plateau. Prior to European settlement, the species probably nested in Tennessee only in grasslands in northwest and north-central Tennessee grasslands.

Changes in farming and other land use practices affecting early seral stage vegetation, particularly old fields, will be the factors that influence the future of Dickcissels in Tennessee. These often ephemeral habitats are shared with few other bird species. As long as early seral stage vegetation is present, which seems particularly assured in West Tennessee, Dickcissels should be with us for a long time, regardless of the prediction of their demise by Fretwell (1986).

Breeding Biology: Despite their local abundance in West Tennessee, relatively few detailed Dickcissel nest records are available from the state. Male Dickcissels are frequently polygynous and take little part in nesting

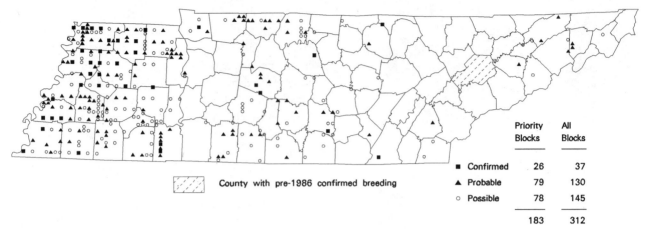

Distribution of the Dickcissel.

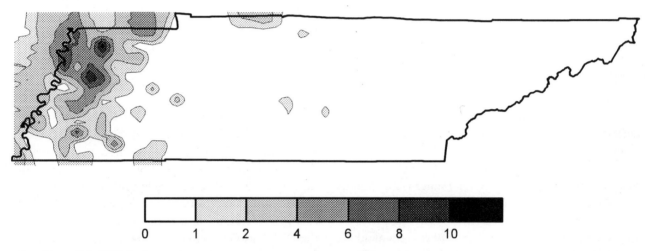

Abundance of the Dickcissel.

activities (Schartz and Zimmerman 1971). The nest is built in dense, low vegetation among herbaceous plants, blackberry tangles, saplings, or shrubs. Average nest height in Tennessee is 36 cm (s.d. = 18 cm, range 15–45 cm, n = 8). Egg laying probably peaks in the second half of May. Females raise only a single brood (Zimmerman 1982), and Tennessee clutches average 4.6 (s.d. = 0.53, range 4–5, n = 7) eggs, which are unmarked blue. Parasitism by the Brown-headed Cowbird, heavy elsewhere (Fretwell 1986), has not been reported in Tennessee, although it likely occurs.—*Douglas B. McNair.*

Eastern Towhee
Pipilo erythropthalmus

The Eastern Towhee, which until recently had the formal English name of Rufous-sided Towhee, is a common permanent resident of Tennessee. It is also one of the most widely distributed birds, nesting from the Mississippi River to the highest elevations of the eastern mountains. Migrants nesting north of Tennessee are also present during the winter. Although found in areas of thick shrubbery, the towhee is conspicuous as it noisily forages by kicking aside dead leaves, frequently calling "tow-hee." The male also sings its distinctive song from conspicuous perches.

Until the 1940s, the towhee was not known to nest in southwest Tennessee, west of a line from Hardin County to eastern Obion County (Coffey 1941b). Following a few breeding season reports of towhees, beginning in 1938, the first southwest Tennessee nest records were in 1945 in Shelby County (Hoyt 1945). Nesting towhees rapidly increased and by the mid-1960s were common throughout West Tennessee (BBS data). The origin of the towhees that spread into southwest Tennessee is unclear; they probably came from the northeast or the southeast. No twentieth-century habitat change facilitating its spread in the state is evident. James and Neal (1986) describe a recent eastward spread of nesting towhees in Arkansas, and Mengel (1965) suggests towhees have become more common in extreme southwest Kentucky during this century.

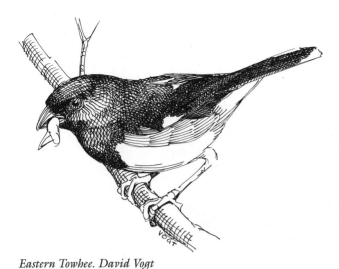

Eastern Towhee. David Vogt

The earliest description of the range of the Eastern Towhee in Tennessee was by Rhoads (1895a), who found it from Nashville eastward and described it as uncommon in Middle Tennessee. Torrey (1896) described it as rather common, but less numerous than expected, in the Chattanooga area. Prior to the increase in early successional habitats resulting from nineteenth-century agricultural clearing, it was probably common in Middle Tennessee cedar glades, Unaka Mountain shrub balds and heaths, dry ridgetop forests with dense, shrubby understories, and canopy openings resulting from blowdowns or other disturbances. Its numbers probably increased with the shrubby growth resulting from agricultural clearing and logging. Tennessee BBS routes show a significant ($p < 0.01$) declining trend in the towhee population of 2.1%/year from 1966 to 1994. Throughout its range, the towhee declined by 1.8%/year ($p < 0.01$) from 1966 to 1993 (Peterjohn, Sauer, and Link 1994). This rangewide decline, most severe in New England, has been described as one of the most dramatic of a nonendangered species and attributed largely to natural forest succession and the resultant decreased area of early successional habitats (Hagan 1993). Cleaner farming techniques and urban development are probably additional factors in Tennessee. The decline has not been uniform across Tennessee, as the towhee population has increased at high elevations of the Smokies following the recent death of mature firs (Alsop and Laughlin 1991).

Atlas workers found the towhee in essentially all of the completed priority blocks and on about 93% of the miniroutes. Its average miniroute relative abundance was 4.0 stops/route. Except for the Mississippi Alluvial Plain, where it averaged 1.7 stops/route, it was most numerous, with physiographic region averages of 4.0 or higher, from the Central Basin westward. Its average density in regions east of the Central Basin was between 3.0 and 4.0.

The Eastern Towhee has been recorded on several census plots from the Central Basin eastward. Kendeigh and Fawver (1981) found the highest density among several forest types in the Great Smokies, 165 pairs/100 ha, to be in recently burned, early successional spruce-fir forest. Densities in relatively closed canopy forest have been generally no more than 12/100 ha (e.g., Howell 1974, Lewis 1983a). Densities in Central Basin cedar forests were 33–49/100 ha (Fowler and Fowler 1983a, 1983b, 1983c), and 16–33/100 ha in recently abandoned agricultural fields (Fowler and Fowler 1984c, 1984d). The towhee population in a deciduous clearcut increased rapidly until peaking at 133/100 ha 8 years after harvest, when the towhee was the most abundant species present (Lewis 1982).

Breeding Biology: Confirming the breeding of the Eastern Towhee was not easy for atlasers, and records of fledglings and adults carrying food accounted for

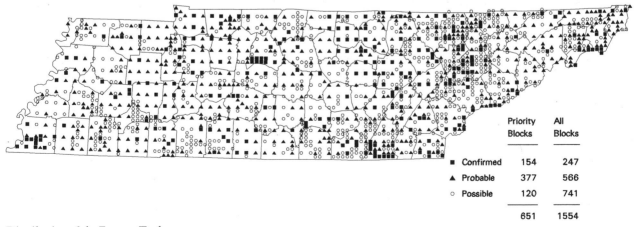

Distribution of the Eastern Towhee.

	Priority Blocks	All Blocks
■ Confirmed	154	247
▲ Probable	377	566
○ Possible	120	741
	651	1554

Confirmed Breeding Species

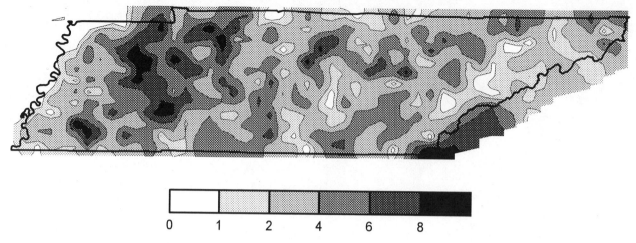

Abundance of the Eastern Towhee.

about three-fourths of the confirmed records. The towhee has a long breeding season in Tennessee, from late March through August, and, as elsewhere (Dickinson 1968), is usually double-brooded. The nest is usually concealed on the ground or in low shrubs.

Towhees are monogamous, and the male displays to the female by flashing the white spots in his wings and tail (Dickinson 1968). The female chooses the nest site and builds the nest. It is bulky and built of strips of bark, leaves, dead grasses, and weed stems and lined with fine grasses and rootlets. Early nests are frequently on the ground in a shallow depression scraped out by the female or in the base of a grass clump. Later nests are in shrubbery or vine-covered saplings. Of 27 ground nests reported in Tennessee, 21 were begun before 15 May. Above-ground nests were up to 4.6 m high, averaging 1.2 m high (s.d. = 0.7, n = 77). More Tennessee nests have been reported in red cedar than any other tree or shrub species.

Egg laying dates show a broad peak from mid-April through mid-May and a lesser peak in mid-June. Clutch size is 2–6 eggs, with 3–5 egg clutches most common, as reported by Dickinson (1968); the average is 3.54 eggs (s.d. = 0.69, n = 95). The female incubates the eggs for 12–13 days and broods the hatched young (Dickinson 1968). Both adults feed the nestlings and fledglings, which leave the nest in 10–12 days. The female starts the second clutch 8–21 days after the young of the first nest have fledged (Dickinson 1968).

Parasitism by the Brown-headed Cowbird is common, both in Tennessee and elsewhere in the towhee's range (Friedmann 1963). Laskey (1945) reported that 7 of 10 towhee nests found at Nashville in 1945 were parasitized, 6 with 1 cowbird egg and 1 with 2 cowbird eggs. At least 10 other parasitized nests have been reported in the state, all containing 1 or 2 cowbird eggs or young, and all begun before the middle of June. The towhee readily accepts cowbird eggs and often successfully raises cowbird young (Friedmann 1963); towhees feeding cowbird fledglings have been observed several times in Tennessee.—*Charles P. Nicholson.*

Bachman's Sparrow
Aimophila aestivalis

The Bachman's Sparrow, although drably plumaged, has one of the most beautiful songs of any sparrow. Its loud, distinctive "see-slip-slip-slip-slip" song, however, has become a rare sound in Tennessee in recent decades. The Bachman's Sparrow is a summer resident, present from April through August and rarely into October (Nicholson 1976, Robinson 1990).

The Bachman's Sparrow is an endemic resident of pinelands of the southeastern United States, traditionally associated with grassy openings in mature pine forests (Dunning and Watts 1990), a habitat shared with the Red-cockaded Woodpecker. In the late nineteenth and early twentieth centuries, the Bachman's Sparrow expanded its range northward from Illinois to Pennsylvania, occupying old fields and eroded hillsides (Brooks 1938a). It was, however, probably fairly common before then in the pinelands and other periodically burned woodlands of prehistoric southwest and eastern Tennessee and was first reported in the state in 1882 in Hamilton County (Fox 1882). Fox (1886) found several in open pines in Roane County in 1884–85, and Rhoads (1895a) described it as numerous on the Cumberland Plateau in Fentress, Scott, and Morgan Counties. As occurred farther north, it apparently soon expanded into old-field habitats throughout much of the state except for the northwest corner and the Unaka Mountains (Nicholson 1976), becoming locally fairly common (e.g., Ganier 1933a, Howell and Monroe 1957).

Bachman's Sparrow. David Vogt

By the 1950s, a decline in Bachman's Sparrow numbers was evident throughout much of Tennessee (Nicholson 1976, Alsop 1979b). This decline occurred throughout much of the sparrow's range and led to its listing as Endangered in Tennessee (TWRA 1975) and, until very recently, as a Category 2 Candidate species for federal listing as Endangered or Threatened (USFWS 1994, 1996). Most Tennessee reports from the 1960s through the early 1980s were in open mature pines, grassy old fields, or edge habitats with scattered large pines, and, particularly on the Cumberland Plateau, young clearcuts replanted with pine, where it was locally common (Nicholson 1976, 1984b; Nicholson pers. obs.).

The Atlas results show that the Bachman's Sparrow is still rare, with its center of abundance in the southern Coastal Plain Uplands physiographic region. Most of the Atlas records were in young pine plantations managed by paper companies; special searches in this habitat increased the number of records in nonpriority blocks. A singing male in northeast Stewart County in late April 1986 may have represented another breeding location, but this site was not checked during the summer and is not included on the Atlas map. The sparrow was recorded at 1 stop on each of 2 miniroutes.

The twentieth-century population dynamics of the Bachman's Sparrow in Tennessee—an early increase followed by a decline and a probable recent, localized increase—are most likely due to changes in the availability of suitable habitat as well as the sparrow's ability to disperse into patches of suitable habitat. Although historically occurring in (at least superficially) a variety of habitats, the Bachman's Sparrow has rather strict habitat requirements, notably a high volume of grasses and forbs, and some scattered trees and shrubs with an open understory on dry, upland sites (Dunning and Watts 1990).

This habitat occurred in the periodically burned, mature, natural pine stands until logging and fire control altered this habitat. The availability of abandoned farmland early in this century provided habitat for the sparrows' early-twentieth-century increase. Because of the infertile, heavily eroded nature of much of this reverting farmland, tree growth was probably retarded, and it supported sparrows for several years. As this habitat eventually reverted to forest and suitable pinelands disappeared, sparrow numbers dropped. Commercial pulpwood plantations, usually of loblolly pine, began providing suitable habitat on the Cumberland Plateau by the early 1970s and on the Western Highland Rim and Coastal Plain Uplands in the 1980s, and most recent Bachman's Sparrow reports have been in this habitat.

The suitability of young pine plantations for Bachman's Sparrows depends on the site preparation methods used before planting. In most Tennessee plantations recently occupied by sparrows, the brush and slash remaining after clearcutting had been bulldozed into windrows and often burned before pines were planted (Simbeck pers. obs.). Most pine plantations only remain suitable for about 3–5 years unless regularly thinned and burned to reduce understory vegetation. Several aspects of this habitat, however, remain poorly known. No Bachman's Sparrows were found during the Atlas period in Cumberland Plateau plantations that appeared superficially suitable (Nicholson pers. obs.); the vegetative differences between these unoccupied areas and plateau plantations occupied in the 1970s as well as recently occupied plantations farther west are largely unknown. Another unknown factor is whether suitable habitat will occur following the harvest of the plantation pines. Because it is unlikely that sufficient suitable mature pine habitat will exist in Tennessee in the future, the long-term survival of the Bachman's Sparrow is probably dependent on young pine plantations. Individual patches of this habitat remain suitable for the sparrow for a relatively short time and it will probably be necessary to provide a sequence of plantations of different ages within a small geographic area to allow dispersal by the sparrows between patches (cf. Dunning and Watts 1990).

Breeding Biology: Bachman's Sparrows have a long nesting season, from late April into mid-August (Haggerty 1988). The nest is built on the ground at the base of a clump of grass, frequently broomsedge, or other low vegetation (Nicholson 1976, Haggerty 1988). It is constructed primarily of grasses, particularly broomsedge and *Panicum* spp., lined with fine grasses or hair, and usually domed with a side entrance. Egg laying begins as early as late April; because of the few nest records, no peak is evident

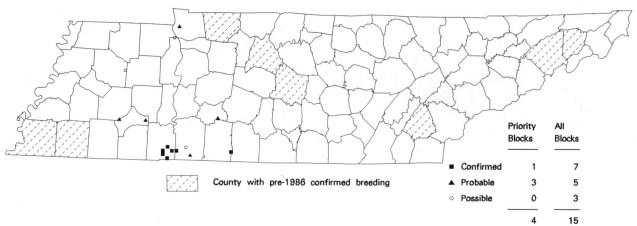

Distribution of the Bachman's Sparrow.

(egg coll. data, Nicholson 1976). Tennessee clutches are equally divided between 4- and 5-egg clutches (n = 10). Clutches of 3–5 eggs, with an average of 4 eggs, have been found elsewhere (McNair 1987a, Haggerty 1988). The pure white eggs are incubated by the female for 12–14 days (Weston 1968, Haggerty 1988). Both adults feed the young, which fledge in 9 days. Renesting following nest failure or fledging of the young is common elsewhere (Haggerty 1988); the large range of egg dates suggests renesting occurs in Tennessee. Cowbird parasitism is uncommon (Friedmann 1963, Haggerty 1988) and not reported in Tennessee.—*Damien J. Simbeck and Charles P. Nicholson.*

Chipping Sparrow
Spizella passerina

The Chipping Sparrow is a fairly common summer resident and a rare winter resident of Tennessee (Robinson 1990). Migrants arrive during March and early April and most depart by late October. It nests at middle to low elevations throughout most of the state in areas of short grass and scattered trees and shrubs, where it often conspicuously feeds on the ground. It is relatively tame and easily identified by its black eye line and reddish brown cap, as well as its rather unmusical trilled song.

The Chipping Sparrow inhabits open, parklike woodlands with little understory and a grassy ground cover, as well as extensive areas of short grass and scattered trees (Stull 1968, James 1971). Typical habitats in Tennessee include wooded suburbs, cemeteries, golf courses, orchards, pastures with scattered trees, and rural roadsides. It also occurs in dry, savannah-like woodlands, both pine- and oak-dominated. Simmers (1983, 1984) recorded 1 pair of Chipping Sparrows on a 27 ha mixed woodland and fields census area; it has not been recorded on other Tennessee plot censuses.

The Chipping Sparrow was probably a locally common bird in the periodically burned oak and pine-oak savannah-like areas of prehistoric Tennessee. It may have been most numerous on the Cumberland Plateau and the barrens of the Eastern Highland Rim, where savannah-like habitat was extensive. Its numbers probably increased with European settlement, as farmyards and pastures increased the amount of suitable habitat. The Chipping Sparrow was first reported in Tennessee by Fox (1886), who found it abundant in Roane County. Langdon (1887) also found it abundant in farmed coves at low elevations of the Smokies. Rhoads (1895a) found it throughout the state except for the highest elevations of Roan Mountain. Ganier (1933a) described it as common in West, Middle, and East Tennessee. In the Unaka Mountains, it has nested as high as 1490 m; a few summer records exist from higher elevations (Stupka 1963).

Atlas workers found the Chipping Sparrow throughout the state in 86% of the completed priority blocks. It was also recorded on almost two-thirds of the miniroutes,

Chipping Sparrow. David Vogt

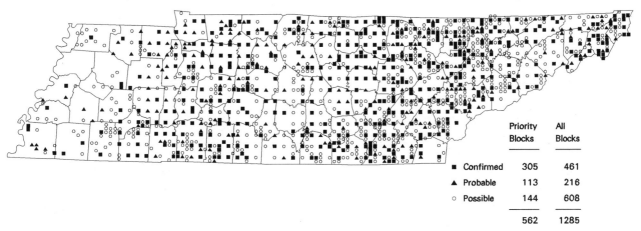

Distribution of the Chipping Sparrow.

at an average relative abundance of 2.7 stops/route. It was most abundant on the north-central Cumberland Plateau in a region with numerous small farms, extensive pasture, and savannah-like woodlands. The Cumberland Plateau also had the highest average abundance, 3.7 stops/route, of any physiographic region. Other regions of abundance were the northern Unaka Mountains, the Ridge and Valley, the Eastern and Western Highland Rims, and the southern Coastal Plain Uplands. The uncommon status of the Chipping Sparrow in the Mississippi Alluvial Plain and Loess Slope has been noted in previous studies (e.g., Whittemore 1937). Its absence from many predominantly forested blocks in the Unaka and Cumberland Mountains, where suitable habitat was lacking, was expected. Its absence from several Middle Tennessee blocks, however, is difficult to explain.

From 1966 to 1994, the Chipping Sparrow population in Tennessee, as measured by BBS routes, decreased by 1.9%/year ($p < 0.05$). Most of this increase occurred from 1966 to 1979, and since then it has shown a slight increasing trend. The reasons for these trends are not clear.

With increasing suburban growth, the area in lawns has increased, at least partially offsetting the loss of fencerow habitat. A similar decrease occurred through 1979 in several states adjoining Tennessee (Robbins, Bystrak, and Geissler 1986). Throughout its range, the Chipping Sparrow population showed no significant overall trend from 1966 to 1993 (Peterjohn, Sauer, and Link 1994); as in Tennessee, it also declined in the first part of this period and then increased from 1978 to 1987 (Robbins et al. 1989).

Confirming the breeding of the Chipping Sparrow was usually fairly easy for Atlas workers; over half of the records in priority blocks were of confirmed breeding. The high rate is due to the species's long nesting season, the conspicuous, noisy behavior of its fledglings, and frequent ease in finding its nest. About half of the confirmed records were of fledglings and about one-quarter were of nests.

Breeding Biology: For the first few weeks after its spring arrival, Chipping Sparrows occur in loose flocks, which break up as males begin territorial defense in early April (Nicholson pers. obs.). The female builds the nest, a

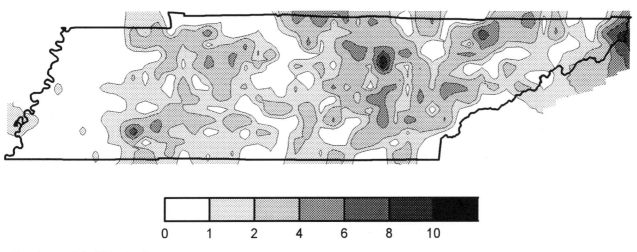

Abundance of the Chipping Sparrow.

Confirmed Breeding Species

compact, well-made cup, in 2–4 days from dead grasses, weed stalks, and rootlets (Stull 1968). The nest is lined with fine rootlets and hair and placed in a dense shrub, vine tangle, on a lower horizontal tree branch, or rarely on the ground. More Tennessee nests have been built in red cedars than any other species. A preference for nesting in conifers has also been noted elsewhere (Stull 1968). Tennessee nests range from 0.46 to 12.2 m above ground, with an average height of 2.1 m (s.d. = 1.9, n = 66); none have been reported on the ground, as occasionally occurs elsewhere.

Egg laying peaks during the last third of April and first week of May. The normal clutch size is 3–5 eggs, with 4 most common and clutches of 2 rare (Stull 1968, Reynolds and Knapton 1984). Tennessee clutches average 3.76 eggs (s.d. = 0.49, n = 79). The eggs are pale blue, with sparse brown to black markings, usually around the larger end of the egg (Stull 1968). The female incubates the eggs, usually for 11 or 12 days (Stull 1968, Reynolds and Knapton 1984). Both adults feed the nestlings, which leave the nest in about 9 days (Reynolds and Knapton 1984). Chipping Sparrows frequently raise 2 broods; Lyle and Tyler (1934) believed them to be triple-brooded. The chronology of Tennessee nest records suggests that second broods are common and third broods rare.

The Chipping Sparrow has been described as one of the most common hosts of the Brown-headed Cowbird (Friedmann 1963). The proportion of parasitized nests, however, varies greatly from 0 to over 50% in different studies (Friedmann 1963, Reynolds and Knapton 1984). It is very rare in Tennessee; only 1 record, of a nest containing a cowbird egg, is available.—*Charles P. Nicholson*.

Field Sparrow
Spizella pusilla

Field Sparrow. David Vogt

The Field Sparrow is a fairly common to common permanent resident found throughout Tennessee. Its easily identified song, a series of plaintive whistles ending in a trill, is a distinctive sound in brushy fields from March through the end of August. Although present throughout the year, a pronounced migration occurs, with an influx of northern birds and movement of some locally breeding birds to the south (Laskey 1934).

Field Sparrows nest throughout the state. Atlas workers found them in 94% of the completed priority blocks, the highest proportion of any sparrow. They were absent from several heavily forested blocks in the Unaka Mountains, a few heavily forested blocks on the Cumberland Plateau, and a few urban and bottomland blocks in extreme West Tennessee. In the Unaka Mountains, Field Sparrows are common in farmed valleys at low to middle elevations. Stupka (1963) reported them nesting on grassy balds at 1300–1500 m in the western Smokies, although Atlas workers did not find them in these areas. The highest elevation at which there is recent evidence of nesting is around 1000 m.

Atlasers recorded Field Sparrows on 575 (83%) of the miniroutes, at an average relative abundance of 4.0 stops/route. They were most numerous in the Central Basin, Eastern Highland Rim, and Ridge and Valley, and least numerous in the Unaka Mountains, Cumberland Mountains, and extreme West Tennessee. Studies elsewhere (Best 1977, Walkinshaw 1968) have shown that the most suitable nest habitat for Field Sparrows is old fields with scattered woody vegetation. Walkinshaw (1968) found an average territory size of about 1.2 ha in Michigan, and Best (1977) found an average territory size of 0.76 ha in Illinois. Field Sparrow densities on Tennessee census plots include 57 pairs/100 ha in a brushy pasture on the Eastern Highland Rim (Simmers 1978), 44/100 ha in a northern Ridge and Valley pasture (Lewis 1975), and 66 and 82/100 ha on 6-year-old abandoned farmland in the Central Basin (Fowler and Fowler 1984c, 1984d).

Although Field Sparrows are presently fairly common to common across the state, their population, as measured by BBS routes, has decreased at the rate of 3.0%/year ($p < 0.01$) from 1966 to 1994. This decline is probably due to the decreasing area of brushy, early successional habitat resulting from cleaner farming methods with improved pastures and reduction of fencerows, the decreasing proportion of young forest area, and urban development. The Tennessee trend is similar to the rangewide 1966–93 decrease of 3.3%/year ($p < 0.01$) (Peterjohn, Sauer, and Link 1994).

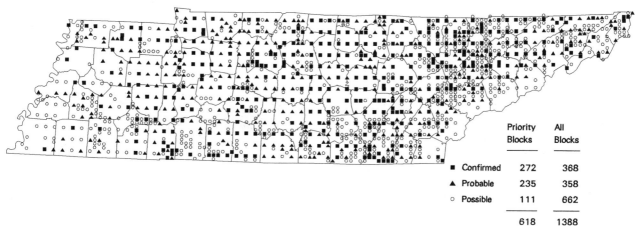

Distribution of the Field Sparrow.

Field Sparrows, however, are still much more common than they were two centuries ago. Prior to the widespread forest clearing during European settlement, Field Sparrows were probably locally common in Central Basin cedar glades, Highland Rim barrens, recently burned forests, and reverting fields near Indian towns. Their numbers probably increased with the nineteenth-century spread of European settlements, as the field edges and brushy pastures provided suitable habitat. Rhoads (1895a) observed Field Sparrows throughout the state except for the summit of Roan Mountain. Their numbers probably peaked in the 1940s when farm abandonment peaked and the proportion of early successional forest was high.

Confirming Field Sparrow breeding was not difficult, although confirmed records only made up about one-quarter of all Atlas records. Over half of the confirmed records were of fledglings and almost one-quarter were of adults carrying food.

Breeding Biology: Male Field Sparrows begin singing regularly and defending territories in late March. The males mate monogamously, often with the first female to arrive on his territory (Walkinshaw 1968, Best 1977). The female chooses the nest site and builds the nest, often closely accompanied by the male. The first nest of the season is usually built in 4 or 5 days, later nests in 2 or 3 days (Walkinshaw 1968). Nests are built in low vegetation and commonly used plants in Tennessee include broomsedge, blackberry, sericea, buckbush, red cedar, and saplings of pines and deciduous trees. Early nests, which are built prior to the leafing out of many trees and shrubs, are often placed low in grass clumps; use of forbs, trees, and shrubs increases later in the season (Walkinshaw 1968, Best 1978). This trend occurs in Tennessee, although use of red cedar is common throughout the season. Average nest height in Tennessee is 0.47 m (s.d. = 0.50 m, range 0–3.7 m, n = 138). Walkinshaw (1968) found an average nest height of 0.21 m, and Best (1978) found an average height of 0.37 m. The nest is made of dead grass stems and leaves and lined with fine grass and hair.

Egg laying begins during the second week of April, peaks during the first half of May, and extends into early

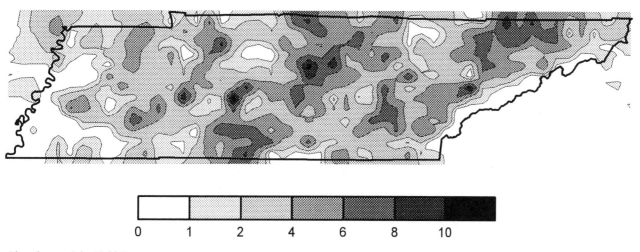

Abundance of the Field Sparrow.

August. Clutch size ranges from 2 to 6 eggs, with 3–5 eggs most common, and decreases as the season progresses (Walkinshaw 1968). The average size of 140 Tennessee clutches is 3.76 eggs (s.d. = 0.55, range 2–5). Walkinshaw (1968) reported an average clutch size of 3.37 eggs; his sample probably includes more late-season clutches than the Tennessee sample. The eggs are creamy white or very pale bluish or greenish with reddish brown spots, usually forming a wreath around the large end.

Incubation begins before the last egg is laid, and the female alone incubates the eggs for an average of 11.6 days (Walkinshaw 1968). Both adults feed the nestlings and remove fecal sacs from the nest. The young leave the nest when 7 or 8 days old and reach independence about 3 weeks later. Nest success is low, with young fledging from only one-third or less of all nests (Nolan 1963, Walkinshaw 1968, Best 1978); predation by snakes is a major cause of nest loss. Field Sparrows readily renest after losing a nest, and the female usually begins laying eggs in the replacement nest 5 days after loss of the previous nest (Walkinshaw 1968). Field Sparrows are also multiple-brooded, and, in the absence of nest losses, probably raise 3 broods per year (Lyle and Tyler 1934).

Field Sparrows are frequently parasitized by Brown-headed Cowbirds. Walkinshaw found 27% of the nests he studied were parasitized; Best (1978) found a 10% parasitism rate and noted parasitism was higher in nests near woodland edge than in more open fields. Only 5 records of cowbird parasitism of Field Sparrows in Tennessee are available, a much lower rate than that reported elsewhere. Field Sparrows frequently desert their nests after cowbirds lay in them (Walkinshaw 1968).—*Charles P. Nicholson.*

Vesper Sparrow
Pooecetes gramineus

The Vesper Sparrow may be best known for its melodious song, which supposedly sounds sweeter in the evening—hence its name. This sparrow nests in northern and western grasslands, reaching its southeastern breeding limit in the mountains of Tennessee and North Carolina (AOU 1983). It is a rare and local summer resident in extreme northeast Tennessee, arriving in March or April and departing by October. Elsewhere in the state it is an uncommon migrant and rare winter resident (Robinson 1990).

The earliest mention of nesting Vesper Sparrows in Tennessee is that of Rhoads (1895a:489), who found them "breeding, but not abundantly" at Johnson City. Two egg sets were collected in Washington County in 1913 and 1940 (Lyle egg coll.). A foray to Shady Valley, Johnson County, in June 1934 found 6 pairs, "two of

Vesper Sparrow. David Vogt

which were feeding young out of the nest" (Ganier and Tyler 1934:23). Wetmore (1939) reported them fairly common there in 1937. A June foray to Roan Mountain in 1936 located 1 bird on the high-elevation grassy balds (Ganier 1936). Herndon (1950b:67) classified it as a "rare summer resident at higher altitudes" in Carter County. Stupka (1963) relates 2 summer records from the Great Smoky Mountains National Park: 9 June 1937 in Cades Cove and 10 June 1937 on Gregory Bald. Finucane (*Migrant* 21:54) reported that this sparrow had been "observed regularly during May thru August for three consecutive years" in the Kingsport area before disappearing in 1950. Both the AOU (1957) checklist and Berger (1968a) mention the Vesper Sparrow breeding at Tate Spring, Grainger County; the source and details of this record, however, could not be determined.

More recent reports of the Vesper Sparrow have continued to come from Shady Valley (e.g., *Migrant* 37:55) and the Roan Mountain balds (Knight pers. obs.), as well as 1 record from Lake Phillip Nelson (*Migrant* 46:91), near Roan Mountain. A 1972 record from Cross Mountain (*Migrant* 43:80), near Shady Valley on the Carter/Johnson County line, was erroneously placed in Campbell County by Robinson (1990). A Vesper Sparrow seen 27 July 1973 at Austin Springs (*Migrant* 44:87), in Washington County, was probably an early migrant or a post-breeding wanderer. In North Carolina, the Vesper Sparrow has nested as far south as Henderson County, on the South Carolina border; most recent records, however, have come from the 3 northernmost counties on the Tennessee border (Simpson 1978a).

During the Atlas, Vesper Sparrows were found in 8 blocks in Johnson County and 1 block in Carter County. All of these were in the Unaka Mountains physiographic

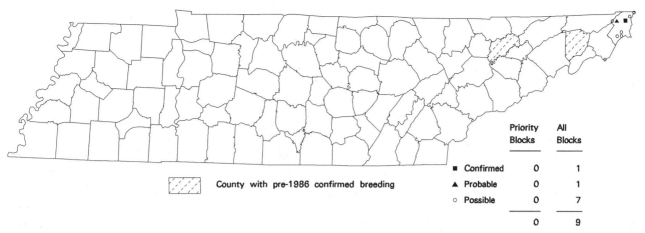

Distribution of the Vesper Sparrow.

region. These records should rather reliably map the current distribution, as this species was intentionally searched for in much of its historic range. One area in Carter County, near Buck Mountain, was not adequately covered, and a Vesper Sparrow was found there after the Atlas period in June 1992 (Knight pers. obs.). Based on the former nesting range as described above, the present breeding range of this species in Tennessee is considerably reduced. It apparently no longer breeds in the Ridge and Valley region.

Most accounts of the abundance of breeding Vesper Sparrows in Tennessee and North Carolina are anecdotal. It was not recorded on the Atlas miniroutes. Simpson (1978a) estimated that a density of about 62 males/100 ha may be reached in parts of northwestern North Carolina, where the species appears to be much more common than in northeastern Tennessee. Whereas early Tennessee accounts called it fairly common (Lyle and Tyler 1934, Wetmore 1939), it is presently uncommon to rare (Eller and Wallace 1984, Knight pers. obs.). Based on the range contraction and apparent population decline, the Vesper Sparrow deserved to be listed as In Need of Management in Tennessee (TWRA 1976). Further evidence of this need comes from the BBS, where this species showed a significant decline in the eastern United States (Robbins, Bystrak, and Geissler 1986). The Vesper Sparrow was also put on National Audubon Society's Blue List for 1977 (Arbib 1976), but downgraded to a "local problem species" for only the northeastern United States in 1982 (Tate and Tate 1982:135). The continued listing as In Need of Management is warranted.

Preferred breeding habitat for Vesper Sparrows in this region can be described as open, short-grass (heavily grazed) fields, often on a moderate to steep slope, with scattered shrubs, fence posts, or rocks for song perches (Mengel 1965, Simpson 1978a, LeGrand and Potter 1980, Knight pers. obs.). Numerous fields in this mountain region have been converted from pasture to Christmas tree farms, which are suitable for only a few years until the trees become too large (Knight pers. obs.). Suitable habitat at high elevations in the Smokies has probably been reduced by the succession of grassy balds to shrubs following the elimination of grazing. Studies elsewhere have shown that Vesper Sparrows require areas of short, dense grass with a high percentage of ground cover (Reed 1986). Compared to Savannah and Grasshopper Sparrows, however, the Vesper occupies areas of sparser grassland (Whitmore 1979). In Johnson County, Tennessee, all 3 species were found in close proximity at some sites, although the Vespers occurred in areas of shorter grass, usually on steeper slopes, and seemed to need elevated song perches (Knight pers. obs.). Atlas records were at elevations of 600 to over 1800 m, while historic sites ranged down to about 335 m at Tate Springs.

Breeding Biology: The Vesper Sparrow places its nest on the ground in a slight depression concealed by adjacent vegetation, usually in an area of low, sparse vegetation (Berger 1968a). The nest is a bulky cup made of grasses and weed stems, with a lining of finer grasses and rootlets. Clutches contain 3–6 eggs, with 4 most common (Berger 1968a). A nest in Shady Valley contained 3 eggs (*Migrant* 37:55), and another in Washington County held 4 (Lyle egg collection); both clutches may have been incomplete. The eggs are creamy white, with various amounts of brown spotting. Incubation is carried out mainly by the female and lasts 12–13 days (Berger 1968a). Both parents tend the young, which fledge in 9–13 days. This species is monogamous and double-brooded. The male may feed the first set of fledglings while the female starts the second nest (Ehrlich, Dobkin, and Wheye 1988). The Vesper Sparrow is an uncommon cowbird host (Friedmann, Kiff, and Rothstein 1977).—*Richard L. Knight.*

Lark Sparrow
Chondestes grammacus

The Lark Sparrow, easily recognized by its striking quail-like head pattern, black breast spot, and white-tipped tail, is a rare, local summer resident in Tennessee (McNair and Nicholson in press, Atlas data). It usually arrives from mid-April to early May and often departs breeding localities by mid-July, exceptionally early for a primarily temperate zone migrant. A few Lark Sparrows have been reported in Tennessee in winter (Robinson 1990).

Lark Sparrows were first reported in Tennessee by Rhoads (1895a), who observed a flock of probable migrants in Obion County, a singing male in Shelby County, and a pair (male collected) at Bellevue, Davidson County, in 1895. Breeding was first confirmed in 1919, near Nashville, and active nests were discovered in 1927 and 1933 in adjacent Rutherford and Wilson Counties (Ganier 1932b, Crook 1935a, *Migrant* 4:22). Lark Sparrows have been regularly observed in the limestone cedar glade region of the inner Central Basin ever since, especially in Davidson, Rutherford, and Wilson Counties, though their numbers have declined (see McNair and Nicholson in press for a historical review of all pre-Atlas breeding records). Lark Sparrows have also been less regularly reported from other areas scattered across the state.

Not surprisingly, Atlas records were concentrated in northeastern Rutherford and southern Wilson Counties. No Lark Sparrows were recorded from Davidson County, probably because of the destruction of large areas of glades by increasing urbanization and the impoundment of Percy Priest Reservoir (Parmer 1985, Quarterman 1989). The only other Lark Sparrow record from the Central Basin was in Marshall County.

Lark Sparrows have undoubtedly nested in the limestone cedar glades of Middle Tennessee since the pre-European settlement period. The landscape and vegetational features to which Lark Sparrows respond are the extensive areas of bare ground (often with rocky outcrops), patchy herbaceous plant cover, and scattered saplings in a xeric summer environment, not only in Tennessee but elsewhere throughout their breeding range (Brooks 1938b, Baepler 1968). They prefer a low density of widely spaced cedar or other shrubs and saplings on breeding territories for use as perch sites and cover for their nests. The heavy logging of cedar in the nineteenth century, resulting in its commercial depletion in much of Middle Tennessee by 1870 (Killebrew 1874), would have created some Lark Sparrow habitat by opening the landscape, and the Lark Sparrow population may have peaked during this period.

A secondary area where Lark Sparrows are concentrated occurs in the southern Coastal Plain Uplands region and the adjacent southwest corner of the Highland Rim. Habitat somewhat similar to the cedar glades occurs near the western arm of the Tennessee River in the Silurian limestone glades centered in Decatur County. Although most of these areas, only a few hectares in size, are probably too small for Lark Sparrows, suitable habitat occurs in this area on upland sites with sparse vegetation and eroded, thin loess or cherty soils. Lark Sparrows were first confirmed breeding in this area in 1956 (McNair and Nicholson in press), and have been sporadically found since, including during Atlas fieldwork. Most of the Lark Sparrow sites in this area, as is true in other parts of the state outside the limestone glades region, have been heavily grazed pastures and cultivated fields. A few sites, however, were large, dry, fallow fields or clearcuts planted to pines. One of the largest populations (at least 10 adults) at a single site yet discovered in Tennessee was found in 1989 in a clearcut planted with small pines and containing scattered herbaceous plants and small thickets (Atlas results).

The few remaining Atlas records of Lark Sparrows are scattered across the state in the northern Loess Plain, the Eastern Highland Rim, the Ridge and Valley, and in the Sequatchie Valley of the Cumberland Plateau. The Ridge and Valley record was a bird observed in late July 1988, a year of unusually severe drought, near Cherokee Reservoir. The only previous Ridge and Valley breeding records were in Sevier and Jefferson Counties in the 1970s (McNair and Nicholson in press). No previous breeding records existed from the Sequatchie Valley, where an adult feeding a juvenile was found in 1987 (*Migrant* 58:146)

Lark Sparrows significantly declined ($p < 0.01$) throughout their range at an average rate of 3.4%/year from

Lark Sparrow. David Vogt

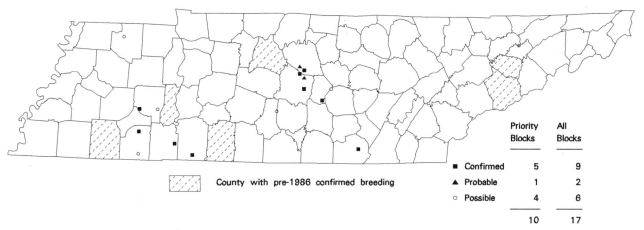

Distribution of the Lark Sparrow.

1966 to 1993 (Peterjohn, Sauer, and Link 1994). East of the Mississippi River, this decline has been evident for over 60 years, and they have receded from much of their former breeding range (Brooks 1938b, Baepler 1968, reports in many state bird journals). Their greatest known concentration is probably in the limestone glades region of Middle Tennessee. Although reliable population size and trend information for the Lark Sparrow is lacking, the inner Central Basin is one of the most rapidly developing parts of Tennessee, and the future of the sparrow there is uncertain. In comparison with the relatively stable, edaphic climax of the limestone glades, sites occupied by Lark Sparrows elsewhere in the state are transitory. Despite the modest improvement in our knowledge of the distribution of Lark Sparrows in Tennessee provided by the Atlas, and the rather wide distribution of breeding records from many areas of the state when both historical and Atlas records are considered, the Lark Sparrow deserves protection and listing as Endangered in Tennessee. It was formerly listed as In Need of Management and recently upgraded to Threatened (TWRA 1976, 1994b).

Breeding Biology: Only 2 of the confirmed Lark Sparrow Atlas records were of nests, and most egg data are from Todd's Rutherford County observations (e.g., DeVore 1975). Unlike in the southwestern United States, where Lark Sparrows often place their nests in shrubs or saplings (Baepler 1968, McNair 1985), all Tennessee nests have been built on the ground, usually in a slight depression at the base of a weedy herbaceous plant, agricultural crop, or shrub that shades the nest (see McNair and Nicholson in press). The mean egg-laying date (corrected for stage of incubation) is 29 May (s.d. = 13 days, range 1 May–24 June). Second broods have been suspected but never confirmed, though renesting is frequent (Baepler 1968, Newman 1970), which probably accounts for the very late nest with young on 20–24 July in Lawrence County (*Migrant* 44:85, NRC). The mean clutch size was 3.9 (s.d. = 0.55, range: 3–5, n = 20), which agrees closely with clutch size data from nearby states (McNair 1985). The female incubates the eggs for 11–12 days, and both adults feed the nestlings, which fledge in 9–10 days (Baepler 1968). Cowbird parasitism has not been reported in Tennessee; it is uncommon to locally heavy elsewhere in the sparrow's range (Newman 1970, Friedmann, Kiff, and Rothstein 1977).—*Douglas B. McNair.*

Savannah Sparrow
Passerculus sandwichensis

The Savannah Sparrow is the most familiar of the grassland sparrows found throughout Tennessee. It is an uncommon to fairly common migrant and winter resident, usually arriving in mid-September and departing by early

Savannah Sparrow. David Vogt

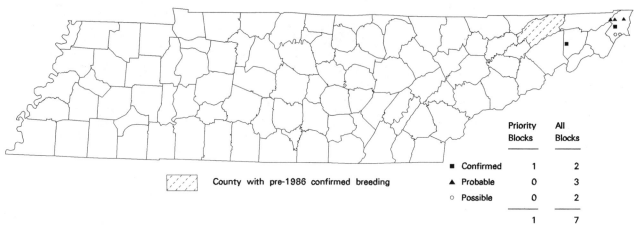

Distribution of the Savannah Sparrow.

May (Robinson 1990). As a breeding bird, however, the Savannah Sparrow has only recently expanded its range southward into the state.

Prior to the Atlas project, only one breeding record and another summer sighting were known from Tennessee. Up to 3 adults were present in a dry Hawkins County hayfield in late June, and a nest with eggs was found there on 2 July 1973 (Alsop 1978). A single bird was observed in Carter County on 13 June 1972 (Knight 1989a). The state's second breeding record was a pair with 2 recently fledged young seen in Washington County on 10 July 1987 (Knight 1989a), during the Atlas period. A single bird was observed in Cocke County on 2 July 1988. In June and July 1991, an incursion of Savannah Sparrows was detected in Johnson County. Ten singing males, at least 3 of them mated, were found in 5 blocks in 4 widely scattered sections of the county (Knight pers. obs.). An intense distraction display on 25 June was the only confirmed breeding evidence obtained there, and a single Savannah Sparrow was found at one miniroute stop.

All breeding season records of Savannah Sparrows have been from the northern portions of the Ridge and Valley and the Unaka Mountain physiographic regions in East Tennessee. Their recent arrival in this area seems to be part of a recent general southward expansion of their breeding range down the Appalachian Mountain chain (Knight 1989a). Rangewide BBS results, however, show a decreasing population trend of 0.8%/year ($p = 0.01$) from 1966 to 1993 (Peterjohn, Sauer, and Link 1994). Three of the 4 Johnson County areas supporting Savannnah Sparrows in 1991 had been surveyed in 1989 or 1990, with none reported. This species could have been overlooked, however, because many southern birders are unfamiliar with its buzzy, insectlike song.

All of the Savannah Sparrows found during the summer in Tennessee have been in open, grassy hayfields or pastures, typical of inland populations in eastern North America (Baird 1968). The dominant species around the 1973 Hawkins County nest was orchard grass (Alsop 1978). The Johnson County sites were in broad, open valleys at elevations of 730–850 m, whereas the sites in the Ridge and Valley region were at 350–450 m. Grasshopper Sparrows were found in all of the same fields and Vesper Sparrows were present at some of the Johnson County sites (Knight pers. obs.).

Breeding Biology: The Savannah Sparrow's nest is a simple cup of fine grasses placed on the ground, often in a slight scrape or depression, and usually well hidden under surrounding vegetation (Potter 1974, Welsh 1975, Alsop 1978). Construction is by the female alone and takes about 4 or 5 days (Welsh 1975). One egg per day is laid and a complete clutch averages 4, with a range of 3–5 (Potter 1974, Welsh 1975, Bedard and LaPointe 1985). The only Tennessee clutch record is of 3 eggs (Alsop 1978). Savannah Sparrows are often double brooded, usually with a slightly smaller average clutch size in the second (Potter 1974, Welsh 1975, Bedard and LaPointe 1985). The eggs are a pale greenish color that can be almost completely obscured by heavy brown blotches (Harrison 1975). Incubation is carried out by the female alone, but both parents feed the young (Potter 1974, Welsh 1975, Dixon 1978). Cowbird parasitism, unreported in Tennessee, varies, for reasons unknown, from heavy to insignificant in other parts of the sparrow's range (Friedmann and Kiff 1985).—*Richard L. Knight.*

Grasshopper Sparrow
Ammodramus savannarum

The Grasshopper Sparrow is aptly named, for it is one of our two breeding species that is restricted to broad grasslands across the length of Tennessee, may feed on grasshopper nymphs during the breeding season, and has a song reminiscent of a grasshopper's stridulations. The Grasshopper Sparrow may have been overlooked in many areas of Tennessee in the past because of its inconspicuous song and behavior, except when males sing from elevated perches; however, Atlas results belie this today. The bird generally arrives on breeding territories from mid- to late April, and breeding is usually not concluded until mid-August. The species is very rare in winter in Tennessee (Robinson 1990, *contra* Alsop 1979b).

Prime natural habitat for the Grasshopper Sparrow is rather open, sparse, dry upland mixed-grass prairie (Smith 1968), though it also occurs in tall-grass prairie. Prior to European settlement, the tall-grass prairies in the northern Gulf Coastal Plain and the Pennyroyal Plateau section of the Western Highland Rim in Tennessee undoubtedly had Grasshopper Sparrows where intrusion of woody plants was minimal (cf. Johnston and Odum 1956, Smith 1968, and others). Natural prairie habitat is now practically nonexistent in Tennessee. Grasshopper Sparrows may likewise have also frequented some larger agricultural fields created by Native Americans or open areas grazed by bison.

After European settlement, the area of grasslands increased dramatically, as probably did the Grasshopper Sparrow population. Detailed habitat information in Tennessee is limited, but studies elsewhere in the eastern United States indicate that breeding Grasshopper Sparrows occupy short- to medium-tall grasslands, preferably intermixed with a high percentage of forbs and patches of bare ground or soil; a low frequency of shrubs can be tolerated (Johnston and Odum 1958, Smith 1968, Wiens 1969, Whitmore 1981). The most frequently occupied habitats in Tennessee are hayfields, lightly grazed pastures, clover and alfalfa fields, and airfields (Alsop 1979b, Atlas results). This sparrow also rarely occurs in maturing winter wheatfields. Although mowing or cultivation is necessary to maintain these habitats, these activities, if carried out during May and June, can substantially reduce nesting success. In the absence of regular maintenance, grasslands become unsuitable habitat after about 4–5 years. The most recent use of a new habitat by Grasshopper Sparrows are human-made grasslands on reclaimed surface mines in the Cumberlands (C. Nicholson pers. comm., Atlas data). Compared to West Virginia, where reclaimed mines are extensively used by Grasshopper Sparrows (Whitmore and Hall 1978, Wray, Strait, and Whitmore 1982), most Tennessee surface mines are unsuitable for this species, and the mined area has greatly decreased in recent years (Nicholson pers. comm.).

Information on the sparrow's population before the mid-twentieth century is very limited. Rhoads (1895a) found it in the Reelfoot Lake area, and Shelby, Davidson,

Grasshopper Sparrow. David Vogt

Hamilton, Knox, Greene, and Washington Counties. Rhoads's visit to the Reelfoot area was in early May, when the Grasshopper Sparrow could still have been migrating; the sparrow was likely nesting at the other locations. Ganier (1933a) described it as fairly common in West and East Tennessee and common in Middle Tennessee. Based on the distribution of grasslands (Bureau of Census 1945) and the few published reports (e.g., Tyler and Lyle 1934), however, Ganier (1933a) probably overestimated its abundance in West and Middle Tennessee and underestimated it in East Tennessee.

Tennessee BBS data from 1966 to 1994 suggest a large, although statistically nonsignificant, decline in the Grasshopper Sparrow population. This decline was significant ($p < 0.01$) at 10.2%/year from 1966 to 1979; since then it has shown a slight increasing trend. Its rangewide population has also declined, by 3.9%/year ($p < 0.01$), from 1966 to 1993 (Peterjohn, Sauer, and Link 1994). An independent assessment of its population in Tennessee was provided by an incomplete analysis of Spring Bird Count data, which suggested a widespread decline had occurred since the early 1960s (Alsop 1979b). A decline in Grasshopper Sparrow populations should not be surprising because the amount of pasturelands peaked in Tennessee in the 1950s. The recent increase is probably due to the species's elimination from many suburban routes during the 1960s and 1970s, largely because of the urban/suburban development of farmland (e.g., see Parmer 1985), and to a gradual increase on rural routes in recent years. Part of the increase has been due to the USDA Conservation Reserve Program, which resulted in the conversion of much cropland to grassland.

The Grasshopper Sparrow was found in 30% of the completed priority Atlas blocks and on 15% of the miniroutes at an average relative abundance of 1.9 stops/route. It

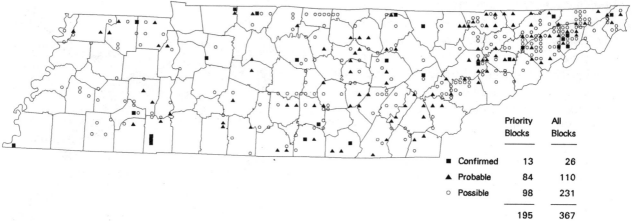

Distribution of the Grasshopper Sparrow.

	Priority Blocks	All Blocks
■ Confirmed	13	26
▲ Probable	84	110
○ Possible	98	231
	195	367

generally increases from west to east across Tennessee at low to mid-elevations, and is most numerous in the Eastern Highland Rim and Ridge and Valley regions. Highest densities are attained in the northeast of the latter region, an area of extensive pastures. The highest number at a single site, however, was on a large Conservation Reserve area in Cumberland County, on the Cumberland Plateau. The Atlas results also suggest a modest increase in West Tennessee, where it was previously rare and local (Alsop 1979b, *contra* Ganier 1933a). In contrast to the rest of the state, the area of farmland has increased in this region during the last 30 years, and the area of Conservation Reserve grasslands is greatest. The increase and expansion of Grasshopper Sparrows into the coastal plain of Tennessee is consistent with the timing of its expansion into other areas of the Southeast coastal plain (cf., McNair 1982, 1984b). The Grasshopper Sparrow is absent from high elevations of the Blue Ridge and rare elsewhere in this heavily forested province. They have nested as high as 885 m at Shady Valley, Tennessee (Wetmore 1939), though they range as high as 1355–1415 m in the Blue Ridge of North Carolina, where they are local and uncommon (Lynch and LeGrand 1989). In Tennessee, they are also generally rare in the heavily forested Cumberland Mountains and Western Highland Rim, as well as in the Mississippi Alluvial Plain where suitable dry grassland habitat is scarce.

Because of its population decline, the Grasshopper Sparrow was listed as Threatened in Tennessee in 1975 (TWRA 1975). Atlas results show that it presently varies from rare to fairly common, is widely distributed over much of the state, and has recently increased in West Tennessee. This information led to its recent status change from Threatened to In Need of Management (TWRA 1994a). This designation is presently justified because of the Grasshopper Sparrow's specialized habitat requirements, and because Conservation Reserve fields will only remain suitable for a few years.

Breeding Biology: Grasshopper Sparrow nests are placed on the ground, usually in a depression and often domed over with grass or forbs. They are well hidden, and only

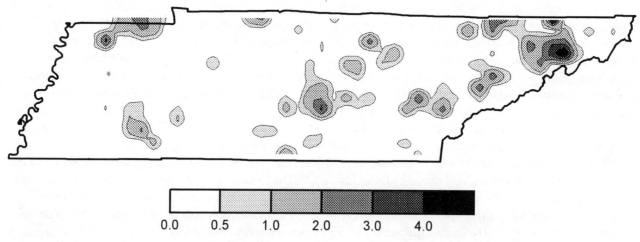

Abundance of the Grasshopper Sparrow.

1 confirmed Atlas record was of a nest. The mean egg-laying date is 26 May (s.d. = 22 days, n = 22). The range extends from 30 April to 13 July for this normally double-brooded species (Smith 1968). The mean clutch size is 4.36 (s.d. = 0.66, range 3–5, n = 22), which closely agrees with clutches elsewhere (Wray, Strait, and Whitmore 1982, McNair 1987a). The female incubates the eggs for 11–12 days, and both adults feed the nestlings, which fledge in about 9 days (Smith 1968). Cowbird parasitism, uncommon to locally common elsewhere (Friedmann, Kiff, and Rothstein 1977), has not been reported in Tennessee.—*Douglas B. McNair.*

Song Sparrow
Melospiza melodia

Song Sparrow. David Vogt

The Song Sparrow is a permanent resident of Tennessee, familiar to many because of its extended January to late summer song period. A large winter population is present throughout the state from October through early April (Robinson 1990). The breeding population, probably nonmigratory, increases in abundance from very rare and local in West Tennessee to abundant in East Tennessee. The Song Sparrow is probably most numerous in thickets along streams and rivers, but also readily occupies brushy field edges and shrubby areas in suburbs and towns. Lewis (1975) recorded 68 pairs/100 ha in a Hawkins County riverside field with hedgerows; no other plot census information from Tennessee is available.

The breeding range of the Song Sparrow has changed dramatically in the last century. Brewster (1886) did not find it in western North Carolina. The first Tennessee breeding season records were by Rhoads (1895a), who found it only at Johnson City and on the lower slopes of Roan Mountain. Wake (1897), using a Knoxville address but not explicitly giving a locality, described it as numerous. In the early 1920s Ganier (1926) found it at the lower and middle elevations of the Smokies. It also spread throughout much of western North Carolina from 1892 to 1910 (Sherman 1910). The sparrow had apparently extended its breeding range into Tennessee by moving southward in the Blue Ridge, occupying brushy stream borders and field edges created by valley farmers, and later moving upslope to occupy cutover forests and balds (Rhoads 1895a, Stupka 1963, Odum and Burleigh 1946). By 1938 it was nesting at the highest elevations of the Smokies (Stupka 1963) and was fairly common at Knoxville (*Migrant* 9:48–50).

Reconstructing the spread of the Song Sparrow in the southern Ridge and Valley and into Middle Tennessee is difficult because of the lack of systematic observations at many locations from the 1930s through 1950s. Singing birds were present at Chattanooga in July 1942 (Coffey 1942c), and Mayfield (1953) reported nesting on the Eastern Highland Rim in Putnam County in 1952 and a pair on the southern Cumberland Plateau in Grundy County in 1953. Stedman (1988b) described the range expansion into Middle and West Tennessee. By 1970, it was nesting at Nashville. Its first appearances in West Tennessee were along the Tennessee River in Benton County in 1975 and along the Mississippi River in Lake County in 1976; at that time it was not reported between Nashville and the western arm of the Tennessee River (Stedman 1988b). At present, Lake and Shelby Counties appear to be the only West Tennessee counties not adjoining the Tennessee River with established populations.

Most of the Middle and West Tennessee locations have been along streams, which likely provided dispersal corridors. Stedman's (1988b) hypothesis that this was a westward movement across Tennessee, however, is questionable. It seems more likely that it spread southward into Middle and West Tennessee from Kentucky and Missouri, as it had done, presumably from Virginia, into East Tennessee. Its spread into western Kentucky, as portrayed by Mengel (1965), may have been from a midwestern source separate from the population in the eastern half of the state. It has also been established in southeast Missouri for some time (Robbins and Easterla 1992). Its first appearances in various parts of Middle and West Tennessee have been at widely scattered points, with few small established populations, arguing against a steady westward spread across the state.

Song Sparrows were found in most Atlas priority blocks from the eastern border west to the northeast corner of the Central Basin, except for those so densely forested that suitable habitat did not exist. It is locally common in the Nashville area, around phosphate settling

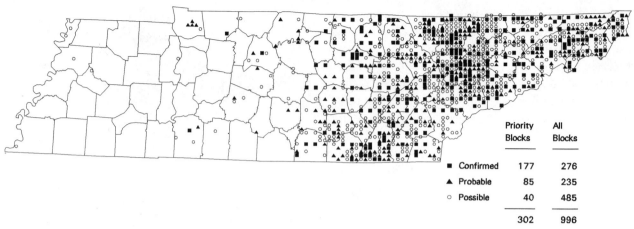

Distribution of the Song Sparrow.

	Priority Blocks	All Blocks
■ Confirmed	177	276
▲ Probable	85	235
○ Possible	40	485
	302	996

ponds in Maury County, and at Cross Creeks NWR. It is rare elsewhere in the Central Basin, Western Highland Rim, and West Tennessee. Few new locations were reported during Atlas work in these areas, suggesting its spread has slowed. Song Sparrows were recorded on slightly over one-third of the miniroutes at an average relative abundance of 5.1 stops/route. Its abundance was greatest to the north and east of Knoxville, where it was first established. By physiographic regions, it was most numerous in the Ridge and Valley, where it was recorded on all but 6 routes and averaged 6.6 stops/route, followed by the Unaka Mountains ($\bar{x}$ = 4.9, on 25 of 43 routes), Cumberland Mountains (4.3, 16/16 routes), and Cumberland Plateau (3.7, 47/71 routes).

Song Sparrow numbers on Tennessee BBS routes increased at the rate of 2.6%/year ($p < 0.01$) from 1966 to 1994. This increase is due to both increases on East Tennessee routes, where it has long been established, as well as its appearance on Middle Tennessee routes. Since 1980, however, its numbers actually decreased ($p < 0.01$) by the rate of 2.1%/year, probably due to the loss of habitat in the east and the slowing of its rate of spread to the southwest. During the Atlas period, it was found on 20 of the 42 BBS routes at averages of 24.5 birds and 17.4 stops/50-stop route. It has not been recently reported on some Middle and West Tennessee routes where it appeared earlier (Stedman 1988b). Its range-wide population decreased by 0.8%/year from 1966 to 1993 ($p < 0.01$) (Peterjohn, Sauer, and Link 1994).

Breeding Biology: The Song Sparrow is strongly territorial and the male defends his territory for over half the year (Nice 1968). It is usually monogamous, although bigamy occasionally occurs. The female usually builds the nest in 3 or 4 days (Nice 1968). The nest, constructed of dead grass and weed stems and lined with fine grass, is usually well concealed and placed either on the ground in a natural depression or among grass clumps or above the ground in thick shrubbery. The proportion of nests built on the ground is highest early in the season, when many plants are not yet leafed out; about one-third of Tennessee nests have been on

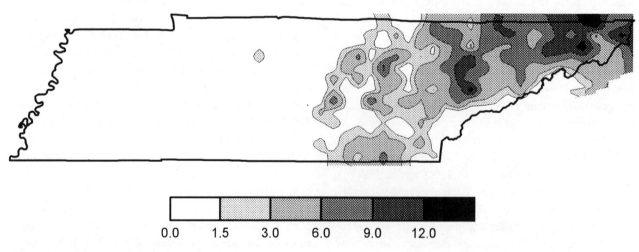

Abundance of the Song Sparrow.

the ground. The average height of above-ground nests is 1.1 m (s.d. = 0.48 m, n = 46), and the maximum 2.4 m. Almost half of the above-ground nests have been in red cedars, and boxwoods and arborvitae are frequently used in suburban areas.

Egg laying begins during the second week of April and peaks during early May. Tennessee clutches average 4.44 eggs (s.d. = 0.57, range 3–5, n = 54), with clutches of 5 more common in April and May than later in the season. The female incubates the eggs for 12 or 13 days. Both adults feed the nestlings, which leave the nest in 10 days, before they are able to fly, and become independent when 28–30 days old (Nice 1968). The species regularly raises 3 and occasionally 4 broods, with the male caring for the fledglings while the female begins the next nest.

In Tennessee, only 2 of about 70 Song Sparrow nests (3%) for which details are available were parasitized by Brown-headed Cowbirds, a very low rate in comparison to the overall rate of 30% throughout its large breeding range (Friedmann, Kiff, and Rothstein 1977). The Song Sparrow frequently successfully raises cowbird young, and a few records exist of adult sparrows feeding cowbird fledglings in Tennessee (Nicholson pers. obs.).
—*Charles P. Nicholson.*

Dark-eyed Junco
Junco hyemalis

The Dark-eyed Junco is probably the most conspicuous and abundant bird nesting at high elevations of the Unaka Mountains. It is easily identified, relatively tame, and popularly known as the "snowbird." Juncos are a permanent resident in the Unakas, although most individuals migrate downslope to winter at lower elevations. Elsewhere in Tennessee, juncos are a common winter resident, usually arriving in early October and departing by mid-April (Robinson 1990). Juncos nesting in the southern Appalachian Mountains were long ago described as a distinct subspecies, the Carolina Junco (*J. hyemalis carolinensis*), on the basis of their larger size, grayer plumage, and darker bill color (Brewster 1886).

The first report of juncos nesting in Tennessee was from the "Unaka Mountains in Southeastern Tennessee" (Ragsdale 1879:239). Since then, nesting juncos have been found throughout the length of the Unaka Mountains (Rhoads 1895a, Tanner 1958), with no apparent historic change in their distribution. Juncos are numerous from the highest elevations down to about 1220 m, and less common at lower elevations (Tanner 1958). Their lowest elevation varies by watershed and

Dark-eyed Junco. David Vogt

probably averages about 915 m. Nesting has been reported as low as 825 m in the Great Smoky Mountains (Tanner 1958, Stukpa 1963).

Although the elevation of several peaks in the Cumberland Mountains exceeds 915 m, nesting juncos have never been found there. They do nest on Big Black Mountain in the Cumberland Mountains of Kentucky, where the maximum elevation is 1265 m. On Big Black Mountain, they occur above 1070 m (Mengel 1965), close to the maximum elevation in the Tennessee Cumberlands. In the Unaka Mountains, atlasers found juncos in all thoroughly worked blocks with maximum elevations above about 1005 m. The population on Big Frog Mountain in Polk County, rather than being isolated as suggested by the Atlas map, is at the northern end of a high ridge extending south into Georgia.

Dark-eyed Juncos occur in all habitat types within their breeding range, including deciduous and coniferous forests and grassy balds. Kendeigh and Fawver (1981) reported breeding densities (averages of several plots censused in 1947–48) of 24 pairs/100 ha in cove forests, 75/100 ha in hemlock-deciduous forests, 58/100 ha in red oak forests, and 140/100 ha in beech gap forests. The highest breeding density, 313/100 ha, was in climax spruce-fir forests. Early successional spruce-fir forests also had high densities, 255/100 ha, while densities in mid-successional spruce-fir forests were lower (145–208/100 ha). The junco was the most abundant species in the beech gap and spruce-fir forests. Alsop and Laughlin (1991) reported a density of 190/100 ha in a virgin spruce-fir forests in 1967. In 1985, after most of the canopy trees had died, they found a density of 165/100 ha.

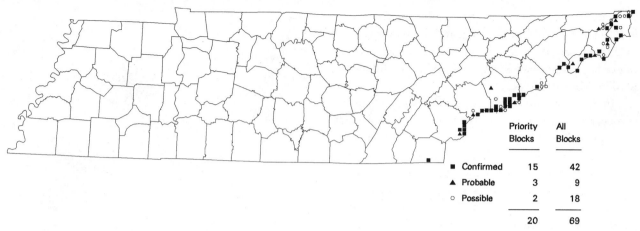

Distribution of the Dark-eyed Junco.

Atlas workers found juncos on 13 miniroutes, at an average relative abundance of 6.3 stops/route. They were most abundant on the highest elevation routes in the Great Smoky Mountains. Information on population trends is limited, as juncoes are not recorded on Tennessee BBS routes. Wilcove (1988) found a significant increase (p = 0.04) in junco numbers between 1947 and 1983 on census plots in the Smokies, possibly due to changes in low-elevation winter habitats.

Breeding Biology: Confirming nesting of the junco was relatively easy for atlasers. Juncos typically raise 2 or 3 broods (Stupka 1963). Their nests are relatively easy to find, and pairs with fledglings are conspicuous. Several studies of the junco's nesting in the southern Appalachians have been published; Tanner's (1958) study in the Great Smoky Mountains is one of the most comprehensive.

Juncos begin returning to their breeding territories in late March. The males defend a relatively small territory that may be surrounded by an undefended area used for feeding by several pairs of birds. Courtship displays, most frequent early in the nesting season, include a display with drooping wings and tail given by the male while singing a soft, jumbled song of trills and warbles (Tanner 1958).

The female builds the nest, which is usually placed on the ground on a sloping bank and concealed by low vegetation. Nests are sometimes placed above ground among the roots of a fallen tree and less frequently in a shrub or evergreen tree as high as 12 m above ground (Tanner 1958). The nest is constructed of moss, grasses, and rootlets; the deep cup is lined with grasses and, less commonly, with moss, rootlets, or hair.

The beginning of egg laying varies with the warmth of the spring and the elevation and is earlier during warm springs and at low altitudes, where it begins in late April. Tanner (1958) observed that the average date of laying the first egg is about 11 days later for each 305 m gain in altitude. At the highest elevations, the first egg may not be laid until late May. First clutches usually contain 4 eggs, less commonly 3 eggs and rarely 5. Tanner (1958) found 4 eggs or young in 71 of 91 nests found from

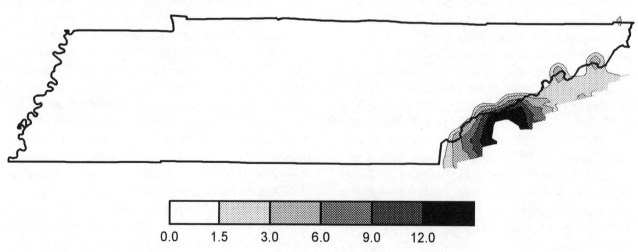

Abundance of the Dark-eyed Junco.

April through June; the remainder had 3 eggs or young. Unpublished records (NRC, egg coll. data) and Stupka (1963) support Tanner's findings. Later clutches tend to be smaller. Tanner (1958) found 3 eggs in 14 of 18 July nests, with 4 eggs in the others; 9 of 16 unpublished July and August records contained 3 eggs, with 4 in the remainder. The eggs, which are quite variable in size and color, have a background of whitish to gray and brown to gray spots or blotches, usually concentrated at the large end.

Incubation begins with the laying of the next-to-last egg and lasts 12 days (Tanner 1958). The female alone incubates the eggs. Both adults feed the nestlings, which leave the nest in 12 or 13 days. The fledglings are fed by the parents for several days, at least until another nest is begun. Tanner (1958) estimated that 35 to 40% of nests he observed fledged young. No records of cowbird parasitism of juncos in Tennessee are available and, because of the rarity of cowbirds in the Smokies and southern Unakas, juncos there are probably rarely parasitized. Juncos in the northern Unakas are probably at least occasionally parasitized. Carolina Juncos are common cowbird hosts in parts of their range (Wolf 1987).
—*Charles P. Nicholson.*

Red-winged Blackbird
Agelaius phoeniceus

The Red-winged Blackbird is a common to abundant permanent Tennessee resident. It nests in marshes, shrub swamps, and grasslands throughout the state. During the winter, Red-wingeds gather with other species of blackbirds into large roosts, which may number in the hundreds of thousands to millions of birds (Meanley and Dolbeer 1978). The winter population is augmented by migrants from farther north; wintering Red-wingeds are less numerous in East Tennessee than elsewhere in the state.

During the breeding season, Red-winged Blackbirds are conspicuous and easily identified. The males sing their loud, easily recognized song from exposed perches, and both males and females are conspicuous in flight. Red-wingeds were one of the species most frequently reported by Atlas workers and were found in 94% of the completed priority blocks. The few completed priority blocks where they were not reported were all heavily wooded blocks with little or no marsh or grassland habitat.

Red-winged Blackbirds were recorded on 86% of the miniroutes, at an average relative abundance of 4.7 stops/route. They were most abundant in the extreme northwest corner of the state, where they were recorded at all stops of several routes. Pitts (1985) noted that the Red-winged Blackbird was the most abundant nesting species on Reelfoot Lake. By physiographic regions, Red-wingeds were most abundant in the Mississippi Alluvial Plain (10.9 stops/route) and the Loess Plain (8.0/route). Their lowest density (2.8/route or less) was in the Cumberlands and the Unaka Mountains. Overall, their abundance correlated positively with the proportion of cropland and negatively with the proportion of forest.

Red-winged Blackbirds have not always been so common in Tennessee. Although they are adapted to nesting in a variety of habitats, prior to the widespread forest clearing in the nineteenth century, their nest habitat was limited to marshes and shrub swamps. Such habitat was then, as at present, most common in West Tennessee. As forests were cleared, they spread into low-lying pastures and hayfields to nest, and grainfields provided additional foraging habitat. In 1895, Rhoads (1895a:487) found Red-wingeds "breeding all over the state, but less abundant than in the Middle States."

Red-winged Blackbird numbers continued to increase during this century, as they shifted to nesting in upland grasslands (e.g., Graber and Graber 1963). The increase was probably most pronounced in West Tennessee due to the widespread drainage and conversion of bottomland forests to farmland. The widespread construction of farmponds throughout the state, many of which develop a fringe of cattails or willows, also increased nesting habitat.

The Red-winged Blackbird is presently one of the most common birds in agricultural landscapes. Their breeding population, as measured by Tennessee BBS routes, does not show a significant overall trend from 1966 to 1994. From 1966 to 1979, however, it increased

Red-winged Blackbird. David Vogt

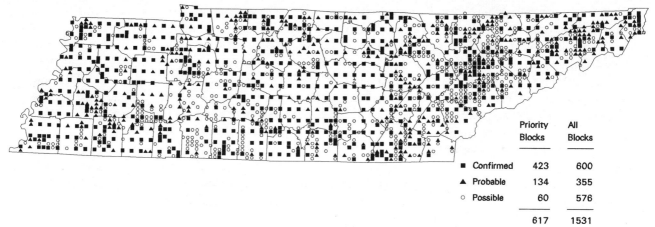

Distribution of the Red-winged Blackbird.

by 4.5%/year (p < 0.01), while since 1980 it decreased by 1.6%/year (p < 0.05). Rangewide, it decreased by 1.1%/year (p < 0.01) from 1966 to 1993 (Peterjohn, Sauer, and Link 1994). Loss of wetlands and grasslands as well as blackbird roost control measures are probably responsible for the declines.

Red-winged Blackbirds are polygynous, with males defending territories in which several females nest. Territories in upland habitats are larger than those in marshes (Yasukawa and Searcy 1995). Information on sex ratios and density in Tennessee is limited. Smith (1967) found an average of 2.0 females/male (n = 5, range 1–3) in a Carter County wet meadow. Fowler and Fowler (1984c, 1984d) reported densities of 66 and 49 males/100 ha in 6-year-old, upland, abandoned agricultural fields in Maury County. Williams (1975b) recorded 128 simultaneously active nests in an 18 ha Blount County marsh.

Breeding Biology: Male Red-winged Blackbirds begin occupying nesting territories in March. Females arrive somewhat later and settle within the male's territory, where some studies have suggested they defend subterritories (Yasukawa and Searcy 1995). Both the male and female may leave the territories to feed. The female chooses the nest site and constructs the nest in 3–6 days (Case and Hewitt 1963). The nest is a fairly large, well-built deep cup with an outer shell of cattail leaves, rushes, or grass stems woven around the supporting vegetation and lined with fine grasses. In marshes, early nests are frequently placed in clumps of cattail, where the dead vegetation from the previous year provides support and concealment (Case and Hewitt 1963, Williams 1975b). Later nests are often placed in bulrushes, shrubs, and small trees, most commonly buttonbush and willow. Nests in upland habitats are frequently built in coarse herbaceous plants such as goldenrod and dock or occasionally placed on the ground (Case and Hewitt 1963, Alsop 1972, Nicholson pers. obs.). The average height of 314 nests in Blount County marshes was 92 cm (s.d. = 37, range 23–243 cm) (Williams 1975b). Nests elsewhere in the state had a similar average height of 96 cm (s.d. = 75, n = 41, range 6–366 cm).

	Priority Blocks	All Blocks
■ Confirmed	423	600
▲ Probable	134	355
○ Possible	60	576
	617	1531

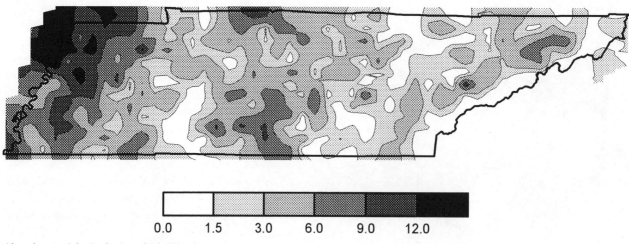

Abundance of the Red-winged Blackbird.

Egg laying in Tennessee peaks near the end of April (Alsop 1972, Williams 1975b). The most common clutch sizes are 3 or 4 eggs. A few clutches of 5 eggs have been reported in Tennessee, as have several clutches of 2 eggs, which may have been incomplete. The average size of apparently complete clutches in Blount County marshes was 3.47 (s.d. = 0.54, n = 53) (Williams 1975b), and of 83 sets in egg collections, 3.72 (s.d. = 0.50). Similar average clutch sizes have been reported elsewhere (e.g., Case and Hewitt 1963). The female incubates the eggs for an average of 11 days and does most of the feeding of the young (Case and Hewitt 1963). The young leave the nest in 10 or 11 days. Females may renest several times after loss of nests early in the season, and a small proportion of them raise second broods (Case and Hewitt 1963, Smith 1967).

Parasitism by Brown-headed Cowbirds, while common in the prairie states, is rare in the East (Friedmann, Kiff, and Rothstein 1977). Alsop (1972) found a cowbird egg in only 1 out of over 3000 Redwing nests, most in Tennessee. Two other incidences of parasitism have been reported in Tennessee (NRC data).—*Charles P. Nicholson.*

Eastern Meadowlark
Sturnella magna

The Eastern Meadowlark is a common, conspicuous permanent resident of grasslands throughout Tennessee. It is easily identified by the contrast of the black V on its breast against its bright yellow underparts, as well as by its call notes, flight pattern, and distinctive whistled song, given from late winter into autumn. The meadowlark is partially migratory and less numerous during the winter.

The Eastern Meadowlark occupies grasslands such as hayfields, pastures, and old fields with sparse shrubs. Fences, isolated trees, and phone wires are frequently used for song perches. It is not particularly sensitive to the density of the grass cover and in this respect is more flexible in its habitat requirements than some other grassland species, such as the Grasshopper Sparrow (Wiens 1969). The meadowlark was probably fairly common on the prairies of prehistoric north-central and northwest Tennessee. Scattered populations probably occurred elsewhere on savannahs maintained through regular burning by Native Americans or natural causes. Wilson (1811a) described it as present in "considerable numbers" in open woodlands with a grassy understory (presumably fire-maintained savannas) in areas occupied by Choctaw and Chickasaw Indians; this description probably applied to parts of southern Tennessee.

Eastern Meadowlark. David Vogt

The Eastern Meadowlark population increased as European settlers cleared woodlands for agricultural uses. A century after European settlement began, Rhoads (1895a) described the meadowlark as uniformly abundant at low elevations across the state. The area of meadowlark habitat in pasture and hayfields peaked about 1959 or 1969, depending on the land classification and sample scheme used; although little quantitative information is available, the meadowlark population probably also peaked during this period. From 1966 to 1994, BBS route results show a decreasing trend of 2.7%/year ($p < 0.01$). This decrease has been pronounced since 1971, and its major cause is probably the decreasing area of grassland. More intensive hayfield management in recent decades, with mowing at least twice a year, has probably reduced nest success, contributing to the decline. The meadowlark's rangewide population decreased by 2.5%/year ($p < 0.01$) from 1966 to 1993 (Peterjohn, Sauer, and Link 1994). The conversion of cropland into permanent grassland in the late 1980s under the Conservation Reserve Program resulted in short-term local increases in meadowlark habitat (Hays and Farmer 1990). In Tennessee, these increases were probably greatest in the Coastal Plain Uplands and Loess Plain regions.

The Eastern Meadowlark was found in 92% of the completed Atlas priority blocks and in a high proportion of nonpriority blocks. The meadowlark occurs throughout the state except for a few heavily forested blocks on the Western Highland Rim, Cumberland Plateau, Cumberland Mountains, and Unaka Mountains. It is common at about 915 m in Johnson County and to about 1370 m on Roan Mountain (Ganier and Tyler 1934,

Confirmed Breeding Species

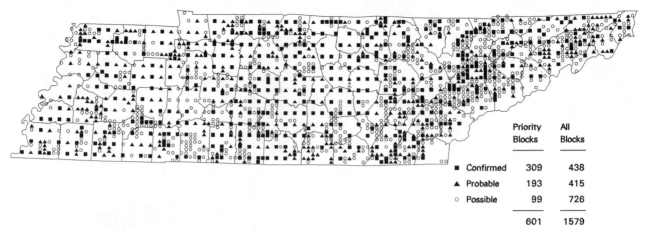

Distribution of the Eastern Meadowlark.

Eller and Wallace 1984, Atlas results). Only occasional vagrant meadowlarks occur on the high-elevation balds of the Smokies (Stupka 1963).

The Eastern Meadowlark was found on 86% of the miniroutes at a comparatively high average relative abundance of 5.5 stops/route. It was most abundant in the Central Basin (7.3 stops/route), Loess Plain (7.1), Eastern Highland Rim (6.3), and Ridge and Valley (5.5) physiographic regions; in each of these regions it was found on over 90% of the routes. Lowest numbers were in the Unaka and Cumberland Mountains. The meadowlark's abundance correlated well with local proportions of farmland except in extreme West Tennessee, where a minor proportion of the predominantly bottomland farmland is in pasture or hay.

Eastern Meadowlark breeding densities have been reported on few census plots. Lewis (1975) found 14 pairs/100 ha in an ungrazed bottomland field with hedgerows, and Fowler and Fowler (1984c, 1984d) found densities of 33 and 49 pairs/100 ha in old fields last cultivated about 6 years earlier. None of these areas appeared to be optimal habitat, which the meadowlark shares with only a few other bird species.

Breeding Biology: Male Eastern Meadowlarks are usually polygynous, with 2 or rarely 3 females nesting within a male's territory (Lanyon 1995). The nest, built by the female, is placed on the ground in an existing depression or a depression excavated by the female. The depression is lined with dried grass, and grasses are usually woven into a canopy over the nest, which has a side opening. The nest cup is lined with fine grasses. Meadowlark nests are very well-concealed, and only about 9% of the confirmed Atlas records were of nests. Most confirmed records were sightings of the conspicuous fledglings and of adults carrying food.

Egg laying begins in early April and continues through July. Clutch size is from 3 to 6 eggs, with 5 egg clutches most common ($\bar{x} = 4.74$, s.d. $= 0.57$, n $= 75$). Similar clutch sizes have been reported elsewhere (Roseberry and Klimstra 1970, Lanyon 1995). Clutch size declines through the nesting season (Roseberry and Klimstra

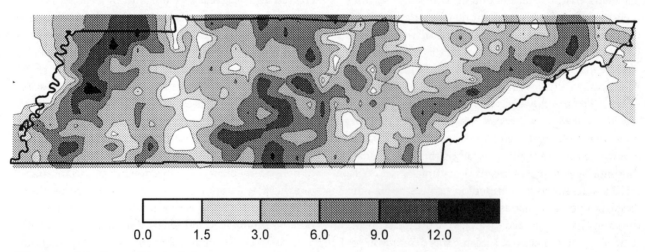

Abundance of the Eastern Meadowlark.

1970), although the difference in Tennessee clutches begun before and after the middle of May is slight. One unusually large clutch of 8 eggs (Vaughn egg coll.), probably produced by more than 1 female, was not included in the above average. Meadowlark eggs are white with brown and purple spots concentrated near the large end. The female incubates the eggs for an average of 14 days, often beginning before the clutch is complete (Roseberry and Klimstra 1970). Both adults feed the nestlings, which fledge in 11 or 12 days. Cowbird parasitism, fairly common in the Midwest (Friedmann, Kiff, and Rothstein 1977), has not been reported in Tennessee.—*Charles P. Nicholson.*

Common Grackle
Quiscalus quiscula

The Common Grackle is one of the most numerous and widespread birds nesting in Tennessee and a locally common winter resident. It occurs in flocks much of the year; during the winter it feeds in farmland and roosts in large aggregations often numbering hundreds of thousands of birds. Wintering birds are most common in the western two-thirds of the state.

Common Grackles nest almost everywhere that feeding areas of short grass, farmland, or wetlands and nesting habitat of dense conifer groves, riparian forest, shrub swamps, or roadside trees and shrubs occur together. In prehistoric Tennessee, they were probably fairly common along the large rivers, in West Tennessee wetlands, and perhaps in the vicinity of Indian villages. Their numbers increased as European settlers cleared the forests, creating additional breeding and foraging habitat. The decrease in deer and turkey populations and virtual extinction of the Passenger Pigeon during the late nineteenth century could also have been beneficial to grackles by reducing competition for acorns, on which the grackle feeds heavily during the fall.

The earliest record of the Common Grackle in Tennessee was by Fox (1886) in Roane County. Rhoads (1895a) found it throughout the state except on the Cumberland Plateau and in the Roan Mountain area. Ganier (1933a) described it as common across the state. Its numbers further increased as the shift to mechanical grain harvesters left more waste grain in fields, increasing its winter food supply (Robbins, Bystrak, and Geissler 1986). This increase continued into the mid-1970s, when it was reversed by a combination of blackbird control efforts and severe winters (Robbins, Bystrak, and Geissler 1986). The Tennessee population, as measured by BBS routes, peaked in 1975. Its overall trend from 1966 to 1994 was a significant ($p < 0.01$) decline of 3.4%/year.

Common Grackle. David Vogt

The rangewide grackle population also showed a significant ($p < 0.01$) declining trend of 1.5%/year from 1966 to 1993; in the east, most of this declined occurred during the latter half of this period (Robbins et al. 1989, Peterjohn, Sauer, and Link 1994).

Atlas workers found the Common Grackle in all but 5% of the completed priority blocks. It was absent from several almost totally forested blocks in the Cumberlands and in the Unaka Mountains. In a few other heavily forested blocks, the only grackle observations were of birds flying high overhead and probably not nesting nearby (Nicholson pers. obs.).

The Common Grackle was also one of the most frequently observed species on Atlas miniroutes. It was found on 88% of the routes at an average density of 5.2 stops/route. Its density was inversely related to the area of forest land and, by physiographic regions, was highest in the Central Basin (7.5 stops/route), followed by the Loess Slope (6.8), Mississippi Alluvial Plain (6.7), and Ridge and Valley (6.0). It was recorded on all routes in the first 3 of these regions. In a survey of West Tennessee forested wetlands, Ford (1990) found the Common Grackle to be one of the most ubiquitous species, present on 55 of 59 plots. It was most abundant in plots with few understory trees and a canopy dominated by cypress or silver maple. The Common Grackle was the second most abundant species along the Duck River in Maury County (Fowler and Fowler 1985).

The proportion of confirmed records was one of the highest of any primarily tree-nesting species. Although the nests are large and often easy to find, they made up only 11% of the confirmed records. Of the confirmed records, 57% were adults carrying fecal sacs or food; these birds are conspicuous as they fly between feeding areas and nests. Most of the remainder of the confirmed records were of fledglings.

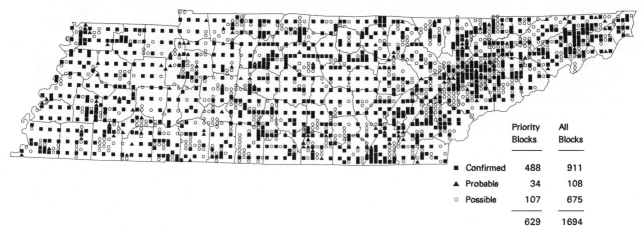

	Priority Blocks	All Blocks
■ Confirmed	488	911
▲ Probable	34	108
○ Possible	107	675
	629	1694

Distribution of the Common Grackle.

Breeding Biology: The large flocks break up in late winter, and grackles occupy breeding areas and begin courtship by mid-March. During courtship, males display to females by fluffing their head and body feathers, holding their tails in a V-shape, and vocalizing (Maxwell 1970). Small groups of males also frequently chase single females. Little territorial defense occurs, and grackles nest as either isolated pairs or in small loose colonies. Some males abandon their mates before the young are fledged to pair and renest with another female (Howe 1976). Females are single-brooded.

The Common Grackle is adaptable in its choice of nest sites. Most Tennessee nests have been high in trees with dense foliage, most frequently red cedars or pines. Other plants used include apple, southern magnolia, hackberry, willow, buttonbush, and small deciduous trees with dense honeysuckle vines. Tree cavities are occasionally used. Vaughn found grackle nests in the hollow tops of cypress stubs 29 and 30.5 m high at Reelfoot Lake; the nests were lined with Great Egret plumes (egg coll. data). Ganier (1950) noted grackles regularly nesting within heron colonies. Nests in trees and shrubs average 7.8 m above ground (s.d. = 4.9 m, range 0.6–30.5 m, n = 75); most are between 4.6 and 9 m. Spero and Pitts (1984) found grackle nests in 20% of Wood Duck boxes in a Kentucky Lake study area. Grackles also occasionally nest in barns and in the steel framework of bridges and electrical substations (*Migrant* 8:36, Nicholson pers. obs.). The large, bulky cup-shaped nest is built by the female of grasses and weed stems, reinforced with mud (Maxwell 1970).

Egg laying begins in early April and peaks during the last half of April. Tennessee clutch size averages 4.7 eggs (s.d. = 0.73, range 3–6, n = 113), with clutches of 5 eggs most frequent, as reported elsewhere in the grackle's range (Maxwell 1970). The eggs are pale green to brown, with brown spots and scrawls. The female incubates the eggs for about 13 days, beginning before the last egg is laid; both adults feed the nestlings, which leave the nest in 12–14 days (Maxwell and Putnam 1972). Fledglings are conspicuous by the first of June, and large flocks of adults and young birds begin forming by late June (Nicholson pers. obs.).—*Charles P. Nicholson.*

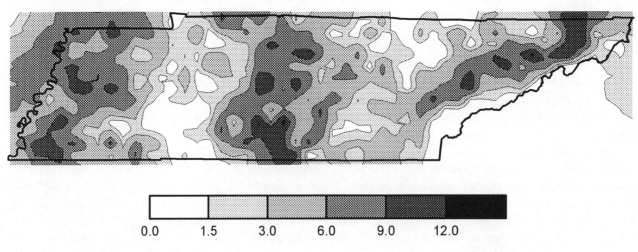

Abundance of the Common Grackle.

Brown-headed Cowbird
Molothrus ater

The Brown-headed Cowbird is an uncommon to common permanent resident. During the winter, it occurs in flocks, primarily in heavily agricultural areas, where it joins large roosts with other blackbirds. It is dispersed throughout most of the state by late March and, as a breeding species, is absent only from some heavily wooded parts of East Tennessee. The Brown-headed Cowbird is the only obligate brood parasite breeding in Tennessee. This habit of laying its eggs in other species's nests is often detrimental to the host species; with declining numbers of many of its host species, the cowbird is attracting increased ornithological attention.

The Brown-headed Cowbird feeds predominantly in areas of short grass (Mayfield 1965), a habit that probably restricted its prehistoric range to the short-grass prairie of central North America and perhaps parts of the tall-grass prairie farther east where bison maintained areas of short grass. It was present on the East Coast by the 1720s, although it is unlikely that it spread there from the prairies as a result of agricultural clearing. By the late 1700s, however, it was common in some northeastern states, probably having spread eastward as early livestock herders spread west of the Alleghenies ahead of settlers (Mayfield 1965).

Nineteenth- and early-twentieth-century agricultural clearing allowed the Brown-headed Cowbird to spread across Tennessee. It was first reported in the state by Fox (1886), who found it uncommon in Roane County in April 1885. The cowbirds observed by Fox were probably migrants breeding north of Tennessee. Rhoads (1895a) observed it with certainty only in the Reelfoot Lake area and concluded that it was rare in summer. The cowbird was found breeding at Nashville in 1918, shortly after resident ornithologists began study there (Monk 1936). The first West Tennessee breeding record was in Shelby County in 1921 (Ganier 1921). Friedmann's (1929) map of the cowbird's breeding range showed it in northern Mississippi, West and Middle Tennessee, and western and central Kentucky. The first East Tennessee breeding records were in 1932 at Knoxville and Johnson City (Woodring 1932). As late as the early 1940s, it was absent from some sparsely settled areas on the Cumberland Plateau (Ganier 1937a, Ganier and Clebsch 1940) where it presently occurs.

The cowbird population has shown a decrease on Tennessee BBS routes of 1.6%/year ($p < 0.10$) from 1966 to 1994. Most of this decrease was from 1966 to 1979, as it has shown a slight increase since then. The range-wide cowbird population also shows a long-term annual

Brown-headed Cowbird. David Vogt

decreasing trend of 0.8%/year ($p < 0.10$) from 1966 to 1993 (Peterjohn, Sauer, and Link 1994). The reasons for this decline are not clear. The decrease in farmland would have reduced suitable foraging habitat, although this is probably locally offset by the spread of mowed lawns and grassy roadsides through suburban development. Blackbird roost control efforts during the 1970s could also have accounted for some of the decline.

Atlas workers found the Brown-headed Cowbird in 97% of the completed priority blocks; it was present essentially statewide except for several blocks in the Unaka Mountains, which were either completely forested or mostly forested and at high elevations. Although its maximum breeding elevation is not well known, it has been found around livestock as high as 1525 m (Ganier and Clebsch 1944). It is either very rare or absent from high elevations distant from farming operations; within the Smokies it rarely occurs far from maintained clearings (Wilcove 1988, Atlas results). The cowbird was recorded on 85% of the miniroutes at an average local abundance of 3.48 stops/route. It showed much local variation in abundance; by physiographic regions, it was least abundant in the Unaka and Cumberland Mountains and most abundant in the Central Basin, Western Highland Rim, Coastal Plain Uplands, and Mississippi Alluvial Plain.

Breeding Biology: In her studies of their behavior at Nashville, Laskey (1950) concluded breeding cowbirds formed monogamous, territorial pairs; other studies have found cowbirds may also be promiscuous, polygynous, or polyandrous (Lowther 1993). Courtship includes bowing displays, often performed by small groups of males toward a single female. In regions of high host nest density,

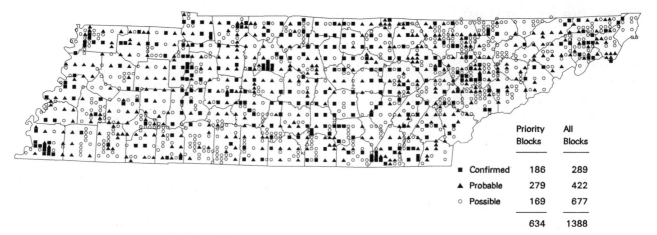

Distribution of the Brown-headed Cowbird.

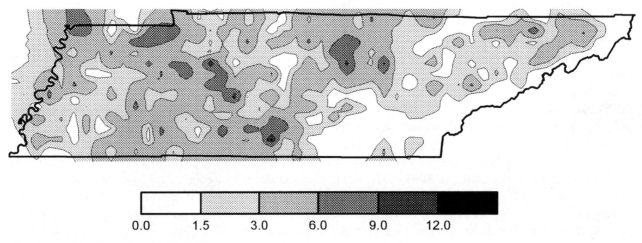

Abundance of the Brown-headed Cowbird.

females maintain territories. They spend much of their time searching for nests by perching motionless for long periods quietly watching other birds building, by walking quietly on the ground in dense woods watching for other birds, and by active, noisy searching of thick vegetation (Norman and Robertson 1975).

Brown-headed Cowbirds lay their eggs in the nests of many different birds. Their eggs or young have been found in the nests of at least 37 species in Tennessee, and cowbirds have been observed feeding fledglings of an additional 7 species (appendix 3). The accounts of these host species give more details on the observed parasitism; among the heavily parasitized species are the Wood Thrush, White-eyed and Red-eyed Vireos, Yellow and Prairie Warblers, Northern Cardinal, Indigo Bunting, and Eastern Towhee.

Individual female cowbirds can lay about 40 eggs over a 2-month laying season (Lowther 1993), which in Tennessee peaks from mid-April through mid-June. Although the proportion of these eggs that are successfully reared is low, they often reduce the reproductive success of their hosts. Several birds parasitized by cowbirds, particularly species of grassland and grassland edge habitats, respond to parasitism by abandoning the nest, burying the cowbird egg in the nest lining, or ejecting the cowbird egg (Mayfield 1965, Friedmann, Kiff, and Rothstein 1977). Other species of eastern forests, thought to be only recently exposed to cowbird parasitism, have no antiparasitism defenses, and cowbirds have been implicated in their population declines (Lowther 1993). Nests near forest openings are more heavily parasitized than in extensive forests, and the increasing fragmentation of our forests by road construction, housing developments, and other land use practices exposes increasing areas to cowbirds.—*Charles P. Nicholson.*

Orchard Oriole
Icterus spurius

The Orchard Oriole is a fairly common summer resident throughout most of Tennessee. It occupies open country, nesting in orchards, wooded fencerows, and scattered

Orchard Oriole. David Vogt

trees in pastures, farmyards, cemeteries, and residential areas. It is well known to many rural Tennesseans, who often mistakenly call it the Baltimore Oriole. Orchard Orioles usually arrive in mid-April and are one of the first songbirds to migrate south, beginning in July. They have usually departed by late August (Robinson 1990).

Atlas workers found the Orchard Oriole in 88% of the completed priority blocks. In the northeastern corner of the state, it was found in the cleared valleys up to an elevation of about 800 m. Because of the scarcity of suitable habitat, it was absent from most of the central and southern Unaka Mountains and parts of the Cumberland Plateau and Mountains. Other gaps in its mapped distribution were probably due to its being overlooked by atlasers. This was most likely in blocks worked late in the season because of the oriole's reduced singing by late June and its early migration.

Orchard Orioles were recorded on 61% of the miniroutes, at an average relative abundance of 2.2 stops/route. The highest abundance was in the southern Central Basin, where it reached 9 stops/route. The reasons for its low numbers in the Ridge and Valley are unclear; the combination of wooded ridges and cleared valleys should have been conducive to higher numbers. Little information on Orchard Oriole densities from breeding bird censuses is available. Lewis (1975) found a density of 5.5 pairs/100 ha in a Hawkins County riverside field with hedgerows. Ford (1990) found an average density of 12 pairs/100 ha on 4 of 59 line transect counts in West Tennessee forested wetlands. Orchard Orioles have occasionally been described as colonial in parts of their range (Orians 1985); this behavior has not been reported in Tennessee.

Based on the distribution of suitable habitat, the Orchard Oriole is presently more abundant in the state than at the time of European settlement. Prior to that time, it was probably restricted to openings maintained by Native Americans and to the naturally maintained openings of glades, prairies, and in the floodplains of large rivers. By the late nineteenth century, the oriole had increased to the point where Rhoads (1985a) described it as abundant statewide. Ganier (1933a) described it as common across the state. It probably reached its peak abundance during this period, when the area of farmland was greatest and the numerous small farms provided abundant suitable habitat.

From 1966 to 1994, the Orchard Oriole population showed a declining trend of 1.5%/year ($p < 0.01$) on Tennessee BBS routes. Increasing urbanization, the decreasing area of farmland, and the trend toward larger farm fields with fewer fencerows have all probably contributed to the declining trend. Throughout its North American range, the Orchard Oriole decreased significantly ($p < 0.01$) by 1.9%/year from 1966 to 1993 (Peterjohn, Sauer, and Link 1994). Because of apparent regional declines, it was listed in the Special Concern category of *American Birds'* 1986 Blue List (Tate 1986). A special protective status is not warranted in Tennessee at this time.

Breeding Biology: Like other North American orioles, male Orchards retain an immature plumage during their first breeding season (Orians 1985). In this plumage, the young male resembles the green-plumaged female except for having a black bib. The function of this immature plumage is not clear (Orians 1985). Yearling males successfully breed, and many pairs containing yearling males were reported by Atlas workers. Orchard Orioles are monogamous and defend a territory containing the nest and foraging area.

The nest is either suspended from the fork of a horizontal tree branch or built in a fork of a more upright limb. It is usually near the outer edge of the tree canopy, which may protect it from predatory mammals unable to climb to the nest. Orchard Orioles frequently nest close to Eastern Kingbirds, often in the same tree (Bent 1958). This relationship was frequently reported by Atlas workers; a tree in Anderson County held active oriole, kingbird, American Robin, and Chipping Sparrow nests within a few meters of one another (Nicholson pers. obs.). The nest, built by the female (Orians 1985), is a cup, somewhat broader than it is deep, woven from blades of grass. The grass, green when the nest is built, dries to a characteristic straw color. Tennessee nests ranged from 1.8 to 15 m above ground, with an average of 5.7 m (s.d. = 3.7, n = 38). Nest heights of from 0.8 to 21 m have been reported elsewhere (Bent 1958). Nests in

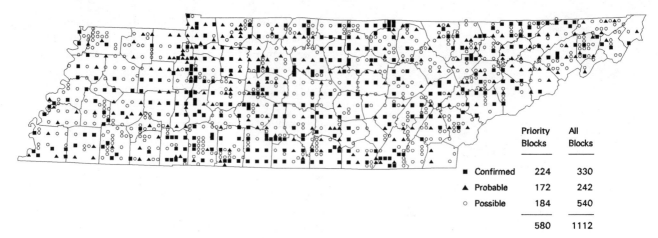

Distribution of the Orchard Oriole.

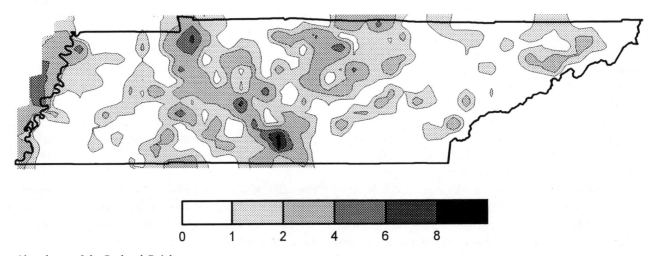

Abundance of the Orchard Oriole.

Tennessee have been built in a variety of tree species, with sycamore, elm, maple, and sweetgum most frequently used. Nests have also been reported in red cedar, loblolly and shortleaf pines, and privet.

Egg laying peaks in late May. Clutches in Tennessee range from 3 to 5 eggs, with an average of 4.14 (s.d. = 0.63, n = 54). Bent (1958) gave 4 to 5 as the most common clutch size. The female alone incubates the eggs for 12–15 days, and may be fed by the male during this period (Bent 1958). The young fledge in 11–14 days; the nestlings and fledglings are fed by both parents (Orians 1985). Flocks form shortly after the young fledge, with the family group often forming the nucleus of these flocks. Fledglings accounted for about one-quarter of the confirmed Atlas records. This relatively low number is probably due to the ease in finding their nests and the early departure of fledglings from the nest territory. Orchard Orioles are single-brooded.

The Orchard Oriole is apparently an uncommon cowbird host in Tennessee. None of 35 egg sets collected primarily in East and Middle Tennessee before 1940 contained cowbird eggs. Atlasers reported at least 2 instances of orioles feeding fledgling cowbirds, and at least one additional record appears in the literature (Monk 1936). The current rate of parasitism is probably higher than these few records suggest, as the contents of few of the recently reported nests were inspected. The Orchard Oriole is a locally common cowbird host elsewhere (Friedmann and Kiff 1985).—*Charles P. Nicholson.*

Baltimore Oriole
Icterus galbula

The Baltimore Oriole is an uncommon summer resident and a fairly common migrant in Tennessee, present from late April through early September (Robinson 1990). During the breeding season, it occurs in tall deciduous trees in open areas and riparian woodlands. From 1973 to 1995, the Baltimore was considered a subspecies of the more widespread Northern Oriole; it is now again considered a separate species (AOU 1995).

The Baltimore Oriole was first reported nesting in

Baltimore Oriole. David Vogt

Tennessee at Nashville in 1886 (Vaughn 1940). Rhoads (1895a) found it throughout the state except at high elevations of the Unaka Mountains. In comparing its status with the Orchard Oriole, Rhoads stated it was "the most abundant, more numerous, indeed, than I ever saw it elsewhere in the United States" (1895a:487). A few years later, Ganier (1917) described it as common in West, Middle, and East Tennessee. If these descriptions are accurate, the Baltimore Oriole has undergone a widespread population decline in Tennessee.

More recent annotated lists, as well as Atlas project results, suggest that the Baltimore Oriole is much less common. Lyle and Tyler (1934), Herndon (1950b), and White (1956) described it as rare in northeastern Tennessee. Howell and Monroe (1957) listed it only as a migrant in Knox County, and Parmer (1985) described it as a rare summer resident in the Nashville area. At Reelfoot Lake, Pindar (1925) considered it common, Whittemore (1937) did not observe it, and it is presently uncommon (Pitts 1985).

The Baltimore Oriole population in much of the state has varied considerably in recent years. From 1966 to 1994, it increased on Tennessee BBS routes by 5.1%/year ($p < 0.05$). Between 1966 and 1991, it was recorded on 30 of the 42 routes. During the Atlas period, it was recorded on 15 routes, only 1 of which was in East Tennessee. Baltimore Oriole numbers on most BBS routes vary greatly from year to year. During the 6-year Atlas period it was found on 15 routes, but it was recorded on only 4 of those routes during more than 2 years. This variability was recognized by Vaughn (1940), who considered the oriole to be nomadic in the Nashville area, rarely nesting at the same site in consecutive years. A few sites in Tennessee, however, such as immediately below Pickwick Dam, Hardin County (Warriner 1974, Atlas results), have been used for many consecutive years.

Throughout its breeding range, the Baltimore Oriole showed no significant, overall population trend from 1966 to 1993 (Peterjohn, Sauer, and Link 1994). From 1978 to 1988, however, it declined significantly ($p < 0.01$) by 2.9%/year in eastern North America (Sauer and Droege 1992).

Atlas workers recorded the Baltimore Oriole in 11% of the completed priority blocks; by physiographic regions, it was most frequently found in the Mississippi Alluvial Plain, followed by the Loess Slope. Most of the records in each of these regions, as well as many records elsewhere in the state, were in riparian areas, either scattered large trees near water or a row of trees between water and fields. Common tree species in this habitat are sycamore, cottonwood, American elm, silver maple, and black willow. A few records were from scattered trees or small groves of trees in open rural areas, distant from water. Few recent reports have been from urban parks or residential areas, habitats more frequently used in the past (e.g., Vaughn 1940).

Atlas miniroute results show little regional difference in Baltimore Oriole numbers. It was recorded on only 19 of the miniroutes, at an average of 1.3 stops/route. The route with the highest count, 5 stops, was in the northwest corner of the state, adjacent to the Mississippi River.

Little census plot information is available for the Baltimore Oriole. Stedman and Stedman (1992) recorded half a territory and fledglings on a 10 ha mature deciduous and pine woodland plot adjoining a Putnam County lake. Ford (1990) found the species on 5 of 59 West Tennessee forest wetland sites at average densities, measured by line transects, of 4 pairs/100 ha. Most of the sites in this survey lacked the forest edge preferred by the Baltimore Oriole (Ford pers. comm.).

Breeding Biology: The Baltimore Oriole begins nest building soon after its spring arrival in Tennessee, and essentially completed nests have been reported during the first week of May. The nest is a deep pouch attached by its rim to the fork of a tree branch near the outer edge of the canopy. It is distinctive in appearance and often persists throughout the winter; used nests accounted for about one-third of the confirmed Atlas records. Tennessee nests average 8.7 m above ground (s.d. = 4.2, n = 27, range 3.7–22.9 m), similar to the average reported by Tyler (1958a). All but 1 nest have been in deciduous trees, with sycamore most frequently used, followed by hackberry and elm. A nest in Johnson County was in a

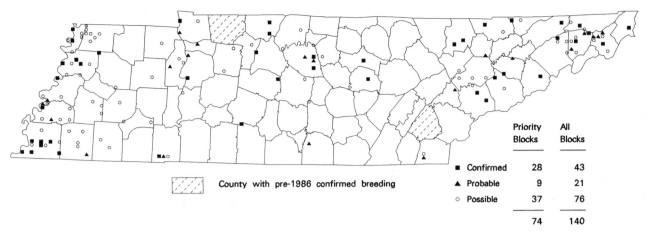

Distribution of the Baltimore Oriole.

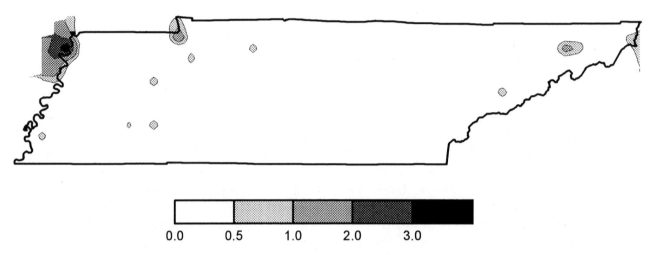

Abundance of the Baltimore Oriole.

hemlock. The nest is woven, primarily by the female, from long strips of milkweed and other plant fibers, grapevine bark, and, when available, string, hair, and bits of rags (Tyler 1958a). At least 2 Tennessee nests have been constructed primarily of monofilament fishing line. The nest is lined with fine grasses, hair, or cotton.

Egg laying peaks about the middle of May. Clutches range from 3 to 5 eggs and average 4.10 (s.d. = 0.74, n = 10). The female incubates the eggs for 12–14 days; both adults feed the nestlings, which fledge in 12–14 days (Tyler 1958a). One brood is raised. Brown-headed Cowbird parasitism of the Baltimore Oriole has not been reported in Tennessee, although a male oriole was observed chasing a female cowbird from its nest in Wilson County on 3 June 1991 (A. T. Tarbell pers. comm.). The Baltimore Oriole typically ejects foreign eggs from its nest (Friedmann, Kiff, and Rothstein 1977).—*Charles P. Nicholson.*

House Finch
Carpodacus mexicanus

The House Finch is a very recent addition to Tennessee's avifauna, having been here since 1972. Its status changed during the 6 years of Atlas project fieldwork as the finches continued to expand their range and increase in numbers; both of these changes are still occurring. Currently the House Finch is locally common in the eastern half of the state, where it is more widespread, and less common in the western half. The state's breeding finches are at least in part permanent residents, as shown by banding results (D. Pitts and M. Pitts pers. comm.). Their numbers are augmented in winter by migrants from the north.

Native to the western United States and Mexico, the House Finch population now inhabiting most of eastern North America grew from small numbers released near New York City in the early 1940s (Elliot and Arbib 1953). Their expansion to the west and south reached explosive proportions in the 1970s and 1980s. The first

House Finch. David Vogt

Tennessee record occurred on 24 March 1972 in Greene County (Holt 1972). Sightings on 17 February 1975 in Davidson County (Parmer 1985) and 4 December 1979 in Shelby County (*Migrant* 51:38) were the first in Middle and West Tennessee. The first summer record was of a bird lingering until 16 June 1978 in Blount County (*Migrant* 49:95). Breeding may have occurred in 1979 in Knox County (*Migrant* 51:95), but was not documented in the state until 1980 when recently fledged young were seen in Sullivan (Laughlin and Phillips 1981) and Knox (*Migrant* 51:95) Counties. The first actual nests were found in 1981 in Washington (Knight 1982) and Sullivan (Phillips 1982) Counties. The rapid expansion of their breeding range within Tennessee is demonstrated by nesting evidence from Hamilton County in 1982 (*Migrant* 53:69), Davidson County in 1984 (Parmer 1985), and Shelby County in 1986 (*Migrant* 57:75). The eastern House Finches have been expanding their range by 2 processes: diffusion, the gradual moving of populations across hospitable terrain, and jump-dispersal, the movement of individuals across great distances over inhospitable terrain (Mundinger and Hope 1982). Expansion down the Ridge and Valley is an example of diffusion, while colonization of areas such as Crossville and Oneida on the Cumberland Plateau typifies jump-dispersal.

During the Atlas fieldwork, House Finches were found statewide and in all physiographic regions, but are localized due to their preference for suburban-like settings. They are most numerous and widely distributed in the Ridge and Valley province, which has much suitable habitat and many observers and was first invaded. Gaps on the map, however, are slightly misleading in some respects. Many cities and small towns, especially those outside priority blocks, were not adequately surveyed, but probably supported House Finches. Also, although a loud and conspicuous singer, some observers may not yet have been familiar with the House Finch's song and, especially, its call notes. The timing of Atlas work in parts of the state, especially West Tennessee, also influenced whether House Finches were detected. The Paris-Henry County area, for example, was primarily worked in 1986–87; had it been worked in later years, more House Finches would probably have been reported. Despite these problems, many blank areas on the map represent genuine gaps in distribution. House Finches do not occur at the higher elevations or in heavily forested areas. They are rather sparse in most agricultural settings except around some houses and barns.

Although not evenly distributed, House Finches are often locally abundant. They are presently one of the most common birds in suburban residential areas in the Ridge and Valley and outnumber House Sparrows in parts of Knoxville and some northwest Tennessee towns (C. Nicholson, D. Pitts pers. comm.). Their abundance was not well reflected on the Atlas miniroutes, where House Finches had an average abundance of 1.8 stops/route on the few routes (n = 44, 6%) recording this species. The location of most priority blocks and, in some cases, the layout of miniroutes were often not favorable for finding this species. The miniroutes do show their relative abundance in the Ridge and Valley. Their rapid increase is quantified by Tennessee BBS routes, which first recorded the species on a Knox County route in 1985; by 1991, it had been found on 12 routes and was increasing at the rate of 47%/year ($p < 0.01$). By 1994, it was found on 16 routes and increasing by 61%/year ($p < 0.01$). The expansion of the eastern North American House Finch population on a broader scale is also well documented by the BBS (e.g., Robbins, Bystrak, and Geissler 1986).

As mentioned above, House Finches have a preference for suburban settings. The grassy lawns and scattered trees and shrubs of residential areas, parks, golf courses, cemeteries, and other landscaped areas provide optimum habitat. In rural areas they may be found around houses and barns. Bird feeders are an important attractant and undoubtedly contributed to their rapid expansion. Once locally established, House Finches often visit and sometimes monopolize feeders year-round, to the occasional chagrin of the feeder operators.

Breeding Biology: The House Finch nest is a well-made cup of grass, weed stems, and a variety of debris, almost always containing something white, such as string or cigarette butts (Harrison 1978, Pitts and Pitts pers. comm.). It is built by the female in a variety of sites around buildings and in shrubs. Many early (March–April)

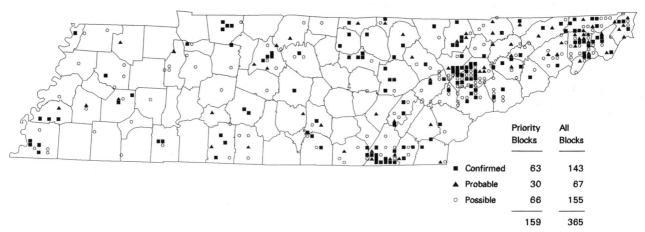

Distribution of the House Finch.

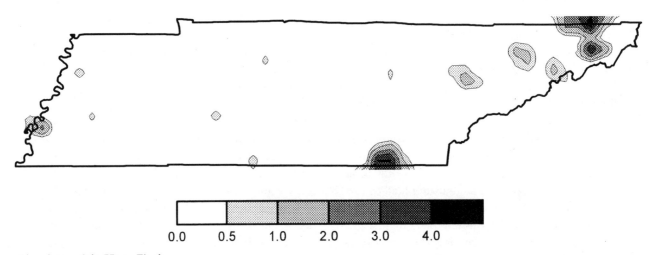

Abundance of the House Finch.

nests in northwest Tennessee are in typical House Sparrow nest sites such as on ledges under metal awnings (Pitts and Pitts pers. comm.). Other Tennessee nest sites are hanging potted plants, horizontal pine limbs, and thick vines on walls and fences. The most frequently reported sites, however, are in densely foliaged, young ornamental conifers such as arborvitae. Nests are occasionally built over an old nest of a House Finch or other species used the previous year. Tennessee nests range from 85 cm to 6.1 m above ground, and most have been from 1.5 to 3 m high. Thirty northwest Tennessee nests averaged 2.0 m high (s.d. = 0.7, n = 30) (Pitts and Pitts pers. comm.). Females occasionally nest very close to one another; 2 active nests in Weakley County were 46 cm apart in the same shrub, and 3 active nests in Obion County were in the same shrub (Nicholson, Pitts and Pitts pers. comm.). Whether these were the products of polygynous males is unknown.

Nest building begins as early as 6 March, and many first clutches are completed by early April. The eggs are pale blue with sparse black dots. Clutch size is 3–5 eggs, with an average of 4.44 (s.d. = 0.61, n = 34). Wootton (1986) reported that clutch size averages larger in the introduced eastern population ($\bar{x}$ = 4.53 ± 0.86) than in the native western population. The female, often fed by the male, incubates the eggs for 12–14 days; both parents tend the young, which fledge in 14–16 days (Harrison 1978). Two or more broods are frequently raised. At least 3 of 42 Tennessee nests contained Brown-headed Cowbird eggs, and the rate of cowbird parasitism is higher in the eastern North American population than in the western (Wootton 1986).—*Richard L. Knight.*

Red Crossbill
Loxia curvirostra

The Red Crossbill is a regular, though somewhat seasonal year-round resident in the southern Appalachian Mountains, especially in the Great Smokies (McNair 1988b). Elsewhere in Tennessee, it is a very rare and irregular visitor, occurring almost every month but most frequently during the winter (Robinson 1990).

Red Crossbill. David Vogt

One of the earliest published bird records from Tennessee is of Red Crossbills, including a small flock containing adults and an immature, in August 1880 at several locations in and near Rugby, Morgan County (Smith 1881, 1883). All but one of Tennessee breeding records are from the Smokies; the other record, discovered during the Atlas period, is from Shelby County. Breeding records of the Red Crossbill throughout the southern Appalachians are not related to irruptive flights (see McNair 1988b), while nesting elsewhere in the Southeast has usually followed an irruptive flight the preceding fall and winter.

Breeding criteria for crossbills are more restrictive than for most passerines. In addition, detection of breeding evidence can be difficult, which is the case for another Cardueline finch, the Pine Siskin (see McNair 1988b). However, the bills of recently fledged juvenile crossbills remain uncrossed and their tails are often less than fully grown until after they have left their birthplaces; they can thus be distinguished from older young with crossed bills whose birthplace is unknown.

Prior to the Atlas period, only 7 breeding records were known from Tennessee or the Tennessee–North Carolina border, all before 1966 (Stupka 1963, McNair 1988b). All these breeding records are based on juveniles seen in the Great Smoky Mountains within the national park or on its border at Gatlinburg from late March to mid-June, either in low-elevation yellow pine forests (400–565 m) or high-elevation (1537–1606 m) fir-spruce forests. Crossbill breeding-season records during the Atlas period are also concentrated in the Great Smokies, though confirmed breeding evidence was lacking here or anywhere else in the Unaka Mountains.

Outside the Unaka Mountains, however, breeding was unexpectedly confirmed during the Atlas period from the opposite end of the state, at Germantown, Shelby County (Peeples 1991, Atlas results). A small flock of Red Crossbills was first observed on 23 May 1991 in loblolly pines full of cones on a golf course. On 26 May, 10 crossbills were present there, including 2 fledglings with uncrossed bills and short tails that begged from and were fed by an adult male. This extralimital breeding record, unlike the only other Gulf Coastal Plain nest record, in Winston County, Mississippi (Warren, Jackson, and Darden 1977), did not follow an irruptive flight. Although a few other crossbills were reported outside of the Unaka Mountains during the Atlas period *(Migrant)*, because of the lack of breeding evidence, these are not included in the Atlas results.

Groth (1988) documented that two discrete call types of crossbills, which he has typified as Call Type 1 and 2, corresponded with differences in size class. Their taxonomic affinities with other populations of Red Crossbills are uncertain (Payne 1987, Groth 1988), though Type 1, the smaller form, may be a southern Appalachian endemic, and Type 2 may be a synonym of *L. c. pusilla*. Groth argued that both size classes of Red Crossbills represent cryptic species because they occur sympatrically without interbreeding. Population data on breeding Red Crossbills of certain call types and size classes in the southern Appalachians are, however, essentially limited to recent information in southwest Virginia and Highlands, North Carolina (Groth 1984, 1988; McNair 1988a, 1988b). No conclusive evidence yet exists that the two size classes breed sympatrically (but see Knox 1990), though both forms pair assortatively with no suggestion of interbreeding. It is also unclear if the proposed cryptic species of Red Crossbills correlate with distinct breeding regimes and habitat, though the smaller-billed form breeds at Highlands, North Carolina, where breeding is timed to correspond with the seeding phenology of white pines (McNair 1988a; see Benkman 1987, 1990; Groth 1988). If the taxonomic affinities of Red Crossbills breeding in the southern Appalachian Mountains are resolved and the arguments of Payne and Groth eventually accepted, the new crossbill species would be the only endemic species of bird in this geographic region. The taxonomic affinities of Red Crossbills breeding elsewhere in the Southeast are unknown except that the few described birds are apparently large-billed forms (Payne 1987, Groth 1988).

Regrettably, the taxonomic complexity of Red Crossbills and the difficulty of reliable field identification of their different forms without special techniques will probably preclude casual ornithologists from making further important contributions, except for documentation of extralimital breeding records and irruptive flights in the

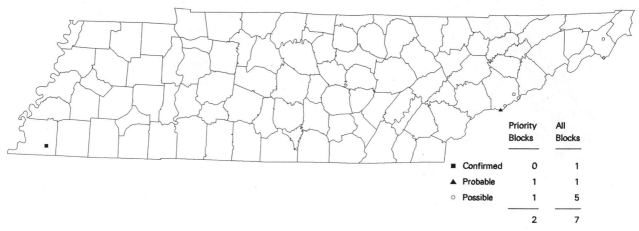

Distribution of the Red Crossbill.

Southeast, which have not occurred since 1976. Our taxonomic knowledge of the forms breeding in the Southeast will be greatly enhanced by recording vocalizations and capturing pairs of known breeding status for measuring and tissue sampling, especially since material evidence of breeding Red Crossbills in Tennessee is limited.

Breeding Biology: As mentioned above, breeding of the Red Crossbill can be difficult to confirm. It may also occur at almost any time of year. In the southeast United States, breeding records of Red Crossbills extend from February through October (see McNair 1988b). One record of a second brood exists from the Southeast (Haggerty 1982). Most nesting in the southern Appalachian Mountains occurs during two periods, from late winter to early spring in most mature coniferous forest types except white pine and from mid-summer through early fall in white pine forests; the extent of late-season breeding on Table Mountain or other yellow pines is unclear. Outside the southern Appalachian Mountains, all breeding records are from yellow pines from February to May, except for one confirmed late summer record at Southern Pines, North Carolina (see McNair 1988b); in this region, there is probably insufficient pine seed available throughout the year to support resident populations of Red Crossbills (Benkman 1987). Throughout the Southeast, all low-elevation breeding has occurred in yellow pines, as high as 1006 m in Table Mountain pines. Otherwise, crossbills breeding in Fraser fir, red spruce, or white pine habitats occur at mid- to high elevations.

The female crossbill builds the bulky nest on a horizontal tree branch, usually far from the trunk (Ehrlich, Dobkin, and Wheye 1988). Clutches of 3 and 4 eggs are most common, and the female incubates them for 12–18 days. Both adults feed the nestlings, which fledge in 15–20 days.—*Douglas. B. McNair.*

American Goldfinch
Carduelis tristis

The American Goldfinch is an uncommon to common permanent resident throughout Tennessee. Migrants from north of the state are present during the winter, when goldfinches frequently visit bird feeders. Little is known of the migratory status of goldfinches nesting in Tennessee.

Goldfinches usually occur in flocks from the fall through late spring. They are most conspicuous in spring, when the males, molted into their bright yellow breeding plumage, begin singing and the flocks feed on dandelion and other seeds in grassy areas or on opening tree buds. The flocks break up, and the birds begin entering breeding condition in early summer (Middleton 1978). Pairs form during this period, although the pair does not settle on a breeding territory until nest construction begins, normally in July. Thus, most Atlas fieldwork was conducted before goldfinches began nesting, and most records were in the possible breeding category. Most of the confirmed records were of used nests, which are easily identified and last well into the winter. A few individual atlasers in East Tennessee reported most of the used nests. Although the confirmed and probable records, mostly of observations of pairs, suggest goldfinches nest statewide, additional confirmed nest records would be valuable to better define the goldfinch's breeding range in West Tennessee.

During the summer, goldfinches usually occur in brushy fields, fencerows, orchards, marshes, and woodland edges, where they feed heavily on seeds of early successional plants, particularly asters. Densities of between 9.9 and 16.4 pairs/100 ha have been reported in relatively uniform goldfinch nesting habitats, including an Eastern Highland Rim brushy pasture (Simmers 1978), a 5- to 7-year-old clearcut in the Unaka Mountains (Lewis 1980, 1981a, 1982), a Central Basin cedar

American Goldfinch. David Vogt

glade (Fowler and Fowler 1983c), and in Central Basin abandoned farmland (Fowler and Fowler 1984c, 1984d). Prior to the widespread clearing of the forests by European settlers, goldfinches were probably much less common in Tennessee and inhabited marshes, fallowed Indian farmlands, and heavily burned areas.

Rhoads (1895a) found American Goldfinches throughout the state and up to 1070 m on Roan Mountain. A few decades later, Ganier (1933a) described them as a rare summer resident in West Tennessee and a common summer resident in Middle and East Tennessee. Goldfinches occur at all elevations of the Smokies during the summer, and nesting has been reported as high as 1768 m elsewhere in the Blue Ridge (Stupka 1963). Atlas miniroute results indicate that they are presently uncommon in West Tennessee and fairly common to common elsewhere in the state. Goldfinches were recorded on about 60% of the miniroutes, at an average relative abundance of 2.3 stop/route. From 1966 to 1994, the early summer goldfinch population, as measured by BBS routes, did not show a significant overall trend. From 1966 to 1979, it decreased by 4.9%/year ($p < 0.01$) and since then increased by 3.8%/year ($p < 0.01$). The early summer goldfinch population showed an overall declining trend of 1.2%/year ($p < 0.01$) throughout North America from 1966 to 1993 (Peterjohn, Sauer, and Link 1994).

Breeding Biology: American Goldfinches begin nesting later than any other bird nesting in Tennessee. Following completion of the extended prenuptial molt, they enter breeding condition in early summer and physiologically peak in July and August (Middleton 1978). Nesting is probably triggered by the flowering of thistles and other asters, which provide nest material and food for the young (Middleton 1979). Although a few atlasers reported goldfinches carrying nest material in mid-June, nest construction appears to peak in mid- to late July. Studies elsewhere (e.g., Middleton 1979) have shown the time at which nesting begins is quite uniform from year to year.

Goldfinches often nest in loose aggregations, and the male defends a small area around the nest (Nickell 1951, Middleton 1979). The female does most of the nest building, often closely accompanied by the male. The nest is a well-built cup woven from fine plant fibers, grasses, and down from thistles or cattails and lined with down, rootlets, and hair. The rim is reinforced with fine bark strips held in place with spider webbing (Nickell 1951). The nest is usually placed in the upright fork of a shrub or sapling, or in the outer canopy of a tree at a forest edge. Most Tennessee nests, for which details are available, have been in saplings, most commonly elm; shrubs used include buttonbush and blackberry. Their height ranged from 0.9 to 7.6 m and averaged 2.3 m (s.d. = 1.6, n = 23). Nests elsewhere have been reported from 0.3 to 18.3 m above ground, and occasionally in herbs such as thistle (Tyler 1968).

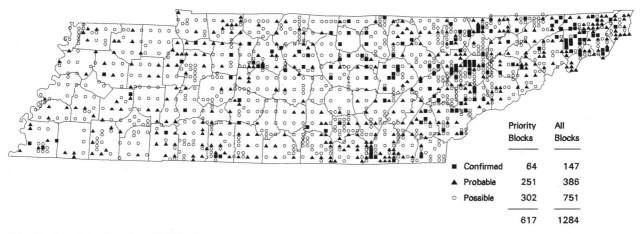

Distribution of the American Goldfinch.

	Priority Blocks	All Blocks
■ Confirmed	64	147
▲ Probable	251	386
○ Possible	302	751
	617	1284

Confirmed Breeding Species

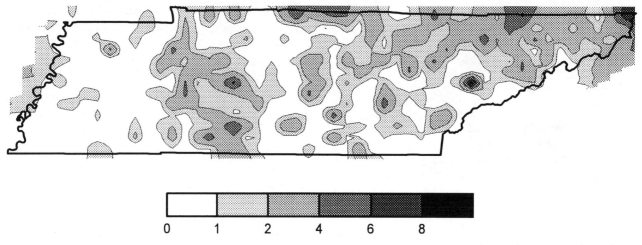

Abundance of the American Goldfinch.

Egg laying peaks near the end of July and extends into September. Goldfinches renest after nest losses, and a small percentage of females attempt to raise second broods (Middleton 1979). The late August and September clutches are probably second or third nest attempts. Clutches of 3–6 eggs have been reported in Tennessee, with the average size being 4.7 (s.d. = 0.81, n = 29). Studies elsewhere have reported clutches of 2–7 eggs, with 5 eggs most common (Berger 1968b, Middleton 1979). The eggs are pale bluish white and normally without markings. The female incubates the eggs for 12–14 days and sometimes begins before the last egg is laid (Berger 1968b). The incubating female, who rarely leaves the nest, is fed by the male who regurgitates food into her mouth (Tyler 1968). Both adults feed the nestlings, which leave the nest in 11–17 days, and most frequently between 13 and 15 days (Stupka 1963, Tyler 1968, Berger 1968b). When renesting, the female often starts the second clutch before the nestlings have fledged from the first nest, leaving the male to care for the young (Middleton 1979).

No records of Brown-headed Cowbird parasitism of goldfinches are available from Tennessee. The number of detailed nest records, however, is small. Elsewhere, cowbird parasitism is rare, as most goldfinches nest later than cowbirds (Friedmann 1963). Cowbird young are rarely, if ever, raised to maturity by goldfinches, as the nestling goldfinches' diet of regurgitated seeds is apparently unsuitable for nestling cowbirds.—*Charles P. Nicholson.*

House Sparrow
Passer domesticus

The House Sparrow is a common to locally abundant permanent resident of Tennessee. It is usually closely associated with human settlements and thus familiar to many Tennesseans.

The House Sparrow is native to Eurasia. It was introduced into North America in an attempt to control insect pests in urban areas, where control by native birds was thought ineffective, and to provide a familiar bird for European immigrants (Robbins 1973). The first release of birds—from England (hence, its other vernacular name of English Sparrow)—was in 1851 in Brooklyn, New York. That release was unsuccessful, but after releases in 1852 and 1853, the species was established in New York. It spread rapidly, aided by legal protection, feeding, and nest boxes. Further releases, of both established stock and imported birds, were made throughout the United States in the next 25 years. The first known releases in Tennessee, of birds transplanted from northeastern populations, were of 4 pairs at Knoxville in 1874 and an unspecified number at about the same time at Memphis (Barrows 1889).

Following its introduction into Tennessee, the House Sparrow spread rapidly and occurred throughout the state by 1886 (Barrows 1889). Torrey (1896a:215) found it "distressingly superabundant" in Chattanooga and its suburbs in 1894. Rhoads (1895a) described it as abundant in all the larger towns, villages, and suburbs he visited across the state. Horse-drawn carriages were then the main mode of transportation, and partially digested seeds in horse manure provided an abundant, year-round food source. Its population probably peaked in the early 1900s just before the horse population decreased following introduction of electric streetcars in the large cities and automobiles throughout the state. This decline has continued in recent years and is probably more pronounced in urban and suburban areas than in the rural countryside. It declined since 1973 on the Christmas Bird Counts in Middle and East Tennessee analyzed by Tanner (1985), which were mostly in urban and suburban areas. Its numbers have not shown a sig-

House Sparrow. David Vogt

nificant trend on Tennessee BBS routes since 1966, which primarily sample rural areas. Throughout North America, it significantly (p < 0.01) declined by 1.8%/year between 1966 and 1993 (Peterjohn, Sauer, and Link 1994).

The House Sparrow was observed in 86% of the completed priority Atlas blocks and found in virtually all blocks with a sizable human population or several farms. Because of its conspicuous behavior and close association with buildings, it is unlikely it was missed in completed blocks. The main areas where it was not found were in extensively forested areas of the Unaka Mountains, Cumberland Mountains, and Cumberland Plateau.

The House Sparrow was recorded on about two-thirds of the Atlas miniroutes, at an average abundance of 2.6 stops/route. By physiographic regions, it was most abundant on the Loess Slope, where it occurred on 90% of the routes at an average of 3.6 stops/route. Next highest were the Eastern Highland Rim (2.9) and Mississippi Alluvial Plain (2.8). During the Atlas period, it was recorded on all but 1 BBS route, at an average of 7.8 stops and 28.3 birds/route.

Breeding Biology: The proportion of confirmed breeding records, 86% in priority blocks and 62% in all blocks, was among the highest of any species. This was due both to the ease in finding House Sparrow nests, which accounted for two-thirds of the confirmed records, and to the bird's long breeding season. Its breeding season regularly extends from March into August; nest building has been observed in January (*Migrant* 13:24). For such an abundant species with a long breeding season, however, relatively few detailed nest records from Tennessee are available; the only published study of its nesting is that of Pitts (1979) in northwest Tennessee.

The House Sparrow usually nests in an artificial or natural cavity. Frequently used sites include holes in the walls and under eaves of buildings, behind wall-mounted signs, in woodpecker holes, and bird boxes. When using bird boxes, it aggressively competes with and often ejects Purple Martins and Eastern Bluebirds. It also frequently occupies Cliff Swallow nests (e.g., Patterson 1966). Within the cavity the male and female House Sparrow construct a partially roofed nest of grasses, straws, and weed stems, lined with feathers and frequently filling the cavity. Domed nests, resembling a large ball of grasses with a hole in the side, are occasionally built in tree branches or similar sites. I once found an open-topped House Sparrow nest in a Barn Swallow nest a few centimeters from the top of a concrete box culvert.

Clutches of 2–7 eggs have been reported in Tennessee, and clutches of 5 are most common. Pitts (1979) found average clutch sizes of 4.42 eggs (s.d. = 1.03, n = 33) in a suburban area and 4.87 (s.d. = 0.89, n = 39) in a rural area. Clutches from Middle and East Tennessee average 4.43 eggs (s.d. = 1.00, n = 28) (egg coll. data). Clutch size elsewhere increases with latitude (Murphy 1978) and Sappington (1977), in the only other detailed study of the species in the Southeast, found an average of 4.2 in Mississippi. The eggs are white to greenish white,

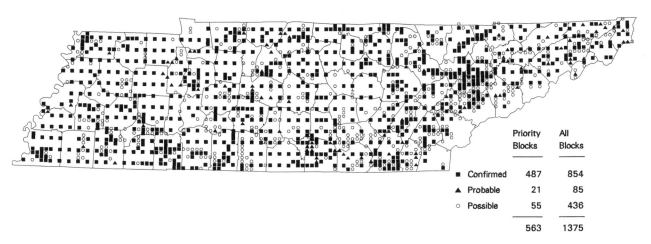

Distribution of the House Sparrow.

	Priority Blocks	All Blocks
■ Confirmed	487	854
▲ Probable	21	85
○ Possible	55	436
	563	1375

Confirmed Breeding Species

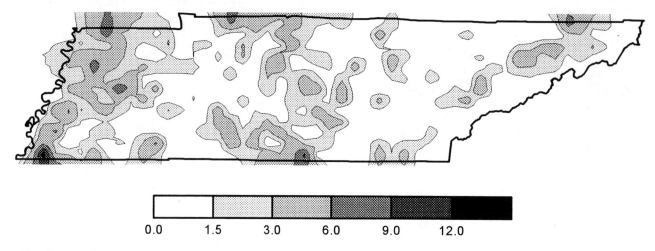

Abundance of the House Sparrow.

with variable gray and brown spots. Most of the incubation is done by the female, and the eggs hatch in about 12 days (Sappington 1977, Pitts 1979)

Both adults, and frequently helpers, feed the young, which fledge in an average of 15 days (Pitts 1979).

House Sparrow pairs lay from 1 to 4 clutches per season, frequently using the same nest cavity. Males typically use the same cavity throughout the nesting season, but occasionally renest with a different female (Sappington 1977).—*Charles P. Nicholson.*

Miscellaneous Species

Possible/Probable Breeding Species, Former Breeding Species, Unsuccessfully Introduced Species, Hybrids

American Bittern
Botaurus lentiginosus

The American Bittern is an uncommon migrant and a very rare summer and winter resident. Confirmed nesting has occurred in 2 counties in south-central Tennessee, most recently in 1976. Summer records exist from a few other sites across the state (Robinson 1990); none were reported during Atlas fieldwork. The American Bittern typically nests in marshes with tall emergent vegetation such as grasses, cattails, and bulrushes (Hancock and Kushlan 1984).

The first nest records of the American Bittern were in 1935, when Ganier and others found nests on 26 May at Goose Pond, Grundy County, and at a small marsh near Morrison, Warren County (Ganier 1935b). At Goose Pond, a 15 ha marsh with water about 60 cm deep and dominated by maidencane grass, were 2 nests, 1 containing an addled egg and the other with 3 nestlings about 5 days old. The nest with young was a flat-topped platform of marsh grass, about 33 cm in diameter and 20 cm above the water. The 2 nests in Warren County each contained small nestlings and were located 15 m apart in an 0.8 ha grassy marsh. No other information describing them is available.

The only other nest records are also from Goose Pond. Todd (1944, egg coll. data) found a nest, containing 4 eggs that had been incubated for 10 days, on 14 May 1939. Dickerson (1977) described a nest found on 22 May 1976 containing 1 egg; a young bird about 2 weeks old was found near the nest. The unhatched egg was still present 2 weeks later. The nest was made of dead grass and other rotting vegetation, anchored in a clump of marsh grass. It was about 50 cm in diameter, 18 cm above water, and well hidden by surrounding vegetation.

American Bittern. Chris Myers

Based on a 28- to 29-day incubation period, starting before clutch completion (Hancock and Kushlan 1984), egg laying in the Tennessee nests occurred between about 9 April and 6 May, when migrating bitterns are usually present (Williams 1975b, Robinson 1990). American Bitterns have been reported, without other evidence of nesting, between late May, presumably after spring migrants have left the state, and July from Reelfoot Lake, Amnicola Marsh in Hamilton County, and other sites in Knox, Sumner, and Davidson Counties (Williams 1975b, Parmer 1985). Some of these reports have been of up to 3 birds. Ganier (1933b), without giving specific dates, noted the bittern's occurrence at Reelfoot Lake during late May and early June and believed it nested there. Pitts (1985) related an unsubstantiated report of bitterns nesting at Reelfoot near the Kentucky border in 1984.

Tennessee is south of the normal breeding range of the American Bittern, which extends to central Missouri, central Ohio, and eastern Virginia (AOU 1983). It has nested in west-central Kentucky (Monroe, Stamm, and Palmer-Ball 1988) and east-central Arkansas (James and Neal 1986), although there are no recent nest records from these areas.—*Charles P. Nicholson.*

Black-bellied Whistling-Duck
Dendrocygna autumnalis

The Black-bellied Whistling-Duck is a large, primarily tropical, gooselike duck whose main breeding range extends southward from the central Texas coast (Bellrose 1976, AOU 1983). It was first recorded in Tennessee in 1978 at the southern end of Reelfoot Lake along the Lake–Obion County line, where a pair was present from at least 1 August through 8 October (Pitts 1982a). Following the disappearance of one of the adults in September, the pair and 8 young were observed by local residents on 7 and 8 October. The birds were not observed on later dates.

Although the Reelfoot whistling-ducks may have been escaped individuals, as suggested in AOU (1983), they gave no indications of being domesticated and rarely allowed approaches closer than about 75 m (Pitts 1982a). Extralimital records of the Black-bellied Whistling-Duck exist as far north as Iowa and Michigan (AOU 1983), and its breeding range has expanded northward in Texas in recent decades (Palmer 1976). Unlike populations farther south, the Texas population is migratory (Bellrose 1976). The whistling-duck usually nests in tree cavities and is adaptable in its use of other sites. It sometimes renests after the successful hatching of the first clutch and clutch initiation occurs as late as mid-August in Texas (Bellrose 1976). Pitts (1982a) estimated the Reelfoot pair began laying about the first of September.

The Black-bellied Whistling-Duck has been reported from Tennessee twice since 1978, in the fall of 1985 (*Migrant* 57:22, 51) and in June of 1988 (*Migrant* 59:123). Each of these observations was in Shelby County. The latter observation was of 4 birds present on 17 and 18 June. Because further details were unavailable, it was not included in the Atlas results.—*Charles P. Nicholson.*

Common Merganser
Mergus merganser

The Common Merganser is an uncommon to rare migrant and winter resident in Tennessee (Robinson 1990). Its breeding range in eastern North America encompasses the boreal forest as far south as central Michigan and upstate New York (Palmer 1976). Although its range is acknowledged to have once extended farther south, and twentieth-century breeding records exist from Virginia and North Carolina, it was not until the report by Kiff (1989) that Tennessee was included in its historic breeding range.

Kiff (1989) described 3 Common Merganser egg sets and gave details of 2 other nests from Tennessee; Knight (1989b) later described another egg set. All 6 records were from Smith County during the years 1896 through 1899; the egg collection dates from 21 April to 13 May. Three nests contained 7 eggs and 3 contained 9 eggs, within the normal range of 7–13 eggs found elsewhere (Palmer 1976). The eggs were either fresh or slightly incubated. All nests were in tree cavities, the typical Common Merganser nest site (Palmer 1976), at an average height of 3.9 m (range 2.2–5.5 m, s.d. = 1.3 m, n = 5). The identity of trees holding 2 nests was not given; 3 others were in cottonwoods and 1 in a beech. One nest was mentioned as occurring on a riverbank. Common Mergansers usually nest near water, typically a stream or river (Palmer 1976). The brood often moves downstream as the young mature. Suitable nest habitat would have occurred in Smith County along the Cumberland and Caney Fork Rivers and their larger tributary streams.

The Smith County records suggest that the Common Merganser once regularly nested in Tennessee. Its full, historic breeding range in the state will never be well known, but it was likely not confined to Smith County. —*Charles P. Nicholson.*

American Swallow-tailed Kite
Elanoides forficatus

The American Swallow-tailed Kite once nested in the Mississippi Valley as far north as Minnesota and up the Ohio River to near Cincinnati (Robertson 1988). Between about 1880 and 1910 its range contracted to the lower Mississippi Valley north to central Mississippi and an isolated population in the northwest Tennessee–southeast Missouri area. Since then its range has continued to contract and it is presently restricted to scattered locations in the southern Atlantic and Gulf Coastal Plain from South Carolina to Florida to Louisiana (Robertson 1988).

The earliest Tennessee record of the American Swallow-tailed Kite is Wilson's observation from along the Duck River in Maury County on 5 May 1810 (Wilson 1812). The next Tennessee record, by Pindar (1889), was of 8 or 9 birds at Woodland Mills, Obion County, on 9 August 1886. Rhoads (1895a) quoted Miles's report of shooting 1 near Brownsville about 1880 and later observing another; according to Miles, it was very rare. Records since then include a small flock on an unspecified date around 1900 near Nashville, an April record, and 7 reports between mid-August and late September (Robinson 1990).

As the above records indicate, no definite Tennessee nest record exists, and all of the records may have been of migrating birds. West Tennessee, however, is within its historic breeding range, and it is very likely that it nested in the state. Within the Mississippi Valley, the kite usually inhabited bottomland timber bordering rivers and lakes (Robertson 1988). Its numbers rapidly declined in the late nineteenth century due to the loss of habitat and shooting of this conspicuous, unwary, gregarious species. Present U.S. populations occur in relatively undisturbed lowland forest near large rivers, and frequently hunt over open country near forest edges (Robertson 1988).

Restoration of a breeding population of the American Swallow-tailed Kite in Tennessee has received little consideration. Recent measures to protect and restore bottomland forests in the Mississippi Alluvial Plain should make this area more attractive for the kite. A hacking project for this species, dependent on the questionable availability of young birds, would be more beneficial than recent efforts to hack the already fairly common Mississippi Kite.—*Charles P. Nicholson.*

Golden Eagle
Aquila chrysaetos

The Golden Eagle is a rare migrant, a rare winter resident, and a very rare summer visitor in Tennessee. For decades it was thought to have formerly nested in Tennessee and other southern Appalachian states (e.g., Palmer 1988, Robinson 1990). Under critical examination (e.g., Lee and Spofford 1990), however, none of the supposed nest records are entirely convincing, and the Golden Eagle is here not considered a former or present native breeding species in Tennessee.

Stupka (1963) quotes Charles Lanman's 1849 account of an eagle's (species not given) nest, with screaming adults nearby, at Alum Cave in the Great Smoky Mountains. Based on the behavior of the birds and the later use of this site by Peregrine Falcons, Stupka (1963) and Lee and Spofford (1990) concluded, probably correctly, that Lanman's observation was of nesting Peregrine Falcons. The earliest specific mention of the Golden Eagle nesting in the southern Appalachians was by Brewster (1886:103), who stated Golden Eagles "were frequently seen, usually in pairs.... They are said to breed on inaccessible cliffs and ledges of the higher mountains" in western North Carolina. LeMoyne (1886) gave a secondhand report of a pair of adult Golden Eagles shot and their 2 eggs collected at a cliff eyrie on Bald Mountain. This nest site was placed in Blount County by Rhoads (1895a) and Ganier (1926), although its location is not explicitly given in LeMoyne's account. Ganier (unpubl. ms., Univ. Tenn. Special Coll., Knoxville) considered this record untenable.

Bent (1937) described 2 Golden Eagle specimens in his personal collection that were taken from a nest on Walden's Ridge in the Cumberland Mountains of Tennessee in 1902 and kept in captivity for over a year. The information on the specimen labels, however, raises doubts about their value as evidence of nesting (Lee and Spofford 1990).

Ganier, in his pursuit of Peregrine Falcon eyries on the Cumberland Plateau, found what he believed were old Golden Eagle nests in Van Buren County and southwest of Jamestown, Fentress County. In Fentress County, Ganier and others observed a "pair" on 31 May 1927 in Buffalo Cove, and in Gwinn Cove the following year observed an "old nest" occupied by Great Horned Owls (Ganier 1937a:27). Ganier reported 3 birds there on 31 May 1930. In Van Buren County, Ganier (*Migrant* 7:47) reported a "picturesque" Golden Eagle nest on a cliff in October 1935; it was occupied by Great Horned Owls the following March. No Golden Eagles were observed during intensive fieldwork at Fall Creek Falls SP, Van Buren County, in June 1940, and in the summary of that work, Ganier and Clebsch (1940:55) stated that the eagle was "probably no longer a summer resident in or near the Park but it has been observed there in earlier years by Ganier and a pair built their nest, which may still be seen, on a protected ledge of Yellow Bluff." The Yellow Bluff site is different from the site of the 1935–36 nest. At no time did Ganier observe a Golden Eagle actually at a nest; the nests he observed could have been built by Red-tailed Hawks.

Although evidence of the Golden Eagle's nesting in Tennessee is meager, there is no doubt that Golden

Eagles once occurred during the summer with some regularity (Stupka 1963, Alsop 1979, Robinson 1990). A large proportion of these records are from the Cumberland Plateau and Unaka Mountains, and many are from the period when livestock, especially sheep, were still raised on remote mountaintop pastures. This livestock probably provided a food supply for the eagles, as typical eagle prey, such as medium-sized mammals, is generally scarce in these areas. Golden Eagles do not begin nesting until at least 4 years old (Palmer 1988), and the summering birds could have been subadult birds, whose plumage can frequently resemble that of adults, from the population nesting in Maine, New York, and eastern Canada (De Smet and James 1987, Lee and Spofford 1990).

Because of its assumed status as formerly nesting, the Golden Eagle is listed as Endangered in Tennessee (TWRA 1975). In light of the poor evidence of nesting, but continued occurrence as a rare winter resident, its status should be downgraded to In Need of Management. In the absence of better evidence of its former nesting, attempts to introduce a breeding population, such as have occurred in North Carolina and Georgia, are unjustified in Tennessee, as are special habitat management efforts for breeding Golden Eagles.—*Charles P. Nicholson.*

Introduced Game Birds

The Tennessee Game and Fish Commission and the Department of Game and Fish, both predecessors of the TWRA, along with private hunting clubs and private landowners, have attempted to establish self-sustaining populations of several species of non-native game birds. Their aims were to provide more huntable species as well as to establish a game bird that would thrive in environments altered by humans to the point where they no longer supported large populations of native game birds. These efforts date back to at least the 1920s; details of many of the first releases are lacking. The program peaked during the 1960s, when brood stock was provided by the U.S. Department of Interior's Foreign Game Bird Investigation Program and large numbers of birds were raised at the Buffalo Springs Game Farm near Morristown. The Tennessee efforts paralleled those of game agencies in many other states.

Introduction efforts concentrated on the Chukar *(Alectoris chukar)*, the Japanese Quail *(Coturnix japonica)* (also called Common or Coturnix Quail, *C. coturnix* in many references), the Red Junglefowl *(Gallus gallus)*, and the Ring-necked Pheasant *(Phasianus colchicus)*. In 1939, 2400 Chukars were released in 24 counties across the state (Solyom 1940). Although some successfully nested their first spring and some survived the first winter, all the populations apparently died out within a few years. About 29,000 Japanese Quail were released in at least 7 counties across the state from 1956 to 1958 (Due and Ruhr 1957, Tennessee State Game and Fish Commission 1959). Banded individuals of this migratory species were recovered several hundred kilometers from the release sites. It apparently failed to establish breeding populations, and the releases were halted in 1959 (Tennessee State Game and Fish Commission 1959). From 1962 to 1966, 566 Red Junglefowl, ancestor of the domestic chicken, were released in Tennessee (Lever 1987). This attempt also quickly failed. Efforts to establish Ring-necked Pheasant populations came closer to achieving success and are described in more detail in the following account.—*Charles P. Nicholson.*

Ring-necked Pheasant
Phasianus colchicus

The effort to establish the Ring-necked Pheasant, native to Asia, was greater than for any other alien game bird. Although it ultimately failed, it probably came closer to achieving success than did the other introductions. The date of the earliest release is unknown; the release of 300 birds in eastern Marion County about 1930 (Ganier 1933c) was probably one of the earlier large releases. At that time Tennessee ornithologists refused to report the pheasant on organized counts because it was not considered established (*Migrant* 3:13). Continued releases met with some short-term success, and pheasants were first reported on the 1951 Christmas Bird Count (*Migrant* 22:66).

The best details on pheasant introductions are from the 1960s (e.g., Hines 1971). From 1961 to 1969, almost 50,000 (the largest number of any state participat-

Ring-necked Pheasant. Chris Myers

ing in the Foreign Game Introduction Program) pheasants were released in 24 counties across the state (Hines 1971, Lever 1987). Between about 500 and 5,000 birds were released in each county, and releases at a particular site were frequently made during 2 to 4 consecutive years. Different subspecies and hybrids between subspecies were stocked to find the strain best suited to local conditions. Spring censuses at release sites generally showed population declines beginning the second spring after the final late summer/fall stocking (Hines 1971). Populations apparently persisted longest at the Old Hickory WMA in Wilson County, at the Hiwassee WMA and Refuge in Meigs County, and in western Greene County (Hines 1971, *Migrant*). Pheasants persisted at Old Hickory until at least 1975 (*Migrant* 46:12) and at Hiwassee until at least 1976 (*Migrant* 47:101).

The eventual failure to establish sustainable populations of the Ring-necked Pheasant in Tennessee and other southern states has been attributed to several causes (Hines 1971). These include the low survival rate of the farm-raised birds, abnormally high cock to hen ratios, illegal hunting, lack of essential soil minerals, and undesirably high humidity and temperature during incubation. Nesting success, however, was apparently not a problem; Hines (1971) noted an average brood size of 5.4 chicks (s.d. = 3.38, n = 91), which he felt adequate.

Attempts to establish pheasant populations have been made throughout much of North America, and in the East sizable populations are established north of about 38° latitude (Johnsgard 1986). Typical pheasant habitat there consists of a combination of grasslands, cultivated cropland, including winter grains, and dense brushy areas. Many of the heavily hunted populations, however, are dependent on the continued release of farm-raised birds, and pheasant numbers have generally declined in recent years throughout much of their American range (Peterjohn and Sauer 1993).

During Atlas fieldwork, Ring-necked Pheasants were reported from 2 blocks: a hen in western Claiborne County and a cock in northern Anderson County. These birds were probably released, either intentionally or accidentally, by local hunters or game bird breeders.—*Charles P. Nicholson.*

Greater Prairie-Chicken
Tympanuchus cupido

The pioneer ornithologist Alexander Wilson, while traveling through south-central Kentucky en route to Nashville in April 1810, found the Greater Prairie-Chicken common in the original prairie or "Barrens" (Wilson 1811a). Historical evidence of the species's occurrence in Tennessee is based on Wilson's secondhand account of one captured a few kilometers from Nashville. Although Wilson (1811a) did not specifically mention observing it in Tennessee, the prairie-chicken has long been considered a part of the original Tennessee avifauna (e.g., Ganier 1933a), and prairies extended from Kentucky southward into north-central and northwest Tennessee. The *"Tetrao"* collected by Michaux near Nashville in June 1795 (Williams 1928) may have been either prairie-chickens or Ruffed Grouse. The Greater Prairie-Chicken was probably extirpated from Tennessee within a decade or two of Wilson's visit.

The Greater Prairie-Chicken inhabits tall-grass prairie, and current, high-quality habitat often consists of a 2 to 1 ratio of native grasslands to cropland (Johnsgard 1983a). Scattered oak woodlands, as occurred on the barrens, are often an important habitat component. In the midwestern states, prairie-chickens often increased with the spread of agriculture until the native grassland area was reduced to the point at which its populations declined (Johnsgard 1983a). Such an increase may have occurred in Tennessee, although its extirpation was more likely due to excessive hunting and trapping than habitat changes. The current range of the Greater Prairie-Chicken is a small fraction of its former range. The few remnant eastern populations, such as in southern Illinois, each number at least a few hundred birds and occupy several square kilometers of managed grasslands (Johnsgard 1983a). Given the scarcity of extensive grasslands in Tennessee, the reintroduction of the Greater Prairie-Chicken seems unlikely. It should, however, be investigated at Fort Campbell in Montgomery County and adjacent Christian County, Kentucky, which probably offers the best potential habitat.—*Charles P. Nicholson.*

Black Rail
Latterallus jamaicensis

The Black Rail, the smallest and most secretive North American rail, is a very rare and perhaps accidental migrant and summer resident in Tennessee. The 5 records of its occurrence date from 1915 to 1983; all are from East Tennessee. Two of the records were during early May, in Jefferson County in 1980 and Knox County in 1983 (Robinson 1990). Another was on an unspecified date in the spring of 1948 in Greene County (Nevius 1964). The first summer record was by N. F. Stokeley, who, while cutting wheat, used a hay fork to collect a Black Rail between 10 and 20 June 1915 near Del Rio, Cocke County (Walker 1935). The only confirmed breeding record was in 1964, when Nevius (1964), while mowing hay on 23 June, observed a pair of adults and 5 downy young in western Greene County. The brood was again observed on 25 and 26 June.

Black Rail. Elizabeth S. Chastain

Within the North American portion of its breeding range, the Black Rail occurs in several disjunct populations on the California coast, along the Atlantic coast from New York south, and in Kansas and several midwestern states (Ripley 1977, AOU 1983). It has also bred in central and western North Carolina and has been recorded during the summer in southwestern Virginia (AOU 1957, Potter, Parnell, and Tuelings 1980, Kain 1987). Much about its distribution and biology, however, is poorly known. Inland nest habitat includes marshes, wet meadows, and dry grasslands. The Greene County brood was in a bottomland hayfield of red clover and grass (Nevius 1964). During most years the field was moist or wet; when the brood was observed, however, the field was unusually dry. The other June record was in a wheatfield, and the May records were in an overgrown, dry hayfield and a wet meadow. The nest is a deep cup woven of grasses and sedges and concealed in thick grass. Clutches number from 6 to 10 eggs (Ripley 1977).

The Black Rail is most readily identified by its "ki-ki-krr" or "ki-ki-do" songs, most frequently sung at night. Systematic nocturnal surveys of wet meadows and grasslands during May and June, using broadcast song recordings, would be useful in better defining its Tennessee range.—*Charles P. Nicholson.*

Passenger Pigeon
Ectopistes migratorius

The extinct Passenger Pigeon, once one of the most abundant birds in the world, occurred in immense flocks throughout the deciduous forests of eastern North America (Bucher 1992). Although scientific and historical treatises contain only a few mentions of its occurrence in Tennessee, those reports, the frequent occurrence of pigeon remains in archaeological deposits (e.g., Chapman 1985), as well as records from adjacent states and place names such as Pigeon River in Cocke County and Pigeon Roost in Humphreys County, suggest that it was an abundant migrant and winter resident. Schorger (1955) compiled migration records of 23–28 April 1797 in East Tennessee and, during the fall, 7 dates between 8 October and 4 November from 1844–83 in West and Middle Tennessee. A thorough review of nineteenth-century newspapers, which frequently reported the arrival and location of pigeon flocks because of their interest to sport and commercial hunters and trappers, would likely produce more records of the pigeon's occurrence.

No definite records of the nesting of the Passenger Pigeon exist for Tennessee, although such nesting probably occurred. Its main breeding range extended south to south-central Kentucky, and there are 2 Mississippi nest records (Schorger 1955). Nesting in small numbers outside of the normal breeding range was apparently not uncommon (Bucher 1992). Rhoads (1895a) quoted Miles's secondhand reports of small numbers of pigeons nesting in the bottoms near Brownsville in the 1840s.

The most detailed account of the Passenger Pigeon in Tennessee is that of Manlove (1933), who described a large roost in the fall of 1870 on Paradise Ridge near Whites Creek, Davidson County. The roost covered several hectares, and the birds dispersed from it each day to feed in beech groves. They remained in the area 2 or 3 weeks until the local nut crop was exhausted. During this period, the pigeons were constantly hunted, both at night in the roost and during the day as the birds flew to and from foraging areas. According to Miles (in Rhoads 1895a), the last large flights near Brownsville were in about 1873; their numbers were then noticeably reduced. The last state report was apparently in the fall of 1893, when 1 of a flock of 8 was killed in Haywood County (Rhoads 1895a). By the end of the decade, the Passenger Pigeon was essentially extinct in the wild (Schorger 1955).

Several causes of the extinction of the Passenger Pigeon have been proposed (Schorger 1955, Blockstein and Tordoff 1985). In a recent review, Bucher (1992) argues that the extinction was the result of the destruction and fragmentation of forests as well as competition for nuts by domestic livestock. These factors reduced the population below the level where the remaining birds could efficiently find food and successfully nest. The minimum viable population of this highly social species was reached when the pigeon was still relatively common, probably in the 1870s. While the intense human predation accelerated this decline, habitat alteration alone would probably have resulted in extinction.—*Charles P. Nicholson.*

Monk Parakeet
Myiopsitta monachus

Feral populations of the Monk Parakeet, a native of southern South America and, during the 1960s, a popular cage bird, were first reported in the United States from New York in 1967 (Neidermyer and Hickey 1977, Lever 1987). Within the next few years, these parakeets were reported from several locations across the country. These populations primarily originated from numerous deliberate and accidental releases of captive birds; some dispersal from established populations may also have occurred.

The first documented Tennessee record was during the winter of 1972–73 in Shelby County (Dinkelspiel 1973). Neidermyer and Hickey (1977) list the occurrence of 2 Monk Parakeets in Middle Tennessee in 1972; further details on this record are not available. Single birds were reported in October 1977 in Knox County and October 1979 in Shelby County (Robinson 1990). At least 1 bird was present at Nashville from December 1979 into May 1980 (Parmer 1985). This bird (or birds) built the large, usually multichambered and communal nest of sticks characteristic of the species; whether breeding occurred is unknown. These nests, unique among parrots, are used as dormitories throughout the year as well as for nesting. The 1972–73 Shelby County bird was observed building a nest with honeysuckle vines in ivy on a black walnut tree (Dinkelspiel 1973).

Because of alarm over the Monk Parakeet's wide distribution, exaggerated reports in the popular press of its population size, and the very real concerns over its potential competition with native birds, damage to crops, and potential for spreading disease, the USFWS embarked on an eradication campaign (officially known as a "retrieval" program) in 1973 (Neidermyer and Hickey 1977). This effort was presumably successful as the number of birds rapidly dropped, and by 1983 the status of remaining wild populations was doubtful (AOU 1983).

The only recent Tennessee report is of a single bird in Greene County on 25 April 1987 (*Migrant* 58:84). Further details of this bird, presumably locally escaped or released, are not available.—*Charles P. Nicholson.*

Carolina Parakeet
Conuropsis carolinensis

The extinct Carolina Parakeet was formerly a common permanent resident, found in several habitats and most numerous in forested river bottoms. The most comprehensive account of its occurrence in Tennessee is that of McKinley (1979). Because of its colorful plumage, occurrence in noisy flocks, and fearless behavior, the Carolina Parakeet was frequently noted by early travelers. The first of these reports was in 1673 by Joliet along the Mississippi River adjacent to Tennessee. Parakeets were reported with some regularity until 1831. The last reports, by Miles (in Rhoads 1895a), were of a flock of 100 birds in the summer of 1874 along the Mississippi River in Lauderdale County and of a single bird in Haywood County in 1876. By the end of the century, the Carolina Parakeet was virtually extinct in the wild (Forshaw 1977).

The Carolina Parakeet occurred throughout much of the eastern United States and was a permanent resident throughout much of its range (Forshaw 1977). The best but still not definitive evidence of Carolina Parakeets nesting in Tennessee is the 1805 letter of William Blount Robertson of Nashville to Benjamin Barton of Philadelphia. In it, Robertson stated that the parakeets laid their eggs and roosted in hollow trees (McKinley 1979). The few descriptions of the parakeet's nesting elsewhere also state that it nested in tree cavities, frequently in sycamores.

Tennessee records with specific dates are during the months of March, April, May, November, and December (McKinley 1979). With one exception, the parakeets were observed along the Mississippi River and the lower Tennessee and Cumberland Rivers. The exception was near the Cumberland River at Flynns Lick, Jackson County. Parakeets also frequented salt licks, as Wilson (1811a) noted in Kentucky and Tennessee.—*Charles P. Nicholson.*

Hermit Thrush
Catharus guttatus

The flutelike song of the Hermit Thrush is often regarded as the most beautiful birdsong in North America. Yet, it is probably unfamiliar to many southern birdwatchers

Hermit Thrush. Elizabeth S. Chastain

because this species breeds in northern and western montane forests and infrequently sings on its southern wintering grounds. Only recently has the Hermit Thrush extended its breeding range into the southern Appalachian Mountains, where it is now a rare and local summer resident. This species is more familiar to Tennesseans as an uncommon to fairly common statewide migrant and winter resident, generally arriving in October and November and departing by early April.

The first breeding season report of the Hermit Thrush in the southern Appalachians was in 1966 on Mount Rogers in southwestern Virginia (Scott 1966). Its population there has since increased, and 2 nests have been found (Scott 1982, P. C. Shelton pers. comm.). A Hermit Thrush was discovered in June 1979 on Roan Mountain (Potter and LeGrand 1980), astride the Tennessee–North Carolina border about 80 km southwest of Mount Rogers. One to 3 singing males were present there during the summers of 1983, 1986, 1991, and 1992 (Knight pers. obs.). Most of these were found on the north slope of Roan High Knob, on the Tennessee side. In western North Carolina, Hermit Thrushes have occurred on Mt. Mitchell since 1983 (LeGrand 1984, McNair 1987c) and on Grandfather Mountain since 1984 (Lee, Audet, and Tarr 1985). One was found on the North Carolina side of the summit of Unaka Mountain in 1992 (H. E. LeGrand Jr. pers. comm.). The 3 southernmost records in the Appalachians are from the Great Smoky Mountains National Park. An undated record is from Peck's Corner, on the state line (McNair 1987c). Three or 4 territorial birds were found in a 20 ha census plot on Mt. Collins in 1986; none were found there in 1990 or 1991 (K. Rabenold unpubl. ms.). A singing bird was found in June 1990 on the Tennessee side of Laurel Top (Knight pers. obs.). Solid breeding evidence is lacking from all of these Tennessee and North Carolina sites, as only territorial males have been located thus far. Confirmation of the breeding of this species is difficult, as evidenced by Atlas results in New York and Michigan, within the Hermit Thrush's main breeding range (Bonney 1988, Winnett-Murray 1991). The 3 blocks on the Atlas map are Roan Mountain, Laurel Top, and Mt. Collins, the only sites with definite Tennessee records.

Abundance and population trend data for the Hermit Thrush in the southern Appalachians are merely anecdotal, not surprising considering the recent and sparse nature of its distribution there. However, the recent southward extension of the Hermit Thrush's summer range would seem to imply an increase. Continental BBS data show an average annual increase of 1.4% (p = 0.02) from 1966 to 1993 (Peterjohn, Sauer, and Link 1994).

The Hermit Thrush is sympatric with 1 or more of the other spotted thrushes over much of its North American breeding range. Coexistence among these potentially competing species has been achieved by separation through selection of specific habitat features (Morse 1971, Noon 1981). Breeding habitat for the Hermit Thrush is generally described as coniferous or mixed deciduous-coniferous forest. A New York study found them associated with edgelike settings within the forest with stands of rather dense, young, mixed growth, such as the margins of lakes or burns and powerline corridors (Dilger 1956). In Vermont, it selected "midsuccessional forests with high canopy cover generated by dense stands of intermediate trees" (Noon 1981:114). Standing dead trees for singing perches seem to be important (Godfrey 1966). The Hermit Thrush is more tolerant of drier habitats than are the other thrushes, but may be found in moist sites as well (Dilger 1956, Morse 1971).

In the southern Appalachians, the Hermit Thrush has been found only in the high-elevation spruce-fir belt (Scott 1982, Lee, Audet, and Tarr 1985, McNair 1987c, Knight pers. obs.). All occurrences have been above 1550 m, with most above 1750 m. On Roan Mountain,

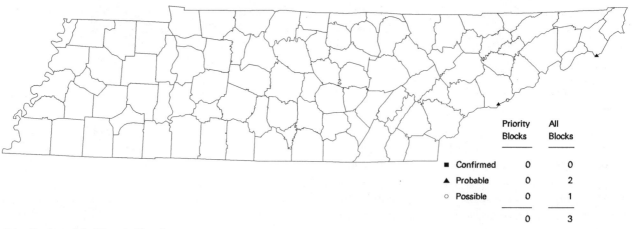

	Priority Blocks	All Blocks
■ Confirmed	0	0
▲ Probable	0	2
○ Possible	0	1
	0	3

Distribution of the Hermit Thrush.

this species has been found most consistently in a stand of living and dead spruce-fir, with a very dense understory of young conifers and other growth (Knight pers. obs.). It was found in an area of dense undergrowth and few large living spruce-fir trees on Mt. Collins (K. Rabenold unpubl. ms.). Much of the Hermit Thrush's southern Appalachian range is shared with the Veery, while the Swainson's Thrush *(C. ustulatus)* is also found on Mount Rogers, Virginia (Scott 1982), and may eventually also move south. Little detailed analysis of the habitat partitioning among these thrushes has been done there (P. C. Shelton pers. comm.), although it seems ripe for study.

The nest of the Hermit Thrush, a bulky cup of twigs and various plant fibers, is built by the female on or near the ground (Harrison 1975). The completed clutch of 3–4 pale blue, unmarked eggs are incubated by the female. Both parents tend the young. In parts of its range, the Hermit Thrush is a regular cowbird host (Friedmann and Kiff 1985); cowbirds are scarce to absent within the high-elevation range of this thrush in the southern Appalachians.—*Richard L. Knight.*

Bell's Vireo
Vireo bellii

The Bell's Vireo is an extremely rare bird in Tennessee and was not found during the Atlas period. Six state records exist, 2 of them most likely of spring migrants, the others during the summer (Robinson 1990). The first record—the only confirmed breeding record—was in June 1935 in Memphis (Coffey 1935). A pair was present from mid-June to early July in an urban lot containing "trees, bushes, vines, and back-yard hedges, along a creek" (Coffey 1935:67, Coffey pers. comm.). On 30 June, Coffey found a nest containing 2 spoiled eggs, located 1.2 m up in an elm bush. This nest was soon abandoned, and, although the pair scolded from other likely nest sites, Coffey and others were unable to find another nest. Bell's Vireos were not found at this site in later years.

The next 2 state records could be considered possible breeders. On 7 July 1946 a singing bird was observed in Natchez Trace State Park, Henderson County (Coffey 1946). On 27 June 1971 a vagrant Bell's Vireo was briefly observed singing on Mud Island, Shelby County, but not found on later visits (*Migrant* 42:68, Coffey pers. comm.). The most recent breeding season record was of a singing bird present from 27 May to 1 June 1976 in southwest Obion County (Pitts 1983). This territorial bird was found in 6 m tall willows along the edge of a slough.

Bell's Vireos are typically found in riparian and brushy second growth, often with White-eyed Vireos (Barlow

Bell's Vireo. Chris Myers

1962). The nest is usually suspended from the fork of a small, pendant branch, in dense leaves at the outer portion of a tree or shrub, usually less than a meter high. The nest is built of bark and grass shreds, held together with silk from spider webs and egg sacs. The 3–4 eggs are incubated for 14 days, and the young leave the nest in 11–12 days (Barlow 1962). A second brood is often raised. Bell's Vireos are heavily parasitized by cowbirds (Barlow 1962, Friedmann, Kiff, and Rothstein 1977), although the cowbird fledging rate is low because the vireo often deserts, removes, or buries the cowbird eggs.

The breeding range of the Bell's Vireo covers much of the central states, extending eastward to eastern Arkansas and Missouri, western Kentucky, southern Indiana, and western Ohio (AOU 1983). In northeastern Arkansas, it is uncommon and occurs regularly (James and Neal 1986). The species extended its breeding range into Indiana in the 1940s, apparently following the prairie peninsula (Barlow 1962). The first Kentucky nest record was in 1980 in McCracken County (Nicholson 1981b). In recent years, the species has been found in additional counties in west-central Kentucky (B. Palmer-Ball Jr. pers. comm). Although there is presently not an established population of the Bell's Vireo in Tennessee, ornithologists should be alert for its occurrence, especially in shrubby riparian habitats in West Tennessee.—*Charles P. Nicholson.*

Brewster's and Lawrence's Warblers
Vermivora pinus x *chrysoptera*

When Blue-winged Warblers interbreed with closely related Golden-winged Warblers, the offspring have a variety of plumage patterns; 2 distinctive types are known as the "Brewster's" and "Lawrence's" Warblers. The more common Brewster's usually resembles a Blue-winged with a plain, whitish throat and underparts, and the rarer Lawrence's resembles a Blue-winged with a Golden-winged's black mask and throat.

Although most Tennessee records of the hybrids have been during migration (Robinson 1990), at least 4 pre-1986 records were of likely local breeders. A male Brewster's singing a Blue-winged-type song was present among several Golden-wingeds on 11 June 1973 in Cumberland County (Alsop 1974), and another Brewster's observed in May and June 1974 in Marion County (Dubke and Dubke 1977, Nicholson 1980a). Other Brewster's Warblers were reported on 30 May 1975 in Unicoi County (*Migrant* 46:91) and on 17 July 1979 in Cheatham County (*Migrant* 50:88).

During the Atlas fieldwork, a Brewster's was observed singing a Golden-winged-type song on 19 May 1990 on abandoned farmland near Ketchen in eastern Scott County (Nicholson pers. obs.). A Blue-winged and 2 Golden-wingeds were singing nearby; by 19 May most migrants of these species have passed through and residents have usually begun nesting. Only Golden-wingeds were observed at this site the following month. Another Brewster's was observed in cutover pines about 8 km southeast of Tullahoma, Franklin County, for at least 2 weeks beginning 16 May 1991 (D. Davidson pers. comm.). Blue-winged Warblers held territories nearby, and the Brewster's aggressively responded to playback of a Blue-winged song. A probable male Brewster's and its apparent Blue-winged mate, both carrying food, were observed on 13 June 1992 a few hundred meters from the 1991 Franklin County site (Davidson pers. comm.).

With the exception of the Cheatham County bird and probably the Franklin County birds, all Brewster's Warblers reported during the breeding season have been within the breeding range of the Golden-winged Warbler. Probable breeding of the Lawrence's Warbler has not yet been reported in Tennessee.

The frequency of the hybrid forms is probably underestimated, as singing Golden-wingeds and Blue-wingeds are often difficult to see (Nicholson pers. obs.), and the songs of the hybrids are not noticeably different from those of the parent species (Morse 1989). Some hybrids are also difficult to identify unless in the hand (Morse 1989). With the increasing frequency of contact between Blue-winged and Golden-winged Warblers north of Tennessee, the frequency of the hybrids in Tennessee during migration should increase. Expansion of the Blue-winged Warbler into the Tennessee breeding range of the Golden-winged Warbler may also result in more hybrids; observers should carefully observe and report their occurrence.—*Charles P. Nicholson.*

Magnolia Warbler
Dendroica magnolia

The Magnolia Warbler is a fairly common spring and fall migrant in Tennessee, present from late April to late May and from late August to mid-October (Robinson 1990). Until the late 1960s, its southern breeding limit was the mountains of West Virginia and adjacent central Virginia (Hall 1983). Since then, however, territorial Magnolias have occurred with increasing frequency in the mountains of southwestern Virginia, western North Carolina, and northeastern Tennessee, and this bird is now a very local probable breeder in Tennessee.

Prior to the late 1960s, there was little evidence that Magnolia Warblers were regular summer residents in the southern Appalachians. Pearson, Brimley, and Brimley (1942) report thirdhand that J. S. Cairns, a late-nineteenth-century naturalist in the Asheville, North Carolina, area, collected the nest and eggs of a Magnolia Warbler. No supporting documentation of this record is available, and it was not mentioned in Cairns's publications. Bent (1953) stated that the Magnolia occasionally bred in the Asheville area, again without supporting documentation and presumably repeating the report of Pearson, Brimley, and Brimley. Herndon (1977) reported, without additional details, single summer records on Roan Mountain, Tennessee–North Carolina, on 4 July 1959 and on 28 July 1962. The first documented records of a breeding population in the southern Appalachians were from southwestern Virginia by Scott (1966), who reported 6 singing males on Beartown Mountain, Russell County, and 2 on Whitetop, Smythe County. Magnolia Warbler populations have since become established in at least 3 sites in southwestern Virginia (Virginia Atlas Project, unpubl. results).

Magnolia Warblers were again observed on Roan Mountain in 1975, when up to 3 singing males, and on 14 July, an adult carrying food, were present on the North Carolina side of the mountain (*Amer. Birds* 29:973, Herndon 1977). Magnolia Warblers have since been observed sporadically on Roan Mountain in both states; a singing male present on 21 June 1988 and not found on later dates was on the North Carolina side of the mountain (*Migrant* 50:91, 53:91, 59:130). Since the early 1970s, singing male Magnolia Warblers have been observed at 5 other sites in western North Carolina as far south as southern Haywood County (Lee 1985, D. S. Lee pers. comm.).

A second Tennessee breeding locality was established with the observation of a singing male on 8 and 17 June 1989 on Unaka Mountain, Unicoi County (*Migrant* 60:110). Additional observations there during 1991 and 1992 revealed the presence of a large popula-

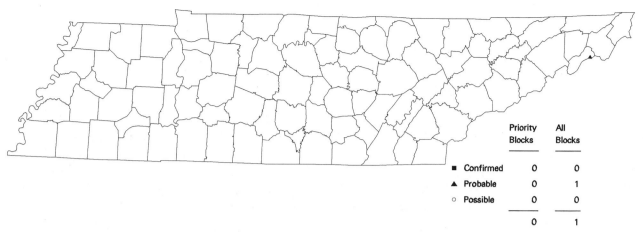

Distribution of the Magnolia Warbler.

tion (*Amer. Birds* 45:116, Nagel pers. obs.). During 1991, Magnolia Warblers were present from 17 May through 12 September, and the highest single-day count was 11 singing males on 17 June (Nagel and F. J. Alsop III pers. obs.). Territorial conflicts between neighboring males were observed on several occasions, but no females or confirmed evidence of breeding were seen. In 1992, observations were continued by R. Mayfield (pers. comm.), and Magnolia Warblers were again abundant throughout the regenerating spruce/shrub habitat. On 2 occasions pairs foraging together were observed, but again there was no evidence of confirmed breeding. In a breeding bird census plot within the Unaka Mountain red spruce forest, Mayfield (1993) found a density of 25 Magnolia males/100 ha concentrated in a corner of the plot where the spruce canopy was broken and mixed with small deciduous trees and shrubs (pers. comm.).

To date, the only reports of Magnolia Warblers in Tennessee during the breeding season are from Roan and Unaka Mountains. The Roan Mountain reports show that a few individuals sporadically occur in the mid-successional spruce/fir forest. The Unaka Mountain population, among the largest in the southern Appalachians, occupies a mid-successional, previously logged and burned forest of red spruce mixed with yellow birch, maples, fire cherry, mountain ash, rhododendrons, and open areas of rock, grass, and ferns. This habitat is very similar to that occupied by Magnolia Warblers in West Virginia (Hall 1983) and Haw Orchard Mountain, southwestern Virginia (Simpson 1976a), suggesting a preference for red spruce. Alternatively, both spruce and fir habitats may be acceptable as long as they consist of small trees with openings of shrubs and blackberries (Lee 1985, Stevens 1976). If this is true, the recent death of mature fir in the Great Smoky Mountains should provide suitable habitat, although summering Magnolia Warblers have yet to be reported there.

The Magnolia Warbler usually nests in small conifers up to 3 m above ground (Bent 1953). The well-hidden, relatively flimsy nest, built by the female, is composed of a variety of plant stems, twigs, and grasses and lined with small, black rootlets. Normal clutch size is 3–5 eggs, incubated by the female for 11–13 days. Both adults feed the nestlings, which fledge in about 10 days.—*Jerry W. Nagel*.

Henslow's Sparrow
Ammodramus henslowii

The Henslow's Sparrow is a very rare spring and fall migrant and winter visitor in Tennessee (Robinson 1990). No confirmed breeding records exist; it is here considered a probable breeder on the basis of singing, apparently territorial males observed in 1957 and, after the atlas fieldwork, in 1994. The first summer observations were of a singing male on 12 and 18 June 1957 in an old field near the Clinch River in northeast Roane County (Howell 1958). The more recent observations were of at least 3 singing males in June and July 1994 near Cheatham Dam, Cheatham County (*Nat. Audubon Soc. Field Notes* 48:953).

The Henslow's Sparrow nests in moist, overgrown fields and meadows with scattered low shrubs or saplings (Graber 1968). The presence of standing dead vegetation from previous seasons is an important habitat component (Zimmerman 1988). The sparrow's main breeding range extends from the north-central and northeastern states south to central Kentucky and northern Virginia (Monroe, Stamm, and Palmer-Bell 1988, Kain 1987). Its population has declined in much of its range, including Virginia (Kain 1987, Robbins, Bystrak, and Geissler 1986), and it was formerly listed as a Category 2 Candidate species for federal listing as Endangered or Threatened (USFWS 1994, 1996). A population was recently discovered on the coastal plain of North Carolina (Lynch and LeGrand 1985). In Tennessee, seemingly suitable habitat is scattered across the state and probably most common on the northern Cumberland Plateau. Tennesseans should become familiar with its "se-lick" song and be aware of its possible occurrence.—*Charles P. Nicholson.*

Bobolink
Dolichonyx oryzivorus

The Bobolink is a fairly common spring migrant, uncommon fall migrant, and very rare summer resident in Tennessee (Robinson 1990). Spring migrants are present from late April to mid-May and fall migrants from late August to early October. The Bobolink usually occurs in lush grasslands, clover, wheat, and alfalfa fields.

The only confirmed nesting of the Bobolink in Tennessee was during 1962 in a meadow in Shady Valley, Johnson County, at 855 m elevation (Dubke 1963). At least 2 adult males and 1 adult female were present at the site, and a nest containing 3 young birds and an unhatched egg was found on 10 June. Three nestlings were again present on 16 June. On 17 June the nest was empty and fledglings were nearby. Only 1 adult male was observed at the site a week later. A pair had been present in Shady Valley the previous year. Bobolinks were again present during early June of 1963 and 1964, but no evidence of nesting was observed (Dubke 1963, *Migrant* 35:66).

Several other June and July observations of Bobolinks exist, primarily from northeast Tennessee. Excluding observations during the first few days of June, which could be of late migrants, possible breeding birds have been reported from Stewart (Robinson 1990), Davidson (*Migrant* 55:71), Jefferson (*Migrant* 51:94–95, 56:114), Hawkins (Phillips 1979a), and Carter Counties (Dubke 1963), and from the Great Smoky Mountains National Park (*Migrant* 47:103). The Hawkins County observations, on 15 July 1977, were of a female that scolded the observers at close range and of a group of 3 males and 4 "female" plumaged birds, possibly young of the year. A pair was reported at this site on 10 May 1978 (*Migrant* 49:70). The Jefferson County observations consist of a male on 14 June 1980 and a "pair on territory" on 2 June 1985. The field was mowed a few days after the 1985 observation and the birds not seen again.

The regular breeding range of the Bobolink extends south to central Illinois, central Ohio, and southern Pennsylvania (AOU 1983). It has nested in north-central Kentucky (Monroe, Stamm, and Palmer-Bell 1988), sporadically in southwest Virginia (Virginia Atlas Project, prelim. results), and in northwest North Carolina (Dubke 1963, Potter, Parnell, and Tuelings 1980).

Based on an average 12-day incubation period and 11-day nestling period (Orians 1985), eggs in the single Tennessee nest were probably laid during the third week of May. Courtship and nest construction would have occurred during the period when spring migrants are normally present. Male Bobolinks are frequently polygamous (Bent 1958). One adult female and 2 adult males were present at the Johnson County nest site (Dubke 1963). No aggression was noted between the 2 males; weak territorial behavior has been described elsewhere (Bent 1958).

The only summer report of a Bobolink during the Atlas period was of a bird on 10 June 1988 in Stewart County; it was considered a late migrant by its observer (Robinson 1990).—*Charles P. Nicholson.*

Western Meadowlark
Sturnella neglecta

The Western Meadowlark is a rare visitor to Tennessee, where it has nested at least once. Most of its occurrences have been between late October and May in Shelby and Lake Counties (Robinson 1990). It was first recorded in the state in 1943 and then was recorded annually from 1950 through 1969. Reports in recent years have been more sporadic. The Western Meadowlark is difficult to separate by plumage characteristics from the much more abundant Eastern Meadowlark. The songs of the 2 species, however, are usually diagnostic.

The only nest record of the Western Meadowlark was at the Shelby County Penal Farm in 1951 (Smith 1951). Two were observed the previous December in the meadow where the nest was later found and at least 3 were seen throughout the winter. A territorial pair remained into the spring. The female was observed carrying food to the nest area on 19 May, and, after intensive searching, the nest with 6 large young was found on 20 May. On 22 May the nest was empty and a fledgling was nearby. Several of the other observations of Western Meadowlarks were in late April and May, when meadowlarks are normally nesting. A male in Knox County during April and May 1959 was suspected of being paired with a female Eastern Meadowlark; no conclusive evidence of their nesting was obtained (Pardue 1959).

During the first half of the twentieth century, the Western Meadowlark expanded its breeding range eastward into Michigan, Indiana, and Ohio (Lanyon 1956). A similar expansion did not occur farther south into eastern Missouri, Arkansas, or Kentucky. Except for the lack of breeding records, the history of the species in Kentucky is very similar to Tennessee, with the first record in the 1940s and a peak in records during the 1950s and 1960s (Mengel 1965, Monroe, Stamm, and

Palmer-Bell 1988). The decreased number of reports in recent years is probably partly due to recent decreases in the breeding population in Michigan and adjacent states (Robbins, Bystrak, and Geissler 1986, Granlund 1991).—*Charles P. Nicholson.*

Pine Siskin
Carduelis pinus

The Pine Siskin is usually an uncommon to fairly common winter resident and a rare summer visitor in Tennessee (Robinson 1990). In the Appalachian Mountains south of Pennsylvania, it is apparently a very rare and erratic breeder (see McNair 1988b, Siebenheller and Siebenheller 1992, Simpson 1993). Most breeding records in the region have occurred following an intermediate or major irruptive flight during the preceding fall and winter (though see Simpson 1993). Pine Siskin winter irruptions into the south-central and southeastern United States have occurred regularly for at least the past 50–100 years and are among the most synchronous of all boreal seed-eating birds (Stupka 1963, Bock and Lepthien 1976b, and others). In Tennessee, siskins are most numerous in the east, especially in the Great Smoky Mountains, where they have been recorded every month of the year (Stupka 1963, Tanner 1985). Consequently, the scarcity of siskin breeding records cannot be attributed to the scarcity of siskins, which occur regularly at moderate and high elevations in the southern Appalachians until late May or early June and occasionally later.

The breeding status of the Pine Siskin is difficult to elucidate (see Yunick 1981). Pine Siskins are Cardueline finches, which are unusual among passerines in that they do not establish territories until after the nest site has been chosen (Newton 1972). Thus, nest building by the female or defense of the small nesting territory is the minimal acceptable evidence to show probable breeding. Courtship feeding, for example, evidence of probable breeding for most passerines, is only evidence of possible breeding for the Pine Siskin. Furthermore, juvenile siskins usually cannot be readily distinguished from immatures in the field, and even juvenile siskins may have dispersed an unknown distance away from their natal localities. Consequently, because only one active nest of the Pine Siskin in the southern Appalachian Mountains has been documented (Simpson 1993), other existing breeding evidence must be interpreted cautiously.

The preponderance of the available evidence suggests that Pine Siskins may breed in fir-spruce forest at high elevations in the southern Appalachian Mountains or, perhaps less frequently, in rural or suburban yards at lower elevations that have coniferous trees used as nest sites and that contain a mixture of coniferous and deciduous trees that provide natural foods in addition to seeds available at feeders (McNair 1988b, Siebenheller and Siebenheller 1992, Simpson 1993). All 5 breeding records in the Unaka Mountains from Tennessee or the Tennessee–North Carolina border occurred prior to the Atlas period at 2 localities, the Great Smoky Mountains National Park and Roan Mountain. Four of these 5 records are from high-elevation spruce-fir forest; the remaining record near Cosby in the Smokies was in unspecified habitat at 825 m. A juvenile female was collected here on 2 July 1937 (Wetmore 1939), but the significance of this record is difficult to interpret without further information. The lack of detailed information also plagues the interpretation of the 3 other breeding records where juvenile or immature siskins were seen, alone or fed by adults, from mid-June to early July. The remaining record, at Roan Mountain, consisted of a female gathering nest material on the anomalous dates of 15–16 July 1975; no nest was located (McNair 1988b). Although Pine Siskins made several irruptive flights during the Atlas period, there was no breeding evidence in the Unaka Mountains of Tennessee.

Significant, however, is an apparent probable breeding record of the Pine Siskin close to the Georgia border at Lookout Mountain, Hamilton County, Tennessee, at 580 m (Caldwell 1991), outside the Blue Ridge Mountains where all other breeding records have occurred. This breeding record followed the prevailing pattern in that it occurred after an irruptive flight of the preceding fall and winter, but it is also confounded by the usual difficulties in interpretation since material evidence is lacking. Despite a convincing description of a juvenile siskin being fed by an adult (possibly a female) at a feeder in late May 1990, no prior nesting activity was noted, though suitable breeding habitat existed nearby. Interestingly, the only late summer record of siskins in Tennessee outside the Unaka Mountains is from nearby Signal Mountain in 1985 (Robinson 1990).

Convincing documentation of confirmation of breeding for the Pine Siskin in Tennessee is still lacking and limited to the report of Simpson (1993) elsewhere in the southern Appalachian Mountains. Detection of breeding evidence is difficult for, among other reasons, siskins feed their young by regurgitation and forage away from the nest. Furthermore, the timing of breeding usually coincides with the presence of large numbers of non-breeding siskins, which may also display some behavior suggestive of breeding. A completely acceptable and definitive breeding record of the Pine Siskin in Tennessee still awaits discovery and convincing documentation.—*Douglas B. McNair.*

Appendix 1. Taxonomic List

The following list contains the vernacular and scientific names of species mentioned in the text. Mammal names are from the *Revised Checklist of North American Mammals North of Mexico, 1991* (Jones et al. 1992). Bird names are from the *Check-List of North American Birds* (AOU 1983) and later supplements. The primary sources for plant names are *Gray's Manual of Botany* (Fernald 1970) and the Soil Conservation Service's *National List of Scientific Plant Names* (SCS 1982).

Mammals

Beaver	*Castor canadensis*
Bison	*Bos bison*
Cat, domestic	*Felis catus*
Deer, white-tailed	*Odocoileus virginianus*
Raccoon, common	*Procyon lotor*

Birds

Anhinga	*Anhinga anhinga*
Bittern, American	*Botaurus lentiginosus*
Bittern, Least	*Ixobrychus exilis*
Blackbird, Red-winged	*Agelaius phoeniceus*
Bluebird, Eastern	*Sialia sialis*
Bobolink	*Dolichonyx oryzivorus*
Bobwhite, Northern	*Colinus virginianus*
Bunting, Indigo	*Passerina cyanea*
Bunting, Painted	*Passerina ciris*
Cardinal, Northern	*Cardinalis cardinalis*
Catbird, Gray	*Dumetella carolinensis*
Chat, Yellow-breasted	*Icteria virens*
Chickadee, Black-capped	*Poecile atricapillus*
Chickadee, Carolina	*Poecile carolinensis*
Chuck-will's-widow	*Caprimulgus carolinensis*
Chukar	*Alectoris chukar*
Coot, American	*Fulica americana*
Cormorant, Double-crested	*Phalacrocorax auritus*
Cowbird, Brown-headed	*Molothrus ater*
Creeper, Brown	*Certhia americana*
Crossbill, Red	*Loxia curvirostra*
Crow, American	*Corvus brachyrhynchos*
Crow, Fish	*Corvus ossifragus*
Cuckoo, Black-billed	*Coccyzus erythropthalmus*
Cuckoo, Yellow-billed	*Coccyzus americanus*
Dickcissel	*Spiza americana*
Dove, Mourning	*Zenaida macroura*
Dove, Rock	*Columba livia*
Duck, American Black	*Anas rubripes*
Duck, Wood	*Aix sponsa*
Eagle, Bald	*Haliaeetus leucocephalus*
Eagle, Golden	*Aquila chrysaetos*
Egret, Cattle	*Bubulcus ibis*
Egret, Great	*Ardea alba*
Egret, Snowy	*Egretta thula*
Falcon, Peregrine	*Falco peregrinus*
Finch, House	*Carpodacus mexicanus*
Flicker, Northern	*Colaptes auratus*
Flycatcher, Acadian	*Empidonax virescens*

Continued on next page

Flycatcher, Alder	*Empidonax alnorum*	Osprey	*Pandion haliaetus*
Flycatcher, Great Crested	*Myiarchus crinitus*	Ovenbird	*Seiurus aurocapillus*
Flycatcher, Least	*Empidonax minimus*	Owl, Barred	*Strix varia*
Flycatcher, Olive-sided	*Contopus cooperi*	Owl, Boreal	*Aegolius funereus*
Flycatcher, Scissor-tailed	*Tyrannus forficatus*	Owl, Barn	*Tyto alba*
Flycatcher, Willow	*Empidonax traillii*	Owl, Great Horned	*Bubo virginianus*
Gallinule, Purple	*Porphyrula martinica*	Owl, Hawk	*Surnia ulula*
Gnatcatcher, Blue-gray	*Polioptila caerulea*	Owl, Northern Saw-whet	*Aegolius acadicus*
Goldfinch, American	*Carduelis tristis*	Parakeet, Carolina	*Conuropsis carolinensis*
Goose, Canada	*Branta canadensis*	Parakeet, Monk	*Myiopsitta monachus*
Grebe, Pied-billed	*Podilymbus podiceps*	Parula, Northern	*Parula americana*
Grackle, Common	*Quiscalus quiscula*	Pelican, White	*Pelecanus erythrorhynchos*
Grosbeak, Blue	*Guiraca caerulea*	Pheasant, Ring-necked	*Phasianus colchicus*
Grosbeak, Pine	*Pinicola enucleator*	Phoebe, Eastern	*Sayornis phoebe*
Grosbeak, Rose-breasted	*Pheuticus ludovicianus*	Pigeon, Passenger	*Ectopistes migratorius*
Grouse, Ruffed	*Bonasa umbellus*	Prairie-Chicken, Greater	*Tympanuchus cupido*
Grouse, Sharp-tailed	*Tympanuchus phasianellus*	Quail, Japanese	*Coturnix japonica*
Hawk, Broad-winged	*Buteo platypterus*	Rail, Black	*Laterallus jamaicensis*
Hawk, Cooper's	*Accipiter cooperii*	Rail, King	*Rallus elegans*
Hawk, Red-shouldered	*Buteo lineatus*	Rail, Virginia	*Rallus limicola*
Hawk, Red-tailed	*Buteo jamaicensis*	Raven, Common	*Corvus corax*
Hawk, Sharp-shinned	*Accipiter striatus*	Redstart, American	*Setophaga ruticilla*
Heron, Great Blue	*Ardea herodias*	Robin, American	*Turdus migratorius*
Heron, Green-backed	*Butorides virescens*	Sandpiper, Spotted	*Actitis macularia*
Heron, Little Blue	*Egretta caerulea*	Sapsucker, Yellow-bellied	*Sphyrapicus varius*
Hummingbird, Ruby-throated	*Archilochus colubris*	Screech-Owl, Eastern	*Otus asio*
Jay, Blue	*Cyanocitta cristata*	Shrike, Loggerhead	*Lanius ludovicianus*
Jay, Gray	*Perisoreus canadensis*	Siskin, Pine	*Carduelis pinus*
Junco, Dark-eyed	*Junco hyemalis*	Sora	*Porzana carolina*
Junglefowl, Red	*Gallus gallus*	Sparrow, Bachman's	*Aimophila aestivalis*
Kestrel, American	*Falco sparverius*	Sparrow, Chipping	*Spizella passerina*
Killdeer	*Charadrius vociferus*	Sparrow, Field	*Spizella pusilla*
Kingbird, Eastern	*Tyrannus tyrannus*	Sparrow, Grasshopper	*Ammodramus savannarum*
Kingfisher, Belted	*Ceryle alcyon*	Sparrow, Henslow's	*Ammodramus henslowii*
Kinglet, Golden-crowned	*Regulus satrapa*	Sparrow, House	*Passer domesticus*
Kite, American Swallow-tailed	*Elanoides forficatus*	Sparrow, Lark	*Chondestes grammacus*
Kite, Black-shouldered	*Elanus leucurus*	Sparrow, Savannah	*Passerculus sandwichensis*
Kite, Mississippi	*Ictinia mississippiensis*	Sparrow, Song	*Melospiza melodia*
Kite, White-tailed	*Elanus leucurus*	Sparrow, Vesper	*Pooecetes gramineus*
Lark, Horned	*Eremophila alpestris*	Starling, European	*Sturnus vulgaris*
Mallard	*Anas platyrhynchos*	Stilt, Black-necked	*Himantopus mexicanus*
Martin, Purple	*Progne subis*	Swallow, Bank	*Riparia riparia*
Meadowlark, Eastern	*Sturnella magna*	Swallow, Barn	*Hirundo rustica*
Meadowlark, Western	*Sturnella neglecta*	Swallow, Cliff	*Petrochelidon pyrrhonota*
Merganser, Common	*Mergus merganser*	Swallow, Northern Rough-winged	*Stelgidopteryx serripennis*
Merganser, Hooded	*Lophodytes cucullatus*	Swallow, Tree	*Tachycineta bicolor*
Mockingbird, Northern	*Mimus polyglottos*	Swan, Trumpeter	*Cygnus buccinator*
Moorhen, Common	*Gallinula chloropus*	Swift, Chimney	*Chaetura pelagica*
Nighthawk, Common	*Chordeiles minor*	Tanager, Scarlet	*Piranga olivacea*
Night-Heron, Black-crowned	*Nycticorax nycticorax*	Tanager, Summer	*Piranga rubra*
Night-Heron, Yellow-crowned	*Nycticorax violaceus*	Teal, Blue-winged	*Anas discors*
Nuthatch, Brown-headed	*Sitta pusilla*	Tern, Least	*Sterna antillarum*
Nuthatch, Red-breasted	*Sitta canadensis*	Thrasher, Brown	*Toxostoma rufum*
Nuthatch, White-breasted	*Sitta carolinensis*	Thrush, Hermit	*Catharus guttatus*
Oriole, Baltimore	*Icterus galbula*	Thrush, Swainson's	*Catharus ustulatus*
Oriole, Orchard	*Icterus spurius*		

Continued on next page

Thrush, Wood	*Hylocichla mustelina*
Titmouse, Tufted	*Baeolophus bicolor*
Towhee, Eastern	*Pipilo erythrophthalmus*
Turkey, Wild	*Meleagris gallopavo*
Veery	*Catharus fuscescens*
Vireo, Bell's	*Vireo bellii*
Vireo, Red-eyed	*Vireo olivaceus*
Vireo, Solitary	*Vireo solitarius*
Vireo, Warbling	*Vireo gilvus*
Vireo, White-eyed	*Vireo griseus*
Vireo, Yellow-throated	*Vireo flavifrons*
Vulture, Black	*Coragyps atratus*
Vulture, Turkey	*Cathartes aura*
Warbler, Black-and-white	*Mniotilta varia*
Warbler, Black-throated Blue	*Dendroica caerulescens*
Warbler, Black-throated Green	*Dendroica virens*
Warbler, Blackburnian	*Dendroica fusca*
Warbler, Blue-winged	*Vermivora pinus*
Brewster's Warbler	*Vermivora pinus* x *chrysoptera*
Lawrence's Warbler	*Vermivora pinus* x *chrysoptera*
Warbler, Canada	*Wilsonia canadensis*
Warbler, Cerulean	*Dendroica cerulea*
Warbler, Chestnut-sided	*Dendroica pensylvanica*
Warbler, Golden-winged	*Vermivora chrysoptera*
Warbler, Hooded	*Wilsonia citrina*
Warbler, Kentucky	*Opornis formosus*
Warbler, Magnolia	*Dendroica magnolia*
Warbler, Nashville	*Vermivora ruficapilla*
Warbler, Pine	*Dendroica pinus*
Warbler, Prairie	*Dendroica discolor*
Warbler, Prothonotary	*Protonotaria citrea*
Warbler, Swainson's	*Limnothlypis swainsonii*
Warbler, Tennessee	*Vermivora peregrina*
Warbler, Worm-eating	*Helmitheros vermivorus*
Warbler, Yellow	*Dendroica petechia*
Warbler, Yellow-throated	*Dendroica dominica*
Waterthrush, Louisiana	*Seiurus motacilla*
Waxwing, Cedar	*Bombycilla cedrorum*
Whip-poor-will	*Caprimulgus vociferus*
Whistling-Duck, Black-bellied	*Dendrocygna autumnalis*
Wood-Pewee, Eastern	*Contopus virens*
Woodcock, American	*Scolopax minor*
Woodpecker, Downy	*Picoides pubescens*
Woodpecker, Hairy	*Picoides villosus*
Woodpecker, Ivory-billed	*Campephilus principalis*
Woodpecker, Pileated	*Dryocopus pileatus*
Woodpecker, Red-bellied	*Melanerpes carolinus*
Woodpecker, Red-cockaded	*Picoides borealis*
Woodpecker, Red-headed	*Melanerpes erythrocephalus*
Wren, Bewick's	*Thryomanes bewickii*
Wren, Carolina	*Thryothorus ludovicianus*
Wren, House	*Troglodytes aedon*
Wren, Winter	*Troglodytes troglodytes*
Yellowthroat, Common	*Geothlypis trichas*

Reptiles

Snake, rat	*Elaphe obsoleta*

Insects

Adelgid, balsam woolly	*Adeleges piceae*
Blow fly, bird	*Protocalliphora deceptor*
Honeybee	*Apis mellifer*
Mantis, praying	family Mantidae
Moth, gypsy	*Lymantria dispar*

Plants

Alder, green	*Alnus crispa* (Dryand. in Ait.) Pursh
Alfalfa	*Medicago sativa* L.
Arborvitae	*Thuja occidentalis* L.
Aspen	*Populus* spp.
Basswood	*Tilia heterophylla* Ventenat
Beautyberry	*Callicarpa americana* L.
Beech, American	*Fagus grandifolia* Ehrh.
Birch, yellow	*Betula alleghaniensis* Britton
Birch, black	*B. lenta* L.
Blackberry	*Rubus* spp.
Blackgum	*Nyssa sylvatica* Marshall
Blueberry	*Vaccinium* spp.
Bluestem, big	*Andropogon gerardii* Vitman
Bluestem, little	*Schizachyrium scoparium* (Michx.) Nash
Boxelder	*Acer negundo* L.
Boxwood	*Buxus sempervirens* L.
Broomsedge	*Andropogon virginicus* L.
Buckeye, yellow	*Aesculus flava* Sonald. ex Hope
Buckeye, Ohio	*A. glabra* Willd.
Buckthorn, Carolina	*Rhamnus caroliniana* Walter
Buffalo-nut	*Pyrularia pubera* Michx.
Bulrush, great	*Scirpus validus* Vahl
Buttonbush	*Cephalanthus occidentalis* L.
Cane	*Arundinaria gigantea* (Walter) Walter ex Muhl.
Cardinal-flower	*Lobelia cardinalis* L.
Cattail	*Typha latifolia* L.
Cedar, red	*Juniperus virginiana* L.
Cherry, black	*Prunus serotina* Ehrh.
Cherry, fire	*P. pensylvanica* L.F.
Chestnut, American	*Castanea dentata* (Marshall) Borkh.
Clover, sweet	*Melilotus* spp.
Clover, red	*Trifolium pratense* L.
Cohosh, blue	*Caulophyllum thalictroides* (L.) Michx.
Coralberry	*Symphoricarpos orbiculatus* Moench
Corn	*Zea mays* L.
Cotton	*Gossypium hirsutum* L.
Crabapple	*Pyrus* spp.
Creeper, Virginia	*Parthenocissus quinquefolia* (L.) Planch.
Cutgrass, giant	*Zizaniopsis miliacea* (Michx.) Doell & Aschers.
Cypress, bald	*Taxodium distichum* (L.) L.C.Rich.
Dock	*Rumex* spp.
Dogwood	*Cornus florida* L.

Continued on next page

Dog-hobble	*Leucothoe fontanesiana* (Steud.) Sleum.	Oak, Nuttall	*Q. nuttallii* E.J.Palmer
Elder, common	*Sambucus canadensis* L.	Oak, overcup	*Q. lyrata* Walter
Elm, American	*Ulmus americana* L.	Oak, pin	*Q. palustris* Muenchh.
Elm, winged	*U. alata* Michx.	Oak, post	*Q. stellata* Wangenh.
Farkleberry	*Vaccinium arboreum* Marshall	Oak, scarlet	*Q. coccinea* Muenchh.
Fern, Christmas	*Polystichum acrostichoides* (Michx.) Schott	Oak, southern red	*Q. falcata* Michx.
Fir, Fraser	*Abies fraseri* (Dougl. ex D.Don) Lindl.	Oak, white	*Q. alba* L.
		Oak, willow	*Q. phellos* L.
Flag, blue	*Iris versicolor* L.	Orange, osage	*Maclura pomifera* (Raf.) C.K.Schneid.
Flag, sweet	*Acorus calamus* L.	Pine, jack	*P. banksiana* Lamb.
Forsythia	*Forsythia suspensa* (Thunb.) Vahl	Pine, loblolly	*P. taeda* L.
Goldenrod	*Solidago* spp.	Pine, pitch	*P. rigida* Mill.
Grape	*Vitis* spp.	Pine, shortleaf	*P. echinata* Mill.
Grass, Indian	*Sorghastrum nutans* (L.) Nash	Pine, Table Mountain	*P. pungens* Lamb.
Grass, Johnson	*Sorghum halapense* (L.) Pers.	Virginia	*Pinus virginiana* Mill.
Grass, orchard	*Dactylis glomerata* L.	Pine, white	*P. strobus* L.
Hackberry	*Celtis occidentalis* L.	Plum, wild	*Prunus* spp.
Hawthorn	*Crataegus* spp.	Pondweed	*Potamogeton epihydrus* Raf.
Hemlock	*Tsuga canadensis* (L.) Carriere *T. caroliniana* Engelm.	Privet	*Ligustrum vulgare* L.
		Rhododendron	*Rhododendron catawbiense* Michx. *R. maximum* L.
Hickory, shagbark	*Carya ovata* (Mill.) K.Koch	Rose, multiflora	*Rosa multiflora* Thunb.
Hickory, water	*C. aquatica* (Michx.F.) Nutt.	Rush, common	*Juncus effusus* L.
Holly, deciduous	*Ilex decidua* Walter	Sericea	*Lespedeza cuneata* (Dumont) G.Don.
Honeysuckle, Japanese	*Lonicera japonica* Thunb.	Silverbell	*Halesia carolina* L.
Huckleberry	*Gaylussacia* spp.	Soybean	*Glycine max* (L.) Merrill
Hydrangea, wild	*Hydrangea arborescens* L.	Spruce, red	*Picea rubens* Sarg.
Ivy, poison	*Toxicodendron radicans* (L.) Kuntze	Squash	*Cucurbita pepo*
Jewelweed	*Impatiens capensis* Meerb. *I. pallida* Nutt.	St. John's-wort	*Hypericum* spp.
		Sugarberry	*Celtis laevigata* Willd.
Knotweed	*Polygonum densiflorum* Meisn.	Sumac	*Rhus* spp.
Laurel, mountain	*Kalmia latifolia* L.	Sweetgum	*Liquidambar styraciflua* L.
Maple, mountain	*Acer spicatum* Lam.	Switchgrass	*Panicum virgatum* L.
Maple, red	*A. rubrum* L.	Sycamore, American	*Platanus occidentalis* L.
Maple, sugar	*A. saccharum* Marshall	Trumpet-creeper	*Campsis radicans* (L.) Seem.
Mountain-ash	*Sorbus americana* Marshall	Tulip-poplar	*Liriodendron tulipifera* L.
Oak, black	*Quercus velutina* Lam.	Tupelo, water	*Nyssa aquatica* L.
Oak, blackjack	*Q. marilandica* Muenchh.	Vetch	*Vicia* spp.
Oak, cherrybark	*Q. falcata* var. *paegodaefolia* Elliott.	Walnut, black	*Juglans nigra* L.
Oak, chestnut	*Q. prinus* L.	Watermilfoil, Eurasian	*Myriophyllum spicatum* L.
Oak, chinquapin	*Q. muhlenbergii* Engelm.	Water-chinquapin	*Nelumbo lutea* (Willd.) Pers.
Oak, northern red	*Q. rubra* L.	Willow	*Salix* spp.

Appendix 2.
Summary of Breeding Chronology

The following information was compiled from all available Tennessee sources, including egg collection records, published accounts, Cornell nest record cards, Atlas observations, and unpublished field notes. The first column lists all species nesting in Tennessee for which a dated breeding observation is available. The second column is the range of dates of complete clutches, either fresh or actively incubated. The third column, the range of dates of nests containing young, is not used for precocial species whose young leave the nest within hours of hatching. The fourth column lists the range of dates of precocial and altricial young that have left the nest and are still dependent on their parents. Multiple ranges of dates are given for a few species that nest throughout much of the year, and in a few instances to show large gaps in the reported ranges of dates. All dates are based on actual observations; none are calculated from incubation, nestling, or fledgling periods. A dash (—) indicates missing data. Dates are given as month/day except when only the month was available; these are noted by enclosing the name of the month in parentheses (e.g. (May)).

Species	Nests with Eggs	Nests with Young	Fledglings
Pied-billed Grebe	4/12–5/20		4/20–9/15
Double-crested Cormorant	4/3–5/29	5/13–5/29	7/15
Anhinga	4/28–5/29	—	—
American Bittern	5/14	5/22–5/30	—
Least Bittern	5/1–7/22	6/18–8/7	—
Great Blue Heron	3/19–5/28	4/27–8/5	—
Great Egret	4/28–5/28	5/13–6/30	—
Snowy Egret	5/17–7/7	7/6	8/11
Little Blue Heron	5/17–7/20	5/17–7/6	6/30
Cattle Egret	6/10–7/7	6/14–9/21	—
Green-backed Heron	4/17–6/26	5/12–7/28	6/11–7/31
Black-crowned Night-Heron	4/20–6/19	5/15–7/20	6/17–8/22
Yellow-crowned Night-Heron	4/3–5/30	5/15–7/11	6/7–7/21
Black-bellied Whistling-Duck	—		10/7–10/8
Canada Goose	4/9–6/3		4/15–8/1
Wood Duck	3/9–6/6		4/15–8/8
American Black Duck	3/1–5/14		5/27
Mallard	3/10–6/1		4/23–8/17

Continued on next page

Appendix 2 Table—*Continued*

Species	Nests with Eggs	Nests with Young	Fledglings
Blue-winged Teal	5/2		5/1–6/30
Hooded Merganser	5/18		4/5–7/13
Common Merganser	4/21–5/13		—
Black Vulture	2/19–7/7	4/3–10/4	6/17–8/24
Turkey Vulture	4/3–5/28	5/13–7/30	7/3–8/2
Osprey	(March–May)	6/2–8/1	
Mississippi Kite	—	6/9–8/26	7/11–9/3
Bald Eagle	—	2/22–6/10	5/26–8/16
Sharp-shinned Hawk	5/2–6/27	6/7–7/28	6/27–8/1
Cooper's Hawk	4/12–5/30	5/27–8/1	6/21–8/1
Red-shouldered Hawk	3/17–4/14	4/14–6/7	5/22–7/6
Broad-winged Hawk	4/18–6/9	5/26–7/8	6/21–7/29
Red-tailed Hawk	2/26–4/28	3/25–7/12	5/25–8/30
American Kestrel	3/21–6/7	5/3–8/15	5/6–8/26
Peregrine Falcon	3/25–5/14	3/29–6/13	4/30–6/17
Ruffed Grouse	4/1–6/1		5/15–8/7
Wild Turkey	4/15–6/4		5/30–8/13
Northern Bobwhite	4/12–9/18		5/1–10/30
Black Rail	—		6/23–6/26
King Rail	3/26–7/4		5/21–7/1
Virginia Rail	4/20–5/8		6/3
Sora	—		4/17–6/8
Purple Gallinule	5/26–6/23		—
Common Moorhen	5/26–6/17		5/24–8/7
American Coot	5/9–6/16		4/27–6/26
Killdeer	2/26–7/31		3/21–8/4
Black-necked Stilt	5/5–7/25		6/1–8/31
Spotted Sandpiper	6/10–6/19		6/4–7/6
American Woodcock	3/3–5/20		3/25–6/4
Least Tern	6/8–8/5	6/22–7/19	7/14
Rock Dove	1/14–6/28	3/24–7/11	6/4–7/10
Mourning Dove	2/13–9/19,	2/16–10/7	2/21–10/13, 12/28–29
Black-billed Cuckoo	4/30–6/19	5/30–6/15	6/15
Yellow-billed Cuckoo	4/30–9/7	5/20–9/15	6/4–9/20
Barn-Owl	1/17–2/18, 3/31–7/2, 12/17	2/11–7/25, 9/11–1/17	3/4–11/20
Eastern Screech-Owl	3/9–5/18	4/13–6/24	5/11–7/22
Great Horned Owl	1/2–3/20	2/9–5/24	3/11–7/30
Barred Owl	3/2–3/30	3/30–7/5	4/30–7/20
Northern Saw-whet Owl	7/1–7/23	8/2–8/29	5/8–5/11, 8/27
Common Nighthawk	5/3–7/14	5/27–7/22	6/23–7/28
Chuck-will's-widow	5/17–6/22	5/28–7/11	6/27–7/11
Whip-poor-will	4/26–7/12	5/10–7/16	6/8–7/12
Chimney Swift	5/17–7/1	6/2–9/1	7/5–8/14
Ruby-throated Hummingbird	5/7–8/2	5/26–8/13	6/14–8/13
Belted Kingfisher	4/7–6/13	5/16–7/1	5/14–7/14
Red-headed Woodpecker	5/15–7/4	5/10–8/24	5/25–9/1
Red-bellied Woodpecker	4/19–7/12	4/30–7/30	5/15–8/8
Yellow-bellied Sapsucker	—	6/10–6/16	6/16–7/16
Downy Woodpecker	4/10–6/7	5/3–7/5	5/17–7/31
Hairy Woodpecker	4/7–5/22	4/26–6/2	5/15–7/18
Red-cockaded Woodpecker	5/20	5/3–6/7	6/1–8/8
Northern Flicker	4/13–6/22	5/2–7/25	5/27–8/7
Pileated Woodpecker	4/4–5/14	4/15–6/23	5/25–9/1
Olive-sided Flycatcher	6/30	7/5	7/18
Eastern Wood-Pewee	5/10–7/17	6/2–7/31	6/4–8/27
Acadian Flycatcher	5/20–7/8	6/3–7/22	6/18–8/16
Alder Flycatcher	—	—	6/24–7/17

Continued on next page

Appendix 2 Table—*Continued*

Species	Nests with Eggs	Nests with Young	Fledglings
Willow Flycatcher	6/2–7/20	6/28–8/3	6/19–8/4
Least Flycatcher	—	6/18–7/14	6/27–7/19, 9/2
Eastern Phoebe	3/6–7/6	4/9–7/24	5/7–8/7
Great Crested Flycatcher	5/2–7/13	5/17–7/15	5/27–7/31
Eastern Kingbird	5/5–7/9	5/26–8/5	5/27–8/22
Scissor-tailed Flycatcher	6/21–6/25	(July)	
Horned Lark	3/7–6/30	4/2–7/9	4/6–7/17
Purple Martin	5/16–6/13	5/11–7/14	6/3–8/26
Tree Swallow	4/30–6/27	5/14–7/20	6/1–7/20
Northern Rough-winged Swallow	4/15–6/19	5/10–7/12	5/16–7/20
Bank Swallow	6/18	5/20–7/8	6/25–7/10
Cliff Swallow	6/5–5/25	5/16–7/15	5/25–7/3
Barn Swallow	5/4–7/30	5/11–8/19	5/25–8/20
Blue Jay	3/30–6/30	4/18–8/7	5/1–8/24
American Crow	3/18–5/16	4/14–6/13	5/5–7/31
Fish Crow	—	6/25	—
Common Raven	3/9	4/5–5/16	5/5–5/20
Black-capped Chickadee	5/30	—	6/27–7/9
Carolina Chickadee	3/16–6/2	3/31–7/1	4/18–7/15
Tufted Titmouse	4/2–6/22	4/21–7/9	5/7–7/25
Red-breasted Nuthatch	6/14–6/23	6/15–6/26	6/7–7/10
White-breasted Nuthatch	3/22–4/30	4/9–5/14	5/5–7/31
Brown-headed Nuthatch	4/17–5/10	5/18	5/25–7/12
Brown Creeper	5/24–6/29	5/14–5/20	5/15–7/31
Carolina Wren	3/21–8/29	4/5–9/4	4/29–9/4
Bewick's Wren	2/28–8/5	4/24–8/10	4/25–8/12
House Wren	5/3–7/4	5/18–8/2	6/2–7/28
Winter Wren	—	6/21	6/14–7/8
Golden-crowned Kinglet	—	6/17–7/7	6/10–7/23
Blue-gray Gnatcatcher	4/6–7/5	5/18–8/5	5/19–8/5
Eastern Bluebird	3/2–8/25	3/21–9/15	3/22–10/15
Veery	5/28–6/21	—	6/9–7/11
Wood Thrush	5/1–7/23	5/23–8/8	5/31–8/9
American Robin	2/26–6/29	4/1–8/16	5/2–8/31
Gray Catbird	4/29–7/29	5/26–8/10	5/28–9/1
Northern Mockingbird	3/17–8/2	4/3–8/26	4/13–8/29
Brown Thrasher	3/23–7/4	4/24–7/31	5/7–8/8
Cedar Waxwing	6/6–8/23	6/3–8/21	6/13–9/10
Loggerhead Shrike	3/30–6/11	4/24–7/11	5/5–8/17
European Starling	3/28–7/10	4/17–7/6	5/5–7/31
White-eyed Vireo	4/24–7/19	5/8–7/10	5/28–8/7
Bell's Vireo	6/30	—	—
Solitary Vireo	5/9–7/17	5/31–7/4	5/28–7/8
Yellow-throated Vireo	4/13–7/24	5/23–6/24	6/1–6/29
Warbling Vireo	5/5–5/29	—	6/5–7/9
Red-eyed Vireo	5/8–7/10	6/7–7/11	6/8–9/1
Blue-winged Warbler	4/7–6/19	5/15–6/14	5/27–6/24
Golden-winged Warbler	5/28–6/9	6/2–6/26	6/5–6/29
Northern Parula	5/5–6/27	6/4–6/24	6/5–7/31
Yellow Warbler	4/28–5/31	5/19–6/30	5/26–7/6
Chestnut-sided Warbler	5/27–6/28	5/29–7/12	6/20–7/7
Black-throated Blue Warbler	5/26–6/26	6/5–6/30	5/31–7/20
Black-throated Green Warbler	5/13–5/17	5/21–6/19	5/22–7/22
Blackburnian Warbler	—	6/27	6/27–7/15
Yellow-throated Warbler	4/5–6/2	6/28	5/24–7/24
Pine Warbler	4/6–6/2	4/23–6/12	5/2–8/19
Prairie Warbler	4/30–7/2	6/5–7/17	5/30–7/19
Cerulean Warbler	—	5/28	6/9–7/28
Black-and-white Warbler	4/23–6/16	5/2–6/11	5/1–7/30
American Redstart	5/12–6/14	5/31–6/6	6/6–7/26

Continued on next page

Appendix 2 Table—*Continued*

Species	Nests with Eggs	Nests with Young	Fledglings
Prothonotary Warbler	4/18–7/13	5/6–7/17	5/16–7/30
Worm-eating Warbler	5/4–6/11	5/26–6/23	5/24–7/12
Swainson's Warbler	5/21–5/30	6/2	7/3–7/5
Ovenbird	5/6–6/21	5/20–7/11	5/26–7/30
Louisiana Waterthrush	4/5–6/7	4/30–6/24	5/12–7/15
Kentucky Warbler	4/25–6/21	5/19–6/26	5/29–7/12
Common Yellowthroat	5/5–6/27	5/27–8/5	6/22–8/5
Hooded Warbler	4/30–6/28	5/25–7/27	5/30–8/7
Canada Warbler	—	—	6/17–7/16
Yellow-breasted Chat	4/22–8/4	5/25–8/12	5/28–7/22
Summer Tanager	5/11–7/5	5/28–8/19	6/8–8/11
Scarlet Tanager	5/10–7/26	5/21–5/31	6/10–7/30
Northern Cardinal	3/28–9/12	4/11–10/1	4/20–10/6
Rose-breasted Grosbeak	5/14–6/2	6/9–6/15	6/9
Blue Grosbeak	5/28–7/10	6/7–8/18	6/5–8/29
Indigo Bunting	5/15–8/28	5/28–8/26	6/4–9/3
Painted Bunting	—	7/17–7/19	7/9–7/25
Dickcissel	5/16–7/29	5/28–6/25	6/22–7/17
Eastern Towhee	3/26–8/30	4/20–9/3	4/29–9/24
Bachman's Sparrow	5/2–7/20	7/1–7/24	6/11–7/27
Chipping Sparrow	4/13–7/19	5/5–8/15	5/15–8/25
Field Sparrow	4/14–8/12	4/30–9/1	5/17–9/1
Vesper Sparrow	5/19–6/5	—	6/5–6/24
Lark Sparrow	5/6–7/2	5/17–7/24	5/24–7/5
Savannah Sparrow	7/2–7/4	7/7–7/10	7/10–7/21
Grasshopper Sparrow	5/3–7/19	5/20–	6/5–8/24
Song Sparrow	4/10–7/23	4/24–8/25	5/4–8/31
Dark-eyed Junco	4/20–8/23	5/19–8/29	6/6–8/14
Bobolink	—	6/10	6/17
Red-winged Blackbird	4/21–7/4	5/9–7/27	5/18–8/7
Eastern Meadowlark	4/13–8/20	5/5–8/31	5/17–8/17
Western Meadowlark	—	5/20	5/22
Common Grackle	4/9–5/29	5/1–7/19	5/7–8/7
Brown-headed Cowbird	4/16–7/18	5/10–6/30	5/12–8/16
Orchard Oriole	5/7–7/9	5/18–7/26	7/3–8/3
Baltimore Oriole	5/10–6/3	5/19–6/30	6/3–7/4
House Finch	3/17–7/9	4/1–7/30	4/14–8/17
Red Crossbill	—	—	3/29–6/14
Pine Siskin	—	—	5/27–7/4
American Goldfinch	7/1–9/19	7/25–10/5	7/28–10/26
House Sparrow	3/25–7/4	4/15–8/1	4/29–8/7

Appendix 3. Brown-headed Cowbird Hosts

The following table lists all known hosts of the Brown-headed Cowbird in Tennessee as well as the type of evidence observed—cowbird egg in the host nest, cowbird nestling in the host nest, or a fledgling cowbird being fed by the host. Additional details are given in the accounts of the individual host species. Information in the table was compiled from all available sources, including egg collection records, published accounts, Cornell nest record cards, Atlas observations, and unpublished field notes.

Host Species	Eggs	Nestlings	Fledglings
Eastern Wood-Pewee			x
Acadian Flycatcher	x	x	x
Eastern Phoebe	x	x	
Carolina Chickadee	x		
Carolina Wren	x		
Bewick's Wren	x	x	x
Blue-gray Gnatcatcher	x		x
Eastern Bluebird	x		
Wood Thrush	x	x	x
Gray Catbird	x		
Brown Thrasher	x		
White-eyed Vireo	x	x	x
Solitary Vireo	x		
Yellow-throated Vireo			x
Warbling Vireo	x		x
Red-eyed Vireo	x	x	x
Blue-winged Warbler			x
Yellow Warbler	x	x	x
Black-throated Green Warbler		x	
Yellow-thr. Warbler			x
Prairie Warbler	x	x	x
Cerulean Warbler			x
Black-and-white Warbler	x		x
American Redstart			x
Prothonotary Warbler	x		
Worm-eating Warbler			x
Ovenbird	x	x	
Louisiana Waterthrush	x	x	x
Kentucky Warbler		x	x
Common Yellowthroat	x		x
Hooded Warbler	x	x	x
Yellow-breasted Chat	x	x	x
Summer Tanager	x	x	x
Scarlet Tanager	x		x
Northern Cardinal	x	x	x
Blue Grosbeak	x		
Indigo Bunting	x	x	x
Eastern Towhee	x	x	x
Chipping Sparrow	x		
Field Sparrow	x	x	x
Song Sparrow	x	x	x
Red-winged Blackbird	x		
Orchard Oriole	x	x	x
House Finch	x		

Literature Cited

Abernathy, B. H. 1955. Jewels in the jewelweeds. Migrant 26:44.

Able, K. P. 1967. Some recent observations from western Kentucky. Kentucky Warbler 43:27–33.

Adams, D. A. 1959. Breeding bird census 9—Fraser's fir forest. Aud. Field Notes 13:464.

Adkisson, C. S. 1990. Accipiters. Pp. 63–69 In Proc. Southeast raptor management symposium and workshop. Natl. Wildl. Fed., Washington, D.C.

———. 1991. Bewick's Wren. Pp. 518–20 In K. Terwilliger, Coordinator. Virginia's endangered species. McDonald and Woodward, Blacksburg, Va.

Adkisson, C. S., and R. N. Conner. 1978. Interspecific vocal imitation in White-eyed Vireos. Auk 95:602–6.

Aldrich, J. W. 1967. Historical background, taxonomy, distribution and present status. Pp. 3–44 In O. H. Hewitt, ed. The Wild Turkey and its management. The Wildlife Society, Wash., D.C.

Aldrich, J. W., and P. Goodrum. 1946. Census 26—Virgin hardwood forest. Audubon Mag. Sec. 2, Audubon Field Notes, pp. 144–45.

Allen, J. A. 1896. Recent literature—Rhoads's list of Tennessee birds. Auk 13:244–45.

Allen, R. W., and M. M. Nice. 1952. A study of the breeding biology of the Purple Martin *(Progne subis)*. Amer. Midl. Nat. 47:606–65.

Allen, T. T. 1961. Notes on the breeding behavior of the Anhinga. Wilson Bull. 73:115–25.

Allison, A. 1907. Notes on the spring birds of Tishomingo County, Mississippi. Auk 24:12–25.

Allred, C. E., and F. M. Fitzgerald. 1939. Forest production areas, Tennessee and United States. Univ. Tenn. Agricul. Exper. Stat. Rural Res. Series Monogr. 93, Knoxville.

Allred, C. E., S. W. Atkins, and F. M. Fitzgerald. 1939. Development of timber industry in Tennessee and United States. Univ. Tenn. Agricul. Exper. Stat. Rural Res. Series Monogr. 92, Knoxville.

Alsop, F. J., III. 1969. Census 21—Virgin spruce-fir forest. Audubon Field Notes 23:716.

———. 1970. King Rails in Knox County. Migrant 41:64–65.

———. 1971a. Traill's Flycatcher nesting in Knox County. Migrant 42:25–26.

———. 1971b. The 1971 foray. Migrant 42:73–81.

———. 1972. Eggshell thickness from Red-winged Blackbird *(Agelaius phoeniceus)* populations with different exposures to DDT. Ph.D. Diss., Univ. of Tennessee, Knoxville.

———. 1974. June records of a Brewster's Warbler and Red Crossbills in Cumberland County, Tennessee. Migrant 45:69–70.

———. 1976. The 1973 foray: Benton County. Migrant 47:81–86.

———. 1978. Savannah Sparrow *(Passerculus sandwichensis)* nest in upper East Tennessee. Migrant 49:1–4.

———. 1979a. Mantids selected as prey by Blue Grosbeaks. Wilson Bull. 91:131–32.

———. 1979b. Population status and management considerations for Tennessee's 13 threatened and endangered bird species. Tenn. Wildl. Res. Agency, Nashville.

———. 1981. The Cliff Swallow *(Petrochelidon pyrrhonota)* in Tennessee. Migrant 52:1–11.

———. 1990. First record of Scissor-tailed Flycatcher in the Great Smoky Mountains National Park. Migrant 61:66.

———. 1991. Birds of the Smokies. Great Smoky Mtns. Nat. Hist. Assoc., Gatlinburg, Tenn.

Alsop, F. J., III, and M. Williams. 1974. The 1972 foray: Lawrence County. Migrant 45:1–9.

Alsop, F. J., III, and T. F. Laughlin. 1991. Changes in the spruce-fir avifauna of Mt. Guyot, Tennessee, 1967–1985. J. Tenn. Acad. Sci. 66:207–9.

American Ornithologists' Union. 1957. Check-list of North American birds, 5th ed. Lord Baltimore Press, Baltimore.

———. 1983. Check-list of North American birds, 6th ed. Allen Press, Lawrence, Kans.

———. 1995. Fortieth supplement to the American Ornithologists' Union Check-list of North American birds. Auk 112:819–30.

Anderson, S. H. 1976. Comparative food habits of Oregon nuthatches. Northwest Sci. 50:213–21.

Anderson, S. H., and H. H. Shugart Jr. 1974. Habitat selection of breeding birds in an East Tennessee forest. Ecology 55:828–37.

Anderson, W. L., and R. D. Duzan. 1978. DDE residues and eggshell thinning in Loggerhead Shrikes. Wilson Bull. 90:215–20.

Andrle, R. F., and J. R. Carroll, eds. 1988. The atlas of breeding birds in New York state. Cornell Univ. Press, Ithaca, N.Y.

Anon. 1933. Late summer nestings. Migrant 4:36–37.

———. 1934. Crow reduction contest. Migrant 5:47.

Apfelbaum, S. I., and P. Seelbach. 1983. Nest tree, habitat selection and productivity of seven North American raptor species based on the Cornell University Nest Record Card Program. Raptor Research 17:97–113.

Arbib, R. 1976. The Blue List for 1977. Am. Birds 30:1031–39.

Armstrong, E. A. 1955. The Wren. Collins, London.

Armstrong, J. T. 1965. Breeding home range in the Nighthawk and other birds; its evolutionary and ecological significance. Ecology 46:619–29.

Artmann, J. W. 1977. Woodcock status report, 1975. U.S. Fish Wildl. Serv. Special Sci. Rep.—Wildlife No. 201.

Askins, R. A., J. F. Lynch, and R. Greenberg. 1990. Population declines in migratory birds in eastern North America. Current Ornithol. 7:1–57.

Audubon, J. J. 1831. Ornithological biography, vol. 1. Edinburgh.

———. 1839. Ornithological biography, vol. 5. Edinburgh.

———. 1929. Journal of John James Audubon made during his trip to New Orleans in 1820–1821. Boston Historical Society, Cambridge, Mass.

Baepler, D. H. 1968. Lark Sparrow. Pp. 886–902 *In* A. C. Bent (O. L. Austin Jr., ed.). Life histories of North American cardinals, grosbeaks, buntings, towhees, finches, sparrows and allies, part 2. U.S. Natl. Mus. Bull. 237.

Bailey, R. W., and K. T. Rinell. 1967. Events in the Turkey year. Pp. 73–91 *In* O. H. Hewitt, ed. The Wild Turkey and its management. The Wildlife Society, Wash., D.C.

Baird, J. 1968. Eastern Savannah Sparrow. Pp. 678–96 *In* A. C. Bent (O. L. Austin Jr., ed.). Life histories of North American cardinals, grosbeaks, buntings, towhees, finches, sparrows and allies, part 2. U.S. Natl. Mus. Bull. 237.

Bamberg, J. 1933. A Woodcock nest at Knoxville. Migrant 4:22–23.

Barclay, F. H. 1957. The natural vegetation of Johnson County, Tennessee, past and present. Ph.D. Diss., Univ. Tennessee, Knoxville.

Barlow, J. C. 1962. Natural history of the Bell's Vireo. Univ. Kansas Publ. Mus. Nat. Hist. 12:241–96.

Barlow, J. C., and J. C. Rice. 1977. Aspects of the comparative behavior of the Red-eyed and Philadelphia Vireos. Can. J. Zool. 55:528–42.

Barrows, W. B. 1889. The English Sparrow in North America. U.S. Dept. Agric. Div. of Economic Ornithology and Mammalogy. Bull. 1.

Bart, J., and J. D. Schoultz. 1984. Reliability of singing bird surveys: changes in observer efficiency with avian density. Auk 101:307–18.

Bart, J., and J. P. Klosiewski. 1989. Use of presence-absence to measure changes in avian density. J. Wildl. Manage. 53:847–52.

Bartsch, P. 1922. Some tern notes. Auk 34:101.

Basham, B. 1969. Brown-headed Nuthatch. Migrant 40:11.

Baskin, J. M., and C. C. Baskin. 1981. The Big Barrens of Kentucky not a part of Transeau's prairie peninsula. Pp. 43–48 *In* R. L. Studkey and K. J. Reese, eds. The prairie peninsula—in the "shadow" of Transeau. Proc. 6th North Amer. Prairie Conf., Ohio Biol. Surv. Note 15.

Bedard, J., and G. LaPointe. 1985. Influence of parental age and season on Savannah Sparrow reproductive success. Condor 87:106–10.

Beddal, B. G. 1963. Range expansion of the Cardinal and other birds in the northeastern states. Wilson Bull. 75:140–56.

Beddow, T. E. 1990. Recovery of the East Tennessee Osprey population. Migrant 61:92–94.

Bednarz, J. C., and J. J. Dinsmore. 1981. Status, habitat use, and management of Red-shouldered Hawks in Iowa. J. Wildl. Manage. 45:236–41.

———. 1982. Nest sites and habitat for Red-shouldered Hawks and Red-tailed Hawks in Iowa. Wilson Bull. 94:31–45.

Bednarz, J. C., D. Klem Jr., L. J. Goodrich, and S. E. Senner. 1990. Migration counts of raptors at Hawk Mountain, Pennsylvania, as indicators of population trends, 1934–1986. Auk 107:96–109.

Belles-Isles, J-C., and J. Picman. 1986. House Wren nest-destroying behavior. Condor 88:190–93.

Bellrose, F. 1938. Notes on birds of the Great Smoky Mountains National Park. Migrant 9:1–4.

Bellrose, F. C. 1976. Ducks, geese and swans of North America. Stackpole Books, Harrisburg, Penn.

———. 1990. The history of Wood Duck management. Pp. 13–20 *In* L. H. Fredrickson et al., eds. Proc. 1988 Wood Duck Symp., St. Louis, Mo.

Belthoff, J. R., and G. Ritchison. 1990. Nest-site selection by Eastern Screech-Owls in central Kentucky. Condor 92:982–90.

Benkman, C. W. 1987. Food profitability and the foraging ecology of crossbills. Ecol. Monogr. 57:251–67.

———. 1990. Intake rates and timing of crossbill reproduction. Auk 107:376–86.

Bent, A. C. 1926. Life histories of North American marsh birds. U.S. Natl. Mus. Bull. 135.

———. 1927. Life histories of North American shore birds. U.S. Natl. Mus. Bull. 142.

———. 1937. Life histories of North American birds of prey, part 1. U.S. Natl. Mus. Bull. 167.

———. 1938. Life histories of North American birds of prey, part 2. U.S. Natl. Mus. Bull. 170.

———. 1939. Life histories of North American woodpeckers. U.S. Natl. Mus. Bull. 174.

———. 1940. Life histories of North American cuckoos, goatsuckers, hummingbirds and their allies. U.S. Natl. Mus. Bull. 176.

———. 1942. Life histories of North American flycatchers, larks, swallows, and their allies. U.S. Natl. Mus. Bull. 179.

———. 1946. Life histories of North American jays, crows and titmice. U.S. Natl. Mus. Bull. 191.

———. 1948. Life histories of North American nuthatches, wrens, thrashers, and their allies. U.S. Natl. Mus. Bull. 195.

———. 1950. Life histories of North American wagtails, shrikes, vireos, and their allies. U.S. Natl. Mus. Bull. 197.

———. 1953. Life histories of North American wood warblers. U.S. Natl. Mus. Bull. 203.

———. 1958. Life histories of North American blackbirds, orioles, tanagers, and allies. U.S. Natl. Mus. Bull. 211.

———. (O. L. Austin Jr., ed.). 1968. Life histories of North American cardinals, grosbeaks, buntings, towhees, finches, sparrows, and allies, part 1. U.S. Natl. Mus. Bull. 237.

Berger, A. J. 1968a. Eastern Vesper Sparrow. pp. 868–82 *In* A. C. Bent (O. L. Austin Jr., ed.). Life histories of North American cardinals, grosbeaks, buntings, towhees, finches, sparrows and allies, part 2. U.S. Natl. Mus. Bull. 237.

———. 1968b. Clutch size, incubation period, and nestling period of the American Goldfinch. Auk 85:494–98.

Berger, D. D., C. R. Sindelar, and K. E. Gamble. 1969. The status of breeding peregrines in the eastern United States. Pp. 165–73 *In* J. J. Hickey, ed. Peregrine Falcon populations: their biology and decline. Univ. Wisconsin Press, Madison.

Bertin, R. I. 1982. The Ruby-throated Hummingbird and its major food plants: ranges, flowering phenology, and migration. Can. J. Zoology 60:210–19.

Best, L. B. 1977. Territory quality and mating success in the Field Sparrow *(Spizella pusilla)*. Condor 79:192–204.

———. 1978. Field Sparrow reproductive success and nesting ecology. Auk 95:9–22.

———. 1986. Conservation tillage: ecological traps for nesting birds. Wildl. Soc. Bull. 14:308–17.

Bierly, M. L. 1978. Brown Creeper nests in Nashville. Migrant 49: 86–87.

———. 1980. Bird finding in Tennessee. Privately published, Nashville, Tenn.

Bird, D. M. 1988. American Kestrel, *Falco sparverius*. Pp. 253–290 *In* R. S. Palmer, ed. Handbook of North American birds, Vol. 5. Yale Univ. Press, New Haven, Conn.

Birdsey, R. A. 1983. Tennessee forest resources. USDA For. Serv. Res. Bull. SO-90.

Blancher, P. J., and R. J. Robertson. 1985. A comparison of Eastern Kingbird breeding biology in lakeshore and upland habitats. Can. J. Zool. 63:2305–12.

Blockstein, D. E., and H. B. Tordoff. 1985. A contemporary look at the extinction of the Passenger Pigeon. Am. Birds 39:845–52.

Bock, C. E., and L. W. Lepthien. 1976a. Changing winter distribution and abundance of the Blue Jay, 1962–1971. Amer. Midl. Nat. 96:232–36.

———. 1976b. Synchronous eruptions of boreal seed-eating birds. Amer. Nat. 110:559–71.

Böhning-Gaese, K., M. L. Taper, and J. H. Brown. 1993. Are declines in North American insectivorous songbirds due to causes on the breeding range? Consv. Biology 7:76–86.

Bonney, T. E., Jr. 1988. Hermit Thrush. Pp. 324–25 *In* R. F. Andrle and J. R. Carroll, eds. The atlas of breeding birds in New York state. Cornell Univ Press, Ithaca, N.Y.

Borror, D. J. 1987. Song in the White-eyed Vireo. Wilson Bull. 99:377–97.

Brackbill, H. 1958. Nesting behavior of the Wood Thrush. Wilson Bull. 70:70–89.

Bradley, R. A. 1980. Vocal and territorial behavior of the White-eyed Vireo. Wilson Bull. 92:302–11.

Braun, E. L. 1950. Deciduous forests of eastern North America. Blakiston Co., Philadelphia.

Breckenridge, W. J. 1956. Measurements of the habitat niche of the Least Flycatcher. Wilson Bull. 68:47–51.

Brewer, R. 1961. Comparative notes on the life history of the Carolina Chickadee. Wilson Bull. 73:348–373.

Brewster, W. 1886. An ornithological reconnaissance in western North Carolina. Auk 3:94–112, 173–79.

Briskie, J. V., and S. G. Sealy. 1987. Polygyny and double-brooding in the Least Flycatcher. Wilson Bull. 99:492–94.

Broley, C. C. 1947. Migration and nesting of Florida Bald Eagles. Wilson Bull. 59:1–20.

Brooks, M. 1938a. Bachman's Sparrow in the north-central portion of its range. Wilson Bull. 50:86–109.

———. 1938b. The eastern Lark Sparrow in the upper Ohio Valley. Cardinal 4:181–200.

———. 1940. The breeding warblers of the central Allegheny Mountain region. Wilson Bull. 52:249–66.

———. 1947. Breeding habitats of certain wood warblers in the unglaciated Appalachian region. Auk 64:291–95.

Brown, C. R., and M. B. Brown. 1986. Ectoparasitism as a cost of coloniality in Cliff Swallows *(Hirundo pyrrhonota)*. Ecology 67:1206–18.

Brown, D. M. 1941. The vegetation of Roan Mountain: a phytosociological and successional study. Ecol. Monogr. 11:64–97.

Brown, R. E., and J. G. Dickson. 1994. Swainson's Warbler *(Limnothlypis swainsonii)*. *In* A. Poole and F. Gill, eds. The birds of North America, No. 126. Acad. Nat. Sci. Philadelphia and Amer. Ornithol. Union., Washington, D.C.

Bruner, S. C., and A. L. Field. 1912. Notes on the birds observed on a trip through the mountains of western North Carolina. Auk 29:368–77.

Brunton, D. H. 1988. Sequential polyandry by a female Killdeer. Wilson Bull. 100:670–72.

Bryson, R. A., and K. F. Hare. 1974. The climates of North America. Pp. 1–47 *In* Bryson and Hare, eds. Climates of North America. Elsevier Scientific Publ. Co., New York.

Bucher, E. H. 1992. The causes of extinction of the Passenger Pigeon. Current Ornithology 9:1–36.

Bump, G., R. W. Darrow, F. C. Edminster, and W. F. Crissey. 1947. The Ruffed Grouse. New York State Cons. Dept., Albany.

Bunn, D. S., A. B. Warburton, and R. D. S. Wilson. 1982. The Barn Owl. Buteo Books, Vermillion, S.Dak.

Burch, M. S. 1982. Nesting ecology of Mourning Doves in Knox and Loudon Counties, Tennessee. M.S. Thesis, Univ. Tennessee, Knoxville.

Bureau of Census. 1925. Census of agriculture, Part II, the southern states. U.S. Dept. Commerce, Bureau of Census, Washington, D.C.

———. 1945. Census of agriculture, Vol. 1, Geographic area series, Part 20—Tennessee. U.S. Dept. Commerce, Bureau of Census, Washington, D.C.

———. 1980. 1978 Census of agriculture, Vol. 1, Geographic area series, Part 42—Tennessee. U.S. Dept. Commerce, Bureau of Census, Washington, D.C.

———. 1989. 1987 Census of agriculture, Vol. 1, Geographic area series, Part 42—Tennessee. U.S. Dept. Commerce, Bureau of Census, Washington, D.C.

Burger, J. 1978. Competition between Cattle Egrets and native North American herons, egrets, and ibises. Condor 80:15–23.

Burleigh, T. D. 1927. Further notes on the breeding birds of northeastern Georgia. Auk 44:229–34.

———. 1935a. Two new birds from the Southern Appalachians. Proc. Biol. Soc. Wash. 48:61–62.

———. 1935b. The present status of the Olive-sided Flycatcher as a breeding bird in western North Carolina and eastern Tennessee. Wilson Bull. 47:165.

———. 1941. Bird life on Mt. Mitchell. Auk 58:334–45.

———. 1958. Georgia birds. Univ. Oklahoma Press, Norman.

Busbee, E. L. 1977. The effects of dieldrin on the behavior of young Loggerhead Shrikes. Auk 94:28–35.

Butler, R. W. 1988. Population dynamics and migration routes of Tree Swallows, *Tachycineta bicolor*, in North America. J. Field Ornithol. 59:395–402.

Butler, T. 1948. Notes on hawks in West Tennessee. Migrant 19:23.

Bystrak, D. 1980. Application of miniroutes to bird population studies. Maryland Birdlife 36:131–38.

Cade, T. J. 1982. The falcons of the world. Cornell Univ. Press, Ithaca, N.Y.

Cade, T. J., J. H. Enderson, C. G. Thelander, and C. M. White, eds. 1988. Peregrine Falcon populations: their management and recovery. Peregrine Fund, Boise, Idaho.

Caldwell, H. R. 1952. Carolina Wren's nest with twelve eggs. Migrant 23:30.

Caldwell, T. 1991. Possible breeding of Pine Siskins at Lookout Mountain, Tennessee. Migrant 62:3–4.

Calhoun, J. B. 1941. Notes on the summer birds of Hardeman and McNairy Counties. J. Tenn. Acad. Sci. 16:293–309.

Campbell, J. M., and J. C. Howell. 1970. Observations of certain birds. Migrant 41:73–75.

Cannings, R. J. 1987. The breeding biology of Northern Saw-whet Owls in southern British Columbia. Pp. 193–98 *In* R. Nero et al., eds. Biology and conservation of northern forest owls: symposium proceedings. USDA For. Serv. Gen. Tech. Rep. RM-142.

Carey, M., and V. Nolan Jr. 1979. Population dynamics of Indigo Buntings and the evolution of avian polygyny. Evolution 33:1180–92.

Case, N. A., and O. H. Hewitt. 1963. Nesting and productivity of the Red-winged Blackbird in relation to habitat. Living Bird 2:7–20.

Catesby, M. 1771. The natural history of Carolina, Florida, and the Bahama islands. London, printed for Benjamin White, at Horace's Head.

Chamberlain-Auger, J. A., P. Auger, and E. Strauss. 1990. Breeding biology of American Crows. Wilson Bull. 102:615–22.

Chapman, J. 1985. Tellico archaeology. Dept. Anthropology, Rep. Investigations No. 43, Univ. Tennessee, Knoxville, and Publ. Anthropology No. 41, Tennessee Valley Authority, Knoxville.

Chapman, J., P. A. Delcourt, P. A. Cridlebaugh, A. B. Shea, and H. R. Delcourt. 1982. Man–land interaction: 10,000 years of American Indian impact on native ecosystems in the lower Little Tennessee River valley, eastern Tennessee. Southeastern Archaeology 1:115–21.

Chester, E. W., and W. H. Ellis. 1989. Plant communities of northwestern Middle Tennessee. J. Tenn. Acad. Sci. 64:75–78.

Christensen, N. L. 1988. Vegetation of the southeastern Coastal Plain. Pp. 317–63 *In* M. G. Barbour and W. D. Billings, eds. North American terrestrial vegetation. Cambridge Univ. Press, N.Y.

Clark, K. L., and R. J. Robertson. 1981. Cowbird parasitism and evolution of anti-parasite strategies in the Yellow Warbler. Wilson Bull. 93:249–58.

Clebsch, A. 1939. The season–Clarksville area. Migrant 10:32.

———. 1942. Swainson's Warbler nesting notes. Migrant 13:45–46.

Clebsch, E. E. C. 1989. Vegetation of the Appalachian mountains of Tennessee east of the Great Valley. J. Tenn. Acad. Sci. 64:79–83.

Coffey, B. B., Jr. 1933. Notes on the Painted Bunting at Memphis. Migrant 4:41–42.

———. 1935. Bell's Vireo at Memphis. Migrant 6:67–68.

———. 1940. Mississippi Kite nesting in a city park. Migrant 11:79.

———. 1941a. Swainson's Warbler in the Memphis area. Migrant 12:30–31.

———. 1941b. Summer range of mid-south Towhees. Migrant 12:51–57.

———. 1942b. Fish Crow at Memphis. Migrant 13:42.

———. 1942c. Chattanooga notes. Migrant 13:48.

———. 1943a. Post-juvenal migration of herons. Bird Banding 14:34–39.

———. 1943b. Phoebe tunnel. Migrant 14:70–72.

———. 1944. Notes on the breeding birds of Natchez Trace State Park. Migrant 15:25–27.

———. 1946. Bell's Vireo and other Natchez Trace notes. Migrant 17:46–47.

———. 1952. Southwestern Tennessee heronries. Migrant 23:45.

———. 1955. Notes on the Blue Grosbeak in the mid-south. Migrant 26:41–42.

———. 1966. Warbler notes from Natchez Trace S. P. and Forest. Migrant 37:14.

———. 1976. The 1975 foray: Lauderdale County. Migrant 47:1–7.

———. 1979. Early records of the Mississippi Kite and a summary. Migrant 50:83–84.

———. 1981. A past Green Heron colony in Memphis. Migrant 52:44.

———. 1985. First state records and nesting of Black-necked Stilts at Memphis, Tennessee. Migrant 56:1–3.

———, ed. 1942a. The wrens of Tennessee. Migrant 13:1–13.

Coffey, B., and L. Coffey. 1980. A West Tennessee foray—June 1979. Migrant 51:12–14.

Coffey, J. W. 1963. A nesting study of the Eastern Phoebe. Migrant 34:41–49.

Coffey, L. C. 1948. Nesting data from Memphis. Migrant 19:11–12.

Coffey, Mrs. B. B., Jr. 1964. Cattle Egret nesting at the Dyersburg heronry. Migrant 35:54.

Coffey, W. 1966. Colonial nesting of the Green Heron. Migrant 37:75.

COHMAP Members. 1988. Climatic changes of the last 18,000 years: observations and model simulations. Science 241:1043–52.

Cole, J. C., and R. W. Dimmick. 1991. Distribution of Ruffed Grouse southeast of the range of quaking aspen. Proc. Annu. Conf. S.E. Assoc. Fish and Wildl. Agencies 45:58–63.

Coleman, J. S., and J. D. Fraser. 1987. Food habits of Black and Turkey Vultures in Pennsylvania and Maryland. J. Wildl. Manage. 51:733–39.

———. 1990. Black and Turkey Vultures. Pp. 78–88 *In* Proc. Southeast raptor management symposium and workshop. Natl. Wildl. Fed., Washington, D.C.

Collins, S. L. 1983. Geographic variation in habitat structure of the Black-throated Green Warbler *(Dendroica virens)*. Auk 100:382–89.

Colvin, B. A. 1985. Common Barn-Owl population decline in Ohio and the relationship to agricultural trends. J. Field. Ornithol. 56:224–35.

Confer, J. L. 1992. Golden-winged Warbler *(Vermivora chrysoptera)*. *In* A. Poole, P. Stettenheim, and F. Gill, eds. The birds of North America, No. 20. Acad. Nat. Sci., Philadelphia, and Amer. Ornithol. Union, Washington, D.C.

Confer, J. L., and K. Knapp. 1981. Golden-winged Warblers and Blue-winged Warblers: the relative success of a habitat specialist and a generalist. Auk 98:108–14.

Conner, R. N., and C. S. Adkisson. 1977. Principal component analysis of woodpecker nesting habitat. Wilson Bull. 89:122–29.

Conner, R. N., R. G. Hooper, H. S. Crawford, and J. S. Mosby. 1975. Woodpecker nesting habitat in cut and uncut woodlands in Virginia. J. Wildl. Manage. 39:144–50.

Conrad, K. F., and R. J. Robertson. 1993. Patterns of parental provisioning by Eastern Phoebes. Condor 95:57–62.

Cook, R. E. 1969. Variation in species density in North American birds. Syst. Zool. 18:63–84.

Cooper, R. J. 1981. Relative abundance of Georgia caprimulgids based on call-counts. Wilson Bull. 93:363–71.

Cope, E. D. 1870. Observations on the fauna of the Southern Alleghanies. Amer. Natur. 4:392–402.

Copeyon, C. K. 1990. A technique for constructing cavities for the Red-cockaded Woodpecker. Wildl. Soc. Bull. 18:303–11.

Corgan, J. X. 1977. Tennessee's early technical and scientific journals. J. Tenn. Acad. Sci. 52:23–26.

Corlew, R. E. 1981. Tennessee, a short history. 2nd ed. Univ. Tenn. Press, Knoxville.

Cornwell, G. W. 1963. Observations of the breeding biology and behavior of a nesting population of Belted Kingfishers. Condor 65:426–31.

Cottrell, S. D., H. H. Prince, and P. I. Padding. 1990. Nest success, duckling survival, and brood habitat selection of Wood Ducks in a Tennessee riverine system. Pp. 191–97 *In* L. H. Fredrickson et al., eds. Proc. 1988 Wood Duck Symp., St. Louis, Mo.

Coues, E. 1897. Characters of *Dendroica caerulescens cairnsi*. Auk 14:96–97.

Cowardin, L. M., V. Carter, F. C. Golet, and E. T. LaRoe. 1979. Classification of wetlands and deepwater habitats of the United States. U.S. Fish Wildl. Serv., Biol. Services Program FWS/OBS-79/31.

Craig, R. J. 1984. Comparative foraging ecology of Louisiana and Northern Waterthrushes. Wilson Bull. 96:173–83.

Criswell, W. G. 1979. Brown Creeper nesting in west Tennessee. Migrant 50:81–82.

Cromer, J. R. 1978. Analysis of eleven years of banding data from a free-flying, resident flock of Canada Geese *(Branta canadensis)* on Old Hickory Lake, Tennessee. M.S. Thesis, Tenn. Tech. Univ., Cookeville.

Crook, C. 1933. Nesting of Blue-winged and Black and White Warblers. Migrant 4:10.

———. 1934. Cowbird notes from Tennessee. Auk 51:384–85.

———. 1935a. Nesting of the Lark Sparrow in central Tennessee. Auk 52:194–95.

———. 1935b. The status of the Black-billed Cuckoo (*Coccyzus erythropthalmus* (Wilson)) in Tennessee. J. Tenn. Acad. Sci. 13:109–19.

Crossner, K. A. 1977. Natural selection and clutch size in the European Starling. Ecology 58:885–92.

Cuthbert, N. I. 1962. The Michigan Audubon Society phoebe study (part II). Jack-Pine Warbler 40:68–83.

Cypert, E. 1949. Three rookeries on Kentucky Lake. Migrant 20:41–42.

———. 1955. Some interesting bird observations on Kentucky Lake. Migrant 26:9–11.

Dahl, T. E. 1990. Wetland losses in the United States 1780's to 1980's. U.S. Fish Wildl. Serv., Washington, D.C.

Daugherty, A. B. 1989. U.S. grazing lands: 1950–1982. U.S. Dept. Agriculture Economic Research Service Stat. Bull. 771.

Davant, M. 1965. A brief history of the T.O.S. chapters. Migrant 36:30–36.

Davis, C. M. 1978. A nesting study of the Brown Creeper. Living Bird 17:237–63.

Davis, D. E. 1959. Observations on territorial behavior of Least Flycatchers. Wilson Bull. 71:73–85.

Davis, W. J. 1982. Territory size in *Megaceryle alcyon* along a stream habitat. Auk 99:353–62.

De Smet, K., and R. D. James. 1987. Golden Eagle. P. 519 *In* M. D. Cadnam, P. F. J. Eagles, and F. M. Helleiner. Atlas of the breeding birds of Ontario. Univ. Waterloo Press, Waterloo, Ontario.

Deaderick, W. H. 1940. Audubon in Tennessee. Migrant 11:59–61.

Defebaugh, J. E. 1906. History of the lumber industry of America, Vol. 1. The American Lumberman, Chicago.

Delcourt, P. A., and H. R. Delcourt. 1981. Vegetation maps for eastern North America: 40,000 YR B.P. to the present. Pp. 123–65 *In* R. C. Romans (ed.), Geobotany, vol II. Plenum Press, New York.

DeSelm, H. R. 1988. The barrens of the Western Highland Rim of Tennessee. Pp. 199–219 *In* L. Snyder, ed. Proc. 1st Ann. Symp. on the Natural History of Lower Tennessee and Cumberland River Valleys. Center for Field Biology of Land Between the Lakes, Austin Peay State Univ., Clarksville, Tenn.

———. 1989. The barrens of Tennessee. J. Tenn. Acad. Sci. 64:89–95.

Devereux, J. G., and J. A. Mosher. 1984. Breeding ecology of Barred Owls in the central Appalachians. Raptor Res. 18:49–58.

DeVore, J. E. 1968a. A nesting study of the King Rail and Least Bittern. Migrant 39:53–58.

———. 1968b. December nesting of the Carolina Wren. Migrant 39:62.

———. 1975. Middle Tennessee ornithological records of the late H. O. Todd, Jr. Migrant 46:25–37.

Dexter, R. W. 1961. Further studies on the nesting of the Common Nighthawk. Bird-banding 32:79–85.

———. 1969. Banding and nesting studies of the Chimney Swift, 1944–1968. Ohio J. Sci. 69:193–213.

———. 1981. Nesting success of Chimney Swifts related to age and the number of adults at the nest, and the subsequent fate of the visitors. J. Field Ornithol. 52:228–32.

Diaz, H. F., and R. G. Quayle. 1980. The climate of the United States since 1895: Spatial and temporal changes. Monthly Weather Review 108:249–66.

Dickerson, K. 1977. American Bittern nest found at Goose Pond. Migrant 48:43.

Dickinson, J. C., Jr. 1968. Rufous-sided Towhee. Pp. 562–79 In A. C. Bent (O. L. Austin Jr., ed.). Life histories of North American cardinals, grosbeaks, buntings, towhees, finches, sparrows, and allies, part 1. U.S. Natl. Mus. Bull 237.

Dickson, R. R. 1975. Climate of the states—Tennessee. Pp. 910–28 In Ruffner, J. A. Climates of the States, Vol. 2. Gale Research Co., Detroit.

Dilger, W. C. 1956. Adaptive modifications and ecological isolating mechanisms in the thrush genera *Catharus* and *Hylocichla*. Wilson Bull. 68:171–99.

Dimmick, R. W. 1971. The influence of controlled burning on nesting patterns of bobwhites in west Tennessee. Proc. Annu. Conf. S.E. Assoc. Game and Fish Comm. 25:149–55.

———. 1974. Populations and reproductive effort among bobwhites in western Tennessee. Proc. Annu. Conf. S.E. Assoc. Game and Fish Comm. 28:594–602.

———. 1992. Northern Bobwhite *(Colinus virginianus)*: Section 4.1.3. U.S. Army Corps of Engineers Wildlife Resources Management Manual, Tech. Rept. EL-92-18, USAE Waterways Exp. Sta. Vicksburg, Miss.

Dimmick, R. W., W. W. Dimmick, and C. Watson. 1980. Red-cockaded Woodpeckers in the Great Smoky Mountains National Park: their status and habitat. Research/Resources Management Report 38, USDI Natl. Park Serv., Southeast Regional Office, Atlanta, Ga.

Dingle, E. von S. 1942. Rough-winged Swallow. Pp. 424–33 In A. C. Bent, ed. Life histories of North American flycatchers, larks, swallows, and their allies. U.S. Natl. Mus. Bull. 179.

Dinkelspiel, H. 1973. Monk Parakeet in Shelby County. Migrant 44:82.

Dixon, C. L. 1978. Breeding biology of the Savannah Sparrow on Kent Island. Auk 95:235–46.

Dobyns, H. F. 1983. Their number become thinned: Native American population dynamics in eastern North America. Univ. Tenn. Press, Knoxville.

Dolbeer, R. A. 1991. Migration patterns of Double-crested Cormorants east of the Rocky Mountains. J. Field Ornithol. 62:83–93.

Douglass, L. E., M. L. Bierly, and K. A. Goodpasture. 1965. Green Herons nest at Basin Spring. Migrant 36:76–80.

Driver, H. E. 1970. The Indians of North America, 2nd ed. Univ. Chicago Press, Chicago.

Droege, S. 1990. The North American Breeding Bird Survey. Pp. 1–4 In J. R. Sauer and S. Droege, eds. Survey designs and statistical methods for the estimation of population trends. U.S. Fish Wildl. Serv., Biol. Rep. 90(1).

Drury, W. H. 1958. Chickadees and nest boxes. Mass Aud. Soc. Bull., Nov., p.1.

Dubke, K. H. 1963. First nesting record of Bobolink in Tennessee. Migrant 34:17–19.

———. 1974. Purple Gallinule nesting at Goose Pond, Grundy County. Migrant 45:94–95.

———. 1982. Fish Crows over Savannah Bay, Hamilton County, Tennessee. Migrant 53:12.

Dubke, K., and L. H. Dubke. 1975. Apparent double nesting of the American Kestrel. Migrant 46:16.

———. 1977. The 1974 foray: Grundy County. Migrant 42:81–85.

Due, L. A., and C. E. Ruhr. 1957. The Coturnix Quail in Tennessee. Migrant 28:48–53.

Duley, L. J. 1979. Life history aspects of the Screech Owl *(Otus asio)* in Tennessee. M.S. Thesis, Univ. of Tennessee, Knoxville.

Dull, C. W., J. D. Ward, H. D. Brown, G. W. Ryan, W. H. Clerke, and R. J. Uhler. 1988. Evaluation of spruce and fir mortality in the southern Appalachian Mountains. USDA For. Serv. South. Reg. Prot. Rep. R8-PR.

Dunning, J. B., Jr., and B. D. Watts. 1990. Regional differences in habitat occupancy by Bachman's Sparrow. Auk 107:463–72.

Duyck, B. E. 1981. Range expansion of nesting Tree Swallows. Chat 45:98–100.

Duyck, B. E., and D. B. McNair. 1990. Brown-headed Nuthatches nest again at Weaverville, Buncombe County, North Carolina. Chat 54:7–9.

Duyck, B. E., D. B. McNair, and C. P. Nicholson. 1991. Dirt-storing behavior by White-breasted Nuthatches. Wilson Bull. 103:308–9.

Eagar, C. 1984. Review of the biology and ecology of the balsam wooly aphid in the Southern Appalachian spruce-fir forests. Pp. 36–50 In P. S. White, ed. The Southern Appalachian spruce-fir ecosystem: its biology and threats. USDI Natl. Park Service Res. Resour. Manage. Rep. Ser-71.

Eastman, J. 1991. Black-billed Cuckoo. Pp. 232–33 In R. Brewer, G. A. McPeek, and R. J. Adams Jr. The atlas of breeding birds of Michigan. Michigan St. Univ. Press, East Lansing.

Eaton, S. W. 1958. A life history study of the Louisiana Waterthrush. Wilson Bull. 70:211–36.

Eckert, A. W. 1981. The wading birds of North America. Doubleday and Co., New York.

Eckhardt, R. C. 1976. Polygyny in the Western Wood-Pewee. Condor 78:561–62.

Eddleman, W. R., F. L. Knopf, B. Meanley, F. A. Reid, and R. Zembal. 1988. Conservation of North American rallids. Wilson Bull. 100:458–75.

Ehrlich, P. R., D. S. Dobkin, and D. Wheye. 1988. The birder's handbook: a field guide to the natural history of North American birds. Fireside, New York.

Eller, G., and G. Wallace. 1984. Birds of Roan Mountain and vicinity. Lee Herndon Chapter, Tenn. Ornithol. Soc., Elizabethton.

Elliot, J. J., and R. Arbib. 1953. Origin and status of the House Finch in the eastern United States. Auk 70:31–37.

Ellis, W. H., and E. W. Chester. 1989. Upland swamps of the Highland Rim of Tennessee. J. Tenn. Acad. Sci. 64:97–101.

Ellison, W. G. 1992. Blue-Gray Gnatcatcher *(Polioptila caerulea)*. In A. Poole, P. Sttenheim, and F. Gill, eds. The birds of North America, No. 23. Acad. Nat. Sci. Philadelphia and Amer. Ornithol. Union., Washington, D.C.

Emlen, J. T., Jr. 1954. Territory, nest building, and pair formation in the Cliff Swallow. Auk 71:16–35.

Erskine, A. J. 1971. Some new perspectives on the breeding ecology of Common Grackles. Wilson Bull. 83:352–70.

Erwin, W. G. 1935. Some nesting habits of the Brown Thrasher. J. Tenn. Acad. Sci. 10:179–204.

Eshbaugh, B. K., and W. H. Eshbaugh. 1979. Removal of fur from a live raccoon by Tufted Titmice. Wilson Bull. 91:328.

Faanes, C. A. 1980. Breeding biology of Eastern Phoebes in northern Wisconsin. Wilson Bull. 92:107–10.

Farrell, L. 1964. Traffic control. Migrant 35:117.

Fawver, B. J. 1950. An analysis of the ecological distribution of breeding bird populations in eastern North America. Ph.D. Diss., Univ. Illinois, Champaign.

Fenneman, N. M. 1938. Physiography of the eastern United States. McGraw-Hill, New York.

Ferguson, J., and S. Ferguson. 1991. Chimney Swifts choose hollow oak tree for nesting site. Migrant 64:97–98.

Ferguson, J. A. 1992. Ben B. Coffey, Jr., elected Fellow in the American Ornithologists' Union. Migrant 63:26–38.

Fernald, M. L. 1970. Gray's manual of botany, 8th edition. D. Van Nostrand, New York.

Ficken, M. S. 1963. Courtship of the American Redstart. Auk 80:307–317.

Ficken, M. S., and R. W. Ficken. 1962. Some aberrant characters of the Yellow-breasted Chat, *Icteria virens*. Auk 79:718–19.

Fitch, F. W. 1950. Life history and ecology of the Scissor-tailed Flycatcher, *Muscivora forficata*. Auk 67:145–68.

Fitch, H. S., and V. R. Fitch. 1955. Observations on the Summer Tanager in northeastern Kansas. Wilson Bull 67:45–54.

Fite, G. C. 1979. Southern agriculture since the Civil War: an overview. Agricul. History 53:3–21.

Fleming, J. H. 1907. Birds observed in Hawkins County, East Tennessee. Wilson Bull. 14:154–57.

Fleming, W. J., and D. R. Petit. 1986. Modified milk carton nest box for studies of Prothonotary Warblers. J. Field Ornithol. 57:313–15.

Fleming, W. J., B. P. Pullin, and D. M. Swineford. 1984. Population trends and environmental contaminants in herons in the Tennessee Valley, 1980–81. Colonial Waterbirds 7:63–73.

Floyd, J. K. 1990. Breeding ecology of Bald Eagles at Cross Creeks National Wildlife Refuge, Dover, Tennessee. M.S. Thesis, Murray State Univ., Murray, Ky.

Ford, R. P. 1987. Summary of recent Brown Creeper observations in West Tennessee. Migrant 58:50–51.

———. 1990. Habitat relationships of breeding birds and winter birds in forested wetlands of West Tennessee. M.S. Thesis, Univ. Tennessee, Knoxville.

———. 1992. New West Tennessee heronries established. Migrant 63:4–5.

Ford, B., and B. Cooper. 1993. Tennessee partners in flight: birds and biodiversity. Tenn. Wildlife 16(5) March/April:5–12.

Ford, R. P. and P. B. Hamel. 1988. The breeding birds of forested habitats of the Central Basin of Tennessee. Pp. 278–94 *In* D. A. Snyder, ed. Proceedings first annual sympoisum on the natural history of lower Tennessee and Cumberland river valleys. Center for Field Biology of Land Between the Lakes, Austin Peay State Univ., Clarksville, Tenn.

Fornari, H. D. 1979. The big change: Cotton to soybeans. Agricul. History 53:245–53.

Forshaw, J. M. 1977. Parrots of the world. T. F. H. Publications, Neptune, New Jersey.

Fowler, D. K., J. R. MacGregor, S. A. Evans, and L. E. Schaaf. 1985. The Common Raven returns to Kentucky. Am. Birds 39:852–53.

Fowler, L. J. 1985. Color phases of the Eastern Screech-Owl in Tennessee. Migrant 56:61–63.

Fowler, L. J., and D. K. Fowler. 1983a. BBC 71—Cedar forest I. Am. Birds 37:72–73.

———. 1983b. BBC 72—Cedar forest II. Am. Birds 37:73.

———. 1983c. BBC 73—Cedar forest III. Am. Birds 37:73.

———. 1984a. BBC 65—Upland hardwood forest I. Am. Birds 38:87.

———. 1984b. BBC 66—Upland hardwood forest II. Am. Birds 38:87.

———. 1984c. BBC 184—Abandoned agricultural lands I. Am. Birds 38:123–24.

———. 1984d. BBC 185—Abandoned agricultural lands II. Am. Birds 38:124.

———. 1985. Breeding birds and vegetation along the Duck River in Middle Tennessee. J. Tenn. Acad. Sci. 60:48–51.

Fowler, L. J., and R. W. Dimmick. 1983. Wildlife use of nest boxes in Eastern Tennessee. Wildl. Soc. Bull. 11:178–81.

Fox, W. H. 1882. Stray notes from Lookout Mountain. Nutt. Bull. 7:191–92.

———. 1886. List of birds found in Roane County, Tennessee, during April, 1884, and March and April, 1885. Auk 3:315–70.

———. 1887. *Vireo solitarius alticola* in Tennessee. Auk 4:164.

Fredrickson, L. H. 1971. Common Gallinule breeding biology and development. Auk 88:914–19.

Fredrickson, L. H., J. M. Anderson, F. M. Kozlik, and R. A. Ryder. 1977. American Coot (*Fulica americana*). Pp. 122–47 *In* G. C. Sanderson. Management of migratory shore and upland game birds in North America. Internat. Assoc. Fish Wildl. Agencies, Washington, D.C.

Freer, V. M. 1979. Factors affecting site tenacity in New York Bank Swallows. Bird-Banding 50:349–57.

Fretwell, S. D. 1986. Distribution and abundance of the Dickcissel. Pages 211–42 *In* R. F. Johnston, ed. Current Ornithology, Vol. 4. Plenum Press, New York.

Friedmann, H. 1929. The cowbirds: a study in the biology of social parasitism. Charles C. Thomas, Springfield, Ill.

———. 1963. Host relations of the parasitic cowbirds. U.S. Natl. Mus. Bull. 233.

———. 1971. Further information on the host relations of the parasitic cowbirds. Auk 88:239–55.

Friedmann, H., and L. F. Kiff. 1985. The parasitic cowbirds and their hosts. Proc. West. Foundation Vert. Zool. 2:225–304.

Friedmann, H., L. F. Kiff, and S. I. Rothstein. 1977. A further contribution to knowledge of the host relations of the parasitic cowbirds. Smithson. Contr. Zool. 235.

Galati, B., and C. B. Galati. 1985. Breeding of the Golden-crowned Kinglet in northern Minnesota. J. Field Ornith. 56:28–40.

Ganier, A. F. 1916a. Organization of the Tennessee Ornithological Association. Wilson Bull. 28:45.

———. 1916b. November bird life at Reelfoot Lake, Tennessee. Wilson Bull. 28:25–30.

———. 1917. Preliminary list of the birds of Tennessee. Tenn. Dept. Game and Fish, Nashville.

———. 1921. Cowbird lays in Prothonotary Warbler's nest. Wilson Bull. 28:146.

———. 1922. Breeding of the Barn Swallow in Tennessee. Wilson Bull. 34:184–85.

———. 1923a. Notes from the Tennessee Cumberlands. Wilson Bull. 35:26–34.

———. 1923b. Nesting of the Sharp-shinned Hawk. Wilson Bull. 35:41–43.

———. 1924. Starlings abundant at Nashville, Tenn. Wilson Bull. 36:31–32.

———. 1926. Summer birds of the Great Smoky Mountains. J. Tenn. Acad. Sci. 1:31–40.

———. 1928. European Starling nesting at Nashville, Tennessee. Wilson Bull. 40:198.

———. 1930. Breeding of the Least Tern on the Mississippi River. Wilson Bull. 42:103–7.

———. 1931a. A list of the birds of the Great Smoky Mountains National Park. Unpubl. manuscript, Univ. Tennessee Library Special Coll., Knoxville.

———. 1931b. Nesting of the Duck Hawk in Tennessee. Wilson Bull. 43:3–8.

———. 1931c. Facts about eagles in Tennessee. J. Tenn. Acad. Sci. 6:49–57.

———. 1931d. Nesting of the Prairie Horned Lark near Nashville. Migrant 2:31.

———. 1932a. Duck Hawks at a Reelfoot heronry. Migrant 3:28–29.

———. 1932b. Lark Sparrow at Nashville. Migrant 3:37.

———. 1933a. A distributional list of the birds of Tennessee. Tenn. Avifauna No. 1, Tenn. Ornithol. Soc., Nashville.

———. 1933b. Water birds of Reelfoot Lake. J. Tenn. Acad. Sci. 8:65–83.

———. 1933c. Two March days at Mullins Cove. Migrant 4:3–6.

———. 1933d. Duck Hawk, etc., on the plateau. Migrant 4:38–39.

———. 1934a. Incubation period of the Killdeer. Wilson Bull. 46:17–19.

———. 1934b. Swainson's Warbler in Tennessee. Migrant 5:11–12.

———. 1934c. The status of the Duck Hawk in the southeast. Auk 51:371–73.

———. 1935a. Spring migration at Athens, Tenn. Migrant 6:2–5.

———. 1935b. Goose Pond and its marsh birds. Migrant 6:22–24.

———. 1935c. The Rock Dove. Migrant 6:93.

———. 1936. Summer birds of Roan Mountain. Migrant 7:83–86.

———. 1937a. Summer birds of Pickett Forest. Migrant 8:24–27.

———. 1937b. The Reelfoot cranetown. Migrant 8:42–43.

———. 1937c. Further notes on a very old Cardinal. Wilson Bull. 49:15–16.

———. 1940a. Notes on Tennessee birds of prey. Migrant 11:1–4.

———. 1940b. Nesting of Swainson's Warbler in Middle Tennessee. Migrant 11:111–12.

———. 1941. Through the seasons with the Cardinal. Migrant 12:1–4.

———. 1946a. Sparrow hawk nests in a cliff. Migrant 17:26.

———. 1946b. Additional records of the Saw-whet Owl. Migrant 17:67–68.

———. 1947. Nesting habits of the Great Horned Owl. Migrant 18:17–24.

———. 1950. A new heronry in northwest Tennessee. Migrant 31:48–49.

———. 1951a. The breeding herons of Tennessee. Migrant 22:1–8.

———. 1951b. Some notes on Bald Eagles. Migrant 22:37–39.

———. 1951c. Cerulean Warblers and Redstarts remove their nests. Migrant 22:43–44.

———. 1952. Purple Gallinules near McMinnville. Migrant 23:46–47.

———. 1953. Observations of the Sycamore Warbler. Migrant 24:22–25.

———. 1954. A new race of the Yellow-bellied Sapsucker. Migrant 25:37–41.

———. 1955. Crested Flycatcher evicts Starling. Migrant 26:65.

———. 1956. Nesting of the Black-throated Blue and Chestnut-sided Warblers. Migrant 27:43–45.

———. 1960. A new heronry in northwest Tennessee. Migrant 31:48–49.

———. 1962. Some nesting records from the Smokies. Migrant 33:1–6.

———. 1964a. The alleged transportation of its eggs or young by the Chuck-will's-widow. Wilson Bull. 76:19–27.

———. 1964b. Some field notes from Reelfoot Lake, Tenn. Migrant 35:30–32.

———. 1965. Ornithological exploration and collecting in Tennessee. Migrant 36:26–29.

———. 1973. The wildlife met by Tennessee's first settlers. Migrant 44:58–74.

Ganier, A. F., and A. Clebsch. 1938. Some June birds of the Great Smokies. Migrant 9:41–45.

———. 1940. Summer birds of Fall Creek State Park. Migrant 11:53–59.

———. 1944. Summer birds of the Unicoi Mountains. Migrant 15:61–65.

———. 1946. Breeding birds of the Unicoi Mountains. Migrant 17:53–59.

Ganier, A. F., and B. P. Tyler. 1934. Summer birds of Shady Valley. Migrant 5:21–23.

Ganier, A. F., and S. A. Weakley. 1936. Nesting of the Cliff Swallow in Tennessee. Migrant 7:29–30, 41–42.

Ganier, A. F., G. R. Mayfield, D. Merritt, and A. C. Webb. 1935. History of the Tennessee Ornithological Society. Migrant 6:41–44.

Gansner, D. A., and O. W. Herrick. 1985. Host preferences of gypsy moth on a new frontier of infestation. USDA For. Serv. Res. Note NE-330.

Garton, A. W. 1973. BBC 32—Upland mixed deciduous forest with strip mines. Am. Birds 27:972–73.

Gattinger, A. 1887. The Tennessee flora; with special reference to the flora of Nashville. Publ. by the author, Nashville.

Gauch, H. G., Jr. 1979. COMPCLUS—A FORTRAN program for rapid initial clustering of large data sets. Cornell Univ., Ithaca, N.Y.

———. 1982. Multivariate analysis in community ecology. Cambridge Univ. Press, Cambridge.

Gawlik, D. E., and K. L. Bildstein. 1990. Reproductive success and nesting habitat of Loggerhead Shrikes in north-central South Carolina. Wilson Bull. 102:37–48.

Geissler, P. H., and J. R. Sauer. 1990. Topics in route-regression analysis. Pp. 54–57 In J. R. Sauer and S. Droege, eds. Survey designs and statistical methods for the estimation of population trends. U.S. Fish Wildl. Serv., Biol. Rep. 90(1).

Geissler, P. H., D. D. Dolton, R. Field, R. A. Coon, H. F. Percival, D. W. Hayne, L. D. Soileau, R. R. George, J. H. Dunks, and S. D. Bunnell. 1987. Mourning Dove nesting: seasonal patterns and effects of September hunting. U.S. Fish Wildl. Serv., Res. Publ. 168.

Gersbacher, E. O. 1939. The heronries at Reelfoot Lake. J. Tenn. Acad. Sci. 14:162–86.

———. 1964. Heronries of Reelfoot Lake—25 years later. J. Tenn. Acad. Sci. 39:15–16.

Gibbs, J. P., and J. Faaborg. 1990. Estimating the viability of Ovenbird and Kentucky Warbler populations in forest fragments. Consv. Biology 4:193–96.

Gill, F. B. 1980. Historical aspects of hybridization between Blue-winged and Golden-winged Warblers. Auk 97:1–18.

Godfrey, W. E. 1966. The birds of Canada. Natl. Mus. Canada Bull. 203.

Goertz, J. W. 1962. An opossum-titmouse incident. Wilson Bull. 74:189–90.

Goetz, R. C., and D. W. Sharp. 1980. The effect of orientation and light intensity on utilization of artificial Wood Duck nest boxes. Proc. Ann. Conf. S.E. Assoc. Fish and Wildl. Agencies 34:591–97.

Goodpasture, K. A. 1949a. Blue-winged Warbler's nest with six eggs. Migrant 20:54.

———. 1949b. A Sycamore Warbler's nest in a pine tree. Migrant 20:55–56.

———. 1950. Horned Lark nesting near Nashville. Migrant 21:37–41.

———. 1951. Cerulean Warbler's nest in Nashville area. Migrant 22:32–33.

———. 1954. Warblers breeding at Basin Springs, Tennessee. Migrant 25:42–45.

———. 1956. Sharp-shinned Hawk's nest with six eggs found in cedar tree. Migrant 27:15.

———. 1968. Summer occurrence of Blue Grosbeaks in middle Tennessee. Migrant 39:1–3.

———. 1977a. Dr. George Morris Curtis a founder of the Tennessee Ornithological Society. Migrant 48:7–10.

———. 1977b. Fall banding at Basin Spring, 1975. Migrant 48:65–69.

Goodpasture, K. A., and F. J. Alsop III. 1972. Traill's Flycatcher nests at Nashville, Tennessee. Migrant 43:81–84.

Goodpasture, K. A., and L. E. Douglass. 1964. Whip-poor-will nests at Basin Spring. Migrant 35:100–101.

Goodwin, D. 1976. Crows of the world. Cornell Univ. Press, Ithaca, N.Y.

———. 1983. Pigeons and doves of the world, 3rd ed. Cornell Univ. Press, Ithaca, N.Y.

Gore, J. F., and C. J. Barstow. 1970. Status of free-flying resident flock of Canada Geese (*Branta canadensis*) in Tennessee. Proc. Ann. Conf. S.E. Assoc. Game & Fish. Comm. 23:101–4.

Gottschalk, K. W., M. J. Twery, and S. I. Smith, eds. 1990. Interagency gypsy moth research review. USDA For. Serv. Gen. Tech. Rep. NE-146.

Graber, J. W. 1968. Western Henslow's Sparrow. Pp. 779–88 In A. C. Bent (O. L. Austin Jr., ed.). Life histories of North American cardinals, grosbeaks, buntings, towhees, finches, sparrows, and allies, part 2. U.S. Natl. Mus. Bull. 237.

Graber, J. W., R. R. Graber, and E. L. Kirk. 1978. Illinois birds: Ciconiiformes. Ill. Natl. Hist. Survey Biol. Notes 109.

Graber, R. R., and J. W. Graber. 1963. A comparative study of bird populations in Illinois, 1906–1909 and 1956–1958. Bull. Ill. Natl. History Survey 28:383–528.

Granlund, J. G. 1991. Western Meadowlark. Pp. 498–99 In R. Brewer, G. A. McPeek, and R. J. Adams Jr. The atlas of breeding birds of Michigan. Michigan State Univ. Press, East Lansing.

Grant, G. S., and T. L. Quay. 1977. Breeding biology of Cliff Swallows in Virginia. Wilson Bull. 89:286–90.

Grazma, A. F. 1967. Responses of brooding nighthawks to a disturbance stimulus. Auk 84:72–86.

Greene, M. A., J. K. Knox, and T. D. Pitts. 1991. Status of the Reelfoot Lake, Tennessee, heron and egret colony: 1990–1991. Migrant 62:89–96.

Greller, A. M. 1988. Deciduous forest. Pp. 287–316 In M. G. Barbour and W. D. Billings, eds. North American terrestrial vegetation. Cambridge Univ. Press, New York.

Griscom, L., and A. Sprunt Jr. 1979. The warblers of America. Doubleday & Company, Inc., Garden City, N.Y.

Gross, A. O. 1940. Eastern Nighthawk. Pp. 209–12 In A. C. Bent. Life histories of North American cuckoos, goatsuckers, hummingbirds, and their allies. U.S. Natl. Mus. Bull. 176.

———. 1942. Cliff Swallow. Pp. 463–84 In A. C. Bent. Life histories of North American flycatchers, larks, swallows, and their allies. U.S. Natl. Mus. Bull. 179.

———. 1948. Eastern House Wren. Pp. 113–41 In A. C. Bent. Life histories of North American nuthatches, wrens, thrashers and their allies. U.S. Natl. Mus. Bull. 195.

———. 1968. Dickcissel. Pp. 158–91 In A. C. Bent (O. L. Austin Jr., ed.). Life histories of North American cardinals, grosbeaks, buntings, towhees, finches, sparrows and allies, part 1. U.S. Natl. Mus. Bull. 237.

Groth, J. G. 1984. Vocalizations and morphology of the Red Crossbill (*Loxia curvirostra* L.) in the southern Appalachians. M.S. Thesis, Virginia Polytechnic Inst. State Univ., Blacksburg.

———. 1988. Resolution of cryptic species in Appalachian Red Crossbills. Condor 90:745–60.

Gudlin, M. J., and R. W. Dimmick. 1984. Habitat utilization by Ruffed Grouse transplanted from Wisconsin to Tennessee. Pp. 75–88 In W. L. Robinson, ed. Ruffed grouse management: state of the art in the early 1980's. Proc. Symp. 45th Midwest Fish and Wildlife Conf., St. Louis, Mo.

Guthrie, M. 1989. A floristic and vegetational overview of Reelfoot Lake. J. Tenn. Acad. Sci. 64:113–16.

Hagan, J. W., III. 1993. Decline of the Rufous-sided Towhee in the eastern United States. Auk 110:863–74.

Haggerty, T. 1982. Confirmation of breeding Red Crossbills in the mountains of North Carolina with notes on nesting behavior. Chat 46:83–86.

Haggerty, T. M. 1988. Aspects of the breeding biology and productivity of Bachman's Sparrow in central Arkansas. Wilson Bull. 100:247–55.

Hale, K. D. 1979. Whip-poor-will foray. Migrant 50:80.

Hale, S. H. 1980. A breeding bird census of boreal forest habitat on Roan Mountain, Mitchell County, North Carolina. M.S. Thesis, East Tenn. State Univ., Johnson City.

Hall, G. A. 1983. West Virginia birds. Carnegie Mus. Nat. Hist. Spec. Publ. No. 7, Pittsburgh.

———. 1989. Birds of the southern Appalachian subalpine forest. Bird Cons. 3:101–17.

Hamel, P. B. 1992. Land manager's guide to birds of the south. The Nature Conservancy, Southeastern Region. Chapel Hill, N.C.

Hamerstrom, F., F. N. Hamerstrom Jr., and J. Hart. 1973. Nest boxes: an effective management tool for Kestrels. J. Wildl. Manage. 37:400–403.

Hamilton, R. B. 1975. Comparative behavior of the American Avocet and the Black-necked Stilt (Recurvirostridae). Ornithol. Monogr. 17.

Hamilton, W. J., III., and M. E. Hamilton. 1965. Breeding characteristics of Yellow-billed Cuckoos in Arizona. Proc. Calif. Acad. Sci. 32:405–32.

Hammer, D. A., and R. M. Hatcher. 1983. Restoring Osprey populations by hacking preflighted young. Pp. 293–97 in D. M. Bird, ed. Biology and management of Bald Eagles and Ospreys. Harpell Press, Ste. Anne de Bellevue, Quebec.

Hammond, J. S., and D. A. Adams. 1986. Breeding bird census 42—Fraser's fir forest. Am. Birds 40:70.

Hancock, J., and J. Kushlan. 1984. The herons handbook. Harper and Row, New York.

Hands, H. H., R. D. Drobney, and M. R. Ryan. 1989. Status of the Cerulean Warbler in the northcentral United States. Missouri Cooperative Fish and Wildlife Research Unit, Columbia, prepared for U.S. Fish and Wildlife Service.

Haney, J. C. 1981. The distribution and life history of the Brown-headed Nuthatch in Tennessee. Migrant 52:77–86.

Hann, H. W. 1937. Life history of the ovenbird in southern Michigan. Wilson Bull. 49:145–237.

Hanson, H. C. 1965. The giant Canada Goose. Southern Ill. Univ. Press, Carbondale.

Haramis, G. M. 1990. The breeding ecology of the Wood Duck: a review. Pp. 45–60 In L. H. Fredrickson et al., eds. Proc. 1988 Wood Duck Symp., St. Louis, Mo.

Harding, K. C. 1931. Nesting habits of the Black-throated Blue Warbler. Auk 48:512–22.

Hardy, C. L. 1991. A comparison of bird communities in loblolly vs. white pine plantations on the Oak Ridge National Environmental Research Park. M.S. Thesis, Univ. Tennessee, Knoxville.

Hardy, J. W. 1957. The Least Tern in the Mississippi Valley. Mich. State Univ. Mus., Biol. Series 1:1–60.

———. 1961. Studies in behavior and phylogeny of certain New World jays (Garrulinae). Univ. Kansas Sci. Bull. 42:13–149.

Harris, P. C. 1984. Scissor-tailed Flycatchers in Meigs County, Tennessee. Migrant 55:66.

Harrison, C. 1978. A field guide to the nests, eggs and nestlings of North American birds. Collins, London.

Harrison, H. H. 1975. A field guide to birds' nests. Houghton Mifflin, Boston.

———. 1984. Wood warblers' world. Simon and Schuster, New York.

Hart, J. F. 1968. Loss and abandonment of cleared farm land in the eastern United States. Annals Assoc. Amer. Geographers 58:417–40.

Hassler, R., and D. Hassler. 1972. Bird finding in Tennessee—Pickett County. Migrant 43:42–43.

Hatcher, R. M. 1991. Computer model projections of Bald Eagle nesting in Tennessee. J. Tenn. Acad. Sci. 66:225–28.

Hays, R. L., and A. H. Farmer. 1990. Effects of the CRP on wildlife habitat: Emergency haying in the midwest and pine plantings in the southeast. Trans. N. A. Wildl. and Natl. Res. Conf. 55:30–39.

Hays, R. L., R. P. Webb, and A. H. Farmer. 1989. Effects of the Conservation Reserve Program on wildlife habitat: results of 1988 monitoring. Trans. N. A. Wildl. and Natl. Resour. Conf. 54:365–76.

Heinrich, B. 1991. Ravens in winter. Summit Books, New York.

Helm, R. M., D. M. Pashley, and P. J. Zwank. 1987. Notes on the nesting of the Common Moorhen and Purple Gallinule in southwest Louisiana. J. Field Ornithol. 58:55–61.

Henderson, C. R. 1985. A Yellow-crowned Night-Heron colony in Rutherford County, Tennessee. Migrant 56:13.

Henny, C. J. 1972. An analysis of the population dynamics of selected avian species. U.S. Fish Wildl. Service Wildl. Res. Rep. 1.

Henny, C. J., and H. M. Wight. 1972. Population ecology and environmental pollution: Red-tailed and Cooper's Hawks. Pp. 229–49 In Population ecology of migratory birds: a symposium. U.S. Dept. Interior, Wildlife Res. Rep. 2.

Henson, J. W. 1990. Aquatic and certain wetland vascular vegetation of Reelfoot Lake, 1920s–1980s. II. Persistent marshes and marsh-swamp transitions. J. Tenn. Acad. Sci. 65:69–74.

Herndon, L. R. 1947. Cliff Swallows nesting in Carter County, Tennessee. Migrant 18:44–45.

———. 1950a. Least Flycatcher nesting near Elizabethton. Migrant 21:49.

———. 1950b. Birds of Carter County, Tennessee. Migrant 21:57–68.

———. 1956. The House Wren in Tennessee. Migrant 27:23–80.

———. 1958. Traill's Flycatcher breeding in Tennessee. Migrant 29:37–42.

———. 1977. Summer visitors on Roan Mountain. Migrant 48:13–14.

Hespenheide, H. A. 1971. Flycatcher habitat selection in the eastern deciduous forest. Auk 88:61–74.

Hickey, J. J. 1942. Eastern population of the Duck Hawk. Auk 59:176–204.

———, ed. 1969. Peregrine Falcon populations: their biology and decline. Univ. Wisconsin Press, Madison.

Hickey, J. J., and D. W. Anderson. 1969. The Peregrine Falcon: life history and population literature. Pp. 3–42 In J. J. Hickey, ed. Peregrine Falcon populations: their biology and decline. Univ. Wisconsin Press, Madison.

Hill, M. O., and H. G. Gauch Jr. 1980. Detrended correspondence analysis, an improved ordination technique. Vegetatio 42:47–58.

Hill, S. R., and J. E. Gates. 1988. Nesting ecology and microhabitat of the Eastern Phoebe in the central Appalachians. Am. Midl. Nat. 120:313–24.

Hines, T. 1971. A final report on Tennessee's efforts to introduce pheasants. Proc. Annu. Conf. S.E. Assoc. Game and Fish Comm. 24:252–68.

Hinkle, C. R. 1989. Forest communities of the Cumberland Plateau of Tennessee. J. Tenn. Acad. Sci. 64:123–29.

Hobson, K. A., and S. G. Sealy. 1989. Female–female aggression in polygynously nesting Yellow Warblers. Wilson Bull. 101:84–86.

Holbrook, H. L., and J. C. Lewis. 1967. Management of the Eastern Turkey in the Southern Appalachian and Cumberland Plateau Region. Pp. 343–70 *In* O. H. Hewitt, ed. The Wild Turkey and its management. The Wildlife Society, Washington, D.C.

Holmes, R. T., C. P. Black, and T. W. Sherry. 1979. Comparative population bioenergetics of three insectivorous passerines in a deciduous forest. Condor 81:9–20.

Holmes, R. T., T. W. Sherry, P. P. Marra, and K. E. Petit. 1992. Multiple brooding and productivity of a neotropical migrant, the Black-throated Blue Warbler *(Dendroica caerulescens)*, in an unfragmented temperate forest. Auk 109:321–33.

Holt, J. G. 1972. House Finches at Greeneville. Migrant 43:87.

Hooper, R. G. 1977. Nesting habitat of Common Ravens in Virginia. Wilson Bull. 89:233–42.

Hooper, R. G., L. J. Niles, R. F. Harlow, and G. W. Woods. 1982. Home ranges of Red-cockaded Woodpeckers in coastal South Carolina. Auk 99:675–82.

Howe, H. F. 1976. Egg size, hatchling asynchrony, sex and brood reduction in the Common Grackle. Ecology 57:1195–1207.

Howell, A. H. 1910. Notes on the summer birds of Kentucky and Tennessee. Auk 27:295–304.

———. 1924. Birds of Alabama. U.S. Dept. Agri. and Alabama Dept. Game and Fish, Montgomery.

Howell, J. C. 1942. Notes on the nesting habits of the American Robin *(Turdus migratorius* L.). Amer. Midl. Nat. 28:529–603.

———. 1951a. The roadside census as a method of measuring bird populations. Auk 68:334–57.

———. 1951b. Some East Tennessee bird observations. Migrant 22:33.

———. 1958. Long-range ecological study of the Oak Ridge area: I. Observations on the summer birds in Melton Valley. Unpubl. Report, ORNL Central Files No. 58-6-14, Oak Ridge, Tenn.

———. 1965. BBC 9—Ridge and valley hardwood forest. Aud. Field Notes 19:593–94.

———. 1971. BBC 16—Ridge and valley hardwood forest. Am. Birds 25:974–75.

———. 1972. BBC 19—Ridge and valley hardwood forest. Am. Birds 26:952.

———. 1973. BBC 31—Ridge and valley hardwood forest. Am. Birds 27:972.

———. 1974. BBC 45—Ridge and valley hardwood forest. Am. Birds 28:1010.

Howell, J. C., and J. M. Campbell. 1972. Observations of Campbell County birds. Migrant 43:1–4.

Howell, J. C., and M. B. Monroe. 1957. The birds of Knox County, Tennessee. J. Tenn. Acad. Sci. 32:247–322.

———. 1958. The birds of Knox County, Tennessee. Migrant 29:17–27.

Howes-Jones, D. 1985. Relationships among song activity, context, and social behavior in the Warbling Vireo. Wilson Bull. 97:4–20.

Hoyt, J. S. Y. 1945. Nesting records of the Towhee at Memphis. Migrant 16:40–41.

Hoyt, S. F. 1953. Incubation and nesting behavior of the Chuck-will's-widow. Wilson Bull. 65:204–5.

———. 1957. The ecology of the Pileated Woodpecker. Ecology 38:247–56.

Hubbard, J. A. 1976. Social organization, dispersion, and population dynamics in a flock of pen-reared wild Canada Geese. Ph.D. Diss., Univ. Tennessee, Knoxville.

Hughes, R. H. 1966. Fire ecology of canebrakes. Proc. Tall Timbers Fire Ecology Conf. 6:149–58.

Hull, B. K. 1990. Behavioral study of the breeding season of the Alder Flycatcher *(Empidonax alnorum)* on Roan Mountain. M.S. Thesis, East Tenn. State Univ., Johnson City, Tenn.

Hull, C. N. 1983. Eastern Phoebe nests at relocated nest site. Jack-Pine Warbler 61:100.

Humphrey, P. S. 1946. Observations at the nest of a Pileated Woodpecker. Migrant 17:43–46.

Hunter, C., ed. 1983. The life and letters of Alexander Wilson. Amer. Philosophical Society, Philadelphia.

Hupp, C. R. 1992. Riparian vegetation recovery patterns following stream channelization: a geomorphic perspective. Ecology 73:1209–26.

Hurley, R. J., and E. C. Franks. 1976. Changes in the breeding ranges of two grassland birds. Auk 93:108–15.

Hursh, C. R. 1948. Local climate in the Copper Basin of Tennessee as modified by the removal of vegetation. U.S. Dept. Agri. Circular No. 774.

Ijams, H. P. 1931. Hawks and owls. Migrant 2:3.

———. 1937. Prothonotary Warblers. Migrant 8:41.

Ijams, H. P., and L. A. Hofferbert. 1934. Nesting records of birds at Athens, Tenn. Migrant 5:1–4.

Imhof, T. A. 1976. Alabama birds, 2nd. ed. Univ. Alabama Press, University, Ala.

Ingold, D. J. 1989. Nesting phenology and competition for nest sites among Red-headed and Red-bellied Woodpeckers and European Starlings. Auk 106:209–17.

———. 1991. Nest-site fidelity in Red-headed and Red-bellied Woodpeckers. Wilson Bull. 103:118–22.

Irwin, O. S. 1959. Sparrow Hawks nesting. Migrant 30:39.

Isler, M. L., and P. R. Isler. 1987. The Tanagers. Smithsonian Inst. Press, Washington, DC.

Jackson, J. A. 1976. A comparison of some aspects of the breeding ecology of Red-headed and Red-bellied Woodpeckers in Kansas. Condor 78:67–76.

———. 1983. Nesting phenology, nest site selection, and reproductive success of Black and Turkey Vultures. Pp. 245–70 *In* S. R. Wilbur and J. A. Jackson, eds. Vulture biology and management. Univ. Calif. Press, Berkeley.

———. 1988a. American Black Vulture. Pp. 11–24 *In* R. S. Palmer, ed. Handbook of North American birds, Vol. 4. Yale Univ. Press, New Haven, Conn.

———. 1988b. Turkey Vulture. Pp. 25–42 *In* R. S. Palmer, ed. Handbook of North American birds, Vol. 4. Yale Univ. Press, New Haven, Conn.

Jackson, J. A., and J. Tate. 1974. An analysis of nest box use by Purple Martins, House Sparrows, and starlings in eastern North America. Wilson Bull. 86:435–49.

Jackson, J. A., and P. G. Burchfield. 1975. Nest-site selection of Barn Swallows in east-central Mississippi. Am. Midl. Nat. 94:503–9.

Jackson, J. A., and R. E. Weeks. 1976. Nesting of the Eastern Phoebe and Barn Swallow in western Alabama. Alabama Birdlife 24:7–9.

Jackson, J. A., D. Werschkul, R. Howell, and T. Darden. 1976. Extension of the known breeding range of the Eastern Phoebe in Mississippi. Mississippi Kite 6:6.

Jacobs, B. 1991. First state nesting record for Black-necked Stilts, *Himantopus mexicanus*. Bluebird 58:7–11.

James, D. A., and J. C. Neal. 1986. Arkansas birds. Univ. Arkansas Press, Fayetteville.

James, F. C. 1971. Ordination of habitat relationships among breeding birds. Wilson Bull. 83:215–36.

James, F. C., and H. H. Shugart Jr. 1974. The phenology of the nesting season of the American Robin *(Turdus migratorius)* in the United States. Condor 76:159–68.

James, F. C., C. E. McCulloch, and L. E. Wolfe. 1990. Methodological issues in the estimation of trends in bird populations with an example: the Pine Warbler. Pp. 84–97 *In* J. R. Sauer and S. Droege, eds. Survey designs and statistical methods for the estimation of avian population trends. U.S. Fish Wildl. Serv., Biol. Rep. 90(1).

James, F. C., D. A. Wiedenfeld, and C. E. McCulloch. 1992. Trends in breeding populations of warblers: declines in the southern highlands and increases in the lowlands. Pp. 43–56 *In* J. M. Hagan III and D. W. Johnston, eds. Ecology and conservation of neotropical migrant landbirds. Smithsonian Inst. Press, Washington, D.C.

James, F. C., R. F. Johnston, N. O. Wamer, G. J. Niemi, and W. J. Boecklen. 1984. The Grinnellian niche of the Wood Thrush. Am. Nat. 124:17–30.

James, R. D. 1976. Foraging behavior and habitat selection of three species of vireos in southern Ontario. Wilson Bull. 88:62–75.

———. 1978. Pairing and nest site selection in Solitary and Yellow-throated Vireos with a description of a ritualized nest building display. Can. J. Zool. 56:1163–69.

———. 1984. Structure, frequency of usage, and apparent learning in the primary song of the Yellow-throated Vireo, with comparative notes on Solitary Vireos (Aves: Vireonidae). Can. J. Zool. 62:468–72.

James, W. K. 1979. Resident waterfowl production and release project progress report, September 1977–September 1978. Tennessee Valley Authority Div. Land and Forest Resources, Norris, Tenn.

Jamison, C., and W. Simpson. 1940. Barn Owls in the Nashville area. Migrant 11:97–98.

Jeter, H. H. 1957. Eastern Phoebe nesting in Louisiana. Wilson Bull. 69:360–61.

Johnsgard, P. A. 1973. Grouse and quails of North America. Univ. Nebraska Press, Lincoln.

———. 1975. North American game birds of upland and shoreline. Univ. Nebraska Press, Lincoln.

———. 1983a. The grouse of the world. Univ. Nebraska Press, Lincoln.

———. 1983b. The hummingbirds of North America. Smithsonian Inst. Press, Washington, D.C.

———. 1986. The pheasants of the world. Oxford Univ. Press, Oxford, U.K.

———. 1988. North American owls. Smithsonian Inst. Press, Washington, D.C.

Johnson, R. R., and J. J. Dinsmore. 1985. Brood-rearing and postbreeding habitat use by Virginia Rails and Soras. Wilson Bull. 97:551–54.

———. 1986. Habitat use by breeding Virginia Rails and Soras. J. Wildl. Manage. 50:387–392.

Johnston, D. W. 1961. The biosystematics of American crows. Univ. Washington Press, Seattle, Wash.

———. 1971. Niche relationships among some deciduous forest flycatchers. Auk 88:796–804.

Johnston, D. W., and E. P. Odum. 1956. Breeding bird populations in relation to plant succession on the piedmont of Georgia. Ecology 37:50–62.

Joliet, L. 1673–77. On the first voyage made by Father Marquette. Reprinted in R. Thwaites. 1896–1901. The Jesuit relations. Vol. 59:149–51. Burrows Bros., Cleveland, Ohio.

Jones, E. L. 1974. Creative disruptions in American agriculture 1620–1820. Agricul. History 48:510–28.

Jones, F. M. 1931. Nesting habits of the Parula Warbler. Migrant 2:20–21.

———. 1933. Notes on Duck Hawk nestings. Migrant 4:43–44.

Jones, J. K., Jr., R. S. Hoffman, D. W. Rice, C. Jones, R. J. Baker, and M. D. Engstrom. 1992. Revised checklist of North American mammals North of Mexico, 1991. Occas. Papers, The Museum, Texas Tech Univ., No. 146.

Jones, L. F. 1979. Evaluation of a Ruffed Grouse restoration attempt on the Western Highland Rim in Tennessee. M.S. Thesis, Univ. Tennessee, Knoxville.

Kahl, R. B., T. S. Baskett, J. A. Ellis, and J. N. Burroughs. 1985. Characteristics of summer habitats of selected nongame birds in Missouri. Univ. Missouri—Columbia Agr. Exp. Sta. Res. Bull. 1056.

Kain, E., ed. 1987. Virginia's birdlife: an annotated checklist. Virginia Soc. Ornithology, Virginia Avifauna 3.

Kalla, P. I. 1979. The distribution, habitat preference, and status of the Mississippi Kite *(Ictinia mississippiensis)* in Tennessee, with reference to other populations. M.S. Thesis, East Tenn. State Univ., Johnson City.

Kalla, P. I., and F. J. Alsop III. 1983. The distribution, habitat preference, and status of the Mississippi Kite in Tennessee. Am. Birds 37:146–49.

Kalla, P. I., and R. W. Dimmick. 1987. Evaluation of a Ruffed Grouse introduction in Tennessee. Proc. Annu. Conf. S.E. Assoc. Fish and Wildlife Agencies. 41:365–72.

Kaufmann, G. W. 1989. Breeding ecology of the Sora, *Porzana carolina*, and the Virginia rail, *Rallus limicola*. Can. Field-Nat. 103:270–82.

Kellogg, R. 1939. Annotated list of Tennessee mammals. Proc. U.S. Natl. Mus. 86:245–303.

Kendeigh, S. C. 1963. Regulation of nesting time and distribution in the House Wren. Wilson Bull. 75:418–27.

Kendeigh, S. C., and B. Fawver. 1981. Breeding bird populations in the Great Smoky Mountains, Tennessee and North Carolina. Wilson Bull. 93:218–42.

Kennamer, R. A., W. F. Harvey IV, and G. R. Hepp. 1988. Notes on Hooded Merganser nests in the coastal plain of South Carolina. Wilson Bull. 100:686–88.

Kessel, B. 1957. A study of the breeding biology of the European Starling *(Sturnus vulgaris* L.) in North America. Amer. Midl. Nat. 58:257–31.

Kiff, L. F. 1989. Historical breeding records of the Common Merganser in the southeastern United States. Wilson Bull. 101:141–43.

Kilham, L. 1959. Early reproductive behavior of flickers. Wilson Bull. 71:323–36.

———. 1960. Courtship and territorial behavior of Hairy Woodpeckers. Auk 77:259–70.

———. 1961. Reproductive behavior of Red-bellied Woodpeckers. Wilson Bull. 73:237–54.

———. 1962. Breeding behavior of Yellow-bellied Sapsuckers. Auk 79:31–43.

———. 1972. Reproductive behavior of White-breasted Nuthatches. II. Courtship. Auk 89:115–19.

———. 1977. Nesting behavior of Yellow-bellied Sapsuckers. Wilson Bull. 89:310–24.

———. 1979. Courtship and the pair-bond of Pileated Woodpeckers. Auk 96:587–94.

———. 1989. The American Crow and the Common Raven. Texas A & M University Press, College Station.

Killebrew, J. B. 1874. Introduction to the resources of Tennessee. Tavel, Eastman and Howell, Nashville, Tenn.

Kinser, G. W., Jr. 1973. Ecology and behavior of the cardinal, *Richmondena cardinalis* (L.) in southern Indiana. Ph.D. Diss., Indiana Univ., Bloomington.

Kirby, R. E. 1990. Wood Duck nonbreeding ecology: fledging to spring migration. Pp. 61–76 *In* L. H. Fredrickson et al., eds. Proc. 1988 Wood Duck Symp., St. Louis, Mo.

Kittle, P. D., and D. C. Patterson. 1990. First nesting record of the Scissor-tailed Flycatcher *(Tyrannus forficatus)* in Alabama. Alabama Birdlife 37:7–8.

Klaas, E. E. 1975. Cowbird parasitism and nesting success in the Eastern Phoebe. Occas. Pap. Univ. Kansas Mus. Nat. Hist. 41.

Klimstra, W. D., and J. L. Roseberry. 1975. Nesting ecology of the bobwhite in southern Illinois. Wildl. Monogr. 41:1–37.

Knight, R. L. 1982. First House Finch nest in Tennessee. Migrant 53:40.

———. 1987. Golden-crowned Kinglet nest on Roan Mountain, North Carolina/Tennessee. Migrant 58:48–49.

———. 1989a. Second Tennessee breeding record of Savannah Sparrow, with comments on its expansion into the southern Appalachians. Migrant 60:69–71.

———. 1989b. Another 19th century Common Merganser egg set collected in Tennessee. Migrant 60:93–94.

Knight, R. L., D. J. Grout, and S. A. Temple. 1987. Nest-defense behavior of the American Crow in urban and rural areas. Condor 89:175–77.

Knopf, F. L., and B. A. Knopf. 1983. Flocking pattern of foraging American Crows in Oklahoma. Wilson Bull. 95:153–55.

Knox, A. G. 1990. The sympatric breeding of Common and Scottish Crossbills *Loxia curvirostra* and *L. scotica* and the evolution of crossbills. Ibis 132:454–66.

Koenig, W. D. 1986. Geographical ecology of clutch size variation in North American woodpeckers. Condor 88:499–504.

Kridelbaugh, A. L. 1983. Nesting ecology of the Loggerhead Shrike in central Missouri. Wilson Bull. 95:303–8.

Kroodsma, D. E. 1980. Winter Wren singing behavior: a pinnacle of song complexity. Condor 82:357–65.

Kroodsma, R. L. 1984. Effect of edge on breeding forest bird species. Wilson Bull. 96:426–36.

Kuchler, A. W. 1964. Potential natural vegetation of the conterminous United States. Amer. Geogr. Soc., Spec. Publ. 36.

Kuerzi, R. G. 1941. Life history studies of the Tree Swallow. Proc. Linn. Soc. N.Y. 52–53:1–52.

Lambert, R. S. 1961. Logging the Great Smokies, 1880–1930. Tenn. Hist. Quarterly 20:350–63.

Langdon, F. W. 1887. August birds of the Chilhowee Mountains, Tennessee. Auk 4:125–33.

Lanyon, W. E. 1956. Ecological aspects of the sympatric distribution of meadowlarks in the north-central states. Ecology 37:98–108.

———. 1995. Eastern Meadowlark *(Sturnella magna)*. *In* A. Poole and F. Gill, eds. The birds of North America, No. 160. Acad. Nat. Sci. Philadelphia and Amer. Ornithol. Union., Washington, D.C.

Larson, D. L., and C. E. Bock. 1986. Eruptions of some North American boreal seedeating birds, 1901–1980. Ibis 128:137–40.

Laskey, A. R. 1934. Eastern Field Sparrow migration at Nashville. Bird-banding 5:172–75.

———. 1939. Bird banding brevities—No. 14. Migrant 10:47–48.

———. 1943. Brown Thrasher parasitized by Cowbird. Migrant 14:57–58.

———. 1944. A study of the Cardinal in Tennessee. Wilson Bull. 56:27–44.

———. 1945. Towhee-cowbird nesting data for 1945. Migrant 16:46.

———. 1946. Some Bewick Wren nesting data. Migrant 17:38–43.

———. 1950. Cowbird behavior. Wilson Bull. 62:157–74.

———. 1956. The Bluebird nest-box project at Nashville, Tennessee. Inland Bird Banding News 28:29–30.

———. 1957. Some Tufted Titmouse life history. Bird-Banding 28:135–45.

———. 1958. Blue Jays at Nashville Tennessee: movements, nesting, age. Bird-banding 29:211–18.

———. 1962. Breeding biology of Mockingbirds. Auk 79:596–606.

———. 1963. A set of seven Mockingbird eggs. Migrant 34:62–63.

———. 1966. Status of Bewick's Wren and House Wren in Nashville. Migrant 37:4–6.

Laughlin, S. B., D. P. Kibbe, and P. F. J. Eagles. 1982. Atlasing the distribution of the breeding birds of North America. Am. Birds 36:6–19.

Laughlin, T. L., and R. A. Phillips. 1981. Probable first nesting of the House Finch in Tennessee. Migrant 52:19.

Lawrence, L. de K. 1953. Nesting life and behaviour of the Red-eyed Vireo. Can. Field-Natur. 67:47–77.

———. 1967. A comparative life history study of four species of woodpeckers. Ornithol. Monogr. No. 5, Allen Press, Lawrence, Kans.

Layne, M. A. 1931. Some nesting habits of the Catbird. M.A. Thesis, Geo. Peabody College for Teachers, Nashville, Tenn.

LeBlanc, D. C., N. S. Nicholas, and S. M. Zedaker. 1992. Prevalence of individual-tree growth decline in red spruce populations of the southern Appalachian Mountains. Can. J. For. Res. 22:905–14.

Leck, C. F., and F. L. Cantor. 1979. Seasonality, clutch size, and hatching success in the Cedar Waxwing. Auk 96:196–98.

Lee, D. S. 1985. Breeding-season records of boreal birds in western North Carolina with additional information on species summering on Grandfather Mountain. Chat 49:85–94.

Lee, D. S., and W. R. Spofford. 1990. Nesting of Golden Eagles in the central and southern Appalachians. Wilson Bull. 102:693–98.

Lee, D. S., D. Audet, and B. Tarr. 1985. Summer bird fauna of North Carolina's Grandfather Mountain. Chat 49:1–14.

Leggett, K. 1968. Heronry at Dyersburg still active. Migrant 39:59.

———. 1970. Heronry at Dyersburg no longer active. Migrant 41:58.

LeGrand, E. 1979. A report on an Alder Flycatcher colony at Roan Mountain with comments on the status of the species in the Southern Appalachians. Chat 43:35–36.

LeGrand, H. E., Jr. 1982. Briefs for the files. Chat 46:21–25.

———. 1984. Briefs for the files. Chat 48:18–26.

———. 1990a. Briefs for the files. Chat 54:65–70.

———. 1990b. Olive-sided Flycatcher. Pp. 36–37 In D. S. Lee and J. F. Parnell, eds. Endangered, threatened, and rare fauna of North Carolina. Part III. A re-evaluation of the birds. Occ. Pap. North Carolina Biol. Surv. 1990–1.

LeGrand, H. E., and E. F. Potter. 1980. Ashe County breeding bird foray—1979. Chat 44:5–13.

Lemon, E. K. 1969. Bewick Wren, host to Brown-headed Cowbird. Can. Field Natur. 83:395–96.

LeMoyne, A. 1886. Notes on some birds of the Great Smoky Mountains. Ornithol. and Oologist 11:115–17, 131–32, 147–48, 163–64, 179–80.

Lennington, S., and T. Mace. 1975. Mate fidelity and nesting site tenacity in the Killdeer. Auk 92:149–51.

Lever, C. 1987. Naturalized birds of the world. John Wiley and Sons, New York.

Lewis, J. C. 1962. The status of Wild Turkeys in Tennessee. Migrant 33:61–62.

Lewis, R. P. 1975a. BBC 84—Field with hedgerows. Am. Birds 29:1115.

———. 1975b. BBC 151—Pasture. Am. Birds 29:1140–41.

———. 1977. BBC 85—Deciduous clearcut. Am. Birds 30:62.

———. 1978. BBC 79—Deciduous clearcut. Am. Birds 32:80.

———. 1979. BBC 85—Deciduous clearcut. Am. Birds 33:78.

———. 1980. BBC 83—Deciduous clearcut. Am. Birds 34:66.

———. 1981a. BBC 106—Deciduous clearcut. Am. Birds 35:76.

———. 1981b. BBC 107—Deciduous clearcut (2-yr.). Am. Birds 35:76.

———. 1982. BBC 103—Deciduous clearcut. Am. Birds 36:78.

———. 1983a. BBC 47—Mixed deciduous forest. Am. Birds 37:66–67.

———. 1983b. BBC 89—Deciduous clearcut. Am. Birds 37:78.

———. 1984a. BBC 64—Mixed deciduous forest. Am. Birds 38:87.

———. 1984b. BBC 92—Deciduous clearcut. Am. Birds 38:95.

———. 1992. BBC 47—Deciduous clearcut. J. Field Ornithol. (supplement) 63:63–64.

Lewis, R., and A. B. Smith. 1975. BBC 90—Deciduous clearcut. Am. Birds 29:1117–18.

Ligon, J. D. 1970. Behavior and breeding biology of the Red-cockaded Woodpecker. Auk 87:255–78.

Link, W. A., and J. R. Sauer. 1994. Estimating equations estimates of trends. Bird Populations 2:23–32.

Lochridge, O. B., and A. R. Lochridge. 1984. Immature Virginia Rail observed in Maury County. Migrant 55:85.

Locke, B. A., R. N. Conner, and J. C. Kroll. 1983. Factors influencing colony site selection by Red-cockaded Woodpeckers. Pp. 46–50 In D. A. Wood, ed. Proceedings Red-cockaded Woodpecker Symposium II. Florida Game and Fresh Water Fish Comm., Tallahassee.

Longcore, J. R., and R. E. Jones. 1969. Reproductive success of the Wood Thrush in a Delaware woodlot. Wilson Bull. 81:396–406.

Lowther, P. E. 1993. Brown-headed Cowbird (*Molothrus ater*). In A. Poole and F. Gill, eds. The birds of North America, No. 47. Acad. Nat. Sci., Philadelphia, and Amer. Ornithol. Union., Washington, D.C.

Ludwig, J. P. 1984. Decline, resurgence and population dynamics of Michigan and Great Lakes Double-crested Cormorants. Jack Pine Warbler 62:92–102.

Lunk, W. A. 1962. The Rough-winged Swallow, *Stelgidopteryx ruficollis* (Vieillot), a study based on its breeding biology in Michigan. Publ. Nuttall Ornithol. Club No. 4.

Lura, R., E. Schell, and G. Wallace. 1979. Nesting Alder Flycatchers in Tennessee. Migrant 50:34–36.

Luther, E. T. 1977. Our restless earth—the geologic regions of Tennessee. Univ. Tenn. Press, Knoxville.

Lyle, R. B., and B. P. Tyler. 1934. The nesting birds of northeastern Tennessee. Migrant 5:49–57.

Lynch, J. M. 1981. Status of the Cerulean Warbler in the Roanoke River basin of North Carolina. Chat 45:29–35.

Lynch, J. M., and H. E. LeGrand Jr. 1985. Breeding-season records of the Henslow's Sparrow in the North Carolina coastal plain. Chat 49:29–35.

———. 1989. Breeding season birds of Long Hope Creek Valley, Watauga and Ashe counties, N.C. Chat 53:29–35.

Lyon, J. A., Jr. 1893. Collecting Black Vulture eggs. Oologist 10:55–56.

Mace, T. R. 1978. Killdeer breeding densities. Wilson Bull. 90:442–43.

Manlove, W. R. 1933. A roost of the Wild Pigeon. Migrant 4:18–19.

Marks, J. S., H. Doremus, and R. J. Cannings. 1989. Polygyny in the Northern Saw-whet Owl. Auk 106:732–34.

Marti, C. D. 1988. The Common Barn-owl. Pp. 535–50 In W. J. Chandler, ed. Audubon Wildlife Report 1988/1989. Academic Press, San Diego.

Marti, C. D., P. W. Wagner, and K. W. Denne. 1979. Nest boxes for the management of Barn Owls. Wildl. Soc. Bull. 7:145–48.

Martin, K., and J. Parker. 1991. Mississippi Kites reborn in Tennessee. Tenn. Conservationist 57(3):5–10.

Martin, T. E., and P. Li. 1992. Life history traits of open- vs. cavity-nesting birds. Ecology 73:579–92.

Martin, W. H. 1989. Forest patterns in the Great Valley of Tennessee. J. Tenn. Acad. Sci. 64:137–43.

Maxwell, G. R., II. 1970. Pair formation, nest building and egg-laying of the Common Grackle in northern Ohio. Ohio J. Sci. 70:284–91.

Maxwell, G. R., II, and L. S. Putnam. 1972. Incubation, care of young, and nest success of the Common Grackle (*Quiscalus quiscula*) in northern Ohio. Auk 89:349–59.

May, D. M. 1991. Forest resources of Tennessee. USDA For. Serv. Res. Bull. SO-160.

Mayfield, G. R. 1953. Song Sparrow extending nesting area westward. Migrant 24:54.

Mayfield, G. R., III. 1993. BBC 58—High altitude red spruce forest. J. Field Ornithol. (supplement) 64:69–70.

Mayfield, G. R., III., and F. J. Alsop III. 1992. First confirmed Tennessee nest record of the Northern Saw-whet Owl. Migrant 63:81–88.

Mayfield, H. 1965. The Brown-headed Cowbird, with old and new hosts. Living Bird 4:13–28.

Mayhew, W. W. 1958. The biology of the Cliff Swallow in California. Condor 60:7–37.

McConnell, J., and O. McConnell. 1983. Breeding birds of the Unicoi Mountains. Chat 47:33–40.

McCormick, J. F., and R. B. Platt. 1980. Recovery of an Appalachian forest following the chestnut blight or Catherine Keever—You were right! Amer. Midl. Nat. 104:264–73.

McEven, J. A., Jr. 1894. A few field notes. Oologist 11:223.

McGarigal, K., and J. D. Fraser. 1984. The effects of forest stand age on owl distribution in southwestern Virginia. J. Wildl. Manage. 48:1393–98.

McGowan, R. W. 1967. A breeding census of the Pileated Woodpecker. Migrant 38:57–59.

McGuiness, J. H. 1989. Nesting and brooding ecology of the Eastern Wild Turkey on Natchez Trace Wildlife Management Area, Tennessee. M.S. Thesis, Tenn. Tech. Univ., Cookeville.

McKinley, D. 1979. A review of the Carolina parakeet in Tennessee. Migrant 50:1–6.

McKinney, G. W., and J. B. Owen. 1989. First evidence of Northern Saw-whet Owls nesting in Tennessee. Migrant 60:5–6.

McKinney, L. E. 1989. Vegetation of the Eastern Highland Rim of Tennessee. J. Tenn. Acad. Sci. 64:145–47.

McLaughlin, J. A. 1926. The original roosting habits of the Chimney Swift. Wilson Bull. 38:36.

McLaughlin, S. B., D. J. Downing, T. J. Blasing, E. R. Cook, and H. S. Adams. 1987. An analysis of climate and competition as contributors to decline of red spruce in high elevation Appalachian forests in the eastern United States. Oecologia 72:487–501.

McNair, D. B. 1982. Grasshopper Sparrows breed in Lowndes County, Mississippi. Alabama Birdlife 29:9–11.

———. 1984a. Nest placement of the Eastern Phoebe under bridges in south-central North Carolina. Oriole 49:1–6.

———. 1984b. Breeding status of the Grasshopper Sparrow in the coastal plain of the Carolinas, with notes on local behavior. Chat 48:1–4.

———. 1984c. Breeding biology of the Fish Crow. Oriole 49:21–32.

———. 1984d. Clutch-size and nest placement in the Brown-headed Nuthatch. Wilson Bull. 96:296–301.

———. 1985. A comparison of oology and nest record card data in evaluating the reproductive biology of Lark Sparrows, *Chondestes grammacus*. Southwestern Naturalist 30:213–24.

———. 1987a. Egg data slips—are they useful for information on egg-laying dates and clutch size? Condor 89:369–76.

———. 1987b. Status and distribution of the Fish Crow in the Carolinas and Georgia. Oriole 52:28–35.

———. 1987c. Recent breeding information on birds in a portion of the Southern Appalachian Mountains. Migrant 58:109–34.

———. 1988a. Red Crossbills breed at Highlands, North Carolina. Migrant 59:45–48.

———. 1988b. Review of breeding records of Red Crossbill and Pine Siskin in the southern Appalachian Mountains and adjacent regions. Migrant 59:105–13.

———. 1990. Eastern Phoebe breeds in the northeast upper coastal plain of South Carolina. Chat 54:59–61.

McNair, D. B., and C. P. Nicholson. In press. Historical breeding-season information of the Lark Sparrow in Tennessee. Migrant.

McNish, E. D. 1923. Nesting of the Blue-winged and Worm-eating Warblers in Tennessee. Wilson Bull. 35:55–56.

Meanley, B. 1963. Nesting ecology and habits of the Dickcissel on the Arkansas Grand Prairie. Wilson Bull. 75:280.

———. 1971. Natural history of the Swainson's Warbler. North American Fauna, No. 69. U.S. Dept. Interior.

———. 1992. King Rail. *In* A. Poole, P. Stettenheim, and F. Gill, eds. The birds of North America, No. 3. American Ornithologists Union, Philadelphia.

Meanley, B., and R. A. Dolbeer. 1978. Source of Common Grackles and Red-winged Blackbirds wintering in Tennessee. Migrant 49:25–28.

Meng, H. K., and R. N. Rosenfield. 1988. Cooper's Hawk. Pp. 320–54 *In* R. S. Palmer, ed. Handbook of North American birds, Vol. 4. Yale Univ. Press, New Haven, Conn.

Mengel, R. M. 1965. The birds of Kentucky. Ornithol. Monogr. 3, Allen Press, Lawrence, Kans.

Meyer, H., and R. R. Nevius. 1943. Some observations on the nesting and development of the Prothonotary Warbler, *Protonotaria citrea*. Migrant 14:31–36.

Middleton, A. L. A. 1978. The annual cycle of the American Goldfinch. Condor 80:401–6.

———. 1979. Influence of age and habitat on reproduction by the American Goldfinch. Ecology 60:418–32.

Miller, N. A., and J. Neiswender. 1989. A plant community study of the Third Chickasaw Bluff, Shelby County, Tennessee. J. Tenn. Acad. Sci. 64:149–54.

Miller, R. A. 1974. The geologic history of Tennessee. Tennessee Div. Geology Bull. 74, Nashville.

Miller, R. A., W. D. Hardeman, and D. S. Fullerton. 1966. Geologic map of Tennessee (4 sheets). Tenn. Div. Geology, Nashville.

Mills, A. M. 1986. The influence of moonlight on the behavior of goatsuckers (Caprimulgidae). Auk 103:370–78.

Mitchell, L. C., and B. A. Millsap. 1990. Buteos and Golden Eagle. Pp. 50–62 *In* Proc. Southeast raptor management symposium and workshop. Natl. Wildl. Fed., Washington, D.C.

Mitchell, L. J., and B. H. Stedman. 1993. BBC 39—Mature maple-beech-birch forest. J. Field Ornith. (supplement):64:57–58.

Monk, H. C. 1929. Bird migration at Nashville. J. Tenn. Acad. Sci. 4:65–77.

———. 1932. Water birds of Radnor Lake. J. Tenn. Acad. Sci. 7:217–32.

———. 1934. Habits of the Warbling Vireo. Migrant 5:33–34.

———. 1936. Cowbird nesting records for Davidson County. Migrant 7:32–33.

———. 1942. A very late Robin brood. Migrant 13:25.

———. 1949. Nesting of the Mourning Dove at Nashville. Migrant 20:1–9.

Monroe, B. L. 1937. A Blue-winged Warbler nest near Nashville. Migrant 8:38–39.

Monroe, B. L., Jr., A. L. Stamm, and B. L. Palmer-Ball Jr. 1988. Annotated checklist of the birds of Kentucky. Commonwealth Printing Co., Louisville, Ky.

Mooney, J. 1972. Myths and sacred formulas of the Cherokees. Reprinted by Charles Elder, Nashville, Tenn.

Moore, W. S. 1995. Northern Flicker *(Colaptes auratus)*. *In* A. Poole and F. Gill, eds. The birds of North America, No. 160. Acad. Nat. Sci. Philadelphia and Amer. Ornithol. Union., Washington, D.C.

Morehouse, E. L., and R. Brewer. 1968. Feeding of nestling and fledgling Eastern Kingbirds. Auk 85:44–54.

Morse, D. H. 1967. Competitive relationships between Parula Warblers and other species during the breeding season. Auk 84:490–502.

———. 1971. Effects of the arrival of a new species upon habitat utilization by two forest thrushes in Maine. Wilson Bull. 83:57–65.

———. 1989. American warblers. Harvard Univ. Press, Cambridge, Mass.

———. 1993. Black-throated Green Warbler (*Dendroica virens*). *In* A. Poole and F. Gill, eds. The birds of North America, No. 55. Acad. Nat. Sci. Philadelphia and Amer. Ornithol. Union., Washington, D.C.

———. 1994. Blackburnian Warbler (*Dendroica fusca*). *In* A. Poole and F. Gill, eds. The birds of North America, No. 102. Acad. Nat. Sci. Philadelphia and Amer. Ornithol. Union., Washington, D.C.

Mosby, H. S. 1973. The changed status of the Wild Turkey over the past three decades. Pp. 71–76 *In* B. C. Sanderson and H. C. Schultz, eds. Wild Turkey management. Missouri Chapter Wildl. Soc., and Univ. Missouri Press, Columbia.

Mosher, J. A., and R. S. Palmer. 1988. Broad-winged Hawk. Pp. 3–33 *In* R. S. Palmer, ed. Handbook of North American birds, Vol 5. Yale Univ. Press, New Haven, Conn.

Mulvihill, R. S. 1992. Golden-crowned Kinglet. Pp. 260–61 *In* D. W. Brauning, ed. Atlas of breeding birds in Pennsylvania. Univ. Pittsburgh Press, Pittsburgh.

Mumford, R. E. 1964. The breeding biology of the Acadian Flycatcher. Univ. Mich. Mus. Zool. Misc. Publ. 125.

Mumford, R. E., and C. E. Keller. 1984. The birds of Indiana. Indiana Univ. Press, Bloomington.

Mundahl, J. T. 1982. Role specialization in the parental and territorial behavior of the Killdeer. Wilson Bull. 94:515–30.

Mundinger, P. C., and S. Hope. 1982. Expansion of the winter range of the House Finch, 1949–1979. Am. Birds 36:347–53.

Murphy, E. C. 1978. Breeding ecology of House Sparrows: Spatial variation. Condor 80:180–93.

Murphy, M. T. 1983. Ecological aspects of the reproductive biology of Eastern Kingbirds: geographic comparisons. Ecology 64:914–28.

———. 1986. Brood parasitism of Eastern Kingbirds by Brown-headed Cowbirds. Auk 103:626–28.

———. 1988. Comparative reproductive biology of kingbirds (*Tyrannus* spp.) in eastern Kansas. Wilson Bull. 100:357–76.

Murphy, M. T., and R. C. Fleischer. 1986. Body size, nest predation, and reproductive patterns in Brown Thrashers and other mimids. Condor 88:446–55.

Murray, G. A. 1976. Geographic variation in the clutch sizes of seven owl species. Auk 93:602–13.

Myers, J. M., and A. S. Johnson. 1978. Bird communities associated with succession and management of loblolly-shortleaf pine forests. Pp. 50–65 *In* Proc. workshop manage. southern for. nongame birds, R. M. DeGraaf (tech. coord.). USDA For. Serv. Gen. Tech. Rep. SE-14.

Neidermyer, W. J., and J. J. Hickey. 1977. The Monk Parakeet in the United States, 1970–1975. Am. Birds 31:273–78.

Nevius, R. R. 1955. Yellow-crowned Night Herons nesting in Greene County. Migrant 26:18.

———. 1964. A Tennessee nesting of the Black Rail. Migrant 35:59–60.

Newman, G. A. 1970. Cowbird parasitism and nesting success of Lark Sparrows in southern Oklahoma. Wilson Bull. 82:304–9.

Newton, I. 1972. Finches. Collins, London.

Nice, M. M. 1957. Nesting success in altricial birds. Auk 74:305–21.

———. 1968. Mississippi Song Sparrow. Pp. 1513–23 *In* A. C. Bent (O. L. Austin Jr., ed.). Life histories of North American cardinals, grosbeaks, buntings, towhees, finches, sparrows, and allies, part 3. U.S. Natl. Mus. Bull. 237(3).

Nicholas, N. S., S. M. Zedaker, C. Eagar, and F. T. Bonner. 1992. Seedling recruitment and stand regeneration in spruce-fir forests of the Great Smoky Mountains. Bull. Torrey Bot. Club 119:289–99.

Nicholls, T. H., and D. W. Warner. 1972. Barred owl habitat use as determined by radio-telemetry. J. Wildl. Manage. 36:213–24.

Nichols, J. D., and F. A. Johnson. 1990. Wood duck population dynamics: a review. Pp. 83–105 *In* L. H. Fredrickson et al., eds. Proc. 1988 Wood Duck Symp., St. Louis, Mo.

Nicholson, C. P. 1976. The Bachman's Sparrow in Tennessee. Migrant 47:53–60.

———. 1977. The Red-cockaded Woodpecker in Tennessee. Migrant 48:53–62.

———. 1979a. BBC 86. Deciduous forest and contour strip mine. Am. Birds 33:78–79.

———. 1979b. BBC 88. Strip mine and deciduous woodlot. Am. Birds 33:79.

———. 1980a. Corrections to the 1974 Grundy Co. foray report. Migrant 51:53.

———. 1980b. Birds of Decatur County. Migrant 51:1–10.

———. 1980c. BBC 84—Deciduous forest and contour strip mine. Am. Birds 34:66.

———. 1980d. BBC 86—Strip mine and deciduous woodlot. Am. Birds 34:67.

———. 1981a. Birds of Fentress County, including the 1979 foray. Migrant 52:53–62.

———. 1981b. Nesting of the Bell's Vireo in Kentucky. Kentucky Warbler 57:77–79.

———. 1982a. The birds of Pickett County, Tennessee. Migrant 53:25–36.

———. 1982b. The Yellow Warbler in West Tennessee. Migrant 53:82–83.

———. 1984a. Cliff Swallows still nesting on Swallow Bluff, Decatur County, Tennessee. Migrant 55:15–16.

———. 1984b. Late spring and summer birds of McNairy County, Tennessee. Migrant 55:29–39.

———. 1986. Alexander Wilson's travels in Tennessee. Migrant 57:1–7.

———. 1987. Notes on high elevation breeding birds of Frozen Head State Natural Area, Tennessee. Migrant 58:39–43.

———. 1991. Geographic patterns in species occurrence of Tennessee's breeding birds. J. Tenn. Acad. Sci. 66:195–98.

———. In press. Nest records of the Black-throated Green Warbler in Tennessee. Migrant.

Nicholson, C. P., and P. B. Hamel. 1986. Tennessee Breeding Bird Atlas handbook. Tenn. Ornithol. Soc., Nashville.

Nicholson, C. P., and T. C. Pitts. 1982. Nesting of the Tree Swallow in Tennessee. Migrant 53:73–80.

Nickell, W. P. 1951. Studies of habitats, territory, and nests of the Eastern Goldfinch. Auk 68:447–70.

———. 1965. Habitats, territory, and nesting of the Catbird. Amer. Midl. Nat. 73:433–78.

Nolan, V., Jr. 1963. Reproductive success of birds in a deciduous scrub habitat. Ecology 44:305–13.

———. 1974. Notes on parental behavior and development of the young in the Wood Thrush. Wilson Bull. 86:144–55.

———. 1978. The ecology and behavior of the Prairie Warbler *Dendroica discolor*. Ornithol. Monogr. 26, Allen Press, Lawrence, Kans.

Nolan, V., Jr., and C. F. Thompson. 1975. The occurrence and significance of anomalous reproductive activities in two North American nonparasitic cuckoos *Coccyzus* spp. Ibis 117:496–503.

Noon, B. R. 1981. The distribution of an avian guild along a temperate elevational gradient: the importance and expression of competition. Ecol. Monogr. 51:105–24.

Norman, R. F., and R. J. Robertson. 1975. Nest searching behavior in the Brown-headed Cowbird. Auk 92:610–11.

Norris, D. J., and W. H. Elder. 1982. Distribution and habitat characteristics of the Painted Bunting in Missouri. Trans., Missouri Acad. Sci. 16:77–83.

Norris, R. A. 1958. Comparative biosystematics and life history of the nuthatches *Sitta pygmaea* and *Sitta pusilla*. Univ. California Publ. Zool. 56:119–300.

Nunley, H. W. 1960. Chuck-will's-widow. Migrant 31:57–58.

Nunnally, L. B., and M. D. Williams. 1977. A Black-crowned Night Heron colony in Sevier County, Tennessee. Migrant 48:42.

Odum, E. P. 1948. Nesting of the Mountain Vireo at Athens, Georgia, conclusive evidence of a southward invasion. Oriole 13:17–20.

Odum, E. P., and D. W. Johnston. 1951. The House Wren breeding in Georgia: An analysis of a range extension. Auk 68:357–66.

Odum, E. P., and T. D. Burleigh. 1946. Southward invasion in Georgia. Auk 63:388–401.

Odum, R. R. 1977. Sora *(Porzana carolina)*. Pp. 57–65 *In* G. C. Sanderson, ed. Management of migratory shore and upland game birds in North America. Internat. Assoc. Fish and Wildl. Agen., Washington, D.C.

O'Halloran, K. A., and R. N. Conner. 1987. Habitat used by Brown-headed Nuthatches. Bull. Texas Ornithol. Soc. 20:7–13.

Orians, G. H. 1985. Blackbirds of the Americas. Univ. Washington Press, Seattle.

Oring, L. W., D. B. Lank, and S. J. Maxson. 1983. Population studies of the polyandrous Spotted Sandpiper. Auk 56:272–85.

Osborn, M. R., F. Llacuna, and M. Linsenbigler. 1992. The Conservation Reserve Program—Enrollment statistics for singup periods 1–11 and fiscal years 1990–1992. U.S.D.A. Economic Research Service, Stat. Bull. 843.

Owen, J. B. 1979a. First Tennessee breeding record of Red-breasted Nuthatch outside of mountains. Migrant 50:36.

———. 1979b. Hummingbird uses nest third year. Migrant 50:81.

Palmer, R. S. 1949. Maine birds. Bull. Mus. Comp. Zoology 102.

———. 1962. Handbook of North American birds, Vol. 1. Yale Univ. Press, New Haven, Conn.

———. 1976. Handbook of North American birds, Vol. 2–3. Yale Univ. Press, New Haven, Conn.

———. 1988. Handbook of North American birds, Vol. 4–5. Yale Univ. Press, New Haven, Conn.

Palmer, T. S. 1954. Biographies of members of the American Ornithologist's Union. Amer. Ornithol. Union, Washington, D.C.; Lord Baltimore Press, Baltimore.

Palmer-Ball, B., Jr. 1991. Current status of colonial nesting waterbirds in Kentucky. J. Tenn. Acad. Sci. 4:211–14.

Pardue, P. S. 1959. A Western Meadowlark in Knox County, Tennessee. Migrant 30:30–31.

Parker, J. W. 1988. Mississippi Kite. Pp. 166–86 *In* R. S. Palmer. Handbook of North American Birds, Vol. 4. Yale Univ. Press, New Haven, Conn.

Parker, J. W., and J. C. Ogden. 1979. The recent history and status of the Mississippi Kite. Am. Birds 33:119–129.

Parker, M. 1991. Season report. Arkansas Audubon Newsletter 36(4):3.

Parmalee, P. W., and W. E. Klippel. 1982. Evidence of a boreal avifauna in Middle Tennessee during the late Pleistocene. Auk 99:365–68.

———. 1991. Seasonal variation in prey of the Barn Owl *(Tyto alba)* in Tennessee. J. Tenn. Acad. Sci. 66:219–24.

Parmelee, D. F. 1959. The breeding behavior of the Painted Bunting in southern Oklahoma. Bird-Banding 30:1–18.

Parmer, H. E. 1985. Birds of the Nashville area, 4th ed. Nashville Chapter, Tenn. Ornithol. Soc.

Patterson, D. E. 1966. Cyclical interaction of Cliff Swallows and House Sparrows. Migrant 37:76.

Pauley, E. F., and E. E. C. Clebsch. 1990. Patterns of *Abies fraseri* regeneration in a Great Smoky Mountains spruce-fir forest. Bull. Torrey Bot. Club 117:375–81.

Payne, R. B. 1982. Ecological consequences of song matching: breeding success and intraspecific song mimicry in Indigo Buntings. Ecology 63:401–11.

———. 1987. Populations and type specimens of a nomadic bird: comments on the North American Crossbills *Loxia pusilla* Gloger 1834 and *Crucirostra minor* Brehm 1845. Occ. Pap. Mus. Zool. Univ. Michigan 714:1–37.

———. 1990. Indigo Bunting. Pp. 1–22, *In* A. Poole, P. Stettenheim, and F. Gill, eds. Birds of North America, No. 4. Amer. Ornithol. Union, Philadelphia.

Peake, R. H., Jr. 1965. Saw-whet Owls in Jackson County, North Carolina. Chat 29:110–11.

Pearson, T. G., ed. 1917. Birds of America, Part 2. The University Society, New York.

Pearson, T. G., C. S. Brimley, and H. H. Brimley. 1942. Birds of North Carolina. North Carolina Dept. Agri., Raleigh.

Peart, D. R., N. S. Nicholas, S. M. Zedaker, M. M. Miller-Weeks, and T. G. Siccama. 1992. Condition and recent trends in high-elevation red spruce populations. Pp. 125–91 *In* C. Eager and M. B. Adams, eds. Ecology and decline of red spruce in the eastern United States. Springer-Verlag, New York.

Peeples, W. R. 1991. Red Crossbills feeding young in Shelby County, Tennessee. Migrant 62:107.

Peterjohn, B. G., and J. R. Sauer. 1993. North American Breeding Bird Survey annual summary 1990–1991. Bird Populations 1:1–15.

Peterjohn, B. G., J. R. Sauer, and W. A. Link. 1994. The 1992 and 1993 summary of the North American Breeding Bird Survey. Bird Populations 2:46–61.

Peters, R. L., and T. E. Lovejoy, eds. 1992. Global warming and biological diversity. Yale Univ. Press, New Haven, Conn.

Petersen, L. 1979. Ecology of Great Horned Owls and Red-tailed Hawks in southeastern Wisconsin. Wisconsin Dept. Nat. Res., Tech. Bull. 11.

Peterson, A. J. 1955. The breeding cycle in the Bank Swallow. Wilson Bull. 67:235–286.

Petit, L. J. 1986. Factors affecting the reproductive success of Prothonotary Warblers *(Protonotaria citrea)* nesting in riverine habitat. M.S. Thesis, Bowling Green State Univ., Bowling Green, Ohio.

———. 1989. Breeding biology of Prothonotary Warblers in riverine habitat in Tennessee. Wilson Bull. 101:51–61.

———. 1991a. Effects of habitat quality on the breeding density, reproductive success, and mating system of Prothonotary Warblers *(Protonotaria citrea)*. Ph.D. Diss., Univ. Arkansas, Fayetteville.

———. 1991b. Experimentally-induced polygyny in a monogamous bird species: Prothonotary Warblers and the polygyny threshold. Behav. Ecol. Sociobiol. 29:177–87.

———. 1991c. Adaptive tolerance of cowbird parasitism by Prothonotary Warblers: a consequence of nest-site limitation? Anim. Behav. 41:425–32.

Petit, L. J., and D. R. Petit. 1988. Use of Red-winged Blackbird nest by a Prothonotary Warbler. Wilson Bull. 100:305–6.

Petit, L. J., W. J. Fleming, K. E. Petit, and D. R. Petit. 1987. Nest box use by Prothonotary Warblers *(Protonotaria citrea)* in riverine habitat. Wilson Bull. 99:485–88.

Petrides, G. A. 1938. A life history study of the Yellow-breasted Chat. Wilson Bull. 50:184–89.

Phillips, A. R. 1986. The known birds of North and Middle America, Part I. Denver.

Phillips, J. C. 1926. Natural history of the ducks, Vol. 4. Houghton Mifflin Co., New York.

Phillips, R. 1982. Nesting of the House Finch at Kingsport, Tennessee. Migrant 53:40–41.

Phillips, R. A. 1979a. Summer record of Dickcissels and Bobolinks in Hawkins County. Migrant 50:17.

———. 1979b. Notes on summer birds of the Canadian zone forest of Roan Mountain. Migrant 50:73–76.

———. 1989. The breeding birds of Meadowview Marsh, Sullivan County, Tennessee. M.S. Thesis, East Tenn. State Univ., Johnson City.

Phillips, R. A., and F. J. Alsop III. 1978. Notes on some adaptive nesting behavior of the Killdeer *(Charadrius vociferus)* in East Tennessee. Migrant 49:73–75.

Pickering, C. F. 1937. A September visit to Reelfoot Lake. Migrant 8:49–50.

———. 1938. Notes from members of the Clarksville chapter. Migrant 9:13.

———. 1941. Interesting days on Reelfoot Lake. Migrant 12:24–26.

Pickwell, G. B. 1931. The Prairie Horned Lark. Trans. St. Louis Acad. Sci. 27.

Pilsbry, H. A., and S. N. Rhoads. 1896. Contributions to the zoology of Tennessee. No. 4, Mollusks. Proc. Acad. Nat. Sci. Phil. 1896:487–506.

Pindar, L. O. 1886. The breeding of *Branta canadensis* at Reelfoot Lake, Tenn. Auk 3:481.

———. 1889. List of the birds of Fulton County, Kentucky. Auk 6:310–16.

———. 1925. Birds of Fulton County, Kentucky. Wilson Bull. 37:77–88, 163–69.

Pinkowski, B. C. 1971. An analysis of banding–recovery data on Eastern Bluebirds banded in Michigan and three neighboring states. Jack-Pine Warbler 49:33–50.

Pitts, T. D. 1972. Nesting of Bank Swallows in Lake County. Migrant 43:48.

———. 1973. Tennessee heron and egret colonies: 1972. Migrant 44:89–93.

———. 1977. Tennessee heron and egret colonies: 1973–1975. Migrant 48:25–29.

———. 1978a. Some nesting habits of Carolina Chickadees. J. Tenn. Acad. Sci. 53:14–16.

———. 1978b. Comparison of American Woodcock courtship activities in Knox and Weakley Counties, Tennessee. Migrant 49:29–30.

———. 1978c. Status of the American Woodcock in Tennessee. Migrant 49:31–36.

———. 1979. Nesting habits of rural and suburban house Sparrows in northwest Tennessee. J. Tenn. Acad. Sci. 54:145–48.

———. 1981. Eastern Bluebird population fluctuations in Tennessee during 1970–1979. Migrant 52:29–37.

———. 1982a. First record of occurrence and possible nesting of Black-bellied Whistling-Duck in Tennessee. Migrant 53:1–3.

———. 1982b. Establishment of a new heron and egret colony at Reelfoot Lake, Tennessee. Migrant 53:63–64.

———. 1982c. Nesting season records of Willow Flycatchers in West Tennessee. Migrant 53:84–85.

———. 1983. Bell's Vireo in Obion County, Tennessee. Migrant 54:38.

———. 1984. Description of American Robin territories in northwest Tennessee. Migrant 55:1–6.

———. 1985. The breeding birds of Reelfoot Lake, Tennessee. Migrant 56:29–41.

Poly, D. M. 1979. Nest site selection in relation to water level and some aspects of hatching success in Giant Canada Geese *(Branta canadensis maxima)*. M.S. Thesis, Middle Tenn. State Univ., Murfreesboro.

Poole, A. F. 1989. Ospreys, a natural and unnatural history. Cambridge Univ. Press, Cambridge.

Portnoy, J. W., and W. E. Dodge. 1979. Red-shouldered Hawk nesting ecology and behavior. Wilson Bull. 91:104–11.

Potter, E. F. 1980. Notes on nesting Yellow-billed Cuckoos. J. Field Ornithol. 51:17–29.

———. 1985. Breeding habits, nestling development, and vocalizations in the Summer Tanager. Chat 49:57–66.

Potter, E. F., and H. E. LeGrand Jr. 1980. Bird finding on Roan Mountain, Mitchell County, N.C. Chat 44:32–36.

Potter, E. F., J. F. Parnell, and R. P. Tuelings. 1980. Birds of the Carolinas. Univ. North Carolina Press, Chapel Hill.

Potter, P. E. 1974. Breeding behavior of Savannah sparrows in southeastern Michigan. Jack-Pine Warbler 52:50–63.

Prescott, D. R. C. 1987. Territorial responses to song playback in allopatric and sympatric populations of Alder *(Empidonax alnorum)* and Willow *(E. traillii)* Flycatchers. Wilson Bull. 99:611–19.

Prescott, K. W. 1965. Studies in the life history of the Scarlet Tanager, *Piranga olivacea*. New Jersy State Mus. Investigations, No. 2.

Preston, C. R., C. S. Harger, and H. E. Harger. 1989. Habitat use and nest-site selection by Red-shouldered Hawks in Arkansas. Southwest. Nat. 34:72–78.

Pullin, B. P. 1983. Great Blue Herons nest on transmission line tower in Henry County, Tennessee. Migrant 54:18.

———. 1985–88. Surveys of wading bird colonies in the Tennessee Valley region. Unpubl. reports, Tennessee Valley Authority Wildlife Res. Dev. Program, Norris, Tenn.

———. 1987. Restoring Great Egrets to the Tennessee Valley. Pp. 11–13 In Proc. 3rd S.E. Nongame and Endangered Species Symposium, Athens, Ga.

———. 1990. Size and trends of wading bird populations in Tennessee during 1977–1988. Migrant 61:95–104.

Purrington, R. D. 1991. The changing seasons: central–southern region. Am. Birds 45:1124–28.

Putnam, L. S. 1949. The life history of the Cedar Waxwing. Wilson Bull. 61:141–82.

Pyle, C. 1984. Pre-park disturbance in the spruce-fir forests of Great Smoky Mountains National Park. Pp. 115–30 in P. S. White, ed. The Southern Appalachian spruce-fir ecosystem: its biology and threats. USDI Natl. Park Service Res. Resour. Manage. Rep. SER-71.

Pyle, C., and M. P. Schafale. 1988. Land use history of three spruce-fir forest sites in southern Appalachia. J. Forest Hist. 32:4–21.

Pyne, S. J. 1982. Fire in America: A cultural history of wildland and rural fire. Princeton Univ. Press, Princeton, N.J.

Quarterman, E. 1989. Structure and dynamics of the limestone cedar glade communities in Tennessee. J. Tenn. Acad. Sci. 64:155–58.

Quinn, M. L. 1992. Should all degraded landscapes be restored? A look at the Appalachian Copper Basin. Land Degradation & Rehabilitation 3:115–34.

Rabenold, K. N. 1978. Foraging strategies, diversity, and seasonality in bird communities of Appalachian spruce-fir forests. Ecol. Monogr. 48:397–424.

———. 1993. Latitudinal gradients in avian species diversity and the role of long-distance migration. Current Ornithology 10:247–74.

Ragsdale, G. H. 1879. Nesting of the Snowbird (*Junco hyemalis*) in eastern Tennessee. Bull. Nuttall Ornithol. Club 4:238–39.

Ratcliffe, D. 1980. The Peregrine Falcon. Buteo Books, Vermillion, S.Dak.

Reed, J. M. 1986. Vegetation structure and Vesper Sparrow territory location. Wilson Bull. 98:144–47.

Reid, M. L. 1992. The Tennessee Biodiversity Program: linking knowledge with the land. Tenn. Conservationist 58(6):15–18.

Reller, A. W. 1972. Aspects of behavioral ecology of Red-headed and Red-bellied Woodpeckers. Am. Midl. Nat. 88:270–90.

Renken, R. B., and E. P. Wiggers. 1989. Forest characteristics related to Pileated Woodpecker territory size in Missouri. Condor 91:642–52.

Reynolds, J. D., and R. W. Knapton. 1984. Nest-site selection and breeding biology of the Chipping Sparrow. Wilson Bull. 96:488–93.

Rhoads, S. N. 1895a. Contributions to the zoology of Tennessee. No. 2, Birds. Proc. Acad. Nat. Sci. Phil. 1895:463–501.

———. 1895b. Contributions to the zoology of Tennessee. No. 1, Reptiles and amphibians. Proc. Acad. Nat. Sci. Phil. 1895:376–407.

———. 1896a. Contributions to the zoology of Tennessee. No. 3, Mammals. Proc. Acad. Nat. Sci. Phil. 1896:175–205.

———. 1896b. Additions to the avifauna of Tennessee. Auk 13:181–82.

Ripley, S. D. 1977. Rails of the world. David R. Godine, Boston.

Ritchison, G. 1981. Breeding biology of the White-breasted Nuthatch. Loon 53:184–87.

———. 1986. The singing behavior of female Northern Cardinals. Condor 88:156–59.

Ritchison, G., P. M. Cavanagh, J. R. Belthoff, and E. J. Sparks. 1988. The singing behavior of Eastern Screech-Owls: seasonal timing and response to playback of conspecific song. Condor 90:648–52.

Robbins, C. S. 1973. Introduction, spread, and present abundance of the House Sparrow in North America. Ornithol. Monogr. 14:3–9, Allen Press, Lawrence, Kans.

Robbins, C. S., D. Bystrak, and P. H. Geissler. 1986. The breeding bird survey: its first fifteen years, 1965–1979. U.S. Fish Wildl. Serv. Res. Publ. 157.

Robbins, C. S., D. K. Dawson, and B. A. Dowell. 1989. Habitat area requirements of breeding forest birds of the middle Atlantic states. Wildl. Monogr. 103.

Robbins, C. S., J. R. Sauer, R. S. Greenburg, and S. Droege. 1989. Population declines in North American birds that migrate to the neotropics. Proc. Natl. Acad. Sci. USA 86:7658–62.

Robbins, C.S., J. W. Fitzpatrick, and P. B. Hamel. 1992. A warbler in trouble: *Dendroica cerulea*. pp. 549–62 in J. M. Hagan III and D. W. Johnston, eds. Ecology and conservation of Neotropical migrant landbirds. Smithsonian Inst. Press, Washington, D.C.

Robbins, M. B., and D. A. Easterla. 1992. Birds of Missouri. Univ. Missouri Press, Columbia.

Robbins, M. B., M. J. Braun, and E. A. Tomey. 1986. Morphological and vocal variation across a contact zone between the chickadees *Parus atricapillus* and *P. carolinensis*. Auk 103:655–66.

Roberts, T. H. 1978. Migration, distribution and breeding of American Woodcock. M.S. Thesis, Univ. Tennessee, Knoxville.

Robertson, K. O. L. 1979. Breeding birds in a virgin mixed mesophytic forest in the western escarpment of the Cumberland Plateau in Tennessee. M.S. Thesis, Austin Peay State Univ., Clarksville, Tenn.

Robertson, W. B., Jr. 1988. American Swallow-tailed Kite. Pp. 109–31 In R. S. Palmer, ed. Handbook of North American birds, Vol. 4. Yale Univ. Press, New Haven, Conn.

Robinson, J. C. 1988. Heron and egret roost discovered near Memphis. Migrant 59:118–19.

———. 1989. A concentration of Bewick's Wrens in Stewart County, Tennessee. Migrant 60:1–3.

——— 1990. An annotated checklist of the birds of Tennessee. Univ. Tenn. Press, Knoxville.

Robinson, J. C., and D. W. Blunk. 1989. The birds of Stewart County, Tennessee. Pp. 70–103 In A. F. Scott, ed. Proc. Contr. Papers Sessions, 2nd Ann. Symp. on the Natural History of Lower Tennessee and Cumberland River Valleys. Center for Field Biology of Land Between the Lakes, Austin Peay State Univ., Clarksville, Tenn.

Roe, F. G. 1970. The North American Buffalo: A critical study of the species in its wild state. 2nd ed. Univ. Toronto Press, Toronto.

Roever, K. 1951. Black-billed Cuckoo nesting near Jackson, Tennessee. Migrant 22:30–31.

Root, R. B. 1969. The behavior and reproductive success of the Blue-gray Gnatcatcher. Condor 71:16–31.

Roseberry, J. W., and W. D. Klimstra. 1970. The nesting ecology and reproductive performance of the Eastern Meadowlark. Wilson Bull. 82:243–67.

Rostlund, E. 1957. The myth of a natural prairie belt in Alabama: an interpretation of historical records. Annals, Assoc. Am. Geographers 47:392–411.

Rothstein, S. I. 1971. High nest density and non-random nest placement in the Cedar Waxwing. Condor 73:483–85.

Rowher, S. A. 1971. Molt and the annual cycle of the Chuck-will's-widow, *Caprimulgus carolinensis*. Auk 88:485–519.

Runkle, J. R. 1982. Patterns of disturbance in some old-growth mesic forests in eastern North America. Ecology 63:1533–46.

———. 1985. Disturbance regimes in temperate forests. Pp. 17–33 *In* S. T. A. Pickett and P. S. White, eds. The ecology of natural disturbance and patch dynamics. Academic Press, Orlando, Florida.

Rusch, D. H., C. D. Ankney, H. Boyd, J. R. Longcore, F. Montalbano III, J. K. Ringleman, and V. D. Stotts. 1989. Population ecology and harvest of the American Black Duck: a review. Wildl. Soc. Bull. 17:379–406.

Samuel, D. E. 1971. The breeding biology of Barn and Cliff Swallows in West Virginia. Wilson Bull. 83:284–301.

Santer, S. 1992. Black-necked Stilt. Pp. 134–35 *In* D. W. Brauning, ed. Atlas of breeding birds in Pennsylvania. Univ. Pittsburg Press, Pittsburg.

Sappington, J. N. 1977. Breeding biology of House Sparrows in north Mississippi. Wilson Bull. 89:300–309.

SAS Institute, Inc. 1987. SAS/STAT guide for personal computers, version 6. SAS Institute, Inc., Cary, N.C.

Satz, R. N. 1979. Tennessee's Indian peoples. Univ. Tennessee Press, Knoxville.

Sauer, J. R. 1990. Route-regression analysis of Scissor-tailed Flycatcher population trends. Pp. 160–63 *In* J. R. Sauer and S. Droege, eds. Survey designs and statistical methods for the estimation of avian population trends. U.S. Fish Wildl. Serv., Biol. Rep. 90(1).

Sauer, J. R., and S. Droege. 1992. Geographic patterns in population trends of neotropical migrants in North America. Pp. 26–42 *In* J. M. Hagan III and D. W. Johnston, eds. Ecology and conservation of neotropical migrant landbirds. Smithsonian Inst. Press, Washington, D.C.

Saunders, R. H. 1973. Some behavioral and demographic characteristics of a bobwhite quail (*Colinus virginianus*) population. M.S. Thesis, Univ. Tennessee, Knoxville.

Savage, H., Jr., and E. J. Savage. 1986. André and Francois André Michaux. Univ. Press of Virginia, Charlottesville.

Savage, T. 1964. Rough-winged swallow nestiing in the Great Smoky Mountains. Migrant 35:51.

Savage, T. 1965. Recent observations on the Saw-whet Owl in G.S.M.N.P. Migrant 36:15–16.

Schallenberg, R. H., and D. A. Ault. 1977. Raw material supply and technological change in the American charcoal iron industry. Technology and Culture 18:436–66.

Schartz, R. L., and J. L. Zimmerman. 1971. The time and energy budget of the male Dickcissel (*Spiza americana*). Condor 73:65–76.

Schorger, A. W. 1952. Introduction of the Domestic Pigeon. Auk 69:462–63.

———. 1953. Obituary—Samuel Nicholson Rhoads. Auk 70:238.

———. 1955. The Passenger Pigeon: Its natural history and extinction. Univ. Wisconsin Press, Madison.

———. 1966. The Wild Turkey: its history and domestication. Univ. Oklahoma Press, Norman.

Schrantz, F. G. 1943. Nest life of the Yellow Warbler. Auk 60:367–87.

Schultz, V. 1953. Status of the Ruffed Grouse in Tennessee. Migrant 24:45–52.

———. 1955. Status of the Wild Turkey in Tennessee. Migrant 26:1–8.

Scott, D. M., J. A. Darley, and A. V. Newsome. 1988. Length of the laying season and clutch size of Gray Catbirds at London, Ontario. J. Field Ornithol. 59:355–60.

Scott, D. M., R. E. Lemon and J. A. Darley. 1987. Relaying interval after nest failure in Gray Catbirds and Northern Cardinals. Wilson Bull. 99:708–12.

Scott, F. R. 1966. Results of Abingdon foray, June 1966. Raven 37:71–76.

———. 1982. The third VSO foray to Mount Rogers—June 1980. Raven 53:3–16.

Scott, J. M., F. Davis, B. Csuti, R. Noss, B. Butterfield, C. Groves, H. Anderson, S. Caicco, F. D'Erchia, T. C. Edwards Jr., J. Ulliman, and R. G. Wright. 1993. Gap analysis: A geographic approach to the protection of biological diversity. Wildl. Monogr. 123.

Sealy, S. G. 1978a. Possible influence of food on egg-laying and clutch size in the Black-billed Cuckoo. Condor 80:103–4.

———. 1978b. Clutch size and nest placement of Pied-billed Grebes in Manitoba. Wilson Bull. 90:301–2.

———. 1985. Erect posture of the young Black-billed Cuckoo: an adaptation for early mobility in a nomadic species. Auk 102:889–892.

Sedgwick, J. A., and F. L. Knopf. 1989. Regionwide polygyny in Willow Flycatchers. Condor 91:473–75.

Sennett, G. B. 1887. Observations in western North Carolina mountains in 1886. Auk 4:240–45.

Shankman, D. 1993. Channel migration and vegetation patterns in the southeastern Coastal Plain. Consv. Biology 7:176–83.

Shanks, R. E. 1954. Climates of the Great Smoky Mountains. Ecology 35:354–61.

———. 1958. Floristic regions of Tennessee. J. Tenn. Acad. Sci. 33:194–210.

Sharp, V., Jr. 1931. Nesting data on middle Tennessee birds, part II. Migrant 2:11–12.

———. 1932. The Pileated Woodpecker. Migrant 3:40–41.

Sharrock, J. T. R. 1976. The atlas of breeding birds in Britain and Ireland. T. and A. D. Poyser, Hertfordshire, England.

Shaver, J. M. 1932. A bibliography of Tennessee ornithology. J. Tenn. Acad. Sci. 7:179–190.

Shaver, J. M., and M. B. Roberts. 1930. Some nesting habits of the Cardinal. J. Tenn. Acad. Sci. 5:157–70.

Sheldon, W. G. 1967. The book of the American Woodcock. Univ. Massachusetts Press, Amherst.

Sherman, F., Jr. 1910. Peculiarities in the distribution of some North Carolina birds. J. Elisha Mitchell Soc. 26:71–75.

Sherrod, S. K. 1983. Behavior of fledgling Peregrines. The Peregrine Fund, Ithaca, N.Y.

Sherry, T. T., and R. T. Holmes. 1992. Population fluctuations in a long-distant Neotropical migrant: Demographic evidence for the importance of breeding season events in the American Redstart. Pp. 431–42 in J. M. Hagan III and D. W. Johnston, eds. Ecology and Conservation of neotropical migrant landbirds. Smithsonian Insti. Press, Washington, D.C.

Shoup, C. S. 1943. Notes from the background of our knowledge of the zoology of Tennessee. J. Tenn. Acad. Sci. 18:126–36.

Shy, E. 1984. Habitat shift and geographical variation in North American tanagers (Thraupinae:*Piranga*). Oecologia 63:281–85.

Sibley, C. G. 1955. Behavioral mimicry in the titmice (Paridae) and certain other birds. Wilson Bull. 67:128–32.

Sibley, C. G., and J. E. Ahlquist. 1990. Phylogeny and classification of birds: a study in molecular evolution. Yale Univ. Press, New Haven, Conn.

Siebenheller, N., and W. A. Siebenheller. 1992. Pine Siskins build nest in Transylvania County, N.C. Chat 56:57–59.

Sights, W. 1943. Hooded Merganser's nest on Reelfoot. Migrant 14:16.

Simmers, R. W., Jr. 1978. BBC 80—Pastures with brush, wooded strips, and scattered trees. Am. Birds 32:80.

———. 1979. BBC 87—Mixed-mesophytic woods, fields, and brush. Am. Birds 33:79.

———. 1980. BBC 85—Mixed-mesophytic woods, fields, and brush. Am. Birds 34:66-67.

———. 1981. BBC 108—Mixed-mesophytic woods, fields, and brush. Am. Birds 35:76.

———. 1982a. BBC 104—Disturbed mixed-mesophytic woodland ravine. Am. Birds 36:78.

———. 1982b. BBC 105—Mixed-mesophytic woods, fields, and brush. Am. Birds 36:78–79.

———. 1983. BBC 90—Mixed-mesophytic woods, fields, and brush. Am. Birds 37:78.

———. 1984. BBC 93. Mixed-mesophytic woods, fields, and brush. Am. Birds 38:95.

———. 1989. BBC 51. Mixed-mesophytic woods, fields, and brush. J. Field Ornith. (supplement) 60:53–54.

———. 1990. BBC 66—Mixed-mesophytic woods and brush. J. Field Ornith. (supplement) 61:67–68.

———. 1991. BBC 71—Mixed mesophytic woods and brushland. J. Field Ornith. (supplement) 62:74.

Simons, T., S. K. Sherrod, M. W. Collopy, and M. A. Jenkins. 1988. Restoring the Bald Eagle. Amer. Scientist 76:253–60.

Simpson, M. B., Jr. 1968. The Saw-whet Owl: breeding distribution in North Carolina. Chat 32:83–89.

———. 1969. Nesting of the Brown-headed Nuthatch in the southern Appalachians. Chat 33:103–4.

———. 1970. In quest of the Saw-whet Owl. Chat 34:59–62.

———. 1972. The Saw-whet Owl population of North Carolina's southern Great Balsam Mountains. Chat 36:39–47.

———. 1976a. Breeding season records of the Magnolia Warbler in Grayson County, Virginia. Raven 47:56.

———. 1976b. Breeding season habitats of the Golden-crowned Kinglet in the southern Blue Ridge Mountains. Chat 40:75–76.

———. 1977. The Black-capped Chickadee in the southern Blue Ridge Mountain province: a review of its ecology and distribution. Chat 41:79–86.

———. 1978a. Breeding season distribution and ecology of the Vesper Sparrow in the southern Blue Ridge Mountain province. Chat 42:1–2.

———. 1978b. Ecological factors contributing to the decline of Bewick's Wren as a breeding species in the southern blue Ridge Mountain province. Chat 42:25–28.

———. 1980. William Brewster's exploration of the Southern Appalachian Mountains: The journal of 1885. North Carolina Hist. Rev. 57:43–77.

———. 1993. Pine Siskin nesting in the Southern Blue Ridge Mountain province. Chat 57:47–49.

Simpson, M. B., Jr., and P. G. Range. 1974. Evidence of the breeding of Saw-whet Owls in western North Carolina. Wilson Bull. 86:173–74.

Simpson, R. C. 1972. A study of bobwhite quail nest initiation dates, clutch sizes and hatch sizes in southwest Georgia. Pp. 199–204 in J. A. Morrison and J. C. Lewis, eds. Proc. First Natl. Bobwhite Quail Symp. Okla State Univ., Stillwater.

Simpson, T. W. 1939. The feeding habits of the coot, Florida gallinule, and least bittern on Reelfoot Lake. J. Tenn. Acad. Sci. 14:110–15.

Slack, R. D. 1976. Nest guarding behavior by male Gray Catbirds. Auk 93:292–300.

Smith, A. B. 1975. BBC 40—Mixed deciduous forest. Am. Birds 29:1097.

———. BBC 49—Maple-gum-hickory forest. Am. Birds 31:47.

Smith, A. B., R. P. Lewis, and F. J. Alsop III. 1975. The nesting of Virginia Rails in Hawkins County. Migrant 46:41–42.

Smith, C. R. 1967. A brief preliminary study of nesting Red-winged Blackbirds. Migrant 38:25–29, 45.

———. 1990. Handbook for atlasing American breeding birds. Vermont Inst. Nat. Sci., Woodstock, Vt.

Smith, D. G., and R. Gilbert. 1984. Eastern Screech-Owl home range and use of suburban habitats in southern Connecticut. J. Field Ornithol. 55:322–29.

Smith, G. P. 1989–90. Surveys of Great Blue Heron colonies in western Tennessee. Unpubl. reports submitted to Tenn. Wildlife Res. Agency, Nashville.

Smith, G. S. 1881. The Red Crossbill (*Loxia curvirostra americana*) in Tennessee. Bull. Nuttall Ornithol. Club 6:56–57.

———. 1883. Crossbills. Ornithol. and Oologist 8:7.

Smith, J. W., and R. B. Renken. 1991. Least Tern nesting habitat in the Mississippi River valley adjacent to Missouri. J. Field Ornithol. 62:497–504.

Smith, J. W., and R. B. Renken. 1993. Reproductive success of Least Terns in the Mississippi River valley. Colonial Waterbirds 16:39–44.

Smith, M. T. 1987. Archaeology of aboriginal culture change in the interior Southeast: Depopulation during the early Historic Period. Univ. Presses of Florida, Gainesville.

Smith, R. D. 1951. Western Meadowlark nesting at Memphis. Migrant 22:21–22.

Smith, R. L. 1968. Grasshopper Sparrow. Pp. 725–45 *In* A. C. Bent (O. L. Austin Jr., ed.). Life histories of North American cardinals, grosbeaks, buntings, towhees, finches, sparrows and allies, part 2. U.S. Natl. Mus. Bull. 237.

Smith, S. D., C. P. Stripling, and J. M. Brannon. 1988. A cultural resource survey of Tennessee's Western Highland Rim iron industry, 1790s–1930s. Tenn. Dept. of Conservation, Div. of Archaeology, Res. Series No. 8, Nashville.

Smith, S. M. 1991. The Black-capped Chickadee: behavioral ecology and natural history. Comstock, Ithaca, New York.

Smith, T. M., and H. H. Shugart. 1987. Territory size variation in the Ovenbird: the role of habitat structure. Ecology 68:695–704.

Smith, W. J., J. Pawlukiewicz, and S. T. Smith. 1978. Kinds of activities correlated with singing patterns of the Yellow-throated Vireo. Anim. Behav. 26:862–84.

Snapp, B. D. 1976. Colonial breeding in the Barn Swallow (Hirundo rustica) and its adaptive significance. Condor 78:471–80.

Soil Conservation Service. 1982. National list of scientific plant names, Vol. 1. USDA Soil Conservation Service, Washington, D.C.

———. 1989. Summary report—1987 Natural Resources Inventory. USDA. Soil Conservation Service, Stat. Bull. 790

Solyom, V. 1940. The Chukar Partridge in Tennessee. Migrant 12:74.

Sordahl, T. A. 1984. Observations on breeding site fidelity and pair formation in American Avocets and Black-necked Stilts. North Amer. Bird Bander 9:8–11.

Soulliere, G. J. 1987. Distinguishing Hooded Merganser and Wood Duck eggs by eggshell thickness. J. Wildl. Manage. 51:534.

Spencer, G. R. 1943. Nesting habits of the Black-billed Cuckoo. Wilson Bull. 55:22.

Spero, V. M., and T. D. Pitts. 1984. Use of Wood Duck nest boxes by Common Grackles. J. Field. Ornithol. 55:482–83.

Spero, V. M., F. G. Dallmieir, R. M. Wheat, and T. D. Pitts. 1983. A nesting study of Wood Ducks on Kentucky Lake, Tennessee. Migrant 54:69–75.

Spofford, W. R. 1941. A day at Reelfoot Lake. Migrant 12:74.

———. 1942a. Nesting of the Peregrine Falcon in Tennessee. Migrant 13:29–31.

———. 1942b. A visit to Cranetown. Migrant 13:41–42.

———. 1943. Peregrines in a west Tennessee swamp. Migrant 14:25–27.

———. 1944. Notes on the Peregrine Falcon. Migrant 15:66–67.

———. 1947a. A successful nesting of the Peregrine Falcon with three adults present. Migrant 18:49–51.

———. 1947b. Another tree-nesting Peregrine Falcon record for Tennessee. Migrant 18:60.

———. 1948a. Some additional notes on the birds of Pickett Forest, Tenn. Migrant 19:12–13.

———. 1948b. Early nesting of Bald Eagle at Reelfoot Lake. Migrant 19:25.

———. 1950. Migratory and non-migratory Peregrine Falcons in Tennessee. Migrant 21:48–49.

———. 1971. The breeding status of the Golden Eagle in the Appalachians. Am. Birds 25:3–7.

Springer, M. E., and J. A. Elder. 1980. Soils of Tennessee. Univ. Tennessee Agricul. Exper. Stat. Bull. 596.

St. Cosmé, J. F. B. 1699–1700. Account of Father J. F. Buisson St. Cosme. reprinted in S. C. Williams. 1928. Early Travels in the Tennessee Country, 1540–1800. Watauga Press, Johnson City, Tenn.

Stafford, S. K., and R. W. Dimmick. 1979. Autumn and winter foods of Ruffed Grouse in the southern Appalachians. J. Wildl. Manage. 93:121–27.

Stalmaster, M. V. 1987. The bald eagle. Universe Books, New York.

Stanford, J. A. 1972. Second broods in bobwhite quail. Pp. 21–27 in J. A. Morrison and J. C. Lewis, eds. Proc. First Natl. Bobwhite Quail Symp., Okla. State Univ., Stillwater.

Stedman, S. J. 1987. Nesting habitat of Willow Flycatcher in Tennessee. Migrant 58:49–50.

———. 1988a. Cattle Egrets nest in Sumner County. Migrant 59:29–30.

———. 1988b. Range expansion and population increase of nesting Song Sparrows in Tennessee. Am. Birds 42:382–84.

Stedman, S. J., and B. H. Stedman. 1992. BBC 82—Mature deciduous-coniferous forest with stream. J. Field Ornith. (supplement) 63:86.

———. 1993. BBC 77—Mature deciduous-coniferous forest with stream. J. Field Ornith. (supplement) 64:81–82.

Stedman, S. J., and D. J. Simbeck. 1988. Northern Rough-winged Swallows build nests in semi-trailers. Migrant 59:51–52.

Stein, R. C. 1963. Isolating mechanisms between populations of Traill's Flycatchers. Proc. Am. Phil. Soc. 107:21–50.

Stevens, C. E. 1976. Notes on summer birds in the Virginia mountains, 1970–1975. Raven 47:35–40.

Stevenson, H. M., and A. Stupka. 1948. The altitudinal limits of certain birds in the mountains of the southeastern states. Migrant 19:33–60.

Stewart, P. A. 1952. Dispersal, breeding behavior, and longevity of banded Barn Owls in North America. Auk 69:227–45.

Stewart, R. E. 1953. A life history study of the Yellowthroat. Wilson Bull. 65:99–115.

Stiehl, R. B. 1985. Brood chronology of the Common Raven. Wilson Bull. 97:78–87.

Stoddard, H. L. 1931. The bobwhite quail: its habits, preservation, and increase. Charles Scribner's Sons, New York.

Stogsdill, E. S. 1983. The nesting chronology of Mourning Doves in Knox County, Tennessee. M.S. Thesis, Univ. Tennessee, Knoxville.

Stone, W. 1918. Recent literature—Preliminary list of the birds of Tennessee. Auk 35:93–94.

Stoner, D. 1939. Parasitism of the English Sparrow on the Northern Cliff Swallow. Wilson Bull. 51:221–22.

Strohmeyer, D. A. 1977. Common Gallinule. Pp. 110–17 In G. C. Sanderson, ed. Management of migratory shore and upland game birds in North America. Internat. Assoc. Fish and Wildl. Agencies, Washington, D.C.

Stull, W. DeM. 1968. Eastern and Canadian Chipping Sparrows. Pp. 1166–84 In A. C. Bent (O. L. Austin Jr., ed.). Life histories of North American cardinals, grosbeaks, buntings, towhees, finches, sparrows and allies, part 2. U.S. Natl. Mus. Bull. 237.

Stupka, A. 1946. Occurrence of the Saw-whet Owl in the Great Smoky Mountains during the breeding season. Migrant 17:60–62.

———. 1963. Notes on the birds of Great Smoky Mountains National Park. Univ. Tennessee Press, Knoxville.

Sturm, L. 1945. A study of the nesting activities of the American Redstart. Auk 62:189–206.

Sudworth, G. B., and J. B. Killebrew. 1897. The forests of Tennessee, their extent, character, and distribution. Nashville, Chattanooga, and St. Louis Railway, Nashville, Tenn.

Sutherland, C. A. 1963. Notes on the behavior of Common Nighthawks in Florida. Living Bird 2:31–39.

Swengel, S. R., and A. B. Swengel. 1987. Study of a Northern Saw-whet Owl population in Sauk County, Wisconsin. Pp. 199–208 In R. Nero et al., eds. Biology and conservation of northern forest owls: symposium proceedings. USDA For. Serv. Gen. Tech. Rep. RM-142.

Swindell, M. 1962. A housing shortage? Migrant 33:73.

Taber, W., and D. W. Johnston. 1968. Indigo Bunting. Pp. 80–111 In A. C. Bent (O. L. Austin Jr. ed.). Life histories of North American cardinals, grosbeaks, buntings, towhees, finches, sparrows, and allies, part 1. U.S. Natl. Mus. Bull. 237(1).

Tanner, J. T. 1942. The Ivory-billed Woodpecker. Natl. Audubon Soc. Research Rep. 1.

———. 1952. Black-capped and Carolina Chickadees in the southern Appalachian Mountains. Auk 69:407–24.

———. 1955. The altitudinal distribution of birds in a part of the Great Smoky Mountains. Migrant 26:37–40.

———. 1957. Sight record of a Saw-whet Owl in the Great Smoky Mountains. Migrant 28:28.

———. 1958. Juncos in the Great Smoky Mountains. Migrant 29:61–65.

———. 1974. Nesting Bank Swallows in Knox County. Migrant 45:53.

———. 1985. An analysis of Christmas Bird Counts in Tennessee. Migrant 56:85–97.

———. 1986. An analysis of spring bird counts in Tennessee. Migrant 57:89–97.

———. 1988. Changing ranges of birds in Tennessee. Migrant 59:73–87.

Tarbell, A. T. 1983. A yearling helper with a Tufted Titmouse brood. J. Field Ornithol. 54:89.

Tate, J., Jr. 1973. Methods and annual sequence of foraging by the sapsucker. Auk 90:840–56.

———. 1986. The Blue List for 1986. Am. Birds 40:227–236.

Tate, J., Jr., and D. J. Tate. 1982. The Blue List for 1982. Am. Birds 36:126–35.

Tautin, J., P. H. Geissler, R. E. Munro, and R. S. Pospahala. 1983. Monitoring the population status of American Woodcock. Trans. N. Am. Wildl. Nat. Res. Conf. 48:376–88.

Taylor, W. K., and M. A. Kershner. 1991. Breeding biology of the Great Crested Flycatcher in central Florida. J. Field. Ornithol. 62:28–39.

Technical Working Group 1993. Tennessee wetlands conservation strategy, draft report. Tennessee Interagency Wetlands Committee, Nashville.

Tennessee Department of Conservation and Tennessee Valley Authority. 1960. Conditions resulting from strip mining for coal in Tennessee. Tenn. Dept. Consv. and Commerce, Nashville, and Tennessee Valley Authority, Knoxville.

Tennessee State Game and Fish Commission. 1959. Coturnix Quail. Migrant 30:58.

Tennessee Valley Authority. 1975. Resident waterfowl production and release project, September 1974–September 1975. Tennessee Valley Authority Div. Forestry, Fish., and Wildl. Dev., Norris, Tenn.

———. 1977. Resident waterfowl production and release project, September 1975–September 1976. Tennessee Valley Authority Div. Forestry, Fish., and Wildl. Dev., Norris, Tenn.

Tennessee Wildlife Resources Agency. 1975. Tennessee Wildlife Resources Commission Proclamation No. 75-15—Endangered or threatened species. Nashville.

———. 1976. Tennessee Wildlife Resources Commission Proclamation No. 76-4—Wildlife in need of management. Nashville.

———. 1981. Action for wildlife—Canada Goose. Tenn. Wildlife 4(4):6–12.

———. 1986. Tennessee Wildlife Resources Proclamation No. 86-29—Wildlife in need of management. Nashville.

———. 1990a. Tennessee implementation plan for wetland and waterfowl conservation. Tennessee Wildl. Res. Agency, Nashville.

———. 1990b. 1990 Wild Turkey report. Tennessee Wildl. Res. Agency Tech. Report No. 90-8, Nashville.

———. 1992. Tennessee Ruffed Grouse/Woodcock hunter survey report, 1991–1992. Tennessee Wildl. Res. Agency Tech. Report No. 92-5, Nashville.

———. 1994a. Tennessee Wildlife Resources Commission Proclamation, No. 94-16—Wildlife in need of management. Nashville.

———. 1994b. Tennessee Wildlife Resources Commission Proclamation, No. 94-17—Endangered or threatened species. Nashville.

———. 1994c. 1994 Wild Turkey report. Tennessee Wildl. Res. Agency Tech. Report No. 94-11, Nashville.

Terres, J. K. 1980. The Audubon Society encyclopedia of North American birds. Alfred A. Knopf, New York.

Teulings, R. P. 1972. Briefs for the files. Chat 36:111–15.

Thompson, C. F., and V. Nolan Jr. 1973. Population biology of the Yellow-breasted Chat (*Icteria virens* L.) in southern Indiana. Ecol. Monogr. 43:145–71.

Thompson, C. W. 1991a. The sequence of molts and plumages in Painted Buntings and implications for theories of delayed plumage maturation. Condor 93:209–35.

———. Is the Painted Bunting actually two species? Problems determining species limits between allopatric populations. Condor 93:987–1000.

Thornthwaite, C. W. 1948. An approach toward a rational classification of climate. Geog. Rev. 38:55–94.

Thwaites, R. G. 1904–7. Early western travels, 1748–1846. A. H. Clark, Cleveland, Ohio.

Titus, K., and J. A. Mosher. 1981. Nest-site habitat selected by woodland hawks in the central Appalachians. Auk 98:270–81.

———. 1987. Selection of nest tree species by Red-shouldered and Broad-winged Hawks in two temperate forest regions. J. Field Ornithol. 58:274–83.

Todd, H. O. 1935. Nesting notes from Murfreesboro. Migrant 6:36.

———. 1936. The cowbird in summer near Murfreesboro. Migrant 7:72.

———. 1938. The Black Vulture in Rutherford County. Migrant 9:23–24.

———. 1944. Some nesting records from Murfreesboro. Migrant 15:21–23.

Torrey, B. 1896a. Spring notes from Tennessee. Riverside Press, Cambridge, Mass.

———. 1896b. Some Tennessee bird notes. Atlantic Monthly 77:198–207.
Tove, M. 1980. First evidence of nesting for the Black-capped Chickadee from North Carolina. Chat 44:1–4.
Trabue, L. O. 1965. A review of Tennessee Christmas counts. Migrant 36:36–37, 42–44.
Transeau, E. N. 1935. The prairie peninsula. Ecology 16:423–37.
Turner, L. J., and D. K. Fowler. 1981a. Breeding Bird Census 25—Oak-maple forest. Am. Birds 35:56.
———. 1981b. Breeding Bird Census 26—Oak-maple forest. Am. Birds 35:56.
Twery, M. J. 1990. Effects of defoliation by gypsy moth. Pp. 27–39 In Gottschalk, K. W., M. J. Twery, and S. I. Smith, eds. Interagency gypsy moth research review. USDA For. Serv. Gen. Tech. Rep. NE-146.
Tyler, B. P. 1922. The Starling in Tennessee. Bird Lore 28:334.
———. 1936. Prairie Horned Lark nesting in N-E Tenn. Migrant 7:50.
———. 1948. Status of the House Wren in northeastern Tennessee. Migrant 19:73–74.
Tyler, B. P., and R. B. Lyle. 1933. Winter birds of northeastern Tennessee. Migrant 4:25–29.
Tyler, W. M. 1940. Chimney Swift. Pp. 271–93 In A. C. Bent. Life histories of North American cuckoos, goatsuckers, hummingbirds, and their allies. U.S. Natl. Mus. Bull. 176.
———. 1942. Tree Swallow. Pp. 384–400 In A. C. Bent. Life histories of North American flycatchers, larks, swallows and their allies. U.S. Natl. Mus. Bull. 170.
———. 1948a. White-breasted Nuthatch. Pp. 1–12 In A. C. Bent. Life histories of North American nuthatches, wrens, thrashers, and their allies. U.S. Natl. Mus. Bull. 195.
———. 1948b. Red-breasted Nuthatch. Pp. 22–35 In A. C. Bent. Life histories of North American nuthatches, wrens, thrashers, and their allies. U.S. Natl. Mus. Bull. 195.
———. 1948c. Brown Creeper. Pp. 56–70 In A. C. Bent. Life histories of North American nuthatches, wrens, thrashers and their allies. U.S. Natl. Mus. Bull. 195.
———. 1950a. Red-eyed Vireo. Pp. 335–48 In A. C. Bent. Life histories of North American wagtails, shrikes, vireos, and their allies. U.S. Natl. Mus. Bull. 197.
———. 1950b. Eastern Warbling Vireo. Pp. 362–373 In A. C. Bent. Life histories of North American wagtails, shrikes, vireos and their allies. U.S. Natl. Mus. Bull. 197.
———. 1953. Black-and-White Warbler. Pp. 5–17 In A. C. Bent. Life histories of North American wood warblers. U.S. Natl. Mus. Bull. 203.
———. 1958a. Baltimore Oriole. pp. 247–70 In A. C. Bent. Life histories of North American blackbirds, orioles, tanagers, and allies. U.S. Natl. Mus. Bull. 211.
———. 1958b. Scarlet Tanager. Pp. 479–91 In A. C. Bent, Life histories of North American blackbirds, orioles, tanagers, and their allies. U.S. Natl. Mus. Bull. 211.
———. 1968. Eastern American Goldfinch. Pp. 447–66 In A. C. Bent (O. L. Austin Jr., ed.). Life histories of North American cardinals, grosbeaks, buntings, towhees, finches, sparrows and allies, part 1. U.S. Natl. Mus. Bull. 237.
U.S. Fish and Wildlife Service. 1985. Red-cockaded Woodpecker recovery plan. U.S. Fish Wildlife Service, Atlanta, Ga.
———. 1989. Southeastern states Bald Eagle recovery plan. U.S. Fish and Wildlife Service, Atlanta, Ga.
———. 1990. Recovery plan for the interior population of the Least Tern *Sterna antillarum*. U.S. Fish and Wildlife Service, Twin Cities, Minn.
———. 1991. Endangered and threatened wildlife and plants; Animal candidate review for listing as endangered or threatened species, proposed rule. Federal Register 56:58804–36.
———. 1992. Endangered and threatened wildlife and plants. 50 CFR 17.11 and 17.12. U.S. Fish and Wildlife Service, Washington, D.C.
———. 1994. Endangered and threatened wildlife and plants; Animal candidate review for listing as endangered or threatened species, proposed rule. Federal Register 59:58982–59028.
———. 1995. Endangered and threatened wildlife and plants; Final rule to reclassify the bald eagle from endangered to threatened in all the lower 48 states. Federal Register 60:36000–10.
———. 1996. Endangered and threatened species; Review of plant and animal taxa that are candidates for listing as endangered or threatened species. Federal Register 61:7596–13.
Van Camp, L. R., and C. J. Henny. 1975. The Screech Owl: its life history and population ecology in northern Ohio. North Amer. Fauna 71.
Van Tyne, J. 1948. Home range and duration of family ties in the Tufted Titmouse. Wilson Bull. 60:121.
Vaughn, H. S. 1932. The Red-tailed hawk. Migrant 3:3–4.
———. 1933. A day in "Cranetown." Migrant 4:13–14.
———. 1940. Hereditary habits. Migrant 11:113.
———. 1943. Nesting of the Red-tailed Hawk. Migrant 14:49–50.
Vickers, B. B., and M. A. Kirby, eds. 1991. Tennessee statistical abstract 1991. Univ. Tenn. Center Business & Econ. Res., Knoxville.
Vissage, J. S., and K. L. Duncan. 1990. Forest statistics for Tennessee counties—1989. USDA For. Serv. Res. Bull. SO-148.
Wake, W. 1893. A narrow escape. Oologist 10:255.
———. 1897. Cardinalis cardinalis. Oologist 14:93–94.
Waldron, M. 1980. Anhinga nesting at Big Hill Pond, McNairy Co. Migrant 51:86.
———. 1981. First nesting of Bank Swallow in Shelby County, Tennessee. Migrant 52:68.
———. 1982. Nest box utilization by Hooded Mergansers at Hatchie National Wildlife Refuge. Migrant 53:13.
———. 1987. Seasonal occurrences of Shelby County, Tennessee birds. Memphis Chapter, Tenn. Ornithological Soc., Memphis, Tenn.
———. 1990. First nesting of Sora in Tennessee. Migrant 61:55.
Waldron, M. G., and D. P. Bean. 1991. The Earth Complex, Memphis, Tennessee. Migrant 62:103–5.
Walker, W. M., Jr. 1935. A collection of birds from Cocke County, Tenn. Migrant 6:48–50.
Walkinshaw, L. H. 1941. The Prothonotary Warbler: a comparison of nesting conditions in Tennessee and Michigan. Wilson Bull. 53:3–21.
———. 1961. The effects of parasitism by the Brown-headed Cowbird on *Empidonax* Flycatchers in Michigan. Auk 78:266–68.
———. 1966a. Summer biology of Traill's Flycatcher. Wilson Bull. 78:31–46.

———. 1966b. Summer observations of the Least Flycatcher in Michigan. Jack-pine Warbler 44:151–68.

———. 1968. Eastern Field Sparrow. Pp. 1217–35 In A. C. Bent (O. L. Austin Jr., ed.). Life histories of North American Cardinals, grosbeaks, buntings, towhees, finches, sparrows, and allies, part 2. U.S. Natl. Mus. Bull. 237.

———. 1979. Prothonotary Warbler. Pp. 156–60 In L. Griscom and A. Sprunt Jr. The Warblers of America. Doubleday and Co., Garden City, N.Y.

Warden, J. C. 1989. Changes in the spruce-fir forest of Roan Mountain in Tennessee over the past fifty years as a result of logging. J. Tenn. Acad. Sci. 64:193–95.

Warren, R. C., J. A. Jackson, and T. L. Darden. 1977. Nesting of the Red Crossbill in Mississippi. Am. Birds 31:1100.

Warriner, B. R. 1945. Some observations on the Blue Grosbeak. Migrant 16:24–26.

Warriner, E. 1974. Northern Orioles at Pickwick Dam. Migrant 45:72.

Watson, R. T., M. C. Zinyowera, and R. H. Moss. 1996. Climate change 1995: impact, adaptation and mitigation of climate change: scientific-technical analyses. Cambridge Univ. Press, New York.

Watts, B. D. 1989. Nest-site characteristics of Yellow-crowned Night-Herons in Virginia. Condor 91:979–83.

Weakley, S. A. 1936. Additional Cliff Swallow colonies. Migrant 7:72–73.

———. 1941. Swifts roosting in hollow tree. Migrant 12:76.

———. 1945. White Pelicans and Cliff Swallows on the Tennessee River. Migrant 16:33.

Webb, A. C. 1900. Some birds and their ways. B. F. Johnson, Richmond, Va.

Weber, W. J. 1975. Notes on Cattle Egret breeding. Auk 92:111–17.

Weeks, H. P., Jr. 1978. Clutch size variation in the Eastern Phoebe in southern Indiana. Auk 95:656–66.

———. 1979. Nesting ecology of the Eastern Phoebe in southern Indiana. Wilson Bull. 91:441–454.

Weigl, P. D., and T. W. Knowles. 1995. Megaherbivores and southern Appalachian grass balds. Growth and Change 26:365–82.

Weise, C. M. 1955. Spotted Sandpiper breeding in Middle Tennessee. Migrant 26:18–19.

Weller, W. M. 1958. Observations on the incubation behavior of a Common Nighthawk. Auk 75:48–59.

———. 1961. Breeding biology of the Least Bittern. Wilson Bull. 73:11–35.

Wells, J. V., and K. J. McGowan. 1991. Range expansion in Fish Crow (Corvus ossifragus): the Ithaca, NY, colony as an example. Kingbird 41:73–82.

Welsh, D. A. 1975. Savannah Sparrow breeding and territoriality on a Nova Scotia dune beach. Auk 92:235–51.

West, A. 1961. Cliff Swallow status in the Chattanooga area. Migrant 32:37–40.

West, Mrs. E. M. 1953. Blue Grosbeaks near Chattanooga. Migrant 24:53–54.

———. 1963. First breeding record of Virginia Rail in Tennessee. Migrant 34:20–21.

Westmoreland, D., L. B. Best, and D. E. Blockstein. 1986. Multiple brooding as a reproductive strategy: time-conserving adaptations in Mourning Doves. Auk 103:196–203.

Weston, F. M. 1968. Bachman's Sparrow. Pp. 956–75 In A. C. Bent (O. L. Austin Jr., ed.). Life Histories of North American cardinals, grosbeaks, buntings, and allies, part 1. U.S. Natl. Mus. Bull. 237.

Wetmore, A. 1939. Notes on the birds of Tennessee. Proc. U.S. Natl. Mus. 86:175–243.

Whitcomb, R. F., C. S. Robbins, J. F. Lynch, B. L. Whitcomb, M. K. Klimkiewicz, and D. Bystrak. 1981. Effects of forest fragmentation on avifauna of the eastern deciduous forest. Pp. 125–205 In R. L. Burgess and D. M. Sharpe, eds. Forest Island Dynamics in Man-dominated Landscapes. Springer Verlag, New York.

White, D., and R. W. Dimmick. 1979a. The distribution of Ruffed Grouse in Tennessee. Tenn Acad. Sci. 54:114–15.

———. 1979b. Survival and habitat use of northern Ruffed Grouse introduced into west Tennessee. Proc. Annu. Conf. S.E. Assoc. Fish and Wildlife Agencies. 32:1–7.

White, J. B. 1956. Birds of Greene County, Tennessee. Migrant 27:3–8.

Whitehead, C. J. 1991. Fisheries, small game, and migratory bird program outputs, FY 1985–1990, with estimated projections to the year 2000. Planning Rept. 91-1. Planning and Federal Aid Div. Tennessee Wildl. Res. Agency, Nashville.

Whitehurst, G. T. 1986. Range extension of the Brown-headed Nuthatch in western North Carolina. Chat 50:19.

Whitmore, R. C. 1979. Temporal variation in the selected habitats of a guild of grassland sparrows. Wilson Bull. 91:592–98.

———. 1981. Structural characteristics of Grasshopper Sparrow habitat. J. Wildl. Manage. 45:811–14.

Whitmore, R. C., and G. A. Hall. 1978. The response of passerine species to a new resource: reclaimed surface mines in West Virginia. Am. Birds 32:6–9.

Whittaker, R. H. 1956. Vegetation of the Great Smoky Mountains. Ecol. Monogr. 26:1–80.

Whittemore, W. L. 1937. Summer birds of Reelfoot Lake. J. Tenn. Acad. Sci. 12:114–12.

Widmann, O. 1895. The Brown Creeper nesting in the cypress swamps of southeast Missouri. Auk 12:350–55.

———. 1907. A preliminary catalog of the birds of Missouri. Trans. Acad. Sci. St. Louis 17:1–288.

Wiens, J. A. 1969. An approach to the study of ecological relationships among grassland birds. Ornithol. Monogr. No. 8, Allen Press, Lawrence, Kans.

Wiggers, E. P., and K. J. Kritz. 1991. Comparison of nesting habitat of coexisting Sharp-shinned and Cooper's Hawks in Missouri. Wilson Bull. 103:568–577.

Wilcove, D. S. 1985. Nest predation in forest tracts and the decline of migratory songbirds. Ecology 66:1211–14.

———. 1988. Changes in the avifauna of the Great Smoky Mountains: 1947–1983. Wilson Bull. 100:256–71.

Williams, M. 1989. Americans and their forests: A historical geography. Cambridge Univ. Press, Cambridge.

Williams, M. D. 1975a. Common Gallinule nesting in East Tennessee. Migrant 46:1–3.

———. 1975b. A survey of the birds of the Alcoa Marshes. M.S. Thesis, Univ. Tennessee, Knoxville.

———. 1976. Nest of Olive-sided Flycatcher in the southern Appalachian Mountains. Migrant 47:69–71.

———. 1977a. First breeding record of the Black Vulture in the Great Smoky Mountains National Park. Migrant 48:11–12.

———. 1977b. The status of the Cherokee Lake heronry in 1976. Migrant 48:95–96.

———. 1977c. A species index to *The Migrant*. Tenn. Ornithol. Soc., Spec. Publ 1. Nashville.

———. 1980a. Recoveries of some raptors banded in Tennessee. Migrant 51:26.

———. 1980b. Notes on the breeding biology and behavior of the Ravens of Peregrine Ridge, Great Smoky Mountains National Park, Tennessee. Migrant 51:77–80.

Williams, M. D., and C. P. Nicholson. 1977. Observations at a Black-crowned Night-Heron nesting colony. Migrant 48:1–6.

Williams, S. C. 1928. Early travels in the Tennessee country, 1540–1800. Watauga Press, Johnson City, Tenn.

Williamson, P. 1971. Feeding ecology of the Red-eyed Vireo *(Vireo olivaceous)* and associated foliage-gleaning birds. Ecol. Monogr. 41:129–52.

Willoughby, E. J., and T. J. Cade. 1964. Breeding behavior of the American Kestrel (Sparrow Hawk). Living Bird 3:75–96.

Wilson, A. 1811a. American ornithology, Vol. III. Bradford and Inskeep, Philadelphia.

———. 1811b. American ornithology, Vol. IV. Bradford and Inskeep, Philadelphia.

———. 1812. American ornithology, Vol. V. Bradford and Inskeep, Philadelphia.

Wilson, L. P. 1959. Chuck-will's-widow nestings. Migrant 30:53–54.

Wing, L. 1943. Spread of the Staring and English Sparrow. Auk 60:74–87.

Winnett-Murray, K. 1991. Hermit Thrush. Pp. 354–55 *In* R. Brewer, G. A. McPeek, and R. J. Adams Jr. The atlas of breeding birds of Michigan. Michigan State Univ. Press, East Lansing.

Witter, J. A., and I. R. Ragenovich. 1986. Regeneration of Fraser fir at Mt. Mitchell, North Carolina, after depredations by the balsam wooly adelgid. Forest Sci. 32:585–94.

Wolf, L. 1987. Host-parasite interactions of Brown-headed Cowbirds and Dark-eyed Juncos in Virginia. Wilson Bull. 99:338–50.

Woodring, G. B. 1932. Cowbirds breeding in East Tennessee. Migrant 3:38.

Woodward, M. 1903. Christmas Bird Census, Knoxville, Tenn. Bird-Lore 5:18.

Wootton, J. T. 1986. Clutch-size differences in western and introduced eastern populations of House Finches: patterns and hypotheses. Wilson Bull. 98:459–62.

Wray, T., II, K. A. Strait, and R. C. Whitmore. 1982. Reproductive success of grassland sparrows on a reclaimed surface mine in West Virginia. Auk 99:157–64.

Wright, A. H. 1915. Early records of the Wild Turkey. IV. Auk 32:207–24.

Yahner, R. H. 1972. BBC 18—Mixed deciduous forest. Am. Birds 26:951–52.

———. 1973. BBC 30—Mixed deciduous forest—strip mine. Am. Birds 27:971.

Yasukawa, K., and W. A. Searcy. 1995. Red-winged Blackbird *(Agelaius phoeniceus)*. *In* A. Poole and F. Gill, eds. The birds of North America, No. 184. Acad. Nat. Sci. Philadelphia and Amer. Ornithol. Union., Washington, D.C.

Yates, I. L., and C. J. Whitehead. 1978. Artificial propogation of the Giant Canada Goose in Tennessee. Proc. Annu. Conf. S.E. Fish Game Agenc. 32:348–355.

Yeatman, H. C. 1974. Sharp-shinned Hawks nesting at Sewanee. Migrant 45:66–67.

Young, H. 1955. Breeding behavior and nesting of the Eastern Robin. Amer. Midl. Nat. 53:329–52.

Yunick, R. P. 1981. Some observations on the breeding status of the Pine Siskin. Kingbird 31:219–25.

Zeleny, L. 1976. The Bluebird. Indiana Univ. Press, Bloomington.

Zimmerman, J. L. 1977. Virginia Rail *(Rallus limicola)*. Pp. 46–56 *In* G. C. Sanderson, ed. Management of migratory shore and upland game birds in North America. Internat. Assoc. Fish & Wildl. Agen., Washington, D.C.

———. 1982. Nesting success of Dickcissels *(Spiza americana)* in preferred and less preferred habitats. Auk 99:292–98.

———. 1988. Breeding season habitat selection by the Henslow's Sparrow *(Ammodramus henslowii)* in Kansas. Wilson Bull. 100:17–24.

Index

The page numbers for the main, formal account of each species are *italic*.

Accipiter cooperii. See Hawk, Cooper's
Accipiter striatus. See Hawk, Sharp-shinned
Actitis macularia. See Sandpiper, Spotted
Aegolius acadicus. See Owl, Northern Saw-whet
Aegolius funereus. See Owl, Boreal
Agelaius phoeniceus. See Blackbird, Red-winged
Aimophila aestivalis. See Sparrow, Bachman's
Aix sponsa. See Duck, Wood
Alectoris chukar. See Chukar
Allison, A., 14
Alsop, F. J., III, 18
Ammodramus henslowii. See Sparrow, Henslow's
Ammodramus savannarum. See Sparrow, Savannah
Anas discors. See Teal, Blue-winged
Anas platyrhynchos. See Mallard
Anas rubripes. See Duck, American Black
Anhinga, 46, 52, *66–67*
Anhinga anhinga. See Anhinga
Appalachian Plateaus Province, physiography, 20–21; vegetation, 27
Aquila chrysaetos. See Eagle, Golden
Archilochus colubris. See Hummingbird, Ruby-throated
Ardea alba. See Egret, Great
Ardea herodias. See Heron, Great Blue
Audubon, J. J., 12, 41

Baeolophus bicolor. See Titmouse, Tufted
Behrend, F. W., 17
Bierly, M. L., 17
Bittern, American, 42, *373–74*
Bittern, Least, 42, 46, 52, *68–69*, 125, 126, 129
Blackbird, Red-winged, 44, 49, 65, *353–55*
Blue Ridge Province, physiography, 19–20; vegetation, 24–26
Bluebird, Eastern, 17, 43, 49, 192, 221, 222, 223, *242–44*, 261, 371
Bobolink, *384*
Bobwhite, Northern, 43, *118–20*
Bombycilla cedrorum. See Waxwing, Cedar
Bonasa umbellus. See Grouse, Ruffed
Botaurus lentiginosus. See Bittern, American
Branta canadensis. See Goose, Canada
Brewster, W., 13
Bubo virginianus. See Owl, Great Horned
Bubulcus ibis. See Egret, Cattle
Bunting, Indigo, 44, 49, 52, 304, *329–31*, 360
Bunting, Painted, 49, 52, *331–32*
Burleigh, T. D., 15
Buteo jamaicensis. See Hawk, Red-tailed
Buteo lineatus. See Hawk, Red-shouldered
Buteo platypterus. See Hawk, Broad-winged
Butorides virescens. See Heron, Green-backed

Calhoun, J. B., 16
Campbell, J. M., 18
Campephilus principalis. See Woodpecker, Ivory-billed
Caprimulgus carolinensis. See Chuck-will's-widow
Caprimulgus vociferus. See Whip-poor-will
Cardinal, Northern, 12, 44, 49, 52, *323–25*, 360
Cardinalis cardinalis. See Cardinal, Northern
Carduelis pinus. See Siskin, Pine
Carduelis tristis. See Goldfinch, American
Carpodacus mexicanus. See Finch, House
Catbird, Gray, 43, *250–52*, 254, 262

Cathartes aura. See Vulture, Turkey
Catharus fuscescens. See Veery
Catharus guttatus. See Thrush, Hermit
Catharus ustulatus. See Thrush, Swainson's
Central Basin. See Interior Low Plateaus Province
Certhia americana. See Creeper, Brown
Ceryle alcyon. See Kingfisher, Belted
Chaetura pelagica. See Swift, Chimney
Charadrius vociferus. See Killdeer
Chat, Yellow-breasted, 44, 52, *317–19*
Chickadee, Black-capped, 52, *217–19*
Chickadee, Carolina, 43, 49, 52, 217, 218, *219–21*, 222
Chlidonias niger. See Tern, Black
Chondestes grammacus. See Sparrow, Lark
Chordeiles minor. See Nighthawk, Common
Chuck-will's-widow, 43, 50, *155–56*, 157
Chukar, *376*
Circus cyaneus. See Harrier, Northern
Clebsch, A., 16
Clebsch, E., 16
Climate, 22–24
Coastal Plain Province, physiography 22; vegetation, 28–29
Coastal Plain Uplands. See Coastal Plain Province
Coccyzus americanus. See Cuckoo, Yellow-billed
Coccyzus erythropthalmus. See Cuckoo, Black-billed
Coffey, B. B., Jr., 16
Coffey, L., 16
Colaptes auratus. See Flicker, Northern
Colinus virginianus. See Bobwhite, Northern
Columba livia. See Dove, Rock
Conservation efforts, 45–47, 50; *see also* individual species accounts

Contopus cooperi. See Flycatcher, Olive-sided
Contopus virens. See Wood-Pewee, Eastern
Conuropsis carolinensis. See Parakeet, Carolina
Coot, American, 49, 52, 125, 126, *128–29*
Cope, E. D., 12
Coragyps atratus. See Vulture, Black
Cormorant, Double-crested, 46, 49, 52, *65–66*, 67, 73
Corvus brachyrhynchos. See Crow, American
Corvus corax. See Raven, Common
Corvus ossifragus. See Crow, Fish
Coturnix japonica. See Quail, Japanese
Cowbird, Brown-headed, 17, 42, 44, 45, 52, 58, 59, *359–60*; hosts of, 395 and *see also* species accounts of potential host species
Crane, Sandhill, 46
Creeper, Brown, 41, 50, *229–30*
Crossbill, Red, 12, 50, 52, *366–68*
Crow, American, 43, 49, 52, *211–13*, 214, 215
Crow, Fish, 42, *213–15*
Cuckoo, Black-billed, 52, *141–43*
Cuckoo, Yellow-billed, 43, 52, 141, 142, *143–45*
Cumberland Mountains. See Appalachian Plateaus Province
Cumberland Plateau. See Appalachian Plateaus Province
Curtis, G. M., 15
Cyanocitta cristata. See Jay, Blue
Cygnus buccinator. See Swan, Trumpeter

Dendrocygna autumnalis. See Whistling-Duck, Black-bellied
Dendroica caerulescens. See Warbler, Black-throated Blue
Dendroica cerulea. See Warbler, Cerulean
Dendroica discolor. See Warbler, Prairie
Dendroica dominica. See Warbler, Yellow-throated
Dendroica fusca. See Warbler, Blackburnian
Dendroica magnolia. See Warbler, Magnolia
Dendroica pensylvanica. See Warbler, Chestnut-sided
Dendroica petechia. See Warbler, Yellow
Dendroica pinus. See Warbler, Pine
Dendroica virens. See Warbler, Black-throated Green
Dickcissel, 44, 45, *332–34*
Dimmick, R. W., 18
Distribution changes, summary of, 41–42; see also individual species accounts
Dolichonyx oryzivorus. See Bobolink
Dove, Mourning, 16, 49, 52, *139–41*
Dove, Rock, 41, 42, *137–39*
Dryocopus pileatus. See Woodpecker, Pileated
Duck, American Black, 52, *88–89*
Duck, Wood, 43, 45, *86–88*, 92, 93, 112, 358
Dumetella carolinensis. See Catbird, Gray

Eagle, Bald, 46, 57, *100–102*
Eagle, Golden, 13, 42, 46, *375–76*
Ectopistes migratorius. See Pigeon, Passenger
Egg collectors and collections, 13, 14, 15, 16, 17, 59
Egret, Cattle, 42, 74, 75, 76, *77–79*, 81, 83

Egret, Great, 46, 67, *71–73*, 76, 358
Egret, Snowy, 46, *73–75*, 76, 78
Egretta caerulea. See Heron, Little Blue
Egretta thula. See Egret, Snowy
Elanoides forficatus. See Kite, American Swallow-tailed
Elanus leucurus. See Kite, Black-shouldered; Kite, White-tailed
Empidonax alnorum. See Flycatcher, Alder
Empidonax minimus. See Flycatcher, Least
Empidonax traillii. See Flycatcher, Willow
Empidonax virescens. See Flycatcher, Acadian
Eremophila alpestris. See Lark, Horned

Falcon, Peregrine, 42, 46, 49, *113–15*, 375
Falco peregrinus. See Falcon, Peregrine
Falco sparverius. See Kestrel, American
Farming practices, 31, 32–35
Farmland area, 31, 33–35
Fawver, B. J., 16
Finch, House, 42, 44, 50, *364–66*
Fleetwood, R. J., 15
Fleming, J. H., 14
Flicker, Northern, 43, *175–77*, 261
Flycatcher, Acadian, 43, 54, *182–84*
Flycatcher, Alder, 42, 52, *184–86*
Flycatcher, Great Crested, *191–93*, 261
Flycatcher, Least, 52, *188–89*
Flycatcher, Olive-sided, 46, 49, *179–80*
Flycatcher, Scissor-tailed, 52, *195–96*
Flycatcher, Willow, 42, 50, 185, *186–87*
Forest, area of, 32, 35, 37, 45; diseases, 26, 37; fires, 31–32, 36, 38
Forestry, 36–37
Fox, W. H., 12–13, 18
Friedmann, H., 18
Fulica americana. See Coot, American

Gallinula chloropus. See Gallinule, Purple
Gallinule, Purple, 42, 46, 49, 50, *125–26*, 129
Gallus gallus. See Junglefowl, Red
Ganier, A. F., 14–15, 18
Geothlypis trichas. See Yellowthroat, Common
Gettys, W. R., 14
Gnatcatcher, Blue-gray, *240–42*
Goldfinch, American, 44, *368–70*
Goodpasture, K. A., 17
Goose, Canada, 42, *84–86*
Grackle, Common, 44, 49, 52, *357–58*
Grebe, Pied-billed, *63–64*
Grosbeak, Blue, 42, 44, 45, *327–29*
Grosbeak, Pine, 30
Grosbeak, Rose-breasted, *325–27*
Grouse, Ruffed, 42, 53, *115–16*, 237
Grouse, Sharp-tailed 30
Grus canadensis. See Crane, Sandhill
Guiraca caerulea. See Grosbeak, Blue

Haliaeetus leucocephalus. See Eagle, Bald
Harrier, Northern, 46
Hawk, Broad-winged, 43, *107–9*
Hawk, Cooper's, 46, 50, *104–5*
Hawk, Red-shouldered, 50, *106–7*
Hawk, Red-tailed, *109–11*, 114, 149, 375
Hawk, Sharp-shinned, 46, 50, *102–4*
Helmitheros vermivorus. See Warbler, Worm-eating

Herndon, L. R., 17
Heron, Great Blue, 42, 43, 65, 66, 67, *69–71*, 72, 76, 82, 83, 84
Heron, Green-backed, 43, *79–81*
Heron, Little Blue, 46, 74, *75–77*, 78
Highland Rim. See Interior Low Plateaus Province
Himantopus mexicanus. See Stilt, Black-necked
Hirundo rustica. See Swallow, Barn
Howell, A. H., 14, 18
Howell, J. C., 17–18
Hughes, H. Y., 14, 15
Hummingbird, Ruby-throated, *160–62*
Hylocichla mustelina. See Thrush, Wood

Icteria virens. See Chat, Yellow-breasted
Icterus galbula. See Oriole, Baltimore
Icterus spurius. See Oriole, Orchard
Ictinia mississippiensis. See Kite, Mississippi
Ijams, H. P., 17
Interior Low Plateaus Province, physiography, 21–22; vegetation, 27–28
Ixobrychus exilis. See Bittern, Least

Jay, Blue, 43, 49, 52, *209–11*
Jay, Gray, 30
Joliet, L., 11
Junco, Dark-eyed, 12, *351–53*
Junco hyemalis. See Junco, Dark-eyed
Junglefowl, Red, 376

Kestrel, American, 43, *111–13*
Killdeer, 43, *129–31*, 134
Kingbird, Eastern, 43, *193–95*, 361
Kingfisher, Belted, 43, *162–64*, 203
Kinglet, Golden-crowned, 52, *238–40*
Kite, American Swallow-tailed, 12, *374–75*
Kite, Mississippi, 46, *99–100*, 375
Kite, White-tailed, 48
Komarek, E. V., 15
Kroodsma, R., 18

Langdon, F. W., 13
Lanius ludovicianus. See Shrike, Loggerhead
Lark, Horned, 43, *196–98*
Laskey, A. R., 16–17
Laterallus jamaicensis. See Rail, Black
LeMoyne, A., 13
Limnothlypis swainsonii. See Warbler, Swainson's
Lingebach, C., 18
Loess Plain. See Coastal Plain Province
Lophodytes cucullatus. See Merganser, Hooded
Loxia curvirostra. See Crossbill, Red
Lyle, R. B., 17

Mallard, *89–90*
Martin, Purple, 43, 112, 192, *198–200*, 202, 261, 371
Mayfield, G. R., 15
Meadowlark, Eastern, 44, 49, *355–57*, 384
Meadowlark, Western, *384–85*
Melanerpes carolinus. See Woodpecker, Red-bellied
Melanerpes erythrocephalus. See Woodpecker, Red-headed

Meleagris gallopavo. See Turkey, Wild
Melospiza melodia. See Sparrow, Song
Merganser, Common, 374
Merganser, Hooded, 52, *92–93*
Mergus merganser. See Merganser, Common
Merritt, D., 14
Michaux, A., 12, 31, 41
Mimus polyglottos. See Mockingbird, Northern
Mining, 38–39
Mississippi Alluvial Plain. See Coastal Plain Province
Mniotilta varia. See Warbler, Black-and-white
Mockingbird, Northern, 17, 43, 49, *252–54,* 262
Molothrus ater. See Cowbird, Brown-headed
Monk, H. C., 16
Monroe, M. B., 18
Moorhen, Common, 42, 46, 52, 125, *126–28,* 129
Myiarchus crinitus. See Flycatcher, Great Crested
Myiopsitta monachus. See Parakeet, Monk

Nicholson, C. P., 18
Nighthawk, Common, 43, 50, *153–55*
Night-Heron, Black-crowned, 42, 46, 73, 74, 76, 77, 78, *81–83,* 84
Night-Heron, Yellow-crowned, 42, *83–84*
Nuthatch, Brown-headed, 42, *227–29*
Nuthatch, Red-breasted, *223–25*
Nuthatch, White-breasted, 43, *225–27*
Nycticorax nycticorax. See Night-Heron, Black-crowned
Nycticorax violaceus. See Night-Heron, Yellow-crowned

Oporornis formosus. See Warbler, Kentucky
Oriole, Baltimore, 44, 361, *362–64*
Oriole, Northern. See Oriole, Baltimore
Oriole, Orchard, 44, 328, *360–62,* 363
Osprey, 42, 46, *97–99*
Otus asio. See Screech-Owl, Eastern
Ovenbird, 43, 54, *305–7*
Owl, Barn, 46, 50, *145–47*
Owl, Barred, 50, *150–51*
Owl, Boreal, 30
Owl, Eastern Screech. See Screech-Owl
Owl, Great Horned, 43, 50, *148–49*
Owl, Hawk, 30
Owl, Northern Saw-whet, 46, 50, 52, *151–53*

Pandion haliaetus. See Osprey
Parakeet, Carolina, 11, 12, 41, 42, *379*
Parakeet, Monk, *379*
Parmer, H. E., 17
Parula americana. See Parula, Northern
Parula, Northern, 43, 45, *276–77*
Passer domesticus. See Sparrow, House
Passerculus sandwichensis. See Sparrow, Savannah
Passerina ciris. See Bunting, Painted
Passerina cyanea. See Bunting, Indigo
Pelecanus erythrorhynchos. See Pelican, White
Pelican, White, 11
Perisoreus canadensis. See Jay, Gray
Perrygo, W. M., 18

Petrochelidon pyrrhonota. See Swallow, Cliff
Phalacrocorax auritus. See Cormorant, Double-crested
Phasianus colchicus. See Pheasant, Ring-necked
Pheasant, Ring-necked, 49, *376–77*
Pheuticus ludovicianus. See Grosbeak, Rose-breasted
Phoebe, Eastern, 42, 43, 49, 52, *189–91,* 209
Picoides borealis. See Woodpecker, Red-cockaded
Picoides pubescens. See Woodpecker, Downy
Picoides villosus. See Woodpecker, Hairy
Pigeon, Passenger, 41, 42, 357, *378*
Pindar, L. O., 13
Pinicola enucleator. See Grosbeak, Pine
Pipilo erythrophthalmus. See Towhee, Eastern
Piranga olivacea. See Tanager, Scarlet
Piranga rubra. See Tanager, Summer
Pitts, T. D., 16
Podilymbus podiceps. See Grebe, Pied-billed
Poecile atricapillus. See Chickadee, Black-capped
Poecile carolinensis. See Chickadee, Carolina
Polioptila caerulea. See Gnatcatcher, Blue-gray
Pooecetes gramineus. See Sparrow, Vesper
Population trends, human, 31, 32, 33; summary of bird, 42–45, 58–61; *see also* individual species accounts
Porphyrula martinica. See Gallinule, Purple
Porzana carolina. See Sora
Prairie-Chicken, Greater, 12, 31, 42, *377*
Progne subis. See Martin, Purple
Protonotaria citrea. See Warbler, Prothonotary

Quail, Japanese, *376*
Quiscalus quiscula. See Grackle, Common

Ragsdale, G. H., 12
Rail, Black, 46, *377–78*
Rail, King, 42, 46, *120–22*
Rail, Virginia, 42, 46, 50, *122–23*
Rallus elegans. See Rail, King
Rallus limicola. See Rail, Virginia
Raven, Common, 12, 42, 46, 212, *215–17*
Redstart, American, 43, *297–99*
Regulus satrapa. See Kinglet, Golden-crowned
Rhoads, S. N., 13–14, 41
Ridge and Valley Province, physiography, 20; vegetation, 26–27
Riparia riparia. See Swallow, Bank
Robin, American, 12, 43, 49, 52, *248–50,* 361
Robinson, J. C., 17

Safford, J. M., 12
Sandpiper, Spotted, 52, *133–34*
Sapsucker, Yellow-bellied, 46, 50, 52, *168–70*
Sayornis phoebe. See Phoebe, Eastern
Schaefer, H. R., 18
Scolopax minor. See Woodcock, American
Screech-Owl, Eastern, 50, *147–48,* 150
Seiurus aurocapillus. See Ovenbird
Seiurus motacilla. See Waterthrush, Louisiana

Seiurus noveboracensis. See Waterthrush, Northern
Sennett, G. B., 13
Setophaga ruticilla. See Redstart, American
Shaver, J. M., 16
Shrike, Loggerhead, 43, 46, *258–60*
Shugart, H. H., Jr., 18
Sialia sialis. See Bluebird, Eastern
Siskin, Pine, 367, *385*
Sitta canadensis. See Nuthatch, Red-breasted
Sitta carolinensis. See Nuthatch, White-breasted
Sitta pusilla. See Nuthatch, Brown-headed
Smith, G. S., 12
Sora, 50, 52, *124–25*
Sparrow, Bachman's, 42, 46, 47, 52, *336–38*
Sparrow, Chipping, 44, *338–40,* 361
Sparrow, Field, 44, *340–42*
Sparrow, Grasshopper, 44, 46, 50, 343, *346–49,* 355
Sparrow, Henslow's, *383–84*
Sparrow, House, 42, 44, 112, 200, 207, 209, 243, 365, *370–72*
Sparrow, Lark, 46, *344–45*
Sparrow, Savannah, 42, 52, 343, *345–46*
Sparrow, Song, 42, 44, 58, *349–51*
Sparrow, Vesper, 46, *342–43*
Species richness trends, 50–52
Sphyrapicus varius. See Sapsucker, Yellow-bellied
Spiza americana. See Dickcissel
Spizella passerina. See Sparrow, Chipping
Spizella pusilla. See Sparrow, Field
St. Cosmé, J. F. B., 11
Starling, European, 41, 42, 43, 49, 112, 164–65, 176, 177, 200, 243, *260–62*
Stelgidopteryx serripennis. See Swallow, Northern Rough-winged
Sterna antillarum. See Tern, Least
Stilt, Black-necked, 42, 52, *131–32*
Strix varia. See Owl, Barred
Stupka, A., 16
Sturnella magna. See Meadowlark, Eastern
Sturnella neglecta. See Meadowlark, Western
Sturnus vulgaris. See Starling, European
Surnia ulula. See Owl, Hawk
Swallow, Bank, 12, 42, 203, *204–5*
Swallow, Barn, 43, 49, *207–9,* 371
Swallow, Cliff, *205–7,* 371
Swallow, Northern Rough-winged, *202–3*
Swallow, Tree, 42, 50, *200–201*
Swan, Trumpeter, 14
Swift, Chimney, 43, 49, 52, *158–60*

Tachycineta bicolor. See Swallow, Tree
Tanager, Scarlet, 44, 53, 319, 320, *321–23*
Tanager, Summer, *319–21,* 322, 328
Tanner, J. T., 16, 17–18
Teal, Blue-winged, 52, *90–92*
Tennessee Ornithological Society, founding of, 14–15
Tern, Black, 136
Tern, Least, 7, 46, *136–37*
Thrasher, Brown, 43, 252, *254–56*
Thrush, Hermit, 49, *379–81*
Thrush, Swainson's, 381
Thrush, Wood, 43, 54, 245, *246–48,* 360
Thryomanes bewickii. See Wren, Bewick's

Thryothorus ludovicianus. See Wren, Carolina
Titmouse, Tufted, 43, 49, 52, 220, *221–23*
Todd, H. O., 17
Torrey, B., 14
Towhee, Eastern, 42, 44, 49, 52, *334–36*, 360
Towhee, Rufous-sided. *See* Towhee, Eastern
Toxostoma rufum. See Thrasher, Brown
Troglodytes aedon. See Wren, House
Troglodytes troglodytes. See Wren, Winter
Troost, G., 12
Turdus migratorius. See Robin, American
Turkey, Wild, 12, 37, 45, 50, *117–18*, 357
Tyler, B. P., 17
Tympanuchus cupido. See Prairie-Chicken, Greater
Tympanuchus phasianellus. See Grouse, Sharp-tailed
Tyrannus forficatus. See Flycatcher, Scissor-tailed
Tyrannus tyrannus. See Kingbird, Eastern
Tyto alba. See Owl, Barn

Unaka Mountains. *See* Blue Ridge Province

Vaughan, H. S., 16
Veery, 53, *244–46*, 282, 381
Vermivora chrysoptera. See Warbler, Golden-winged
Vermivora chrysoptera x pinus. See Warbler, "Brewster's;" Warbler, "Lawrence's"
Vermivora peregrina. See Warbler, Tennessee
Vermivora pinus. See Warbler, Blue-winged
Vermivora ruficapilla. See Warbler, Nashville
Vireo bellii. See Vireo, Bell's
Vireo flavifrons. See Vireo, Yellow-throated
Vireo gilvus. See Vireo, Warbling

Vireo griseus. See Vireo, White-eyed
Vireo olivaceus. See Vireo, Red-eyed
Vireo solitarius. See Vireo, Solitary
Vireo, Bell's, *381*
Vireo, Red-eyed, 43, 45, 264, 266, *270–71*, 360
Vireo, Solitary, 53, *264–66*
Vireo, Warbling, *268–69*
Vireo, White-eyed, 43, *262–64*, 360, 381
Vireo, Yellow-throated, *266–68*
Vulture, Black, *93–95*
Vulture, Turkey, 43, 94, *95–97*

Waldron, M. G., 16
Walker, W. M., 16
Warbler, Black-and-white, 54, *295–97*
Warbler, Blackburnian, 52, *285–87*
Warbler, Black-throated Blue, *282–83*
Warbler, Black-throated Green, *284–85*
Warbler, Blue-winged, *272–73*, 274, 275, *381–82*
Warbler, "Brewster's," *381–82*
Warbler, "Lawrence's," *381–82*
Warbler, Canada, *315–17*
Warbler, Cerulean, 43, 46, *293–95*
Warbler, Chestnut-sided, *280–81*
Warbler, Golden-winged, 42, 272, *274–76*, *381–82*
Warbler, Hooded, 44, *313–15*
Warbler, Kentucky, 43, *309–11*
Warbler, Magnolia, 49, 50, *382–83*
Warbler, Nashville, 12
Warbler, Pine, 42, 43, *289–91*
Warbler, Prairie, 43, *291–93*, 360
Warbler, Prothonotary, 201, *299–301*
Warbler, Swainson's, 14, 46, *303–5*
Warbler, Tennessee, 12

Warbler, Worm-eating, 43, 54, *301–3*
Warbler, Yellow, 43, *278–80*, 360
Warbler, Yellow-throated, 43, *287–89*
Waterthrush, Louisiana, 43, *307–9*
Waterthrush, Northern, 308
Waxwing, Cedar, 7, 43, 50, *256–58*
Webb, A. C., 14, 15
Wetlands, losses of, 39–40; types, 29, 39; vegetation, 29
Wetmore, A., 18
Whip-poor-will, 43, 50, 155, *156–58*
Whistling-Duck, Black-bellied, *374*
Whittemore, W. L., 16
Williams, M. D., 17
Wilson, A., 12, 31, 41
Wilsonia canadensis. See Warbler, Canada
Wilsonia citrina. See Warbler, Hooded
Woodcock, American, 50, *134–36*
Woodpecker, Downy, 166, *170–71*, 220, 221
Woodpecker, Hairy, 54, *172–73*
Woodpecker, Ivory-billed, 12, 41
Woodpecker, Pileated, 54, 88, *177–79*
Woodpecker, Red-bellied, *166–68*, 170, 261
Woodpecker, Red-cockaded, 14, 42, 46, 47, 50, 52, *173–75*, 227, 228, 336
Woodpecker, Red-headed, 43, *164–66*, 261
Wood-Pewee, Eastern, *180–82*
Woodward, M., 14
Wren, Bewick's, 42, 43, 46, *232–34*, 236
Wren, Carolina, 43, 49, 52, *230–32*
Wren, House, 42, 233, *234–36*
Wren, Winter, 53, *236–38*

Yellowthroat, Common, 49, 52, 284, *311–13*

Zenaida macroura. See Dove, Mourning

Atlas of the Breeding Birds of Tennessee was designed and typeset on a Macintosh computer system using PageMaker software and set in Galliard. This book was designed and composed by Sheila Hart and printed and bound by Thomson-Shore, Inc. The recycled paper used in this book is designed for an effective life of at least three hundred years.